PHYSIOLOGY AND BIOPHYSICS

AN AMERICAN TEXT-BOOK OF PHYSIOLOGY
edited by William H. Howell, Ph.D., M.D.
W. B. SAUNDERS AND COMPANY   1896

A TEXT-BOOK OF PHYSIOLOGY FOR MEDICAL STUDENTS AND PHYSICIANS
written by William H. Howell, Ph.D., M.D., LL.D.
W. B. SAUNDERS AND COMPANY   1905

HOWELL'S TEXTBOOK OF PHYSIOLOGY 15th edition
edited by John F. Fulton, M.D.
W. B. SAUNDERS COMPANY   1946

A TEXTBOOK OF PHYSIOLOGY 16th and 17th editions
edited by John F. Fulton, M.D.
W. B. SAUNDERS COMPANY   1949 AND 1955

MEDICAL PHYSIOLOGY AND BIOPHYSICS 18th edition
edited by Theodore C. Ruch, Ph.D., and John F. Fulton, M.D.
W. B. SAUNDERS COMPANY   1960

PHYSIOLOGY AND BIOPHYSICS 19th edition
edited by Theodore C. Ruch, Ph.D., and Harry D. Patton, Ph.D., M.D.
W. B. SAUNDERS COMPANY   1965

PHYSIOLOGY AND BIOPHYSICS 20th edition
Volume III: Digestion, Metabolism, Endocrine Function and Reproduction
edited by Theodore C. Ruch, Ph.D., and Harry D. Patton, Ph.D., M.D.
W. B. SAUNDERS COMPANY   1973

EDITED BY

THEODORE C. RUCH, Ph.D.

Professor of Physiology and Biophysics,
Core Staff, Regional Primate Research Center,
University of Washington School of Medicine

and

HARRY D. PATTON, Ph.D., M.D.

Chairman of the Department and Professor of
Physiology and Biophysics,
University of Washington School of Medicine

ASSOCIATE EDITOR

ALLEN M. SCHER, Ph.D.

Professor of Physiology and Biophysics,
University of Washington School of Medicine

TWENTIETH EDITION

HOWELL-FULTON

PHYSIOLOGY
AND
BIOPHYSICS

CIRCULATION,
RESPIRATION
AND
FLUID BALANCE

W. B. SAUNDERS COMPANY

PHILADELPHIA · LONDON · TORONTO

W. B. Saunders Company: West Washington Square
Philadelphia, PA 19105

1 St. Anne's Road
Eastbourne, East Sussex BN21 3UN, England

1 Goldthorne Avenue
Toronto, Ontario M8Z 5T9, Canada

Physiology and Biophysics — Volume II   ISBN 0-7216-7818-1

Print No.:   9   8   7   6   5   4

CONTRIBUTORS

ROSEMARY BIGGS, B.Sc., Ph.D., M.D., M.R.C.P., M.A. University Lecturer in Haematology, Oxford University; Director, Oxford Haemophilia Centre, Churchill Hospital, Headington, Oxford, England.

ARTHUR C. BROWN, Ph.D. Professor, Department of Physiology and Biophysics, University of Washington School of Medicine.

MARSHAL H. CALEY, M.D. (Deceased) Late Holder of PHS Fellowship, Department of Physiology and Biophysics, University of Washington School of Medicine.

LOUIS G. D'ALECY, D.M.D., Ph.D. Assistant Professor of Physiology, University of Michigan School of Medicine.

ERIC O. FEIGL, M.D. Professor, Department of Physiology and Biophysics, University of Washington School of Medicine.

JACOB HILDEBRANDT, Ph.D. Senior Investigator, Virginia Mason Research Center, Seattle; Associate Professor, Department of Physiology and Biophysics and Department of Medicine, University of Washington.

THOMAS F. HORNBEIN, M.D. Professor, Departments of Anesthesiology and Physiology and Biophysics, Vice-Chairman, Department of Anesthesiology, University of Washington School of Medicine.

GLENN L. KERRICK, Ph.D. Assistant Professor, Department of Physiology and Biophysics, University of Washington School of Medicine.

ALAN KOCH, Ph.D. Associate Professor of Zoophysiology, Washington State University.

CLAUDE LENFANT, M.D. Director, Division of Lung Diseases, National Heart and Lung Institute, National Institutes of Health, Bethesda, Maryland.

LORING B. ROWELL, Ph.D. Professor, Department of Physiology and Biophysics, and Adjunct Professor of Medicine, University of Washington School of Medicine.

THEODORE C. RUCH, Ph.D. Professor, Department of Physiology and Biophysics, University of Washington School of Medicine; Core Staff, Regional Primate Research Center, University of Washington.

ALLEN M. SCHER, Ph.D. Professor, Department of Physiology and Biophysics, University of Washington School of Medicine.

SØREN C. SØRENSEN, M.D. Research Associate, Institute of Medical Physiology A, Copenhagen, Denmark.

ROBERT B. STEPHENSON Predoctoral Fellow, Department of Physiology and Biophysics, University of Washington School of Medicine.

CURT A. WIEDERHIELM, Ph.D. Professor, Department of Physiology and Biophysics, University of Washington School of Medicine.

DIXON M. WOODBURY, Ph.D. Professor and Chairman, Department of Pharmacology, University of Utah College of Medicine.

J. WALTER WOODBURY, Ph.D. Professor of Physiology, University of Utah College of Medicine.

ALLAN C. YOUNG, Ph.D. Professor, Department of Physiology and Biophysics, University of Washington School of Medicine.

PREFACE

The Howell-Fulton *Textbook of Physiology and Biophysics* is now in its seventy-seventh year of continuous publication, a life span not equaled by any other physiology textbook in any language. This edition is being published in three volumes (Vol. III was published in March 1973), and each volume will deal with a coherent group of topics. The division into three volumes is partially dictated by the enormous growth of physiological knowledge. More importantly, a three-volume work is better adapted to medical curricula arranged according to organ systems, and no less useful to graduate students and young faculty members in physiology and related subjects.

The 28 chapters in Volume II relate to circulation, respiration and fluid balance, a meaningful triad. Of these 28 chapters 17 are completely new, having been written by new contributors to this textbook. The remaining 11 chapters have been rewritten or revised extensively. For the first time considerable emphasis has been placed on the circulation to specific regions or tissues, *i.e.*, muscle, splanchnic, renal, coronary and cerebral circulation. This change recognizes that there are limits to the generalizations that can be made about the circulation as a whole. However, a view of the woods precedes that of the trees. The capillary segment of the circulation has been treated more quantitatively in keeping with the growing interest in the microcirculation. Textbooks do not ordinarily contain original concepts but such will be found, for example in the chapter on pH, written from a rigorous point of view.

Full titles of journal articles are given for the first time in this edition. The abbreviations of journal titles are based upon the most complete and least parochial system in respect to disciplines. Unfortunately, this reference (*World Medical Periodicals*, 3rd edition, H. A. Clegg, ed.) is no longer available, partly through lack of support of an international effort by biological editors of this country.

Much of the responsibility for this volume was assumed by Dr. Patton. However, the senior editor, author of a single chapter, would like to thank those who critically reviewed his chapter, Dr. Saul Boyarsky and Dr. William Bradley, but takes full responsibility for the content. In addition, we would like to thank Ms. Paula Ann Karlberg and Ms. Pamela Stokes for their efforts in coordinating and editing this volume and Mrs. Helen Halsey for the illustrations.

T. C. RUCH

H. D. PATTON

Seattle, 1974

CONTENTS

CHAPTER 19

COAGULATION OF BLOOD

Rosemary Biggs

SECTION II.   RESPIRATION

CHAPTER 20

ANATOMY AND PHYSICS OF RESPIRATION

J. Hildebrandt

CHAPTER 21

GAS TRANSPORT AND GAS EXCHANGE

Claude Lenfant

CHAPTER 22

NEURAL CONTROL OF RESPIRATION

Allan C. Young

VOLUME II

PHYSIOLOGY
AND BIOPHYSICS

CIRCULATION OF BLOOD AND LYMPH

CHAPTER 1

GENERAL CHARACTERISTICS OF THE CARDIOVASCULAR SYSTEM

by ALLEN M. SCHER

DIFFUSION

EVOLUTIONARY CHANGES AND THE CIRCULATION
  Respiratory Adaptations
  Circulatory Fluids
  Further Developments

THE MAMMALIAN CARDIOVASCULAR SYSTEM
  The Systemic Circulation
  The Pulmonary Circulation

ORGANIZATION OF CARDIOVASCULAR CHAPTERS

This chapter introduces Volume II, *Circulation, Respiration* and *Fluid Balance.* As a general introduction, the role and importance of diffusion in cellular activity and the limitations that diffusion may impose on the size and activity of organisms will be discussed. The evolution of the cardiovascular system will be briefly considered as a sequence of developments which circumvent the limitations of diffusion.

Finally, the mammalian cardiovascular system will be briefly described and, following this introduction, a guide will be presented to the chapters which discuss mammalian cardiovascular physiology.

DIFFUSION

The unicellular organism, floating in a sea or pond, must find the necessities for its metabolism and thus its survival at its external cellular membrane. Short-term survival is dependent on the availability of oxygen and other soluble materials which participate in the basic cellular metabolic reactions. There is, of course, a two-way exchange, and the cell must lose metabolic end products to the medium. Usually the aqueous environment undergoes a continuous convective movement which mixes and distributes the dissolved metabolites and assures replenishment of the fluid at the

1

cell surface. The surface of the fluid is exposed to the oxygen of the air, and here the concentration of dissolved oxygen is highest and the organism's chance of finding adequate oxygen is greatest. Movement of material from the cell surface through the cell membrane can be likened to diffusion through submicroscopic pores or slits. Diffusion in aqueous solutions and across membranes involves movement of materials from areas of high concentration to areas of low concentration. The amount of diffusion across the membrane depends on the concentration difference across the membrane and on the mobility of the diffusing ions and molecules (mobility depends on the mass and the size of the dissolved particles) through the membrane. The interested reader can find a detailed discussion of diffusion elsewhere (Chap. 24, Vol. II).[2, 3] Availability of soluble materials in the external medium may, as indicated, impose no limitations to continued metabolism by the cell, particularly near the surface of the medium; diffusion of material through the cell wall is virtually never a limiting process in cellular survival.

Diffusion is a remarkably rapid process over short distances and a remarkably slow one over longer distances. Estimations of speed of diffusion by Hill,[2] based on physics of particle movement in aqueous media, indicate that a deoxygenated nerve fiber 0.7 μ thick, placed in an oxygen-rich aqueous environment, will attain 90 per cent saturation with oxygen in 5.4 msec. For a fiber 0.7 mm thick (1000 times as thick, but still small) the same degree of saturation would require 54 sec. Hill[2] utilized figures for the oxygen consumption and lactic acid production of contracting frog muscle to compute the relationship between the diameter of a muscle fiber bundle and the amount of activity it could tolerate without exhausting available oxygen or accumulating excess lactic acid. Assuming that such a muscle bundle underwent 8 twitch contractions per sec (a very minimal frequency of contraction), the fiber could have a thickness of about 0.5 mm. A more active fiber than this would have to be thinner. Thus, diffusion limits both the size and activity of tissue.

The rate of diffusion is proportional to the surface area of a cell or cellular mass. The metabolism is more nearly proportional to the cell volume. For spherical cells, the surface area increases as the radius squared, but volume increases as the radius cubed. The smaller a cell the larger its ratio of surface area to volume. Just as diffusion limits the size to which any single cell or muscle mass can grow, it obviously limits the size to which any organism or cellular aggregation can grow if diffusion across a surface is the only method through which the interior of a cell mass can gain oxygen or lose metabolites (the exact requirement for exchange is dependent also on the activity of the cell, mass or organism).

EVOLUTIONARY CHANGES AND THE CIRCULATION

Through evolutionary changes, organisms with low metabolic rates, adrift in aqueous environments and dependent for their survival on availability of materials in these environments, were ultimately succeeded by large animals with high metabolic rates, moving freely on land and with major capabilities to adapt to changes in environment and even to control and alter that environment. Such development is accompanied by increased metabolic activity and must be accompanied by an increased capability for transport of dissolved oxygen into the cells and of metabolites from the cells.

The cardiovascular changes accompanying these evolutionary changes are surprisingly varied. They are well described in several texts and monographs.[1, 4, 5] A detailed consideration is impossible here, and the following discussion is intended to indicate the general rather than specific progression of circulatory adaptations which favor increased oxygen delivery. The term "oxygen delivery" is used as a general term for a variety of changes which increase the oxygen available to the organism (and which increase the movement of metabolic end products out of the organism into the environment).

The line of evolutionary development, here as in other areas, is not characterized by a clear, stepwise movement from the primitive to the advanced. Some features of the mammalian cardiovascular system appear in early invertebrates; some appear and are lost only to reappear later. As pointed out by Martin and Johansen,[4] organisms which evolved separately (and often millions of years apart) in the direction of increasing size and mobility may show strikingly similar evolutionary changes. On the other hand, there may be vast differences between primitive and advanced members of the same phylum. To simplify the following discussion, we will separately discuss changes in the form of the circulatory system, development of respiratory systems and development of specialized circulatory fluids.

A first type of adaptation, seen in simple but multicellular aqueous organisms, involves movement of the external fluid by the cells. This movement supplements the convection which brings oxygenated fluid near the cell surface. At times the organism itself moves to accomplish this convection; at times cilia or similar structures are employed.

In the sponges (porifera or pore-bearers) (Fig. 1–2) the organism takes the form of a branched network of cells, each attached to one or more others. The sponges are considered to be a side branch on the evolutionary tree, but their adaptation is interesting. Channels in the meshwork allow the external fluid to be brought into close contact with each cell. Mixing of the fluid by the movement of the entire mass of tissue increases the availability of oxygen. The single-celled organism has an intracellular organization in which mitochondria, vacuoles, psuedopods and cilia perform specialized functions; the sponges show a higher type of specialization in which particular cells may function as specialized feeding cells, supporting cells, collar cells (which move fluid with their flagella), reproductive cells, etc. Some cells are unspecialized and can develop into the other types. The sponges illustrate a high degree of organization and specialization and show capability beyond that of single cells.

The coelenterates (Fig. 1–3) have a cylindrical form with an internal body cavity filled with fluid. The cavity serves both a digestive and a circulatory function. The name coelenterate means hollow digestive cavity. The hydra, a simple coelenterate, has two layers of cells – endoderm and ectoderm. The level of organization is higher than that of the sponges and may be considered to be on a tissue level. Groups of cells act together as a tissue. The organism has muscle tissue, connective tissue, cell groups with tentacles which can poison passing organisms, glands and nervous tissue.

The internal cavity of the coelenterates, the coelom, becomes extensively branched in some members of this phylum, as in the jellyfish (Fig. 1–4, ctenophore). The movement of the muscles of the body wall propels the contained fluid within the branches and causes exchange with the external medium. The level of organization is not far above that of the hydra, but the hydra lives a relatively fixed life attached to a stone or leaf, moving slowly, while the jellyfish swims by muscular movement and has a more advanced nervous and muscular mechanism. Extensive branching of the gastrovascular cavity and consequent increase in surface area probably facilitate the more complex activity of these coelenterates.

In the nematodes, or roundworms, the digestive and coelomic spaces are separate (Fig. 1–5), a change which anticipates a separate circulatory system. The organism, by this time, contains groups of tissues which make up specialized organs – the digestive system has a mouth, intestine and anus. The nervous system is more complex, and the reproductive system shows sexual differentiation.

The circulatory system of the earthworm, an annelid (Fig. 1–6), represents several major steps forward. A dorsal blood vessel lies on the digestive tube and can rhythmically contract, moving the blood forward. Transverse segmental vessels run from it in each body segment to a ventral vessel below the digestive system, and some branches of the segmental vessels pass to the intestine, to other internal organs and to the body wall. There are valves in some vessels to prevent backflow. The vessels branch extensively, finally forming a network of microscopic capillaries. These latter have walls one cell thick which facilitate extensive exchange of gases, wastes, nutrients, etc., across the wall. Capillaries are so extensive that they bring blood into close contact with each cell. Specialization of internal organs is again more advanced. Respiratory exchange takes place at the skin. The excretory, reproductive and nervous systems are far more advanced. The blood contains dissolved hemoglobin which increases its oxygen-carrying capacity.

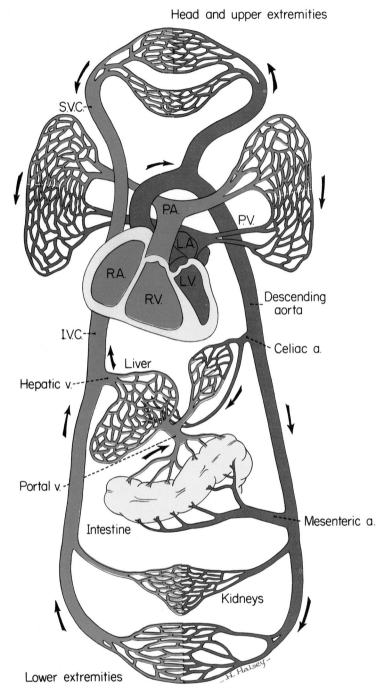

Figure 1–1 A diagram of the heart and blood vessels. Arterial blood is indicated in red, venous blood in blue. Direction of blood flow is indicated by the black arrows. R.A. = right atrium, R.V. = right ventricle, L.A. = left atrium, and L.V. = left ventricle. P.A. = pulmonary artery, P.V. = pulmonary vein, S.V.C. = superior vena cava, and I.V.C. = inferior vena cava. Blood flows into the right atrium from the two venae cavae. From there it flows into the right ventricle, and thence into the pulmonary artery where it is oxygenated by the lungs. It returns through the pulmonary vein to the left atrium and left ventricle. From the left ventricle, the blood enters the aorta and runs to the various parallel beds of the systemic circulation. Arterial blood loses oxygen in the various organs, and then returns to the two venae cavae from the peripheral veins.

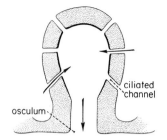

Figure 1–2 Porifera. In the sponges, fluid enters the colony through the lateral ciliated channels and is excreted through the osculum (mouth). Fluid is moved by flagellated cells and by the movement of the entire organism.

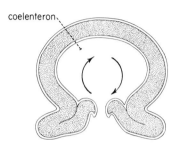

Figure 1–3 Coelenterate. A simple coelenterate has a combined gastrovascular cavity, or coelenteron, with a single opening to the outside. Fluid is circulated by rhythmic contractions of the entire organism and by movement of tentacles.

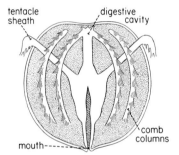

Figure 1–4 Ctenophore. A more highly developed coelenterate, the jellyfish, has a highly branched system for the distribution of the contents of the gastrovascular cavity. The actual cavity is reduced in size, but surface area is increased.

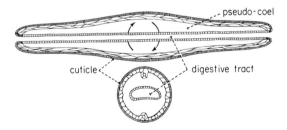

Figure 1–5 Nematode. The nematode, or roundworm, shows separate digestive and coelomic cavities. There are no specific circulatory or respiratory organs. The pseudo-coel is interposed between the digestive tract and the other tissues and is similar to the coelenteron of the more primitive forms, but does not communicate directly with the external medium.

Figure 1–6 Annelid. The annelids are segmented worms, whose bodies contain essentially similar ring-like segments. The circulatory system is closed. The dorsal blood vessel above the digestive tube can contract rhythmically, moving the blood, and gives off lateral branches to the various organs in each somite. Valves in the dorsal vessels and in the contractile vessel serve to prevent backflow. An extensive network of capillaries is present. The blood consists of a fluid plasma containing corpuscles with a dissolved respiratory pigment, hemoglobin. The skin serves as a respiratory organ.

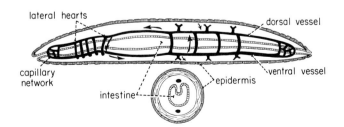

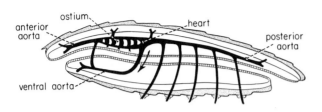

Figure 1–7 Arthropod. The arthropods, segmented like the annelids, are more advanced but have a simpler circulatory system. The circulatory system is open. The dorsal heart pumps blood into the arteries and through the various organs and tissues. From there it returns to the heart via the open circulatory spaces. The blood re-enters the heart through the ostia. The arteries contain valves which prevent backflow. Exchange with the external fluid occurs in the gills. The blood contains a respiratory pigment, hemocyanin.

The "open" circulatory system, as seen in arthropods (Fig. 1–7), is simpler. A contractile portion of the blood vessel pumps fluid—called hemolymph or blood—from the coelomic cavity into arteries and arterioles. The fluid then flows into sinuses or tissue spaces which are part of the coelom. In these spaces the blood bathes the various organs. From these spaces it moves slowly back to large cavities around the heart. From these it moves into the heart through openings in the sides of that organ. Since the blood is only partially confined to the circulatory system, this is called an "open" circulation. This system seems less advanced than the closed system seen in the annelids, but in aquatic arthropods the accompanying respiratory apparatus now includes gills, a specialized invagination of the body wall with a large surface area. The oxygenated external fluid lies on one side of this and the blood lies on the other, making respiratory exchange easier. Other systems (nervous, reproductive) show increased specialization. The blood contains a respiratory pigment, hemocyanin.

A more complicated closed circulatory system, like that seen in mammals, appears in chordates (Fig. 1–8). Here the circulatory fluid is clearly separate, and there is an extensive set of distributing vessels, including arteries, capillaries and sometimes sinuses.

The heart has evolved as a muscular organ, pumping blood through the systemic and pulmonary circuits. The gill circulation and the systemic circulation are in a semi-series connection. Blood flows from the heart into the gills, and some of the blood then returns to the heart. Most of the blood from the gills passes via the dorsal aorta into the various systemic organs, returning from the capillaries into the heart. Directionality of blood flow is assured by valves in the veins. The blood contains red corpuscles incorporating hemoglobin to increase oxygen-carrying capacity.

Respiratory Adaptations. The development of a system of tubes for moving fluid through the body (as contrasted to exchange of fluid with an external medium) would serve no purpose if the fluid could not be oxygenated. In some primitive systems the movement of blood increases exchange with the external fluid, but there is no specific oxygenating organ. Once the system becomes closed, and once there is a true circulation in which fluid moves around a vascular loop and returns to the same place without appreciable exchange with the external environment, specialized adaptations are necessary for exchange with the environment. In some cases the skin serves as an oxygenating organ. In the earthworm, capillaries run extensively to the

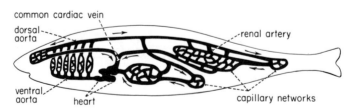

Figure 1–8 Chordate. In the chordates the circulatory system includes a well-developed muscular heart, which lies ventral to the digestive tract. Each contraction propels the blood through a closed system comprised of arteries, capillaries, and veins. The blood moves from the heart into the gills, from which most of the blood then perfuses the various vascular beds of the systemic circulation. The blood then returns to the heart through the venous system. Note that the gills are in series with the systemic circulation, whereas in later developments in the vertebrates the gills constitute a separate circulation.

skin and return oxygenated blood to the central circulation. Even where other exchange mechanisms exist, as they do in frogs, gas exchange through the skin may constitute a sizable portion of the total exchange, and excretion of carbon dioxide by this route may be close to that through the lungs. Since dissolved gases move most rapidly in an aqueous medium, the skin is most effective as a gas exchange organ when it is wet.

Further capability accompanies the development of gills. These are respiratory organs, often with a high surface area to mass ratio, through which blood flows. They are exposed on one side to the blood and on the other side to the external, hopefully oxygenated fluid. The movement of the blood through gills may be aided by the beating of cilia or by other means. Sometimes special appendages move fluid past the gills; sometimes the gills themselves move or fluid is moved past the gill surfaces by swimming movements.

The lungs are a more advanced respiratory structure. Here the principle of compartmentation is extensively applied through the development of small (air) spaces surrounded by blood in capillaries, so that within a small volume there may be a very extensive surface area. Some early lungs are filled with water, but air lungs are more common. The frog has an elementary type of lung in which there is not much compartmentation; it is like a simple bag with air on one side and circulatory fluid on the other side. Early forms of lungs which do not expand or contract furnish increased area for diffusion. In more advanced forms the lungs are actively ventilated by the organism. Sometimes the organism swallows air; sometimes there is movement of the chest cavity. The frog forces air into his chest with movements of the mouth; in mammals, increasing chest size by movement of respiratory muscles draws air into the lungs. There is now a blood pump and an air pump. This makes for a most efficient transfer of oxygen from the air to the cells.

Circulatory Fluids. Oxygen dissolved in aqueous fluids is a relatively meager source of oxygen for metabolism, since the solubility is not high. A number of organic pigments, particularly several containing iron such as hemoglobin, have the capability to combine with large amounts of oxygen, and if these pigments are added to circulatory fluids the ability to transport oxygen is markedly increased. Some pigments of this sort are seen in coelomic fluids of earthworms. As the circulatory system develops, these fluids tend to be concentrated in specialized cells within the blood. Although it is discussed last here, the utilization of specialized oxygen-binding pigments is probably a more primary adaptation than the development of a circulatory system or of a respiratory system.

The blood plasma and the interstitial fluid are similar in composition, although not always in osmolarity, to the primordial sea water in which life evolved and to the coelomic fluid of the primitive forms. Blood is vastly improved over these as an oxygen carrier. The thick integument (plus the reabsorptive ability of the kidneys) prevents the loss of fluid to the external environment. Animals now carry their own extracellular fluid, its composition and osmolarity regulated by heart, lungs, kidneys and liver.

Further Developments. In the invertebrates and lower forms, all the units which make up the mammalian circulatory system have appeared in at least elementary form. In further development, the heart develops as a single, stronger pump, nervous control of the heart and blood vessels emerges, and a diagram of the system would now resemble a diagram of the system seen in mammals. However, in many invertebrate forms the circulatory system operates at a much lower pressure than the system in mammals. The systems in higher invertebrates are intermediate between the open circulatory systems of the lower animals, which function near atmospheric pressure, and the mammalian system, which functions at a mean arterial pressure near 100 mm Hg.

The independence and mobility of terrestrial animals are made possible by the development of the blood as a transporting fluid, by the extensive ramifications of the capillaries which bring oxygenated blood into virtually direct contact with the extracellular space surrounding each cell, by the lungs whose pulmonary capillaries replenish the oxygen of the blood, and by the cardiovascular system of pumps and tubes. Although all of these are important, and although adequate function would be impossible without each of them, the lungs make it uniquely possible to move freely out of the aqueous medium and onto land, and the high pressure arterial system seems to bestow on man and other mammals some of the unique capabilities which they display to move freely and adapt to changes in the external world.

THE MAMMALIAN CARDIOVASCULAR SYSTEM

The mammalian system with which we are concerned consists of a four-chambered heart of which one atrium and one ventricle, the left, are in the systemic circulation, whereas the right atrium and ventricle are part of the pulmonary circulation (Fig. 1–1).

The Systemic Circulation. Arterial blood from the left ventricle is pumped into the thick-walled rigid pressure reservoir, the aorta. Thence it goes to a system of parallel distributing arteries which bring it to the organs of the systemic circulation. In the organs of the systemic circulation, the blood next passes through arteriolar vessels whose caliber can be altered by nervous or metabolic control. Changes in arteriolar caliber can regulate the overall pressure and flow in the systemic circuit and/or shift the blood from one organ system to another. Each organ exhibits a characteristic arteriolar control pattern. The arterioles lead into the capillaries where the ultimate exchange between the blood and the extracellular spaces takes place. In the capillaries, blood loses some of the oxygen which was taken up in the lungs, and the oxygen passes through capillary walls into the extracellular space. Other metabolites are also transported from the blood to the extracellular space in the capillaries. The cells are nourished by exchange with this extracellular fluid, and some cellular metabolic end products and other materials pass into the blood from the extracellular spaces. The blood, containing less oxygen and now called venous blood, enters the venules and moves to the veins, which function as conduits and a storage system for the blood. The major veins unite to form the two venae cavae. From these the blood then enters the right atrium.

The Pulmonary Circulation. Venous blood flows from the right atrium to the right ventricle, which pumps the blood into the pulmonary artery, whence it goes directly into smaller arteries and from these into the pulmonary capillaries. Here, loss of carbon dioxide and uptake of oxygen occur by diffusion across the lungs into the alveolar air. From this point the oxygenated blood enters the pulmonary veins, which return it to the left atrium. The pulmonary circulation does not have the extensive neural control of arteriolar caliber seen in the systemic circulation. This undoubtedly reflects its unitary function. In the systemic circulation there may be a need to shift blood as organs change their activity. Obviously this is not true of the pulmonary circulation. From the pulmonary capillaries, blood flows into pulmonary veins, then into the left atrium and returns to the starting point, the left ventricle. A simplified representation of the circulatory system as a physical array of tubes and pumps is shown in Figure 1–9.

ORGANIZATION OF CARDIOVASCULAR CHAPTERS

As indicated above, our cardiovascular system is closed and dual. As shown in Figure 1–9, we can separate the two pumps and diagram the two systems in series. Clearly, such a system has neither beginning nor end. If we cut the line of blood flow completely across at any level, flow will soon cease everywhere. The circularity of the system has scientific, clinical and educational effects.

The fact that the cardiovascular system is closed and circular gives us both freedom and problems in organizing a text. For instance, we can start with a discussion of capillaries or of the heart and follow the course of the blood through the system and arrive back at the starting point. Almost any order of treatment is thus possible, and yet any subject treated early will be physiologically affected by those treated later. The organization of Volume II is described below.

This chapter is followed by Chapter 2, *Physics of the Cardiovascular System*, which is devoted to those physical factors important for a study of cardiovascular physiology, and by a related chapter (3) which discusses the *Measurement of Blood Pressure and Blood Flow* made to characterize the cardiovascular system and the techniques which are used to make these measurements.

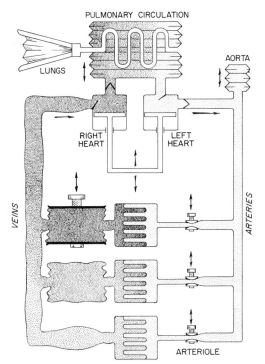

Figure 1–9 Representation of the cardiovascular system as an assemblage of pumps and tubes. Blood enters the right heart from the veins, and is pumped into the lungs by the right heart. In the lungs, the blood is oxygenated and from there it passes into the left heart. The left heart pump, which is coupled to the right heart, pushes the blood into the aorta. The aorta has attached to it a stiff diaphragm, which serves as a pressure reservoir. Blood flows from the arteries into the various systemic vessels through the arterioles, vessels of alterable caliber. In the various organs, shown as square boxes, the blood loses oxygen and acquires carbon dioxide, going from a light color to a dark color. The venous system includes several different types of compliant vessels (some encapsulated, some with a large capacity and some with a small capacity). Blood in the venous system has a mixed oxygen and carbon dioxide content, which is determined by the various organs. The blood returns again to the right heart through the venous system.

Several chapters then lead into a detailed discussion of the heart. These include Chapter 4, *Cardiac Muscle Mechanics*, and Chapter 5, *Electrical Characteristics of the Cardiac Cell*. These two chapters characterize the unique muscle of which the heart is composed. They are followed by Chapter 6, *Electrocardiogram*, which discusses the electrical activity of the heart and the origin of the electrocardiogram. Chapter 7, *Mechanical Events of the Cardiac Cycle*, discusses the sequence of electrical events, changes in pressure, flow, valve opening and closure, and heart sounds which occur in the heart chambers. These four chapters make it possible to understand the behavior of the heart and the sequence of contraction of the various chambers, opening and closure of the valves, etc.

The emphasis then shifts from the heart to the peripheral circulation. Chapter 8, *The Arterial System*, discusses and indicates the functions of the larger vessels and the manner in which the arterial vessels act to regulate the cardiovascular system. This is followed by Chapter 9, *The Capillaries, Veins and Lymphatics*, which deals largely with the microcirculation, exchange of material between blood and tissue, and the return of blood to the heart.

The control of the cardiovascular system is discussed in two chapters: Chapter 10, *Control of Arterial Blood Pressure*, and Chapter 11, *Control of Cardiac Output.*

A number of separate chapters consider special characteristics: Chapter 12, *The Cutaneous Circulation;* Chapter 13, *Circulation to Skeletal Muscle*; Chapter 14, *The Splanchnic Circulation*; Chapter 15, *The Renal Circulation*; Chapter 16, *The Coronary Circulation*; and Chapter 17, *The Cerebral Circulation*. Finally, two chapters discuss *The Pulmonary Circulation* (18) and the *Coagulation of Blood* (19).

The remaining chapters (20 through 29) are divided into Section II, *Respiration;* Section III, *Biophysics of Transport Across Membranes;* and Section IV, *Kidney Function and Body Fluids.*

This chapter indicates how evolutionary forces may have shaped the cardiovascular system to favor the development of large and active organisms. The chapters listed above indicate how the highly developed mammalian cardiovascular system functions to meet the demands imposed upon it by a variety of external and internal factors.

REFERENCES

1. Florey, E. *An introduction to general and comparative animal physiology*. Philadelphia, W. B. Saunders Co., 1966.
2. Hill, A. V. The diffusion of oxygen and lactic acid through tissues. *Proc. Roy. Soc.*, 1928, *B104*, 39–96.
3. Jacobs, M. H. *Diffusion processes*. New York, Springer-Verlag, 1967.
4. Martin, A. W., and Johansen, K. Adaptations of the circulation in invertebrate animals. *Handb. Physiol.*, 1965, sec. 2, vol. III, 2545–2581.
5. Prosser, C. L., and Brown, F. A., Jr. *Comparative animal physiology*, 2nd ed. Philadelphia, W. B. Saunders Co., 1961.

CHAPTER 2   PHYSICS OF THE CARDIOVASCULAR SYSTEM

by ERIC O. FEIGL

The cardiovascular system is a complex network of branched tubes (blood vessels) supplied by a pump (the heart). It is natural to begin the study of the circulation with a review of the physical principles which apply to the motion of liquids in rigid cylindrical tubes, since the same general principles apply to the more complicated physiological situation. Our intention is to make the physical principles relevant to the reader by presenting the physics and its cardiovascular manifestation together in the same section. For example, the problem of blood pressure measurement with changes in posture is presented together with hydrostatics.

FLUIDS. A fluid is a material that flows. It is a substance that deforms to a shearing force but does not tend to return to its original form when the stress is removed as a solid does. There are no elastic forces in liquids which tend to restore their original unstressed shape. Fluids are found in two forms, gases and liquids. Gases are readily compressed while liquids are not. A gas expands to entirely fill the container that encloses it. We are all aware that liquids in a gravitational field tend to flow downhill, and that they occupy the bottom of a container, assuming the shape of the vessel they are held in.

PRESSURE AND HYDROSTATICS

Force in a liquid system is manifest as pressure. In the cardiovascular system the effects of pressure may be observed in two major ways: first, the pressure differences move blood through the blood vessels; and second, pressure can distend the cardiac chambers and arteries and veins.

Pressure is expressed as force per unit area, and the appropriate basic unit is dyne/cm².* Pressure is often given in terms of the height of a liquid column in a U-tube (Fig. 2–1). The first measurement of arterial pressure was made by Stephen Hales in 1733. He attached a vertical glass tube with a brass fitting to the femoral artery of a horse and observed that the blood in the tube rose to a height of 8 feet 3 inches above the level of the left ventricle. The pressure of 8 feet 3 inches of blood is equal to 188 mm of mercury. Arterial pressure is usually expressed in millimeters of mercury (mm Hg). Mercury is chosen for its high density,

---

*Biologists sometimes use the less rigorous units of grams-force per cm² for pressure. A gram is a unit of mass, not force, and gram-force means the force that 1 gram exerts in the gravitational field (1 gram-force = 1 gram times the gravitational acceleration).

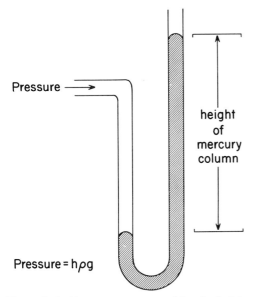

Figure 2–1 Pressure is measured by the height of the mercury column in a U-tube manometer. The pressure is equal to the height of the mercury column (h) times the density of mercury (ρ) times the gravitational constant (g).

which makes it possible to measure arterial pressure in a relatively short tube.

The pressure in dyne/cm² may be obtained from a liquid height (h) by knowing the density of the liquid (ρ) and the gravitational acceleration (g)

$$dyne/cm^2 = h \cdot \rho \cdot g$$
$$dyne/cm^2 = cm \cdot gm/cm^3 \cdot cm/sec^2$$

(remembering the definition of dyne = $gm \cdot cm/sec^2$).

The pressure at a given depth in a liquid is equal in all directions. The shape of the vessel does not alter this: The pressure h cm below the surface of the liquid is equal to

$P = h \cdot \rho \cdot g$ whether the vessel has straight, sloping or curved sides (Fig. 2–2).

Physiologists and physicians use the ambient atmospheric pressure as the zero reference point when measuring pressures in the cardiovascular system. Thus a blood pressure of 90 mm Hg means the pressure is 90 mm Hg above atmospheric pressure.

The hydrostatic pressure caused by the height of a liquid column is important in physiology when posture is considered. The arterial transmural pressure in the feet of a person lying down is essentially the same as that at the root of the aorta. When an individual stands, the additional pressure caused by the column of blood in the arterial tree above the feet is added to the arterial pressure generated by the heart (Fig. 2–3). If the added height of the fluid column is 130 cm, then the additional transmural pressure will be 100 mm Hg. (The density of mercury is approximately 13 times that of blood.) This points up the necessity of an additional convention in measuring cardiovascular pressures—a zero reference level. Unless otherwise specified, the zero reference pressure level for the cardiovascular system is the position of the heart. Arterial diaphragm manometers (see Chap. 4, Vol. II) are usually placed at heart level. The sphygmomanometer cuff (see Chap. 4, Vol. II) measures transmural pressure, and it is important that the artery under the cuff be at the level of the heart or that a correction be made. The usual convention is to measure blood pressure in the brachial artery above the elbow—approximately at heart level when the patient is seated. Pressure measurements in the legs are usually made with the patient lying down so that the vessel is approximately at cardiac level.

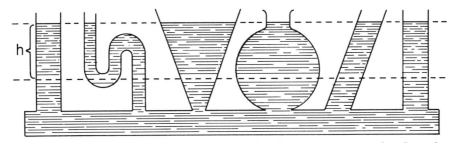

Figure 2–2 The pressure at a given depth (h) in the liquid is the same no matter what shape the vessel. This is intuitively confirmed by the observation that the surface liquid level is the same in all the chambers.

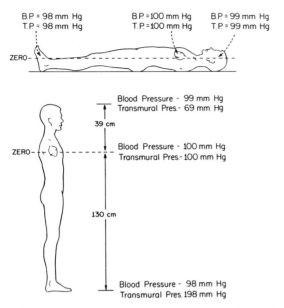

Figure 2–3 The transmural pressure across the wall of a blood vessel is altered by the height of the column of blood above it. In this diagrammatic representation the mean blood pressure at the level of the heart is 100 mm Hg, 99 mm Hg in the head and 98 mm Hg in the feet. With the subject lying down, the head, heart and feet are all at the same level, and the transmural pressure equals the blood pressure which is referenced to the cardiac level. When the subject is standing, the transmural pressure is greater below the heart and less above the heart.

FLOW AND FLOW VELOCITY

The blood flow to an organ is the blood delivered by the arteries to that organ to supply its metabolism. Blood flow in the cardiovascular system is expressed in terms of volume of blood per time. For example, the cardiac output in resting man is about 5 liters per min.

Blood flow *velocity* (linear velocity) is the displacement in time of a particle of blood and is usually expressed in centimeters per second. The average flow velocity times the cross-sectional area of the tube equals the flow: $F = \bar{v} \cdot A$. For the case in which an incompressible liquid (such as blood) flows in an unbranched tube of variable cross-section, the flow will be equal at all cross-sections but the average velocity will vary according to the relation, $\bar{v} = F/A$ (Fig. 2–4).

The cardiovascular system is a branched network of tubes in which the characteristic branching pattern is that the total cross-sectional area of the branches is larger than the cross-sectional area of the parent trunk. This means that the total cross-sectional area of the cardiovascular system increases as the aorta branches to arteries, arterioles and capillaries. Correspondingly, the average blood flow velocity in the small vessels is less than in large arteries or veins (Fig. 2–5).

RELATION BETWEEN PRESSURE AND VELOCITY

Flowing liquids have kinetic energy because of their mass. The law of conservation of energy applied to flowing fluids is described by Bernoulli's equation. For simplicity, consider a liquid without viscosity flowing in a horizontal tube. This permits us to neglect energy dissipated as heat (viscous frictional losses) and changes in energy associated with raising or lowering the liquid in the gravitational field. The energy of the flowing liquid will be given by the sum of the potential energy (pressure) and the kinetic energy ($1/2\ \rho\ v^2$). Bernoulli's equation states that the total energy will be constant.

$$P + 1/2\ \rho v^2 = \text{constant}$$

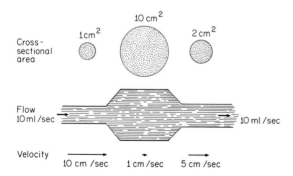

Figure 2–4 Flow of an incompressible liquid in a rigid tube with a variable cross-sectional area. The total flow is 10 ml per sec and is the same for each segment of the tube. The linear flow velocity must be greater in narrow segments to maintain a constant flow of 10 ml per sec.

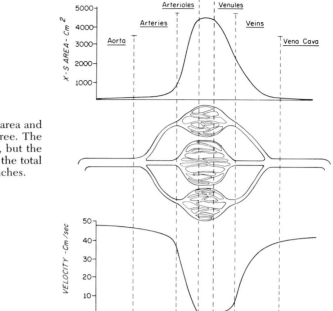

Figure 2–5 The estimated cross-sectional area and linear blood velocity for the cardiovascular tree. The total flow is the same throughout the system, but the linear velocity in a segment is dependent on the total cross-sectional area of the many parallel branches.

where P = pressure; ρ = density; v = velocity. An equivalent statement of Bernoulli's equation is that the energy will be equal at two different points in a tube.

$$P_1 + 1/2\ \rho v_1{}^2 = P_2 + 1/2\ \rho v_2{}^2$$

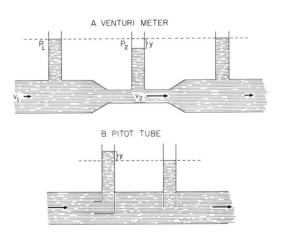

Figure 2–6 The flow of an incompressible liquid without viscosity is illustrated in a Venturi meter and a Pitot tube. A, The linear velocity is greater in the narrow segment, which results in a greater proportion of the fluid energy in the kinetic term $(1/2mv^2)$ and less in the side pressure $(P_2 < P_1)$. B, The curved Pitot tube stops the flowing liquid, thus converting kinetic energy to pressure; end-on pressure is greater than lateral pressure, as indicated by the difference (y) in the two columns.

This relation explains the Venturi meter shown in Figure 2–6A. The fluid velocity and thus the kinetic energy are greater in the constricted segment of the tube, and since the total energy is constant the side pressure will be less.

The pressures in the Pitot tube shown in Figure 2–6B are also readily understood from the Bernoulli equation. With the curved pitot tube facing the liquid stream, the fluid velocity is stopped at its entrance and kinetic energy is converted to pressure. Thus end pressure facing the flow is greater than side pressure.

The problem of end and side pressures has practical application when measuring arterial pressure with a catheter. If the catheter has an end hole and is facing the flow, the kinetic energy of the flowing blood will add to the measured pressure. Blood flow velocity is greatest in the ascending aorta where the difference between end and side hole catheter pressures is a few millimeters of mercury. The effect is less at other sites in the cardiovascular system.

VISCOSITY

Viscosity may be thought of as the internal friction of a fluid. Because of vis-

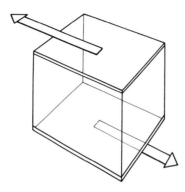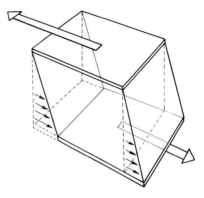

Figure 2–7 A small cube of material shown undistorted on the left is subjected to a shear stress which results in a shear strain shown on the right. The distortion can be envisioned as slippage between many parallel lamellae.

cosity, a force must be exerted to move one layer of fluid past another. Laminar flow of a liquid is analogous to shearing strain in a solid (Fig. 2–7). Ideal liquids do not manifest elastic restoring forces when deformed, and it can be observed that force necessary to shear a liquid is proportional to the shear rate or velocity. Viscosity is analogous to friction, and the energy used to overcome viscosity is degraded to heat. Viscosity is the proportionality constant between shearing force per unit area (the shear stress) and the rate of shear and has the units dyne-sec/cm^2. A viscosity of 1 dyne-sec/cm^2 is called a poise. Water at 20° C has a viscosity of 0.01 poise or 1.0 centipoise (cp). The viscosity of blood is 2 to 4 times the viscosity of water. Fluids which manifest a constant viscosity with varying shear rates are said to be "ideal" or "Newtonian." As discussed below, the viscosity of blood is shear rate dependent and is not constant, and this is related to the fact that blood is a suspension of red cells in plasma.

RELATION BETWEEN PRESSURE AND FLOW

A pressure gradient is needed to make a viscous liquid flow. With non-turbulent laminar flow the flow rate is proportional to the pressure gradient, which may be defined as the pressure drop over a given length ($\Delta P/\Delta L$). For a tube with uniform radius the pressure gradient may be determined from the inlet pressure minus the outlet pressure divided by the tube length (Fig. 2–8). The simple relation between pressure head and tube length shown in Figure 2–8A to E does not include the effects of tube radius. The resistance to fluid flow is strongly dependent on the tube radius as discussed below. In Figure 2–8F, the pressure gradient across the narrow tube segment is greater than it is across the larger tube segment.

A mean pressure gradient may be observed in the vascular system from arteries to capillaries to veins and is shown schematically in Figure 2–9. The vascular tree has many parallel branches for blood flow and is not strictly comparable to the single tube shown in Figure 2–8. However, the principle that the pressure gradient produces flow is valid for the vascular system. The most important point in Figure 2–9 is that the pressure gradient is steepest across the arterioles. This is the major site of resistance in the vascular tree. Arterioles are not the vessels with the smallest radius (the capillaries are clearly the narrowest vessels); rather, arterioles are narrow compared to the vessels that feed them, and there are few arterioles compared to the many capillaries that follow them. Thus in an "average" way the arterioles represent the dominant site of vascular resistance.

RESISTANCE, IMPEDANCE AND THE POISEUILLE EQUATION

RESISTANCE. Hemodynamic resistance is simply the hindrance to the flow of viscous blood provided by the vessels. The

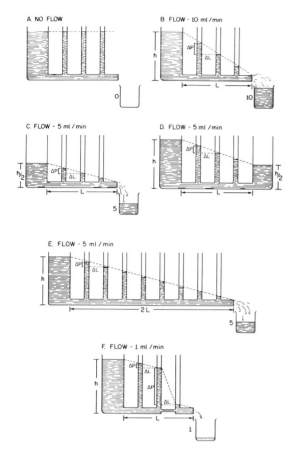

Figure 2–8 The relation between pressure gradient ($\Delta P/\Delta L$) and flow is illustrated for a viscous liquid. *A,* When the end of the tube is closed off, there is no flow and no pressure gradient. *B,* For a pressure head h over a length L, the flow is 10 ml per min. *C,* When the height of the pressure reservoir is halved and the length of the tube is constant the pressure gradient and the flow are also halved. *D,* The pressure gradient is halved by increasing the outflow pressure rather than decreasing the inflow pressure. However, the result is the same, for a halving of the pressure gradient the flow is reduced by one-half. *E,* The pressure gradient is reduced to half by doubling the tube length with the original reservoir height. *F,* The importance of the tube radius in determining the pressure gradient and limiting flow is illustrated.

resistance to flow in a tube is primarily determined by the radius. The smaller the radius, the greater the resistance to flow.

Resistance is expressed in the relationship between pressure gradient ($P_1 - P_2$) and flow (F).

$$R = \frac{P_1 - P_2}{F}$$

This is equivalent to Ohm's law in an electrical circuit (E = IR). The unit of fluid resistance is dyne-sec/cm^5 in the centimeter-gram-second system. The hydrodynamic resistance of a tube is a function of its dimensions and also the viscosity of the fluid flowing through it as shown in the Poiseuille equation.

POISEUILLE EQUATION. Poiseuille was a 19th-century French physician who studied liquid flow in small tubes and experimentally discovered that the flow, for a given pressure gradient, is proportional to the fourth power of the tube radius and inversely proportional to fluid viscosity. The unit of viscosity (poise) is named in his honor. The Poiseuille equation for ideal fluids (constant viscosity) with laminar, smooth, steady (parabolic) flow through a cylindrical tube may be derived as given below. The derivation is presented because of the importance of vascular radius in controlling blood flow and to identify the physical forces involved in viscous flow.

Consider a cylindrical tube of length ℓ and internal radius r_i in which a viscous fluid is moving because of a pressure difference $P_1 - P_2$ (Fig. 2–10). When the flow is steady, the system will be in equilibrium so that the force caused

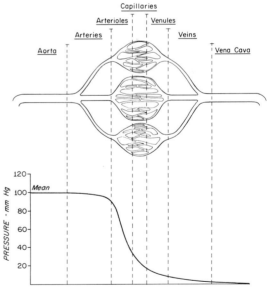

Figure 2–9 Mean pressure in the various vascular segments. The pressure drop from aorta to major arteries is only a few millimeters of mercury. The steep drop in pressure across the arterioles demonstrates that this is the major site of vascular resistance. Although individual capillaries are narrower than arterioles, there are many more of them so that the total resistance of the capillary bed is less than the arteriolar bed.

Steady Laminar Flow

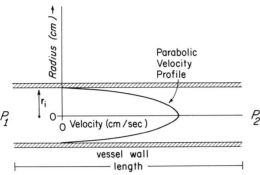

Figure 2-10 A plot of a parabolic velocity profile superimposed on a vessel. In ideal laminar flow the velocity at the inner wall of the tube (r_i) is zero and is maximum at the center of the tube. Flow is shown from left to right with a pressure gradient $P_1 > P_2$. The average velocity will be half the maximal velocity found at the central axis of the tube.

by the pressure is balanced by the shearing viscous force in the fluid. The force caused by pressure will be the pressure difference times the area over which the pressure is manifest.

$$F_{(Press.)} = (P_1 - P_2)\, \pi\, r^2 \qquad (1)$$

The retarding viscous force will be equal to the viscosity η (eta) times the surface area of the cylinder of fluid ($2\,\pi\,r\,\ell$) times the velocity gradient $\left(\dfrac{dv}{dr}\right)$

$$F_{(Vis)} = 2\,\pi\,r\,\ell\,\eta\,\frac{dv}{dr} \qquad (2)$$

Setting the pressure and viscous forces equal and opposite

$$(P_1 - P_2)\, \pi\, r^2 = -2\,\pi\,r\,\ell\,\eta\,\frac{dv}{dr} \qquad (3)$$

Rearranging for the velocity gradient

$$\frac{dv}{dr} = -\frac{r\,(P_1 - P_2)}{2\,\ell\,\eta} \qquad (4)$$

Integrating both sides gives the velocity at radius r (v_r)

$$v_r = -\frac{r^2\,(P_1 - P_2)}{4\,\ell\,\eta} + C \qquad (5)$$

where C is the constant of integration. Evaluation of C requires knowledge or an assumption about the boundary conditions of the tube. The usual assumption is that the layer of liquid adjacent to the wall is stationary. That is, at the inside tube radius (r_i) the fluid does not slip over the wall ($v = 0$). Rearranging for C at r_i and $v = 0$, we have

$$C = \frac{r_i^2\,(P_1 - P_2)}{4\,\ell\,\eta} \qquad (6)$$

so that equation (5) becomes

$$v_r = -\frac{r^2\,(P_1 - P_2)}{4\,\ell\,\eta} + \frac{r_i^2\,(P_1 - P_2)}{4\,\ell\,\eta} \qquad (7)$$

$$v_r = \frac{P_1 - P_2}{4\,\ell\,\eta}\,(r_i^2 - r^2)$$

This is the equation for a parabola; thus the velocity profile is parabolic.

The maximum velocity at the central axis of the tube (r = 0) will be given by

$$v_{max} = \frac{(P_1 - P_2)\,r_i^2}{4\,\ell\,\eta} \qquad (8)$$

The average velocity will be half this value.

To obtain the volume flow rate through the tube, another integration is needed. Consider the volume flow rate through an annulus of radius r and width dr and flow velocity v_r

$$\text{annulus volume flow rate} = 2\,\pi\,r\cdot dr\cdot v_r$$

The total volume flow (F) through a cylindrical tube will be the integral from the center of the tube (r = 0) to the inside edge of the tube (r_i)

$$F = 2\,\pi \int_0^{r_i} r\cdot v_r\cdot dr \qquad (9)$$

v_r is given in equation (7) so that

$$F = 2\,\pi \int_0^{r_i} r\,\frac{P_1 - P_2}{4\,\ell\,\eta}\,(r_i^2 - r^2)\,dr \qquad (10)$$

which yields

$$\text{FLOW} = (P_1 - P_2)\,r_i^4\,\frac{\pi}{8\,\ell\,\eta} \qquad (11)$$

This is Poiseuille's equation, which states that the flow through a cylindrical tube is proportional to the pressure gradient and the fourth power of the radius and inversely proportional to the length of the tube and the fluid viscosity. It should be recognized that the radius to the fourth power has immense effects. For example, doubling the radius increases flow sixteen times!

The appropriate centimeter-gram-second units will be F = cm³/sec, P = dynes/cm², r and ℓ = cm and η the viscosity in poise (dyne sec/cm²).

The assumptions for the Poiseuille equation should be remembered. They are:

(i) A long rigid cylindrical tube where the length of tube is long compared to the radius.

(ii) Ideal fluid of constant viscosity where the viscosity is not a function of the shear rate.

(iii) Steady laminar flow that is not pulsatile and not turbulent.

(iv) The fluid velocity at the edge of the tube is zero.

Blood is a suspension of cells and has a viscosity which is shear rate dependent. The flow in the arterial system is pulsatile and may be turbulent. For these reasons the Poiseuille equation must only be considered a useful approximation for the arterial system. However, the great importance of the tube radius is demonstrated. Physiological regulation of the circulation occurs when vascular smooth muscle changes the radius of the arteries and arterioles.

Resistance may be obtained by rearranging the Poiseuille equation to give

$$R = \frac{P_1 - P_2}{F} = \frac{1}{r_i^4} \ell \, \eta \, \frac{8}{\pi}$$

This emphasizes that resistance is a function not only of the tube radius and length but also of the viscosity of the fluid flowing through it. The same tube will manifest a different pressure-flow relation (resistance) for fluids of different viscosities.

The velocity profile of aortic blood flow has been determined[10] and has a flat profile which varies from moment to moment in the cardiac cycle (Fig. 2–11). The velocity profile is flat rather than parabolic for several reasons which will not be discussed in detail but which include: (i) the blood vessels are not long uniform unbranched rigid tubes, (ii) blood has complicated rheological properties (vide infra), and (iii) the flow is pulsatile rather than steady.

PERIPHERAL RESISTANCE UNITS. In physiological circumstances arterial resistance is calculated by measuring pressure and flow. This has led to the definition of a peripheral resistance unit (PRU) which is the proportionality constant between pressure (mm Hg) and flow (ml per min) as physiologists usually measure them. Thus,

$$PRU = \frac{mm \ Hg}{ml/min}$$

Because the same organ varies in size between different individuals and species, it is sometimes convenient to normalize resistance by organ weight to give PRU per 100 grams of tissue. Peripheral resistance is calculated using mean pressure and mean flow and is analogous to DC resistance in an electrical circuit.

SERIES AND PARALLEL RESISTANCE. The resistance of a system of tubes arranged in series will be the total of the individual

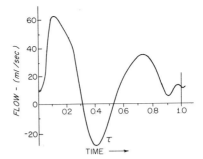
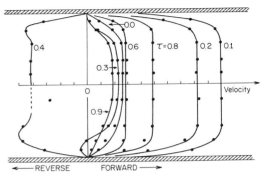

Figure 2–11 The flow (ml per sec) and the flow velocity (cm per sec) profile from a pig's descending thoracic aorta are illustrated. The velocity profile across the vessel is shown at intervals of one-tenth of the cardiac cycle. Note the blunt velocity profile. The example was chosen for a case with large backward flow during a part of the cycle. (After Ling et al., Circulat. Res., 1968, 23, 789–801.)

resistances as illustrated in Figure 2–12. The addition of another resistor in series will always increase the total network resistance.

The resistance of a system of tubes arranged in parallel will be proportional to the reciprocal of the resistances as shown in Figure 2–13. The addition of another resistor in parallel will always decrease the total network resistance.

When dealing with parallel circuits it is convenient to use conductance which is the reciprocal of resistance (C = 1/R or C = F/P).

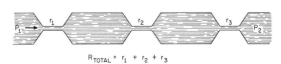

Figure 2–12 Hydraulic resistors in series. The total resistance is the sum of the individual resistance. Adding an additional resistor in series always increases the total resistance.

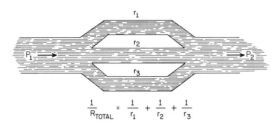

$$\frac{1}{R_{TOTAL}} = \frac{1}{r_1} + \frac{1}{r_2} + \frac{1}{r_3}$$

Figure 2–13 Hydraulic resistors in parallel. The reciprocal of the total resistance is given by adding the reciprocals of the individual resistors. Adding an additional resistor in parallel will always decrease the total resistance.

The total network conductance for the example given in Figure 2–13 can be obtained by adding the conductances which are in parallel, *e.g.*, C = c_1 + c_2 + c_3, which is exactly equivalent to $1/R = 1/r_1 + 1/r_2 + 1/r_3$. The various organs of the systemic circulation are arranged in parallel. The total conductance is the sum of all the separate conductances.

IMPEDANCE. Since the arterial system has elasticity, viscosity and mass, the hindrance to blood flow will be frequency dependent. A careful analysis requires more than a simple steady (DC) resistance calculation. A complex hemodynamic impedance may be determined which is analogous to electrical impedance.[13, 14] The aortic input impedance varies with the cardiovascular state. Typical values of aortic impedance for frequencies from 0.5 to 15 Hz are about 5 per cent of the DC (zero frequency) resistance with a negative phase angle (Fig. 2–14). The negative phase angle

indicates the capacitive property of the arterial tree.[14]

TURBULENCE

In streamlined flow, the layers of fluid slip smoothly over one another, and fluid particles remain in a particular streamline as shown in Figure 2–15. When turbulence occurs the fluid is filled with eddies and whirls, and the fluid particles move from one region of the tube to another in an irregular fashion. This violent mixing of the fluid consumes energy; thus turbulent flow requires a greater pressure gradient than streamline flow.

Under the conditions specified for the Poiseuille equation (steady flow, etc.) the critical velocity for turbulence may be predicted using the non-dimensional Reynolds number

$$\text{Reynolds number} = \frac{\bar{v}D\rho}{\eta}$$

where $\bar{v}$ is the average flow velocity across the tube, D is the internal diameter of the tube, ρ the fluid density, and η the fluid viscosity. When the Reynolds number is expressed in centimeter-gram-second units, it is found that the critical value for turbulence is approximately 2000 for many fluids, including blood.[5] Reynolds numbers of over 30,000 have been achieved without turbulence with special precautions to

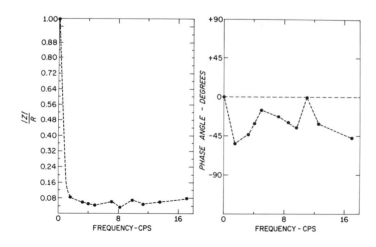

Figure 2–14 The input impedance to aorta in dogs. The value is normalized with respect to the DC mean resistance. Mean resistance is large compared to the phasic components. The negative phase angles indicate the capacitive properties of the arterial tree. (From Patel *et al.*, *J. appl. Physiol.*, 1963, *18*, 134–140.)

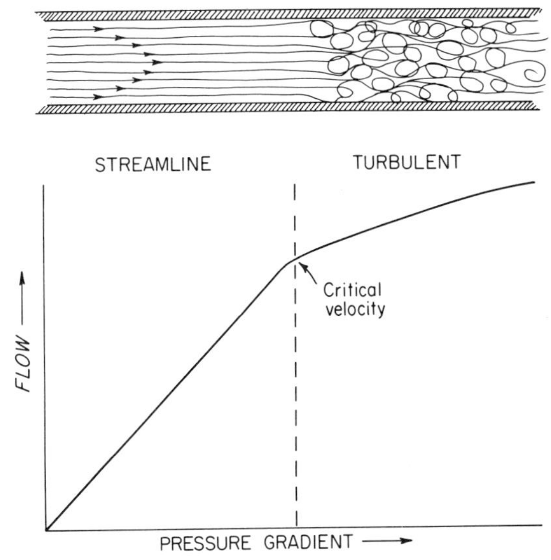

Figure 2–15 The linear relation between pressure gradient and flow for a liquid with constant viscosity is illustrated. Beyond a critical velocity, turbulence begins and the relation between pressure gradient and flow is no longer linear.

have a gently rounded tube entrance and flow from a very still reservoir. Thus a Reynolds number of 2000 should be considered a lower limit, below which turbulence is unlikely in tubes with steady flow.

In practice turbulence begins with slight variations in pressure or flow or at constrictions or irregularities of the tube. The Reynolds number predicts that turbulence will occur in tubes with large radii at lower velocities than tubes with small radii. However, if a constant flow goes through a constriction, the velocity will increase in relation to the area ($F = \bar{v}A$) which is related to radius squared so that there will be an increase in the Reynolds number.

Reynolds numbers have been calculated for the vascular system and give values above 2000 for the proximal aorta and values below 2000 for more distal sites.[11] Other evidence indicates that there is disturbed or turbulent flow in the ascending aorta,[7, 10] with relatively non-turbulent flow beyond the aortic arch. Local turbulence in the form of eddies may occur at arterial branchings — a condition which may be exacerbated by arteriosclerotic disease.

BLOOD VISCOSITY

Thus far this chapter has dealt with the physics of an ideal liquid, such as water, which manifests a constant viscosity over a wide range of shear rates. Blood, because it is a suspension of cells in plasma containing large protein molecules, does not have a constant viscosity at different shear rates. The viscous properties of blood are complex and incompletely understood. Our intention is to introduce the relevant factors without detailed analysis. Serious students should consult reviews and monographs on the subject.[2, 4, 9, 12, 17]

YIELD SHEAR STRESS. Blood, unlike water or plasma, does not flow at low levels of applied shearing force. A definite shear stress must be applied before blood will begin to flow and exhibit the properties of a viscous liquid. The critical shear stress necessary to initiate flow is called the yield shear stress. The presence of a yield stress for blood appears to be dependent on the interaction of red cells and plasma fibrinogen (native fibrinogen, not fibrin). Blood with abnormally low fibrinogen or a low hematocrit exhibits a low yield shear stress.[12]

HEMATOCRIT. The concentration of red cells in blood alters the viscosity of the fluid. The hematocrit is the percentage volume of blood occupied by red blood cells. Polycythemia (an increase in hematocrit) increases blood viscosity more in large tubes than in small tubes which approximate the size of the vessels of the microcirculation (Fig. 2–16). When blood passes through

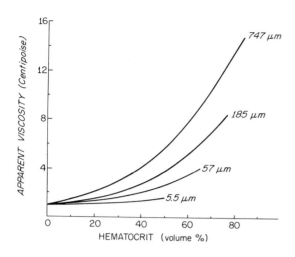

Figure 2–16 The effect of blood hematocrit on viscosity is illustrated in tubes with radii of 5.5, 57, 185, and 747 μm. The viscosity of blood is elevated at higher hematocrits. (After Haynes, *Trans. Soc. Rheol.*, 1961, 5, 85–101.)

capillaries of 4 to 7 μm diameter, where red cells must be deformed, the resistance to flow may be more related to the viscosity of plasma and the deformability of red cells[15, 16] than to the cell concentration.

ERYTHROCYTE DEFORMABILITY. In the living circulation, the biconcave disc red cells, with a diameter of approximately 7 μm, squeeze through capillaries of 4 μm to 8 μm diameter. In cine films of the microcirculation, red cells may be observed folding up in the fashion of partially closed umbrellas as they squeeze through the narrow capillaries. The red cell resistance to deformation is a factor in the overall "viscosity" of blood flow. Blood containing erythrocytes which have been hardened with formaldehyde shows increased resist-

ance to flow through small pores, compared with normal red blood cells.[3, 8]

SHEAR RATE. An "ideal" or "Newtonian" fluid is one where the viscosity is a constant, and is independent of how rapidly the fluid is sheared. Whole blood viscosity is clearly shear rate dependent—the viscosity decreasing with increasing shear rates (Fig. 2–17). Plasma without red cells behaves as a Newtonian fluid.[12]

The anomalous (non-ideal) viscosity of whole blood is thought to be related to the axial streaming of red cells. At slow rates of flow, red cells tend to be dispersed across the entire width of the tube, turning and colliding frequently. At higher (but still non-turbulent) flow rates, the red cells stream in the center of the tube, leaving a

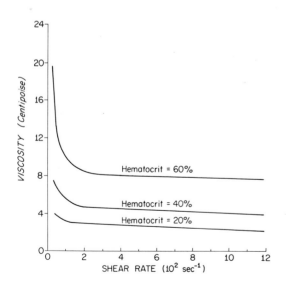

Figure 2–17 The viscosity of human blood as a function of shear rate at three different hematocrits. Blood viscosity is shear rate dependent and thus is a non-ideal (non-Newtonian) liquid. (After Haynes, *Trans. Soc. Rheol.*, 1961, 5, 85–101.)

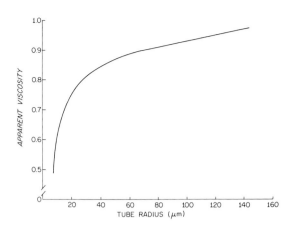

Figure 2–18 The Fåhraeus-Lindquist effect of tube diameter on blood viscosity is illustrated. The apparent viscosity of blood is less in small tubes of arteriolar and capillary size. (After Bayliss, In: *Deformation and Flow in Biological Systems*, Amsterdam, North-Holland Pub. Co., 1952.)

relatively cell-free plasma sleeve near the margin of the vessel. The axial streaming of red cells is due to the velocity gradient across the tube. The faster flow velocity near the center of the tube imparts a spin to a suspended red cell which makes it migrate toward the center of the tube. The effect is somewhat similar to making a baseball "curve" by giving it a spin when it is thrown. The steeper velocity gradient with higher flow rates magnifies this effect until it becomes limited by the increasing frequency of cell collisions in the axial stream. The development of red cell axial streaming as flow rate increases is associated with a progressive decrease in apparent viscosity, until a maximal axial orientation is reached. At still higher shear rates, blood manifests a constant viscosity.

TUBE DIAMETER. The apparent viscosity of blood decreases with decreasing tube diameter in tubes smaller than 200 μm diameter. This is the Fåhraeus-Lindquist[6] effect which is due to suspended red cells in blood (Fig. 2–18). The explanation is uncertain, although similar phenomena are observed in suspensions of clay and in paints. One explanation is that when the tube size is not much larger than the particle size, the fluid flows in discrete, relatively unsheared columns behind the particles, rather than in an infinite number of lamellae distributed across the tube. A mathematical model has been derived which sums up the several columns across the tube to predict the flow characteristics of suspensions in small tubes. The phenomenon of a decreased viscosity of suspensions in small tubes is also called the sigma effect,

because the mathematical model makes a summation across the tube.[1]

TEMPERATURE. The viscosity of many liquids, including water, is temperature dependent, the viscosity decreasing with increasing temperature. The temperature coefficient for the viscosity of blood is the same as that for water in the range 10 to 37° C. The yield stress of blood is not dependent on temperature (15 to 37° C) in the majority of human subjects.[12]

REFERENCES

1. Bayliss, L. E. Rheology of blood and lymph. In: *Deformation and flow in biological systems*, A. Frey-Wyssling, ed. Amsterdam, North-Holland Publ. Co., 1952.
2. Bayliss, L. E. The rheology of blood, *Handb. Physiol.*, 1962, sec. 2, vol. I, 137–150.
3. Braasch, D. Red cell deformability and capillary blood flow. *Physiol. Rev.*, 1971, *51*, 679–701.
4. Copley, A. L., ed. *First International Conference on Hemorheology.* Oxford, Pergamon Press, 1968.
5. Coulter, N. A., Jr., and Pappenheimer, J. R. Development of turbulence in flowing blood. *Amer. J. Physiol.*, 1949, *159*, 401–408.
6. Fåhraeus, R., and Lindquist, T. The viscosity of the blood in narrow capillary tubes. *Amer. J. Physiol.*, 1931, *96*, 562–568.
7. Freis, E. D., and Heath, W. C. Hydrodynamics of aortic blood flow. *Circulat. Res.*, 1964, *14*, 105–116.
8. Ham, T. H., Dunn, R. F., Sayre, R. W., and Murphy, J. R. Physical properties of red cells as related to effects *in vivo*. I. Increased rigidity of erythrocytes as measured by viscosity of cells altered by chemical fixation, sickling and hypertonicity. *Blood*, 1968, *32*, 847–861.
9. Haynes, R. H. The rheology of blood. *Trans. Soc. Rheol.*, 1961, *5*, 85–101.
10. Ling, S. C., Atabek, H. B., Fry, D. L., Patel, D. J.,

and Janicki, J. S. Application of heated-film velocity and shear probes to hemodynamic studies. *Circulat. Res.*, 1968, *23*, 789–801.

11. McDonald, D. A. *Blood flow in arteries.* Baltimore, Williams & Wilkins Co., 1960.

12. Merrill, E. W. Rheology of blood. *Physiol. Rev.*, 1969, *49*, 863–888.

13. O'Rourke, M. F., and Taylor, M. G. Input impedance of the systemic circulation. *Circulat. Res.*, 1967, *20*, 365–380.

14. Patel, D. J., de Freitas, F. M., and Fry, D. L. Hydraulic input impedance to aorta and pulmonary artery in dogs. *J. appl. Physiol.*, 1963, *18*, 134–140.

15. Prothero, J. W., and Burton, A. C. The physics of blood flow in capillaries. II. The capillary resistance to flow. *Biophys. J.*, 1962, *2*, 199–211.

16. Prothero, J. W., and Burton, A. C. The physics of blood flow in capillaries. III. The pressure required to deform erythrocytes in acid-citrate-dextrose. *Biophys. J.*, 1962, *2*, 213–222.

17. Whitmore, R. L. *Rheology of the circulation.* Oxford, Pergamon Press, 1968.

CHAPTER 3 MEASUREMENT OF BLOOD PRESSURE AND BLOOD FLOW

by ERIC O. FEIGL

BLOOD PRESSURE MEASUREMENT

Arterial pressure is expressed relative to ambient atmospheric pressure. That is, atmospheric pressure is taken as the zero pressure for arterial measurements. The anatomic reference point for arterial pressure is the level of the heart. Arterial pressure measurements should be corrected or measured at the level of the heart lest the height of the blood column above or below the heart decrease or add to the observed pressure.

Direct Arterial Blood Pressure Measurement. The arterial pressure may be measured directly by coupling the inside of an artery to a mercury U-tube manometer. A mercury manometer has a large inertia in the mercury column which severely limits the frequency response. A mercury manometer may be used to observe slow changes in arterial pressure but is inadequate for recording dynamic events, including systolic and diastolic pressures.

Accurate dynamic recording of arterial pressure requires a stiff diaphragm transducer with amplification. The principle of such a transducer is that the arterial pressure causes a slight motion of a metal diaphragm which is proportional to the pressure. In the most common transducer, re-sistance wires are attached to the back of the diaphragm so that two wires are lengthened and two wires are shortened with diaphragm movement. The resistance wires form part of an electronic bridge circuit where the change in bridge current is amplified to give an electronic signal for recording. These devices are often called strain gauge transducers because the resistance wires are strain gauges. Pressure is transduced to a deformation, or strain, which is transduced to an electrical signal proportional to the pressure.

The stiffer the diaphragm, the smaller the volume displacement of fluid for a given

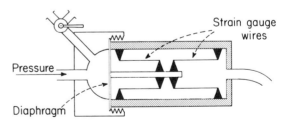

Figure 3–1 Construction of a resistance strain gauge manometer. Pressure in the dome causes the diaphragm to move to the right, which slightly stretches two of the strain gauge wires and compresses the other two. The four strain gauge wires form the four resistors of a Wheatstone bridge in an electronic circuit (not shown).

23

change in pressure. A small volume displacement improves the frequency response of the instrument but decreases the sensitivity. The overall frequency response of a blood pressure recording will depend on the catheter coupling the artery to the transducer, the viscosity and density of the fluid, the transducer, the bridge amplifier and the recorder. The usual limitation of frequency response is in the catheter manometer combination. The longer and narrower the catheter, the greater will be the frictional damping. Air bubbles in the saline which fill the catheter and manometer severely limit the frequency response, as will clotted blood. The problems of adequate pressure recording have been extensively analyzed by Fry.[3]

Indirect Arterial Pressure Measurement. Human arterial pressure is usually indirectly estimated because direct measurement requires arterial puncture. A sphygmomanometer is used to determine arterial pressure in the limbs—usually the arm. The principle of the technique is that a pneumatic cuff is inflated encircling the arm so that the cuff pressure is transmitted through the tissue to compress the brachial artery. A stethoscope is used to listen to the artery distal to the cuff; the cuff is pumped up above systolic pressure, and then slowly lowered; distinctive Korotkov sounds are heard when the cuff pressure equals systolic and diastolic pressure.

When cuff pressure exceeds systolic pressure, no sounds are heard distal to the cuff. When the cuff pressure is just below systolic pressure, an intermittent surge of blood flows past the cuff with each systole. This can be heard with a stethoscope as a sharp tapping sound. As the pressure in the cuff is reduced further, the systolic sound becomes louder until the diastolic pressure is reached when the Korotkov sounds become muffled and then disappear. Whether the point of muffling or the point of disappearance is the better indicator of the diastolic pressure is unknown. Usually the muffling point is 5 to 10 mm Hg above the disappearance pressure. In many patients with aortic regurgitation and in some normal subjects, especially after exercise, the Korotkov sounds do not disappear at any pressure down to zero.

The American Heart Association[5] recommends that three numbers be recorded when measuring human blood pressure— the systolic pressure, when the first Korotkov sound is heard, and both the point of muffling and the point of disappearance, e.g., 120/75–70.

BLOOD FLOW MEASUREMENT

Indicator Dilution. In the Stewart-Hamilton method[8] the cardiac output can be measured by injecting a known amount of indicator (usually the dye indocyanine green) in the right side of the heart and recording the indicator concentration as a function of time in a major artery. The principle of the method is based on the

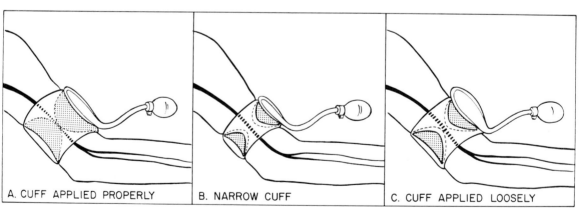

A. CUFF APPLIED PROPERLY B. NARROW CUFF C. CUFF APPLIED LOOSELY

Figure 3–2 The sphygmomanometer cuff must be large enough and applied snugly to transmit cuff pressure to the brachial artery. Shaded area indicates how cuff pressure is transmitted through tissue. (After Rushmer, *Cardiovascular dynamics.* Philadelphia, W. B. Saunders Co., 1970.)

conservation of material in a two port system.

To explain the idea, let us first determine a volume of stationary fluid by indicator dilution, followed by a description of the volume per time (flow rate) of a moving fluid. The volume of liquid in a beaker may be determined by introducing a known quantity of indicator, mixing, and measuring the concentration of the indicator. If the indicator is neither gained nor lost, then the total quantity before and after mixing will be the same (conservation of material). The quantity (Q) of indicator will be equal to the mixed concentration (C) times the volume (V) of dilution

$$Q = V \cdot C$$
$$\text{moles} = \text{liter} \cdot \text{moles/liter} \qquad (1)$$

rearranging for volume

$$V = \frac{Q}{C} \qquad (2)$$

Now consider liquid flowing in a pipe with an indicator injection site upstream and a sampling site downstream. If a known quantity of indicator is injected as a slug, then the volume rate of flow will determine the concentration at the sampling site. The total quantity of indicator (Q) will be equal to the integral of the instantaneous flow rate (F) times instantaneous concentration (C) from the time of injection until all the indicator has passed the sampling site.

$$Q = \int_0^\infty FC \, dt \qquad (3)$$

$$\text{moles} = \int_0^\infty \frac{\text{liter}}{\text{min}} \cdot \frac{\text{moles}}{\text{liter}} \cdot \text{min}$$

If we assume the flow is constant, then it can be brought out of the integral giving

$$Q = F \int_0^\infty C \, dt \qquad (4)$$

rearranging for flow

$$F = \frac{Q}{\int_0^\infty C \, dt} \qquad (5)$$

Thus the cardiac output may be determined if a known quantity of indicator is injected into the right side of the heart and the area under the concentration versus time curve from an arterial site is known.

The assumptions made in this method are as follows:

(i) One inflow and one outflow port exist for the system.

(ii) The indicator represents native fluid. That is, the indicator is well mixed and contained in the vascular tree. The indicator is not lost or gained in passage through the lung and must not be a drug which alters the flow.

(iii) The volume of fluid in the system is constant over the measurement period.

(iv) The flow rate is constant. The flow may be pulsatile, providing that the period of the pulse is short compared to the measurement period and the mean flow is constant.

(v) The frequency distribution of transit times from injection site to sampling site is constant (stationarity). Not only must the mean flow be constant, but the flow in the multiple paths between injection and sampling site must be constant.

(vi) No recirculation. If the indicator recirculates, it will be sampled more than once.

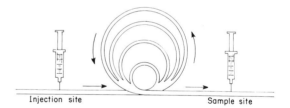

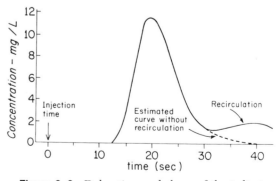

Figure 3–3 Delay time and shape of the indicator dilution curve at the sampling site are functions of the rate of flow in the multiple paths from injection site to sample site. (After Fox and Wood, *Medical physics*, Vol. 3. Chicago, Year Book Publishers, 1960.)

In the practice of measuring cardiac output, these conditions are reasonably well met except for recirculation. Indicator clearly recirculates in the cardiovascular system and recirculation is observable as a second peak on the dye dilution curve. Integration of the dye dilution curve without recirculation is critical to the method so that the curves are corrected to subtract recirculation. (This is done by extrapolating the initial portion of the downward slope before recirculation.) The down slope is usually fitted and extrapolated assuming an exponential decay. This may be done graphically with semilog paper or with a computer. After a corrected curve is found, integration for the area under the curve may be obtained using graphical or computer methods. The problem of recirculation can be minimized by injecting the dye into the right side of the heart and sampling in the ascending aorta. This shortens the transit time without lengthening the recirculation time, permitting better separation of the initial curve from the recirculation curve. Much work has been done on the selection of a proper indicator and validation of the indicator dilution technique.[8] Prospective users should understand the assumptions and limitations of the method.

Fick Principle. The Fick principle is based on the conservation of material in a three port system and is the basis for blood flow measurement in many organs. To understand the Fick principle, consider the hypothetical organ shown in Figure 3-4. Full boxes of apples enter the organ and partially full boxes leave the organ. The organ excretes 12 apples per min through a third port without creating or destroying apples. What is the flow through the organ in boxes per minute? Most persons intuitively arrive at the correct solution of 4 boxes per min. Formally, the extraction rate $\left(\frac{dQ}{dt}\right)$ is divided by the input output concentration difference $(C_{in} - C_{out})$ to determine the flow rate (F).

$$F = \frac{dQ/dt}{C_{in} - C_{out}} \qquad (6)$$

in fluid units:

$$\frac{\text{liter}}{\text{min}} = \frac{\dfrac{\text{moles}}{\text{min}}}{\dfrac{\text{moles}}{\text{liter}}}$$

The assumptions for determining blood flow by the Fick principle are:

(i) Steady state. Both flow (F) and extraction $\left(\frac{dQ}{dt}\right)$ must be constant over the period of measurement.

(ii) The indicator (Q) must be conserved. It must not be metabolized or produced by the organ, although it may be metabolized elsewhere in the body.

(iii) The capacity of the system for the indicator must be constant. This means the amount of blood in the organ must be a constant with no other changing depots for the indicator.

The Fick principle may be applied when the third port is used to add an indicator to blood instead of extract a material from the blood stream. Variations of the Fick principle are used to measure renal blood flow, cerebral blood flow, and other circulations. These modifications are presented in the chapters on the regional circulations.

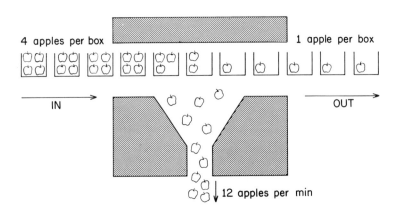

4 apples per box   1 apple per box

IN   OUT

12 apples per min

Figure 3–4 The Fick principle is illustrated in an organ which extracts apples. The problem is to determine the flow of boxes per minute through the organ when the inflow-outflow concentration difference is 3 apples per box and the extraction rate is 12 apples per min.

DIRECT FICK DETERMINATION OF CARDIAC OUTPUT. The Fick principle may be used to determine the blood flow through the lung by using oxygen as an indicator. Three measurements are needed: (i) the concentration of oxygen in the blood entering the lung (pulmonary artery); (ii) the concentration of oxygen in the blood leaving the lung (pulmonary vein or any systemic artery); and (iii) the rate of oxygen consumption of the body. The method is called direct because the blood gas concentrations are determined by direct measurement.

In practice a mixed venous sample is obtained by right heart catheterization. A representative sample must be drawn from the right ventricle or preferably the pulmonary artery. Since a negligible amount of oxygen is lost from the pulmonary veins and left side of the heart and large arteries, the arterial sample may be drawn from any suitable artery. The rate of oxygen consumption is measured with a respiratory spirometer.

The problems with the method are concerned with how well the assumptions of steady state and a constant lung oxygen capacity are met. The oxygen consumption is usually determined over a 3 to 5 min period to obtain a reliable value. During this period the oxygen consumption and the cardiac output must remain constant. A serious potential source of error is that the lung volume at the beginning and end of the measurement period must be constant. Since a spirometer measures the rate of entry of oxygen into the lung rather than oxygen entry into the blood, a change in lung capacity will erroneously be measured as a change in oxygen consumption.

INDIRECT FICK DETERMINATION OF CARDIAC OUTPUT. Arterial puncture and right heart catheterization are avoided in this method. Carbon dioxide excretion by the lung is used in the Fick equation. Arterial and mixed venous carbon dioxide concentrations are indirectly estimated from respiratory gas measurements. Arterial carbon dioxide concentration is estimated from an end expiratory alveolar sample. Mixed venous carbon dioxide is estimated by rebreathing into a closed bag. With rebreathing the carbon dioxide in the bag will come into equilibrium with the venous blood in the lung. The rebreathing is done in an interrupted manner so that the blood level of carbon dioxide is not increased. The indirect Fick method has become more useful and accurate with the introduction of methods for instantaneous gas analysis.[2]

Indicator Washout. An application of the Fick principle is found in the use of an indicator washout curve for determining blood flow.[4, 7] The technique usually employs a radioactive indicator so that the total quantity of indicator in an organ can be determined by counting. A bolus of radioactive material is introduced into the organ either by arterial injection or directly into a tissue such as muscle. The indicator may be in the interstitial or intracellular compartment or both before being washed out by the circulation. A radioactivity counter over the tissue is used to record the rate of disappearance or washout of the indicator from the organ. The more rapid the blood flow, the more rapid the indicator washout.

Beginning with the Fick principle equation (7)

$$F = \frac{\frac{dQ}{dt}}{C_{in} - C_{out}} \qquad (7)$$

we recognize that dQ/dt is the rate of disappearance of the indicator which will be recorded by the counter. Rearranging for dQ/dt

$$\frac{dQ}{dt} = F (C_{in} - C_{out}) \qquad (8)$$

The technique is usually used under circumstances where C_{in} is zero. That is, the arterial concentration is zero after the initial bolus is given. Frequently the indicator is a radioactive gas such as krypton or xenon which is expired from the lung. Thus indicator removed from the test organ travels to the lung via the venous circulation and is lost from the blood in the lung, giving a systemic arterial concentration of essentially zero. Setting $C_{in} = 0$ we have

$$\frac{dQ}{dt} = -FC_{out} \qquad (9)$$

Assuming that the removal of the indicator from the tissue is flow limited but not diffusion limited, then the partition coefficient may be used to estimate C_{out} (the ve-

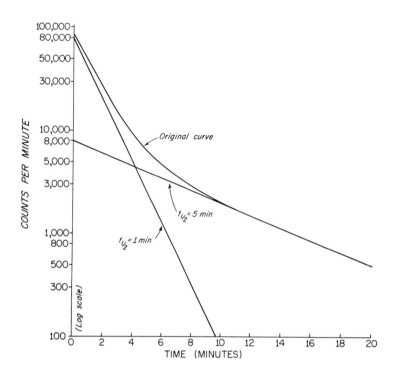

Figure 3–5 An indicator washout curve for an ideal two-compartment system. The line for the compartment with the long half time is found by fitting a straight line in semilogarithmic coordinates to the tail of the curve. The line for the compartment with the short half time is found by subtracting the long half time line from the original curve.

nous concentration). The partition coefficient (λ) is the ratio of solubility of the indicator in the tissue to that in blood. This constant may be determined by a separate experiment where the concentration of indicator in blood and tissue is determined when the two phases are in equilibrium. Thus the partition coefficient is defined.

$$\lambda = \frac{C_{Tissue}}{C_{blood}} \text{ (not saturated)} \quad (10)$$

If no diffusion limitation exists between blood and tissue, then tissue and venous blood will be in equilibrium so that C_{out} is available

$$C_{out} = \frac{C_{Tissue}}{\lambda} \quad (11)$$

and by definition the concentration in tissue is the quantity of indicator (Q) divided by the volume of tissue (V_T).

$$C_T = \frac{Q}{V_T} \quad (12)$$

Substituting (12) into (11) for C_T gives

$$C_{out} = \frac{Q}{V_T} \cdot \frac{1}{\lambda} \quad (13)$$

Substituting (13) into (9) for C_{out} gives

$$\frac{dQ}{dt} = -F \cdot \frac{Q}{V_T \lambda} \quad (14)$$

rearrange for Q on the left

$$\frac{1}{Q} \cdot dQ = \frac{-F}{V_T \lambda} \cdot dt \quad (15)$$

This may be integrated over time

$$\int_{Q_o}^{Q_t} \frac{1}{Q} \cdot dQ = \frac{-F}{V_T \lambda} \int_o^t dt \quad (16)$$

at t = 0
  Q = Q_o and t = 0
at t
  Q = Q_t and t = t

giving

$$\ln \frac{Q_t}{Q_o} = \frac{-F}{V_T \lambda} \cdot t \quad (17)$$

Rearrange for flow (F) per volume of tissue (V_T)

$$\frac{F}{V_T} = \frac{-\ln \frac{Q_t}{Q_o} \lambda}{t} \quad (18)$$

An arbitrary value of $\frac{Q_t}{Q_o}$ must be chosen to find the proper t from the experimental record. Customarily, the time ($t_{1/2}$) is chosen when half the original quantity of indicator (Q_o) is washed out. Setting $\frac{Q_t}{Q_o} = 0.5$ and using this in equation (18) gives

$$\frac{F}{V_T} = \frac{-(\ln 0.5) \lambda}{t_{1/2}} \qquad (19)$$

The natural logarithm of $0.5 = -.6931$ and the equation is often given

$$\frac{F}{V_T} = \frac{.6931 \lambda}{t_{1/2}} \qquad (20)$$

which states that the flow per volume of tissue may be obtained from the half time of the washout curve and the partition coefficient.

Most vascular beds do not show a monoexponential washout curve as would be predicted by equation (20) but a more complex curve. This is assumed to be an indication of inhomogeneity of the vascular bed and that more than one flow rate to different compartments of the vascular bed are present. For example, gray and white matter in the brain or fat in a skeletal muscle vascular bed have different flow rates. The further assumption is made that these components have exponenetial washout curves which are additive for the total curve which gives a modification of equation (18):

$$\frac{F}{V_T} = -\frac{\ln(Q_{t_1}/Q_{o_1})\lambda_1}{t_1} - \frac{\ln(Q_{t_2}/Q_{o_2})\lambda_2}{t_2} -$$
$$- \frac{\ln(Q_{t_3}/Q_{o_3})\lambda_3}{t_3} \cdots \qquad (21)$$

where the numbered subscripts indicate the various flow compartments in the vascular bed.

In practice the radioactive decay rate (indicating the total quantity of indicator) is plotted versus time on semilog paper. The tail of the curve is fitted with a straight line and extrapolated back to time zero. The values of this compartment's washout are then subtracted from the original curve and the remainder replotted. The tail is again fitted and subtracted for each recognizable compartment. The flow per volume of tissue for each compartment is calculated from equation (20), using the half time and partition coefficient for that compartment.

The assumptions in the indicator washout method are:

(i) The flow must be constant over the entire period of measurement, and the capacity for indicator in the tissue is constant (steady state).

(ii) The flow in each of the compartments is constant during the measurement period (stationarity).

(iii) The washout is a true exponential function described by a straight line on a semilog plot.

(iv) Washout from the tissue is only flow limited and not limited by diffusion into the capillary from the tissue.

(v) The partition coefficient is a true constant and concentrations of indicator in blood and tissue are below saturation. True equilibrium exists between tissue and venous blood (related to iv above).

(vi) The only exit for the indicator is via the venous circulation.

(vii) Arterial inflow concentration of indicator is nil during the washout period.

(viii) Indicator is not carried by the circulation or by diffusion from one compartment to another. That is, the compartments are in parallel.

(ix) The correct number of homogeneous compartments is identified by the curve peeling procedure. This is a difficult point, because some arbitrary rule must be used to fit a line to the tail of the curve.

(x) The indicator is not a drug which alters flow.

If some of these conditions are not met, corrections may be applied from a mathematical model of the system. However, the model requires detailed information about the system which may be more difficult to obtain than a blood flow measurement by a different technique.

Venous Occlusion Plethysmograph. The blood flow into a limb may be determined by momentarily occluding the venous outflow and recording the increase in limb volume during the occlusion. The principle is that the limb volume will swell at the rate that arterial blood flows into it during the first moments after venous occlusion. The assumption is that the arterial inflow

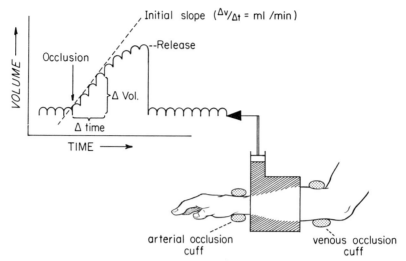

Figure 3–6 A water-filled venous occlusion plethysmograph shown diagrammatically. Blood flow distal to the plethysmograph is prevented by an arterial tourniquet. Sudden venous occlusion is accomplished with a pneumatic cuff inflated to a pressure above venous pressure but below arterial pressure. The initial slope of the volume record gives the rate of arterial inflow (ml/min). The inflow decreases with continued engorgement of the limb so that the occlusion cuff must be released after a few seconds. The maneuver may be repeated several times a minute if the occlusion period is only a few seconds.

is unaltered by venous occlusion with a pneumatic cuff inflated to a pressure above venous pressure but below arterial diastolic pressure. The method is non-invasive and non-traumatic so that repeated measurements may be made in unanesthetized humans. Continuous blood flow is not measured but rather a series of flow determinations, one for each venous occlusion. Care must be used to avoid artifacts from inflating the venous occlusion cuff or motion of the limb in and out of the plethysmograph. The technique has been extensively used to study the circulation in human extremities, and much of our knowledge of muscle and cutaneous blood flow has been obtained with a plethysmograph.[6]

Electromagnetic Flowmeter. A fundamental property of electromagnetism is that an electric current is generated when a conductor is moved through a magnetic field. This principle is used to measure blood flow. The vessel is placed in a magnetic field. The blood, which is a conductor, flows through the field creating an electromagnetic force which is proportional to the flow rate.

The idea is most easily understood with a constant magnetic field. If a constant magnetic field is applied from north to south across a vessel which has blood flowing in it from east to west, then a current will flow

through the vessel from top to bottom (right hand rule) as shown in Figure 3–7. Electrodes are placed on the top and bottom of the vessel to pick up the electrical signal for amplification and recording. The signal is proportional to the number of lines of magnetic force cut per unit time by the moving conductor. Since the magnetic field is constant, the signal will be proportional to the flow rate of the liquid conductor. The method is dependent on the ions in blood acting as conductors.

In practice an alternating current is used to excite the magnet rather than a direct current as described above. The advantages of an alternating field are (i) better rejection of unwanted signals such as the electrocardiogram; (ii) avoidance of electrode polarization; and (iii) the engineering benefit of ac input coupling.

An electromotive force will be generated whenever a conductor cuts a magnetic line of force whether the conductor moves through the magnetic field or the field moves past the conductor. A generator such as an electromagnetic flowmeter is a device where mechanical motion of the conductor or field generates a current. Conversely, a transformer is a device with two mechanically stable conductors. An alternating current in the transformer's primary conductor with its attendant electromagnetic field

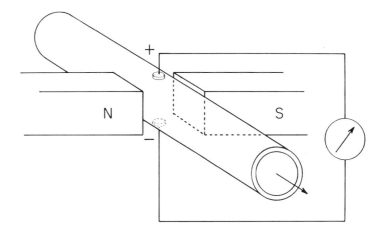

Figure 3–7 Diagram of the electromagnetic flowmeter. When a conducting fluid (with ions) flows through the tube in a magnetic field, an electromagnetic force will be generated perpendicular to the field and the direction of flow.

induces an electromagnetic force in the secondary conductor by building and collapsing the alternating field. In effect the moving lines of magnetic force cut the stationary conductor. The induced electromotive force in the secondary will be proportional to the *rate* that lines of magnetic force cut through it. In the configuration of an electromagnetic flowmeter, the transformer's electromotive force will be greatest when the rate of change of the magnetic field is greatest. Conversely, the electromotive force generated by the moving blood will be greatest when the magnetic field is strongest (most magnetic lines per cm^2) at a time when the rate of change of the magnetic field may be zero.

If the electromagnet is energized by a sine wave alternating current, the flow signal will be greatest when the magnetic field is at its height and the rate of change of the field will be at its minimum. Ninety degrees later the field will be at a minimum and the rate of change will be at a maximum. Hence the flow signal and transformer signal are 90° out of phase or in quadrature (one quadrant of a full 360° cycle = 90°). In practice the flow signal is sampled for a brief period near the peak and rejected for the rest of the half cycle. This is called gating.

Currently available electromagnetic flowmeters have sine wave, square wave or pulsed magnet excitation. Square wave excitation with an alternating square wave is analogous to turning on briefly a direct current magnet and then turning it on with reverse polarity. The signal is gated to sample during the latter portion of the square wave to avoid the transient resulting

from rapidly activating the magnet. Pulsed excitation is a square wave with a 50 per cent duty cycle. A positive square pulse is used for a fourth of the cycle followed by a quarter cycle at zero, followed by a fourth cycle negative square pulse, followed by a final quarter cycle at zero before repeating. Pulsed excitation is analogous to square wave excitation with the advantage that undesirable transducer heating is halved.

The advantages of the electromagnetic flowmeter are that it is sensitive to the flow rate in the magnetic field and virtually insensitive to the shape of the velocity profile in the vessel. A great advantage is that the flow transducer may be placed around the outside of the vessel with the vessel wall acting as a conductor. The most serious disadvantage of the electromagnetic flowmeter is that the vessel must be repeatedly occluded to determine the zero flow. Electromagnetic flowmeters with an electronic nonocclusive zero are under development.

Ultrasonic Doppler Shift Flowmeter. The velocity of an object may be determined by a Doppler frequency shift. A common example of a Doppler shift is the change in pitch (frequency) of a moving train whistle which is higher when approaching than when stationary or receding. If the stationary frequency of the whistle and the speed of sound are known, the velocity of the train with respect to the observer may be calculated from the Doppler equation.

$$\Delta f = \frac{f_0 \, V \, Cos \, \theta}{C} \qquad (22)$$

where Δf is the Doppler shift in frequency

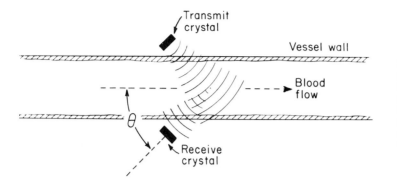

Figure 3–8 The ultrasonic Doppler shift flowmeter shown diagrammatically. Sound is generated at the transmit crystal and reflected from red cells in the blood to the receive crystal. If blood is flowing, the received frequency will be shifted as a function of the red cell velocity.

shift, f_o the frequency when stationary, V the relative velocity of source and observer, C the velocity of sound, and θ the angle between the direction of motion and the observer of the Doppler shift.

To measure the velocity of blood flow, the blood must be made a sonic source. This is done by reflecting sound from the red blood cells and then observing the Doppler shift of the reflected sound. Under these conditions the Doppler shift will be doubled. The Doppler shift will be manifest once from generator to reflecting site and once again from reflecting site to receiver, giving

$$\Delta f = \frac{2f_o\ V\ \mathrm{Cos}\ \theta}{C} \qquad (23)$$

If the transducer is constructed with a constant frequency sonic generator (f_o) and a fixed geometry (θ) and the velocity of sound in blood and tissue is a constant (C), then the Doppler shift (Δf) will be in direct proportion to the velocity (V).

The advantage of the Doppler shift flowmeter is that zero flow is easily determined because it is zero frequency shift from the generator. The transducer may be placed around the vessel and is free from the problem of electrode polarization found with electromagnetic flowmeters. The major limitation of the method is that the velocity of reflecting red cells is measured, not the flow rate in liters per minute. This means that if the red cells in a vessel have different velocities, different Doppler shifts will be observed. Thus the instrument is sensitive to changes in the velocity profile in the vessel. If the velocity profile is unchanging, an "average" Doppler shift

will be recorded which may be calibrated with flowing blood *in situ*. If the vessel pulsates, changing its cross-sectional area with pressure, the average velocity may change without a change in flow rate. An additional problem with the electronic circuitry of some Doppler shift meters is the inability to detect antegrade from retrograde flow. The circuitry is such that the frequency shift is measured but not its sign. Red cell velocity is detected in either direction, but the recording does not indicate if the flow is forward or backward.

REFERENCES

1. Cappelen, C., Jr., ed. *New findings in blood flowmetry.* Universitetsforlaget, Oslo, Aas & Wahls Boktrykkeri, 1968.
2. Ferguson, R. J., Faulkner, J. A., Julius, S., and Conway, J. Comparison of cardiac output determined by CO_2 rebreathing and dye-dilution methods. *J. appl. Physiol.*, 1968, 25, 450–454.
3. Fry, D. L. Physiologic recording by modern instruments with particular reference to pressure recording. *Physiol. Rev.*, 1960, 40, 753–788.
4. Kety, S. S. The theory and applications of the exchange of inert gas at the lungs and tissues. *Pharmacol. Rev.*, 1951, 3, 1–41.
5. Kirkendall, W. M., Burton, A. C., Epstein, F. H., and Freis, E. D. Recommendations for human blood pressure determination by sphygmomanometers. American Heart Association (EM34 rev), 1967.
6. Shepherd, J. T. *Physiology of the circulation in human limbs in health and disease.* Philadelphia, W. B. Saunders Co., 1963.
7. Thorburn, G. D., Kopald, H. H., Herd, J. A., Hollenberg, M., O'Morchoe, C. C., and Barger, A. C. Intrarenal distribution of nutrient blood flow determined with Krypton[85] in the unanesthetized dog. *Circulat. Res.*, 1963, 13, 290–307.
8. Wood, E. H., ed. Symposium on use of indicator-dilution technics in the study of the circulation. *Circulat. Res.*, 1962, 10 (No. 3, Part 2), 377–581.

CHAPTER 4 CARDIAC MUSCLE MECHANICS

by ERIC O. FEIGL

This chapter describes the mechanical properties of cardiac muscle cells. Here the myocardium is treated as a tissue, whereas in succeeding chapters the heart is considered as an intact organ and as part of the cardiovascular system. The intrinsic properties of the myocardium given in this chapter underlie cardiac performance, which is modulated by hemodynamic, neural and hormonal mechanisms.

EXPERIMENTAL PREPARATIONS

Isolated cardiac muscle tissue, such as a strip of atrium or a papillary muscle, can be studied while the muscle contracts in several different mechanical arrangements (Fig. 4–1):

ISOMETRIC CONTRACTION (isos = same, metron = measure). The muscle is held at a constant fixed length throughout the contraction. The isometric apparatus is usually adjustable so that muscle length can be changed between beats—although muscle length is constant for any given beat.

ISOTONIC FREELOADED CONTRACTION (isos = same, tonos = tension). The muscle is attached to a first class lever where muscle shortening lifts the load on the other side of the fulcrum. In a true freeloaded isotonic contraction the tensions during rest and contraction are equal.

ISOTONIC AFTERLOADED CONTRACTION. An adjustable stop is added to the isotonic muscle lever so that the preload (tension during rest) and afterload (tension during contraction) are different. In an isotonic afterloaded contraction, the initial portion of the contraction is isometric until the tension generated by the muscle equals the afterload when the muscle shortens at constant tension. Since freeloaded isotonic contractions (equal preload and afterload) are infrequently studied in cardiac muscle, isotonic afterloaded contractions are often called simply "isotonic."

AUXOTONIC CONTRACTION (auxano = to grow, tonos = tension). The muscle is arranged to pull against a spring so that the greater the shortening, the greater the tension. In an auxotonic contraction, external shortening is against an increasing load.

AUXOTONIC AFTERLOADED CONTRACTION. A stop is added to the auxotonic apparatus so that the diastolic tension is in-

33

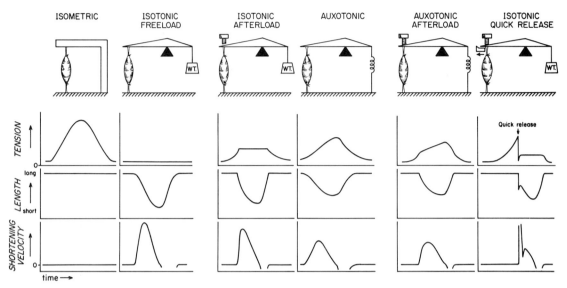

Figure 4–1 The various experimental preparations for studying isolated cardiac muscle are schematically represented. The most frequently used are isometric and isotonic afterloaded preparations. Auxotonic afterloaded contractions most closely resemble the mechanical loading in the intact heart. Isotonic quick-release preparations are used to study muscle elastic properties and the active state.

dependently adjustable. The initial phase of an auxotonic afterloaded contraction is isometric until the muscle tension equals the spring afterload tension, at which point external shortening begins. With continued shortening, tension increases until relaxation begins. This is the type of contraction the intact ventricle makes: The diastolic filling pressure is low. Systole begins with an isovolumic (constant volume) period until ventricular pressure exceeds aortic pressure, thus opening the aortic valve. Ejection (external shortening) begins, increasing the pressure still further.

QUICK RELEASE TO ISOTONIC CONTRACTION. A latch is added to the apparatus which may be instantaneously released during the contraction. This latch holds the muscle isometrically until release, whereupon the muscle quickly snaps to a shorter length, determined by the isotonic afterload. After the quick release transient, the muscle shortens, lifting the isotonic load. The release may be timed to occur at any moment during the beat. Thus the muscle is held at constant length until quick release to isotonic conditions. Quick-release contractions are used to determine the series elastic stiffness and in measurements of the active state.

LENGTH-TENSION RELATION

A fundamental property of cardiac muscle is the length vs. isometric tension relation. Very simply, the greater the initial muscle length at the beginning of contraction (up to a point), the more forceful the contraction will be. This is readily observed in an isolated papillary muscle preparation, where a thin papillary muscle from a cat or rabbit is studied under isometric conditions in an organ bath. The result of such an experiment is shown in Figure 4–2. Various isometric lengths are shown on the abscissa, and tension or force on the ordinate.

Three tensions should be recognized in Figure 4–2. The first is *resting tension*. Resting, unactivated, noncontracting cardiac muscle has elastic properties, *viz.*, the more it is stretched, the greater the tension, like a nonlinear spring. The second tension is *peak isometric tension*. This is the maximum force the muscle develops during a contraction. As can be seen in Figure 4–2, the peak isometric tension varies with length. The third tension is *developed tension*. This is the difference between resting and peak tension. Developed tension is particularly important for cardiac muscle because the body is dependent on

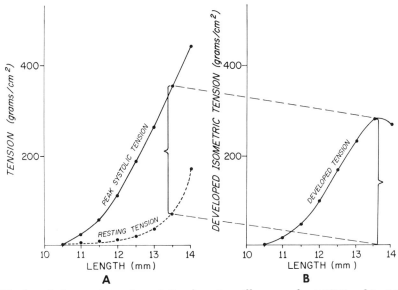

Figure 4–2 The length–isometric tension relation for cat papillary muscle at 25° C and 2 mM calcium ion concentration. The resting and peak isometric active tension as a function of muscle length are plotted in A. The difference between resting and peak tension is the developed tension which is plotted in B. The optimal length for this muscle was 13.5 mm. (After Sonnenblick, *Amer. J. Physiol.*, 1962, *202*, 931–939.)

the force the heart generates in transition from diastole to systole. Developed tension in Figure 4–2A is replotted in Figure 4–2B to show the length-tension relation.

The basic isometric length-tension relation shown in Figure 4–2B is understood according to the sliding filament theory of muscular contraction (see Chap. 6, Vol. I). Briefly, cardiac muscle is composed of interdigitating thick and thin filaments. The thick filaments have an "active" zone at their ends where crossbridges are formed to the thin filaments. The amount of force that can be generated between thick and thin sliding filaments depends on the number of crossbridges which can be formed. The greater the overlap between the crossbridge zone on the thick filament and thin filament, the greater the contractile force. A contractile force plateau would be expected when a thin filament completely overlapped the "active" crossbridge zone of the thick filament. Such a plateau has been observed in careful experiments in skeletal muscle.[20, 21] Contractile tension will not remain constant with shorter and shorter muscle lengths because of other restraints with short muscles. Thin filaments will begin overlapping each other at a sarcomere spacing (Z line to Z line) of less than 2.0 μm, and thin filaments will begin overlapping

the crossbridges on the opposite end of the thick filament at a sarcomere spacing of about 1.8 μm. This double overlap interferes with contraction, and finally, as the Z lines begin to abut on the stiff thick filaments, very little shortening or external force generation is possible.

Intact cardiac muscle gives a smooth curve of tension-vs.-length without a plateau. This rounding is a result of the distribution of the lengths of the sarcomere. An additional factor may be that not all the sarcomeres are arranged in parallel. A recent study[19] has shown a wavy pattern of cardiac muscle fibers in isolated papillary muscles. The extent to which wavy or buckled sarcomeres contribute to the performance of intact hearts is unknown.

The force of contraction is dependent on the muscle (sarcomere) length because of the extent of overlap between thick and thin filaments.[22, 41] The relation between distension of the intact heart and contractile vigor was first observed for cardiac muscle by Otto Frank[16] and documented in the English literature by Ernest Starling.[11] The Frank-Starling relation, or Starling's "law of the heart," is a statement of the length-tension property of cardiac muscle. That is, the longer the initial resting length of cardiac muscle, the greater the strength of contraction in the ensuing beat. This is true up to some critical length beyond which contractile force declines.

In summary, the length-tension property of cardiac muscle relates diastolic length to generated peak isometric tension. The sliding filament theory of muscle contraction explains this by the amount of overlap between thin filaments and the active (crossbridge) area of the thick filaments. This length-tension relation is the basis for Starling's "law of the heart."

INTERVAL-STRENGTH RELATION

The interval between beats is a major determinant of the force of cardiac muscle contraction (Fig. 4-3). This phenomenon is sometimes called the staircase, or Treppe (= staircase in German) phenomenon, because the strength of successive beats changes in a stepwise manner when the heart rate is altered. The experimental record may resemble a staircase, as in Figure 4-4.

The relation between the strength of contraction and the frequency of activation is not a simple one, because two factors seem to be in operation: (i) Frequent activation with brief intervals between beats causes increased strength of contraction. (ii) A long rest between beats augments the force of the next beat. Furthermore, these effects can accumulate over more than one inter-

val so that a sudden change in beat frequency results in a stepwise change in beat strength from one beat to the next until a new steady state is reached. The cellular mechanism responsible for the force-frequency relation is unknown, although an alteration in intracellular calcium ion concentration is probably involved. (See Excitation-Contraction Coupling.)

In the absence of a known mechanism, the phenomenon has been described in terms of hypothetical models. Blinks and Koch-Weser[3, 27] present a model which uses the theoretical notions of Positive Inotropic* Effect of Activation (PIEA) and Negative Inotropic Effect of Activation (NIEA). NIEA both manifests and decays more rapidly than PIEA. The force-frequency relation varies for different cardiac preparations and is not the same for atrium and ventricle.

Bowditch originally demonstrated part of the interval-strength relation which showed a stepwise increase in force with an increase in stimulation frequency, as in Figure 4-4. This observation has been called a Bowditch, or ascending, staircase. Woodworth demonstrated that pauses increase the strength of contraction and that stepwise decreases in contraction may occur with a change in beat frequency. Such changes have been called a Woodworth, or descending, staircase. The various staircase

---

*Inotropic refers to the strength of fiber shortening.

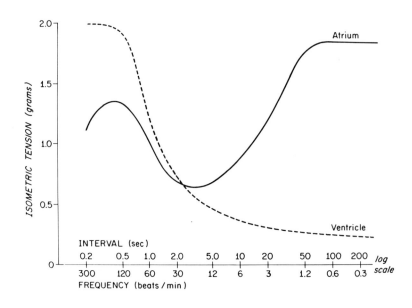

Figure 4-3 The steady state interval-strength relation for isolated kitten atrium and ventricle at 38° C. (After Koch-Weser et al., Pharmacol. Rev., 1963, 15, 601–652.)

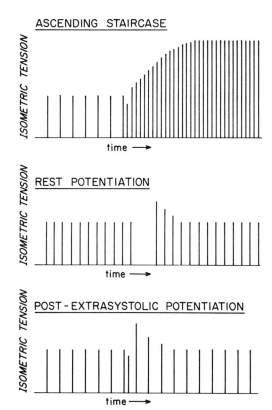

Figure 4-4 Changes in the strength of contraction of cardiac muscle resulting from altering the beat frequency. These are examples of transitional or non-steady state changes in contraction strength from changing the beat frequency.

phenomena represent the transition phase from one beat frequency to another. The interval-strength relation is dependent on the species of animal, tissue type and beat frequency, and the transition from one frequency to another can be quite complex. For example, the transition from a long pause to a rapid beat frequency can show several downward steps after the first beat, followed by an upward staircase.

Post-extrasystolic potentiation (Fig. 4-4) is a special case of the force-frequency relation. In an intact heart a prolonged diastole following an extrasystole may also augment post-extrasystolic potentiation by greater diastolic filling.

Confusion has resulted from trying to characterize cardiac tissue by the various staircase phenomena it exhibits. The same preparation can demonstrate one type of staircase in the transition from a low-to-medium beat frequency and a very different staircase in the transition from a medium-

to-high beat frequency. Such behavior can often be predicted with a knowledge of the steady state interval-strength relation plus a model such as the PIEA-NIEA theory of Blinks and Koch-Weser.

In summary, frequency of the heart beat is a major determinant of the strength of contraction. This is an intrinsic property of cardiac muscle that is not dependent on innervation or hormones. The cellular mechanism of the interval-strength relation is unknown, although a good description of the effect may be obtained with a hypothetical model.

FORCE-VELOCITY RELATION

The shortening velocity of cardiac muscle is inversely related to load.[1, 35, 36] The relation between peak shortening velocity and isotonic load is given in Figure 4-5. (Data

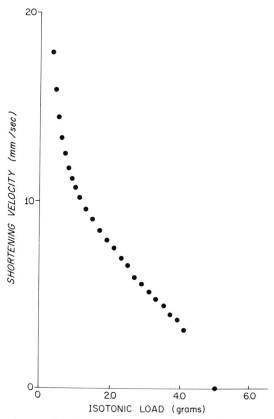

Figure 4-5 The peak shortening velocity vs. isotonic afterload of isolated cat papillary muscle at 36° C. The inverse relation between shortening velocity and load is one of the fundamental properties of cardiac muscle.

for this curve were obtained from an isolated cat papillary muscle). The muscle was arranged in the organ bath for isotonic afterloaded contractions. The shortening velocity plotted is the peak shortening velocity observed with a given load starting from a fixed initial length. Hill[24] proposed that the relationship between shortening velocity and load for skeletal muscle has the form of a rectangular hyperbola and fit his data with the following equation for a hyperbola,

$$(P + a)(V + b) = K$$

where P is muscle tension (or load or force), V is shortening velocity, K is a constant and a and b are the fitting constants. This has been called the "characteristic equation" for muscle.

However, it is now uncertain whether the cardiac force-velocity relation is a hyperbola. All investigators agree on the general inverse relation between shortening velocity and load.[1] Sonnenblick[35, 36] found that cardiac force-velocity data are best described by a hyperbola. However, Noble et al.[30] observed a hyperbolic curve for cardiac muscle only at long muscle lengths, and doubt whether a hyperbolic relation for the contractile element exists. Brady[4] found that cardiac force-velocity data are badly fit by a hyperbola. A good description of the relation between load and shortening velocity is still needed for a clear understanding of cardiac mechanics. Alterations in the force-velocity curve have been used to describe changes in the vigor of cardiac contraction. This is discussed in the section on contractility.

In summary, the force-velocity relation is simply the curve relating shortening velocity to the isotonic afterload. Shortening velocity is rapid for small loads and slow with heavy loads.

ELASTIC PROPERTIES

In addition to the contractile properties described above, cardiac muscle also has passive elastic properties. Resting, nonactivated cardiac muscle behaves as a nonlinear spring. Resting muscle elongates with increasing tension applied by the filling pressure of the heart or the preload in an isolated muscle apparatus. With in-

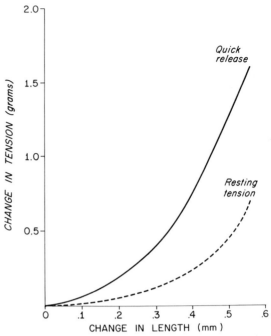

Figure 4–6 The elastic length-tension curves for resting and active elastic properties of isolated papillary muscle. The elastic stiffness of the muscle is given by the slope of the curve. Active muscle (quick release) is stiffer than resting muscle. (After Brady, *Physiologist*, 1967, *10*, 75–86.)

creasing applied tensions, the proportionate lengthening is less.[6, 23, 31, 37] That is, diastolic muscle is stiffer (deforms less) at high tensions than at low tensions (Fig. 4–6).

A recurrent question in cardiac physiology is the possibility of a variable diastolic compliance, or elasticity. The diastolic length-tension relation of cardiac muscle determines the initial muscle length and thus the force of cardiac contraction (length-tension relation). The initial length is determined by the preload during diastole. If cardiac diastolic extensibility were a physiologic variable, then different initial muscle lengths would be observed with the same preload. Variable diastolic extensibility implies partial or residual contractile activity during diastole—this has been termed *tonus*. Cardiac tonus has been proposed a number of times but has not been substantiated. The current concept is that resting extensibility is not altered by neural, hormonal or other physiologic mechanisms.[3, 7, 15, 25]

Active muscle has elastic properties which are most easily observed by quickly releasing the tension during an isometric contraction. When the active tension is rapidly released to a predetermined isotonic level, the muscle quickly shortens. This resembles the rapid snap of a stretched spring when the tension holding it is quickly released. The shortening velocity of the quick release far exceeds any shortening velocity the muscle can develop by contractile shortening. This implies that the contractile machinery of the muscle is in series with an undamped spring.

The three-element model of cardiac muscle is a simplifying description of the muscle which incorporates the resting and active elastic properties of the muscle. The model consists of two springs and a contractile element. One spring, the parallel elastic element, predominantly represents the elastic behavior of resting muscle. The other spring, the series elastic element, predominantly represents the elastic behavior of active cardiac muscle. These elastic elements are so named because they are in parallel and series with the contractile element. All the contractile properties of the muscle (length-tension, force-velocity, etc.) are assigned to the contractile element portion of the model.

Two configurations of the three elements in the model are possible, as shown in Figure 4–7. In Model I, or the Maxwell model, the parallel elastic element is parallel to both the contractile element and the series elastic element. In Model II, or the Voigt model, the parallel elastic element is parallel only to the contractile element. Models I and II have similar properties, and both are reasonable but incomplete analogs of cardiac muscle. Although both

Model I and Model II are functional analogs of muscle behavior, no part of either model corresponds to any anatomical part of the muscle. This cannot be stressed too strongly. Although a noncontractile tendon at the end of the muscle will make the series elastic component more compliant, the series elastic element does not represent a tendon.

The assumptions necessary for the three-element model are as follows:

(i) The elastic elements are nonlinear, becoming stiffer with greater stretch.

(ii) The contractile element is freely extensible at rest. This means that there is no tension across the contractile element at rest. The resting tension of the muscle is borne by the elastic elements.

(iii) The model is valid for extensions and shortenings, but not for compressive stress. Real muscles bend and fold if longitudinal compression is attempted.

INTERPRETATION OF THE ISOMETRIC BEAT USING THE THREE-ELEMENT MODEL. As seen in Figure 4–8, tension in a cardiac isometric twitch rises smoothly to a peak and then relaxes smoothly with time. At rest between beats the diastolic preload is borne entirely by the eleastic elements of the muscle. With activation the contractile element begins to shorten and develop tension. Tension does not rise abruptly because the contractile element must shorten to develop tension. Internal (contractile element) shortening is rapid at first because the contractile element has a small load as a result of its just beginning to pull on the series elastic element. As contractile element shortening proceeds, the series elastic element lengthens and tension rises. Because

Figure 4–7 Two alternate arrangements of the three-element elastic model of cardiac muscle. The elastic elements are passive nonlinear springs. The contractile element is freely extensible at rest and manifests all the active properties of muscle during contraction.

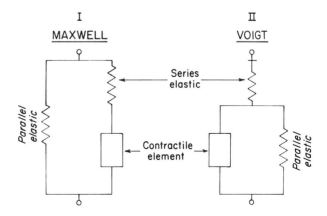

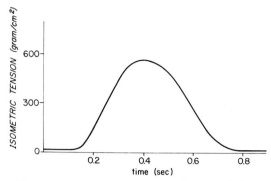

Figure 4–8 Isometric tension developed by an isolated cat papillary muscle at 36° C. The external length of the muscle is held constant (isometric). Tension is developed by internal shortening of the contractile element stretching out the series elastic element.

of the force-velocity relation (low shortening velocity with high loads), contractile element shortening velocity will slow down with increasing tension. This process continues until relaxation begins. The rise and fall of tension in an isometric cardiac beat are also smooth because the onset and turning off of mechanical activity are gradual. (See Active State.)

The three-element model permits a more exact description of myocardial function, because the intrinsic mechanical properties, such as the length-tension and the force-velocity relations, may be redefined for the contractile element rather than the whole muscle. This means that the contractile behavior of the muscle may be rigorously described without including noncontractile elastic properties.

In summary, cardiac tissue exhibits nonlinear elastic behavior at rest that does not vary physiologically. Active myocardium manifests different nonlinear elastic properties from resting muscle. These are most simply understood in terms of a three-element model. The three-element model separates the passive elastic properties and the active contractile properties. In the model the passive properties are in the parallel and series elastic elements, and all the active properties reside in the contractile elements.

ACTIVE STATE

The active state of muscle is a concept which describes the intensity with which

muscle is turned on during activation. In cardiac muscle the concept is used to describe the speed of onset and duration of mechanical activity. Cardiac muscle manifests two clear states—rest and activity. The transition from rest to activity is important to understand. Cardiac muscle does not become mechanically active at a rapid rate as though a fast switch had been thrown. Rather, mechanical activity (not to be confused with the electrical action potential) has a slow onset and decay with each beat. Active state is measured by the ability of the muscle to lift or bear a load. In practice, a special experimental maneuver is required to estimate the active state because of the series elastic property of cardiac muscle. Active state measurements have been made by quick stretch, quick release to an isotonic load and quick release to zero isometric tension.[5, 39] The methods have the disadvantage that the experimental quick stretch or quick release may disrupt the active state. Brady has provided evidence that the active state is more easily disrupted late in the beat than early in the beat.[5]

The best quantitative measurements of cardiac active state have been made with the quick release-to-isotonic load method. In this method the muscle is held isometric until a predetermined time during the beat, whereupon the muscle is quickly released to a fixed isotonic load. After the quick shortening of the series elastic element, the contractile element begins to shorten and lifts the isotonic load. The maximum isotonic shortening velocity is then recorded. The rationale of the method is that the time of onset of isotonic shortening can be experimentally determined during the twitch. Since the isotonic load is the same for each release, the initial shortening velocity for a constant load and an approximately constant length can be determined at different times during the beat. The initial peak-shortening velocity after the quick release is then used as a relative index of the intensity of the active state, as shown in Figure 4–9. The active state in cardiac muscle has a slow onset compared to skeletal muscle. The peak active state occurs at about half the time it takes to develop full isometric tension in the same muscle. The descending limb of the active state curve measurement is less reliable because quick releases later in the beat tend to decrease

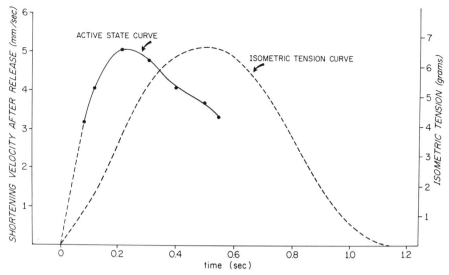

Figure 4–9 The time course of the active state of cat papillary muscle at 22° C. The active state was determined by the shortening velocity after quick release at different times to a constant isotonic afterload. The isometric tension curve is given for comparison. (After Brady, *J. Physiol.* [*Lond.*], 1966, *184*, 560–580. The last portion of the isometric curve has been extrapolated beyond the curve of the original figure.)

the active state.[5, 8] Other methods of measuring the time course give similar results.[5]

The active state should not be confused with the strength of contraction. Strength of contraction as measured by peak isometric tension is dependent upon the active state but is not the same. For example, the active state might have a rapid onset and a high peak intensity but be so brief in duration that a very small isometric tension is developed before relaxation begins. The active state may be considered as the mechanical potential of the muscle to shorten or bear a load at a moment in time. The actual tension or shortening generated by the muscle depends on how it is loaded as well as the intensity and duration of the active state.

Although the active state is usually the ability to shorten or bear a load, shortening and load bearing are not the same nor do they have exactly the same time course during a cardiac beat. Shortening against a light load requires fewer active cross-bridges than full load bearing, and the onset of activity for shortening is probably more rapid than for isometric load bearing.

In summary, the active state is a description of the moment-to-moment contractile potential of the muscle during a beat. This is a useful concept which separates the intrinsic contractile property from the external performance of cardiac muscle. The active state of cardiac muscle has a slow onset and decay during activity, reaching a peak before the height of isometric tension.

EXCITATION-CONTRACTION COUPLING

Mammalian cardiac muscle cells are invested by a cell membrane which has tubular projections extending down into the center of the cell. These T (transverse) tubules are open to the extracellular space and enter the muscle at the level of the Z lines. The sarcoplasmic reticulum is a second intracellular membrane system and invests the myocardial fibrils extending into clefts in the fibril. The plexiform sarcoplasmic reticulum is continuous across Z lines. Specialized subsarcolemmal cisternae of the sarcoplasmic reticulum make intimate contact with the surface membrane and the T tubules[14, 29] (Fig. 4–10).

Excitation of cardiac muscle is equivalent to an action potential propagated along the outer cell membrane. This excitation is carried inward through the T tubules to the center of the muscle. It is assumed that the close apposition of subsarcolemmal cisternae to the surface membrane and the T tubule permits these structures to excite

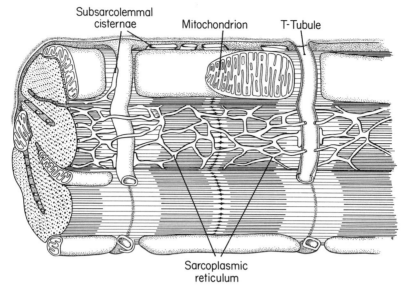

Figure 4–10   Microscopic structure of mammalian cardiac muscle. The mass of myofilaments is partially divided by clefts which contain sarcoplasmic reticulum. The reticulum is continuous across the Z line and T tubule. Small saccular expansions of the reticulum, called subsarcolemmal cisternae, are in close contact with T tubules or with the sarcolemma. The T tubule is continuous with the outer cell membrane. (After Bloom and Fawcett, *A textbook of histology.* Philadelphia, W. B. Saunders Co., 1969.)

the sarcoplasmic reticulum to release calcium ions.

The interaction between myosin (thick filaments) and actin (thin filaments) is physiologically regulated by the concentration of intracellular calcium ions. In the presence of a magnesium ATP complex, actin and myosin split the terminal phosphate from ATP to cause shortening, a conversion of chemical energy into mechanical energy.[26] Calcium ·ion sensitivity is conferred on this reaction by troponin and tropomyosin which are on the thin filament.[12] An increase of the concentration of intracellular calcium ions from about 2×10^{-7} M to 10^{-5} M results in activation of the contractile proteins, and contraction occurs. Relaxation occurs when the ion concentration is lowered below about 2×10^{-7} M. Thus concentration of intracellular calcium ions is understood to be the trigger or mediator of contraction.

The sarcoplasmic reticulum is capable of reversibly binding calcium ions at the appropriate concentrations. During activation, the reticulum and surface membrane are understood to be the sources of calcium ions and to be a sink for calcium during relaxation. Calcium ions enter myocardial cells with activation, and the amount entering is related to contractile vigor.[28]

In summary, electrical excitation at the outer cell membrane is transmitted to the interior of mammalian cardiac muscle by T tubules. Subsarcolemmal cisternae of the sarcoplasmic reticulum are thought to be activated from the surface cell membrane and through the T tubules. Calcium ions are released from the sarcoplasmic reticulum, thus activating the reaction between actin and myosin, resulting in contraction. Relaxation is thought to occur as a result of the rebinding of calcium ions by the sarcoplasmic reticulum.

CONTRACTILITY

Myocardial contractility is the term used to express the vigor of cardiac contraction. Although the term eludes exact definition, it represents a very important concept. The strength of contraction of skeletal muscle is modulated by recruitment: When a greater contractile effort is required of skeletal muscle, a greater number of fibers is activated by the innervating nerves. However, this is not the case in cardiac muscle; every myocardial fiber is activated with each heart beat, and the strength of contraction is modulated by an alteration of contractility. Physiologically, the autonomic nervous system modulates contractility by releasing the neurotransmitters, norepinephrine and acetylcholine.

Intuitively, contractility is the intrinsic property that makes contraction more vigorous or more forceful. Operationally, contractility is the fundamental attribute that increases with norepinephrine and calcium ion concentration and decreases in heart failure. In a sense, contractility is the mechanical reflection of the biochemical status of the muscle.

A clear distinction between contractility and cardiac performance must be made. Cardiac performance may be badly impaired by faulty heart valves limiting cardiac output, but the contractility of the muscle fiber may be excellent. Cardiac performance is an indication of how well the heart is pumping blood and serving the organism. Heart performance is dependent on intrinsic contractility but involves other factors, such as valve function and adequate blood volume, whereas contractility is solely a property of myocardial tissue.

Most observers define contractility independently of length. That is, the changes in contractile vigor or strength resulting from a change in the initial muscle length by virtue of the length-tension relation are not considered alterations in contractility. The positive inotropic action of norepinephrine or the decrease in contractility with heart failure does not result from altered sarcomere lengths. The change in contractile strength resulting from alterations in beat frequency (interval-strength relation) is also considered a change in contractility. A good definition of contractility has not yet been achieved, but a number of approaches to the problem are presented below.

Isometric Tension. A simple and usually reliable indication of altered contractility is the peak isometric tension developed in a cardiac beat. This must be observed at a constant initial length to avoid changes in the peak tension owing to the length-tension relation. In the sliding filament theory of muscle contraction, isometric tension is a reflection of the number of cross-bridges formed between thick and thin filaments, which is probably proportional to the number of calcium ions available for activation. A major problem with using isometric or isovolumic cardiac preparations

is that they are not coupled to the rest of the circulatory system. External shortening and thus cardiac output are then by definition zero, and the organism must be maintained by an artificial circuit. Therefore reflexes in the intact circulation are lost.

Peak isometric tension is a function of the active state duration. Under isometric conditions, internal shortening of the contractile element and lengthening of the series elastic element must occur for tension to be developed. Shortening of the contractile element takes time; thus isometric tension will lag behind active state, as shown in Figure 4–9. If the active state is shortened, peak isometric tension may decrease because time for tension development is inadequate. Thus an alteration of the duration of the active state without a change in the intensity of the active state will modify isometric tension or cardiac performance.

Brady[7] advocates tension measurement with the length of the contractile element fixed to avoid the problems of internal shortening. The logic is to measure first the series elastic element behavior and then pull on the muscle during systole so that the series elastic element is pulled out at just the correct rate to hold the contractile element length constant. This means that the muscle is no longer isometric but is elongated to hold the length of the contractile element constant throughout the beat. The resultant tension recorded is thus a representation of the active state in respect to time. The peak tension will not be altered by changes in the duration of the active state, because no internal shortening occurs.

This approach has a number of disadvantages. The cardiac muscle is even further removed from intact circulation than in the case of isometric contraction. This method requires that either a Maxwell or a Voigt three-element model be used for determining the amount of pull. The method must be used at a constant length because of the length-tension relation, but evidence shows that pulling on the muscle alters contractility—which is what is being measured.[7, 8]

V_{max}. V_{max} is the theoretical maximum shortening velocity of the contractile element at zero load. On the force-velocity plot this is the intercept point on the velocity axis (ordinate). V_{max} is a reflection of the rapidity of the conversion of chemical to mechanical energy when no mechanical hindrance or loading of the system occurs. The rate of myosin ATPase activity has been found to correlate well with the maximum shortening velocity for a number of muscles.[2] V_{max} may be a mechanical representation of myosin ATPase activity, but the rate of ATPase activity is not synonymous with contractility. V_{max} is a theoretical value which cannot be measured directly; some load is on the contractile element even if just the load of the muscle itself. This means that V_{max} is obtainable only by extrapolation from a set of points on a force-velocity curve. This extrapolation presents several difficulties. The form of the force-velocity curve must be known in order to be extrapolated to zero load. As noted above, the form of the force-velocity curve is not established.

If V_{max} is to be used as an index of contractility, a major assumption is that V_{max} is independent of length, which means that the unloaded shortening velocity does not vary with muscle length as isometric tension does. Shortening velocity at any finite load is clearly length dependent, and a critical part of the extrapolation to zero load is the correction for series and parallel elastic properties, using a Maxwell or Voigt model.[23, 33, 40]

Sonnenblick[36] examined V_{max} in cat papillary muscle, using a hyperbolic force velocity curve with no elastic correction, and found V_{max} independent of length. However, because of the difficulties given above, this has not been confirmed in other laboratories,[4, 30] and separate studies have found that V_{max} does vary with muscle length at both short[8] and long (near optimum)[32] muscle lengths.

The relation between V_{max} and the active state is unclear. V_{max} may vary from moment to moment with the active state, or could conceivably be independent of the time course of the active state. Actual force-velocity measurements with real loads are, of course, time dependent. To remove time dependence from V_{max} extrapolated from such time-varying curves would be difficult.

Three-Dimensional Representation. Fry[17, 18] proposed that a surface in the three-dimensional coordinates of length, load and shortening velocity could be used to characterize the state of cardiac muscle. The logic

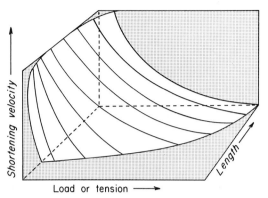

Figure 4–11 The three-dimensional surface characterizing myocardial state. The surface combines the property of length-tension with that of force-velocity. Active muscle traverses a path across this surface as determined by the three coordinates—muscle length, shortening velocity and load. This scheme does not include the effects of time or previous shortening on the active state. (After Fry, *Fed. Proc.*, 1962, *21*, 991–993.)

of this scheme is to conceive of a force-velocity relation with the addition of instantaneous muscle length (Fig. 4–11). This scheme incorporates the force-velocity and length-tension relations of cardiac muscle. A point on the surface represents the muscle at a given instantaneous length and load, with the resulting shortening velocity. Active muscle would trace a path in time on the surface, moving through successive lengths with shortening. An isotonic contraction would be a straight line on the surface, an auxotonic contraction a curved line.

An increase in contractility would result in an upward shift in the surface. That is, for a given instantaneous length and load, the muscle would manifest a greater shortening velocity. The collection of points from a muscle with higher contractility would form a new surface above the surface of lower contractility. The force-velocity-length surface is appealing because it unifies two important relational properties of cardiac muscle (force-velocity and length-tension) in a single scheme. Sonnenblick[38] has adopted three-dimensional analysis and has demonstrated an upward shift in the surface with increased contractility.

An assumption in three-dimensional analysis is that instantaneous muscle length, rather than initial muscle length, is important. However, a much more significant problem is that changes in the active state with time are not accounted for. A three-dimensional surface would be valid if the active state rose to a constant plateau during systole. This is clearly not the case. However, the force-velocity-length analysis could be expanded to four dimensions by adding time. This would then be a motion picture of the force-velocity-length surface rising and falling with the active state during the beat. A scheme with four-dimensional complexity is unwieldy and may be beyond a useful simplification.

An additional problem in cardiac muscle is that shortening abbreviates the active state.[13] The active state for a beat persists longer for an isometric beat than when shortening occurs. This is readily observable from superimposed isotonic and isometric tension records from the same muscle (Fig. 4–12). An instantaneous point in force, length and time does not uniquely determine the shortening velocity. The previous history of shortening also determines the state of the muscle. Thus, not even four dimensions are adequate for a rigorous definition of muscle state. If a four-dimensional analysis is unwieldy, an analysis of five dimensions is truly awkward.

Brutsaert and co-workers[8, 9a, 10] have examined the three-dimensional representa-

tion of cardiac muscle, using phase display in a plane where muscle length is one dimension and shortening velocity is the other. They find the three-dimensional surface of muscle state insensitive to time and the effects of muscle shortening. This implies that the active state rapidly rises to a constant plateau and that shortening does not alter the active state. Why the length-velocity plane analysis gives results contradictory to other methods is unknown.

Work. The amount of shortening times the load lifted equals the external work done by a muscle. For an intact heart the stroke volume times the aortic pressure equals the stroke work. Characterizing contractility by myocardial work is deceptively simple. The work accomplished in a beat is a function of the initial length as given in a Sarnoff-Berglund ventricular function curve[34] (Fig. 4–13). This curve relates initial ventricular size (often given as end diastolic pressure) to the resulting stroke work in an isolated supported heart preparation. The analogous curve for isolated cardiac muscle would show work as a function of initial length. However, work is also a function of load. This is intuitively clear if one considers three points on a load vs. work curve (Fig. 4–14). If load is zero, then any amount of shorten-

Figure 4–12 Shortening abbreviates the active state. Four isotonic afterloaded contractions (1-4) and one isometric contraction (5) are shown from the same isolated cat papillary muscle at 36° C. All contractions begin with the same preload and initial length. The light isotonic afterload (1) permits greater shortening which abbreviates the duration of the beat. If shortening did not abbreviate the duration of the active state, the duration of all the curves would be the same.

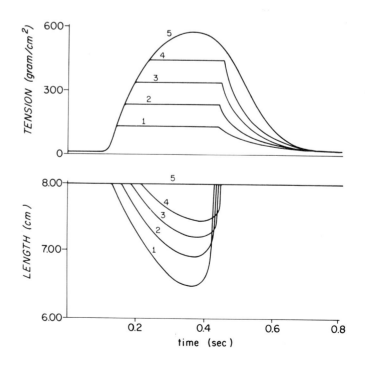

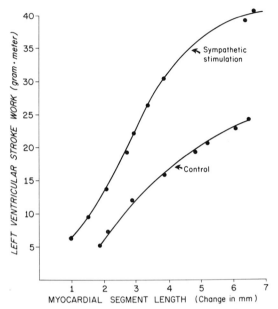

Figure 4–13 Ventricular function curves showing the relation between ventricular work and myocardial fiber length (a manifestation of the length-tension property of cardiac muscle). Increased cardiac work results from sympathetic stimulation. Ventricular function curves of this type fail to account for alterations in cardiac work which result from changes in afterload. (From Sarnoff *et al.*, *Circulat. Res.*, 1960, *8*, 1108–1122. By permission of the American Heart Association, Inc.)

ing times zero load equals zero work. If the afterload exceeds the peak isometric tension, then there will be no external shortening and no work. Thus work will be zero when the load is zero or when the load is equal to or greater than isometric tension. However, cardiac muscle does shorten against intermediate loads and does perform external work. The complete curve is given in Figure 4–14. The conclusion is that a change in work can result from a change in afterload without any change in contractility, invalidating the ventricular function curve as a measure of contractility.

A way around this difficulty is to examine the muscle with a fixed initial length and a fixed afterload and measure the extent of shortening. However, the problems of the time course of the active state or the effects of shortening on active state are not avoided.

Power and dP/dt. A number of indices for contractility have been proposed, based on the idea that time rates might be good

reflections of the myocardial state. Power is the rate of doing work and suffers from all the disadvantages of a work calculation. The dP/dt method records the peak rate of pressure rise in the ventricle during the isovolumic period. The maximum rate of pressure rise is dependent on initial muscle length, the series elastic stiffness and the magnitude of the afterload. Peak dP/dt usually occurs about the time of aortic valve opening, and aortic diastolic pressure in large part determines dP/dt.[42] Power and dP/dt measurements often follow contractility changes because positive inotropic sympathometric agents produce a more rapid onset of the active state. They do not add precision toward a definition of contractility.

In summary, contractility is the intrinsic property of cardiac muscle which describes the vigor of contraction. Contractility is a very important concept because it represents what is wrong with a failing heart and is the target of much cardiac therapy. Contractility is the intrinsic cardiac muscle property which determines the vigor of contraction independent of external loading, valvular function, filling pressure and the like. Cardiac performance is the function of the heart within the cardiovascular system and therefore includes load, valvular function and the like. Contractility has not been quantitatively defined because of the

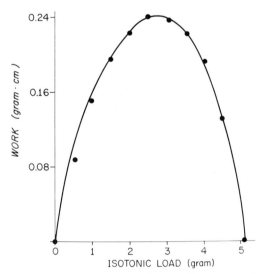

Figure 4–14 The external work (force times distance) performed by an isolated cat papillary muscle at 36° C. The muscle contracted with an isotonic afterload. Cardiac work is critically dependent on load.

complexities of cardiac mechanics. The two extremes of the force-velocity curve, maximum unloaded shortening velocity (V_{max}) and peak isometric tension (P_0) are under investigation for a definition of contractility. Both approaches have practical, technical and theoretical difficulties, primarily because of the time course of the active state in cardiac muscle. Contractility is probably inappropriately defined in terms of a single variable such as V_{max} or P_0 or even a surface in three dimensions (shortening velocity, load and length). The idea of contractility encompasses changes in V_{max} and P_0 but also the time course of the active state. Thus the contractility should be characterized by measurements of the force-velocity relation and the active state. Work and power are poor indicators of contractility because they are very dependent on the external loading of the heart.

REFERENCES

1. Abbott, B. C., and Mommaerts, W. F. H. M. A study of inotropic mechanisms in the papillary muscle preparation. *J. gen. Physiol.,* 1959, *42,* 533–551.
2. Bárány, M. ATPase activity of myosin correlated with speed of muscle shortening. *J. gen. Physiol.,* 1967, *50,* 197–218.
3. Blinks, J. R., and Koch-Weser, J. Physical factors in the analysis of the actions of drugs on myocardial contractility. *Pharmacol. Rev.,* 1963, *15,* 531–599.
4. Brady, A. J. Time and displacement dependence of cardiac contractility: problems in defining the active state and force-velocity relations. *Fed. Proc.,* 1965, *24,* 1410–1420.
5. Brady, A. J. Onset of contractility in cardiac muscle. *J. Physiol. (Lond.),* 1966, *184,* 560–580.
6. Brady, A. J. The three element model of muscle mechanics: its applicability to cardiac muscle. *Physiologist,* 1967, *10,* 75–86.
7. Brady, A. J. Active state in cardiac muscle. *Physiol. Rev.,* 1968, *48,* 570–600.
8. Brutsaert, D. L., Claes, V. A., and Sonnenblick, E. H. Effects of abrupt load alterations on force-velocity-length and time relations during isotonic contractions of heart muscle: load clamping. *J. Physiol. (Lond.),* 1971, *216,* 319–330.
9. Brutsaert, D. L., Claes, V. A., and Sonnenblick, E. H. Velocity of shortening of unloaded heart muscle and the length-tension relation. *Circulat. Res.,* 1971, *28,* 63–75.
9a. Brutsaert, D. L., Parmley, W. W., and Sonnenblick, E. H. Effects of various inotropic interventions on the dynamic properties of the contractile elements in heart muscle of the cat. *Circulat. Res.,* 1970, *27,* 513–522.
10. Brutsaert, D. L., and Sonnenblick, E. H. Force-velocity-length-time relations of the contractile elements in heart muscle of the cat. *Circulat. Res.,* 1969, *24,* 137–149.
11. Chapman, C. B., and Mitchell, J. H. *Starling on the heart.* London, Dawsons of Pall Mall, 1965.
12. Ebashi, S., Endo, M., and Ohtsuki, I. Control of muscle contraction. *Quart. Rev. Biophys.,* 1969, *2,* 351–384.
13. Edman, K. A. P., and Nilsson, E. The mechanical parameters of myocardial contraction studied at a constant length of the contractile element. *Acta physiol. scand.,* 1968, *72,* 205–219.
14. Fawcett, D. W., and McNutt, N. S. The ultrastructure of the cat myocardium. I. Ventricular papillary muscle. *J. Cell Biol.,* 1969, *42,* 1–45.
15. Feigl, E. O. Effects of stimulation frequency on myocardial extensibility. *Circulat. Res.,* 1967, *20,* 447–458.
16. Frank, O. Zur Dynamik des Herzmuskels. *Z. Biol.,* 1895, *32,* 370–447. Translated by C. B. Chapman and E. Wasserman in *Amer. Heart J.,* 1959, *58,* 282–317, 467–478.
17. Fry, D. L. Discussion. *Fed. Proc.,* 1962, *21,* 991–993.
18. Fry, D. L., Griggs, D. M., Jr., and Greenfield, J. C., Jr. Myocardial mechanics: tension-velocity-length relationships of heart muscle. *Circulat. Res.,* 1964, *14,* 73–85.
19. Gay, W. A., Jr., and Johnson, E. A. An anatomical evaluation of the myocardial length-tension diagram. *Circulat. Res.,* 1967, *21,* 33–43.
20. Gordon, A. M., Huxley, A. F., and Julian, F. J. Tension development in highly stretched vertebrate muscle fibres. *J. Physiol. (Lond.),* 1966, *184,* 143–169.
21. Gordon, A. M., Huxley, A. F., and Julian, F. J. The variation in isometric tension with sarcomere length in vertebrate muscle fibres. *J. Physiol. (Lond.),* 1966, *184,* 170–192.
22. Grimm, A. F., and Whitehorn, W. V. Myocardial length-tension sarcomere relationships. *Amer. J. Physiol.,* 1968, *214,* 1378–1387.
23. Hefner, L. L., and Bowen, T. E., Jr. Elastic components of cat papillary muscle. *Amer. J. Physiol.,* 1967, *212,* 1221–1227.
24. Hill, A. V. The heat of shortening and the dynamics constants of muscle. *Proc. roy. Soc.,* 1938, *B126,* 136–195.
25. Jewell, B. R., and Blinks, J. R. Drugs and the mechanical properties of heart muscle. *Ann. Rev. Pharmacol.,* 1968, *8,* 113–130.
26. Katz, A. M. Contractile proteins of the heart. *Physiol. Rev.,* 1970, *50,* 63–158.
27. Koch-Weser, J., and Blinks, J. R. The influence of the interval between beats on myocardial contractility. *Pharmacol. Rev.,* 1963, *15,* 601–652.
28. Langer, G. A. Ion fluxes in cardiac excitation and contraction and their relation to myocardial contractility. *Physiol. Rev.,* 1968, *48,* 708–757.
29. McNutt, N. S., and Fawcett, D. W. The ultrastructure of the cat myocardium. II. Atrial muscle. *J. Cell Biol.,* 1969, *42,* 46–67.
30. Noble, M. I. M., Bowen, T. E., and Hefner, L. L. Force-velocity relationship of cat cardiac muscle, studied by isotonic and quick-release techniques. *Circulat. Res.,* 1969, *24,* 821–833.
31. Parmley, W. W., and Sonnenblick, E. H. Series

elasticity in heart muscle. Its relation to contractile element velocity and proposed muscle models. *Circulat. Res.*, 1967, *20*, 112–123.

32. Parmley, W. W., Chuck, L., and Sonnenblick, E. H. The relation of V_{max} to different models of cardiac muscle. *Circulat. Res.*, 1972, *30*, 34–43.

33. Pollack, G. H. Maximum velocity as an index of contractility in cardiac muscle. *Circulat. Res.*, 1970, *26*, 111–127.

34. Sarnoff, S. J., and Mitchell, J. H. The control of the function of the heart. *Handb. Physiol.*, 1962, sec. 1, vol. I, 489–532.

35. Sonnenblick, E. H. Force-velocity relations in mammalian heart muscle. *Amer. J. Physiol.*, 1962, *202*, 931–939.

36. Sonnenblick, E. H. Implications of muscle mechanics in the heart. *Fed. Proc.*, 1962, *21*, 975–990.

37. Sonnenblick, E. H. Series elastic and contractile elements in heart muscle: changes in muscle length. *Amer. J. Physiol.*, 1964, *207*, 1330–1338.

38. Sonnenblick, E. H. Instantaneous force-velocity-length determinants in the contraction of heart muscle. *Circulat. Res.*, 1965, *16*, 441–451.

39. Sonnenblick, E. H. Active state in heart muscle. Its delayed onset and modification by inotropic agents. *J. gen. Physiol.*, 1967, *50* (3), 661–676.

40. Sonnenblick, E. H. Contractility of cardiac muscle. *Circulat. Res.*, 1970, *27*, 479–481.

41. Spiro, D., and Sonnenblick, E. H. Comparison of the ultrastructural basis of the contractile process in heart and skeletal muscle. *Circulat. Res.*, 1964, *15*, 14–37.

42. Wildenthal, K., Mierzwiak, D. S., and Mitchell, J. H. Effect of sudden changes in aortic pressure on left ventricular dp/dt. *Amer. J. Physiol.*, 1969, *216*, 185–190.

CHAPTER 5 ELECTRICAL CHARACTERISTICS OF THE CARDIAC CELL

by ALLEN M. SCHER *and* W. GLENN L. KERRICK

Earlier chapters of this text (Chaps. 1 and 3, Vol. I) have discussed in detail the origins of transmembrane potentials in living cells and the ionic mechanisms underlying the origin and propagation of the action potential in nerve and muscle. This chapter will discuss the origin of the resting membrane potential and action potentials of cardiac muscle cells, and will present related material concerning the conduction of action potentials from one cardiac cell to another. The discussion will, of necessity, repeat some material from the earlier chapters, which can be consulted for more detail.

TRANSMEMBRANE IONIC CONCENTRATIONS IN CARDIAC CELLS

The membrane of the resting cell is a barrier to the diffusion of materials into or out of the cell. Diffusion of ions across the membrane is far slower than diffusion in an aqueous medium, but the cell membrane does more than impose a passive barrier to diffusion. Through both "active" and passive properties of the membrane, ionic concentration and voltage differences are maintained across cell membranes. Extracellular fluid contains high concentrations of sodium (Na^+) and chloride (Cl^-) ions and low concentrations of potassium (K^+) ions. Intracellular fluid contains low concentrations of sodium and chloride ions and high concentrations of potassium ion (Table 5–1). In Table 5–1 the transmembrane concentration and potential differences for these ions and others are shown for skeletal muscle, which is similar to cardiac muscle. Figure 5–1 shows this distribution diagrammatically. The observed concentration differences are established and maintained by active and passive properties of the membrane. The metabolic processes transport sodium out of the cell and potassium into the cell. This coupled sodium-potassium transport system is loosely referred to as the

Table 5–1 *Approximate Steady State Ion Concentrations and Potentials in Mammalian Muscle Cells and Interstitial Fluid* °

INTERSTITIAL FLUID		INTRACELLULAR FLUID		$\xi ion = $	
[Ion]		[Ion]		$\dfrac{[Ion]o}{[Ion]i}$ $\dfrac{61}{Z}$ log $\dfrac{[Ion]o}{[Ion]i}$ (mV)	
μM per cm^3		μM per cm^3			
Cations		Cations			
Na$^+$	145	Na$^+$	12	12.1	66
K$^+$	4	K$^+$	155	1/39	−97
H$^+$	3.8×10^{-8}	H$^+$	13×10^{-8}	1/3.4	−32
pH	7.43	pH	6.9		
Others	5				
Anions		Anions			
Cl$^-$	120	Cl$^-$	4†	30	−90
HCO$_3^-$	27	HCO$_3^-$	8	3.4	−32
Others	7	A$^-$	155		
Potential	0		−90 mV	1/30	−90

The left hand columns give the approximate concentrations of the more important ions in the interstitial fluid of mammals. Intracellular concentrations are estimated from chemical analysis of a known weight of tissue and a measurement of the fraction of tissue water which is in the interstitial space. The total amount of any ion in the interstitial fluid is then obtained by the product of the interstitial concentration and the fractional volume. This amount is subtracted from the total of ion in the tissue sample to give the amount of ion in the intracellular water. Intracellular concentration is the ratio of the amount of ion to the amount of water in the cells. The middle columns show the concentrations of the more important ions in the intracellular water of mammalian skeletal muscle. Although intracellular concentrations vary considerably from tissue to tissue, the electrolyte pattern of muscle is fairly representative. To summarize, interstitial fluid has high concentrations of Na$^+$ and Cl$^-$; the interstitial fluid has high concentrations of K$^+$ and the largely unknown organic anions (A$^-$). The right hand column indicates the transmembrane potential predicted by the Nernst equation from the distribution of each ion. (From Woodbury in: *Physiology and biophysics*, 19th ed., T. C. Ruch and H. D. Patton, eds. Philadelphia, W. B. Saunders Co., 1965.)

°Vertical double line represents membrane.
†Calculated from membrane potential using the Nernst equation for a univalent anion, *i.e.*, Z = −1.

"sodium-potassium pump." This transport of ions is coupled to enzymatic processes and metabolic energy utilization. Detailed discussion of the transport mechanisms can be found elsewhere (Chap. 24, Vol. II).

TRANSMEMBRANE IONIC DISTRIBUTION: ACTIVE TRANSPORT, CONCENTRATION AND VOLTAGE EFFECTS

The distribution of ions across biological membranes is determined by active transport, referred to above, by selective permeability of the membrane to certain ions, and by certain passive mechanisms which arise from transmembrane ionic concentration and electrical potential differences. A molecule (or ion) which is found at differing concentrations on the two sides of a membrane will, if the molecule or ion can pass through the membrane, move across it to eliminate the concentration difference. The ease with which this movement is accomplished is a function of the *permeability* of the membrane to the substance in question. Ions which are in solution on two sides of a permeable membrane will move across the membrane owing to transmembrane concentration differences and/or electrical forces (electrical field caused by an electrical potential difference across the membrane). For example, positive ions will move to the more negative side of the membrane. Any ion or molecule will move (diffuse) across a membrane so as to eliminate transmembrane concentration differences. If a permeable membrane is bathed on both sides with solutions of identical ionic composition and an electrical potential is then applied, an ion will initially move across the membrane because of electrical forces. This ionic movement will establish a transmembrane concentration difference of the ion,

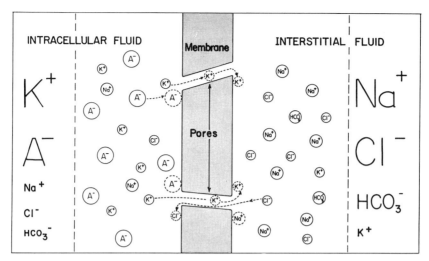

Figure 5–1 Development of transmembrane voltage by an ion concentration gradient. Diagram of an intracellular fluid–membrane–interstitial fluid system. Hypothetical membrane is pierced by pores of such size that K^+ and Cl^- can move through them easily, Na^+ with difficulty and A^- not at all. Sizes of symbols in left- and right-hand columns indicate relative concentrations of ions in fluids bathing the membrane. Dashed arrows and circles show paths taken by K^+, A^-, Na^+ and Cl^- as a K^+ or Cl^- travels through a pore. Penetration of the pore by a K^+ or Cl^- follows a collision between the K^+ or Cl^- and water molecules (not shown), giving the K^+ or Cl^- the necessary kinetic energy and proper direction. An A^- or Na^+ unable to cross the membrane is left behind when a K^+ or Cl^-, respectively, diffuses through a pore. Because K^+ is more concentrated on left than on right, more K^+ diffuses from left to right than from right to left, and conversely for Cl^-. Therefore right-hand border of membrane becomes positively charged (K^+, Na^+) and left-hand negatively charged (Cl^-, A^-). Fluids away from the membrane are electrically neutral because of attraction between + and − charges. Charges separated by membrane stay near it because of their attraction. (From Woodbury, in *Physiology and biophysics*, 19th ed., T. C. Ruch and H. D. Patton, eds. Philadelphia, W. B. Saunders Co., 1965.)

which will cause more ions to diffuse across the membrane toward the side of its lowest concentration. The ionic movement caused by the concentration difference is in a direction opposite to that which would be expected from the electrical forces. Ultimately, at equilibrium the movement of the ion caused by the concentration difference will be equal and opposite to the movement caused by the electrical potential difference, and no net movement of ions across the membrane will occur.

The aforementioned example shows that the transmembrane concentration difference can be established by a transmembrane potential difference. In similar fashion, transmembrane concentration differences can produce transmembrane potential differences. The establishment of a potential difference caused by a concentration difference across a cell membrane can be illustrated by considering the hypothetical situation in which a concentration difference of a salt across the cell membrane is suddenly created, no transmembrane potential exists at first and the membrane is permeable to

only one ion. Initially, the ion will move from the side of the membrane where its concentration is highest to the side of the lowest concentration. This movement will transfer electrical charges associated with the ion from one side of the membrane to the other, and this separation of charge across the membrane will establish a transmembrane electrical potential difference. This transmembrane potential will exert a force on the ion in the opposite direction to ion movement caused by the concentration gradient, and ultimately the resulting fluxes of the ion will be equal and opposite. In the real resting cardiac muscle cell, the concentration difference of the ion is established by the "sodium-potassium pump," and the ion, analogous to that ion in the example just given, is K^+, because the resting membrane is relatively more permeable to K^+ than to Na^+. Other ions which are not actively pumped across the membrane and to which the membrane is permeable (Cl^-) will ultimately distribute themselves, as described in the previous example, so that the fluxes of each ion caused by electrical and chemical gradients are equal and opposite. Details of the ionic movement and establishment of the transmembrane potential are described in Chapter 1, Volume I.

The Nernst Equation. The quantitative relationship (for passive ion distribution) between the equilibrium transmembrane concentration difference and the equilibrium transmembrane voltage difference is stated by the Nernst equation. This equation applies when there is no net movement of the ions across the membrane caused by electrochemical gradients. The relationship is as follows:

$$\mathscr{E}_m = RT/FZ_x \log_e [X]_o/[X]_i \quad (1)$$
$$\text{(Nernst equation)}$$

where $\mathscr{E}_m$ is the electrical potential across the membrane, R the universal gas constant, T the absolute temperature, F the Faraday (number of coulombs per mole of charge), Z_x the valence of the ion X, $\log_e$ indicates natural logarithms (to the base e, 2.718) and $[X]_o$ and $[X]_i$ indicate concentrations of the ion X outside and inside the membrane. A less complicated equation derived from (1) for a monovalent ion at 37° C is as follows:

$$\mathscr{E}_m = 61 \log_{10} [X]_o/[X]_i \quad (2)$$

Here we have converted to base 10 logarithms, and $\mathscr{E}_m$ is in millivolts. The constant, 61, includes the terms R, T, F, Z_x and the change in logarithm base. Note again that this equation applies to a monovalent ion which is distributed at equilibrium across the membrane.

THE RESTING POTENTIAL

Selective Permeability to K$^+$. The resting cardiac cell, in common with other excitable cells (Chap. 1, Vol. I), displays a property which is important for the generation of a resting and action potential. The cell membrane displays a selective permeability to ions. This permits some ions to cross the membrane relatively freely, while restricting the movement of others. Four ions are important for our consideration (Fig. 5–1). These are sodium, potassium, chloride and an unknown anion which is present intracellularly in a sufficient amount to make positive and negative ions equal in the cells and extracellular fluid. Other ions are present but are less important. Of

these four ions, the resting membrane is preferentially permeable to potassium and chloride. Permeability to these ions is not large, but it is far greater than that to sodium or to the unidentified anion. As described previously, K$^+$ is actively transported across the membrane by the sodium-potassium pump, whereas chloride is not. Pumping activity establishes a potassium and a sodium concentration gradient. Owing to the greater membrane permeability to K$^+$, more actively transported K$^+$ than Na$^+$ moves back across the membrane carrying its associated charge. This movement establishes a voltage gradient across the membrane which minimizes the net movement of K$^+$ caused by K$^+$ concentration difference. There is thus an excess of positive charge on the outside of the cell and an excess of negative charge on the inside. Cl$^-$ is not actively pumped, and its distribution across the membrane is a passive response to the transmembrane voltage gradient.

As in other excitable tissues (Chap. 1, Vol. I), the number of ions which are separated across the cell membrane to produce the resting potential is a very small fraction of the total number of ions present in the cell. According to Woodbury's calculations, a separation of about one in a million is sufficient to establish the transmembrane electrical potential. This concentration difference is not measurable by chemical means. Nevertheless, a sizable transmembrane electrical potential difference is established. The movement of ions through the membrane appears to be via channels which are highly selective for particular ions. Permeability is determined by the size of the hydrated ions, features of their geometry and associated charge, and the membrane geometry and charge distribution. Selective permeability is a feature of the resting cell, and we will later see that membranes can undergo changes in their ionic permeability so that other channels open or close with resultant changes in transmembrane current and voltage.

If we return to the Nernst equation (2), the concentration of potassium outside the cell is approximately 4 mM. The concentration of potassium inside is approximately 155 mM. Utilizing the equation, we would expect the potassium equilibrium potential to be approximately −97 mv:

$$\mathscr{E}_{K^+} = 61 \log_{10} 4/155 = -97 \text{ mv} \quad (3)$$

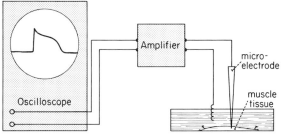

Figure 5–2 At the right of the figure a strip of muscle tissue is immersed in a bath of physiological fluid. A hollow glass ultramicroelectrode is shown penetrating the tissue mass. A ground electrode makes contact with the bath. The potential difference between the ultramicroelectrode and the ground electrode is amplified and displayed on an oscilloscope. The horizontal axis is a time base. The potential shown is similar to that recorded by an ultramicroelectrode which penetrates a cardiac muscle cell.

We would also expect the transmembrane electrical potential to be approximately −97 mv (inside negative with respect to outside the cell), because the membrane is much more permeable to K⁺ than Na⁺. The resting potential can be measured by inserting into the cell a hollow glass ultramicroelectrode filled with a concentrated salt solution (Fig. 5–2). The electrode serves as a conducting contact to the inside of the cell and is used to record potential difference between the inside of the cell and an extracellular reference point. For the cardiac cell the resting potential is −90 mv, and the inside of the cell is negative with respect to the outside. This is obviously

quite close to the calculated potassium resting potential of −97 mv. Since Cl⁻ is not actively transported, and since the membrane is permeable to Cl⁻, it is distributed passively according to the Nernst equation (Table 5–1). Sodium ion, unlike potassium and chloride, is not distributed according to the Nernst equation, and thus is not in equilibrium. The large sodium concentration gradient across the membrane, with a high extracellular sodium concentration, and the negative membrane potential would tend to drive sodium into the cell. From the Nernst equation, we can calculate the sodium equilibrium potential from Table 5–1 to be +66 mv. The membrane potential would be positive, rather than negative, in the resting state if the permeability to sodium ion were much greater than to other ions. If the external potassium is altered in the normal resting cell, the resting potential will vary with the logarithm of the transmembrane potassium concentration ratio,[24, 28] as predicted by the Nernst equation (Fig. 5–3). There are some deviations from a linear relationship which reflect the fact that the membrane is slightly permeable to sodium. These deviations show up at low K⁺ concentrations, where Na⁺ concentration differences have a greater effect.

The establishment of a resting potential by pumping of sodium, coupled with pumping of potassium, is not confined to excitable tissues or to animal tissues but seems (with variations in the actual concentrations) to be a universal characteristic of the living

Figure 5–3 The resting membrane potential (vertical axis) of cardiac (auricular) cells changes as shown when the external K⁺ concentration (horizontal axis) is altered (solid triangles). The dotted line indicates the changes which would be predicted from the normal intracellular concentration according to the Nernst equation. The discrepancies between the curves are considered to be due to the fact that the fibers are permeable to other ions; these are particularly important at low external K⁺ concentrations. (After Trautwein and Dudel, *Pflügers Arch. ges. Physiol.*, 1958, 266, 324–334.)

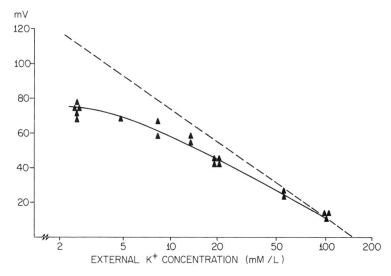

cell. The ionic concentration differences and resultant resting potential are roughly similar for nerve, skeletal muscle and cardiac muscle cells. There are some differences in the exact magnitude of transmembrane ionic concentration and resting potentials in various cells. Further, this description applies to those types of cardiac muscle cells – atrial cells, ventricular cells and Purkinje fibers – which do not normally exhibit pacemaker behavior.

THE ACTION POTENTIAL

Changes in Selective Permeability; Conductance. Although all cells utilize ionic pumping mechanisms and display transmembrane ionic concentration differences, excitable cells, *i.e.*, nerve and muscle cells, have an additional and unique property. When the transmembrane voltage is altered, the membrane of excitable cells undergoes a change in its permeability and thus its conductance to the various ions which are separated across the cell membrane. These conductance changes are due to the opening and closing of various ionic channels in the membrane. As the membrane conductance changes, the transmembrane voltage will be determined by the conductance and by the transmembrane concentration differences.

When a charged particle, *i.e.*, an ion, moves across the membrane, its movement is accompanied by a transfer of electrical charge (current) across the membrane. The term "conductance" indicates how easily current flows across a membrane for a given voltage difference. For ions, permeability and conductance are proportional and conductance is a measure of ionic permeability. Current flow, associated with ion movement, is far easier to measure than chemical indications of permeability, such as concentration differences. The changes in membrane conductance, brought about by changes in transmembrane voltage, are usually transient. The onset of an increase in membrane conductance is referred to as "activation," and a decrease in conductance as "inactivation." Activation and inactivation depend on voltage or time, or both. We have indicated that in the resting cell the transmembrane voltage is determined by the potassium ion but that, owing to a low

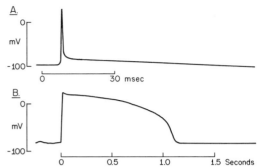

Figure 5–4 Action potentials of frog skeletal muscle fiber (A) and of single cardiac muscle fiber (B) from frog ventricle. The ordinate shows transmembrane potential, and the abscissa, time. Note that skeletal muscle action potential is a few msec in duration while cardiac action potential is near 1 sec in duration.

permeability and thus low conductance to sodium, the sodium concentration difference does not influence the resting membrane potential. As sodium permeability increases, so that the membrane becomes more permeable to sodium relative to potassium, sodium current will tend to depolarize the membrane toward the sodium equilibrium potential (+66 mv). When the membrane's permeability for the different ions changes, the transmembrane voltage will be determined by the permeabilities and by the transmembrane ionic concentration differences.

Action potentials are associated with information transmission in nerve cells and are necessary precursors of mechanical activity in skeletal, smooth and cardiac muscle cells (Fig. 5–4). The action potential is the change in transmembrane voltage determined by a sequence of voltage and time-dependent permeability changes to various ions. It is ideally measured with the ultramicroelectrode and is then referred to as an intracellular potential (Fig. 5–4).

Action Potential Shapes of Cardiac Cells. The depolarization of the heart begins in the pacemaker cells of the sinus node, proceeds through the cells of the atrium, encounters the specialized cells of the atrioventricular node, and passes through these into the common bundle. Thence it moves into the right and left conducting bundles. The ventricular terminations of the conducting bundles (Purkinje fibers) bring the electrical activity to the ventricular cells. Details of activation of the heart are presented in another chapter (Chap. 6, Vol. II). The cells of the sinus node and atrioventricular node are small (3.5 to 5 μ diameter),[26] are embedded in connective tissue and conduct impulses very slowly. The cells of the common bundle, right and left bundles and the Purkinje cells are larger (20 μ)[26] and appear to conduct impulses very rapidly (about 2 m per sec). Atrial cells are medium-sized (12 μ),[26] conducting impulses at about the rate of 0.8 m per sec. Ventricular cells (11 μ) are smaller than Purkinje cells in human hearts and conduct impulses at a velocity of about 0.4 m per sec. The so-called specialized tissues include the sinus and AV nodes, the conducting bundles and the Purkinje fibers.

There are thus four major types of cardiac cells: (i) nodal cells, from both the sinus node and AV node (these often function as pacemakers—see below); (ii) atrial cells; (iii) Purkinje cells (which can act as pacemakers); and (iv) ventricular cells. Each of these cell types has a unique intracellular action potential shape[5] (Fig. 5-5), and yet all have in common certain aspects of action potential shapes. The action potentials of nonpacemaker cells all feature a relatively rapid initial depolarization, usually from approximately −90 to +30 mv. The fibers tend to remain depolarized for several hundred milliseconds, and then repolarize slowly. Some fibers show a maintained "plateau" (voltage maintained near zero for periods up to several hundred milliseconds) after initial depolarization.

The action potentials of atrial cells show a rapid depolarization. The voltage at the peak of the action potential reaches +20 to +30 mv, and this is followed by a slow return to the resting level, which takes about 200 msec, and during which time the voltage changes at a relatively constant rate. Ventricular cells show the same rapid depolarization, a rapid partial repolarization to about zero potential, after which the cells remain for some time at a "plateau" level (near zero potential) for about 200 msec, and a final phase of repolarization which returns the cells to resting potential over about 200 msec. Cells of the Purkinje system have an action potential similar to that of ventricular cells but with a more marked early repolarization "notch," a more sloping "plateau" and a longer duration. Pacemaker cells of the sinus node show a spontaneous slow depolarization to threshold potential, followed by a more rapid phase of depolarization to a positive peak potential (inside of the cell positive relative to the outside). The rate of rapid depolarization and the size of the action potential are less than those found in other cardiac cells. AV nodal cells have action potentials similar to pacemaker cells, except that the initial slow depolarization to threshold is much slower.

Study of the Action Potential. As indi-

Figure 5–5 Intracellular action potential shapes at different sites of the heart. Pacemaker type (*A*) is characteristic of sinus node and AV node. Atrial action potential (*B*) has neither diastolic depolarization of (*A*) nor extended plateau seen in ventricular cells (*D*). Purkinje fiber action potential (*C*) has longer plateau than ventricular action potential (*D*). Voltage and time calibration above. Zero transmembrane voltage is indicated on each action potential. Action potential shapes after Brooks *et al.*, *Excitability of the heart*. New York, Grune & Stratton, 1955.

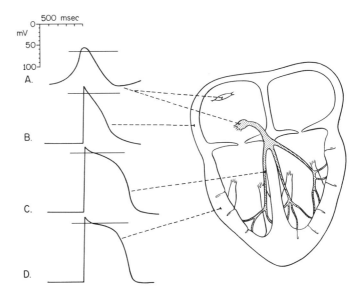

cated, the cardiac action potential is caused by a series of changes in membrane conductance which are initiated by a decrease in the resting potential to the threshold potential. Two problems exist: (i) to describe the sequential changes in membrane permeability in the cardiac cells and the resultant changes in transmembrane potential, and (ii) to indicate the molecular mechanisms by which these changes occur. This second problem is not well understood and will not be discussed here. Concerning the first problem, as indicated elsewhere (Chap. 1, Vol. I), three experimental procedures have been utilized to study membrane properties: (i) alter the transmembrane ionic concentration (usually by changing the extracellular concentration) and determine the effects on action and resting potentials, (ii) measure the ionic fluxes across the membrane utilizing radioactive ions and (iii) *clamp* (hold) the transmembrane voltage to a constant level and record the resultant transmembrane current changes, utilizing an intracellular electrode or some equivalent procedure. To illustrate the first procedure, consider a cell which at some time is permeable to a single ion—for example, K^+. If that ion's transmembrane concentration is altered, the transmembrane potential will be altered. Earlier we discussed this mechanism in connection with the resting potential.

The second procedure, use of radioactive isotopes, will demonstrate whether or not an ion moves across the membrane (radioactive ions placed extracellularly will be diluted in concentration if there is exchange with cellular ions, etc.); *e.g.*, separation of ions across the membrane may or may not be accompanied by movement of a particular ion during an action potential.

The third of these procedures, voltage clamping, has been most useful in determining the changes in membrane conductance which underlie the action potential.[5] The membrane conductance during the action potential varies with transmembrane voltage and with time. The voltage clamp allows the investigator to control the membrane voltage. In one type of study, two microelectrodes are placed within the same cell with one electrode used to record the transmembrane voltage; the second furnishes current to hold (clamp) that voltage at some level determined by the experimenter. Usually the membrane is brought rapidly from rest to a new transmembrane voltage, and the changes in current as a function of time after this step are measured. This current is proportional to the membrane conductance, and is therefore a measure of conductance changes in the time following the change in voltage. The procedure may be supplemented by altering the extracellular concentration of ions, or by using drugs which block conductance increases of specific ions. As a further example, voltage clamp depolarization to the Na^+ equilibrium potential ($+66$ mv) will eliminate Na^+ current (as determined by the Nernst equation), and allow the investigator to study time and voltage dependent conductance changes of the membrane to ions other than Na^+.

INITIAL STAGE OF DEPOLARIZATION: THRESHOLD. When a current is applied to the resting cell through an intracellular electrode, in such a fashion as to depolarize the cell (positive terminal inside the cell), several things happen. First, K^+ will move out of and Cl^- will move into the cell, since the ion flux caused by electrical forces on these ions will be less than that caused by the concentration gradient. The current carried by K^+ and Cl^- will tend to repolarize the membrane to the resting level. The lowered potential leads to a second change: the membrane increases its conductance to Na^+. The extent of the increase in Na^+ permeability is dependent on the degree of depolarization. If the current applied is small, the outward ionic current carried by K^+ and Cl^- will be greater than the ionic current carried inward by Na^+. Termination of current application will be followed by a return of the membrane to the resting potential ($I_{Cl^-} + I_{K^+} > I_{Na^+}$). (In this discussion, we are neglecting transient currents which charge membrane capacitance. These are treated elsewhere [Chap. 3, Vol. I].) If the applied current is somewhat larger, so that the resting membrane potential is decreased by about 30 mv, the voltage-dependent increase in Na^+ permeability will cause sodium inward current to exceed the outward current of K^+ and Cl^-, and the cell will depolarize. The sodium conductance increases, and this increased Na^+ conductance leads to movement of Na^+ ions into the cell, which depolarizes the membrane still further (*i.e.*, brings it closer to the Na⁺ equilibrium potential). This decrease in voltage again further increases the permeability to sodium, which again lowers the membrane potential. The maximal change in membrane permeability to sodium is substantial; during rest the Na⁺ permeability is 1/50 of the K⁺ permeability. At the peak of this initial phase of depolarization, the permeability to Na⁺ increases 500 times so that it greatly exceeds the K⁺ permeability. The electrical resistance of the membrane decreases markedly.[29] The regenerative changes producing rapid depolarization are referred to as the Hodgkin cycle (Chap. 1, Vol. I). The phenomena near threshold have an explosive character owing to the regenerative nature of the Na⁺ permeability change, and the threshold— which looks like a discontinuous event— arises out of smooth and continuous voltage-dependent changes in permeability.

Since the membrane at the peak of depolarization is preferentially permeable to Na⁺, the transmembrane potential is predominantly influenced by the transmembrane Na⁺ concentration difference. The transmembrane potential reaches a peak of about +30 mv; the sodium equilibrium potential is greater than this, approximately +66 mv. Thus the membrane, although predominantly permeable to Na⁺ at this time, is also permeable to other ions.

Much evidence supports the importance of Na⁺ in this phase. (i) When the external sodium concentration is altered, the peak height of the action potential (or overshoot beyond zero) is (roughly) linearly related to the logarithm of the external sodium concentration (Fig. 5–6).[3, 7] (ii) Weidmann found that the initial rate of rise of the action potential, which is a measure of the number of ions available to carry current, varied with external sodium concentration.[29] The more sodium available, the faster the regenerative Hodgkin cycle. (iii) The drug tetrodotoxin, which blocks Na⁺ channels, decreases both the magnitude of the peak of the action potential and the rate of change of voltage during this period.[9] The kinetics of these permeability changes during this phase of the action potential are similar to those found in nerve and skeletal muscle.

Later Phases of the Action Potential. Most of what has been said about the initial phase of the action potential is widely accepted and experimentally validated, and

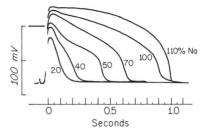

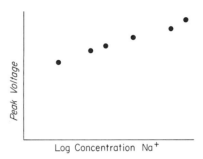

Figure 5–6   *Top:* Superposition of action potentials from frog ventricular fibers perfused with solutions containing different concentrations of Na⁺. Note the change in peak voltage during the spike. Resting potential does not change. *Bottom:* A plot of peak voltage against the logarithm of the Na⁺ concentration from the top figure. (After Brady and Woodbury, *Ann. N.Y. Acad. Sci.,* 1957, 65, 687–692.)

is similar to what occurs in other excitable tissues. Conclusions about the remainder of the action potential are less firmly established and more controversial. The uncertainty about these later phases is largely due to a very substantial difficulty in performing clear-cut experiments.[14] In this discussion we will not distinguish between the various types of cardiac cells. Most of the available information about cardiac cells comes from study of Purkinje fibers.

INACTIVATION OF Na⁺ CONDUCTANCE. The Na⁺ conductance change which leads to rapid depolarization would, if it continued, bring the cell to a new transmembrane potential from which it could not deviate. As indicated, once such conductance changes are initiated, a counterprocess, Na⁺ inactivation (closing of the Na⁺ channels), commences.

In nerve and skeletal muscle, Na⁺ permeability decreases rapidly to near the resting level. In contrast, cardiac muscle Na⁺ permeability does not return to the resting level. Na⁺ permeability remains at an intermediate level for several hundred milliseconds following an initial rapid decrease in Na⁺⁺ permeability. The rapid fall in potential after the peak of the depolarization spike reflects this rapid decrease in Na^+ permeability.

CHLORIDE CURRENT. The initial rapid repolarization from the peak of the depolarization spike, seen in Purkinje fibers, is in part the consequence of a rise in inward chloride current (Cl^- moving out of the cell).[8] Some inactivation of sodium, as described above, also causes a decrease in depolarizing current. Therefore the membrane tends to repolarize.

CURRENTS DURING THE PLATEAU. A "plateau" (*i.e.,* a period during which the membrane potential is relatively constant) characterizes Purkinje cells and ventricular cells. Other cardiac cells show a prolonged action potential duration, and it seems possible that they undergo similar ionic changes during the period between depolarization and final repolarization. Currents during the plateau period have been studied by many investigators.[5, 6, 19] During the plateau the membrane potential is close to zero and the sodium permeability is greater than at rest. In addition, the data from voltage clamp experiments indicate that the conductance to potassium *decreases* immediately after the beginning of the action potential and stays at a lower-than-resting level during the plateau period. (This is in contrast to nerve, where potassium conductance begins to *increase* immediately after the initial rapid depolarization, and where this increased K^+ conductance aids in terminating the action potential.) The K^+ channel which appears to "cut off" at this time is called K_1, and it has a rectifying property to be discussed. In addition to the Na^+ and K^+ conductance changes, there are important effects of Ca^{++} and of membrane rectification during the plateau.

CALCIUM CURRENT. Over the last several years, considerable debate has concerned a slow inward current in myocardial fibers and the role of calcium in this current. Early observations[16] indicated that changing Ca^{++} concentration could alter currents during the plateau, but since Ca^{++} alters membrane properties such evidence was not sufficient to demonstrate a role in the normal action potential. In addition, voltage clamp experiments which were directed at this question were difficult. Recent data,[6, 15, 20] however, indicate that a slowly developing, and slowly inactivated,

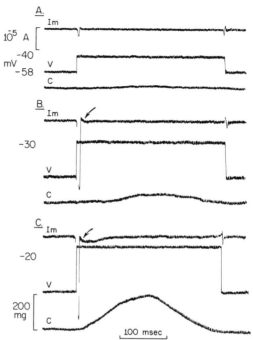

Figure 5–7 Effects of voltage clamping (sucrose gap technique) in cat papillary muscle to three different voltage levels. *Top trace:* Applied current (Im), an indication of conductance changes. *Middle trace:* Transmembrane (clamped) voltage (V). *Bottom trace:* Developed tension (C). The small depolarization in *A*, from −58 to −40 mv, evokes a sodium spike seen slightly to the right of the voltage step in the top curve in *A*. (The current record has been shifted slightly to the right to make the figure clearer.) Note that the trace in *A* is relatively flat, which indicates no conductance change beyond the initial sodium spike, and that there is very little contraction seen in the bottom curve. In *B*, the depolarizing step is larger and takes the membrane to −30 mv. Note that in the top trace there is a slight "sag" of the current (slow inward current) which extends beyond the initial sodium spike, and that the lower trace shows a slight development of tension. In *C*, the depolarization is larger, taking the membrane to −20 mv. The slow inward current is more apparent in the top trace and the tension is larger than 200 mg. (After New and Trautwein, *Pflügers Arch. ges. Physiol.,* 1972, 334, 24–38.)

calcium current exists during early depolarization. This slow Ca^{++} current is not large, but it can be separated from the Na^+ current. It is of importance in maintaining the plateau, and it is of major importance in cardiac function, because increases in intracellular Ca^{++} concentration are necessary for cardiac muscle contraction. Depolarization of the cardiac cell to the level at which this slow inward current is activated appears to be necessary for tension development and modulation (Fig. 5–7). This inward Ca^{++} current appears to be necessary to move Ca^{++} to the sites where it is stored intracellularly, and to be responsible for triggering the release of Ca^{++} from these stores to initiate contraction. The importance of these Ca^{++} stores and of Ca^{++} is discussed in Chapters 4 and 13, Volume II.

RECTIFICATION. During the plateau the membrane displays rectification. (Current moves across the membrane more easily in one direction than in the other.) The cardiac cell membrane shows anomalous (or inward) rectification of K^+ current. K^+ current is able to move across the membrane from outside of the cell to inside easier than in an outward direction. Since this rectification is in the opposite direction from that which would be expected for a cell with a large internal K^+ concentration relative to the external concentration, it is called anomalous.[17] During the plateau phase of the action potential, the electrochemical gradient causes K^+ current to flow out of the cell, which tends to repolarize the cell. However, because of anomalous rectification this K^+ current is very much reduced, and thus this repolarizing K^+ current does not exceed the depolarizing effects of Na^+ and Ca^{++} currents. The change in transmembrane potential is small during the plateau of the action potential, because the depolarizing currents of Na^+ and Ca^{++} are essentially equal and opposite to the repolarizing currents of K^+ and Cl^-. Explanation of rectification requires the use of complicated physical models which will not be discussed in detail.[1]

In a simple model of rectification, the electrical resistance (conductance) of the membrane will vary with the number of available ions to carry current. Outward K^+ current will be facilitated by the large number of K^+ ions inside the cell; but inward current will be more limited owing to the lower external K^+ concentration. This example of rectification, seen as a result of K^+ concentration differences, should favor outward current. As indicated, the observed (anomalous) rectification during the plateau opposes outward current. One suggested model for this is that some ("competitive") positively charged substance is moved into the K^+ channels in the membrane whenever an outward K^+ current commences. This substance, perhaps owing to its size, closes off the K^+ channel for one direction of K^+ movement, but does not obstruct it if the direction of K^+ movement is reversed.

To summarize the events during the plateau: (i) There is a substantial Na^+ permeability, although it is less than during initial depolarization. (ii) There is a Ca^{++} current of probable importance in electromechanical coupling. (iii) K^+ conductance appears to be lower than at rest owing to anomalous rectification (the decrease of i_{K_1} tends to keep the potential steady during the plateau). All three of these events are responsible for the plateau of the action potential.

Final Repolarization. Repolarization is completed by an outward current, in a channel referred to by Noble as X_1. It is mainly, but not exclusively, a potassium channel. Voltage clamp studies show a current reversal at about -75 mv during the repolarization period, whereas the K^+ equilibrium potential is near -100 mv at this time. This repolarization (X_1) current shows some inward-going rectification. The sodium and calcium currents, described previously, terminate during the later part of the plateau.[20] After the cessation of this current (X_1), the cardiac cell is at the resting potential.

The generation of the cardiac action potential is a complicated process.[10] There appear to be five separate conductance changes, one each for Na^+, Cl^- and Ca^{++}, and two for K^+, one of which (X_1) is not specific for K^+ alone. At the present stage of our knowledge, the situation appears to be even more complicated than the above. The Na^+ current change appears to have two components; there is a third K^+ current (i_{K_S}) seen in pacemaker cells (to be described below) and another current (X_2) caused by permeability changes which increase too slowly to affect the action potential. Each of the K^+ conductance changes appears to involve a separate channel in the membrane and, as indicated, the selectivity is such that more than one ion determines the transmembrane potential at most instants during the action potential. Figure 5–8 and Table 5–2 show permeability and current changes during the action potential. Most of the evidence for these currents which describe the action potential comes from voltage clamp experiments. It is difficult to properly apply a voltage clamp to cardiac muscle for a variety of reasons.[13] It is thus possible that a simpler (or unhappily a more complicated) picture of the generation of the cardiac action potential may emerge from future studies. However, it seems well established that Na^+ current is responsible for depolarization, whereas outward K^+ and Cl^- currents and inactiva-

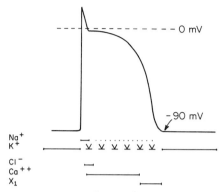

Figure 5–8 Sequence of permeability changes during the action potential of the Purkinje fiber. A typical action potential is shown above with duration of 300 msec and a voltage change from −90 to approximately +30 mv. Below the action potential, five different lines show changes in permeability of the membrane to a specific ion or ions during the action potential. The permeability of the membrane to sodium, which is low at rest (Na$^+$ line), increases markedly at the beginning of the action potential (solid line labeled Na$^+$) and then goes down to a level which is higher than the resting level, but lower than the maximal (dotted line). The membrane permeability to potassium, which is high at rest (solid line labeled K$^+$), becomes low and shows rectification, indicated by the rectifier symbols (V), during most of the action potential, and then the membrane returns to the initial high potassium permeability at the end of the action potential. A short chloride membrane permeability increase (solid line labeled Cl$^-$) is seen during the "notch" after initial depolarization. Calcium permeability increases at about the time of the "notch" and is high during most of the action potential (solid line labeled Ca$^+$). A membrane channel, known as X$_1$, increases its current during the terminal portion of the action potential (solid line labeled X$_1$) and causes final repolarization, then decreases (no line) following the end of the action potential. This is largely a potassium permeability change, but it also includes other ions. (After Fozzard and Gibbons, *Amer. J. Cardiol.*, 1973, *31*, 182–192.)

tion of inward Na$^+$ current are responsible for repolarization of cardiac muscle.

PACEMAKER POTENTIALS

The heart is made up of unspecialized atrial and ventricular cells, and of special-ized cells of the sinus node, the atrioventricular node, the conducting bundles and Purkinje fibers. All these specialized cells can act as pacemakers; that is, they can spontaneously depolarize. However, in normal circumstances, pacemaker activity occurs in the SA node. The major studies have been conducted in Purkinje cells

Table 5–2 *Contribution of Various Ionic Channels to Phases of the Cardiac Action Potential*

						PHASES			
ION	DIRECTION	CHANNEL	Upstroke	Rapid Repolarization	Notch	Plateau	Rapid Repolarization	Diastolic Depolarization
Sodium	Inward	i$_{Na}$(fast)	On	Off	...	...	...	...
		i$_{Na}$(slow)	...	...	On	Off	...	...
Chloride	Inward	i$_{Cl}$	...	On	Off	...	...	...
Calcium	Inward	i$_{Ca}$	...	...	On	...	Off	...
Potassium	Outward	i$_{K_1}$	Off	...	...	...	On	...
		i$_{X_1}$	...	...	...	...	On	Off
		i$_{X_2}$	...	...	(Too slow to participate)			...
		i$_{K_s}$	...	...	...	On	...	Off

This table indicates most of the permeability changes which have been described for cardiac muscle cells, including pacemaker cells. Initial inward sodium current produces the rapid upstroke and is rapidly inactivated during rapid repolarization. The second sodium current, referred to here as slow, is about one-third as large as the rapid inward current, and tends to persist during the notch and plateau. The chloride current is important during the rapid repolarization and the notch. Calcium current, which begins during the notch and goes off during final repolarization, is important in excitation-contraction coupling. K$_1$ decreases during the rapid upstroke and remains decreased until final repolarization. X$_1$, which is partially but not entirely potassium, is turned on during final repolarization and is the major control of that repolarization. X$_2$, a very slow current (too slow to participate in the action potential), has not been described above. The current known as K$_s$ is a potassium current important in pacemaker cells, and which increases slowly from the time of the plateau of the action potential to a short time following final repolarization, and then decreases during repolarization and is responsible for the pacemaker potential (slow depolarization of resting potential to threshold) in cardiac cells. (From Fozzard and Gibbons, *Amer. J. Cardiol.*, 1973, *31*, 182–192.)

acting as pacemakers. Pacemaker cells have a characteristic action potential (Fig. 5–5). Following repolarization, they do not remain at a potential near the normal K^+ equilibrium potential, but rather they reach a maximal repolarized level (less than the level seen in nonpacemaker cells) and then gradually depolarize toward the threshold potential, reach that potential and then undergo rapid depolarization similar to that seen in other cells (caused by Na^+ permeability changes). The sinus node's rate of diastolic depolarization is 15 to 60 mv per sec, and the Purkinje cell's is 5 to 40 mv per sec. In the intact heart, that cell which depolarizes first to the threshold level (at which sodium conductance begins to regeneratively increase) will be the pacemaker for the entire heart. We will therefore assume that pacemaker activity occurs by a similar process in all these specialized cardiac cells.

Pacemaker activity occurs because of two special properties of the pacemaker cells. First, they are more permeable to Na^+ than other cardiac cells. Although nonpacemaker cells have a "background" Na^+ current, it is comparatively small. If the external Na^+ of nonpacemaker cells is altered, little or no change in resting potential is observed owing to this low Na^+ conductance. In pacemaker cells, a much larger change in membrane potential can be seen.[25] The high Na^+ conductance would tend to keep the cells at a potential which is closer to zero than the K^+ equilibrium potential. Second, and more important, in pacemaker cells during the diastolic period there is a time-dependent *decrease* in permeability to $K^{+18, 27}$ and in the current carried by K^+, so that the membrane tends to approach the threshold (the membrane moves toward the Na^+ equilibrium potential and away from the K^+ equilibrium potential). This process brings the membrane finally to the threshold where the regenerative increase in Na^+ conductance occurs (Hodgkin cycle), bringing about complete depolarization or an action potential. The voltage-dependent K^+ conductance change involves a separate slow K^+ conductance, which increases during the plateau and is fully developed[18] at the end of the plateau (referred to by Noble as i_{K_s}). This causes the cell to repolarize. This current contrasts markedly with the previously described K^+ currents in nonpacemaker cells. If extracellular Na^+ concentration is decreased, or

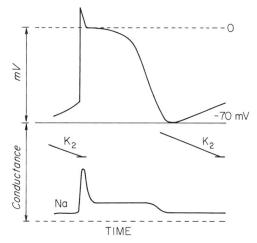

Figure 5–9 This figure diagrams the permeability changes characteristic of pacemaker cells. The action potential is from a Purkinje cell acting as a pacemaker and is similar to that of the Purkinje fiber shown in Figure 5–8, except that the maximal diastolic voltage is −70 mv (inside negative) as compared to −90 mv in the non-pacemaker cell. In addition, there is a slow diastolic depolarization to threshold which causes the cell to fire spontaneously. The sodium permeability of these cells, in the period before rapid depolarization (line labeled Na), is slightly higher than the resting Na^+ permeability of non-pacemaking Purkinje fibers (dotted line). They then show a conventional sodium action potential caused by a sudden increase in membrane sodium permeability, followed by a decrease in membrane sodium permeability to a value higher than the initial value (value at the resting potential) for most of the action potential. Na^+ permeability then returns to the initial value during final repolarization. In addition to the high sodium permeability there is an ionic channel, known and labeled as K_2, which is largely a potassium channel. The conductance of this channel decreases during the diastolic period, and this decrease in potassium permeability, coupled with the high sodium permeability, allows the membrane voltage to slowly increase (slow depolarization). When the cell potential reaches threshold an action potential will occur. (Modified from Fozzard and Gibbons, *Amer. J. Cardiol.*, 1973, *31*, 182–192.)

K^+ concentration is increased, pacemaker activity will be suppressed. Figures 5–8 and 5–9 and Table 5–2 indicate current changes found in pacemaker cells during the action potential.

Autonomic Effects on Pacemaker Potentials. The heart rate is altered by the sympathetic and vagal fibers which run to the pacemaker region. Heart rate may be altered either by a change in the rate at which pacemaker cells depolarize to threshold, or by a change in the threshold.

SYMPATHETIC EFFECTS. Sympathetic stimulation releases catecholamines, probably epinephrine and norepinephrine, which increase the slope of the diastolic pacemaker potential. The membrane reaches threshold more rapidly, and the frequency of pacemaker action potentials increases (Fig. 5–10). There is no effect on the maximal membrane potential, or on the *threshold* for generation of the action potential. As the slope increases, the cell moves more rapidly from the maximally attained diastolic (repolarized) membrane potential to the threshold, and consequently the heart rate increases. Similar effects are seen with the administered epinephrine and norepinephrine. Hauswirth *et al.*[11] found that if epinephrine is present during the application of a voltage clamp to a Purkinje fiber, more depolarizing current flows during the diastolic period, and thus the membrane potential moves toward the threshold more rapidly than during the normal action potential. This effect has not been explained.

VAGAL EFFECTS. During vagal stimulation, acetylcholine is released from vagal terminations. A cell from the sinus venosus shows an increased transmembrane potential difference during maximal repolarization (increased "resting" potential) and a slower rate of diastolic depolarization (Fig. 5–11). Hutter and Trautwein[12] showed that vagal stimulation decreased the duration of the action potential. The membrane potential reached a larger negative level than during a control beat, and the rate of slow diastolic depolarization (which was increased, as has been seen, by catecholamines) decreased. All these changes can

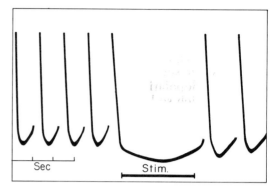

Figure 5–11 Effects of vagal stimulation on the pacemaker potential. The ordinate shows transmembrane potential, and the abscissa, time. Stimulus applied to the vagus nerve, as indicated by the heavy line, caused an increase in the "diastolic" transmembrane potential, decreased the rate of depolarization of the membrane to threshold and shortened the action potential duration. (From Hecht, *Ann. N. Y. Acad. Sci.*, 1965, *127*, 49–83.)

result from increased K^+ conductance. These effects were seen in a number of types of pacemakers.

This effect of acetylcholine on resting membrane potential is due to an increase in the conductance of the membrane to potassium during the diastolic period.[12] The current is referred to as i_{K_s} in Table 5–2.[18] The membrane potential thus moves closer to the potassium equilibrium potential than it normally does, and *hyperpolarization* ensues. Remember that in pacemaker cells, as compared to other cardiac cells, there is a greater "resting" inward sodium current which moves the resting membrane potential away from the potassium equilibrium potential, because K^+ current is reduced. If potassium permeability is increased during diastole by vagal stimulation or acetylcholine, the membrane potential moves toward the potassium equilibrium potential. If the resting membrane potential is above the K^+ equilibrium potential, acetylcholine will cause the membrane to hyperpolarize toward the K^+ equilibrium potential.

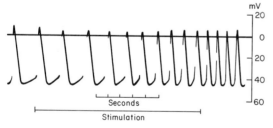

Figure 5–10 Effects of sympathetic nerve stimulation on the transmembrane potential of a frog's sinus venosus (pacemaker) cell. The vago-sympathetic trunk was stimulated at 20 imp/sec during the period indicated by the lower black line. All parasympathetic effects had been blocked by atropine. Ordinate shows transmembrane potential in millivolts; abscissa shows time. Sympathetic stimulation increases the rate of voltage change during the period before the rapid (Na^+) depolarization, and thus increases heart rate. Effects of stimulation are slow in onset and last beyond the end of stimulation. (After Hutter and Trautwein, *J. gen. Physiol.*, 1956, *39*, 715–733.)

CELL-TO-CELL CONDUCTION IN CARDIAC MUSCLE

Since the earliest studies of cardiac muscle, it has been known that electrical or mechanical stimulation at a point within a mass of heart muscle usually leads to depolarization of the entire muscle mass. For instance, if a point in the atrium is stimulated, a wave of depolarization travels away from this point to depolarize the entire atrial mass. Generally, such a wave passes through the atrioventricular conducting system into the ventricles. Atrioventricular conduction may, however, be impaired if the specialized conduction tissue is not functioning normally. Similarly, when a ventricular point is stimulated, a wave of depolarization passes through the entire ventricle (retrograde conduction from ventricle to atrium does not always follow stimulation within the ventricle). Conduction thus seems to be syncytial (*i.e.*, large masses of myocardium behave like a network of electrically interconnected fibers, or like a single cell) and when a wave of depolarization is started at any point in the heart, it proceeds throughout most of the network or "cell."

Light microscopy reveals no serious anatomic objection to this network analogy. Electron micrographs, however, reveal that the myocardium is anatomically divided into units by "cellular" membranes (Fig. 5–12).[22] The intercalated discs, which are perpendicular to the long axes of the fibers, are thought to be the electrical connection between the cells. At times, the separate cell boundaries in the intercalated disc region come into close contact, so that the extracellular space is virtually eliminated. These regions where membranes come into close apposition are known as tight junctions, nexuses or close junctions. They appear to provide an anatomical basis for syncytial conduction.

The physiologic mechanism of syncytial conduction has been a subject of controversy, but a number of experimental results indicate that cell-to-cell conduction is electrical. In studies by Woodbury and Crill,[31] current was injected into a single cell of a trabecula of the rat atrium. The spread of this current was mapped with a second intracellular electrode. Current injected into one cell was found to spread into adjacent cells with surprising effectiveness. In studies by Weidmann,[30] the diffusion of potassium from one cell to another was found to be so rapid that it appeared that the cells were chemically interconnected, as if no cellular boundary existed. (Also, G. H. Pollack [personal communi-

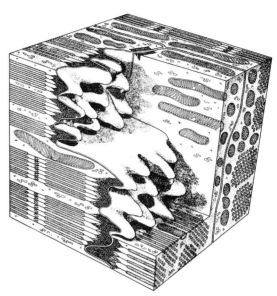

Figure 5–12 Reconstruction by Sjöstrand of submicroscopic anatomy of cardiac muscle. Long narrow structures running from left to right in anterior aspect are myofibrils. Vertical and horizontal double membranes divide the mass of muscle into "units" in these planes. Cutaway section shows continuation of these vertical and horizontal membranes into the intercalated discs, the many tonguelike processes which extend from left to right. These discs form a boundary between "domains" of muscle and constitute justification for discarding the term "syncytial" insofar as anatomy is concerned. Insofar as conduction is concerned, however, the cardiac mass does seem to be syncytial or ephaptic.

cation] has shown that large ions, such as fluorescein dye [M. W. 332], diffuse freely between cells.)

Finally, in studies by Barr and co-workers,[2] conduction was blocked in a strip of muscle by surrounding a section with nonconducting sucrose. Conduction could be restored by connecting a wire across the area of blocked conduction. These studies have been contested,[23] but the bulk of the evidence supports the idea that adjacent cells are electrically connected so that depolarization of one cell will cause sufficient current flow into an adjacent cell to guarantee its depolarization.

REFERENCES

1. Adrian, R. H. Rectification in muscle membrane. *Progr. Biophys. & molecular Biol.*, 1969, *19* (2), 341–369.
2. Barr, L., Dewey, M. M., and Berger, W. Propagation of action potentials and the structure of the

nexus in cardiac muscle. *J. gen. Physiol.*, 1968, *48*, 797–823.

3. Brady, A. J., and Woodbury, J. W. Effects of sodium and potassium on repolarization in frog ventricular fibers. *Ann. N. Y. Acad. Sci.*, 1957, *65*, 687–692.

4. Brooks, C. McC., Hoffman, B. F., Suckling, E. E., and Orias, O. *Excitability of the heart.* New York, Grune & Stratton, 1955.

5. Deck, K. A., Kern, R., and Trautwein, W. Voltage clamp technique in mammalian cardiac fibers. *Pflügers Arch. ges. Physiol.*, 1964, *280*, 50–62.

6. Deck, K. A., and Trautwein, W. Ionic currents in cardiac excitation. *Pflügers Arch. ges. Physiol.*, 1964, *280*, 63–80.

7. Draper, M. H., and Weidmann, S. Cardiac resting and action potentials recorded with an intracellular electrode. *J. Physiol. (Lond.)*, 1951, *115*, 74–94.

8. Dudel, J., Peper, K., Rüdel, R., and Trautwein, W. The dynamic chloride component of membrane current in Purkinje fibers. *Pflügers Arch. ges. Physiol.*, 1967, *295*, 197–212.

9. Dudel, J. Peper, K., Rüdel, R., and Trautwein, W. The effect of tetrodotoxin on the membrane current in cardiac muscle (Purkinje fibers). *Pflügers Arch. ges. Physiol.*, 1967, *295*, 213–226.

10. Fozzard, H. A., and Gibbons, W. R. Action potential and contraction of heart muscle. *Amer. J. Cardiol.*, 1973, *31*, 182–192.

11. Hauswirth, O., Noble, D., and Tsien, R. W. Adrenaline: Mechanism of action on the pacemaker potential in cardiac Purkinje fibers. *Science*, 1968, *162*, 916–917.

12. Hutter, O. F., and Trautwein, W. Vagal and sympathetic effects on the pacemaker fibers in the sinus venosus of the heart. *J. gen. Physiol.*, 1956, *39*, 715–733.

13. Johnson, E. A., and Lieberman, M. Heart: excitation and contraction. *Ann. Rev. Physiol.*, 1971, *33*, 479–532.

14. Johnson, E. A., and Sommer, J. R. A strand of cardiac muscle. Its ultrastructure and the electrophysiological implications of its geometry. *J. Cell Biol.*, 1967, *33*, 103–129.

15. New, W., and Trautwein, W. The ionic nature of slow inward current and its relation to contraction. *Pflügers Arch. ges. Physiol.*, 1972, *334*, 24–38.

16. Niedergerke, R., and Orkand, R. K. The dual effect of calcium on the action potential of the frog's heart. *J. Physiol. (Lond.)*, 1966, *184*, 291–311.

17. Noble, D. Electrical properties of cardiac muscle attributable to inward going (anomalous) recti-

fication. *J. cell. comp. Physiol.*, 1965, *66* (Suppl. 2), 127–136.

18. Noble, D., and Tsien, R. W. The kinetics and rectifier properties of the slow potassium current in cardiac Purkinje fibres. *J. Physiol. (Lond.)*, 1968, *195*, 185–214.

19. Noble, D., and Tsien, R. W. Outward membrane currents activated in the plateau range of potentials in cardiac Purkinje fibres. *J. Physiol. (Lond.)*, 1969, *200*, 205–231.

20. Noble, D., and Tsien, R. W. Reconstruction of the repolarization process in cardiac Purkinje fibers based on voltage clamp measurements of membrane current. *J. Physiol. (Lond.)*, 1969, *200*, 233–254.

21. Reuter, H. The dependence of slow inward current in Purkinje fibers on the extracellular calcium-concentration. *J. Physiol. (Lond.)*, 1967, *192*, 479–492.

22. Sjöstrand, F. S., Andersson-Cedergren, E., and Dewey, M. M. The ultrastructure of the intercalated discs of frog, mouse, and guinea pig cardiac muscle. *J. Ultrastr. Res.*, 1958, *1*, 271–287.

23. Tarr, M., and Sperelakis, N. Weak electronic interaction between contiguous cardiac cells. *Amer. J. Physiol.*, 1964, *207*, 691–700.

24. Trautwein, W. Zum Mechanismus der Membranwirkung der Acetylcholin an der Herzmuskelfaser. *Pflügers Arch. ges. Physiol.*, 1958, *266*, 324–334.

25. Trautwein, W., and Kassebaum, D. G. On the mechanism of spontaneous impulse generation in the pacemaker of the heart. *J. gen. Physiol.*, 1961, *45*, 317–330.

26. Truex, R. C. Comparative anatomy and functional consideration of the cardiac conduction system. In: *The specialized tissues of the heart*, A. P. de Carvalho, W. C. de Mello, and B. F. Hoffman, eds. Amsterdam, Elsevier Publishing Co., 1961.

27. Vassalle, M. Analysis of cardiac pacemaker potential using a "voltage clamp" technique. *Amer. J. Physiol.*, 1966, *210*, 1335–1341.

28. Weidmann, S. *Elektrophysiolgie der Herzmuskelfaser.* Bern, Hans Huber, 1956.

29. Weidmann, S. Resting and action potentials of cardiac muscle. *Ann. N. Y. Acad. Sci.*, 1957, *65*, 663–678.

30. Weidmann, S. The diffusion of radiopotassium across intercalated discs of mammalian cardiac muscle. *J. Physiol. (Lond.)*, 1966, *187*, 323–342.

31. Woodbury, J. W., and Crill, W. E. On the problem of impulse conduction in the atrium. In: *Nervous inhibition*, E. Florey, ed. New York, Pergamon Press, 1961.

CHAPTER 6 ELECTROCARDIOGRAM

by ALLEN M. SCHER

INTRODUCTION

At about the turn of the twentieth century Willem Einthoven,[18, 19] a Dutch physician, designed and constructed a galvanometer which made it possible routinely to record voltages produced at the body surface by electrical activity of the heart. A record of these voltage changes in time is referred to as an electrocardiogram. From Einthoven's time until the late 1940's, electrocardiography developed as a useful clinical tool; the gross relationships between the electrocardiogram and the electrical activity of the heart were elucidated, and some foundations of scientific electrocardiography were established by Lewis and co-workers[34, 35, 36]

in England and by Wilson *et al.*[66, 67] in the United States, among others. Recording techniques also improved. Einthoven's original galvanometer, which required photographic processing of records, was replaced by vacuum tube and then transistor recording systems that produced records instantaneously. Although not always as rapid in frequency response as the photographic galvanometer, these devices proved to be excellent for electrocardiographic recording.

P, QRS, T. The normal electrocardiogram for a single heart beat consists of three main complexes which, as indicated in the last chapter, are designated the P wave, QRS complex and T wave. If electrodes are

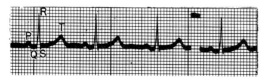

Figure 6–1 Normal lead II electrocardiogram. Initial, low, rounded deflection about 1 mm high and 2 mm long is the P wave. Second deflection, about 1 mm wide and 10 mm high shows a rapid rise and fall and is called the QRS complex. Third, peaked deflection, about 3.5 mm high and 6 mm long is the T wave. Here sequence is repeated three times. Standardization at right, 1 mV. Small black vertical lines are 40 msec apart; larger lines (five spaces) are 200 msec apart; heaviest lines are 1 sec apart. (From Winsor, ed., *Electrocardiographic test book.* New York, American Heart Association, 1956.)

connected to the right arm and left leg of a normal adult and the difference in potential between these extremities is recorded in the fashion described later in this chapter, curves of voltage-in-time like those appearing in Figure 6–1 will result. The initial upright, low, rounded deflection, the *P wave*, results from the depolarization of atrial muscle. The second deflection, the *QRS complex*, is produced by depolarization of the ventricular myocardium. The final deflection, the *T wave*, results from the return of the ventricular muscle to the polarized or resting state. At times, the T wave is followed by a small *U wave* which seems, like the T wave, to be produced by ventricular repolarization. These potentials result *only* from *electrical* changes within the heart.

An understanding of the shape of any electrocardiographic complex—and of the origin of the normal (or abnormal) electrocardiogram—depends on three types of information. (i) What electrical changes take place in single cells as they pass from the resting to the depolarized (or active) state and then return to the resting state? (ii) How do cells in a conducting medium produce potentials at recording electrodes within, or at, the boundaries of that medium? Analysis of current flow in such volume conductors should encompass conductors of the shape and conductivity of the human torso. (iii) What are successive positions of the boundary between resting and active cells during depolarization and repolarization of the heart? This means that we must know the exact pathway of depolarization

and of repolarization of the muscular mass surrounding the cardiac chambers.

Early analysis in the fields of electricity and magnetism developed a theoretical basis for the second factor (volume conduction). This theory was competently applied to some problems of electrocardiography by Craib[11] and Wilson *et al.*[67] Little was known about the first (cellular activity) and third (pathway of activity). Hence our understanding of the electrocardiographic complex was severely limited through most of the 1950's. However, during the last 15 years our knowledge of these two factors has substantially increased, and the outlook is now improved for a quantitative understanding of the normal (or abnormal) electrocardiogram. Advances in knowledge have resulted from studies along the following lines:

In 1946 Graham and Gerard[25] described a hollow glass ultramicroelectrode so small (about 1 μ in diameter) that it could be inserted into a single cell. Ling and Gerard[37] utilized this electrode to record the potentials in skeletal muscle cells, and this marked the beginning of a study of electrochemical events within excitable cells. The ultramicroelectrode and other techniques which permit the same type of measurement have been used to study many types of excitable cells.[15, 16] Electrochemical events in nerve, muscle and cardiac muscle are discussed in Chapters 3 and 6, Volume I, and Chapter 5, Volume II.

About 1950 a number of laboratories[56, 57, 60, 61] began gathering information about the sequence of excitation of the atrial and ventricular musculature. They utilized small, usually multiterminal, metal *extracellular* electrodes which were inserted into the myocardium to record the time of activity at many sites. In the early 1950's there appeared to be substantial promise of furthering our understanding of electrochemical events within the cardiac cells and of conduction within the heart, and through these two of furthering our knowledge of the origin of the electrocardiogram. The substantial progress in these areas has been accompanied by an increased understanding of the underlying physiological events which produced the electrocardiogram and by a sharpening of the use of the electrocardiogram as a clinical tool.

The term "cardiac electrophysiology" has

been used to describe a large area of research extending from the study of electrochemical events in single cells up to the clinical use of the electrocardiogram. Between these two are several other areas of study, particularly the study of pacemaker activity, of conduction from cell to cell, of the pathway of cardiac excitation and of the action of ions and hormones on cardiac cells. Some of these are discussed elsewhere in this textbook (Chap. 5, Vol. II).

In the preceding chapter we have discussed the cellular events which are important to an understanding of the electrical activity of the heart and to the origin of the electrocardiogram. In the present chapter we will cover some aspects of volume conduction, the normal depolarization and repolarization sequence of the heart, the origin of the normal electrocardiogram and some abnormalities of cardiac activity, and we will briefly discuss recording techniques.

Volume Conduction. The instantaneous potential produced at a point in a homogeneous and infinite volume conductor by a boundary between active and resting cells is a function of (i) the number of charges per unit area across the boundary, (ii) the solid angle subtended by the boundary at the point and (iii) the resistivity of the medium. This is shown in the formula:

$$\mathcal{E}_p = K_1 K_2\ \Omega\ \phi$$

in which $\mathcal{E}_p$ denotes the potential at a point, K_1 is a constant for tissue resistivity, K_2 a geometric constant, Ω the solid angle subtended at the point by the boundary, and ϕ is a voltage proportional to the charge density per unit solid angle across the boundary; ϕ is normally constant and is about 100 mV (overshoot minus resting potential) where the solid angle is 4π steradians (see Chap. 3, Vol. I).

To understand why a potential of a certain magnitude exists at a particular point on the body surface at some instant during the electrocardiogram (*i.e.*, to apply this formula), we must know the position and geometry of the boundary at that instant, the dipole moment (number of charges multiplied by distance between poles) per unit area across the boundary and the corrections necessary to compensate for the shape and resistivity of the torso (Fig. 6–2). The

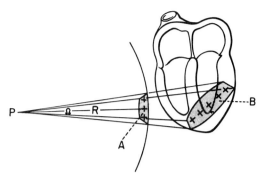

Figure 6–2 Hypothetical boundary between active and resting muscle within heart. Potential at point P is a function of the dipole moment per unit area across this boundary and of the solid angle subtended at P by the boundary. A is the projection of B on sphere of radius R. Solid angle is A/R^2. B is boundary between resting and active tissue.

specific resistivity of blood and lung is 160 ohm-cm and 2000 ohm-cm respectively. Heart muscle averages 400 ohm-cm but is anisotropic, *i.e.*, resistance has one value along one axis of the tissue and another in a perpendicular direction.[49] These measurements indicate that tissue resistance may have to be taken into account when a quantitative model of the electrocardiogram is developed.

A cell or aggregation of cells in the resting state or otherwise uniformly polarized or depolarized can give rise to no potential at an external recording electrode (Fig. 6–3; see also Fig. 4–20, Chap. 4, Vol. I). But if a cell is partially depolarized, an electrode which faces the *resting* portion of the cell will record a positive potential (Fig. 4–20,

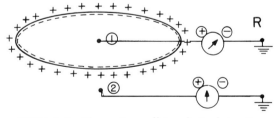

Figure 6–3 The resting cell is polarized; *i.e.*, there are charges distributed evenly around its membrane, negative charges on the inside and positive charges on the outside. If we record the potential between electrode *1*, inside the cell, and an external ground reference electrode at R, the potential inside will be negative. However, an electrode outside the cell, as at *2*, is equally influenced by the positive and negative charges from the resting cell and therefore records no potential.

Chap. 4, Vol. I). This would occur if a wave of depolarization were moving from left to right or if the cell were repolarizing from right to left (Fig. 6–4).

An alternative to the "solid angle" analysis is to consider an active cell or aggregation of cells as a "dipole." A dipole consists of equal numbers of positive and negative charges separated by an infinitesimal distance. The potential at a point in an infinite homogeneous medium produced by such a dipole is proportional to the dipole moment (m) (number of charges per pole multiplied by the distance between the poles), and to the cosine of the angle between the recording point and midpoint of the line joining the two poles, and inversely proportional to the square of the distance (r) between the recording point and the dipole:

$$\mathcal{E} = \frac{m \cos \theta}{r^2}.$$

In practice, the dipole concept is applied semiquantitatively or qualitatively. An aggregation of charges such as a boundary in the heart is approximated by a single arrow with its head pointing in the direction of the positive charges. The boundary between resting and active cells is thus approximated by a single arrow (Fig. 6–4). If such an arrow is drawn for an instant during the period of depolarization, the head will point in the direction of the movement of the wave front. In some recent electrocardiographic techniques, the entire heart has been considered as a single dipolar generator (vide infra).

Anatomic Features. The specialized tissue of the heart traditionally consists of the sinus node and the atrioventricular conduction system, including the ventricular Purkinje fibers. Recent histological studies by James,[31, 32] discussed below, indicate that there are three anatomically distinguishable conducting pathways in the atria between the sinus node and the A-V node. The sinus node (Chap. 7, Vol. II) lies at the junction of the superior vena cava with the right atrium and is the site of origin of impulses within the heart. The atrioventricular conduction system is composed of the A-V node, the common bundle and the right and left bundles. The A-V node lies in the interatrial septum just above the

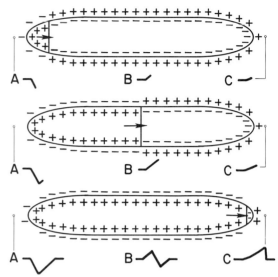

Figure 6–4 As a wave of activity moves from the left-hand side to the right-hand side of the cell, as shown, the potentials recorded at A, B, and C are as indicated. Note that initially the electrode at A views the negative side of the depolarized section on the left-hand side of the cell and also views the negative (inside) portion of the resting portion of the cell (on the right). This process continues, so that A records only a negative potential as the cell goes from the resting to the depolarized state. The electrode at B is initially closest to the positive charges across the resting portion of the membrane at the right and to the positive charges across the depolarized portion of the membrane at the left. It thus records an initial positive potential. However, when the electrode passes it, it records a negative potential. The electrode at C at the right-hand side of the cell records only positive (approaching) activity as the cell depolarizes.

tricuspid valve. It gives rise to the common bundle, which passes into the basal interventricular septum, giving rise to the right and left bundles. These course along the septum toward the apex, the right bundle as a single strand, the left usually dividing into two or more strands. The bundles terminate in the middle regions of the septum and give rise to numerous strands of Purkinje tissue that cover much of the ventricular endocardium in the dog and man and penetrate widely into the ventricular myocardial mass in ungulates. (Purkinje fibers can be observed grossly in the ungulates and less clearly in dog and man.) Purkinje fibers do not deeply penetrate the canine or human myocardium but terminate near the endocardium.[16] The sinus node and each of the constituents of the atrioventricular conduction system differ somewhat in their histologic appearance from the cells of the ordinary atrial and ventricular myocardium.[65]

The atrial walls can be considered as two triangular sections of a sphere, curved to join superiorly to the interatrial septum and superiorly, posteriorly and anteriorly to each other. Inferiorly, the walls and septum join the fibrous ring to which are attached the corresponding ventricular structures.

The following structural details are important in considering ventricular electrical activity: (i) The right bundle merges with the ventricular musculature near the right anterior papillary muscle, where it divides to send strands of Purkinje fibers to the endocardium of the free wall from this location. (ii) The left bundle usually splits into anterior and posterior divisions, which run, respectively, toward the anterior and posterior papillary muscles on the left and give rise to numerous false tendons (Purkinje strands) that cross the left cavity and merge with ordinary myocardium near the endocardium. (iii) The right wall is normally thin, generally no more than 3 or 4 mm thick; the left wall is up to 15 mm thick. (In infants the two ventricular walls are about equally thick.) (iv) The crescent-shaped right wall can be considered a "cap" on the thick-walled cylindrical structure composed of left wall and septum. (v) The endocardial Purkinje network is more widespread in the central and apical portions of the wall and septum bilaterally. This network is sparse or nonexistent in the basal septum.

EXCITATION OF THE HEART

Our knowledge of the pathway of cardiac excitation has been obtained primarily from animal experiments, mostly on dogs. The various components of the canine electrocardiogram last about one-half as long as do those of the human cycle; results obtained in dogs are applicable to man.[1, 4, 5, 30, 61, 66] This has been clearly demonstrated in the studies by Roos et al.[47] and Durrer et al.[17] on excitation of the perfused human heart.

The Cardiac Pacemaker; Excitation of Atrium. One property of cardiac tissue is automaticity—the ability to beat rhythmically without external stimuli (Chap. 6,

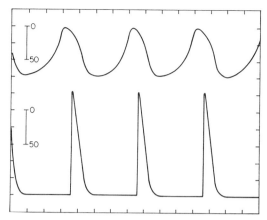

Figure 6–5 *Upper trace,* Potentials recorded by ultramicroelectrode in the pacemaker region. *Lower trace,* Potentials recorded by second ultramicroelectrode in normal atrial tissue. Note diastolic prepotential of smaller amplitude in record from pacemaker; also, notice differences in shapes of potentials in rabbit. (After West *et al., J. Pharmacol.,* 1956, *117,* 245–252.)

Vol. I, and Chap. 7, Vol. II). The cells with the most rapid inherent rhythm are called "pacemaker cells" and control the cardiac rhythm. In cold-blooded animals, pacemaker activity seems to be possible for all parts of the heart, but in intact warm-blooded animals, pacemaker activity is normally confined to the S-A node and the A-V node (Fig. 6–5). However, with even minor departures from normal physiology, extrasystolic (*i.e.,* abnormal) beats may originate at both atrial and ventricular sites. Normally, the dominant pacemaker is the S-A node. If the S-A node fails or is abnormally slowed, the A-V node usually determines the heart rate.

The process by which the S-A and A-V nodes generate impulses is not entirely clear. These cells differ from ordinary myocardium in being more permeable to sodium when at rest.[64] Because of an apparent decrease in potassium conductance accompanied by a sodium leakage current, the pacemaker cells gradually depolarize to threshold at the end of each cycle (see Chap. 5, Vol. II).

Electrical activity in the atrium commences in the S-A node and spreads in a pattern like that produced by dropping a stone into still water. The elliptical shape of the area of initial depolarization probably results from nearly simultaneous pacemaker activity at many points in the S-A node.

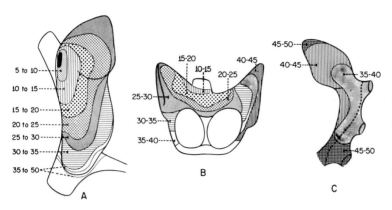

Figure 6–6 Pathway and mode of atrial activation. A, Right atrium and right atrial appendage viewed from right. Activity begins in sinus node *(black)* and progresses toward borders of atrium. B, Activation of atria viewed from anterior aspect. C, Activation of left atrium and appendage. Shading shows areas activated within each 5 msec period. Duration of P wave was 50 msec. (After Puech, *L'activité électrique auriculaire.* Paris, Masson et Cie, 1916.)

Plots of atrial excitation by Lewis *et al.*,[35] by Puech[46] and by Paes de Carvalho *et al.*[43] agree closely (Figs. 6–6 and 6–8). From the region of the S-A node the wave of atrial depolarization proceeds at a velocity of slightly less than 1 m per sec toward the borders of the two atria and of the interatrial septum. The right atrium, being nearer the sinus node, is normally excited before the left.

As indicated briefly earlier in this chapter, anatomical evidence has been presented[31, 32] for an atrial conduction system. This system appears to be composed of cells which are larger than ordinary atrial fibers and have anatomical features similar to those of ventricular Purkinje fibers. James[31, 32] has indicated that there are three "connecting pathways" between the sinus node and the A-V node in the human. Similar findings were reported by Merideth and Titus.[40] From a physiological point of view the evidence for a special atrial conduction pathway(s) is controversial. Many of the intracellular records from cells in the supposed pathway resemble records from nodal cells which conduct slowly and thus could not facilitate internodal conduction,[43] although some fibers with Purkinje-like action potentials have been found.[29] Further, plots of atrial excitation in the rabbit,[52] dog,[24,43,46] horse[27] and human[17] cast some doubt on the function of these pathways. Experiments in which the supposed pathways were cut[22] do not appear conclusive at present. We must therefore await final word on these pathways, although evidence that they resemble the ventricular conduction system is at present negative.

If we consider the shape of the atrium and its position in the body, we may think of normal atrial excitation as consisting of three divergent waves moving inferiorly from the S-A node toward the atrial borders. Initially the direction of activity is to the right and anteriorly; later activity is di-

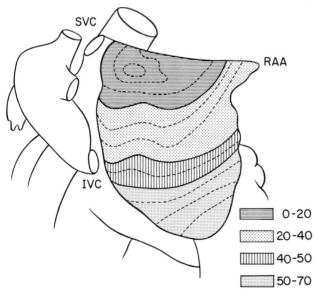

Figure 6–7 The sequence of atrial activation in the perfused human heart as determined by experiments of Durrer *et al.*[17] This figure shows the right atrial free wall and appendage. The sinus node is approximately in the middle of the top, horizontally shaded portion. We have only indicated those areas activated between 5 and 20 msec, 25 and 40 msec, etc., after the earliest activity as indicated on the lower right. Note that the wave of activity proceeds from the elliptical region of the sinus node outward, with what appears to be a relatively constant velocity, arriving latest at the tip of the appendage. As in Figure 6–6, similar, roughly parallel waves of activity move roughly simultaneously in the intra-atrial septum and left atrial wall. (After Durrer *et al., Circulation*, 1970, *41*, 899–912.)

rected leftward and posteriorly. The "atrial conduction system"[29, 31, 32] or, more probably, the direction of atrial fibers controls the sequence of atrial depolarization. These findings in animals have been supplemented by recent experiments on human hearts by Durrer et al.[17] (Fig. 6–7).

Passage of the Impulse Through the A-V Node. The period from the beginning of the P wave to the beginning of the QRS complex is referred to as the P-R interval. It usually has a duration of 0.12 to 0.2 sec (average 0.16 sec) in man and 0.08 sec in the dog. The potentials generated by the A-V node and the Purkinje fibers are far too small to influence electrodes at the body surface or to be recorded by extracellular electrodes that are farther than a few millimeters from these tissues. Recently, however, it has become possible to record potentials from the cells in and near the A-V node with both intracellular and extracellular electrodes.

The atrial musculature in the region near the A-V node is depolarized when atrial

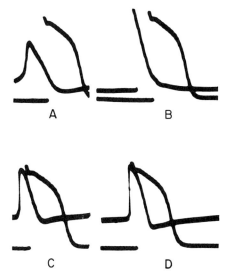

Figure 6–9 Intracellular records from sinus node, atrial muscle, A-V nodal region and common bundle. For timing purposes, large potential which has lowest base line is repeated. It is recorded near the common bundle and has a duration of about 150 msec. *A*, Smaller potential which begins earlier is from sinus nodal region. It shows a diastolic prepotential and slow rate of depolarization with lack of overshoot. *B*, Earlier potential, which has a resting potential slightly lower than that of common bundle and which depolarizes to about the same extent, is from ordinary atrial muscle. Atrial potential occurs somewhat later than potential from sinus node. *C*, Earlier potential, from upper A-V node, shows a smaller amplitude and a small diastolic prepotential after rapid repolarization. *D*, A similar potential with a diastolic prepotential is seen occurring somewhat later and closer in time to the depolarization of the common bundle. This potential is from the mid A-V node. (From Hoffman *et al.*, *Circulat. Res.*, 1959, 7, 11–18.)

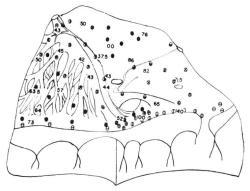

Figure 6–8 A view from the interior of the rabbit's right atrium. Trabeculated area is at the left. Numbers indicate the instant of activity in milliseconds after depolarization of the sinus node. Note that activity spreads approximately radially from the sinus node, which is in the upper central region of the figure. The shapes of action potentials recorded are indicated by the various symbols in the figure. Note the region of the common bundle along the atrioventricular margin on the right (circle with vertical line). Immediately upstream from this (circle with cross) are cells from an intermediate region along the A-V conduction pathway, which the experimenters termed nodobundle cells. Above these (black circle with horizontal white line) are cells which the researchers considered to be the first link in the A-V transmission system and which they termed atrial-nodal cells. Note further that these atrial-nodal cells extend from the trabecular region into the A-V nodal region. (From Paes de Carvalho *et al.*, *Amer. J. Physiol.*, 1959, *196*, 483–488.)

depolarization is about two-thirds complete (Fig. 6–8). A large part of the time between atrial firing and firing of the common bundle is consumed while the electrical wave passes through cells in the A-V nodal region.[28, 56] The cells in this region differ in electrical characteristics from either atrial or bundle cells, resembling instead cells of the sinus node (Fig. 6–9). A-V nodal cells have a diastolic prepotential, a slower rate of initial depolarization than other cardiac cells and a smaller action potential.[42]

An extracellular electrode in the A-V nodal region of a canine or bovine heart[45, 56] records no clear potentials for 5 to 15 msec after the depolarization of the S-A node (Fig. 6–10). Then the upper A-V node displays a negative-going potential which develops slowly (over 10 to 15 msec), remains at its maximal negative value for

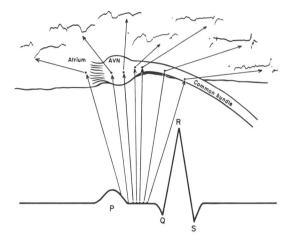

Figure 6–10 Potentials recorded extracellularly at seven sites near A-V nodal and common bundle region. Potential at far left is recorded from atrial muscle upstream from A-V node. As can be seen, it occurs during downstroke of P wave. Second potential from left is from head or upper end of A-V node. It shows an atrial potential followed by a large negative-going A-V nodal potential. Third potential from left is recorded in middle of A-V node. It shows a positive-negative atrial potential followed by a rapid negative-going common bundle potential. Farther downstream, common bundle potentials show more positivity, becoming positive-negative at the far right. As can be seen, a large part of interval between end of P wave and beginning of QRS complex is occupied by events in the A-V nodal region. (From Scher *et al.*, *Circulat. Res.*, 1959, 7, 54–61.)

some time, and returns more rapidly toward zero. The potential in the center of the node is positive-negative, and that at the lower end of the node is positive and is terminated by a negative-going common bundle potential. These potentials within the A-V node are similar to those recorded from a propagated wave in a muscle strip, and we may conclude that conduction *within* the A-V node qualitatively resembles conduction elsewhere in the heart. If the atrium is stimulated at a very rapid rate, conduction time between atrium and ventricle gradually increases, and then complete block occurs (Fig. 6–11). During this period of prolonged conduction (first degree block), the A-V nodal potential occurs later than normal (Fig. 6–11), but its configuration is not altered. This indicates that the block is between the atrium and the A-V node, and it appears that cells immediately *above* the node are the most susceptible to A-V block.

Because of this susceptibility to incomplete and complete conduction block,

the region of the A-V node is said to have the lowest safety factor found in the atrioventricular conduction system. The region lying immediately upstream from the A-V node is composed of very fine muscle fibers interspersed with connective tissue. The slow conduction velocity (0.05 m per sec) and the low safety factor here may result from some property of the muscle fibers (their small size and their particular ionic permeabilities) or from their geometric relation to the A-V node or both.

Within the A-V node the conduction velocity (about 0.1 m per sec) is somewhat faster than in the region immediately above the node. Once the A-V node is excited, the

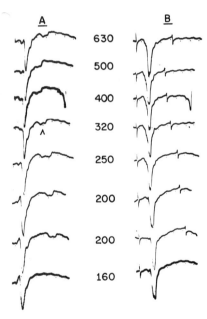

Figure 6–11 Effects of changing frequency of atrial stimulation on potentials recorded from A-V node and common bundle. In column A a large negative potential due to depolarization of the atria is followed by a smaller negative potential (marked by arrow in fourth record from the top), produced by electrical activity of the A-V node. In column B, recorded from a different site, the atrial potential is followed by a much more rapid negative-going potential produced by the depolarization of the common bundle. As the interval between stimuli decreases progressively from 600 to 160 per sec, the nodal potential at first occurs closer in time to the atrial potential, but with progressively shorter intervals it occurs progressively later, but without changing shape, which indicates that the slowing of conduction ("block") is above the node. At an interstimulus interval of 160 msec, the A-V nodal potential no longer appears, and the common bundle is not depolarized (complete block). (From Scher *et al.*, *Circulat. Res.*, 1959, 7, 54–61.)

impulse passes to the Purkinje fibers of the common bundle and the right and left bundles, and thence to the ventricles. Recently potentials from the atrioventricular bundles in man[12, 13] have been recorded with catheter electrodes. This procedure produces records similar to those seen in Figure 6–11. An example of such potentials is presented later in the discussion of atrioventricular block. The conduction velocity in the Purkinje fibers of the bundles is much higher, up to 2.0 m per sec, and reflects their large size. The conduction system is syncytial and normally responds in an all-or-none fashion to atrial activation. Mendez and Moe have presented evidence for a dual conduction system.[39]

Ventricular Activation. Ventricular activation has been plotted in detail in the dog, and some studies have been conducted on the rhesus monkey and on man.[14, 17, 47] When applied to humans, figures for the time of activity in the dog should be multiplied by about 2.5, because ventricular depolarization in the dog requires only 35 msec compared to about 80 msec in the human.

Lewis and Rothschild in 1915[36] noted that the time required for the impulse to travel from a point of stimulation on the ventricular surface to another surface point was not altered if the epicardial muscle between these points was cut. They concluded that the impulse travels slowly from the point of stimulation to the endocardium, moves rapidly over the endocardial surface, and then travels slowly from the inside to the outside at the recording point. They further deduced that the endocardial Purkinje network conducts more rapidly than mural myocardium and that the impulse normally travels outward in the wall. Ventricular depolarization in dog and man has been plotted with extracellular electrode assemblies consisting of several recording terminals along a central shaft[5, 16, 57, 60] (Figs. 6–12, 6–13, and 6–14).

The spread of activity in a coronal section of a dog heart is shown in Figure 6–13. The section shown did not contain the first and last points activated in the ventricles. In Figures 6–12 and 6–13 it can be seen that the general direction of excitation in the right and left walls is from within out. The process is not entirely uniform; the direction of activation may reverse, particularly

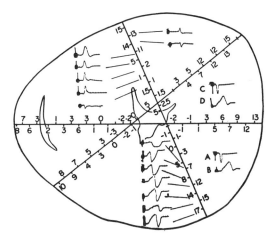

Figure 6–12 Cross-section of dog's heart near its apex showing potentials recorded at various sites. Unipolar records appear below horizontal line and bipolar records above. Unipolar records near cavity show mostly a negative (downward) potential; those near surface of heart show a positive potential followed by a negative potential. Bipolar records are generally upright, indicating that wave of activity moves from inside out in posterior left wall. Instant of activity at a point is indicated by peak of bipolar record. Numbers indicate time of local activity as measured from the bipolar records.

Time reference potential A and electrocardiogram B were taken simultaneously with unipolar records; C and D were simultaneous with bipolar records. Last points to be activated in wall are near epicardium; first points near endocardium. Latest point in section shown is in center of septum along horizontal electrode. (From Scher and Young, *Circulat. Res.*, 1956, *4*, 461–469.)

in the regions near the endocardium and under the papillary muscles and trabeculations. The major portion of the septum is excited by waves moving toward its center from both endocardial surfaces; however, the activation of the basal septum is predominantly from the left.

The average velocity of conduction through the ventricular muscle is about 0.3 m per sec. Calculation of this velocity requires use of a three-dimensional plot of activity (Fig. 6–14) and measurement of the time and distance between two successive positions of the wavefront. If endocardial velocity is ascertained by stimulating an endocardial point and recording the resultant spread of activity, the velocity is about 1 m per sec.[55] These general aspects of ventricular excitation are widely accepted.[2, 5, 16, 38]

ACTIVATION OF MURAL MYOCARDIUM. The manner in which activity spreads near

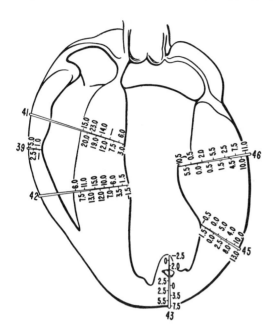

Figure 6–13 Coronal section of dog's heart showing time of activation at many points along six electrodes. In wall, activity generally moves from inside out. There are, however, several reversals of directions. Under papillary muscle along electrode 46, wave of activity moved toward endocardium and toward epicardium. Septum is excited by waves moving centrally from both septal surfaces. Latest points activated are in middle of basal septum.

the endocardium is difficult to establish. However, the general movement in the walls is from within out, although there are reversals of direction near the endocardium. In the right wall, activity always moves from the inside to the outside.

SEPTAL ACTIVATION. A number of investigators have reached similar conclusions, as follows:[2, 5, 14, 55, 57] The Purkinje fibers on the left give off numerous branches to the endocardium of the middle and apical septum. On the right, the Purkinje fibers have less extensive terminations. Consequently, as the QRS complex begins, a size-

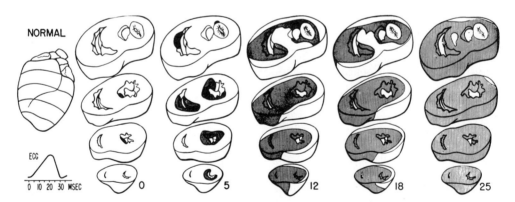

Figure 6–14 Pathway of normal ventricular excitation in dog as discerned by noting extent of depolarization at 0, 5, 12, 18 and 25 msec after beginning of QRS complex. Small drawing of heart indicates positions of planes in which records were taken. Lead II electrocardiogram is labeled to indicate total duration of electrical activity.

At 0 msec, small amount of muscle bordering left cavity is active. Apparently this volume of muscle is too small to give a deflection in peripheral electrocardiogram at this amplification. At 5 msec after beginning of QRS, an incomplete and irregular cone of activity surrounds left cavity, mostly on septal aspect, and a smaller cone surrounds right cavity. By 12 msec after beginning of QRS, these two cones have united in lower three sections and have joined slightly in upper section. Heart now contains a cone of depolarized muscle within an incomplete cone of muscle which is still in resting state. Notice breakthrough of electrical activity anteriorly on right. This leaves activity in posterior and leftward portion of ventricles unopposed. This pattern of excitation continues during next 6 msec. Picture at 18 msec is generally unchanged, although amount of muscle depolarized is, of course, larger; fraction of posterior and left portions in resting state has become smaller. At 25 msec after beginning of QRS complex, only a small amount of muscle in posterior and lateral portion of left wall and of basal septum remains to be excited.

able amount of septal tissue is excited on the left and there is a smaller excitation wave on the right. On both sides of the septum, activity proceeds toward the center of the septum with a preponderance of left-to-right activity. Sodi-Pallares and co-workers[38, 60] at one time postulated a barrier to conduction in the central septum, but this has not been confirmed. In the basal septum there is little Purkinje tissue; this region is excited very late in QRS and mainly from left to right.

DETAILS OF VENTRICULAR ACTIVATION (Fig. 6–14).[5, 55, 57] Since, as stated above, septal activity is at first primarily left to right, ventricular activity likewise usually begins earlier on the left and proceeds toward the right. On the right the earliest activity occurs at the septal termination of the right bundle in the region of the anterior papillary muscle of the right ventricle. Even when activity does not begin earlier on the left, more tissue on that side is activated early in ventricular depolarization. The smaller and usually later activity on the right is directed from right to left. The resultant of these opposing forces is directed to the right.

Immediately after the regions near the septal terminations of the bundles are activated, the impulse spreads very rapidly over a large portion of the endocardium near the apex of the heart and in the central region on both sides. This rapid activation is possible because the conduction system ramifies extensively along the endocardium on both sides. We may liken the intraventricular conduction to a tree, with the impulses starting near the trunk. Although the speed of propagation is only 1 m per sec, the impulse reaches the peripheral branches and excites the subendocardial myocardium at many places almost simultaneously. Within a short period, most of the central and apical endocardium on both sides is activated, and the impulse can then proceed in only one general direction, from endocardium to epicardium.

In the dog, the rapid excitation of the endocardium produces, on both sides, incomplete cones of depolarized muscle which extend through both walls and the septum. At 5 msec, as at 10, after the beginning of the QRS complex, these cones are growing by movement of the advancing wave outward in the walls and toward the center of the septum. Electrocardiographically, a consequence of the double invasion of the septum is that the septal forces tend to cancel one another. At 12 msec after the start of the QRS, the cones have united in the septum and have broken through to the surface of the thinner right ventricle. As a consequence of the breakthrough to the anterior right epicardial surface, there is no longer a boundary between active and resting tissue in this region, and the boundary on the left and posterior parts of the heart is now unopposed. At 18 msec after QRS has begun, invasion of the left and right central ventricular surface is complete. At this time, a thin slice of tissue extending from base to apex in the lateral and posterior left wall remains in the resting state, as does a portion of the basal septum. At the end of 25 msec, a small region in the posterior left wall and another in the basal septum remain unexcited. Activity in these regions, directed from apex to base, continues until the end of the QRS complex. The regions which are depolarized last lie in the basal septum bordering the atrium.

Depolarization of the Human Heart. Heroic efforts in the laboratory of Durrer[17] have produced information about the excitation of the human heart (studied at surgery or, more successfully, during perfusion after removal from individuals who have died accidentally). These data indicate that the pattern of excitation is extremely similar to that described in the dog[17, 47] (Figs. 6–15 and 6–16).

The differences between the dog and man are as follows: (i) The intraventricular septum seems to have more left-to-right excitation from the left bundle early in activation in man. (ii) Also, there seems to be more excitation early in the left lateral wall during ventricular excitation in man. (iii) The area which is excited early near the right septum by the right bundle appears to be relatively small; combined with the increased left septal activity, it gives a larger left-to-right component in the septal activation. This is due both to more extensive ramifications of the left bundle and less extensive ramifications of the right bundle. It is not clear if the basal septum in man is activated very late in QRS, as in the dog, but it does appear to be true that the basal left and right walls are activated very late. There appears to be less extensive

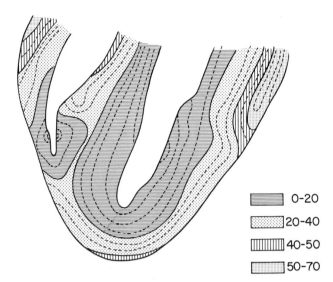

0-20
20-40
40-50
50-70

Figure 6–15 Excitation of the ventricle seen in a sagittal section in the human heart. The shading indicates the sequence of depolarization. The dotted lines within each shading indicate time of depolarization after the earliest activity; lines at 5 msec intervals are found within the larger 20 msec intervals. Note that activity in the septum seems to begin at about the middle of the left septal surface and to move from left to right. The left basal portion of the septum is excited at about 25 msec and the right septal surface in the upper portion of the septum between 45 and 50 msec. There is also some early activity in the apical region of the right septum, extending into the apical wall. In the left and right walls, the activity proceeds from inside out and from apex to base. The earliest activity in the left wall is in the central region, and depolarization spreads from this area so that there is a continuous wave of activity in the left wall and septum. The center of the left wall (midway from endocardium to epicardium) is excited at about 25 msec by the wave of activation moving from endocardium to epicardium. There is also activation from apex to base in the lateral basal left wall. (After Durrer *et al., Circulation,* 1970, *41,* 899–912.)

left Purkinje fiber activity in the right wall and more in the left than in the dog.

Atrial and Ventricular Repolarization. From the duration of the action potential, it is apparent that repolarization of the atrium in the dog and man normally occurs during the depolarization of the ventricles. There is an isoelectric period between the end of

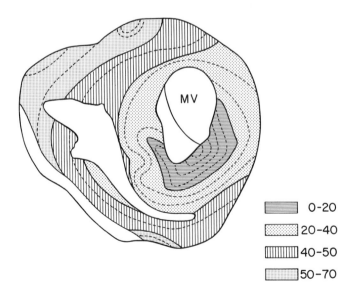

MV

0-20
20-40
40-50
50-70

Figure 6–16 Activation of the left wall of the human heart seen in a coronal section near the base of the heart. Note that the earliest activity occurs around the left cavity in this section, extending somewhat onto the septal surface of the cavity, and that depolarization moves toward the center of the septum and toward the free wall. Note, also, that the last areas depolarized are in the posterior wall of the heart, predominantly on the right but also on the left, with some late activity also seen anteriorly. (After Durrer *et al., Circulation,* 1970, *41,* 899–912.)

atrial depolarization and the beginning of ventricular depolarization during the plateau of the intracellular action potential. During this time, portions of the atrioventricular conduction system are depolarizing. Repolarization of the atrium probably progresses in a direction similar to that followed by atrial depolarization.

However, the wave of ventricular repolarization, as reflected in the electrocardiogram, tends to move oppositely to the wave of depolarization. The duration of the action potential in ventricular cells is such as to indicate that repolarization occurs during the T wave of the electrocardiogram. Both atrial and ventricular depolarization will be considered in greater detail later in the chapter.

ORIGIN OF THE ELECTROCARDIOGRAM

Electrocardiographic Recording Apparatus and Conventions. APPARATUS. Einthoven's string galvanometer,[18, 19] which made electrocardiographic recording possible, has been supplanted by the direct-writing galvanometer employing vacuum tube or transistor amplifiers. In this device, the small voltages at the body surface are amplified to produce currents that can drive a large galvanometer, which in turn moves a hot stylus across heat-sensitive paper (Fig. 6–17).

The usual amplifier has a "push-pull" or balanced input to make it less sensitive to the alternating currents that may be picked up in any location near conventional wiring. Alternating current is picked up either through a capacitive coupling between a lead-in to the amplifier and

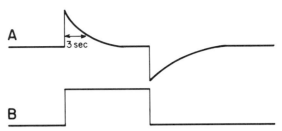

Figure 6–18 When a square wave input *(B)* is applied to the input of a conventional electrocardiographic recorder, the response will be as shown in *A*. Because of input condensers, the amplifier does not respond to a maintained ("DC") input. The 3-sec time constant is sufficiently long that electrocardiographic potential changes are faithfully recorded.

a wire which carries AC, or because alternating magnetic fields cut across the electrocardiographic recording leads. The alternating current is, however, picked up nearly equally by both input leads of the push-pull circuit. Since the amplifier measures the differences in potential between these leads, the alternating current in the leads will tend to cancel.

Electrocardiographic amplifiers are also condenser-coupled. That is, each lead is interrupted by a condenser (Fig. 6–17). Condenser coupling has two purposes: (i) Maintained differences in voltage, *i.e.*, "direct currents," at the two recording leads do not pass condensers. Such voltages may be produced by electrolytic processes (identical to those in batteries) involving the electrodes, electrode paste, perspiration, etc. Voltages resulting from these effects are generally much larger than the electrocardiographic voltages and, since they will usually change in time, much rebalancing would be necessary were they not eliminated. (ii) Condenser coupling also makes it easier to construct a drift-free amplifier because the amplifier may be condenser-coupled between stages. This feature minimizes changes in vacuum tube or transistor performance attributable to temperature, aging, etc. Condenser coupling for electrocardiographic recording has a time constant of 3 sec; *i.e.*, if a constant voltage is applied between the two input terminals, the pen will return about two-thirds of the way to its zero position in 3 sec (Fig. 6–18). This time constant makes it possible to record the slower components of the ECG and does not affect the rapid components. The European standard is 2 sec.

CONVENTIONS. Electrocardiographic convention specifies that 1 mV input to the amplifier shall produce a 1 cm deflection of the pen. Recording paper is moved at 25 mm per sec. The conventional bipolar limb leads, designated leads I, II, and III, record differences between the right arm, left arm

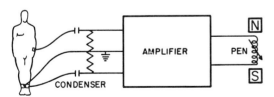

Figure 6–17 Conventional electronic amplifier-pen recorder. Voltages recorded from left arm and leg of patient lead to input tubes of amplifier through a resistance-condenser input. There is also a ground connection, and input is "balanced." Output of amplifier is sufficient to drive a large galvanometer, which moves a recording stylus across a paper. In this way, a record is instantaneously available.

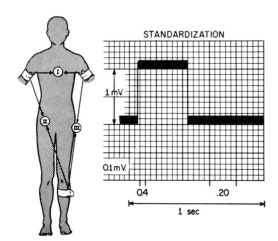

Figure 6–19 Electrocardiographic limb lead connections and conventions for sensitivity and paper speed. Conventionally, 1 mV equals 1 cm vertically; 1 mm horizontally equals 40 msec. Over-all horizontal speed is 25 mm per sec. Conventional bipolar limb leads, designated *I, II* and *III,* are recorded as indicated by the Roman numerals.

and left leg (Fig. 6–19). We can think of the extremities merely as lead wires connected to the body; *i.e.,* the potentials are not altered if the electrodes are moved along the extremities; for convenience the wrists and ankles are used as recording points. Electrocardiographic electrodes are fabricated of corrosion-resistant metal. They are coated with a film of conducting paste and are applied lightly but firmly to the body with rubber straps. It is important that the electrical contact with the body be good, and for this purpose the paste may be rubbed into the skin. The right leg is used as a ground connection. The amplifier is connected to these leads as indicated. In lead I, positivity of the left arm (or negativity of the right arm) produces an upward deflection. In leads II and III, positivity of the leg or negativity of the appropriate upper extremity produces an upward deflection. These conventions were originally established by Einthoven, who arranged them so that the major deflection in each lead would be upright in a normal subject.

In the present technique for recording limb and precordial "unipolar" leads,[3] the negative terminal of the amplifier (the one which will produce an upward pen deflection if its potential is negative) is connected, through a resistive network (Fig. 6–20), to

all three extremities. This arrangement constitutes Wilson's "central terminal." The positive electrode is designated the "exploring electrode"; it is connected to the limb electrodes individually and moved through several precordial positions, designated V_1 through V_6 (Fig. 6–21). The purpose of the unipolar leads is to record predominantly the potential at the point where the exploring electrode is placed. Wilson's central terminal is considered to be approximately a "zero terminal"; that is, it approximates the potential at some point which, because of its symmetry with respect to the voltages produced by the heart or because of its great distance from the heart, is not influenced by the voltages produced by the heart. The idea that a true zero potential can be found on the body surface is theoretically unsound and has been criticized;[23] practically, however, the Wilson terminal, or a modified version thereof, seems to be adequate.[3]

The unipolar limb leads are designated VR, VL and VF. The letter R indicates that the exploring electrode is connected to the right arm, L to the left arm, and F to the left foot. The V indicates that the Wilson terminal is the reference. Potentials so recorded may be slightly smaller than desirable, and for this reason a system of augmented unipolar extremity leads, originally described by Goldberger and designated as aVR, aVL, etc., is more commonly used. In this system the potential difference is recorded between one extremity and the other two, both of which are connected directly to each other (Fig. 6–20). Goldberger's leads are wired into most electrocardiographic recorders.

HEART POSITION. The shape of an electrocardiographic complex recorded at the body surface is determined by the pattern of activation within the heart and the position of the heart and of the recording electrode. In discussing the origin of *human* electrocardiographic complexes, we shall transpose patterns of cardiac activity measured in animals, primarily the dog, to the human heart as it lies in the human thorax. The dog heart lies vertically in the chest, with its major axis parallel to the long axis of the chest. The human heart lies quite differently. The right chambers tend to be anteriorly placed, the left posteriorly. The interatrial and interventricular septa are

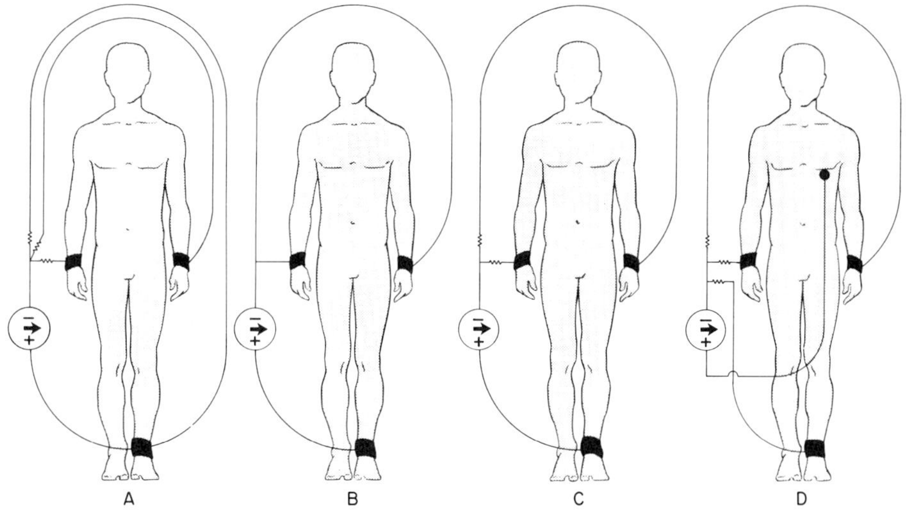

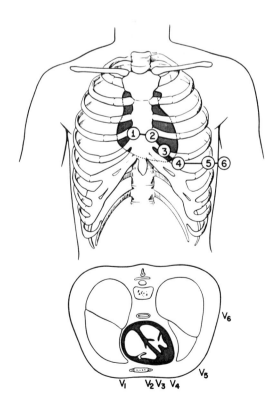

Figure 6–20 Various types of connection used in electrocardiographic recording of unipolar leads.

A, Recording of unipolar leads according to convention established by Wilson. All three extremities are connected by a resistive network; resistors are equal, usually 5000 ohms or larger. This network is connected to negative terminal of amplifier and is called the Wilson central terminal. Exploring or positive electrode is either connected to each extremity, as shown (for unipolar limb lead recording), or placed at precordial positions shown in Figure 6–21.

B, Modified unipolar *limb* lead recording system devised by Goldberger. Potential at one limb electrode, connected to positive terminal of amplifier, is recorded against potential of other two, connected together *without* resistors.

C, Form of unipolar limb recording devised by Wilson wherein potential at one extremity is compared with potentials at other two connected *with* resistors. System shown in C has been generally replaced by system shown in B for unipolar limb lead recording.

D, System for recording unipolar chest leads with Wilson central terminal.

Figure 6–21 Positions of unipolar precordial (chest) leads as routinely recorded in electrocardiography. V_1 is immediately to right of sternum at fourth intercostal space. V_2 is just to left of sternum in fourth intercostal space. V_4, in fifth intercostal space, is in midclavicular line. V_3 is between V_2 and V_4. V_5 is in fifth intercostal space in anterior axillary line. V_6, in fifth intercostal space, is at midaxillary line. The two portions of figure indicate vertical and horizontal positions of these leads.

nearly parallel to the anterior chest wall, etc. (Fig. 6–21). These factors will be considered in the discussion of electrocardiographic complexes which follows.

Origin of the P Wave. The two atrial chambers can be considered as equivalent to three roughly parallel sheets of muscle, lying in the chest in the position shown in Figure 6–22. We can think of normal atrial excitation as consisting of three waves moving inferiorly and to the left from the sinus node. Initially the wave moves anteriorly as the right atrium is depolarized earlier. During the latter half of atrial depolarization, there is activity moving posteriorly in the left atrium. A plane can be imagined drawn through the middle of the atrial mass and extending from the left shoulder down to the level of the sixth interspace beneath the right axilla. Along this plane the potentials caused by atrial excitation would be initially positive, then negative (Fig. 6–23). This would be the "zero line" (equal positive and negative potentials) for atrial depolarization. A recording electrode placed above this, when paired with the Wilson central terminal, would record a predominantly negative potential. All the normal precordial leads are positive during atrial depolarization. Lead VR, the unipolar right arm lead, shows a negative P wave, as do certain esophageal leads which are at times recorded, particularly to diagnose atrial abnormalities. The P wave, in general, has a smooth and rounded contour with little notching or peaking and has an average duration of 90 msec in man and an amplitude of less than 0.25 mV.

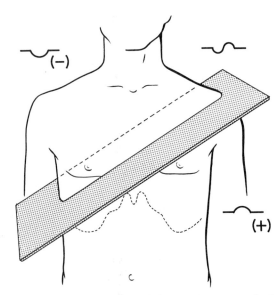

Figure 6–23 Diagrams of shapes of P waves which would be recorded at various places at body surface. Since general direction of atrial activation is from right arm toward left leg, electrodes on upper part of body will see a negative potential during atrial activation; those on the lower part will see a positive potential. There will be a plane, as indicated on drawing, where an electrode would record both positive and negative activity.

THE SILENT PERIOD. As mentioned previously, the initial phase of depolarization in cardiac cells is followed by a plateau, during which the membrane potential changes little. During this period of "electrical systole," all cells are in nearly the same electrical state, and virtually no current flows from one region to another. Consequently, no changes of potential are seen in extracellular leads until the rapid phase of repolarization terminates this plateau and produces the repolarization complex. There is thus a period between the end of the P wave and the beginning of the QRS complex when the normal electrocardiogram shows a flat baseline. During this period, cells of the A-V conduction system are depolarizing, but the number of cells is so small that they produce no potentials in electrocardiographic leads. Electrocardiographic effects of repolarization of the atria will be discussed later.

Ventricular Activation and the QRS Complex. In man, the right ventricle lies anteriorly and the left posteriorly. The septum is tilted slightly forward apically, and the base-to-apex axis of the ventricle is often nearly parallel to the diaphragm (Fig. 6–22).

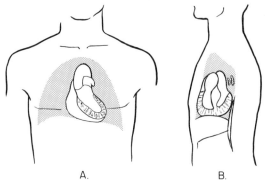

Figure 6–22 Position of heart with respect to body surface in man. Right ventricle is anterior and left ventricle is posterior when viewed from left side. When viewed from front, left ventricle is tilted to left and upward.

The human ventricular electrocardiogram can be closely approximated if we transpose the dog's activation pattern to the human thorax and take into account the differences in QRS duration. As indicated, recent studies of human ventricular activation[17, 47] suggest that the process is similar in dog and man.

As indicated, the initial phase of ventricular activity is usually directed from left to right in the septum and results from earlier or greater initial left-to-right activity or both. This activity, transposed to the human heart (Fig. 6–24), would produce a wave directed to the right, toward the head (since the left side of the septum lies caudally) and slightly anteriorly. This wave will produce an initial negative deflection in leads I, II and III, which accounts for the Q wave. For the leads on the precordium, the picture is also clear. The leads on the right side of the chest (V_1 and V_2) face the positive side of wavefront and record an upward deflection, while those on the far left (V_5 and V_6) record a negative deflection.

Immediately after invasion of the septum begins (Fig. 6–25), rapid conduction through the Purkinje system results in an irregular pattern of inside-out spread in the walls; the transition from the first phase of activity to this second and major phase of

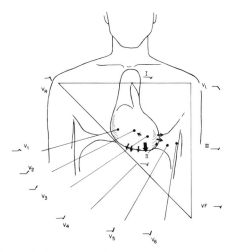

Figure 6–25 When about one-quarter of QRS interval has passed, activity is proceeding from left to right in septum, and activity from inside out in wall has begun. Total activity is such that potentials are near zero in all limb leads, both bipolar and unipolar, and in V_1 and V_6. Other leads on chest are positive because activity proceeds toward the apex and free left wall.

ventricular activity is smooth. Within the septum, left-to-right activity predominates slightly. Arrows drawn perpendicular to the advancing wavefront depict the instantaneous vector of depolarization. At 12 msec after the beginning of QRS in man (by extrapolation from the dog), the average direction of these arrows indicates a pattern of activity directed slightly forward, to the right, and from base to apex. Such a pattern will result in positive deflections in leads II and III and little or no deflection in lead I, which may be either positive or negative at this time. The potentials in the leads on the anterior chest surface will differ slightly from those occurring during the earlier phase, because the leads on the right will now "see"* both approaching (left to right) and receding (base to apex) activity. The approaching activity will be in the right wall and on the left septal surface, the receding activity in the left wall and on the right septal surface. At this time, there may be little or no potential in these leads and a positive deflection in V_3.

At about 35 to 40 msec after the onset of QRS (Fig. 6–26), union of the two separate masses of activated tissue has pro-

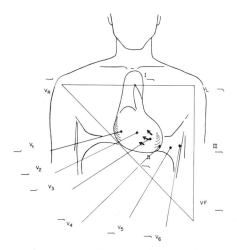

Figure 6–24 Mean direction of activity during earliest portion of QRS transposed from canine to human heart. First activity goes from left to right in septum. Because of position of septum in human chest, this results in negative deflection in all bipolar limb leads, positive deflection in VR and in leads on right side of precordium (V_1 through V_4) and negative deflection in V_5, V_6, VL and VF.

*What is meant by an electrode "seeing" is explained in Chapter 3, Volume I.

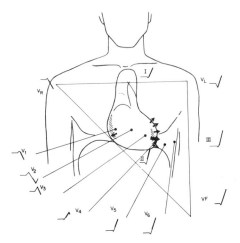

Figure 6–26 At about middle of QRS interval, breakthrough of activity to anterior right ventricle has left forces moving to left posteriorly relatively unopposed. The result is a negative deflection in lead VR and positive deflections in all other limb leads. The leads on far right of chest (V_1 and V_2) now see negative activity; potential at V_3 is near zero, and potentials at V_4, V_5 and V_6 are positive.

duced strong forces directed posteriorly, to the left and inferiorly. The breakthrough of activity to the anterior right wall has greatly reduced the left-to-right component, and over-all activity is directed apically, posteriorly and to the left. The major source of potentials is in the apical, lateral and anterior left wall. Some opposing inside-out activity persists in the basal right wall. At this time, positive deflections will appear in all standard limb leads, and the leads on the left side of the chest will "see" approaching activity. The continuation of this pattern results in the eventual disappearance of the wave front anteriorly, on the right, and in the central and apical portions of the heart; *i.e.*, all the myocardium in these regions has fully depolarized. Overlying precordial leads will therefore record negative potentials.

The over-all pattern of activity immediately following the above, *i.e.*, about midway through QRS, is a continuation of the movement toward the thin slice of lateral posterior left ventricle which remains in the resting state and a smaller movement toward the basal septum. Depolarization reaches the apex of the heart on the right, but some muscle in the apical region of the left ventricle remains to be depolarized. The net result is a wave moving posteriorly,

leftward and slightly toward the apex. Again, the limb leads will be positive. The chest leads except V_5 and V_6 (on the far left) will, however, show negativity.

Finally, after depolarization of the apical regions is complete (60 msec after the onset of QRS), *i.e.*, for the last quarter of the QRS complex, a wave moves toward the base of the left ventricle, particularly posteriorly and from apex to base in the septum. This wave is relatively ineffective in causing potentials in lead I, although leads II and III should show a negative potential; the potentials in the chest leads will be small but generally negative (Fig. 6–27).

In Figure 6–14 it can be seen that the process exhibits a great amount of symmetry around the longitudinal axis of the heart. At various instants, activity is proceeding in opposite directions in the lateral walls or in the septum or in both. This symmetry of depolarization leads to "cancellation" of much of the cardiac electrical activity as recorded from the body surface. It has been estimated that the recorded potentials are 5 to 10 per cent of what might be expected if there were no cancellation.[53] Any condition which alters the sequence of ventricular depolarization in a manner to reduce this cancellation will, of course, produce an increase in the magnitude of the potentials recorded in one or more leads. This is true

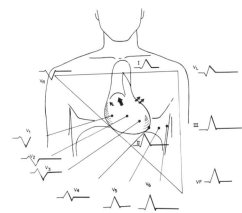

Figure 6–27 During terminal portion of QRS complex, activity is directed to left and posteriorly in basal left ventricle and basally in upper septum. This condition results in potentials which are small in all leads. Deflection in VR is positive; deflections in all other limb leads are negative. Potentials are now returned to zero from negative peak in V_1 and V_2 and from positive peak in V_4, V_5 and V_6. This activity results in slight negative potentials in V_3, V_4, V_5 and V_6.

of bundle branch block and of many types of infarction, and also of cases in which the impulse arises in abnormal sites.

Summary of Ventricular Depolarization. The excitation of the ventricles is here divided into four phases. Any such division is arbitrary, because the process is smooth and continuous and its phases blend imperceptibly into one another. The pattern of excitation is produced by the distribution of the Purkinje fibers to the ventricles. The electrocardiographic potentials resulting from it will depend both on changes in details of the Purkinje fiber anatomy and on the position of the heart in the chest.

During the first phase of ventricular depolarization, the wave of activity proceeds toward the center of the septum from the endocardium bilaterally in the middle and apical portions of the septum. Predominance of rightward activity is due to the fact that Purkinje fibers are distributed more widely on the left septal surface. Because of the position of the heart within the torso, this leads to a wave of depolarization which is directed rightward, superiorly and anteriorly in the torso.

Left-to-right depolarization of the septum continues, but owing to the extensive ramifications of the Purkinje fibers on the apical regions of the mural endocardium, a wave of depolarization moves from inside-out in the walls bilaterally. This begins the second phase of ventricular activation. Within the torso, activity is now directed inferiorly and slightly to the right, without much movement either anteriorly or posteriorly.

Phase three continues phase two, but the activity moving toward the right and anteriorly terminates as the thin right wall is completely depolarized. However, activity continues to move from endocardium to epicardium in the apical and central regions of the thick left ventricular wall. Because the left ventricle lies posteriorly in the human thorax, activity during this phase is directed posteriorly and toward the left leg. This activity produces the major deflection (R wave) of the normal QRS complex.

Finally, most of the apical regions of the ventricle are completely depolarized, and the wave of depolarization still moves toward those regions of the basal septum and left ventricular wall which are farthest re-moved from the terminations of the ventricular Purkinje system. During this period of time, the wave moves toward the base of the heart in the basal septum and posterior basal left ventricular wall. During this time, the direction of depolarization within the torso is predominantly cephalad and posterior.

Duration of QRS is about 80 msec in man. Voltages are of the order of 1 mV but vary with the recording lead. A normal 12-lead electrocardiogram is shown in Figure 6–28.

Repolarization of Atria and Ventricles. As previously noted, the passage of a wave of depolarization along the cell can be likened to the passage of electrical negativity along the outer surface of the cell membrane. In nerve, depolarization is virtually instantaneously followed by repolarization which tends to make any recorded potential a mixture of the two effects (Fig. 4–21a, Chap. 4, Vol. I). Because of the long duration of the action potential in cardiac muscle, depolarization and repolarization are almost always separated in time and produce their effects quite separately. Repolarization can be likened to passage of a wave of positivity along the cell membrane. Thus, we would expect that, if cardiac muscle cells had symmetrical boundaries during depolarization and repolarization, and if a depolarizing wave moving along a cell were followed a few hundred milliseconds later by a repolarizing wave, the potential produced by repolarization would have electrical polarity opposite that produced by depolarization. The pathways of atrial and ventricular repolarization have not been successfully traced in mammalian hearts because the injury potentials produced by extracellular electrodes obscure the potentials caused by repolarization. Intracellular electrodes cannot be used for this purpose either, because, although they work effectively with surface cells, they cannot easily give information about the repolarization of deeper myocardial cells. Furthermore, repolarization potentials are changed by almost any type of experimental intervention (opening the chest, etc.). Therefore, to get any hint of the direction of normal repolarization, we must use whatever information is available; most of this information, particularly in humans, comes from the body surface electrocardiogram.

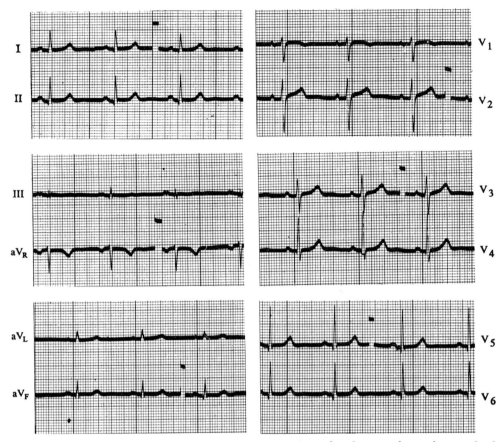

Figure 6-28 A normal 12 lead electrocardiogram. (From Winsor, ed., *Electrocardiographic test book*. New York, American Heart Association, 1956.)

The Ta Wave. Atrial repolarization in the dog and in man normally occurs during the depolarization of the ventricles, and the repolarization potential is concealed by the much larger potentials of the ventricles. There is thus an isoelectric period between the end of atrial depolarization and the beginning of ventricular depolarization, although, as has been discussed, portions of the atrioventricular (A-V) conduction system are depolarizing during this time. Infrequently, the ventricular potentials do not conceal the atrial repolarization potential (referred to as the Ta wave), and it may be seen as a very small mirror image of the P wave. Thus, it is probable, although there is no direct evidence to support this contention, that repolarization of the atrium progresses in a direction similar to that followed by depolarization. The small size of the repolarization complex reflects the fact that, during repolarization, activity in some portions of the atrium cancels activity of other portions. Also, the repolarization

deflection occurs during the terminal (most rapid) phase of cellular repolarization, and during this phase the total voltage change is much less than that during depolarization.

The T Wave. In most electrocardiographic leads *in man*, the T wave has the same electrical polarity as the QRS complex, *i.e.*, is usually upright when the QRS complex is upright and vice versa. This indicates that repolarization does not follow the same pathway as depolarization, and indeed that the pathways tend to be opposite. It is important to consider whether repolarization is electrically propagated, as is depolarization. Although repolarization of a fiber can be induced by appropriate stimulation (*i.e.*, by stimuli causing the *inside* of the fiber to become negative) and can propagate through a fiber, the current flowing during repolarization is less than 1 per cent of that flowing during depolarization. Such a small current probably cannot initiate a propagated wave.

If repolarization is not propagated, we

may wonder why the configuration of the T wave is consistent under normal conditions. Several factors have been thought to control the sequence of repolarization; among them are temperature and pressure. According to one theory, the pressure differential within the walls favors initiation of repolarization in the outer layers, and repolarization occurs later near the endocardium.[36] Potentials within the right and left cavities of the human heart are negative during repolarization. Sodi-Pallares[60] interpreted these findings as indicating that the T wave normally results from spread of repolarization from the outside to the inside of the left wall. He believed further that electrical forces from other portions of the ventricles cancel one another and that the right wall and the septum are electrically silent during repolarization, i.e., have no clear-cut direction of repolarization but repolarize at random. Available data do not allow complete acceptance of any theory concerning ventricular repolarization, although the normal "pathway" of repolarization is apparently independent of, although statistically generally opposite to, the pathway of depolarization (note the T waves in Fig. 6–28).

ABNORMALITIES OF THE ELECTROCARDIOGRAM

Arrhythmias. An "arrhythmia" is a disturbance of the heart rate, the cardiac rhythm, or the sequence in which the chambers are excited. The arrhythmias have two main causes: disorders of impulse formation and disorders of conduction. In the first case, the impulse may be generated in the normal site (the S-A node) but at an abnormal rate, or some other portion of the myocardium may function as pacemaker. Origination of the heart beat in the A-V node is abnormal, even though this tissue normally has the ability to generate impulses. Arrhythmias also may result from pacemaker activity in the branches of the conduction system or within unspecialized tissues — sites which do not normally generate impulses.

SINUS RHYTHM. *Normal sinus rhythm* indicates that the pacemaker is within the S-A node and that the heart rate is within normal limits. In the condition called *sinus arrhythmia* (Fig. 6–29), common in children

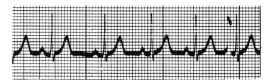

Figure 6–29 Sinus arrhythmia; lead II of a child. Intervals between successive QRS complexes from left to right are 680, 600, 600 and 520 msec in duration respectively. Heart rates would be 88, 100, 100 and 115 beats per min. (From Winsor, ed., *Electrocardiographic test book.* New York, American Heart Association, 1956.)

and young adults, the heart rate changes phasically during the respiratory cycle. Sinus arrhythmia is not pathologic and is most commonly seen when the person is resting; it may disappear when the heart rate increases. The electrocardiographic complexes will have a normal configuration.

In *sinus bradycardia* (Fig. 6–30) the rate at which the S-A node is producing impulses is subnormal; the heart rate is below 60 beats per min but rarely below 40 beats per min. Conversely, the heart rate is more than 100 beats per min in *sinus tachycardia* (Fig. 6–31) because the sinus node is producing impulses at an accelerated rate. The electrocardiographic complex is of normal shape in either of these conditions, and these definitions are somewhat arbitrary. In infants, the normal resting pulse rate is frequently 120 beats per min or more. When the rate exceeds 160 beats per min, the condition is usually *atrial tachycardia*, in which case the pacemaker is atrial. In paroxysmal atrial tachycardia the reversion to sinus rhythm is abrupt, and during the paroxysm the rhythm is very regular. Paroxysmal tachycardia of sinus origin is rare; the pacemaker is usually an atrial focus outside the sinus node. As will be discussed, this condition can produce A-V block. The condition is at times controlled by pressure on the carotid sinus or on the eyeballs, procedures

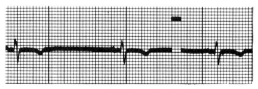

Figure 6–30 Sinus bradycardia; lead III in an athlete. Interval between QRS complexes is 1400 msec (34 msec × 40 msec per mm). Also shows some sinus arrhythmia. Heart rate is 43 beats per min. (From Winsor, ed., *Electrocardiographic test book.* New York, American Heart Association, 1956.)

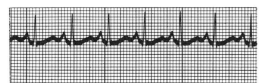

Figure 6-31 Sinus tachycardia in a 13-year-old child (lead II). Heart rate is 130 per min; pacemaker is in sinus node, since each QRS complex is preceded by a P wave at a normal interval. In children of this age, heart rate should not exceed 109 beats per min. S-T segments are considered to be normal in view of rapid rate. (From Winsor, ed., *Electrocardiographic test book*. New York, American Heart Association, 1956.)

activating receptors which can reflexly decrease the heart rate. Ventricular tachycardia and atrial flutter and fibrillation will be discussed later.

Sinus arrest (Fig. 6–32) is a rare condition in which the sinus node does not initiate impulses. It usually results from treatment of heart disease with drugs. Often, the P waves occur less and less frequently and then disappear.

In the condition known as *wandering pacemaker*, there are minor changes in the shape of the P wave and changes in the P-R interval, yet the heart rate is not greatly disturbed. It is thought that this condition arises from movement of the pacemaker. A computer analysis by Brody *et al.*[7] indicates that there may be a shift between two (or more) atrial pacemakers with concomitant changes in P-wave shape in normal persons. The P wave changes seen are small and would not be noted in ordinary clinical examination.

ATRIOVENTRICULAR NODAL RHYTHM. As indicated previously, the A-V node has the second highest rhythmicity of the specialized cardiac tissues. Thus, if activity of the sinus node becomes depressed, or if the rhythmicity of the A-V node is in-

creased, the latter may take over the task of initiating impulses and become the cardiac pacemaker. Conduction will progress normally to the ventricular myocardium and often, but less frequently, will also move in a retrograde direction to the atrium.

EXTRASYSTOLES. All cardiac tissue seems to be capable of generating impulses. Slight injury, anoxia, mechanical trauma or friction apparently can increase this tendency, so that a nonspecialized part of the myocardium may become, either continuously or sporadically, a pacemaker for the heart. Such beats are referred to as "ectopic." Abnormal impulses may be formed in the atrium outside the sinus node (atrial premature beats), in the A-V node (nodal premature beats), in the conducting bundles, or in the ventricular musculature (ventricular ectopic beats). A second class of extrasystoles arises in the sinus node as an interpolated premature beat.

Atrial Premature Contractions (Fig. 6–33). Alteration in the site of the atrial pacemaker may result from failure of the sinus node or from increased excitability of some other locus. If the atrial pacemaker is not in the S-A node, an abnormal P wave will result, usually followed by normal QRS and T complexes. Should the pacemaker be appreciably closer to or farther from the A-V node than the sinus node, the P-R interval will also be prolonged or shortened. The P-R interval will be prolonged if an atrial extrasystole arrives at the A-V node during its relatively refractory period. As indicated, the A-V node has a low safety factor and cannot conduct at a high frequency. If the atrial extrasystole arrives during the absolutely refractory period of the A-V

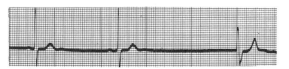

Figure 6-32 Sinus arrest induced by pressure in carotid sinus in a normal subject (lead V_3). Two normal beats at a slow rate (33 beats per min, interval 1.8 sec) followed by a ventricular premature contraction not preceded by a P wave (vagal escape). (From Winsor, ed., *Electrocardiographic test book*. New York, American Heart Association, 1956.)

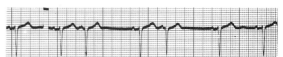

Figure 6-33 Two atrial premature contractions (lead V_1). First two complete complexes are followed by a premature beat, which is followed by a compensatory pause. Fourth complex is followed by a second premature contraction. Fifth beat is again delayed by a compensatory pause; last complex is at a normal interval. Two premature complexes bear a fixed relationship to preceding beats. (From Winsor, ed., *Electrocardiographic test book*. New York, American Heart Association, 1956.)

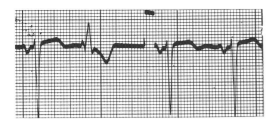

Figure 6–34 Second complex is followed by a ventricular premature beat. Note absence of P wave before this QRS complex, abnormal shape of QRS, inversion of T wave. (T wave has opposite polarity to QRS, while normally it has the same polarity.) (From Winsor, ed., *Electrocardiographic test book.* New York, American Heart Association, 1956.)

node, the abnormal P wave is not followed by a QRS complex.

Nodal Premature Beats. Although the A-V node at times acts as the pacemaker for the heart or for the ventricles, extrasystoles seldom arise within the A-V node.

Ventricular Extrasystoles (Fig. 6–34). An irritable focus within the ventricle may cause regular or irregular ventricular extrasystoles. When the ectopic beat originates within the Purkinje system, the ventricular complex may be normal (origin of beat within the common bundle) or abnormal (origin of beat below the common bundle). When a beat originates in the ventricular myocardium, the QRS complex will be abnormal in shape and duration. The origin of an extrasystole can often be deduced from the shape of the QRS complexes recorded from the body surface. For example, if all chest leads show a marked increase in positivity, the extrasystole must originate in the posterior portion of the heart.

In general, the QRS complex will be prolonged, because conduction over the Purkinje system is not following the normal path. If the analogy between the Purkinje system and a tree is recalled, it will be seen that an impulse originating near a peripheral branch of the tree must be conducted along that branch until it intercepts the normal conduction pathway near the trunk. Even though the ectopic impulse is conducted along the endocardium at normal Purkinje velocity, about 1 m per sec, the time required for depolarization of the ventricles will be increased. With prolongation of the interval between the beginning and the end of ventricular depolarization, those regions which depolarize first tend to repolarize first, and the polarity of the T wave will usually be opposite that of the QRS complex.

At times, a ventricular extrasystole arises from a single focus which, for unknown reasons, discharges rhythmically to produce a wave with a fixed shape and often with a constant relationship to the normal QRS complex. Such a beat is referred to as a *coupled beat.* It may occur after every normal beat or, often, after every third, fourth or fifth normal beat. The resulting rhythm is referred to as bigeminy, trigeminy, etc. When an ectopic beat originates below the A-V node, the atria will often beat independently, because the safety factor for retrograde conduction is lower. An ectopic beat may be conducted in a retrograde direction and may depolarize either the A-V node or, if conducted to the atria, the sinus node. If the S-A node is thus depolarized, it cannot generate a normal impulse until repolarized. Consequently, the interval between the extrasystole and the next normal sinus beat will be abnormally long, and the ventricles will fill to a degree greater than normal. The stroke volume of the beat following the pause will be higher than normal. The combination of the pause and the large stroke volume of the succeeding beat is often perceived by the individual as a "skipped beat" and a definite thump.

Paroxysmal Ventricular Tachycardia. At times, an abnormal ventricular focus produces a maintained tachycardia. It frequently leads to ventricular fibrillation (see below), a fatal rhythm.

ARRHYTHMIAS INVOLVING GREATLY INCREASED HEART RATE; FLUTTER AND FIBRILLATION. *Paroxysmal tachycardia* (Fig. 6–35) is an episode of rapid beating with a sudden onset and termination. At rates of

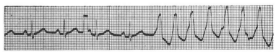

Figure 6–35 Onset of an episode of paroxysmal ventricular tachycardia. In this condition, interval between QRS complexes is somewhat irregular. Individual QRS complexes appear somewhat different, perhaps as a result of superimposition of P waves. Although heart rate (150) is somewhat slow for tachycardia, complexes are typical of this condition, as is inverted T wave. (From Winsor, ed., *Electrocardiographic test book.* New York, American Heart Association, 1956.)

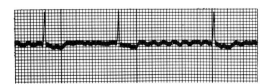

Figure 6–36 Atrial flutter and fibrillation (lead IV). Fine oscillations with duration of 160 msec (4 mm) are flutter. Periods when baseline is "silent" are probably periods of fibrillation. (From Winsor, ed., *Electrocardiographic test book*. New York, American Heart Association, 1956.)

300 beats per min or more the disorder becomes *flutter*. In atrial flutter, the P waves merge, giving a rapid, sawtoothed atrial complex. Usually, the atria depolarize at a rate so rapid that the A-V conduction system cannot adequately respond. Therefore, not every P wave is followed by a QRS complex (second degree A-V block).

Atrial fibrillation (Fig. 6–36), which may develop from atrial flutter, is identified by an irregular baseline of fine small oscillations, at rates greater than 500 beats per min. The ventricular rate is rapid but irregular, because the A-V node responds irregularly to the atrial impulse. The atrial contraction is ineffective.

Ventricular fibrillation is similar to atrial fibrillation (Fig. 6–36), but, whereas atrial fibrillation may continue for years, fibrillation of the ventricles is generally a terminal event. The chaotic contraction pattern is not adequate for propulsion of the blood into the circulation.

Sir Thomas Lewis[34] thought that atrial fibrillation results from a circus movement of the wave of excitation, *i.e.*, that the conduction process is altered in such a fashion that a wave may "catch its own tail." Models of this condition have been created in animal experiments.[41] If a portion of the tissue around the superior vena cava is damaged by being clamped in a forceps for a few seconds, a low safety factor may result. An impulse started at one side of the damaged region may proceed around the ring, return to the original site, and be conducted slowly through the damaged region and again move around the ring.[48] The term "circus movement" is obviously appropriate. Several alternatives to Lewis' theory have been offered.[58] In one of these, a rapidly firing focus is considered the cause of arrhythmia. In another, multiple ectopic

foci are thought to discharge constantly and to produce chaotic excitation.

Certain aspects of fibrillation suggested that some orderly sequence of events must lead to fibrillation. In the first place, a certain volume of tissue is necessary for fibrillation to occur. The cat's atrium cannot be made to fibrillate by electrical stimulation, and the cat's ventricle will fibrillate only transiently and recovers spontaneously. The dog's atrium will generally fibrillate transiently and recover spontaneously, whereas, if the dog's ventricle begins to fibrillate, heroic measures are required to end the condition.

Cooling the heart tends to increase the incidence of fibrillation, and addition of hormones and certain ions (potassium) to bathing solutions increases the tendency to fibrillation.

An important relevant point is the clinical observation that the Q-T interval shortens as the heart rate increases. This phenomenon has been extensively investigated by Carmeliet and Lacquet,[9] using the intracellular electrode. If a cardiac cell is stimulated near the end of repolarization, the next beat will show a decreased duration of the action potential and a slightly decreased rate of initial depolarization. If the stimulation is repeated before the end of the shortened complex, the action potential will again shorten and the rate of initial depolarization will again decrease. The decreased slope of initial depolarization leads to slowing of the conduction velocity. At the same time, the length of time that the muscle is depolarized becomes shorter. If the action potential becomes continuously shorter at the same time the velocity becomes continuously slower, the length of a depolarized segment may decrease. When this decrease occurs, it is possible for a wave to "catch its own tail" (at normal action potential duration and conduction velocities this is not possible). Once this has happened, the process can be repeated. Each successive completion of the circuit can result in a slower conduction velocity, a decreased action potential duration, and a shorter pathway for the circus wave until, finally, it has a very short pathway. At this time, there could be a large number of circus pathways on the myocardium, possibly undergoing continuous change.

Records obtained with both intracellular and extracellular electrodes indicate that once fibrillation has been established the situation is one of complete chaos. Multiple ectopic foci could cause the waves; however, during electrically induced atrial fibrillation, Sano and Scher[51] found that there is initially a simple rapidly firing focus, although circus-like waves which change continuously may exist later.

A common cause of sudden death, ventricular fibrillation, is encountered in cardiac surgery and following accidental contact with high voltage. It can often be terminated by proper electrical stimulation or, at times, by solutions of unphysiologic ionic composition. The cardiac surgeon often prefers to work on a fibrillating heart during cardiac bypass, and routinely fibrillates and defibrillates the heart.

CONDUCTION BLOCK. Conduction block may occur almost anywhere between the sinus node and the ventricles, although the most likely site is at the A-V node. S-A block, as previously noted, is rare.

Atrioventricular Block. Several disorders of A-V conduction can be observed in animal experiments. Stimulating the sinus node at progressively increasing frequencies produces at first a slight *decrease* in conduction time. With more rapid stimulation, the A-V interval increases (Fig. 6–37). As indicated, the A-V node, with its low safety factor, is the structure most susceptible to conduction block. In *first degree block* (Fig. 6–38), the P-R interval is prolonged; this prolongation is sometimes seen in patients with normal heart rates. The disorder in some of these patients is analogous to the prolongation of the P-R interval induced in experimental animals by stimulation of the atrium at progressively increasing rates. An increase in the heart rate of a normal person results in a *decrease* in P-R interval (over the range of 70 to 130 beats per min [Table 6–1]) as long as the heart rate is not faster than 2 beats per sec. This reponse of normal man and experimental animals is attributable to unknown alterations in the physiologic condition of the cells as the heart rate increases. Thus, it would seem that first degree block occurring in man at lower heart rates follows depression of the A-V node by anoxia or some similar factor which makes it impossible for the node to conduct at a normal velocity, even though impulses are arriving at a normal frequency. (In some otherwise normal individuals, A-V block is present from birth.) When, as in paroxysmal tachycardia, impulses arrive at the A-V node at an abnormally high rate, the node may be unable to conduct normally. First degree block or a more serious block may then be seen.

In *second degree block* (Fig. 6–39), atrial excitation does not always lead to ventricu-

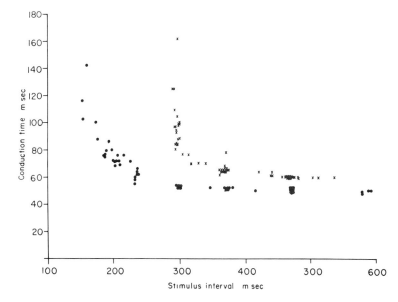

Figure 6–37 Stimuli delivered to S-A node at intervals indicated lead to conduction times between S-A and A-V nodes shown by dots. Retrograde conduction is indicated by crosses. When stimuli are delivered to A-V node, conduction time is from A-V to S-A node.

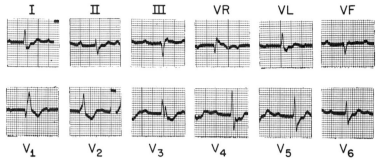

Figure 6-38 First degree A-V block in a heart with several other abnormalities. P-R interval, measuring 320 msec, is very much prolonged; at this heart rate, upper normal limit is 200 msec. That this is first degree A-V block is indicated by QRS complex following every P wave. It should be noted further that there are negative S-T segment shifts in leads V_1, V_2 and V_3. Late activity coming toward right in these leads suggests right bundle branch block. This subject shows a first degree A-V block and a right bundle branch block. (From Winsor, ed., *Electrocardiographic test book.* New York, American Heart Association, 1956.)

lar excitation; *i.e.,* some P waves are not followed by QRS complexes. This condition is an extension of that seen in first degree block and is observed in the dog when the heart rate is artificially raised above 5 beats per sec. When only alternate P waves are followed by a ventricular complex, the condition is referred to as 2:1 block. Other ratios commonly seen are 3:1, 3:2, 4:1, etc. As indicated earlier, records are now obtained from the human heart through the use of catheter electrodes.[12] An example of block between atrium and His's bundle is shown in Figure 6–40.[13]

In second degree A-V block, a *Wenckebach phenomenon* may also occur. In this condition the sinus node generates impulses at a constant rate, but the P-R interval grows progressively longer during several beats until there is an atrial complex which is not followed by a ventricular complex. The next atrial complex is followed by a QRS complex and there is a short P-R interval; the interval again grows progressively longer, and the phenomenon is

repeated. In experimental animals, the Wenckebach phenomenon can be duplicated with high-frequency stimulation at a rate almost sufficient to cause 2:1 A-V block. Although the exact mechanism is not known, it would appear that the cells between the atrial and the A-V node are functional but close to failure. Consequently, with each successive beat, A-V conduction is slower until, finally, it is completely blocked. When complete block occurs, the cells have a long time in which to recover, so the beat after complete A-V block has a short P-R interval.

Third degree or complete A-V block (Fig. 6–41) is a condition in which the A-V node is entirely unable to conduct impulses. A pacemaker within the A-V node or in the ventricle controls the ventricular beat, which is independent of and slower than the atrial beat.

BUNDLE BRANCH BLOCK. Bundle branch block results from failure of transmission in either the right or left conduction bundles or in their terminal ramifications.

Table 6-1 *Upper Limits of the Normal P-R Interval**

HEART RATE	BELOW 70	71–90	91–110	111–130	ABOVE 130
Large adults	0.21	0.20	0.19	0.18	0.17
Small adults	0.20	0.19	0.18	0.17	0.16
Children, ages 14 to 17	0.19	0.18	0.17	0.16	0.15
Children, ages 7 to 13	0.18	0.17	0.16	0.15	0.14
Children, ages 1½ to 6	0.17	0.165	0.155	0.145	0.135
Children, ages 0 to 1½	0.16	0.15	0.145	0.135	0.125

*From Ashman, R. and Hull, E.: *Essentials of electrocardiography*, 2nd ed. New York, The Macmillan Co., 1945.

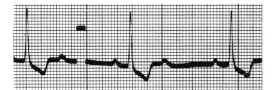

Figure 6–39 Two-to-one A-V block. In lead II, record shows three QRS complexes and four P waves; second and fourth P waves are not followed by QRS complexes. Other two P waves are followed by QRS complexes at about a normal interval. Disease is suggested, although by no means proved, by inverted T wave. (From Winsor, ed., *Electrocardiographic test book.* New York, American Heart Association, 1956.)

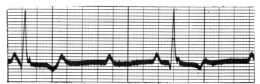

Figure 6–41 Complete A-V block. Atria and ventricles are beating independently. Atrial rate is 90 beats per min; ventricular rate is 29 beats per min. Atrial pacemaker appears to be in sinus node, and, although it cannot be seen from this record, ventricular pacemaker appears to be near base of ventricles. (This rate is somewhat slower than would be expected were pacemaker in A-V node, and tracings are somewhat prolonged—a finding also indicating that ventricular pacemaker is probably not in A-V node.) (From Winsor, ed., *Electrocardiographic test book.* New York, American Heart Association, 1956.)

The usual cause is probably myocardial damage from infarction or fibrosis from long-standing cardiac disease, although right bundle branch block may occur in normal young persons. The term "complete bundle branch block" is an arbitrary designation for beats originating in the A-V node but having a QRS duration of over 120 msec in man. The pattern of ventricular excitation in complete left bundle branch block in the dog is shown in Figure 6–42; Figure 6–43 indicates the changes resulting from right bundle branch block. As might be

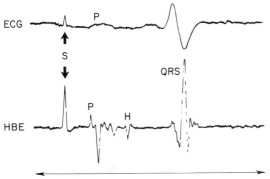

Figure 6–40 A record taken by a catheter electrode within the human heart. The catheter was percutaneously introduced into an antecubital vein and positioned in the region of the right atrium near the His bundle and sinus node. Atrial stimulation was used to control the heart rate. The top record (ECG) shows the conventional electrocardiogram with stimulus artifact followed by upright P wave, QRS complex and T wave. The lower record (HBE) is considered to be recorded from the His bundle. The stimulus artifact is followed by a multiphasic potential from the atrium labeled P. This is then followed by a potential labeled H from the His bundle, and a mostly negative multiphasic potential from the ventricle beneath the QRS complex. Note the similarity to Figure 6–11. (After Damato *et al., Circulation*, 1969, 39, 435–447.)

expected, after the main bundle is interrupted, the normal double envelopment of the septum is replaced by one-way activation from the unblocked side, and activation of the free wall begins at the sites first reached by spread of depolarization across the septum. The wave of excitation utilizes the endocardial Purkinje fibers and travels across the endocardium on the blocked side at about 1 m per sec.[4, 20]

Prolongation of the QRS complex in bundle branch block results both from the longer time required to activate the septum and from the longer time required to activate the blocked free wall. The change in the activation of the free wall during right bundle branch block in the dog is shown in Figure 6–43. Normally, it requires about 18 msec to activate the right mural endocardium, and a large central area is activated within a few milliseconds by the branching Purkinje system. After block, the impulse reaches the wall at the inferior and posterior junctions of the wall and the septum and spreads anteriorly and superiorly. The smooth progression of the wave is altered as it breaks through the septum superiorly. The total time required to activate the right free wall after block is 35 msec.[20] Similar changes in mural activation are seen in left bundle branch block.

In the dog, complete right bundle branch block doubles the duration of QRS; complete left bundle branch block increases it two and one-half times. Comparable durations of the QRS in man would be 160 and 200 msec for complete right and complete left block, respectively. A clinical diagnosis

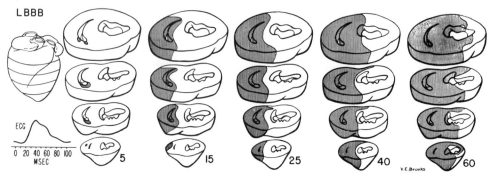

Figure 6-42 Ventricular depolarization after left bundle branch block. This figure should be compared with Figure 6–14, which shows a normal depolarization pattern. Shaded area represents portion of myocardium depolarized up to particular instant indicated at bottom of column, and this is compared with lead II electrocardiogram. Note that activity begins around right cavity, proceeds gradually across septum as depolarization of right free wall is completed, and has reached approximately center of septum at 25 msec after beginning of QRS. Activity continues across septum through 40 msec, and, even at 60 msec after beginning of QRS, lateral left ventricle is not completely excited. Note that both septal activation and activation of left wall are altered by bundle branch block. Increased time required to excite septum and left wall accounts for prolongation of depolarization in complete left bundle branch blocks. This figure, like Figure 6–14, represents activity in the dog heart, in which the duration of QRS is 40 msec or less. (Becker *et al., Amer. Heart J.,* 1958, 55, 547–556.)

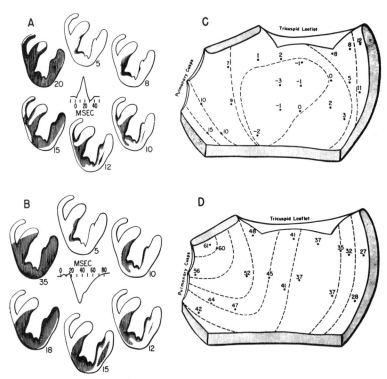

Figure 6-43 Pattern of ventricular depolarization before and after right bundle branch block. *A* and *B,* Sagittal sections through right and left ventricles showing pathway as measured by 9 multipolar insertions. Small figures show position in heart of depolarization wave at various stages of depolarization. *A,* Normal depolarization; normal lead II QRS is shown at center. *B,* Pattern of ventricular depolarization during right bundle branch block (same insertion as in *A*). Lead II QRS is typical of canine right bundle branch block.

C and *D,* Pattern of activation of right mural endocardium as viewed from inside right cavity. Shaded areas indicate junction of right wall and septum. Numbers indicate time of depolarization in milliseconds after onset of QRS. Dotted lines approximate wavefront position at 5 msec intervals. *C,* Normal depolarization. *D,* Pattern of activation after right bundle branch block. (From Erickson *et al., Circulat. Res.,* 1957, 5, 5–10.)

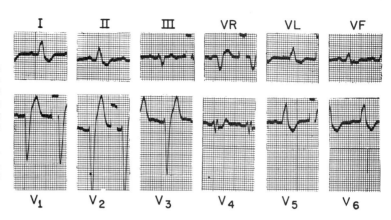

Figure 6-44 Clinical complete left bundle branch block. QRS complexes are 130 msec long. Sinus rhythm. Leads on right of chest see only receding activity, those on left only approaching activity. T waves have opposite polarity to QRS. (From Winsor, ed., *Electrocardiographic test book.* New York, American Heart Association, 1956.)

of complete block is based on far less prolongation of the QRS complex, *i.e.*, 120 msec or more. In complete right bundle branch block we would expect that the right wall would be the last portion of the heart to be activated, a situation which would produce late positive deflections in V_2 and V_2 and aVR. Grant[26] believed that these conditions are met only rarely in clinical examinations and concluded that truly complete right bundle branch block is extremely rare. Complete left bundle branch block (Fig. 6-44) is more common and is accompanied by clear signs of right-to-left activation of the septum and left ventricle. In bundle branch block, as in ventricular ectopic beats, those portions of the ventricular myocardium which depolarize first tend to repolarize first and, similarly, the last areas to fire are the last to recover. Consequently, the T wave tends to become a mirror image of the QRS complex; leads in which the QRS is upright show a downward T wave, etc.

As indicated above, cutting the right or left bundle produces a substantial increase in the duration of the QRS complex. In the dog, the duration may double. In human electrocardiography, the term "incomplete bundle branch block" is applied to conditions where the QRS complex lasts up to 120 msec—approximately a 50 per cent prolongation. Prolongation is thus less than would be expected from cutting a conducting bundle. The possibility exists that this abnormality represents damage to branches of the Purkinje system. Some data indicate that it is more likely due to thickening of the ventricular walls. A study by Boineau *et al.*[5] indicates that in atrial septal

defect in dogs, which resembles incomplete block, endocardial conduction is normal, although QRS is prolonged. The implication from this is that "incomplete block" is due to increased wall thickness. On the other hand, Schamroth[54] has described a patient who showed "progressive" changes, exhibiting varying degrees of prolongation of the QRS complex at different times. This finding would seem to point to the possibility that partial block of the conduction system can occur.

MYOCARDIAL INJURY; ISCHEMIA AND INFARCTION. *Phase I.* If a region of myocardium is partially deprived of oxygen, the first change observed electrocardiographically is an alternation of the T wave. Apparently the region of ischemia cannot repolarize normally. Possibly the ischemic region remains depolarized while adjacent regions have returned to the resting state. An overlying electrode will therefore record a negative T wave that is usually larger than normal. A similar change in the T wave is seen during recovery from an infarction. It should be noted that the T wave is a most labile portion of the electrocardiogram and less reliable for diagnostic purposes than other portions of the ventricular complex. Changes similar to those resulting from ischemia arise also from benign causes. If the entire heart is uniformly deprived of oxygen, T wave changes may be widespread.

Phase II. When a blood vessel supplying the ventricular myocardium is completely occluded by a thrombus or deposition of atheromatous plaques in the vessel wall, and when no collateral circulation exists, the cells previously supplied by this

AT REST   AFTER DEPOLARIZATION

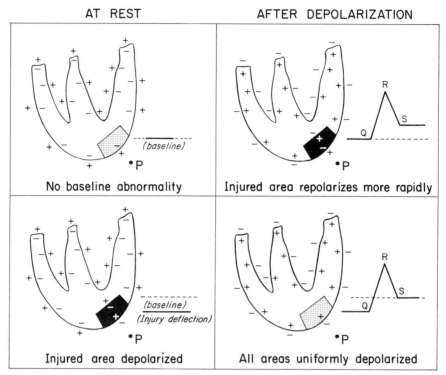

No baseline abnormality   Injured area repolarizes more rapidly

Injured area depolarized   All areas uniformly depolarized

Figure 6–45 Two modes of production of S-T segment shifts following myocardial ischemia. In the top two drawings, the ischemic area—indicated by the light shading—is identical to the remainder of the muscle at rest. The recorded potential at point P is thus on the zero baseline, which is indicated by the dotted line. However, during repolarization the injured area (black) repolarizes more rapidly, as shown in the top right figure. The recording point P thus sees positive charges from the injured area, whereas the rest of the heart shows negative charges, *i.e.*, it is depolarized.

In the bottom two drawings, the injured area is depolarized at rest. There is thus an injury deflection, the potential recorded at P being below baseline. However, when depolarization is complete, the injured area is identical to the rest of the heart and thus is at zero potential.

As shown, both these changes produce an S-T segment which is elevated above the "T-Q" period (before the Q wave). Actually, both these changes are found in ischemic muscle.

vessel will be completely deprived of oxygen. A complicated series of events will ensue, *all* of them causing the same change in the relationship between the S-T and T-Q segments (Fig. 6–45).

The first change which takes place is a shortening of the intracellular action potential. This occurs a few seconds after the tying of a ligature in an experimental animal, as has been demonstrated by several investigators.[50, 63] The injured cells depolarize normally but repolarize more rapidly than do adjacent normal cells. For this reason, during the period of repolarization, *i.e.*, during the S-T segment of the electrocardiogram, current flows from the injured cells to the adjacent normal cells (because, by definition, current flow is from positive to negative). This flow

then leads to a change in the S-T segment, which becomes elevated in unipolar leads facing the area of the infarct. This elevation is a primary change in the S-T segment; it is transient, however, and recovery from this phase occurs within a few minutes. While this change is still in effect (and during the period of recovery), a second change takes place: a decrease in the resting potential of the injured cells. Since the resting potential of the injured cells is now lower than that of the adjacent uninjured cells, current flows from the normal cells into the injured ones during electrical diastole. (This current is frequently referred to as the "current of injury.") This condition produces a depression of the T-Q segment of the electrocardiogram in unipolar electrodes facing the area of injury. At first this depression adds to the true S-T segment shift mentioned previously, but it continues after the initial change disappears.

The input capacitors of the electrocardiographic recorder prevent its use to discriminate between a shift in the T-Q segment and a shift in the S-T segment. In the clinical literature, both of these changes are referred to as "S-T segment elevation" in electrodes facing the injury. (Electrodes facing the rear of the injury record the opposite changes.) At present, there seems to be no need to discriminate among these various causes of "S-T segment elevation" during acute and chronic occlusion of the vessels, and such discrimination would present overwhelming technical problems.

Phase III. After the initial phases of ischemia and injury have disappeared, the S-T segment and T wave may return to normal. The diagnosis of such chronic infarcts is a difficult problem for the practitioner and may be important in determining whether a patient should be treated or should limit his activity. The major problem exists with regard to the QRS complex. In many cases, a sizable portion of the myocardium will have been replaced by scar tissue which is, of course, electrically silent. If the conduction system has not been impaired by the infarction, the duration of the QRS complex may be normal; but if a large amount of myocardium is missing, the complex will be changed, *i.e.*, it will lack the potentials previously contributed by the infarcted region. An electrode which faces the infarcted region and which previously recorded approaching (positive) activity from that region will record less positivity than normal, or it may record a negative deflection. If the area of infarction is in a part of the heart which is normally depolarized early, this increased negativity may either produce an initial negative deflection or increase the magnitude of a negative deflection which would normally occur in the lead facing the region. An initial negative deflection is referred to as a Q wave, and an *abnormal Q wave* is the most common diagnostic sign of chronic infarction. Q waves are not abnormal *per se*: the abnormality is frequently definable only in terms of magnitude or duration of the wave in a particular lead. Some portions of the body surface normally show Q waves. Al-

though this sign is useful, if it is the sole criterion of infarction, a diagnosis obviously can be made only when the infarction lies in regions which are depolarized early in QRS.

Some large infarcts may damage the conduction system, thus causing a prolongation of the QRS complex. It may be important to differentiate such prolongation from that seen in bundle branch block, because the latter may be present and innocuous in an otherwise healthy heart. The value of a control electrocardiogram taken before any reason exists to suspect myocardial damage should be apparent.

It is interesting that, during the period after infarction, a fixed relationship between changes in the S-T segment and T wave is often observed: those leads which show elevated S-T segments show a negative T wave. Although the mechanism has not been directly determined, a possible explanation is available by extrapolation from the observations of Durrer *et al.*[15] and Conrad *et al.*[10] Since depolarization is delayed in some cells in the infarcted region, might not these cells also repolarize late? Late depolarization would elevate the S-T segment and late repolarization would cause a negative T wave over the region. That is, the wave of repolarization would approach a lead over the infarction so slowly as to give a large negative deflection. (Remember that approaching repolarization produces a negative potential.)

ANALYSIS OF THE ELECTROCARDIOGRAM

The foregoing discussion is intended as a physiologic consideration and not as a textbook of electrocardiography. Indeed, a perusal of a complete electrocardiographic textbook will show that only a small fraction of the usual topics have been considered here; and the interpretation of clinical electrocardiograms obviously requires both much more familiarity with pathologic changes and much practical experience. Nevertheless, we shall briefly list here a sequence of procedures to be used in the evaluation of the electrocardiogram. This procedure will probably be sufficient only for the diagnosis of the common arrhythmias which have been listed in this chapter and at times will indicate that other electrocardiographic abnormalities exist, although the exact nature of the lesion may not be apparent.

The conventional electrocardiogram contains the tracings from three unipolar limb leads, the three bipolar limb leads and the six unipolar chest leads. The first procedure is to make sure that these tracings are technically above reproach and to make sure that all the leads have been recorded. At this time the standardization record should be examined to see whether a deflection of exactly 1 cm has been produced by 1 mV standardization. Each time that the standardization is repeated this accuracy should be checked again. Of course, if the amplitude of the standardization is slightly less or more than 1 cm, a correction is easily applied. Rarely, with damaged electronic components, the standardization may exhibit gross overshoot or slurring; if these occur, they are a sign that the instrument is in need of repair. Each electrocardiographic lead recorded should contain several beats. The tracing for each lead should have a baseline that does not drift up and down too widely, and there should be no artifact caused by loose electrodes, 60 cycle interference or muscle tremor.

Second, the cardiac rate should be determined by noting the interval between successive beats. At standard speed each millimeter of the electrocardiographic record equals 40 msec; 25 mm equals 1 sec. Computation of the heart rate is thus quite simple. Electrocardiographic textbooks contain tables which enable one to measure the interval in millimeters and read off the rate directly; calibrated rulers are also available. The ventricular and atrial rates must be determined separately if the chambers depolarized independently. It should be noted whether the rate is regular or irregular; if it is irregular, the number of beats within 10, 20 or 30 sec must be noted to determine the number of beats per min. The irregularity may itself be significant. It should next be noted whether the relationships of the P, QRS and T waves are constant or variable. The P-R interval (from the beginning of the P wave to the beginning of QRS) should be measured to see if it is normal, longer than normal or irregular. Changes in P-R interval result from abnormal activity of the conduction system. The duration of QRS and the Q-T interval (from the beginning of QRS to the end of the T wave) should also be measured. Changes in QRS duration occur in bundle branch block, after infarction, and during extrasystolic beats. The Q-T interval is altered by abnormal concentration of ions which can occur for a variety of reasons. Since these parameters are related to rate and age, tables must be consulted for normal limits.

The next step is to examine the complexes to see whether their shapes and durations are normal. We have previously considered conditions in which the P wave might have an abnormal shape. Several conditions are listed among the arrhythmias in which the QRS complex would be definitely abnormal in shape or prolonged or both. Electrocardiographic diagnosis of ventricular infarction, as can be seen above, rests on alterations in the T wave, the S-T segment or the T wave and QRS complex or combinations of such alterations. It is common practice to determine the electrical axis of the heart (see below), because much diagnosis of electrocardiographic abnormality depends on this determination.

The aforementioned procedures make it possible to detect many arrhythmias; however, it should be borne in mind that bundle branch block, particularly, cannot be diagnosed by analysis of a single tracing. Neither can the origin of many extrasystoles be adequately determined in this way. In cases of myocardial infarction, the problem is more complicated, and, although the example given perhaps may seem clear to the student, there are many types of infarcts and there is no substitute for experience in diagnosis.

Calculation of the Electrical Axis. At any instant during QRS, the electromotive force developed by the ventricles can be considered to have an average magnitude and direction closely related to the ventricular activation process. To plot accurately such a "vector" in the plane of the limb leads, we need two simultaneous recordings. In practice, a mean vector for all of the QRS complex is plotted according to a procedure developed by Einthoven. For this purpose, it is assumed that (i) the extremities form an equilateral triangle in the frontal plane, (ii) the heart lies at the center of this triangle, (iii) the mean polarity of a lead (i.e., the net area or mean voltage multiplied by time) can be determined from the difference between the positive and negative peaks of the QRS complex, and (iv) the body is a homogeneous conductor.

These assumptions contain certain inaccuracies; nevertheless, the electrical axis has empirical usefulness. The algebraic sum of the positive and negative peaks in a lead is measured in millimeters and is plotted along the proper side of the triangle, as shown in Figure 6–46. A perpendicular is then drawn at the termination of this line. The procedure is repeated for a second lead. The line joining the center of the triangle and the intersection of the two perpendiculars is the mean electrical axis. In most normal persons the axis falls between 0° and +90°. If the mean electrical axis lies at an angle greater than +90°, there is said to be *right axis deviation.* If the axis falls in the negative portion of the circle, *left axis deviation* is present. Again, a table of normal and abnormal values should be

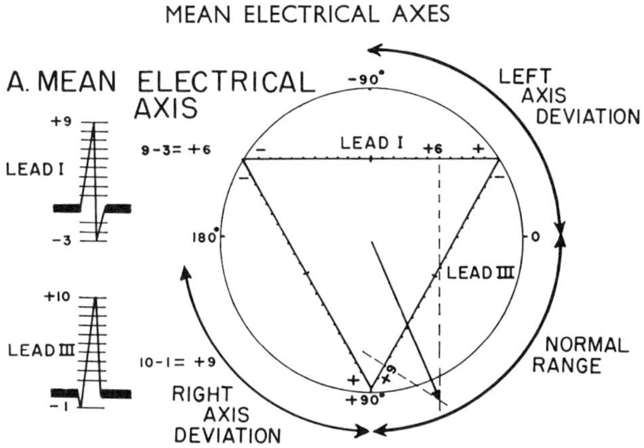

MEAN ELECTRICAL AXES

Figure 6–46   A, The mean electrical axis is computed from 2 of the 3 standard limb leads (*e.g.*, leads I and III). The sum of the downward deflections is subtracted from the sum of the upward deflections. For example, the vertical height of the R wave above the baseline is measured in millimeters (+9 mm in lead I). The total amplitude of the downward deflections (−3 mm in lead I) is added algebraically to the height of the R wave (+9) and leaves a net value of +6. At a point 6 units toward the plus sign on the lead I line of the triangle, a perpendicular is erected. The net amplitude of upward and downward deflections in lead III is +9 (+10 − 1). A perpendicular erected 9 units toward the plus sign on lead III is extended to intersect the perpendicular from lead I. An arrow drawn from the center of the triangle to the intersection of these two perpendicular lines is the *mean electrical axis*. (From Rushmer, *Cardiovascular dynamics*. Philadelphia, W. B. Saunders Co., 1961.)

consulted in borderline cases. A procedure for determining the mean electrical axis by inspection from a *hexaxial* reference system involving unipolar and bipolar limb leads is also often used.[26] Changes in electrical axis often indicate hypertrophy of one or both chambers or abnormal heart position or both. They also occur after infarction.

Vectorcardiography.   Much of our present understanding of electrocardiographic recording stems from the pioneering work of F. N. Wilson and his associates[66, 67] who developed our present system of recording and interpreting unipolar chest leads. He also pioneered in the recording of *vectorcardiograms*, which are at present widely considered as a type of recording which may partially supplant conventional scalar recordings. A vectorial recording plots voltage in one lead (on one axis) against voltage in a second lead (on a second recording axis). Conventional scalar recordings show changes in one lead as a function of time. The major competition between these systems concerns the recording of the QRS complex.

To emphasize the difference between these recording methods, we may regard the conventional scalar recordings as based on the theory that ventricular depolarization produces some purely "local" potentials

on the body surface; that is, each chest lead records in part the unique activity of the immediately underlying myocardium, and adjacent chest leads do *not* merely record different views of the same phenomenon.

Vectorcardiography, on the other hand, is based on the assumption that the heart is electrically so distant from the body surface that the potentials on that surface (or at least the important potentials on the body surface) can be considered to arise from a single fixed-location dipole within the chest. We may consider such a dipole to be a special case of the more general condition when three, and only three, independent current generators are connected to any array of electrodes on or within the body. If three such generators deliver current to the body, the potential at a given point on the body surface will be a linear function of the instantaneous current of each of these generators. In the case of the cardiac dipole, its X, Y and Z components can each be considered a single instantaneous current generator, and the voltage at a particular body surface point can be considered the resultant of the instantaneous current in the X, Y and Z directions. Actually, at present, no basis exists for differentiating between the dipolar and the more general three-function system. In either case, vectorcardiography is based on the assumption that the voltages recorded on the body surface are indistinguishable from those which would be produced by a dipole within the thorax. Further, the vectorcardiographic technique involves the assumption that all the (important) information that can possibly be derived about ventricular depolarization is contained in three leads which can be recorded from four or more body surface recording points. (Note that a lead in such a case might be recorded from an array of more than two electrodes.) Recording more than three independent leads may add no new information.

Schmitt[59] and Frank[23] devised the widely used lead placement systems for vectorcardiography. They have further presented evidence that a dipolar interpretation of cardiac electrical activity gives most of the information derived from our present scalar recording techniques. A study of electrocardiographic techniques through factor analysis has indicated that most of the "information" on the body surface can be ascribed to three internal generators. It has, however, been clearly shown by Taccardi[62] that the body surface potential distribution is at times not compatible with what would be expected from a dipolar generator in a homogeneous conductor. Other objections have been voiced by Brody et al.[6] and by Evans et al.[21] A search continues, particularly by Brody, for recording systems which will bring some of this nondipolar information to light.

Vectorcardiographic display employs a cathode ray oscilloscope. A single such record is taken by connecting one electrocardiographic lead to the horizontal amplifier of the oscilloscope and another lead to the vertical amplifiers. The leads used are selected to show the anterior-posterior, head-to-foot and left-to-right components of the heart's electrical activity. There are many competing schemes for making the connections to the body surface. Figure 6–47 shows the pathway of ventricular activity and resultant scalar potentials on the chest in the horizontal plane (front-to-back and left-to-right leads). Figure 6–48 shows a similar vectorial "loop" on a cathode ray tube from an anterior-posterior and a left-to-right lead.

Grant[26] has devised a system of electrocardiographic interpretation in which vectorial arrows are constructed by using the

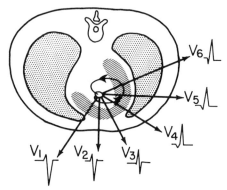

Figure 6–47 A representation of a horizontal cross-section through the chest, showing the locations of the lungs and heart. The figure in the center indicates the sequential directions of the wave of ventricular depolarization within the ventricle. Note the arrows. Initial direction in this is anteriorly and to the right. In the middle of QRS there is a large lateral, posterior, leftward direction of activity, and finally a direction of activity toward the base of the heart, equivalent to the terminal activity in the septum. Outside the thoracic outline are shown the potentials which would be recorded from this "vectorial" representation of activity on the chest surface. It can be seen that the V leads are merely representations of this summed activity.

conventional twelve-lead scalar electrocardiogram. Diagnosis in this system depends on the relationship of the arrows.

Unfortunately, even if the three-function or vectorcardiographic approach is accepted, the next step is not clear. The vectorcardiographic recording is in some respects more difficult technically than the recording of scalar leads. The fact that only three factors (i.e., three independent voltage changes in time) are involved in ventricular depolarization does not necessarily indicate that these should be recorded vectorially. It is thus possible that recording of three scalar leads will be the method of choice at some future date. The vectorcardiographers themselves do not agree on the recording system to be used. At present, there are several systems of vectorcardiographic recording, each having its proponents. The vast majority of clinical electrocardiograms consist of scalar leads exclusively, although an increasing number of cardiologists utilize the principles of vector analysis to interpret them and attempt to relate the information to normal depolarization pathways. On the other hand, some research in electrocardiography[33, 62] is being conducted on the use of total body surface maps. In these maps, the body surface potential distribution for

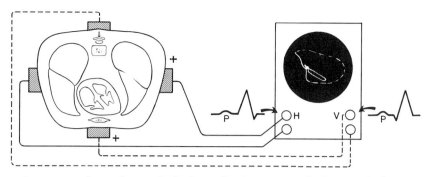

Figure 6–48 The vectorcardiographic method of recording a tracing in the horizontal plane employs two sets of electrodes, one recording left-to-right potential differences, the other recording anterior-posterior potential differences. On the right of the figure are shown the potentials that would be recorded on each of these two leads, and on the oscilloscope face is shown a loop which would be described by the potentials shown. Vectorcardiograms can be "constructed" from scalar leads or can be recorded in this fashion with a cathode ray oscillograph.

many instances in time is presented (usually derived from 60 to 240 separate electrocardiographic recordings) as a series of maps or field plots. Whether any future system of recording will increase the usefulness of electrocardiography remains to be seen.

Computer Analysis of the Electrocardiogram and Vectorcardiogram.[8, 44] The high-speed digital computer is being widely examined as a tool for both the researcher and the diagnostician. Computers have been used in electrocardiographic studies in a number of ways, including attempts to duplicate the analysis that might be made by a competent physician and attempts to discover new diagnostic data. Computer programs exist which rapidly produce the measurements (heart rate, duration of intervals and complexes, electrical axis, etc.) routinely made by the physician. Other computer programs indicate the probability of a particular cardiac abnormality.

REFERENCES

1. Abilskov, J. A., Wilkinson, R. S., Vincent, W. A., and Cohen, W. An experimental study of the electrocardiographic effects of localized myocardial lesions. *Amer. J. Cardiol.*, 1961, 8, 485–492.
2. Amer, N. S., Stuckey, J. H., Hoffman, B. F., Cappelletti, R. R., and Domingo, R. T. Activation of the interventricular septal myocardium studied during cardiopulmonary bypass. *Amer. Heart J.*, 1960, 59, 224–237.
3. Bayley, R. H. Exploratory lead systems and "zero potentials." *Ann. N.Y. Acad. Sci.*, 1957, 65, 1110–1126.
4. Becker, R. A., Scher, A. M., and Erickson, R. V.

Ventricular excitation in experimental left bundle branch block. *Amer. Heart J.*, 1958, 55, 547–556.
5. Boineau, J. P., Spach, M. S., and Ayers, C. R. Genesis of the electrocardiogram in atrial septal defect. *Amer. Heart J.*, 1964, 68, 637–651.
6. Brody, D. A., Bradshaw, J. C., and Evans, J. W. A theoretical basis for determining heart lead relationships of the equivalent cardiac multipole. *IRE Trans. Biomed. Electronics*, 1961, BME8, 139–143.
7. Brody, D. A., Woolsey, M. D., and Arzbaecher, R. C. Application of computer techniques to the detection and analysis of spontaneous P-wave variations. *Circulation*, 1967, 36, 359–371.
8. Caceres, C. A. Electrocardiographic analysis by a computer system. *Arch. intern. Med.*, 1963, 111, 196–202.
9. Carmeliet, E., and Lacquet, L. Durée du potentiel d'action ventriculaire de grenouille en fonction de la fréquence. Influence des variations ioniques de potassium et sodium. *Arch. int. Physiol.*, 1958, 66, 1–21.
10. Conrad, L. L., Cuddy, T. E., and Bayley, R. H. Activation of the ischemic ventricle and acute peri-infarction block in experimental coronary occlusion. *Circulat. Res.*, 1959, 7, 555–563.
11. Craib, W. H. A study of the electrical field surrounding active heart muscle. *Heart*, 1927, 14, 71–109.
12. Damato, A. N., Lau, S. H., Berkowitz, W. D., Rosen, K. M., and Lisi, K. R. Recording of specialized conducting fibers (A-V nodal, His bundle, and right bundle branch) in man using an electrode catheter technic. *Circulation*, 1969, 39, 435–447.
13. Damato, A. N., Lau, S. H., Helfant, R., Stein, E., Patton, R. D., Scherlag, B. J., and Berkowitz, W. D. A study of heart block in man using His bundle recordings. *Circulation*, 1969, 39, 297–305.
14. Durrer, D., Roos, J. P., and Buller, J. The spread of excitation in canine and human heart. In: *Electrophysiology of the heart*, B. Taccardi, ed. New York, Pergamon Press, 1965.
15. Durrer, D., van Lier, A. A. W., van Dam, R. T.,

Jonkman, E., and David, G. The intramural electrocardiogram of myocardial infarction. *Res. Comm., Congr. mond. Cardiol.*, 1958, 3, 355–356.

16. Durrer, D., and Van Der Tweel, L. H. Excitation of the left ventricular wall of the dog and goat. *Ann. N.Y. Acad. Sci.*, 1957, 65, 779–802.

17. Durrer, D., van Dam, R. T., Freud, G. E., Janse, M. J., Meijler, F. L., and Arzbaecher, R. C. Total excitation of the isolated human heart. *Circulation*, 1970, 41, 899–912.

18. Einthoven, W. Ein neues Galvanometer. *Annal. Physik*, 1903, 12, 1059–1071.

19. Einthoven, W. Le télécardiogramme. *Arch. int. Physiol.*, 1906, 4, 132–164.

20. Erickson, R. V., Scher, A. M., and Becker, R. A. Ventricular excitation in experimental bundle branch block. *Circulat. Res.*, 1957, 5, 5–10.

21. Evans, J. W., Erb, B. D., and Brody, D. A. Comparative proximity and remoteness characteristics of conventional electrocardiographic leads. *Amer. Heart. J.*, 1961, 61, 615–621.

22. Eyster, J. A. E., and Meek, W. J. Experiments on the origin and conduction of the cardiac impulse. VI. Conduction of the excitation from the sino-auricular node to the right auricle and auriculoventricular node. *Arch. intern. Med.*, 1916, 18, 775–799.

23. Frank, E. Spread of current in volume conductors of finite extent. *Ann. N.Y. Acad. Sci.*, 1957, 65, 980–1002.

24. Goodman, D., Van Der Steen, A. B. M., and van Dam, R. T. Endocardial and epicardial activation pathways of the canine right atrium. *Amer. J. Physiol.*, 1971, 220, 1–11.

25. Graham, J., and Gerard, R. W. Membrane potentials and excitation of impaled single muscle fibers. *J. cell. comp. Physiol.*, 1946, 28, 99–117.

26. Grant, R. P. *Clinical electrocardiography.* New York, McGraw-Hill Book Co., 1957.

27. Hamlin, R. L., Smetzer, D. L., Senta, R., and Smith, C. R. Atrial activation paths and P waves in horses. *Amer. J. Physiol.*, 1970, 219, 306–313.

28. Hoffman, B. F., Paes de Carvalho, A., Mello, W. C., and Cranefield, P. F. Electrical activity of single fibers of the atrioventricular node. *Circulat. Res.*, 1959, 7, 11–18.

29. Hogan, P. M., and Davis, L. D. Evidence for specialized fibers in the canine right atrium. *Circulat. Res.*, 1968, 23, 387–396.

30. Jacobson, E. D., Rush, S., Zinberg, S., and Abildskov, J. A. The effect of infarction on magnitiude and orientation of electrical events in the heart. *Amer. Heart J.*, 1959, 58, 863–872.

31. James, T. N. The specialized conducting tissue of the atria. In: *Mechanisms and therapy of cardiac arrhythmias*, L. S. Dreifus and W. Likoff eds. New York, Grune & Stratton, 1966.

32. James, T. N. The connecting pathways between the sinus node and A-V node and between the right and left atrium in the human heart. *Amer. Heart J.*, 1963, 66, 498–508.

33. Karsh, R. B., Spach, M. S., and Barr, R. C. Interpretation of isopotential surface maps in patients with ostium primum and secundum atrial defects. *Circulation*, 1970, 41, 913–933.

34. Lewis, T. *The mechanism and graphic registra-tion of the heart beat.* London, Shaw and Sons, 1925.

35. Lewis, T., Meakins, J., and White, P. D. The excitatory process in the dog's heart. Part I. The auricles. *Phil. Trans.*, 1915, B205, 375–420.

36. Lewis, T., and Rothschild, M. A. The excitatory process in the dog's heart. Part II. The ventricles. *Phil. Trans.*, 1915, B206, 181–226.

37. Ling, G., and Gerard, R. W. The normal membrane potential of frog sartorius fibers. *J. cell. comp. Physiol.*, 1949, 34, 383–396.

38. Medrano, G. A., Bisteni, A., Brancato, R. W., Pileggi, F., and Sodi-Pallares, D. The activation of the interventricular septum in the dog's heart under normal conditions and in the bundle-branch block. *Ann. N. Y. Acad. Sci.*, 1957, 65, 804–817.

39. Mendez, D., and Moe, G. K. Demonstration of a dual A-V nodal conduction system in the isolated rabbit heart. *Circulat. Res.*, 1966, 19, 378–393.

40. Merideth, J., and Titus, J. L. The anatomic atrial connections between sinus and A-V node. *Circulation*, 1968, 37, 566–579.

41. Mines, G. R. On dynamic equilibrium in the heart. *J. Physiol. (Lond.)*, 1913, 46, 349–383.

42. Paes de Carvalho, A., and Langan, W. B. Influence of extracellular potassium levels on atrioventricular transmission. *Amer. J. Physiol.*, 1963, 205, 375–381.

43. Paes de Carvalho, A., De Mello, W. C., and Hoffman, B. F. Electrophysiological evidence for specialized fiber types in rabbit atrium. *Amer. J. Physiol.*, 1959, 196, 483–488.

44. Pipberger, H. V. Use of computers in interpretation of electrocardiograms. *Circulat. Res.*, 1962, 11, 555–562.

45. Pruitt, R. D., and Essex, H. E. Potential changes attending the excitation process in the atrioventricular conduction of bovine and canine hearts. *Circulat. Res.*, 1960, 8, 149–174.

46. Puech, P. *L'activité électrique auriculare.* Paris, Masson et Cie, 1916.

47. Roos, J. P., van Dam, R. T., and Durrer, D. Epicardial and intramural excitation of normal hearts in six patients 50 years of age and older. *Brit. Heart J.*, 1968, 30, 630–637.

48. Rosenblueth, A., and Garcia-Ramos, J. Studies on flutter and fibrillation; influence of artificial obstacles on experimental auricular flutter. *Amer. Heart. J.*, 1947, 33, 677–684.

49. Rush, S., Abildskov, J. A., and McFee, R. Resistivity of body tissues at low frequencies. *Circulat. Res.*, 1963, 12, 40–50.

50. Samson, W. E., and Scher, A. M. Mechanism of S-T segment alteration during acute myocardial injury. *Circulat. Res.*, 1960, 8, 780–787.

51. Sano, T., and Scher, A. M. Multiple recording during electrically induced atrial fibrillation. *Circulat. Res.*, 1964, 14, 117–125.

52. Sano, T., Suzuki, F., and Tsuchihashi, H. Function of potential bypass tracts for atrioventricular conduction. *Circulation*, 1970, 41, 413–422.

53. Schaefer, H. The general order of excitation and of recovery. *Ann. N. Y. Acad. Sci.*, 1957, 65, 743–766.

54. Schamroth, L., and Bradlow, B. A. Incomplete

left bundle branch block. *Brit. Heart J.*, 1964, 26, 285–288.

55. Scher, A. M. Excitation of the heart. *Handb. Physiol.*, 1962, sec. 2, vol. 1, 287–322.

56. Scher, A. M., Rodriguez, M. I., Liikane, J., and Young, A. C. The mechanism of atrioventricular conduction. *Circulat. Res.*, 1959, 7, 54–61.

57. Scher, A. M., and Young, A. C. Ventricular depolarization and the genesis of QRS. *Ann. N. Y. Acad. Sci.*, 1957, 65, 768–778.

58. Scherf, D., and Schott, A. *Extrasystoles and allied arrhythmias.* New York, Grune & Stratton, 1953.

59. Schmitt, O. H. Lead vectors and transfer impedance. *Ann. N. Y. Acad. Sci.*, 1957, 65, 1092–1109.

60. Sodi-Pallares, D. *New bases of electrocardiography.* St. Louis, C. V. Mosby Co., 1956.

61. Sodi-Pallares, D., Brancato, R. W., Pileggi, F., Medrano, G. A., Bisteni, A., and Barbato, E. The ventricular activation and the vector cardiographic curve. *Amer. Heart J.*, 1957, 54, 498–510.

62. Taccardi, B. Distribution of heart potentials on the thoracic surface of normal human subjects. *Circulat. Res.*, 1963, 12, 341–352.

63. Trautwein, W., and Dudel, J. Aktionspotential und Kontraktion des Herzmuskels im Sauerstoffmangel. *Pflügers Arch. ges. Physiol.*, 1956, 263, 23–32.

64. Trautwein, W., and Kassebaum, D. G. On the mechanism of spontaneous impulse generation in the pacemaker of the heart. *J. gen. Physiol.*, 1961, 45, 317–330.

65. Truex, R. C. Recent observations on the human cardiac conduction system, with special considerations of the atrioventricular node and bundle. In: *Electrophysiology of the heart*, B. Taccardi, ed. New York, Pergamon Press, 1965.

66. Wilson, F. N., Johnston, F. D., Rosenbaum, F. F., Erlanger, H., Kosmann, C. E., Hecht, H. H., Cotrim, N., Menezes de Oliveira, R., Scarsi, R., and Barker, P. S. Precordial electrocardiogram. *Amer. Heart J.*, 1944, 27, 19–85.

67. Wilson, F. N., Rosenbaum, F. F., and Johnston, F. D. Interpretation of ventricular complex of electrocardiogram. *Advanc. intern. Med.*, 1947, 2, 1–63.

CHAPTER 7 MECHANICAL EVENTS OF THE CARDIAC CYCLE

by ALLEN M. SCHER

As stated elsewhere in this text, survival of large multicellular organisms depends on the constant replenishment of the extracellular fluid with materials rich in oxygen and metabolic substrate and on the removal of metabolic wastes. Although cellular survival depends on the exchange of materials across the walls of capillaries, the basic energy is supplied by the pumping action of the heart. In many types of circulatory distress the ultimate cause of incapacity or death is the lack of transfer of material across capillaries; in a high percentage of such cases the deficiency is not in the capillaries but in the performance of the heart. Thus during the routine physical examination, the physician spends a large percentage of time investigating the function of the heart. Before the advent of certain newer diagnostic procedures, he depended to a large extent on information supplied by his own sense organs. Even now, the careful physician examines the veins for signs of enlargement and high venous pressure. Placing his stethoscope in many places over the thorax, he carefully examines the chest for abnormal impact which may be imparted to it by the heart. Such examination supplements the use of newer diagnostic tools.

In interpreting the information which he receives from his own sense organs and newer measuring devices, the physician must know whether or not a particular sign or symptom is within the normal range. In addition, the scientific physician should know what sequence of physiologic phenomena produces the information he receives and should have some understanding of what physiologic derangements lead to abnormal events. For such an examination of the cardiovascular system, it is essential that the physician know the normal sequence of mechanical and electrical events which take place in the cardiac chambers.

Functional Anatomy. As indicated in Chapter 4, Volume II, cardiac muscle is intermediate between striated and smooth muscle. This intermediacy is both structural and functional. Cardiac muscle is cross-

102

striated like skeletal muscle but, like smooth muscle, it is involuntary, *i.e.*, is not under the control of the will. Functionally, if not anatomically, cardiac muscle is a syncytium like smooth muscle. That is, the units of the muscle are functionally interconnected, so that, once electrical depolarization is initiated in some unit, it continues through all units of the syncytium (see Chap. 4, Vol. II). Striated muscle is able to develop a large amount of tension in a short time, but will fatigue; in contrast, smooth muscle contracts slowly but is capable of exerting moderate amounts of tension for long periods of time. In this respect, cardiac muscle is closer to skeletal muscle, because the heart can contract rapidly, although not as rapidly as skeletal muscle. Some smooth muscle (arterioles, sphincters) is continuously and tonically contracted. Skeletal muscle may contract for long periods, but then usually can rest for long periods. The heart is unique in its lifelong, continuous, rhythmic contraction and relaxation. The heart is composed of a large number of separate muscle bundles which can be separated only with difficulty. Individual bundles have been named, but are virtually indistinguishable.[9]

The fibrous skeleton of the heart is a supporting framework which includes the valves and to which the muscle masses are attached.[11] The fibrous skeleton surrounds the tricuspid and mitral valve orifices, the atria and ventricles being attached about these openings. The fibrous skeleton also surrounds the smaller aortic and pulmonary valve orifices where the aorta and the pulmonary artery insert. Figure 7–1 indicates this general structure.

In addition to the "unspecialized" cardiac muscle which makes up the walls of the atria and ventricle, there is specialized conduction tissue, usually histologically and anatomically separable from the ordinary myocardium (Fig. 7–1). It consists of the *sino-atrial* (sino-auricular) node, which is the mammalian analog of the sinus venosus, the A-V node, the common bundle, the right and left bundles and the ventricular Purkinje* fibers. As discussed in Chapter

---

*Named for Johannes Evangelista Purkinje (1787), a Bohemian physiologist. Pronounced "pur-kin'-je."

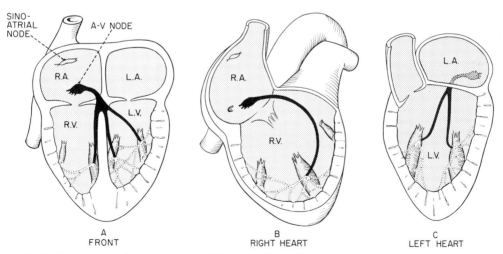

Figure 7–1 The interior of the heart viewed from the "front" (a view showing both cavities), from the right and from the left. Right and left atria and ventricles are identified. The specialized conduction tissue shown in black is of modified muscle fibers with specialized properties which generally lie within the unspecialized muscle mass and are clearly seen histologically. The specialized regions include the sino-atrial node, which lies at the junction of the superior vena cava and the right atrium; the atrioventricular (A-V) node, which lies in the interatrial septum just above the tricuspid valve, and which gives rise to the common bundle which leads to the single right bundle and to the two-branched left bundle. The bundles run into the middle and apical portions of the two septal surfaces on the endocardial surfaces of the septum. From the terminations of the bundles, Purkinje fibers arise and run in part across the cavities and in part along the endocardium to the free ventricular walls and to the papillary muscles. Some of these are the only portion of the specialized tissue not imbedded in the muscle mass. Note the greater thickness of the left ventricular wall and the more extensive distribution of Purkinje fibers to the left septal surface.

5, Volume II, there is also anatomical evidence for specialized conduction pathways or tissue in the atria, although the physiological evidence is subject to dispute. The sino-atrial node (S-A node), or the node of Keith and Flack, is the pacemaker of the mammalian heart. A section of Keith and Flack's description of this node follows:

"Our search for a well-differentiated system of fibres within the sinus, which might serve as a basis for the inception of the cardiac rhythm, has led us to attach importance to this peculiar musculature surrounding the artery at the sino-auricular junction. In the human heart, the fibres are striated, fusiform, with well-marked elóngated nuclei, plexiform in arrangement, and embedded in densely packed connective tissue — in fact, of closely similar structure to the Knoten (A-V node). The amount of this musculature varies, depending upon how much of the sinus has remained of the primitive type; but in the neighborhood of the taenia terminalis there is always some of this primitive tissue found. Macroscopically, the fibres resemble those of the a.-v. bundle in being paler than the surrounding musculature, *i.e.*, in being of the white variety. . . ."[4]

The S-A node consists of small, closely interlaced cells interspersed with connective tissue. The region is easily recognized microscopically with low magnification.

In the interatrial septum, near the coronary sinus, there is a second mass of specialized conduction tissue, the atrio-ventricular node (A-V node), or node of Tawara. The A-V node, as indicated, closely resembles the S-A node. The A-V node gives rise to the common bundle, which passes through the fibrous skeleton. The A-V node and the Purkinje tissue constitute a bridge for electrical conduction of the impulse from the atria to the ventricles.

The Purkinje tissue, a system of specialized muscle fibers, begins at the common bundle and continues through the right and left bundles. Its network of branching fibers covers much of the endocardium. Purkinje tissue has anatomic and electrical characteristics somewhat different from ordinary myocardium. In general, the fibers of the common bundle and the right and left bundles are larger, have clearer cytoplasm and contain more glycogen (indicated by glycogen-specific stains). These fibers conduct more rapidly than ordinary cardiac muscle, and in some species, particularly

the ungulates, are surrounded by a substantial connective tissue sheath; in ungulates, the Purkinje tissue penetrates the walls of the ventricles. This tissue, like the rest of the myocardium, is functionally syncytial. Electrically, the differences are less marked, the intracellular action potential differing somewhat from that of ordinary myocardial fibers (Chap. 4, Vol. II).

Electrical Precursors of Mechanical Activity. The contraction and relaxation of cardiac muscles results from the cyclic electrical depolarization and repolarization of the membranes of the cardiac muscle units. The normal sequence of electrical events in single skeletal and cardiac muscle cells has been discussed previously (see Chap. 4, Vol. II), and the electrical events within the entire cardiac mass will be discussed in detail in a later chapter. It is necessary, however, to summarize the sequence of electrical changes in the heart as an introduction to the consideration of cardiac mechanical events.

CARDIAC PACEMAKER. Certain tissues of the heart have the ability to act as pacemakers, *i.e.*, to depolarize spontaneously. Localization of the cardiac pacemaker cells in cold-blooded animals was anticipated in some very early experiments performed by Harvey in 1628. He found that isolated small bits of cardiac tissue continued to beat rhythmically and that pieces of atrium had a higher inherent rate than did pieces of ventricular muscle. Several other kinds of experimental evidence can be produced.

If ligatures are tied on the frog's heart (i) between the sinus venosus and the atria and (ii) between the atria and the ventricles (Stannius' ligatures), the pre-ligature heart rate is maintained only above the ligature, because the pacemaker is in the sinus venosus. In the mammalian heart, the S-A node is the pacemaker. Warming or cooling the pacemaker in the heart of a cold-blooded or a warm-blooded animal will change the rate of the entire heart, whereas changes of temperature of other regions will not.

The pacemaker at any instant is that portion of the heart with the highest rate. Techniques for electrical mapping of pacemaker sites have been suggested by Lewis.[7] Recently, intracellular recording has been used to find pacemaker cells, which indeed have unique characteristics (see Fig. 5–5, Chap. 5, Vol. II).[13] In the embryonic development of the chick heart, pacemaker activity begins in the precursor of ventricular muscle and moves to the atrial muscle and then to the sinus venosus.

SPREAD OF ACTIVITY. Once electrical activity is initiated in the atrial pacemaker, it spreads through both atrial walls and through the interatrial septum. The wave of excitation spreads concentrically from the S-A node at a rate slightly less than 1 m per sec. Depolarization of the entire human atrium requires about 80 msec. It produces an electrical event recorded at the body surface, called the "P wave" of the body surface electrocardiogram (Fig. 7–2). The electrically excited state in the atrium continues for approximately 150 msec. Atrial repolarization occurs during the depolarization of the ventricles.

The sole muscular connection between the atrium and the ventricle normally consists of the A-V node, the common bundle and the right and left conducting bundles. The velocity of conduction in the A-V node is very low (about 0.1 m per sec). The conduction velocity in the bundles is about 2 m per sec. In the body surface electrocardiogram, there is a period between the end of the excitation of the atrium and the beginning of the excitation of the ventricles when no potential changes are recorded. After the electrical wave has traveled down the right and left bundles, it rapidly traverses the Purkinje fibers, which are widely distributed to the endocardium on both sides. Consequently, excitation is distributed quite synchronously to most of the mural and septal endocardium. The velocity of conduction along the endocardium is about 1 m per sec.

The electrical wave then travels through the ventricular muscle, generally from endocardium to epicardium at about 0.3 m per sec. Ventricular excitation produces an electrical potential at the body surface, the "QRS complex" (Fig. 6–1, Chap. 6, Vol. II). In man, about 80 msec are required for all the ventricular muscle to become electrically excited. During the period when the ventricles are depolarizing, the atria are repolarizing, i.e., returning to the resting state. The ventricular cells remain depolarized for about 300 msec, the range being from slightly above 200 msec to slightly below 500 msec. In the electrocardiogram, the "T wave" signals ventricular repolarization. Cardiac cells begin to contract about 10 msec after they become depolarized and remain in the contracted state while they remain depolarized. Thus, the recurring electrical events lead to a rhythmic contraction of the cardiac muscle which pumps the blood.

Mechanical Characteristics of the Heart. The atria may be likened to a single, irregularly shaped, thin-walled cone split into two chambers by the interatrial septum. The chambers have little ability to do work. The right atrium produces a pressure differential of 5 to 6 mm Hg as it contracts; the left atrium, a differential of 7 to 8 mm Hg.

The right ventricle, which pumps blood returned from the systemic circulation into the pulmonary artery, is a crescent-shaped chamber which sits atop the wall of the interventricular septum. This ventricle appears suited to the accommodation and ejection of large and variable amounts of blood with minimal myocardial shortening. The shape does not appear suited to the development of high pressures. Right *intraventricular* systolic pressure is about 25 mm Hg.* The shape of the left ventricle

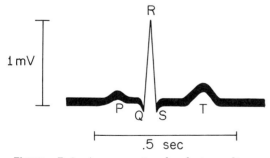

Figure 7–2 A conventional electrocardiogram showing the approximate duration, amplitude and designations of the potentials. The P wave results from atrial depolarization, the QRS complex from ventricular depolarization and the T wave from ventricular repolarization. Discussion in text.

*In discussion of pressures in the cardiovascular system, the values given are the pressure in excess of atmospheric pressure. Systolic pressure is the highest pressure produced by contraction. If used without qualification, "systolic pressure" refers to the peak pressure in the aorta. Diastolic pressure is the lowest pressure reached during ventricular relaxation. At times the terms "electrical systole" and "electrical diastole" are used loosely to designate the period of the cardiac action potential and the period between action potentials, respectively.

may be compared to a cylinder with a small cone at the end. This ventricle is in effect a pressure pump[11] (that is, it tends to eject a constant stroke volume against pressures which vary widely), and its function is to pump oxygenated blood, returned from the lungs, into the aorta. Left ventricular systolic pressure is approximately 120 mm Hg. Diastolic pressure within all four chambers is close to zero.

Valves. The atrium and ventricle on each side are separated by valves which move in response to pressure-induced flow changes. If the pressure in the atria is higher than that in the ventricles, the atrioventricular (A-V) valves will open and blood will enter the ventricles. Conversely, if the ventricular pressure is higher, a slight backflow will occur and the valves will close. The aortic and pulmonary valves function in a similar manner. If the pressure in the aorta is higher than that in the left ventricle, the aortic valves will be closed; if ventricular pressure is higher, the valves will open and blood will flow from the ventricle into the aorta.

Aorta and Pulmonary Artery. Some of the blood ejected into the aorta and pulmonary artery during systole distends the elastic walls of these vessels, storing potential energy. This stored potential energy is released as blood flow during diastole. The aortic and pulmonary arterial pressures rise to a peak during the contraction of the ventricle, but the pressures in these vessels do not fall to zero between beats. Peak systolic pressure is normally about 120 mm Hg in the aorta and 25 mm Hg in the pulmonary artery. The diastolic pressure in the aorta is about 80 mm Hg; that in the pulmonary artery is about 7 mm Hg.

Measurements. The events of the cardiac cycle consist of a number of physical changes. The electrocardiogram has been considered briefly and is discussed further in the preceding chapter. The mechanical contraction which results from activation of the cardiac muscle produces cyclic pressure changes in the chambers of the heart, in the aorta and in the veins. These pressure changes were recorded in anesthetized animals during the last four decades by Wiggers *et al.*[14] Most of the measurements have been repeated in humans and in intact dogs with essentially similar results.[2] A discussion of pressure recording techniques is presented in Chapter 10, Volume II.

In the classic studies, the combined volume of two ventricles was recorded by placing both of them in a glass container (cardiometer) which was closed by a rubber ring around the atrioventricular groove. The container was connected to a recording tambour, *i.e.*, a rubber diaphragm which moved a pen or mirror. This system could be used only on experimental animals with opened chests. Recently designed transducers[11] permit the estimation of ventricular volume from measurements of ventricular diameter or circumference. Further, as indicated in Chapter 10, Volume II, it is now possible to measure instantaneous aortic flow and to compute stroke volume in animals in either short-term or long-term experiments with flowmeters of either the electromagnetic or ultrasonic type.

EVENTS OF THE CARDIAC CYCLE

Because electrical events precede mechanical events, the electrocardiogram is a key to the mechanical events of the cycle. The events will be described vertically— that is, we will describe the changes in all variables in one phase of the cycle, then discuss the next portion of the cycle, and so forth.

It is a general rule that where two chambers are directly interconnected, *i.e.*, when the valves are open, the pressures will be identical with a small pressure gradient in the direction of flow. The phases of contraction and relaxation are named in a fashion which describes the activity of the ventricle. Figure 7–3 shows the events of the cardiac cycle; the vertical lines mark the beginning of the successive phases.

Diastasis (Fig. 7–3A). The cycle begins at the end of diastole. The pressure in the aorta is falling, owing to the "runoff" of blood into the peripheral vessels. Volume and pressure in the atria and ventricles are rising slightly, because the venous pressure exceeds the pressure within the chambers. The atrioventricular valves have long been open. No potentials are recorded in the electrocardiogram, and no sounds are heard stethoscopically. The period extending from the end of the rapid filling phase in one cycle to the atrial contraction in the next cycle is known as the period of *diastasis.* Diastasis is of variable duration and depends on the heart rate. At slow heart rates there is a period of stasis, *i.e.*, volumes and pressures in the heart are virtually unchanging. The P wave of the electrocardiogram occurs during this phase.

Figure 7–3 Events of the cardiac cycle. This diagram portrays the changes in aortic pressure, ventricular pressure, atrial pressure, and ventricular volume and aortic flow through a cardiac cycle. It also shows the conventional electrocardiogram and the heart sounds. The phases of the cycle are separated by vertical lines. The phases are (A) diastasis atrial contraction, (B) isovolumetric ventricular contraction, (C) maximal ejection, (D) reduced ejection, (E) isovolumetric relaxation of ventricle, (F) rapid filling, and (G) period of diastasis which merges with (A). The aortic flow curve shows flow at the origin of the aorta. The volume curve shows volume of blood in the ventricle. See text for detailed description. This figure is further supplemented by Figures 7–4 and 7–5. (Modified from Wiggers, *Circulatory dynamics.* New York, Grune & Stratton, 1952.)

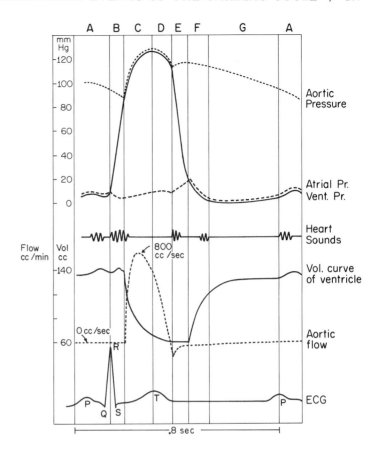

Atrial Contraction. Slightly after the beginning of the P wave (during the diastolic period) the atria commence their contraction. This contraction leads to a surprisingly slight rise in the intra-atrial pressure (the pressure change is about 5 mm Hg). The change develops over about 0.1 sec. The ventricular volume and pressure increase slightly owing to the atrial ejection of blood (ventricular pressure increases by about 3 mm Hg). The major portion of ventricular filling occurs during the period of diastasis (see below). Atrial contraction has generally been considered as a relatively trivial contributor to the filling of the ventricles. It now appears that, if the heart rate is slow enough so that there is a true period of diastasis, atrial contraction produces a rise in ventricular pressure and the A-V valves often close because of atrial ejection, so that this rise in pressure is maintained. That is, the atria eject blood into the ventricles, this blood starts to flow back into the atria and the valves close. If the heart rate is rapid, as in many acute experiments

on animals, ventricular pressure records show no clear atrial component.[3] It has been estimated that the atria contribute 20 per cent or less to ventricular filling. During this time, the pressure in the aorta continues to decrease as blood flows into the arterioles. A very faint atrial vibration, not normally perceived as a sound, occurs at this time. The ventricles begin to depolarize during this period, as shown by the beginning of the QRS complex.

Ventricular Isometric (Isovolumetric) Contraction (Fig. 7–3*B*). Ventricular contraction begins shortly after the onset of the ventricular electrocardiographic complex (QRS). The first period of ventricular contraction is called the "isometric phase." At the beginning of ventricular contraction, the A-V valves may be open or closed. If they are open, they close almost immediately after ventricular contraction begins. The aortic and pulmonary valves are, of course, also closed. Since fluid is incompressible, this is by definition an "isovolumetric" or "isometric" phase of contraction;

i.e., the volume of blood in the ventricles is constant while the pressure is rising rapidly.

At the beginning of ventricular contraction, aortic pressure is about 80 mm Hg and ventricular pressure is only slightly above atmospheric pressure. The ventricles change dimensions as the muscle fibers contract, but no blood is ejected into the arteries and none flows retrograde into the atria once the A-V valves close. Pressures in the arteries and the atria are thus not directly affected even though ventricular pressure rises steeply. This change in shape can produce a slight increase in *apparent* ventricular volume despite the fact that both inflow and outflow valves are closed.

RAPID EJECTION (Fig. 7–3C). When the pressure in the left ventricle exceeds the pressure in the aorta, the aortic valves open. Since there is now a large orifice between the aorta and the ventricle, the two form virtually a single chamber; pressure curves measured in the two regions follow one another closely. Blood flows rapidly from the ventricle into the aorta. During this period of maximal ejection, the ventricular volume decreases sharply. Since the original description of the events of the cardiac cycle by Wiggers and co-workers, recordings of aortic flow records have become commonplace (Fig. 7–3). It appears from these records that the ventricular volume curve recorded with a cardiometer (Fig. 7–3) does not accurately represent ejection (or flow).

Aortic flowmeters show that flow reaches its peak velocity in about 0.10 sec, whereas pressure reaches its maximum in about 0.18 sec. The flow curve is asymmetrical, with rapid initial ejection followed by a slower return to zero. By the time pressure reaches its peak, flow may have declined to two-thirds or one-half of its maximal value. Atrial pressure falls below venous pressure, and the atria begin filling at this time. At the end of the period of maximal ejection, the beginning of ventricular repolarization is signaled by the onset of the T wave.

The period of increasing rate of flow (*i.e.*, positively accelerating flow) is short. Although flow from ventricle to aorta continues, the velocity decreases (Figs. 7–3D and 7–4). Rate of change of velocity goes to zero and then becomes negative. Spencer has pointed out that, while the flow velocity is decreasing, the aortic pressure

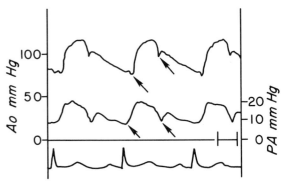

Figure 7–4 Aortic (upper curve) and pulmonary arterial (lower curve) pressures in the human. Electrocardiogram below. Comparison of first arrows on upper and lower curves shows that pulmonary pressure rises first and that pulmonary valves open before aortic valves. Second arrows indicate valve closure, which is later on the right. (After Braunwald *et al., Circulat. Res.*, 1956, *4*, 100–107.)

may be slightly *higher* than the ventricular pressure.[12] On the average, ventricular pressure exceeds aortic pressure only during the initial 45 per cent of the ejection phase (Figs. 7–3 and 7–4). The simple "ohmic" equation, relating pressure to flow and resistance, is $\Delta P = \dot{Q}R$, where Q indicates volume and $\dot{Q}$ is the instantaneous rate of flow. All pressure drops in this relationship are frictional. For a more exact description, an additional term is required involving the inertance (L) of the blood:

$$\Delta P = \dot{Q}R + (\rho l/A)(d\dot{Q}/dt)$$

The term $(\rho l/A)$ is the inertance, L: in this term, ρ is the density of the blood, A the area of the segment, and l the segment length over which measurement is made; $d\dot{Q}/dt$ is the instantaneous rate of change of flow.

Using Spencer's data, let us assume no $\dot{Q}R$ pressure fall: after reaching its peak, the flow rate falls from about 9000 cm^3 per sec to zero in about 0.1 sec. The acceleration (negative) is thus $-90,000/60 = -1500$ cm^3 per sec per sec. If we assume an area of 4.0 cm^2 and a density of 1, with l = 4 cm, the inertance term has the value of $\frac{1 \times 4 \times 1500}{4} = -1500$ dynes per cm^2 = 1500/980 cm H_2O = about -1.5 mm Hg. Aortic pressure thus can exceed ventricular pressure by about 1.5 mm.

DECREASED EJECTION (Fig. 7–3D). Following the initial period of rapid ejection, the rate of outflow from the ventricle decreases markedly and there is a period of reduced ejection. The ventricular volume curve starts to level off, and ventricular and aortic pressures begin to fall. Decreased ejection results because the fibers have reached a shorter length, are contracting isotonically and can no longer contract forcefully. The ventricles appear to exert their maximal effort during the initial phase of ejection. It is also possible that there is some influence of the end of depolarization.

The venous pressure continues to be greater than atrial pressure; the atria continue to fill. Electrically this period is marked by the end of the T wave; *i.e.*, ventricular repolarization becomes complete.

Isometric (Isovolumetric) Relaxation (Fig. 7–3E). When the ventricular ejection falls to zero, the left ventricular pressure falls below the pressures in the aorta and pulmonary artery. The aortic and pulmonary artery valves therefore close. The ventricular pressure continues to fall rapidly as the ventricles relax. The A-V valves remain closed while the ventricular pressure exceeds atrial pressure. This is the period of isometric relaxation. Because the valves at both ends of the ventricles are closed, the amount of fluid contained in the ventricles obviously cannot change. The term "isovolumetric relaxation" is applied to this period.

Rapid Ventricular Filling (Fig. 7–3F). The isometric relaxation phase ends when the ventricular pressure falls below pressure in the atria; the A-V valves open, and a phase of rapid ventricular filling begins. During all this period, flow of blood from the aorta to the peripheral arteries continues and the aortic pressure falls slowly. It has recently been claimed that ventricular diastolic suction contributes to the ventricular filling.[1] Apparently the ventricle is able to do work filling itself with blood, *i.e.*, the fact that the ventricle is empty but relaxed makes the atrioventricular pressure difference greater than the difference between the atrial and intrathoracic pressures.

The phase of rapid inflow is followed by a variable phase of slow filling during which filling is much less rapid. Filling is limited, too, because the ventricle has come close to a maximum diastolic size, which for a given cycle length is determined by the atrial pressure (although it may be changed by nervous, hormonal and other factors). The period of diastasis ends the cycle and is terminated by atrial systole, which begins the next cycle (this is where we came in;

Table 7–1

	MAN	DOG
Isometric contraction	.05	.05
Maximum ejection	.09	.12
Reduced ejection	.13	.10
Total systole	.27	.27
Protodiastole°	.04	.02
Isometric relaxation	.08	.05
Rapid inflow	.11	.06
Diastasis	.19	.29
Atrial systole	.11	.11
Total diastole	.53	.53

°The protodiastolic period, which has not been discussed, is the period after the ventricles cease to eject and before the aortic valves close.

Figs 7–3A and G). The durations of the various phases in seconds, taken from Wiggers, are shown in Table 7–1.[14]

Pressure changes in the human heart have been recorded during cardiac surgery[2] (Fig. 7–4). In general the results are similar to those previously described. Additionally, a complete picture would include pressure on the right and left sides of the heart. These are shown in Figure 7–5. Right atrial contraction precedes left atrial contraction by about 20 msec. Right ventricular ejection also begins slightly earlier than does ejection from the left. Further, mitral valve closure follows tricuspid closure and the pulmonic valves open before and close after the aortic valves. Pressures in the several chambers and vessels are shown below and in Figure 7–4.

HEART SOUNDS

The mechanical events of the cardiac cycle produce sounds which may be heard at the body surface. Sound-producing events include oscillations of the blood, movements of the heart walls, blood vessels and valves and turbulence of blood flow. These sounds may be heard by placing the ear on the chest wall or by using a stethoscope. They may be amplified and recorded with a microphone and recording system. The various sounds are heard with differing intensities at various locations on the body surface. In certain experimental procedures, microphones or catheters

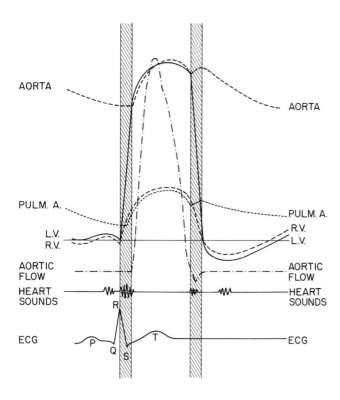

Figure 7–5 An expanded version of the events of the cardiac cycle (Fig. 7–3). Isovolumetric contraction and relaxation of left ventricle shown in cross-hatched areas. It also shows pressure in the pulmonary artery and right ventricle and a curve of aortic flow as recorded just beyond the aortic valves. The pulmonary artery flow curve shown is very similar to the curve in the aorta. It shows an initial period of very rapid flow, followed by a period of lower flow and return to zero flow. As will be seen, valves open and close at slightly different times on right and left, but otherwise the flow curves are nearly identical. Heart sounds and the ECG are as in Figure 7–3. Note also that ventricular pressure exceeds aortic pressure only during the first portion of systole.

connected to microphones have been placed within the chambers of the heart. For details, the reviews of McKusick[8] and Lewis[6] should be consulted. Sounds are shown in Figures 7–2, 7–3, 7–5 and 7–6.

Causes of Specific Heart Sounds. During diastole blood flows smoothly from the atrium into the ventricle and from the aorta into the peripheral vascular beds. This smooth flow of blood produces no audible vibrations.

First Heart Sound. The first heart sound occurs at the termination of atrial contraction and the onset of contractions of the ventricles. Before ventricular contraction, the A-V valves (mitral and tricuspid)

Table 7–2 *Commonly Used Pressures in the Various Cardiac Chambers and Major Vessels in Millimeters of Mercury**

Right atrium	3	(0–5)
Right ventricle	24/4	(17–32/1–7)
Pulmonary artery	24/9	(17–32/4–13)
Left atrium	8	(2–12)
Left ventricle	130/9	(90–140/5–12)
Aorta	125/70	(90–140/60–90)

*Mean pressures are shown for the atria. In other chambers, the slash indicates systolic/diastolic pressures, figures in brackets show range of normal values.

are open and, as ventricular pressure rises, blood moves toward the atrium. As indicated, the contraction of the atria may also move blood into the ventricle and may play a part in mitral (or tricuspid?) valve closure. The atria may thus contribute to this sound. As a result of this movement, the atrioventricular valves close. The initial movement of the blood, the closure of the valves and the resulting abrupt cessation of movement of blood into the atrium produce sounds which are part of the *physiologic first sound.*

Valve closure is generally considered the major contributor to the first sound;[5] however, the continued increase in ventricular pressure moves blood into the great vessels, and the distension of these vessels by the increased pressure may produce vibrations which are part of the first heart sound. Thus the great vessels, the valves, the blood and the ventricular walls may all be vibrating interdependently. A further component of the first sound may result from turbulence in the flow of blood through the great vessels (Chap. 2, Vol. II). In many normal persons, the mitral and tricuspid valves apparently close slightly asynchronously. If this asynchrony is marked, a *split first sound* will be heard.

If the P-R interval is long (0.2 sec or

more), and ventricular contraction is delayed, the valve leaflets will move close together before ventricular contraction occurs. In this case, the valves will not travel far to close, and the first sound will be unusually faint. If the P-R interval is short, valve closure will be abrupt and the first sound will be loud. If the atrial pressure on either side is abnormally high, the valve leaflets may remain widely open and then close abruptly during ventricular contraction. The first sound will be abnormally loud in this case. Increase in thickness of a valve by growth of connective tissue will also increase the sound intensity.

SECOND HEART SOUND. Once the flow of blood from the ventricle into the aorta has been established and the valves are open, no sounds will be heard until the onset of the relaxation phases, unless the blood flows turbulently. When the ventricular pressure falls below the aortic (or pulmonary arterial) pressure, a slight backflow will close the aortic and pulmonary valves. The slight backflow of blood toward the ventricles is followed by rapid cessation of the movement as the valves close. This recoil initiates movements of the ventricular chambers and of the stretched valve cusps. As with the first sound, various components of the perceived sound have been considered to be produced by movements of specific structures. Again, valve movement is considered of prime importance.[5] If the pressure in the pulmonary artery or aorta is abnormally high, the closure of the valves may be exceptionally rapid and the sounds very loud.

Right ventricular systole terminates after left ventricular systole. Consequently, it is possible for the right and left ventricular components of the second sound to be separated or split. This splitting is more marked in inspiration and, usually, during right bundle branch block; it is possibly due to increased right ventricular filling. Concerning the respiratory variation, filling is increased during inspiration because the transmural pressure is then greater. Contributions made by the left ventricle to the second sound are considered to be greater than those of the right ventricle. Figure 7–6 shows normal first and second sounds and shows relations of the sounds to aortic, ventricular and atrial pressures and to the opening and closing of atrio-

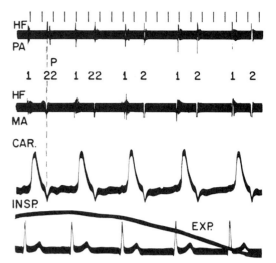

Figure 7–6 Phonocardiograms correlated with carotid pulse, respiration and electrocardiogram. Phonocardiograms were obtained with a high frequency (*HF*) recording phonocardiograph from pulmonary (*PA*) and mitral (*MA*) areas of the chest. The numbers under the top tracing indicate the first and second sounds. Where the number 2 is printed once, the second sound is not split. Where the number 2 is repeated, the second sound is split. Below heart sound records is indirect carotid pulse recorded with a device which is sensitive to changes in pressure but not to absolute pressure. Below this are indicated the phases of respiration and at the bottom conventional lead II of the ECG. Note that first sound is always simultaneous with upstroke of carotid pulse.

Second sound in pulmonary area has two components during inspiration but only one during expiration. During inspiration first component is synchronous with downward notch in carotid pulse, coinciding with closure of aortic valves. Second component of second sound is perceived in pulmonary but not in mitral area and occurs after closure of aortic valves. These observations indicate that second component is of pulmonary origin. As recorded in mitral area, second sound correlates only with aortic valve closure; in pulmonary area, closure of pulmonary as well as aortic valves is indicated and resultant sounds may be fused or separate. Time intervals are 40 msec. (After Leatham, *Lancet*, 1958, *2*, 703–708.)

ventricular and semilunar valves. Figure 7–7 also shows a comparison of events in the left and right heart—atrial and ventricular contraction, tricuspid, mitral and aortic valve opening and closing and the sounds.

THIRD HEART SOUND. The third and fourth heart sounds are known as "diastolic" sounds, in contrast to the first and second or "systolic" sounds. Their origin is less perfectly understood. As the ventricles relax, their internal pressure drops below the pressure in the atrium. The atrioventricu-

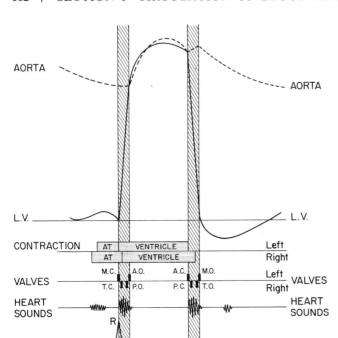

Figure 7–7 Events of the cardiac cycle, valve movement and heart sounds. Aortic and ventricular pressures, contraction of the atria and the ECG are also shown. The stippled black line labeled "contraction" shows the sequence of contraction of atria (AT) and ventricles for the left (above line) and right heart (below line). The line marked "valves" shows the opening and closure of mitral (M), aortic (A), tricuspid (T) and pulmonary (P) valves. Also shown are the first and second heart sounds and the smaller third and atrial sounds.

lar valves will then open, and blood will move into the relaxed ventricular chambers. Movement of blood into the ventricle produces vibrations of the chamber walls at about the time that the rapid filling phase terminates.

ATRIAL SOUND; FOURTH HEART SOUND. The contraction of the atrium moves blood through the partially open atrioventricular valves into the well distended ventricles. This movement gives rise to a vibration of low frequency and amplitude preceding the first heart sound. The third and fourth sounds are generally inaudible.

AUDIBILITY OF HEART SOUNDS. Lewis et al[6] have devised a procedure for intracardiac phonocardiography in man. In the right heart—the region which they studied—the first and third heart sounds were loudest in the right ventricle. The second sound was loudest in the pulmonary artery; the fourth was loudest in the right atrium. These results are compatible with the concepts of the origins of the sounds given above and indicate that movement of the blood within the chambers is probably necessary for transmission of the heart sounds.

With a stethoscope, only two distinct sounds are normally heard. The first heart sound is of lower pitch, more booming and longer. Its frequency content is apparently between 30 and 100 Hz, whereas the second heart sound contains higher frequencies. (Remarks about the frequency of these sounds are misleading unless one specifies where and how they are recorded.) The third sound, when present, is not loud; the atrial sound is rarely heard unless some form of amplification is used. The first heart sound has a duration of between 50 and 100 msec. The second sound lasts from 25 to 50 msec.

The heart sounds have their maximum intensities at different locations; these depend on the site of origin of the sound and on the way in which the fluids of the body conduct the sound to the surface (Fig. 7–8). When valvular defects produce murmurs and turbulence, it is usually possible to identify the defective valve from the location of the sound. The pulmonary valve produces sounds in the pulmonary area at the third (and second) left intercostal space in the left parasternal line. The aortic valve produces sounds which are of maximum intensity at the right of the sternum in the second right intercostal space. The sounds

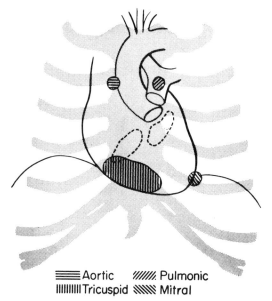

≡≡≡Aortic ▨ Pulmonic
ⅢⅢⅢTricuspid ▧▧ Mitral

Figure 7–8 Locations on body surface at which sounds from particular valve regions are best perceived.

of the tricuspid valve are loudest at the right sternal border in the fourth intercostal space; those from the mitral valve are heard best near the apex of the heart. The sounds from the pulmonary and A-V valve of the right heart tend to be of maximal amplitude near the underlying valvular positions. However, the sounds from the aortic and A-V valve of the left heart valves are not heard best over the valve rings, probably because the sounds are transmitted more successfully through a liquid medium than directly through the lungs. The physician usually moves his stethoscope from one area to another, noting where the sounds are loudest and picking out components from each valve in seeking to assess the sound.

Characteristics of Systems Used in Detecting the Heart Sounds. When sound reaches the body surface, it causes that surface to move, and this movement may be perceived by the physician placing his ear directly in contact with the thorax. In such a situation, he will perceive the sound through the gas, solids and liquid of his outer, middle and inner ear. The sensation which he receives may not at all represent the actual events at the surface of the thorax because the audibility curve (Chaps. 19 and 20, Vol. I) shows that the auditory sys-

tem cannot perceive sounds below 20 imp/sec. Even when it perceives low-frequency sounds, the auditory system is not linear—*i.e.*, the perceived intensity of sound at different frequencies is not proportional to the actual amplitude of the vibrations (Chaps. 19 and 20, Vol. I). Heart sounds are generally between 20 and 200 imp/sec. In this range, low frequency sounds are minimized by the nonlinearity of the ear. The stethoscope is used as a convenience in listening to heart sounds; it does not amplify sounds but actually distorts them, changing their characteristics and generally accentuating the nonlinearity of the auditory system. Various types of stethoscopes have different effects on sounds; for certain purposes one type may be better than another. For example, a stethoscope of the diaphragm type selectively attentuates the low frequencies, whereas the bell type attenuates the high frequencies.

PHONOCARDIOGRAPHY. Phonocardiography directly records the heart sounds so that a visual record is obtained. Such records supplement auscultation, *i.e.*, the use of the stethoscope to hear the heart sounds. The major problem concerns the differences between behavior of the ear and the responses of the conventional microphones, amplifiers and recorders—both optical and direct—which have been used. The perception by the ear of vibrations in the audible range (20 to 1500 Hz) is nonlinear. To date, the microphones available for phonocardiography have been either linear in their response or nonlinear in a manner different from the nonlinearity of the ear. The physician who wishes to see a phonocardiogram which accurately represents the perception of the sound will be disappointed if the phonocardiographic system is one which faithfully reproduces the sounds. A discussion of techniques can be found in McKusick *et al.*[8]

Abnormalities. SPLITTING OF HEART SOUNDS. In a number of circumstances a dissociation of the component vibrations of a sound takes place, dividing it into separate parts, each of which is heard or recorded as a separate entity. As far as the first sound is concerned, this may be a purely physiologic event, and frequently can be recorded as an exaggerated separation of the two major components, the isometric phase and the ejection phase components; this may often occur at the end of expiration (Fig. 7–6).

Splitting of the second sound proceeds from an entirely different cause, namely, asynchronous closure of the aortic and pulmonary valves. The phenomenon occurs in normal subjects, but is more frequent during right bundle branch block when the interval between right and left ventricular contraction is abnormally prolonged (Fig. 7–9). The second sound associated with some ventricular extrasystoles (*i.e.*, beats which originate in abnormal sites and during which more time than normal is required for ventricular depolarization) may also be split for the same reason. In left bundle branch block the natural asynchrony may be masked.

OPENING SNAP OF MITRAL VALVES. In mitral stenosis (pathologic narrowing of

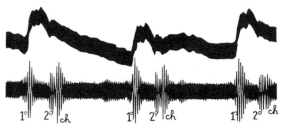

Figure 7–10 Opening snap of mitral valve in mitral stenosis. Snap *(ch)* represents an increase to an audible stage of vibrations normally present but inaudible. *Above:* Central arterial pulse. *Below:* Phonocardiogram. (From Orias and Braun-Menéndez, *The heart sounds in normal and pathological conditions*, 1st ed. London, Oxford University Press, 1939.)

the mitral orifice) a third sound is heard, and, when it is recorded phonocardiographically, it is found to be coincident in time with the opening of the A-V valves. It represents an abnormal intensification of a component of the normal second heart sound, which because of attenuation in the stethoscope-ear combination is not normally heard (Fig. 7–10).

GALLOP RHYTHMS. When a loud third sound is heard in a rapidly beating heart, the resulting triple rhythm has a cadence resembling the sound of a galloping horse. Such gallop rhythms are heard most frequently in cases of serious cardiac disease, and it is presumed that the abnormal intensification of the third or the atrial sound is related in some way to an altered ventricular response to rapid filling or atrial systole. A systolic gallop may occasionally be heard when there is marked splitting of the first heart sound.

MURMURS. When fluid flows slowly through a smooth tube of uniform diameter, no sound may be heard through a stethoscope placed on the tube. If, however, the velocity of flow is greatly increased or the viscosity of the fluid reduced, flow is no longer smooth but becomes turbulent; *i.e.*, eddy currents are set up, and these produce vibrations which may be audible. The velocity at which turbulence begins is greatly diminished by annular expansions or constrictions of the tube or inequalities of its surface. These factors therefore favor the development of murmurs.

In the normal heart the critical velocities for turbulent flow are not quite attained in

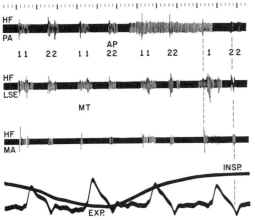

Figure 7–9 Splitting of first and second sounds sometimes occurs in complete right bundle branch block. Closure of tricuspid valve (*T*) [seen in second record taken at left sternal edge (*LSE*)] and of pulmonary valve *(P)* (seen in upper record from pulmonary region) are both delayed, although mitral and aortic valve closures, indicated by *M* and *A*, are at approximately the normal time. Delayed tricuspid component of first sound at lower left sternal edge is not altered by respiration, but splitting of second sound in record from pulmonary area is greater during inspiration, because closure of pulmonary valve is delayed owing to increased filling during inspiration. First component of split second sound is synchronous with dicrotic notch in carotid tracing (there is some delay in carotid recording system). Pulmonary nature of second component is indicated by its great intensity at pulmonary area. HF indicates that this is a high frequency recording. Splitting of sounds as indicated in Figure 7–6. (From Leatham, *Pediat. Clin. N. Amer.*, 1958, 839–870.)

average circumstances. In strenuous exercise and in other conditions in which the velocity of flow is augmented, systolic murmurs may appear in normal hearts. Increased velocity of blood flow during the ejection phase, with resulting turbulence, is probably also responsible for the systolic murmurs that may appear in anemia and in thyrotoxicosis. The murmurs of a patent ductus arteriosus (Fig. 7–11), the hum over arteriovenous aneurysms, the Korotkov sounds (sounds heard during auscultatory determination of blood pressure) and Duroziez' sign (systolic and diastolic murmurs heard over the femoral artery in aortic incompetence and modified by the degree of pressure exerted by the stethoscope) are other examples of sounds produced by turbulence at constrictions in smooth tubes.

Abnormal narrowing of an orifice such as occurs in mitral or aortic stenosis will both increase flow velocity and lower the velocity at which turbulence occurs. These effects account in the main for the murmurs heard in these conditions, although the roughening of the walls of the orifice by scarring and partial destruction of the valves undoubtedly also contribute in some measure. In mitral stenosis the murmurs will be diastolic, occupying typically the periods of rapid filling and atrial systole,

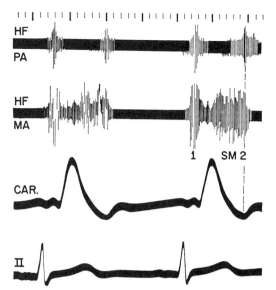

Figure 7–12 Heart sounds in a mild case of regurgitation from left atrium to ventricle. The murmur (SM2), as seen in mitral area, extends throughout systole, although its intensity diminishes during rising phase of carotid pulse. Note relative absence of this murmur in pulmonary region and tendency for murmur to blend with and virtually obscure second sound. Abbreviations as in Figure 7–9. (From Leatham, *Pediat. Clin. N. Amer.*, 1958, 839–870.)

although the murmur may be continuous throughout diastole owing to overlapping of the two phases. The typical murmur of aortic stenosis occurs in systole during the phase of ejection.

It is understandable that in the presence of valvular incompetence in addition to stenosis, murmurs may be heard during systole in mitral valvular disease and during diastole in aortic valvular disease, owing in part to the regurgitation of blood through the narrow orifice (Fig. 7–12).

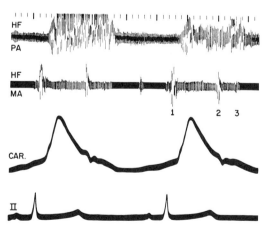

Figure 7–11 In patent ductus arteriosus blood flows continuously from the high pressure aorta to the low pressure pulmonary artery. Resultant murmur is loud, often increases in intensity late in systole, and differs in its characteristics from murmurs which result from regurgitation. Abbreviations as in Figure 7–9. (From Leatham, *Pediat. Clin. N. Amer.*, 1958, 839–870.)

REFERENCES

1. Bloom, W. L., and Ferris, E. B. Negative ventricular diastolic pressure in beating heart studied *in vitro* and *in vivo. Proc. Soc. exp. Biol. (N.Y.)*, 1956, *93*, 451–454.
2. Braunwald, E., Fishman, A. P., and Cournand, A. Time relationship of dynamic events in the cardiac chambers, pulmonary artery and aorta in man. *Circulat. Res.*, 1956, *4*, 100–107.
3. Brockman, S. K. Dynamic function of atrial contraction in regulation of cardiac performance. *Amer. J. Physiol.*, 1963, *204*, 597–603.
4. Keith, A., and Flack, M. The form and nature of

the muscular connections between the primary divisions of the vertebrate heart. *J. Anat. (Lond.)*, 1907, *41*, 172–189.

5. Leatham, A. Auscultation of the heart. *Lancet*, 1958, *2*, 703–708; 757–766.
6. Lewis, D. H. Phonocardiography. *Handb. Physiol.*, 1962, sec. 2, vol. 1, 695–732.
7. Lewis, T. *The mechanism and graphic registration of the heart beat*, 3rd ed. London, Shaw and Sons, Ltd., 1925.
8. McKusick, V. A., Talbot, S. A., Webb, G. N., and Battersby, E. J. Technical aspects of the study of cardiovascular sound. *Handb. Physiol.*, 1962, sec. 2, vol. 1, 681–694.
9. Robb, J. S. The structure of the mammalian auricle. *Med. Wom. J.*, 1934, *41*, 143–152.
10. Rushmer, R. F. Initial phase of ventricular systole: asynchronous contraction. *Amer. J. Physiol.*, 1956, *184*, 188–194.
11. Rushmer, R. F. *Cardiac diagnosis*. Philadelphia, W. B. Saunders Co., 1955.
12. Spencer, M. P., and Greiss, F. C. Dynamics of ventricular ejection. *Circulat. Res.*, 1962, *10*, 274–279.
13. Weidmann, S. *Elektrophysiologie der Herzmuskelfaser*. Bern, Hans Huber, 1956.
14. Wiggers, C. J. *Circulatory dynamics*. New York, Grune & Stratton, 1952.

CHAPTER 8   THE ARTERIAL SYSTEM

by ERIC O. FEIGL

The arterial system is a branching system of conduits for blood flow to the various tissues of the body. Since the arterial tree is distensible, it acts as an elastic reservoir or capacitor for left ventricular output, maintaining pressure and flow during diastole when the heart is not ejecting blood. Small arteries and arterioles are the sites of greatest hemodynamic resistance in the cardiovascular circuit.

The distribution of blood flow to the various organs is primarily determined by vasomotion of small arteries and arterioles. This chapter introduces the general characteristics of the arterial system. The physiological significance of this material is more completely developed in subsequent chapters on the regional circulations and on cardiovascular regulation.

ARTERIAL PRESSURE PULSE

A recording of the arterial pressure pulse has an ascending limb and a descending limb. Occasionally a small irregularity called the anacrotic notch can be observed in the ascending period. The descending

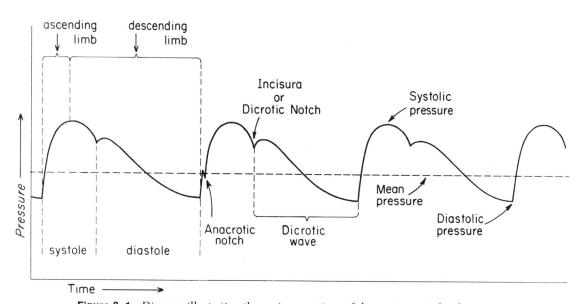

Figure 8–1 Diagram illustrating the various portions of the aortic arterial pulse wave.

117

limb of the aortic pulse always contains a dicrotic notch or incisura which results from aortic valve closure at the end of ejection. The interval following the incisura before the next ascending limb is referred to as the diastolic run-off period. A pressure undulation, often observed in this period, is named the dicrotic wave. The peak of the arterial pressure pulse is the systolic blood pressure; the nadir is the diastolic pressure. The difference between systolic and diastolic pressure is the pulse pressure. The mean pressure is the average pressure over the entire period (the pressure-time integral from the beginning of the pulse to the beginning of the next pulse divided by the period). The mean pressure is most accurately determined by electronic averaging of a pressure pulse (see Chap. 3,

Vol. II); but in the brachial artery, it can be approximated by the formula:

$$\text{mean pressure} = \text{diastolic pressure} + \tfrac{1}{3}\text{ pulse pressure.}$$

The fraction of the pulse pressure added to the diastolic pressure is different at the various sites in the vascular tree because the pulse amplitude increases peripherally (see below).

The arterial pressure wave travels down the aorta with a velocity of approximately 5 m per sec. This is much more rapid than the flow wave, which has a speed less than 100 cm per sec. The pressure pulse wave velocity is a function of elastic stiffness, radius and thickness of the vessel wall and density of the blood.[10, 19]

A complex wave such as the aortic pres-

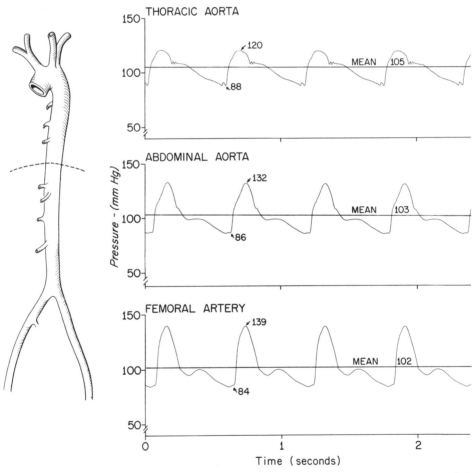

Figure 8–2 Simultaneous blood pressure recordings from the thoracic and abdominal aorta and the femoral artery of a dog. The wave form changes as it progresses down the arterial tree. Systolic and pulse pressures are greater at peripheral sites, whereas mean pressure falls slightly. Mean pressures were determined by electronic averaging.

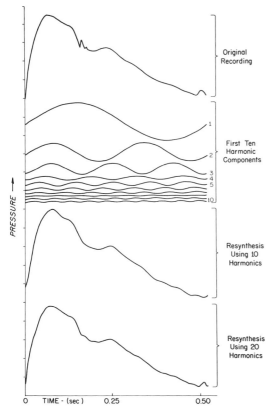

Figure 8–3 Fourier analysis of an aortic blood pressure wave. The original wave is shown above with the first 10 harmonics. The sine wave harmonics are shown with the appropriate amplitudes and phase angles. Resynthesized waves using the first 10 and first 20 harmonics are shown below.

sure wave may be considered mathematically as being composed of a number of sine waves of various amplitudes and frequencies.[2] This analysis of a periodic complex wave shape into a series of sine waves is called Fourier or harmonic analysis. The construction of a complex wave by adding sine waves of various amplitudes and frequencies is Fourier synthesis.

The arterial pulse changes shape and amplitude as it travels down the arterial tree. In general, the pulse pressure becomes higher, the sharp incisura is lost, the dicrotic waves becomes apparent and the mean pressure falls slightly as the pressure wave travels from the aortic arch to the femoral artery.

Partial pressure wave reflection will occur at any sharp discontinuity in the arterial tree. Each branching of the arterial tree is a discontinuity, although the reflection coeffi-

cient for most branchings is probably small. The reflection coefficient is a number varying between 0 and 1.0 which represents the extent to which the incident wave is reflected. A right angle closed end to a vessel would have a high reflection coefficient (close to 1.0). Thus the pressure at a given point in the aorta is the result of an antegrade pressure pulse from the heart and multiple small retrograde waves from distal reflecting sites.

The transformation of the pressure pulse was formerly believed to be due to a standing wave in the aorta with a node in the region of the diaphragm. The most compelling reason that this is unlikely is that a reflecting site must be at an exact multiple of half the wave length of the traveling wave for a standing wave to form. The arterial system has multiple reflecting sites which are clearly not at half wave length distances.

The blood and vascular wall have viscous properties which attentuate high frequencies. This results in the damping out of high frequency components (rapid oscillations) in the pressure pulse. The highest frequency oscillations in the pressure pulse at the root of the aorta occur during the incisura. As the pressure wave travels from the ascending aorta down the vascular tree, the sharp dicrotic notch is blunted and disappears before the wave reaches the femoral artery.

Harmonic dispersion will occur in systems, such as the arterial tree, which have viscosity and mass.[14] The "effective stiffness" of the arterial wall is greater for high frequency harmonics than for low frequency harmonics so that the pulse wave velocity is greater for higher frequencies. The faster wave velocity for higher frequencies results in the aortic pressure wave becoming dispersed in distance and time along the arterial tree, thus changing the shape of the pressure wave.

An additional factor contributing to the distortion of the pressure pulse as it travels out the arterial tree is the elastic nonuniformity of the vascular wall. The peripheral vessels are stiffer than the proximal vessels, and this increase in vascular elastic modulus[3, 4, 13] at more distal sites contributes to the increased pulse pressure in peripheral arteries.[16a]

It is at first startling to find that the systolic blood pressure is higher in the brachial

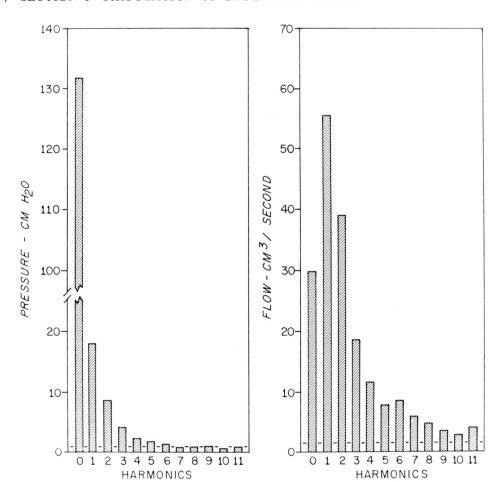

Figure 8–4 Harmonic content of the pressure and flow waves from the ascending aorta in dogs obtained by Fourier analysis. The zero harmonic is the DC or mean term; the first harmonic has the frequency of the heart rate. The dashed line indicates the approximate noise level in the recording and analysis. (From Patel *et al.*, *J. appl. Physiol.*, 1963, *18*, 134–150.)

or femoral artery than in the aorta. This is not evidence for a propulsive action of arterial smooth muscle contraction with each heart beat. Nor is the law of conservation of energy contradicted; rather, the aortic pressure wave is altered by reflection, damping and dispersion which can readily account for increased peripheral systolic pressures. The energy imparted to the arterial tree with each heart beat is in the form of pressure *and* flow. Since pressure can be transformed into flow and flow into pressure, the energy in the system cannot be accounted for by pressure alone. The systolic pressure often rises from the root of the aorta to the brachial and femoral arteries and falls again farther out in the arterial tree. Mean pressure falls slightly from aorta to femoral artery.

In summary, the arterial pressure pulse has a characteristic shape which is modified as it travels out the arterial tree. The peak of the pressure wave is the systolic pressure; the trough is the diastolic pressure. The altered shape and height of the pressure wave as it travels are due to reflection, damping and dispersion.

ARTERIAL ELASTIC PROPERTIES

Large arteries act as elastic reservoirs or capacitors, distending during cardiac systole and narrowing during diastole. The magnitude of the distension of the aorta with each pressure wave is determined by the mechanical properties of the aortic wall. Although the vessel wall has mass and vis-

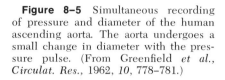

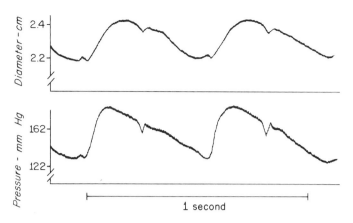

1 second

Figure 8–5 Simultaneous recording of pressure and diameter of the human ascending aorta. The aorta undergoes a small change in diameter with the pressure pulse. (From Greenfield *et al.*, *Circulat. Res.*, 1962, *10*, 778–781.)

cosity, the dominant characteristic is elasticity.[4, 8, 12, 13] A recording of the aortic diameter during a heart beat closely resembles the pressure wave (Fig. 8–5). The diameter of the aorta changes about 5 per cent during the cardiac cycle.

With each beat, the heart delivers potential and kinetic energy to the arterial system. The kinetic energy is in the flowing blood and its acceleration during ejection from the heart. The potential energy is manifest as the increase in aortic pressure with each heart beat. The blood pressure does not fall to zero in the period between cardiac contractions, because the arterial tree acts as an elastic reservoir in series with the peripheral resistance. This capacitive property of the arterial tree together with the action of the aortic valve allows a portion of the ventricular stroke volume to be stored during systole and thus to maintain flow during diastole. When the arterial tree becomes stiffer with aging, the pulse pressure becomes greater—evidence of the loss of capacitive function. In older literature, the elastic reservoir property of the arterial tree was referred to as a Windkessel (German for wind kettle), after the pneumatic tank capacitor used with some hydraulic lines. The capacitor of the arterial system is not a partially filled chamber where air is compressed as in a Windkessel, but is the arterial vessels which distend with blood.

Arteries exhibit nonlinear elastic properties, becoming stiffer with increasing distension.[3] The vessel wall is composed of connective tissue and vascular smooth muscle, both contributing to the elastic stiffness of the vessel. Vascular smooth muscle contraction decreases vessel radius, and this may be interpreted as an increase in elastic stiffness when plotted as a function of vessel size or as a decrease in stiffness when plotted as a function of distending pressure.[6]

In summary, the elastic properties of arteries result in the capacitive or Windkessel function of the arterial tree. Arterial stiffness increases with age, decreasing the capacitive function of the arterial tree which results in a greater pulse pressure.

AUTOREGULATION

Autoregulation is said to be present when blood flow remains relatively constant in the presence of a varying perfusion pressure (inflow pressure minus outflow pressure). Autoregulation is lost below a minimum pressure and above a maximum pressure. Below and above these pressures, a change in perfusion pressure results in a change in blood flow as expected in a passive set of tubes. In the autoregulatory range, however, blood flow remains relatively constant despite changes in perfusion pressure. From an operational point of view, autoregulation may be defined as flattening of the flow versus pressure curve in an isolated organ (Fig. 8–6).

Flow regulation in the intact circulation is subject to autonomic and hormonal control; but as the name implies, autoregulation is a vascular bed characteristic which is independent of extrinsic control mechanisms. Autoregulation (a term sometimes inappropriately used to describe other types of constant flow regulation), can

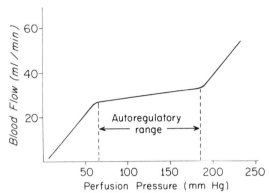

Figure 8–6 A schematic representation of autoregulation. The relative constancy of flow between 60 and 190 mm Hg perfusion pressure would not be observed in a simple tube.

be proved only in a preparation without extrinsic neural and hormonal control. There are three hypotheses as to how autoregulation occurs: (i) myogenic, (ii) metabolic and (iii) tissue pressure.[9]

(i) The *myogenic hypothesis* is frequently called the Bayliss hypothesis. Bayliss rapidly inflated an excised artery and visually observed a contraction or spasm of the artery. Quick stretching is a stimulus for contraction for many smooth muscles, including vascular smooth muscle. The hypothesis is that the stretch imposed by an increased arterial pressure stimulates arterial and arteriolar smooth muscle to contract, and this vasoconstriction reduces the flow. This hypothesis is not completely convincing as a means of flow regulation, because pressure and not flow is involved in the regulatory scheme. Folkow[7] favors a modified myogenic hypothesis of autoregulation on the basis that precapillary sphincters respond to transmural pressure. The precapillary sphincters will close more frequently when stretched by increased pressure so that the average resistance to flow would be increased.

(ii) The *metabolic hypothesis* of autoregulation postulates that a relatively constant flow is maintained because the vascular smooth contraction is controlled by tissue levels of either a metabolic substrate or a metabolite. Blood flow is linked to tissue metabolism in a number of vascular beds. Large increases in tissue metabolism result in local vasodilation by a local mechanism (see Metabolic Control, below).

Real as metabolic control is, the metabolic hypothesis does not appear to explain all forms of autoregulation. The most evident counterexample is the renal circulation. The kidney has a very large blood flow, far in excess of that required for its metabolism, because the bulk of flow is used to form the glomerular filtrate. Yet the kidney exhibits very strong autoregulation. Renal autoregulation is difficult to explain on the basis of a critical level of a metabolite or metabolic substrate.

A further reservation concerning the metabolic hypothesis is that autoregulation appears to occur at a higher perfusion vs. metabolism ratio than is exhibited by metabolic control in other circumstances. In moderate exercise with less than maximum blood flow, the muscle venous oxygen content falls much below the level observed during autoregulation in resting muscle. Using the venous oxygen content as a rough index of the match between blood flow and metabolism implies that autoregulation occurs at a higher blood flow to metabolism ratio at rest than during activity. Therefore the regulatory scheme includes more than a simple matching of flow to metabolism in these two circumstances.

(iii) The *tissue pressure hypothesis* of autoregulation is based on the observation that an increase in perfusion pressure can result in an increase in tissue pressure. The amount of fluid entering the extracellular tissue space by ultrafiltration through the capillaries is dependent on capillary blood pressure (see Chap. 9, Vol. II). In general, an increase in perfusion pressure augments capillary pressure and fluid movement into the extracellular tissue space. If the tissue is encapsulated in a fairly rigid container, an increase in tissue fluid volume will raise the interstitial pressure in the tissue. This increased tissue pressure will compress small veins and venules, increasing the resistance to blood flow through them. An artificial system of porous tubes in series with collapsible rubber tubing enclosed in a rigid case exhibits "autoregulation" by this mechanism.[15]

Increased small vein resistances have been demonstrated in the isolated kidney during autoregulation, and a tissue pressure mechanism probably accounts for at least part of renal autoregulation. The kidney is unusual because it has a rather stiff capsule which permits a rapid rise in tissue pressure with an increased extracellular fluid volume. A tissue pressure mechanism for autoregulation in intestine or muscle is less satisfactory, because these tissues do not have a stiff capsule and tissue pressure rises only slightly with extracellular fluid accumulation.

In summary, autoregulation is a characteristic of isolated vascular beds which maintains a relatively constant blood flow in the face of an altered perfusion pressure. The mechanism of autoregulation is not settled. A tissue pressure mechanism is likely in the kidney, but not in muscle or intestine. Metabolic autoregulation is reasonable for muscle and intestine, but unlikely in the kidney. A myogenic mechanism is possible, but has a number of difficulties. A possible compromise is that all three mechanisms are involved in varying proportions in different tissues. Such a compromise is tempting but may only be a cover for our ignorance. A unified theory of autoregulation is still needed.

METABOLIC CONTROL OF THE ARTERIAL CIRCULATION

Metabolic circulatory control adjusts the blood flow to the metabolic rate of the organ. It exists because tissue metabolically required substances and/or metabolites are intrinsically linked to the smooth muscle of vessels in the tissue. Metabolic control increases blood flow when metabolic requirements increase and may or may not account for autoregulation. Two examples of metabolic control are (i) the marked increase in skeletal muscle blood flow with exercise, and (ii) the increase in coronary blood flow when the heart rate increases.

The postulated mechanism of metabolic control is that a metabolically required substance such as oxygen or a metabolite such as carbon dioxide acts directly or through some chemical intermediate on vascular smooth muscle. In the case of carbon dioxide, an inadequate tissue blood flow would result in an increase in carbon dioxide in the extracellular fluid which surrounds the vascular wall. This would cause the vascular smooth muscle to relax, increasing the vessel radius and blood flow. The increased flow would carry away the excess carbon dioxide, lowering the tissue level and the stimulus to vasodilation. The analogous scheme where an inadequate supply of a required metabolic material causes vasodilation is postulated for substances such as oxygen.

The apparent link between metabolism and vasomotion is different in various tissues.[11] In the heart, the link is tissue oxygen tension; the heart is relatively insensitive to the accumulation of carbon dioxide. In contrast, the cerebral circulation vasodilates greatly when carbon dioxide levels increase, but shows limited vasomotion with altered oxygen tension. Further, no single metabolically required substance or metabolite has been found which accounts for exercise hyperemia in skeletal muscle. The separate details of metabolic control for each tissue are given in the chapters on the regional circulations. The concept of metabolic control is introduced at this point to provide a complete picture of the several types of arterial control.

In most organs, if the blood flow is interrupted by occluding the artery for a brief period, marked vasodilation will be observed. This phenomenon is called reactive hyperemia and is a manifestation of metabolic control. During the period of arterial occlusion, either the metabolic substrate is depleted or metabolite concentration is built up. After the occlusion is released, the reactive hyperemia "pays back" part of the blood flow debt.

In summary, metabolic control is a coupling between the metabolic rate of a tissue and the blood flow through it, independent of nerves and reflexes. The metabolically required substance or metabolite responsible for vasomotion varies in different organs.

NEURAL CONTROL OF THE ARTERIAL CIRCULATION

Arterial vessels are innervated by the two divisions of the autonomic nervous system. The sympathetic division innervates the entire arterial tree. The coronary and the cerebral circulations are additionally innervated by the parasympathetic divisions. The sympathetic (thoracolumbar outflow) and parasympathetic (craniosacral outflow) are simply anatomical designations (*cf.* Chap. 11, Vol. I) and not based on the transmitter involved. Similarly, these designations are not concerned with the response that the nerve elicits, because stimulation of sympathetic nerves may cause either smooth muscle contractions (*e.g.,* arterial vasoconstriction) or smooth muscle relaxation (*e.g.,* bronchial dilation). Sympathetic postganglionic fibers usually are "adrenergic" and release norepinephrine to the effector cell (*e.g.,* the heart) or maybe "cholinergic" and release acetylcholine to the effector cell (*e.g.,* sweat glands). Parasympathetic fibers release acetylcholine at effector cells (*e.g.,* cardiac pacemaker).

Sympathetic Adrenergic Fibers. Classification of effects of autonomic nerves (and autonomic drugs) has been greatly aided by the concept that there are "receptors" in the autonomic target organs (cardiac muscle, arterioles, etc.). The postulate is that a specific receptor on the effector cells responds to a specific neurotransmitter. Receptors are hypothetical constructs which categorize neurotransmitter or humoral effects and are classified by the groups of drugs which block them.[18] It should be emphasized that no receptor has been identified or "seen."

The concept of specific receptors in the membranes of effector cells explains why some drugs selectively block autonomic responses in some tissue and not others, even when the neurotransmitter is the same. One group of compounds, called *alpha-receptor blocking agents,* selectively block sympathetic vasoconstriction, leaving sympathetic influences on the heart intact. Examples of alpha-blocking agents are phenoxybenzamine or phentolamine. *Beta-blocking agents,* e.g., propranolol, block sympathetically mediated cardiac acceleration and increased cardiac contractility without affecting vasoconstriction. Both the cardiac effects and vasoconstriction are mediated by norepinephrine released by sympathetic fibers, but the differential effects of the blocking agents indicate the existence of different postjunctional receptors into two tissues.

Sympathomimetic drugs differ in their ability to activate alpha- and beta-receptors.[1] For example, isoproterenol is considered a "pure" beta-receptor agonist with no alpha-receptor activity, whereas phenylephrine is considered a "pure" alpha-receptor activator. Most other sympathomimetic drugs, however, have both alpha- and beta-receptor actions in various proportions.

In general, sympathetic nerve activation causes arterial vasoconstriction in all vascular beds. This is blocked by alpha-receptor blocking agents. Norepinephrine is released from sympathetic nerve endings, activating vascular smooth muscle alpha-receptors.[5] An injection of the hormone epinephrine results in vasoconstriction in most vascular beds. The arterial tree in skeletal muscle dilates at low concentrations of epinephrine and vasoconstricts with larger doses of injected epinephrine. The vasoconstriction is blocked by alpha-receptor blocking agents which convert the response to pure vasodilation. The vasodilation may be blocked by beta-receptor blocking agents. Thus the response of the skeletal muscle arterial bed to injected epinephrine is interpreted as activation of beta-receptor vasodilation at low doses and alpha-receptor vasoconstriction dominating at higher doses.

Sympathetic Cholinergic Fibers. Small arterial vessels in skeletal muscle are unique because they are innervated by an additional group of sympathetic fibers which release acetylcholine at their nerve endings. Acetylcholine causes active vasodilation, and this effect may be blocked by the cholinergic blocking agent, atropine.

We are now in a position to understand the experiment shown in Figure 8–7. The sympathetic nerve to a skeletal muscle vascular bed is electrically stimulated. Intense vasoconstriction results from norepinephrine activation of alpha-receptors, which is the dominant effect. If this is blocked with an alpha-receptor blocking agent, sympathetic nerve stimulation will result in a large vasodilation owing to sympathetic cholinergic fibers which may be blocked by atropine. After atropine, sympathetic nerve stimulation still results in a modest active vasodilation of beta-receptors by catecholamine activation.[17] Atropine and alpha-receptor blockage have "unmasked" beta-receptor active vasodilation. This active vasodilation may be blocked by beta-receptor blocking agents.

In skeletal muscle, the dominant effect of sympathetic trunk nerve stimulation is vasoconstriction. Sympathetic cholinergic vasodilation may be selectively activated in a defense alarm reaction (see Chap. 13,

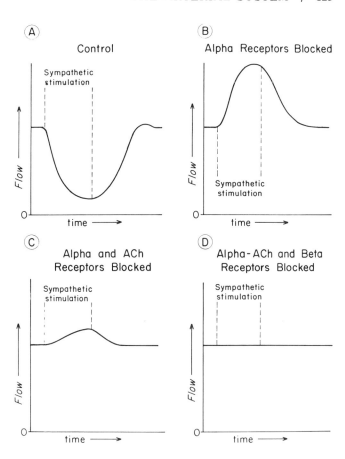

Figure 8–7 The effects of sympathetic nerve stimulation on blood flow in skeletal muscle shown diagrammatically to illustrate the various sympathetic receptors. *A,* Sympathetic stimulation results in vasoconstriction of the alpha-receptor type. *B,* Selective blockade of the alpha-receptors unmasks sympathetic vasodilation. *C,* The bulk of sympathetic vasodilation is gone after blocking the acetylcholine receptors with atropine. A small sympathetic beta-receptor vasodilation still remains. *D,* When all receptors are blocked sympathetic nerve stimulation no longer results in vasomotion.

Vol. II). Beta-receptor vasodilation in skeletal muscle may be observed pharmacologically by injecting isoproterenol without prior alpha-receptor blockade.

In summary, the sympathetic division of the autonomic nervous system innervates the arterial tree throughout the body. In most organs, sympathetic nerve activity results in alpha-receptor vasoconstriction. In skeletal muscle, sympathetic beta-receptor vasodilation and sympathetic cholinergic vasodilation can be demonstrated under special circumstances. The arteries of the brain and heart are also innervated by the parasympathetic division of the autonomic nervous system. Neural control varies from organ to organ. The details are given in Chapters 12 to 18, Volume II.

ARTERIAL TONE AND RANGE OF VASCULAR CONTROL

Basal Arterial Tone. The degree of vascular contraction found in a resting tissue without autonomic influence is called the basal arterial tone. Arterial and arteriolar smooth muscle will maintain a steady degree of contraction without sympathetic nerve action in a tissue which is resting or at basal metabolism. The concept is not exact because of the difficulties in defining basal metabolism. The basal state for skeletal muscle is at rest where metabolism is low and the metabolic control of arterial resistance is slight; for skin, the basal state is at neutral temperature without sensible sweating. The idea of a basal state is less clear in an organ such as the heart which is always active. Basal arterial tone is the conceptual reference point to which various vasomotor influences are added—either vasoconstriction or vasodilation (Fig. 8–8).

The level of basal tone in part determines the variation in vascular resistance that is possible in an organ. For example, basal tone is high in resting skeletal muscle and the vascular bed is capable of tremendous active vasodilation during exercise owing to a metabolic mechanism. In contrast,

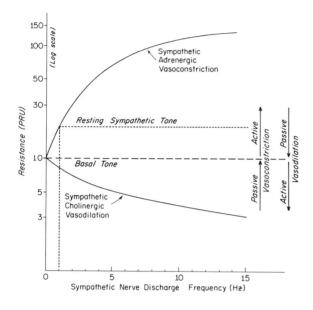

Figure 8–8 Diagram of sympathetic vasoconstriction and sympathetic cholinergic vasodilation as found in skeletal muscle. Basal tone is the vascular resistance in resting skeletal muscle with no neural influence. Active vasoconstriction and dilation are changes in resistance away from basal tone. Passive vasomotion is toward basal tone. Resting sympathetic tone is illustrated at a frequency of 1 Hz.

kidney function is essentially maximum in the basal state so that an increase in blood flow above "basal" is unlikely.

Resting Sympathetic Tone. A regular tonic discharge of sympathetic fibers that maintains a steady vasoconstriction in most vascular beds occurs in a resting supine man. The degree of vascular constriction which is added to basal tone by tonic (steady) sympathetic nerve firing to the vascular smooth muscle when the general level of cardiovascular stress is minimal is the resting sympathetic tone. Resting sympathetic vasoconstriction permits vasodilation by the withdrawal of sympathetic tone and is a significant mode of vasodilation in many vascular beds. The presence or absence of resting sympathetic tone in the vascular bed of an organ and its magnitude are important to understanding the autonomic control of the circulation. The renal vasculature manifests no resting sympathetic tone and is thus incapable of decreasing its vascular resistance by sympathetic withdrawal. In contrast, the abdominal viscera and skeletal muscle vascular beds do manifest resting sympathetic tone.

Active vasoconstriction is a term used to indicate an increase in vascular resistance capable of increasing resistance above basal tone by the addition of some active factor, *e.g.*, sympathetic vasoconstriction. If a vascular bed which is widely dilated by vigorous tissue metabolism returns toward basal tone because of a decrease in metabolism, this would not be active vasoconstriction (Fig. 8–8). *Active vasodilation* is a decrease in vascular resistance capable of decreasing resistance below basal tone owing to an active factor, as might result from an increase in tissue metabolism or stimulation of a vasodilator nerve fiber. Withdrawal of sympathetic vasoconstriction is not active vasodilation (Fig. 8–8).

Passive vasodilation is in general the opposite of active vasoconstriction; that is, a decrease in vascular resistance toward basal tone such as would occur with the withdrawal of sympathetic constrictor tone. Passive vasodilation is also used to express the idea of the increased vessel diameter which results from an increased distending pressure across the vessel wall. Since arteries are distensible tubes, an increase in intra-arterial pressure will increase the internal radius and thus decrease the resistance to flow. *Passive vasoconstriction* is an increase in vascular resistance toward basal tone such as might occur when an active tissue lowers its metabolism. The decrease in arterial radius that results from a fall in transmural arterial distending pressure may also be named passive vasoconstriction.

Differential Control. The responsiveness of the various vascular beds to sympathetic nerve stimulation is different (Fig.

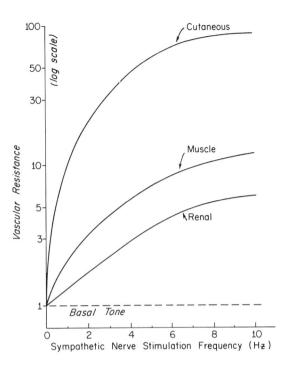

Figure 8–9 The relative magnitude of sympathetic vasoconstriction in cutaneous, muscle and renal vascular beds. The cutaneous blood vessels show the greatest response to sympathetic nerve stimulation. (After Celander, *Acta physiol. scand.*, 1954, *32*, 1–132.)

8–9). In the basal state, skin and muscle vascular beds show a much greater increase in vascular resistance than renal vessels for a given increment in sympathetic nerve discharge rate. This difference may be due to the density of sympathetic innervation, the extent of smooth muscle in the vessel, or the sensitivity of the vascular smooth muscle to the sympathetic transmitter. The sensitivity of a vascular bed to sympathetic control is dependent on the metabolic state of the organ involved. Arterial vessels in skeletal muscle are much more responsive to sympathetic stimulation at rest than during exercise.[16] The mechanism is understood as a competition between metabolic vasodilation with exercise and sympathetic vasoconstriction.

In summary, basal arterial tone is the steady, vascular smooth muscle constriction manifest under basal conditions, independent of neural or hormonal mechanisms. Resting sympathetic tone is the level of steady vascular constriction which results from tonic sympathetic nerve discharge to the vessels under resting conditions. Active vasoconstriction is an increase in vascular resistance capable of increasing resistance above basal tone, whereas passive vasoconstriction is an increase in vascu-

lar resistance toward basal tone from a previously vasodilated state. The level of basal tone and of resting sympathetic tone and the sensitivity of the vascular bed to sympathetic nerve discharge all combine to provide for the differential control of blood flow to the various organs of the body.

REFERENCES

1. Ahlquist, R. P. A study of the adrenotropic receptors. *Amer. J. Physiol.*, 1948, *153*, 586–600.
2. Attinger, E. O., ed. *Pulsatile blood flow.* (Proceedings of the First International Symposium on Pulsatile Blood Flow.) New York, McGraw-Hill Book Co., 1964.
3. Bergel, D. H. The static elastic properties of the arterial wall. *J. Physiol. (Lond.)*, 1961, *156*, 445–457.
4. Bergel, D. H. The dynamic elastic properties of the arterial wall. *J. Physiol. (Lond.)*, 1961, *156*, 458–469.
5. Celander, O. The range of control exercised by 'sympathico-adrenal system.' A quantitative study on blood vessels and other smooth muscle effectors in the cat. *Acta physiol. scand.*, 1954, *32*, 1–132.
6. Dobrin, P. B., and Rovick, A. A. Influence of vascular smooth muscle on contractile mechanics and elasticity of arteries. *Amer. J. Physiol.*, 1969, *217*, 1644–1651.
7. Folkow, B. Description of the myogenic hypothesis. *Circulat. Res.*, 1964, *15*, I-279–I-287.
8. Greenfield, J. C., Jr., and Patel, D. J. Relation

between pressure and diameter in the ascending aorta of man. *Circulat. Res.*, 1962, *10*, 778–781.

9. Johnson, P. C., ed. Autoregulation of blood flow. *Circulat. Res.*, 1964, *15*, 1–291.

10. McDonald, D. A. *Blood flow in arteries.* Baltimore, Williams & Wilkins Company, 1960.

11. Mellander, S., and Johansson, B. Control of resistance, exchange, and capacitance functions in the peripheral circulation. *Pharmacol. Rev.*, 1968, *20*, 117–196.

12. Patel, D. J., Mallos, A. J., and Fry, D. L. Aortic mechanics in the living dog. *J. appl. Physiol.*, 1961, *16*, 293–299.

13. Peterson, L. H., Jensen, R. E., and Parnell, J. Mechanical properties of arteries in vivo. *Circulat. Res.*, 1960, *8*, 622–639.

14. Peterson, L. H., and Shepard, R. B. Some relationships of blood pressure to the cardiovascular system. *Surg. Clin. N. Amer.*, 1955, *35*, 1613–1628.

15. Rodbard, S. Autoregulation in encapsulated, passive, soft-walled vessels. *Amer. Heart J.*, 1963, *65*, 648–655.

16. Rowlands, D. J., and Donald, D. E. Sympathetic vasoconstrictive responses during exercise- or drug-induced vasodilation. *Circulat. Res.*, 1968, *23*, 45–60.

16a. Taylor, M. G. Wave-travel in a non-unit uniform transmission line, in relation to pulses in arteries. *Phys. Med. Biol.*, 1965, *10*, 539–550.

17. Viveros, O. H., Garlick, D. G., and Renkin, E. M. Sympathetic beta adrenergic vasodilatation skeletal muscle of the dog. *Amer. J. Physiol.*, 1968, *215*, 1218–1225.

18. Waud, D. R. Pharmacological receptors. *Pharmacol. Rev.*, 1968, *20*, 49–88.

19. Womersley, J. R. An elastic tube theory of pulse transmission and oscillatory flow in mammalian arteries. *WADC Technical Report TR 56-614.* Aeronautical Research Laboratory, January 1957.

CHAPTER 9   THE CAPILLARIES, VEINS AND LYMPHATICS

by CURT A. WIEDERHIELM

CAPILLARIES

The purpose of circulation—the transport of nutrients and oxygen from the blood into the tissue and removal of CO_2 and waste products—is fulfilled in the terminal arborization of the vascular tree, the capillaries. Other segments of the terminal vascular bed, incorporated within the term "the microvascular bed," also include vessels which regulate blood flow to organs and others which serve as blood volume reservoirs. The subunit of circulation, the microvascular bed, is composed of blood vessels smaller than 100 μ in diameter. A rational subdivision of the blood vessels comprising the microvascular bed was given by Chambers and Zweifach[3] based on histomorphologic criteria (see Fig. 9–1). The *terminal arteries* have a wall composed of an inner layer of endothelium, an internal elastic lamina and a surrounding sheath of at least two continuous layers of vascular smooth muscle cells. The *arterioles*, the next generation of branches, are distinguished by a single continuous layer of vascular smooth muscle. The succeeding generation of branches derived from the arterioles, the *metarterioles*, have a single discontinuous layer of vascular smooth muscle cells. The sympathetic vasoconstrictor innervation in most instances extends as far as arterioles, but occasionally it is found in metarterioles. The capillary networks take origin from the metarterioles, and in many tissues their origin is surrounded by a cushion of vascular smooth muscle cells, the so-called *precapillary sphincters.* Metarterioles and precapillary sphincters are usually not innervated, but appear to be mainly under local control; *i.e.*, waste products released from tissues, lack of oxygen or possibly other factors may lead to dilation of these vessels, resulting in an increase in blood flow and oxygen supply. Characteristically, these vessels exhibit an alternating vasoconstriction and vasodilation, *rhythmic vasomotion*, in contrast with arterioles which exhibit irregular changes in caliber. The time course of vasomotion may depend on local

129

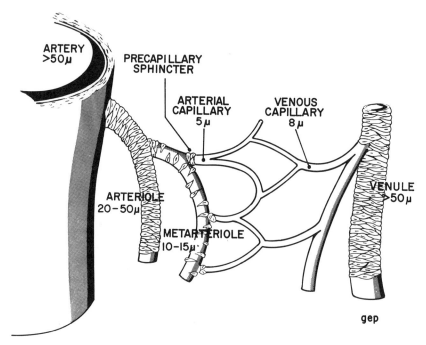

Figure 9–1 Anatomic classification of vascular bed.

stimuli such that accumulation of waste products may induce a prolonged period of flow through the metarterioles and pre-capillary sphincter while the subsequent period of closure is shortened.

The *capillaries* consist of a single layer of endothelial cells and a basement membrane, surrounded by a fine reinforcing network of reticular collagen fibers, and average 4 to 10 μ in diameter. The capillaries form richly anastomotic networks which have extremely large surface areas, and thus are particularly well suited for carrying out the exchange function. The individual capillaries in the network can extend for several hundred microns. Capillaries are divided into arterial and venous capillaries, depending on their proximity to the metarteriole or the draining collecting venules. As discussed below, significant differences may exist in the exchange properties of the arterial and venous capillaries. The venous capillary network forms confluent channels, which eventually become heavily invested by connective tissue, the nonmuscular venules. Farther along, vascular smooth muscle cells appear in the walls of the muscular venules. With increasing diameter, several such vascular smooth muscle layers are found in the *veins*.

Veins and muscular venules generally receive a scant sympathetic vasoconstrictor nerve supply. The major function of veins and venules is to transport the blood from the capillary bed back to the heart. In addition, some vascular beds may also serve as volume reservoirs. In emergencies, such as hemorrhage, sympathetic venoconstriction mobilizes the blood volume stored in the venules and veins.

The microvascular bed can be divided into three segments, each serving a different function: (i) the resistance vessels, composed of arterioles, metarterioles and pre-capillary sphincters, which serve to control the blood flow to the organ; (ii) the exchange vessels, *i.e.*, the capillaries; and (iii) the reservoir vessels which include venules and veins. Among the resistance vessels, changes in resistance to flow by sympathetic vasoconstrictor fiber discharge appear mainly to serve the organism as a whole, whereas the metarterioles and pre-capillary sphincters respond to the immediate needs of the local tissue.

Ultrastructure of Capillary Walls. The capillary walls have many of the properties of a selectively permeable membrane, *i.e.*, they permit rapid transport of certain molecules, whereas others are transported at

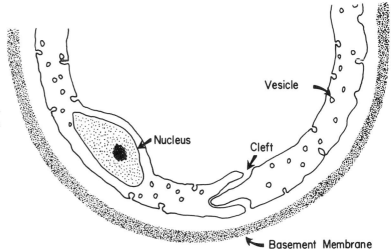

Figure 9–2 Schematic representation of a capillary section from electron micrographs.

considerably slower rates, or even prevented from entering the tissues surrounding the capillary. The basis for this selective property becomes apparent when one considers the ultrastructure of a "typical" capillary wall, as indicated schematically from electron micrographs in Figure 9–2. In their normal distended state, endothelial cells have a thickness ranging from 0.1 to 3 μ (near the nucleus). Between adjacent cells is a cleft, the intercellular space, which is approximately 100 Å wide. Virtually every cleft has a constricted region which is in the order of 40 Å wide.[5] Obviously, molecules with diameters larger than 40 Å cannot pass through these interstices. The intercellular clefts are generally filled with an amorphous material. Within the cytoplasm of the endothelial cells are numerous small vesicles, which continually coalesce with both the inner and outer cell membranes. Transport across the capillary wall may take place through Brownian motion of these pinocytotic vesicles. The diameter of these vesicles is in the order of 100 Å, and thus could make possible transcapillary transport of molecules that cannot pass through the intercellular spaces. The whole capillary is surrounded by a basement membrane composed of fine reticular collagen fibers and high molecular weight substances, called mucopolysaccharides.

However, capillary endothelium in various organs shows special adaptation to the function of that organ. For instance in glands, where relatively large amounts of water are filtered, the capillary endothelium is modified. It shows a large number of fenestrations, i.e., areas where the inner and outer capillary cell membrane appear to fuse, leaving virtually no cytoplasm to interfere with the transport across the endothelial cell (Fig. 9–3A). In the glomerulus of the kidney, a type of fenestrated endothelial cell is found, in which the fenestrae are not closed by cell membranes. The diameter of the fenestrae averages 100 Å, and they thus offer no resistance to transport of most plasma proteins. The glomerular capillary basement membrane is, however, unusually thick, and serves as an effective barrier to all plasma proteins. On the other hand, in other organs such as the liver and bone marrow the intercellular clefts between the endothelial cells are not uniform but show large defects (Fig. 9–3B). The basement membrane is discontinuous, permitting transport of the largest molecules present in blood as well as cellular elements. Since the endothelium offers little or no resistance to movement of plasma protein molecules, the interstitium surrounding the so-called liver sinusoids has a protein concentration almost identical to that of plasma, and lymph derived from the liver usually has a protein content of 5 to 6 per cent. The capillaries within the central nervous system show still another kind of specialized structure, in which the 40 Å spaces are almost nonexistent, that is, the membranes of adjacent endothelial cells appear to fuse. This is reflected in

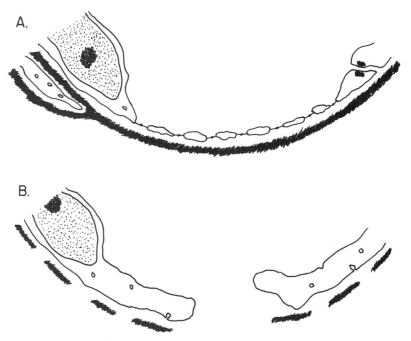

Figure 9–3 *A*, Typical arrangement of endothelium cell junctions in glands. *B*, Defects between endothelium cells as found in liver and bone marrow capillaries.

absence of transport or larger molecules such as plasma proteins. Indeed, the spaces between the cells are so small that even certain relatively low molecular weight antibiotics cannot penetrate the capillary endothelium. This constitutes the "blood-brain barrier."

Capillary Permeability. The permeability of a membrane is conveniently expressed as the mass flow rate per unit time per unit of driving force, and is usually expressed for water in terms of cm³ per min per mm Hg. The permeability of capillary membranes to water is several hundred times larger than that of a cell membrane, an expected finding because the capillary cell membranes are composed of a hydrophobic layer of lipoproteins. Therefore only a small fraction of the total exchange of water between blood and tissues occurs through the endothelial cells, whereas most occurs through the intercellular clefts. In contrast, other materials which are lipid soluble, *e.g.*, O_2, CO_2 and many anesthetic agents, penetrate the capillary cell membranes with little hindrance. For these substances almost the total cellular area of the capillary endothelium is available for rapid exchange between the blood and tissues. This exchange is facilitated because the velocity of the blood flow through the capillary is considerably slower than any other portion of the vascular system (1 to 10 mm per sec). This allows a longer period of time for the erythrocytes to be in contact with the capillary endothelium, thus prolonging the time available for transport of gases. Since transport rates by diffusion over small distances (0.1 to 0.3 μ) are extremely rapid, the same molecule may traverse the capillary walls several times during a single capillary passage.

By contrast, the transcapillary exchange of water and water-soluble materials, *e.g.*, the ionic components of the blood plasma, Na, Cl, K, Ca and others, occurs at slower rates. However, the surface area of the intercellular clefts is sufficiently large so that diffusion exchange still occurs at rapid rates. It should be emphasized that the major route of transport between the blood and the tissues for these substances is through the intercellular clefts, and only relatively small amounts traverse the endothelial cell membranes either by diffusion or pinocytosis. Larger molecular species such as plasma proteins, *i.e.*, albumin, globulins and fibrinogen, penetrate at rates considerably slower than that of water. Some controversy centers on the transport mechanisms of the macromolecules between the blood and the tissues. One such mechanism has already been discussed, transport by pinocytotic vesicles. An alternative is that larger defects may exist in certain areas of capillary endothelium, the so-called large pore system; these pores may reach diameters ranging up to 350 Å. These defects are found mainly in the venous capillaries. Current experimental data, however, tend to favor the hypothesis that pinocytotic transport is the dominant mechanism for macromolecular transport. The movement of plasma proteins, particularly the alpha-globulins, into the tissues is more than accidental, because these substances through immune responses are important to the body's defense against infection. If the globulins were restricted to the circulation, this defense would by necessity exclude tissues in which the defense reaction is most urgently needed.

Cellular elements such as leukocytes and lymphocytes penetrate the capillary wall into the tissues. Earlier, these cells were believed to penetrate the capillary wall through the large pore system mentioned above. Recently, these cells have been shown to migrate through the intercellular clefts, or even through the endothelial cytoplasm by ameboid motion, actually dissecting their way through the endothelial cell and the basement membrane. This will be described in more detail below in a section on pathophysiology of the capillary bed.

Capillary Water Balance and the Starling-Landis Hypothesis. The total body fluid volume is maintained within relatively narrow limits by the kidneys and their associated hormonal control systems. The total body fluid volume may be divided into three compartments: (i) the plasma volume, *i.e.*, the volume of blood plasma contained within the circulatory system; (ii) the intracellular fluid volume contained within the cells, including the erythrocytes of the blood; and (iii) the interstitial space volume, *i.e.*, fluid contained in the space between the cells in the tissues. The plasma volume and the interstitial volume are also maintained within relatively narrow limits, and the partition of fluid between the circula-

tion and the interstitial space appears to be relatively constant. The major mechanism whereby this precise partition of fluid between the circulation and the interstitial space is maintained resides at the capillary level. Four major factors are involved in this partition mechanism: (i) the plasma oncotic pressure, *i.e.*, the osmotic pressure contributed by the plasma proteins, (ii) the hydrostatic pressures in various segments of the capillary bed, (iii) the oncotic (colloidal) pressure of the tissue fluid within the interstitial space and (iv) the hydrostatic pressure within the interstitial space.

OSMOTIC PRESSURE. The term "osmotic pressure" may best be defined in operational terms. If a chamber is divided into two compartments by means of a semipermeable membrane, *i.e.*, a membrane permeable to water but not to solutes, and the chamber is filled with water, the hydrostatic pressure on both sides of the membrane will be identical. When one side of the chamber is filled with water and the other side with a 0.9 per cent saline solution (osmotically equivalent to the concentration of electrolytes and other solutes in the blood), water will move across the membrane from the water compartment into the saline compartment because of the lower "physicochemical potential" of the water containing the solute (Fig. 9–4A). Were this system closed, a hydrostatic pressure would be generated within the saline compartment, and at a certain equilibrium value fluid transport would be prevented by the hydrostatic pressure (about 16 atmospheres). This pressure, which is just sufficient to prevent water movement across the semipermeable membrane, is defined as the osmotic pressure. However, this is not what is found in the capillary walls. The capillaries permit relatively free exchange of water and small molecules, whereas plasma proteins are prevented from entering the interstitial space because of the 40 Å junctions in the intercellular clefts. A similar situation can be simulated by using a selectively permeable membrane in the aforementioned chamber which has a pore size in the order of 40 Å. In these circumstances, electrolytes and other smaller solutes would be free to disperse on both sides of the membrane, whereas albumin would be selectively retained on one side (Fig. 9–4B). If

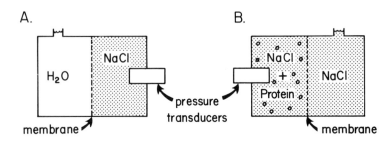

Figure 9–4 *A*, Osmotic pressure determination using a membrane impermeable to sodium chloride but permeable to water. *B*, Osmotic pressure experiment using membrane permeable to both water and sodium chloride but not to protein molecules.

one compartment contains 0.9 per cent saline and the other compartment 0.9 per cent saline to which 7 per cent plasma protein has been added, flux of water through the membrane into the protein-containing compartment will again occur. Owing to the high molecular weight of the plasma proteins the hydrostatic pressure at equilibrium would be considerably smaller than that in the case of the ideal semipermeable membrane referred to earlier (25 mm Hg). The osmotic pressure will now be due only to the plasma proteins, *i.e.*, the colloid osmotic pressure of the solution, also called the oncotic pressure. The relationship between the osmotic pressure and solute concentration is given by van't Hoff's law:

$$\pi = CRT \qquad (1)$$

where π = osmotic pressure expressed in atmospheres, C = the concentration of the solutes expressed in moles per liter, R = universal gas constant and T = absolute temperature. For small molecules in dilute solutions the osmotic pressures are linearly related to the concentration. For larger molecules, however, the relationship becomes markedly nonlinear, and more complex mathematical descriptions are necessary.

ONCOTIC PRESSURES OF PLASMA PROTEINS. Human blood plasma protein contains 51 per cent albumin, 17 per cent globulin and 4 per cent fibrinogen, the remaining 28 per cent made up of a number of other proteins. Albumin, because of its high concentration and its relatively low molecular weight, is the major determinant of the plasma oncotic pressure, which depends on the number of particles. The total concentration of proteins in human and most mammal plasma is approximately 7 per cent; protein concentrations in cold-blooded animals and birds are generally much lower

than in mammals, yielding oncotic pressures from 5 to 10 mm Hg. The first studies of human oncotic pressures were published by Starling in 1896.[13] With relatively crude instrumentation he obtained values which were higher than modern estimates, although later measurements with improved osmometers yielded values in the same general range as the capillary hydrostatic pressure (1896). He postulated that the oncotic pressure of the plasma proteins counteracted the hydrostatic capillary pressures in such a manner that a continuous filtration of water into the tissue did not occur.

CAPILLARY HYDROSTATIC PRESSURES. In 1930 Landis, penetrating capillaries in the fingernail bed of man, for the first time achieved direct measurements of capillary hydrostatic pressures.[7] Similar observations were also carried out in other species and in other vascular beds. Pressure measurements were obtained by inserting micropipets into the capillaries with a micromanipulator. An ingenious nulling system, using the influx and efflux of dye from the tip as pressure in the pipet was varied, measured the capillary pressure. He obtained an average value for arterial capillary pressure of 32 mm Hg and in venous capillaries 15 mm Hg. Near the mid-point, values of about 25 mm Hg were found. Essentially similar values have been found in recent studies of the skin in the wing web of intact unanesthetized bats (Fig. 9–5). The agreement is surprising because vigorous vasomotion of precapillary sphincters within the bat wing causes the arterial capillary pressure to fall to venous level during the closed phase, whereas the pressures approximate those in the feeding metarteriole (55 mm Hg) during the open phase. In both instances measurements were obtained under resting, "normal" conditions. A considerably larger variability in capil-

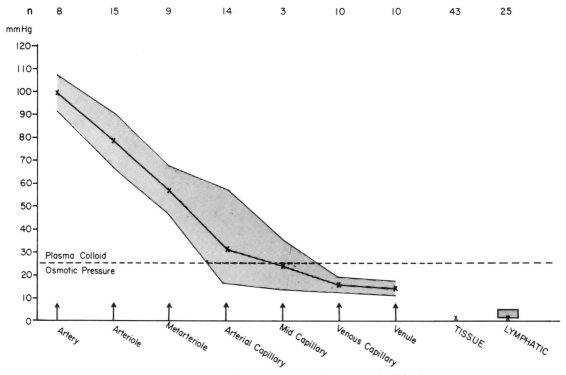

Figure 9-5 Microvascular pressures in the wing web of the bat.

lary pressures may be observed under such conditions as sympathetic vasoconstriction, vasodilation or postural changes.

INTERSTITIAL FLUID ONCOTIC PRESSURES. The composition of tissue fluid is unknown, because of the difficulty in withdrawing samples under normal conditions. This is probably due to the fact that interstitial fluid is not a free fluid pool but is immobilized by relatively dense fibrous structures interspersed with a gel-like amorphous structure, the ground substance. In edema, where the interstitial fluid has been expanded by excessive filtration, tissue fluid protein concentrations ranging from 0.1 to 0.4 per cent have been found. Similarly by elevating the venous pressure and thus producing excessive filtration, the lymphatic protein concentration (which presumably reflects tissue fluid composition) has an average value ranging from 0.2 to 0.4 per cent.[8] This, however, is considerably lower than the values found for normal lymph, which generally range from 3 to 4 per cent.[12] The question about the tissue fluid protein concentration thus is far from resolved, but in most instances the

oncotic pressure of the tissue fluid is assumed to be negligible.

TISSUE HYDROSTATIC PRESSURES. Earlier attempts at measuring tissue hydrostatic pressure (by inserting a hypodermic needle subcutaneously and injecting a small volume of fluid) were frustrated because the distension of the tissues by the injected fluid volume introduced artifacts; the measurements obtained in this fashion were generally too high. Perhaps the most careful study of tissue hydrostatic pressures was that carried out by McMaster, who inserted a 30-gauge (300 μ O.D.) needle under microscopic observation into the skin.[10] Particular care was exercised to avoid injury to capillaries and lymphatics. Saline was infused at very slow rates through the needle, and the pressure was estimated by means of a conventional pressure transducer connected to the needle. The values found averaged about 2.5 mm Hg in skin and subcutaneous tissues and 4.5 mm Hg in muscle. Even these very carefully conducted studies were criticized on the grounds that a 30-gauge needle is very large compared to the blood vessels in the skin, and would cause tissue injury. McMaster's measurements have, however, recently been confirmed in a series of measurements of tissue pressures in the bat wing, in which glass micropipets with tip diameters less than 1 μ were used for pressure measurements; significant tissue injury was thus avoided.[15] This study yielded a value for tissue pressure of 1.8 mm Hg, which is close to McMaster's average.

THE STARLING-LANDIS HYPOTHESIS. While Starling was the first investigator to point out the importance of the oncotic pressure of blood plasma in preventing excessive translocation of fluid from the circulation into the tissues, his work was subsequently supplemented by the study of Landis.[7] When Starling conducted his classic studies no direct measurements of capillary pressures were available and only unreliable indirect techniques were used. Landis' measurements indicated that the pressures in the arterial and venous capillaries bracketed the plasma colloid osmotic pressure (Fig. 9–6). From these values filtration (movement of fluid from blood into the tissue) or reabsorption (back flow of fluid from the tissues into the capillary lumen) can be evaluated. A simple mathematical expression may be written which describes the fluid movement through the capillary wall:

$$F = k(P_t - \Pi_c - P_t + \Pi_t) \qquad (2)$$

where F = net movement of fluid across the capillary endothelium, k = a constant, including the surface area and permeability of the capillary membrane, P_t = hydrostatic capillary pressure, Π_c = oncotic pressure of blood plasma, P_t = hydrostatic tissue pressure, Π_t = tissue oncotic pressure. The hydrostatic tissue pressure and tissue oncotic pressure are assumed to be approximately equal; they will therefore cancel out, and equation (2) simplifies to F = k($P_c - \Pi_c$). The amount of fluid to be reabsorbed would thus mainly be determined by the hydrostatic capillary pressure and the plasma oncotic pressure. The pressure gradient along the capillaries is essentially linear between 32 and 15 mm Hg. At the arterial capillary the hydrostatic pressure exceeds the osmotic pressure by 7 mm Hg, and filtration occurs. Conversely, at the venous end of the capillary, the capillary hydrostatic pressure is approximately 10 mm Hg lower than the plasma oncotic pressure and reabsorption occurs (Fig. 9–6). In the intermediate regions of the capillary, a gradual decline in filtration and reabsorption takes place. Note that the sections under the two triangles between the hydrostatic and oncotic pressures have approximately equal areas, indicating that the filtration in the arterial capillary and reabsorption in the venous capillary will be approximately equal. This is essentially the basic statement of the Starling-Landis hypothesis of capillary fluid balance. The amount of water filtered in the whole body (excluding the kidneys) has been estimated to be 20 liters per day. Of these 20, 16 to 18 liters is reabsorbed. The balance of 2 to 4 liters per day is removed from the tissues by the lymphatics. These values are small compared with the cardiac output, approximately 8000 to 9000 liters per day.[9] Thus the effect of filtration on the concentration of plasma proteins within the capillary is negligible, because only a very small fraction of the total blood flow through organs is filtered into the tissues (excepting the kidney).

LYMPHATIC CAPILLARIES. Since lymphatic capillaries are intimately involved in the fluid exchange between blood and tissues, as well as in protein transport, they will be discussed in this section rather than in a later one on lymphatics.

The lymphatic capillaries show distinct morphologic differences from the capillaries in the cardiovascular system. They usually terminate in blind, flattened sacs within the connective tissue and may be as wide as 30 μ, whereas their thickness in the normal state may be only 2 to 10 μ. The endothelium of the lymphatic capillary and

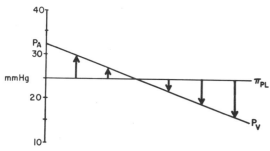

Figure 9–6 Filtration and reabsorption along a capillary. π_{PL} = Osmotic pressure of blood plasma; P_A = hydrostatic pressure in arterial capillary; P_V = hydrostatic pressure in venous capillary.

the blood capillaries show some similarities, although lymphatic endothelial cells are usually larger and thinner. Their cytoplasm also contains pinocytotic vesicles. Distinct differences are also apparent, a major one being the large defects between lymphatic endothelial cells which may reach a size of 5 μ. The basement membrane is discontinuous and in many places totally absent.[6] The lymphatic capillaries are thus in intimate contact with the interstitial space, permitting free entry of fluid, small and large molecules as well as cellular elements. In infections, bacteria and leukocytes with phagocytized bacteria are generally transported into the lymphatics and removed from the tissues. In certain species, contractile elements may be found scattered intermittently in the wall of the lymphatic capillaries, giving rise to spontaneous slow rhythmic contractions, which propel the lymph toward larger lymphatic trunks. Hydrostatic pressures in the lymphatic capillaries have been measured recently, and are not significantly different from those in the tissue (1.8 mm Hg). These findings support the earlier morphologic description, i.e., that the interior of lymphatic capillaries is in relatively free communication with the interstitial space. Lymphatic capillaries are widely distributed in the connective tissue, in many instances forming anastomotic connections similar to capillary networks. The lymphatic capillaries thus have a close contact with a major part of the interstitial space, facilitating the transport of materials from the interstitium into the lymphatic system. Because plasma proteins are continuously entering the interstitial space from the cir-culation and because it has recently been demonstrated that labeled protein injected into the tissue is not returned to the blood, in time an accumulation of protein would occur within the tissues. The removal of protein from the interstitial space is one of the major roles of the lymphatics.

CAPILLARY ENVIRONMENT. The capillary exists in a complex environment, the extravascular connective tissue. Viewed with the light microscope, stained connective tissue appears to be composed of fibers which course through a ground substance (Fig. 9-7). Fibers are classified on the basis of morphologic and chemical criteria as collagen, reticular and elastic fibers. Collagen and reticular fibers are chemically related proteins, whereas elastin has a different chemical structure. The ground substance is amorphous, is optically homogeneous and stains pink or red with appropriate staining agents. The capillary basement membrane stains as a dense main line and may represent a condensation of components of the ground substance. The ground substance comprises a heterogeneous group of materials; the most important ones are produced by local synthesis in fibroblast cells, whereas others are products of transcapillary transport from the blood. Thus about 50 per cent of the whole body protein pool is found in the interstitial space. Many components belong to the mucopolysaccharide groups and have been isolated in the past 30 years. Among these are hyaluronic acid, chondroitin sulfate, keratosulfate and heparitin sulfate. Of particular interest is hyaluronic acid because of its high molecular weight (M.W. 1 million to 10 million) and its molecular conforma-

Figure 9-7 Diagram of relations of connective tissue ground substance (light stipple) to cells, fibers and a blood vessel. The heavier stipple represents basement membrane of epithelium, muscle and endothelium of a small blood vessel capillary. (After Gersh and Catchpole, *Perspect. Biol. Med.*, 1960, 3, 282–319.)

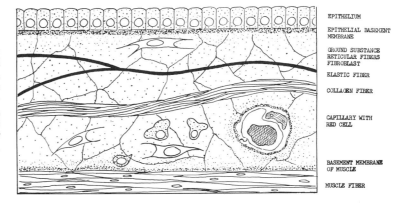

EPITHELIUM

EPITHELIAL BASEMENT MEMBRANE

GROUND SUBSTANCE
RETICULAR FIBERS
FIBROBLAST

ELASTIC FIBER

COLLAGEN FIBER

CAPILLARY WITH RED CELL

BASEMENT MEMBRANE OF MUSCLE

MUSCLE FIBER

tion. Owing to its very large dimensions it forms an entangled meshwork within the ground substance which exhibits certain gel-like properties such as swelling and shrinkage. Hyaluronic acid is perhaps mainly responsible for the relatively slow transport rates of intermediate size molecules such as sucrose within the interstitial space. It also exhibits significant water-binding and electrolyte-binding properties, which has led to the postulate that the interstitial space basically should be considered as a two-phase system. By special preparation for electron micrography, the seemingly amorphous ground substance of muscle has been resolved to a structure consisting of fluid with vacuoles 600 to 1200 Å in diameter (the free fluid phase) surrounded by denser walls (the gel-like colloid phase). These two phases are thought to exist in a dynamic equilibrium where changes in composition, such as protein concentration of the free fluid phase, would produce compensatory swelling or shrinking of the gel-like colloid phase.[2] Changes in the relative proportion of the two phases may be initiated by many different stimuli and are thought to be the basis of reversible changes in the physical characteristics of the ground substance.

Pathophysiology of the Capillary Bed. EDEMA FORMATION. The term edema refers to a pathologic state in which an excessive, clinically observable accumulation of fluid occurs in the interstitial space. Many disturbances may induce edema, the most important of which are (i) reduction of plasma oncotic pressure; (ii) elevation of venous pressure; (iii) a general increase in capillary permeability of fluid and proteins; (iv) interference with removal of protein from the interstitial space by blockage of the lymphatics; and (v) an increased content of mucopolysaccharides, particularly hyaluronic acid, in the interstitial space. The basic mechanisms for edema formation in the first two cases are shown in Figure 9–8. Panel A shows the normal condition, with arterial and venous mean capillary hydrostatic pressures of 32 and 15 mm Hg respectively and a plasma oncotic pressure of 25 mm Hg. As previously stated, filtration and reabsorption within the whole capillary are then approximately equal, and no significant change in the interstitial fluid volume would be expected. Obviously, this

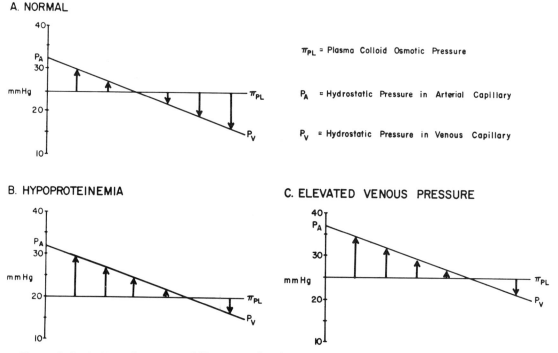

Figure 9–8 *A*, Normal patterns of filtration and reabsorption. *B*, altered pattern following decreased plasma oncotic pressure. *C*, Similar response to elevated pressure.

equilibrium of filtration and reabsorption is unstable, and minor changes in hydrostatic or plasma oncotic pressures, 5 to 10 mm Hg, might give rise to an imbalance in this process.

Panel B shows what would happen if the plasma oncotic pressure were lowered from 25 to 20 mm Hg. Under these conditions, the filtration would extend throughout a larger part of the capillary and be substantially increased, whereas the reabsorption process would be proportionately decreased. This will lead to an excessive filtration and consequent fluid accumulation within the interstitial space. This condition, termed hypoproteinemia, occurs in disease states such as the nephrotic syndrome, which involves the kidneys and leads to excessive loss of protein in the urine. Similarly hypoproteinemia may be induced accidentally if the thoracic duct (the major lymphatic channel for the circulation) is severed during intrathoracic surgery.

Panel C shows the result of an increase in the venous capillary pressure by 5 mm Hg. In maintaining the flow through the capillary, this increase would be reflected back to the arterial capillaries and produce a similar increase. Essentially the same conditions described previously would prevail, and excessive filtration would result. This condition occurs typically in cardiac edema of heart failure, in which venous pressure is typically elevated 5 to 10 mm Hg or more.

The most common cause of the third mechanism of edema formation is the inflammatory state, which will be discussed more fully in the section on inflammatory response.

The fourth cause of edema is due to the lymphatics failing to remove protein rapidly enough from the interstitial space and is termed lymphedema. This form of edema is relatively rare, perhaps because large lymphatic trunks have numerous collateral connections, and the trunks tend to regenerate rapidly after blockade. After infection by a tropical, microscopic parasite (filariasis), the lymphatics of the legs or scrotum are frequently completely occluded and gross lymphedema results. This is perhaps the most striking form of edema, because the amount of protein accumulating in the tissues over prolonged periods of time is

unlimited. In lymphedema the volume of the involved leg can increase to three times its normal volume, a condition termed elephantiasis owing to its resemblance to an elephant's leg.

The fifth mechanism, excessive accumulation of mucopolysaccharides in the interstitial space, characteristically involves a diffuse edema of the skin and is caused by the affinity of water to mucopolysaccharides. This condition occurs mainly in hormonal disturbances in which insufficient amounts of thyroid hormone are secreted and is termed myxedema.

INFLAMMATORY RESPONSE. Inflammatory responses occur in virtually every organ of the body in response to tissue injury. Inflammation is most prominent in the skin, and the following discussion will be centered on the skin. Inflammation occurs in response to virtually every form of tissue injury, including trauma, thermal injury and infection. In the case of infection, the tissue injury reaction is usually caused by release of bacterial toxins. In the skin the inflammatory response consists of a generalized vasodilation of the resistance vessels, reflected in a reddening of the skin, and sometimes extending beyond the margin of injury. Later, swelling of the injured area occurs owing to increased capillary permeability.

The inflammatory response appears to be mainly induced by activity of a specialized cell, found in the connective tissue of the skin and other organs, the mast cell. The mast cell is a heavily granulated cell which may serve other functions besides being the primary factor in inflammation. In tissue injury, these cells tend to release their granules, which in part are composed of histamine and serotonin,[1] both powerful vasodilators which act on the terminal arteries, arterioles, metarterioles and precapillary sphincters. They thus give rise to a large increase in capillary blood flow which accounts for the reddening of the skin. Histamine also has a direct effect on the capillary endothelium, which becomes considerably more permeable to water and small molecule solutes as well as to protein. This movement of fluid and protein into the tissues accounts for the swelling associated with inflammation. When the inflamed area is observed under the microscope, characteristic changes are seen in the behavior of leukocytes. Normally the leukocytes tend to adhere to the endothelial cells lining the capillary. In the inflammatory state they become firmly attached to the endothelium, particularly in venules. New leukocytes continuously enter the inflamed area from the blood moving rapidly through the capillary and tend to accumulate in inflamed areas. After adhering to the endothelium, the leukocytes, by an ameboid motion, dissect their way through the interstices between endothelial cells, which are less tightly adherent because of the histamine-induced increase in capillary permeability. Exceedingly large numbers of leukocytes may be found in inflamed tissues. Edema formation makes the ground substance less dense and thus permits relatively free movement of the leukocytes.

At the onset of normal tissue repair, the leukocytes and excessive interstitial plasma proteins are taken up by the lymphatic capillaries and returned to the circulation. The capillary permeability returns to normal as does the capillary blood flow. In severe tissue injury, however, blood clots may form in the small ramifications of the arterial tree, and if widely distributed, blood flow to the tissue will be interrupted, leading to death of the tissues and subsequent scar formation.

A Critical View on Capillary Fluid Balance. Experimental data indicate that edema generally does not develop until venous pressures are elevated or plasma oncotic pressure is lowered by at least 10 to 15 mm Hg. Thus a protective mechanism appears to limit transcapillary shifts of fluid for a range of 10 mm Hg around the normal capillary hydrostatic and oncotic pressures. The Starling-Landis hypothesis fails to explain this feature of fluid balance at the capillary level.

The traditional model of fluid balance implies three unstated assumptions: (i) the permeability to water and solutes is uniform along the capillary; (ii) the surface areas of the arterial and venous ends of the capillary are identical; (iii) hydrostatic and colloid osmotic pressures of tissue fluid are negligible compared to corresponding blood values. In this section the validity of these assumptions will be examined in view of recent experimental data.

The first assumption was challenged as early as 1930, when it was demonstrated that water-soluble dyes escape more rapidly from venous capillaries and venules than from arterial capillaries in mammalian skin and muscle.[11] Also, in recent studies an increased permeability of water has been shown to exist in venous capillaries.[4] The second assumption in the traditional model, namely that the surface areas available for filtration and reabsorption are similar, may also be challenged. By careful measurements of dimensions and numbers of arterial and venous capillaries in the bat wing, the venous capillary surface area has been shown to be as much as six times larger than that of the arterial capillaries.[14] Similar measurements in the skin of man indicate that the surface area of the dermal capillaries can be approximately four times larger than that of the arterial capillary. If both permeability and the surface area of the venous capillaries are higher than those of the arterial capillaries, reabsorption would be favored, leading to a relative dehydration of the tissue.

The third assumption, that tissue fluid oncotic and hydrostatic pressures are identical, is also open to challenge. Other osmotically active molecules, e.g., hyaluronate, are present in the interstitial space, and by themselves could yield an oncotic pressure of 4 mm Hg. In addition, if the tissue fluid concentration is identical to that of the average capillary filtration (0.2 per cent), some active mechanism must be invoked to explain elevation of protein concentration from that in the interstitial space to reach the relatively high values in the lymphatic capillaries (3.5 per cent). Because lymphatic capillaries are in free communication with the interstitial space, the site of such a mechanism is difficult to pinpoint.

If all these various factors are included in a computer analysis, a solution can be obtained for the tissue oncotic pressure required for equilibrium of filtration and reabsorption under these more complex conditions.[16] This analysis indicates that a tissue oncotic pressure in the order of 10 mm Hg, equivalent to a protein concentration of 3.5 per cent, would be required to keep filtration and reabsorption in equilibrium. Additional experimental data, obtained by osmometry in the subcutaneous tissue of the rabbit, yield comparable values. Thus, although the basic statements of the Starling-Landis hypothesis are true, namely, that the hydrostatic and oncotic pressure of blood plasma and interstitial fluid are in equilibrium, the actual mechanism is considerably more complex than originally postulated. If the effective protein concentration within the interstitium is approximately 3.5 per cent, it would form an important first line of defense against edema formation as originally pointed out by Starling.[13] If capillary hydrostatic pressures are elevated, a slight increase in filtration of fluid would occur, diluting the tissue fluid proteins and leading to a decrease in tissue fluid oncotic pressure. This dilution would proceed until a new point of equilibrium was established, in which the reduction of interstitial oncotic pressure would compensate for the capillary pressure increase with a relatively small change in tissue volume. The computer study indicates that this compensatory range extends over approximately 10 mm Hg, similar to that described at the beginning of this chapter. (For details of the material discussed in this section, see ref. 16.)

VEINS

Venous Blood Flow and Pressure. As the blood leaves the capillaries it enters the veins, a system of converging vessels serving both as a conduit and as a low pressure reservoir of large and variable capacity. From this reservoir the blood is fed to the right heart and then to the lungs.

The walls of the veins make them well suited for this storage function at low pressures. Their thin, elastic walls are sparsely covered with vascular smooth muscle. The thin, flaccid walls of empty veins are flattened and assume an elliptic cross-section so that they yield to very small increments of internal pressure. A rise of 1 mm Hg may increase the capacity of a vein threefold; at this pressure the cross-section of the veins becomes circular. A rise of 10 mm Hg increases the capacity about sixfold by passive distension of the connective tissue elements and the vascular smooth muscle within the wall. Beyond this level the venous wall becomes progressively stiffer, and the increment of capacity increase for each

unit rise in pressure falls rapidly. The capacity of the vein may be altered by the muscular elements in the walls; these elements respond to nervous, hormonal and chemical stimuli. In general, veins are more sparsely innervated than arteries; considerable variability, however, is found in different vascular beds; thus skeletal muscle veins appear to lack innervation. By activating sympathetic fiber endings in innervated venous beds, blood stored within them may be mobilized.

The major veins generally run parallel with the major arteries and their branches, but show a rich anastomotic connection with other veins. Thus the subcutaneous veins in the forearm form an anastomotic network, which also communicates freely with deeper veins in the underlying skeletal muscle.

In the recumbent man, the major driving force for propulsion of the blood from the capillary bed back to the heart is the arterial pressure, transmitted through the capillary network. Since the veins generally have a considerably larger cross-sectional area than the corresponding arteries, the pressure drop across the venous bed is relatively low. Thus the venous pressure in the recumbent individual averages 13 mm Hg at the dorsum of the foot, whereas the right atrial pressure of the heart averages 3 mm Hg. Changes in blood flow and environmental factors may, however, drastically alter the venous pressure (see next section). The veins, particularly in the extremities, have bicuspid valves which first appear in veins approximately 1 mm in diameter. These valves restrict blood flow only in the direction of the heart. Some veins, such as the intrathoracic, intra-abdominal and cerebral veins, usually do not have valves. Because of their large diameters, the veins and venules contain the largest fraction of the blood volume, approximately 75 per cent, as compared with approximately 20 per cent in the arterial tree and 5 per cent in the capillaries.

In the intrathoracic section of the venae cavae, the transmural pressure is higher because the intrathoracic pressure is slightly subatmospheric, ranging from −3 to −7 mm Hg, with the most negative values obtained during inspiration. The elevated transmural pressure at the peak of inspiration tends to distend the veins, thus facilitating the influx of venous blood into the thorax and the right atrium; expiration tends to decrease the inflow. The pressure on the intrathoracic and intra-abdominal veins can be increased considerably if the respiratory and abdominal muscles contract to produce a forced expiration against a closed glottis (40 to 50 mm Hg or higher), as during the expiratory effort when lifting a heavy weight or straining at stool. Under these conditions the intrathoracic pressure rises sufficiently to collapse the venae cavae, temporarily increasing the inflow into the right atrium because reflux into the extremities is prevented by venous valves. Later, inflow of blood into the atrium is prevented by the venous collapse. If maintained for prolonged periods of time, a reduction in left ventricular ejection will occur with a consequent fall in arterial blood pressure, which, if sufficiently pronounced, can induce loss of consciousness.

Effects of Postural Changes on the Venous Pressure and Flow. Thus far only the principal factors influencing the venous pressures and flow in recumbent, relaxed individuals have been discussed. When considering the effects of a change in posture, it is useful to consider a system of fluid-filled tubes with rigid walls, in which the pressures throughout the system are the same when the system is horizontal. When this system of tubes is placed vertically it remains filled with fluid throughout;

the pressure in the lower parts is elevated owing to the long hydrostatic columns in the systems. However, when an elastic distensible system of tubes is shifted from the horizontal to the vertical position, the fluid accumulates in the lower parts because of hydrostatic pressure. The vascular bed is such a system of elastic tubes; when a person moves from a recumbent to a vertical position, blood tends to accumulate in the vessels of the dependent portions, *i.e.*, the veins and venules of the foot and leg. As the veins in the lower extremities are filled by inflow from the capillary bed, a continuous hydrostatic column develops from the heart level to the foot. In these circumstances the venous valves remain open, because flow is essentially unidirectional. Thus the pressure in the veins of the foot will be equal to the weight of the hydrostatic column of blood between the foot and the level on the atrium. The venous pressures at various levels in the leg veins and abdominal aorta are illustrated in Figure 9–8A. In the arterial system also, the weight of the hydrostatic column in the arterial tree must be added to the mean aortic pressure at heart level (Fig. 9–9A). The venous pressure in the foot would thus be approximately 80 mm Hg, whereas the arterial pressure is 180 mm Hg. The net perfusion pressure across the capillary bed remains unchanged, that is, about 100 mm Hg. In these conditions excessive capil-

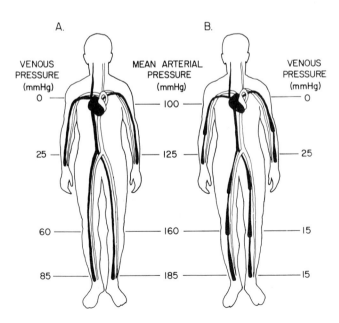

Figure 9–9 *A*, Formation of continuous venous column of blood in erect man (heavy lines). *B*, Interruption of hydrostatic column by muscular exercise (bulges).

lary filtration and edema could occur in the foot, but compensatory mechanisms prevent this. (A substantial fraction of the blood volume [500 cc] may, however, be stored in the legs as a result of distension of the dependent veins.) Occasionally, large amounts of blood may be sequestered in the legs and impair the return of the blood to the right heart. This is ultimately reflected in a decrease in left cardiac output and falling arterial blood pressure which may be sufficient to induce unconsciousness. This situation is encountered on parade grounds where soldiers are required to stand at attention for long periods of time.

The blood trapped in the veins of the lower extremities may be mobilized by skeletal muscular activity, either by exercise or by isometric contraction of the leg muscles. Measurements of the tissue pressure in contracting skeletal muscle have yielded values as high as 85 mm Hg. Since this pressure is generally higher than the distending pressure of the veins, the contracting skeletal muscle would compress the vein and eject its content toward the heart. After only seven steps the venous pressure at the ankle level has returned to the normal level of 10 to 15 mm Hg, because the continuous hydrostatic column extending from the heart to the foot has now been broken up, and the venous valves prevent the reflux of blood into the extremities (Fig. 9–9B). This ability of contracting skeletal muscle to propel the blood toward the heart has been termed the "muscle pump." Although the muscle pump acts only on the deep vessels within the skeletal muscle, the distended cutaneous veins will drain into the deeper vessels through the anastomotic connections in which reflux is prevented by valves. It is noteworthy that the arterial pressure at the ankle level still remains at 190 mm Hg and that the effective perfusion pressure is increased to approximately 180 mm Hg. This elevation of perfusion pressure may serve to increase blood flow required by exercising skeletal muscle.

Veins located above the heart level behave oppositely from those below heart level. Because the right atrial pressure is about 3 mm Hg, it can support only a hydrostatic column of 4 cm H_2O; if the veins were rigid, pressure in the neck veins (the jugular veins) would be negative. Since they are flaccid, atmospheric pressure would collapse the extrathoracic veins above the heart level up to their entrance in the cranial cavity. Within the cranial cavity the situation is more complex, because the cerebrospinal fluid column, extending from the cranium to the end of the spinal canal, also constitutes a hydrostatic column, generating a negative counter pressure approximately equal in amount to the net distending pressure within the cerebral veins. Since these vessels are contained within the rigid cranium, they remain patent and relatively independent of postural changes.

LYMPHATICS

The networks of lymphatic capillaries ultimately join together in small and finally large lymphatic vessels and trunks. Histologically the lymphatics are similar to veins: lined by endothelium and surrounded initially by discontinuous layers of smooth muscle cells which, in the larger trunks, become continuous. The lymphatic vessels and trunks, like the veins, have valves consisting of either a single or a double cusp. These valves restrict movement of lymph through the lymphatics to the direction of the thoracic duct. Some tissues appear to lack lymphatics, e.g., bone marrow, lung alveoli, cartilage endothelium, splenic pulp and probably the central nervous system. On the other hand, they are abundant in the dermis and in the connective tissue of the genitourinary, respiratory and gastrointestinal tracts. The large lymphatic channels are interrupted by lymph nodes, widely distributed throughout the body, concentrating in groups throughout the prevertebral regions, the mesentery and the connective tissues of axilla and groin. Generally several lymph vessels converge in the capsule of the node. Inside the node they break up into numerous sinuses lined by lymphoid cells. The sinuses pervade the node and reunite into one or more lymph channels carrying lymph from the node. Ultimately all lymphatics converge into two main trunks, the thoracic and the right lymphatic duct, which empty into the junctions of the subclavian and internal jugular veins on the left and right side, respectively. The thoracic duct drains both lower extremities, the pelvis, abdominal cavity, left thorax, pleura, left head and neck and upper extremity. The right lymphatics serve the right head and neck, thorax and upper extremity.

Composition and Origin of Lymph. Ordinarily the volume of lymph represents the difference between capillary filtration and reabsorption; approximately 2 to 4 liters of lymph leaves the tissues and returns to circulation daily. A large fraction of lymph flow through the thoracic duct is derived from the liver (30 to 50 per cent) through the highly permeable liver sinusoids. This is also reflected in the relatively high protein content of thoracic duct lymph, ranging from 4 to 5 per cent.

In the fasting animal, lymph is a transparent liquid, usually slightly yellowish in color because of red cell content. Shortly after a meal it appears milky because of minute globules of fat which, in the intestinal lymphatics, are seen as fine white lines passing from the mucosa to the mesentery. On reaching the blood the fat globules are discharged as chylomicrons, having a diameter of 1 μ or less. The compositions of lymph and blood plasma are similar in all constituents except proteins, which in the lymph from most organs are approximately one-half that in plasma. Nevertheless all plasma proteins are present in lymph, indicating that the capillary wall is relatively permeable to the blood proteins, even those of the largest molecular size, such as fibrinogen. In addition to solutes, lymph contains particulate matter escaped from the capillaries. Bacteria, blood cells and their breakdown products are scavenged by lymphatics and may be phagocytized wholly or in part in the lymph nodes. In its passage through the lymph nodes, the lymph acquires an important population of cells. Peripheral lymph contains a total of only a few hundred white cells per mm³, whereas thoracic duct lymph may contain 8000 to 12,000 cells, which are mostly lymphocytes.

Lymphatic Pressures and Flow. Pressures in the peripheral lymphatic vessels are low; values around 1 to 2 mm Hg are usually found. This is essentially similar to the tissue pressures found in the subcutaneous connective tissue. Smooth muscle within the wall of lymphatic vessels and trunks may elevate lymphatic pressures to 5 to 10 mm Hg during their rhythmic contraction. This contraction of lymphatic trunks is synchronous along segments extending between lymphatic valves, and serves to propel the lymph from one segment into the next. In addition to this *active* mode of lymph transport, *passive* transport may occur by compression and massage of lymphatics by skeletal muscle, or other surrounding tissues, while the valves prevent backflow. The larger lymphatic vessels are innervated, but stimulation of their nerves has not yielded conclusive evidence of altered lymph flow.

The lymph flow is also increased by any procedure or agent which increases the rate of filtration from the capillaries, *e.g.*, raising venous pressure, reducing plasma oncotic pressure or inflammation. A general increase in systemic arterial pressure has little effect on lymph flow; however, as would be expected, a decrease in arterial pressure leads to diminution or stoppage of flow.

Lymphatic Tissues. The lymph nodes are flattened, rounded bodies, 1 to 25 mm in diameter. They consist of a capsule of dense collagen fibers from which radiate branching trabeculae, and provide a framework for the node. Between the trabeculae is a finer network of reticular fibrils. This network supports the cells of the nodes, mainly primitive reticular cells and fixed macrophages, that form the walls of lymphatic sinuses through which the lymph is flowing. By mitotic division, reticulocytes may produce either lymphocytes of different sizes or plasma cells. Lymphocytes enter the lymphatic sinuses and eventually the vascular system, whereas the plasma cells remain in the lymph nodes.

In addition to the lymph nodes, aggregations of lymphatic tissue occur in the spleen and in the thymus gland. Small amounts exist in the bone marrow and respiratory, genitourinary and alimentary tracts. Lymphoid tissues constitute about 1 per cent of total body weight.

The lymph node can effectively arrest small particles by phagocytosis by macrophages attached to the sinus walls. Pyogenic bacteria of sufficient virulence may be retained but not destroyed by the node. In this event they multiply and cause swollen inflamed nodes. Living tumor cells similarly lodge and proliferate to form metastasis within lymph nodes.

Lymph nodes have long been associated

with formation of antibodies which are believed to be produced mainly in the plasma cells. Destruction of lymphoid tissues by whole body irradiation or cortisone treatment abolishes immune reactions and leads to a state of immunologic tolerance.

REFERENCES

1. Bloom, G. E. Structural and biochemical characteristics of mast cells. In: *The inflammatory process.* B. W. Zweifach, L. Grant, and R. T. McCluskey, eds. New York, Academic Press, 1965.
2. Bondareff, W. Submicroscopic morphology of connective tissue ground substance with particular regard to fibrillogenesis and aging. *Gerontologia,* 1957, *1*, 222–233.
3. Chambers, R., and Zweifach, B. W. Functional activity of the blood capillary bed, with special reference to visceral tissue. *Ann. N.Y. Acad. Sci.,* 1946, *46*, 683–694.
4. Intaglietta, M. Evidence for a gradient of permeability in frog mesenteric capillaries. *Bibl. anat.* (Basel), 1967, *9*, 465–468.
5. Karnovsky, M. J. The ultrastructural basis of capillary permeability, studied with peroxidase as a tracer. *J. Cell Biol.,* 1967, *35*, 213–236.
6. Kato, F. The fine structure of the lymphatics and the passage of china ink particles through their walls. *Nagoya med. J.,* 1966, *12*, 221–246.
7. Landis, E. M. Micro-injection studies of capillary blood pressure in human skin. *Heart,* 1930, *15*, 209–228.
8. Landis, E. M., Jonas, L., Angevine, M., and Erb, W. The passage of fluid and protein through the human capillary wall during venous congestion. *J. clin. Invest.,* 1932, *11*, 717–734.
9. Landis, E. M., and Pappenheimer, J. R. Exchange of substances through the capillary walls. *Handb. Physiol.,* 1963, sec. 2, vol. II, 961–1034.
10. McMaster, P. D. The pressure and interstitial resistance prevailing in the normal and edematous skin of animals and man. *J. exp. Med.,* 1946, *84*, 473–494.
11. Rous, P., Gilding, H. P., and Smith, F. The gradient of vascular permeability. *J. exp. Med.,* 1930, *51*, 807–830.
12. Rusznyák, O., Földi, M., and Szabó, G. Composition of lymph. In: *Lymphatics and lymph circulation.* New York, Pergamon Press, Ltd., 1960.
13. Starling, E. H. On the absorption of fluids from the connective tissue spaces. *J. Physiol. (Lond.),* 1896, *19*, 312–326.
14. Wiedeman, M. P. Dimensions of blood vessels from distributing artery to collecting vein. *Circulat. Res.,* 1963, *12*, 375–378.
15. Wiederhielm, C. A. The interstitial space and lymphatic pressures in the bat wing. In: *The pulmonary circulation and interstitial space.* A. P. Fishman and H. H. Hecht, eds. Chicago, University of Chicago Press, 1969.
16. Wiederhielm, C. A. Dynamics of transcapillary fluid exchange. *J. gen. Physiol.,* 1968, *52*, 29–63.

CHAPTER 10 CONTROL OF ARTERIAL BLOOD PRESSURE

by ALLEN M. SCHER

INTRODUCTION

Human arterial blood pressure (and insofar as we know, blood pressure in other mammals) stays within narrow limits over a wide variety of bodily postures and states. Indeed, under normal conditions only massive changes in bodily activity cause a substantial alteration in the arterial blood pressure. The mean blood pressure rises only slightly in exercise, despite a greatly increased energy expenditure. With substantial blood loss, the blood pressure can fall, but here the compensatory mechanisms, which operate to maintain the normal pressure and which are described in this chapter, are functioning maximally. The necessity for blood pressure control is related to the maintenance of a constant internal environment and is necessary for cellular survival. It was pointed out during the discussion of motor functions of the nervous system that vertebrate organisms tend to—in fact, are forced for survival to— maintain a relatively constant internal environment. This requires control of the composition of the extracellular fluid. For the physiologist, Claude Bernard's picture of the regulated *milieu interne*[10] or Cannon's *homeostasis*[19] conceptualizes this control of extracellular fluid composition. The importance of pressure regulation here is that, when the pressure is maintained, the blood circulates adequately and the plasma is continually transported to the extracellular space after oxygenation in the lungs, addition of absorbed substances from the gastrointestinal organs and removal of metabolic end products in the kidney. Diffusion from the extracellular space to

146

this constantly replenished plasma (as discussed in Chap. 26, Vol. II) maintains the proper environment for the cells. In the mammalian circulatory system a high pressure is maintained in a central aortic reservoir. Changes in the resistance of arterioles alter the flow of blood out of this reservoir into the various vascular beds.

A certain level of pressure is necessary to maintain an adequate blood supply to the brain and to the heart. In a conventional view, which may be questioned below (Chap. 17, Vol. I), these most essential organs probably have little vasoconstrictor potential. Even if sympathetic "tone" is increased throughout the body, with a rise in arterial pressure, the brain and heart will have a nearly normal vascular resistance. They thus have a certain priority or protection insofar as perfusion with blood is concerned.

Elsewhere, this textbook considers regulation through feedback control in the case of the stretch reflex as it functions to main-

tain the upright posture (Chap. 8, Vol. I). The most important known reflexes involving the circulatory system control the arterial blood pressure. The blood pressure is altered by reflexes initiated by receptors (pressoreceptors) sensitive to *pressure* (or to a function of pressure). Blood pressure is also strongly influenced by receptors (chemoreceptors) sensitive to the chemical composition of the blood. These two classes of reflexes are the major subjects of this chapter, which also includes details of a variety of mechanisms of a local or central nature which can change the blood pressure. A review by Korner discusses much of this material.[53]

THE PRESSORECEPTOR REFLEXES

A diagram of the anatomic units involved in blood pressure regulation is presented in Figure 10–1. Receptors are located in two general sites. Those which are most

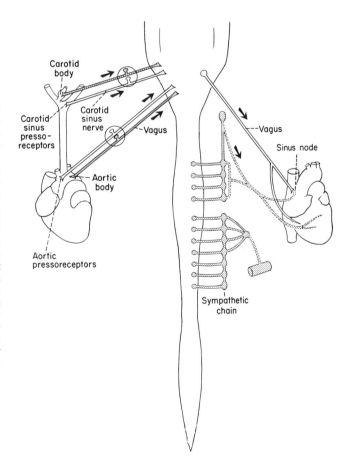

Figure 10–1 The medullary blood pressure control system. Sensory elements are shown on the left of the figure. The pressure receptors are the carotid sinus pressoreceptors and the receptors of the aortic arch. Chemoreceptors (shaded lines) are the carotid and aortic bodies. The aortic receptors give rise to afferent nerve fibers which run in the vagus nerve (and depressor nerve). The carotid receptors send impulses to the central nervous system via the nerve of Hering, which joins the glossopharyngeal trunk.

Effectors are shown on the right. The vagus nerve alters the heart rate through its effect on the sino-atrial node and may also change the strength of cardiac contraction. Sympathetic fibers (shaded lines) alter heart rate and strength of cardiac contraction. Sympathetic fibers run to the sino-atrial node (pacemaker) and to the ventricular musculature. Sympathetic fibers also alter pressure by their widespread innervation of the arterioles.

discretely localized, the carotid sinus receptors, are at the bifurcation of the carotid artery in the neck. Another group of receptors, the aortic receptors, are found in the general region of the arch of the aorta and the origin of the subclavian arteries (Fig. 10–2). There are receptors scattered along the blood vessels between these two sites.[36, 44] Detailed anatomic and physiologic characteristics of these receptors will be presented later. The receptors change their frequency of firing with changes in the arterial blood pressure; i.e., the higher the pressure, the greater the firing frequency, and the lower the pressure, the smaller the number of impulses sent to the nervous system per unit time. Impulses generated by the receptors in the carotid sinus travel in fibers of the nerve of Hering (or carotid sinus nerve), which joins the glossopharyngeal nerve. Afferent impulses from the receptors in the region of the aortic arch travel in sensory fibers accompanying the vagus nerve. At normal arterial pressure some receptor units are active. As the pressure rises, these receptors produce impulses at a higher frequency and other receptors begin to fire. This increase in the number of active receptors and in the frequency of impulses generated by individual receptors continues with further increase in pressure. Conversely, if the pressure decreases, the frequency of impulses from an active receptor decreases, and some receptors cease to fire.

The afferent nerve fibers from the receptors enter the brain stem with the glossopharyngeal and vagus nerves and travel to the medulla oblongata, entering the "vasomotor center" or "cardiac center" or both. Here, through one or more synaptic connections, impulses from the pressoreceptors alter the discharge in the appropriate motor nerves from the sympathetic nervous system and from the vagal branches of the parasympathetic nervous system. The reflex connections are such that when the arterial pressure rises (and the frequency of impulses in the pressoreceptor nerves increases), activity decreases in the sympathetic nervous system and increases in the vagal fibers.

Tonic activity is considered to be present in both the sympathetic and parasympathetic outflow from the brain stem. In the case of the sympathetic nervous system, decreased activity in the fibers running to

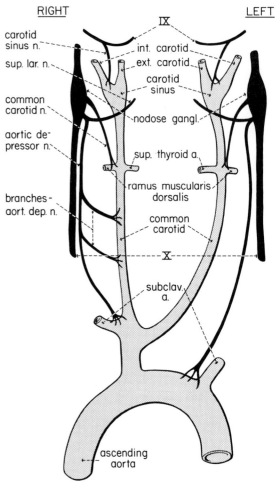

Figure 10–2 Position and nerve supply of pressoreceptors and chemoreceptors of the carotid and aortic regions in the cat. Blood vessels are shaded and nerve fibers are black. The aortic receptors lie near the origin of the subclavian artery bilaterally. Fibers from the receptors join the vagus (X) nerves. Other similar pressoreceptors originate as indicated along the common carotid artery, and fibers from these also join the vagus nerves. The aortic depressor nerve, although drawn separately, is clearly separate only in the rabbit and sometimes in the cat. In other species it is within the vagal sheath. The carotid pressoreceptors lie in the carotid bifurcation, and the carotid sinus nerves run from them into the glossopharyngeal (IX) nerve. (Note that there are chemoreceptor afferent fibers which run in the same nerves as the fibers from the pressoreceptors in the aortic and carotid regions.)

the heart (which might reflexly arise from increased arterial pressure) tends to decrease the heart rate. Decreased activity in fibers which innervate peripheral blood vessels allows the caliber of the vessels to increase, thereby decreasing the peripheral resistance. A further effect of decreased sympathetic activity to the heart is a decline in the vigor of cardiac contraction (Chap. 11, Vol. II). The increase in vagal firing occasioned by an increase in arterial blood pressure reduces the heart rate, synergistically with the decreased sympathetic activity. It may also decrease the strength of cardiac contraction.[25]

When arterial blood pressure falls, activity is decreased in the afferent fibers which run from the pressoreceptors to the brain stem. This leads to a reflex increase in sympathetic motor activity and a decrease in vagal motor activity. The increased sympathetic activity tends to increase the heart rate, the strength of cardiac contraction and the degree of constriction of the peripheral blood vessels. The decrease in vagal activity increases the heart rate.

The qualitative description of the reflexes is as follows: When the pressoreceptor impulses increase, vagal discharge to the heart increases reflexly, slowing the heart. Concomitantly, sympathetic firing to the heart and blood vessels decreases. Decreased sympathetic discharge to the heart also leads to a decrease in the strength of ventricular contraction. The decreased discharge to the vasoconstrictor nerve fibers allows the arterial pressure to fall. Thus the blood pressure tends to remain at a normal level.

The existence of the carotid sinus reflex was classically demonstrated by Heymans.[43] He ingeniously arranged two dogs so that the carotid arteries of one, the recipient (including his carotid receptors), were perfused by a second dog, the donor. Blood was returned to the donor from the recipient's jugular vein. The nervous connections between the recipient's head and the rest of its circulatory system were intact. When the blood pressure of the donor animal was raised through drugs or otherwise, the pressure in the recipient's carotid arteries also increased (Fig. 10–3). This produced a reflex fall in the systemic blood pressure of the recipient. The carotid receptors of the recipient were exposed to the increased pressure, and produced this reflex change in the recipient's blood pressure. The nervous connections to the recipient's effector organs were, of course, intact. A schematic diagram of the blood pressure control system is seen in Figure 10–4. Treatments of various aspects of blood pressure control in terms of engineering control theory have appeared recently.[30, 39, 49, 53, 71, 87]

Fragmentary pieces of evidence indicated the existence of the blood pressure control reflexes long before they were clearly elucidated.[79] In 1799 a Welsh physician, Parry,[62] reported an effect of the reflexes (although he did not understand the mechanism): "In patients whose hearts have been beating with undue quickness and force I have often, in a few seconds, retarded their motion many pulsations in a minute by strong pressure on the carotid arteries." A similar observation was reported to the Royal Society in London in 1862 by Waller,[84] who felt that the results were due to irritation of the vagus nerve. In 1866 de Cyon and Ludwig[21] noted that stimulation of the aortic nerve (depressor) in the rabbit produced a reflex fall in heart rate and in blood pressure. This nerve appears to contain, largely or exclusively, fibers from the aortic pressoreceptors.

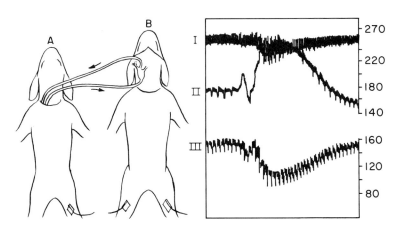

Figure 10–3 The preparation used by Heymans[43] to study pressoreceptor reflexes. A recipient dog (B) is arranged so that its carotid arteries are perfused by the donor dog (A). Blood is returned to the donor from the recipient's jugular vein. In this experiment, the blood pressure of the donor animal (II) was raised by an injection of epinephrine. This, of course, raised the pressure in the carotid arteries of the recipient. This led to a marked fall in pressure (III) in the major portion of the recipient's circulation, which was perfused normally. The recipient's heart rate also fell (I).

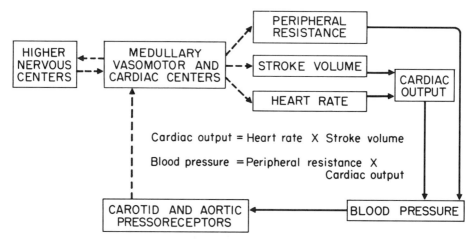

Figure 10–4 Blood pressure is sensed by the pressoreceptors, which send impulses to the medulla. The medullary centers send impulses to motor fibers of the sympathetic and parasympathetic nervous system. When pressure at the receptor rises or falls, it is reflexly corrected by an alteration of heart rate, stroke volume and peripheral resistance. Dotted lines indicate neural connections.

In 1925 Hering noted that mechanical stimulation of the carotid sinus region in a patient produced a reflex bradycardia.[42] Hering later stimulated the nerves from the receptors in animals, with results similar to those reported by Ludwig and de Cyon.[21] Because of technical difficulties in duplicating Hering's experiments, a long period of time followed when the reflexes were a subject for controversy. In 1929 Heymans[43] performed the ingenious experiments previously cited, ending the confusion over these reflexes.

Characteristics of Pressoreceptors and Pressoreceptor Reflexes. The carotid sinus can be grossly observed as a dilatation of the internal carotid artery at its origin. This region has rich sensory innervation from the sinus nerve, and the wall of the vessel appears extremely thin.[20, 44] The aortic arch receptors cannot be grossly observed but can be seen in histologic section.

The pressoreceptors, on histologic examination, appear to be the terminal ramifications of undifferentiated nerve fibers which branch extensively in the adventitia and media of the carotid sinus and aortic arch regions (Fig. 10–5). When the vessel increases in diameter because of increased blood pressure, the receptors are stretched. Through this deformation of the wall and of the fine nerve terminations the receptors transform pressure into nerve impulses. They are not, therefore, truly pressoreceptors but are indicators of length or deformation of a nerve termination within a vessel segment. de Castro[20] found about 700 mye-

linated fibers in the carotid sinus nerve of the cat. Most of these were smaller than 5 μ in diameter. Paintal[61] measured a conduction velocity of 12 to 53 m per sec in the aortic depressor nerve of the rabbit. This agrees with the measured fiber diameter (Chap. 3, Vol. I).

The functional characteristics of the pressoreceptors have been studied by several investigators.[17, 27, 31, 54] The carotid sinus receptors are better adapted to these studies because the cut carotid sinus nerve contains predominantly sensory (pressoreceptor and chemoreceptor) fibers and be-

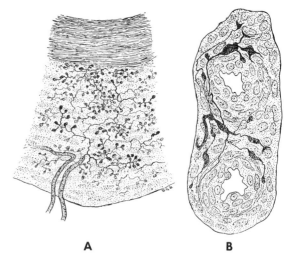

A   B

Figure 10–5 Diagram showing characteristics of afferent endings in (A) carotid sinus and (B) carotid body. (After de Castro.)

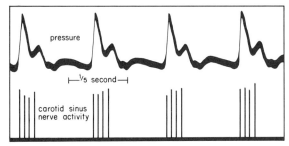

Figure 10-6 Blood pressure above; impulses in the carotid sinus nerve below. Impulses occur during the rise in pressure early in each beat. (After Bronk and Stella, *Amer. J. Physiol.*, 1935, *110*, 708-714.)

cause the carotid bifurcation can be isolated and the pressure therein controlled. The sinus nerve is carefully dissected, and a few fibers or a single fiber are carefully separated from the main nerve trunk. These fine fibers are laid on electrodes connected to an amplifier and recording oscillograph. The response to normal blood pressure changes can then be observed (Fig. 10-6). If responses to controlled pressure changes are to be studied, the carotid sinus is converted to a closed sinus; that is, the major branches of the internal and external carotid arteries are tied. Careful surgical technique is required to avoid damage to the sinus nerve. The pressure in the carotid

sinus can then be altered, using various hydraulic systems. The receptor response can, for convenience, be divided into two main portions. The first of these is the static response, *i.e.*, the response when the rate of change of pressure is zero. The second is a response to phasic pressure changes.

STATIC RESPONSE. The following description comes from the studies of Bronk and Stella,[17] Landgren,[54] Green,[36] Spickler and Kezdi[76] and Franz *et al.*[31] Below a certain threshold pressure, the receptor (single fiber) does not fire. This threshold is variable but seems to be as low as 30 mm Hg for some fibers.[54] At the threshold pressure the receptor begins to fire at some minimal frequency (Fig. 10-7). (Firing does not go to "zero frequency"; the earliest activity at pressure threshold may be at 20 imp/sec or higher.) As the mean pressure at the receptor is raised past the threshold, the firing rate increases roughly in proportion to the increase in pressure. Eventually, at a pressure of about 220 mm Hg or higher, the firing rate reaches a maximum for a particular fiber. The receptor region contains fibers of various sizes, pressure thresholds and sensitivities. Landgren[54] found that the small nerve fibers, which are not easy to isolate and from which recording is difficult, have approximately the same characteristics

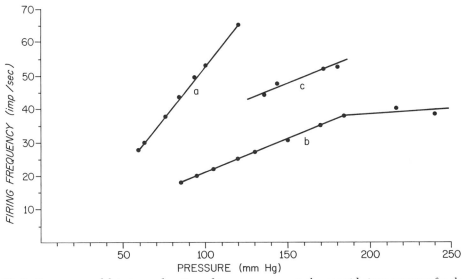

Figure 10-7 Frequency of firing as a function of input pressure at the carotid sinus receptor for three single fibers from the carotid sinus. Note that in each case there is a pressure threshold and a minimal firing frequency, and that the response is approximately linear. Fiber B tends to show a maximal firing frequency (limiting). (After Franz *et al.*, *J. appl. Physiol.*, 1971, *30*, 527-535.)

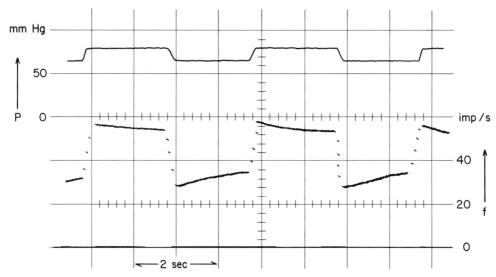

Figure 10–8 Effects of step changes in pressure at the carotid sinus (upper trace) on a firing frequency in a single pressoreceptor fiber in the rabbit. With a step change in pressure, the pressure rises rapidly to a new level and then decreases to a maintained (static) level. Note that in addition to this initial overshoot there is also an undershoot after a step decrease in pressure. (After Franz *et al.*, *J. appl. Physiol.*, 1971, *30*, 527–535.)

as the larger fibers, more commonly examined experimentally.

PHASIC RESPONSE. Phasic changes in pressure are, of course, normally imposed on the receptor by the beating of the heart. In experiments on the pressoreceptors, phasic inputs are often generated by external hydraulic systems. With phasic inputs, the receptors fire most rapidly when the pressure is increasing and when the rate of change of pressure is greatest. When the rate of change of pressure becomes negative, *i.e.*, when pressure is falling, the receptor often ceases to fire, even when the pressure is above the threshold.[17, 31, 54, 76] These responses are clearly seen when square waves or sine waves are imposed on the receptors (see Fig. 10–8). The receptor is thus not only a mean pressure level sensing device, but is also, like other mechanoreceptors, sensitive to the rate of change of pressure (Chap. 8, Vol. I).

RECEPTOR EQUATION. From the work cited above,[73] the simplest equation which approximates receptor performance seems to be the following:

$$F_0 + K_3 dF_0/dt = K_1(P_I - P_T) + K_2 dP_I/dt$$

where F_0 is the receptor frequency, P_I the arterial pressure, P_T the threshold pressure and dF_0/dt and dP_I/dt the derivatives of frequency and pressure, respectively. That this receptor

equation is justified is seen most clearly when square waves of pressure are applied to the receptor. Here the response to a step increase in pressure is initially a marked increase in firing ("overshoot"), followed by a gradual fall to a new firing frequency higher than that seen at the original pressure (Fig. 10–8). The necessity for the term in dP_I/dt is seen in the initial overshoot. The dF_0/dt term is needed because the rate of firing continues to change after the input has become stationary. At a mean pressure just below threshold, the receptor will not fire; however, if pulsations are added to the pressure *without* changing the mean level, the receptor will fire during the positive-going portion of these pulsations. The receptor thus fires, although the mean level has not changed (Fig. 10–9). This nonlinear receptor property apparently accounts for a type of "rectification" seen in the carotid sinus reflex (see below). A more realistic description of receptor function requires higher derivative terms.[31] As indicated, F_0 has a threshold or minimal firing frequency. The receptor does not go smoothly to zero frequency.

Several investigators have claimed that the sensitivity of the receptor can be reflexly set. According to this view, the sympathetic fibers terminate near the receptor. This sympathetic activity changes the diameter of the vessel or the sensitivity of the receptor, so that the firing is different at a given pressure than it would be without sympathetic activity. There is some disagreement concerning the existence of this reflex alteration in receptor sensitivity.

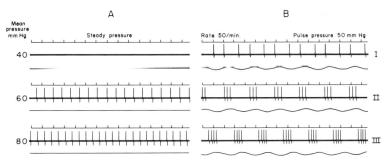

Figure 10-9 On the left, the response of a single fiber of the common carotid nerve to various maintained pressure levels. Note that at 40 mm Hg the nerve does not fire. In the records on the right, mean input pressure is the same as on the left, but sinusoidal pressure variations are also imposed on the receptor. At 40 mm Hg the nerve is now active; at higher frequencies the firing occurs mostly when the pressure is increasing. (After Green in *Reflexogenic areas of the cardiovascular system*, Heymans and Neil, eds. London, J. & A. Churchill, Ltd., 1958.)

Medullary Vasomotor and Cardiac Control Regions. Impulses from the carotid and aortic receptors travel in the ninth and tenth nerves to a region of the medulla oblongata where the major integration of cardiovascular responses is considered to take place. The importance of this region has been investigated by three methods: (i) progressive brain stem sectioning, (ii) electrical stimulation of the brain stem and (iii) recording of electrical activity of the region. If the brain stem is sectioned transversely at the level of the upper pons[2] (Fig. 10–10), the blood pressure is well maintained, as are the basic blood pressure-controlling reflexes. When the brain stem is sectioned below the upper pons, the pressure falls, and this fall is greater the more caudal the section. Finally, when the brain stem is sectioned at the first cervical segment, pressure is markedly reduced and reflexes cannot be elicited from the pressoreceptors.

Stimulation experiments have been conducted by Ranson and Billingsley[68] and by Alexander,[2] Lindgren,[57] Uvnäs[80] and others. The blood pressure can be altered by stimulation in a region extending from the middle of the pons to the obex. Stimulation of certain regions of the lateral reticular formation in the rostral two-thirds of the medulla increases blood pressure. This region is therefore referred to as the "pressor" center. Stimulation of a more central and caudal "depressor center" produces a fall in blood pressure (Fig. 10–9). These regions overlap somewhat in the anteroposterior direction and laterally. Evidence from animal experiments indicates that the pressor center is normally "tonically" active, *i.e.*, that it constantly discharges impulses to the preganglionic vasoconstrictor neurons of the periphery; these tonic impulses tend to maintain some degree of constriction in the blood vessels and to increase the heart rate. That the depressor center is also tonically active is indicated by the existence of a tonic vagal discharge to the heart. (Vagotomy leads to an increase in heart rate.)

A section of the brain stem is, of course, a gross alteration of structure, interrupting many tracts concerned with functions other than those being directly examined. Stimulation of local medullary regions produces clearer results, but these are difficult to evaluate because one cannot assume that punctate stimulation bears a real relationship to what normally takes place. The sectioning and stimulation experiments both indicate that control of cardiovascular functions is widely distributed in the medullary region rather than discretely localized.

Improved understanding of the blood pressure control system could be achieved if we could exactly describe the "input-output" functions of the medullary region. Chains of neurons of unknown length lie between the pressoreceptor fibers and the sympathetic and vagal motor fibers, and ideally we might understand the function of each. As an approximation to this ideal, we have information of several sorts. (i) Some recordings have been made from central neurons which alter firing rates when arterial pressure changes or when pressoreceptor nerves are stimulated (Figs. 10–10 and 10–11). (ii) Some recordings have

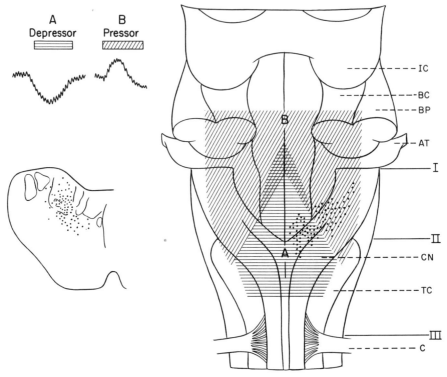

Figure 10–10 Location of areas in the medulla which participate in the control of blood pressure. *AT*, Auditory tubercle; *BC*, brachium conjunctiva; *BP*, brachium ponti; *CN*, cuneate nucleus; *IC*, inferior colliculus; *TC*, tuberculum cinereum; *C*, first cervical nerve. Cross section shown at left is at level indicated by I. Horizontal shadings indicate the areas where electrical stimulation produces a fall in blood pressure (depressor response). Stimulation of diagonally shaded areas increases blood pressure (pressor response). Note that there is an area of overlap between these two. The typical responses from stimulation in these areas are shown in curves A and B. Stimulation of region A produces a fall in pressure, whereas stimulation of region B produces a rise in pressure. The dots on the right-hand side of the large drawing and on the left-hand side of the small cross-section show areas in which electrical records have been made from nerve cells which fire in phase with normal blood pressure or change their firing patterns when the blood pressure is altered. (After Alexander, *J.Neurophysiol.*, 1946, 9, 205–217; Ranson and Billingsley, *Amer. J. Physiol.*, 1941, *134*, 359–383; and Humphrey, in *Baroreceptors and hypertension*, P. Kezdi, ed. Oxford, Pergamon Press, 1967.)

been made of the firing of pressoreceptor fibers and motor nerves and/or of the relationship between arterial pressure changes and firing of the motor nerves.[35, 48, 50]

Results of successful recordings *within the medullary region* are not yet conclusive in establishing the nature of central blood pressure control (Figs. 10–10 and 10–11). Fibers have been found which respond to arterial pressure and which alter their activity when pressure changes.[40] In some studies such fibers change firing levels with changes in arterial pressure but the fibers do not show a cardiovascular rhythm. Some of the fibers studied do not seem to be directly involved in blood pressure control but may be concerned in the relationships between blood pressure control and other functions.[40, 46]

Recordings have been made of the firing in sympathetic[50] (Figs. 10–12 and 10–13) and vagal motor[48] (Fig. 10–14) fibers. In most of the studies, arterial pressure has been simultaneously recorded and, in some, pressoreceptor firing has also been recorded either simultaneously or during identical maneuvers. The relationship between baroreceptor input and sympathetic motor discharge is approximately linear and inverse (Figs. 10–12 and 10–13), *i.e.*, sympathetic fibers decrease firing frequency as arterial pressure and baroreceptor discharge increase.[35, 50] The experiments of Kezdi and Geller[50] show some time delay between input and output, and the motor nerve fibers are capable of following pressure changes up to 2 Hz without attenuation[40, 72] (Figs. 10–12 and 10–13).

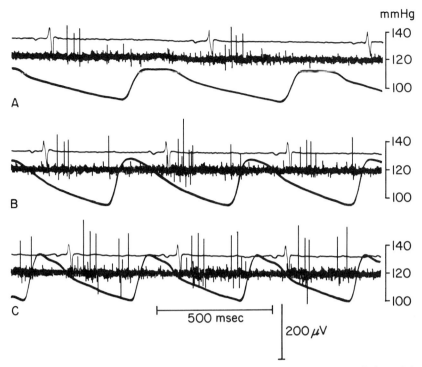

Figure 10–11 Records of one type of nerve impulse recorded in the rhombencephalon of the cat associated with cardiovascular events. A, A control record obtained from the electrocardiogram; B, an electrode in the medulla; C, and the femoral arterial pressure. In this control record, several nerve impulses can be seen in synchrony with the heart beat, approximately during the T wave of the electrocardiogram and half way between the peak of systole and the end of diastole.

During B and C, an infusion of epinephrine was given. The firing of the neurons seen in the control record is augmented sporadically as the arterial pressure rises, and a second "group" of nerve impulses is seen later in diastole. This second group might arise from atrial activity or from the increased pressure. All the impulses seem to be recorded from nerve fibers one or more synapses removed from the receptors. (After Hellner and von Baumgarten, *Pflügers Arch. ges. Physiol.*, 1961, *273*, 223–234.)

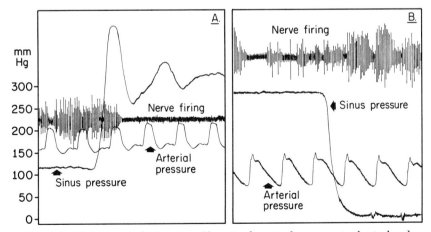

Figure 10–12 Response of the sympathetic motor fibers to change of pressure in the isolated carotid sinus. In A, on the left, a sudden increase in pressure at the carotid sinus has little effect on arterial pressure, but after a lag of about 200 msec the firing in the sympathetic motor fiber ceases. In B, on the right, a change in carotid sinus pressure from approximately 250 mm Hg to 0 mm Hg causes, after a somewhat longer lag, a marked increase in firing of the sympathetic fibers. (After Kezdi and Geller, *Amer. J. Physiol.*, 1968, *214*, 427–435.)

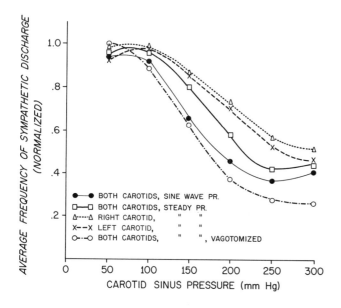

Figure 10–13 Frequency of firing of post-ganglionic sympathetic fibers in response to changes in carotid sinus pressure. The curves have been normalized to make the highest frequency equal to 1 in each case. Note that the decreased frequency of firing is greatest in the vagotomized animal when both carotids are subjected to the increase in pressure, responses are slightly less when the input pressure is sinusoidal and the responses are further decreased when only one carotid is subjected to steady pressure at varying levels. (After Kezdi and Geller, *Amer. J. Physiol.*, 1968, *214*, 427–435.)

There is less information about vagal firing, but in the records available the reflex relationship between sinus pressure and motor nerve firing appears to be approximately linear, positive rather than inverse, and to involve a time delay and the same frequency-following characteristics as sympathetic discharge[48] (Fig. 10–14).

CENTERS. As indicated, the medullary regions are often referred to as centers,

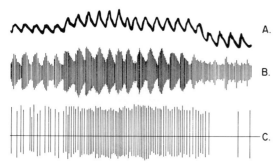

Figure 10–14 Effects of pressure on pressorecep-tors and vagal firing. When the pressure in the carotid sinus is changed, as in curve A, the firing of a single fiber from the baroreceptor changes, as in curve B, increasing as the pressure is raised and decreasing as the pressure falls. The firing of a vagal motor fiber to the heart is shown in C. The frequency is low at the left when the arterial pressure and the firing of the baroreceptors are lowest; it increases when these two increase, and when the pressure falls at the right side of curve A, the vagal firing temporarily stops and then resumes. (Adapted from Katona and Barnett, *Ann. N.Y. Acad. Sci.*, 1969, *156*, 779–786; and Katona *et al.*, *Amer. J. Physiol.*, 1970, *218*, 1030–1037.)

identified by such names as vasomotor center, cardiac center, vasoconstrictor center, vasodilator center, cardio-accelerator center and cardio-inhibitory center. These names suggest the existence of a specific organization in the medulla, perhaps like a telephone switchboard with separate controls for each function. Apparently, the medullary regions have many connections to the "input" and "output" cells of the blood pressure control system and to inter-neurons, but the specificity of the organization is not yet established. Connections occur in this region between cells concerned with cardiovascular regulation and those concerned with other forms of autonomic regulation and with respiration. Also, the neurons which descend from the higher centers and which influence the cardio-vascular system synapse here with those cells concerned with basic autonomic regulation. The higher centers, including the hypothalamus and the cortex, apparently contribute to regulation of the circulatory system through their effects on the medullary neurons.[67, 86] These facts justify the use of the term "center" but should not be taken to indicate a specific localized type of control. As will be discussed later, the medullary regions are sensitive to certain metabolites, and there are also peripheral chemoreceptors, both of which can alter blood pressure. Peiss[64] has presented the theory that our present concept of medul-

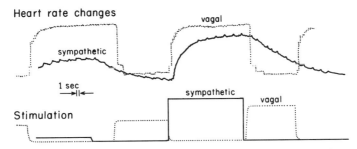

Figure 10–15 The lower record shows the duration of stimulation of the sympathetic (solid line) and parasympathetic (dotted line) fibers to the heart. The curves above show the effects on heart rate of the two types of stimulation. The curves have been superimposed to facilitate comparison. Note that the cardiac slowing in response to vagal stimulation is accomplished very rapidly—less than 1 sec is required for the response to be virtually complete. The "off" response when vagal stimulation ceases is slightly slower. In contrast, the heart rate increase in response to sympathetic stimulation may not be fully complete in 20 sec, and the slowing of the rate when sympathetic stimulation ceases is even slower. (After Warner and Cox, *J. appl. Physiol.*, 1962, *17*, 349–355.)

lary areas as primary integrators of input and output signals may be incorrect and that primary control may be more widely distributed from spinal to cortical centers.

Autonomic Neuroeffectors. The vagal fibers originate in the dorsal motor nucleus of the vagus and descend in the vagal trunk to ganglia in the cardiac plexus and possibly in the atrial wall. The most important distribution of these endings is to the sinoatrial node and to the A-V node. Sympathetic outflow (described in detail in Chap. 11, Vol. I), originates in the intermediolateral gray column of Th 1 to Th 5 and passes through the ventral roots to the postganglionic neurons in the upper thoracic and cervical sympathetic ganglia. The cardiac nerves originate here and run to the cardiac plexus and then to the heart, most importantly to the sinus node and to the ventricular myocardium. As previously mentioned, there is an extensive sympathetic innervation of the blood vessels in most of the organs perfused by the systemic circulation. In an earlier chapter (Chap. 8, Vol. II), the transmitters were discussed; a review by Korner[53] covers this material in depth.

As shown in Figures 10–12 and 10–14, the sympathetic and vagal motor fibers respond quite rapidly to changes in carotid sinus pressure. If we lump these responses of the motor nerves with the responses of the neuroeffector junction and the smooth muscle of the blood pressure, the responses are far slower. The overall responses have been evaluated in part by observing the entire reflex and in part by stimulating motor nerves and observing the results.

The most quantitative studies of the cardiovascular innervation are those of Wang and Borison[85] and of Warner and Cox (Fig. 10–15).[87] The sympathetic effects on resistance have been studied by Scher and Young.[73] Parasympathetic innervation of the heart acts rapidly. Comparisons of sympathetic and parasympathetic effects by Wang and Borison indicate that heart rate can be altered by the vagus nerve within the interval of one heart beat, and a change may be complete within three or four heart beats (*i.e.*, in about as many seconds). Effects of the sympathetic motor nerves on the blood vessels and on the heart are much slower, and a complete response to a change in motor nerve activity may require 20 sec. The control of heart rate by the vagus is rapid and can provide virtually instantaneous regulation, but the slow response of the heart and blood vessels to sympathetic activation makes it appear that these cannot function when extremely rapid adaptation must occur.[73, 87]

THE INTEGRATED REFLEX

A conceptually simple technique utilized by Koch,[52] Scher and Young[73] and Levison *et al.*[56] consists of controlling the pressure in the carotid arteries (and thus the pressure affecting the receptors) through a hydraulic system and noting the effects on systemic pressure as pressure at the receptor is changed. In some of these experiments the vagi are cut to denervate the aortic arch receptors, permitting the carotid sinus receptors to be studied separately.

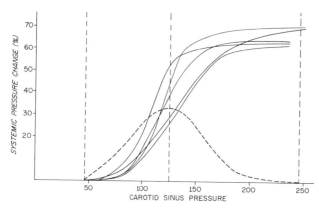

Figure 10–16 Changes in systemic pressure resulting from changes in carotid sinus pressure in the dog. Solid lines show systemic pressure changes plotted as percentage of the resting pressure. Dotted line is the slope of one of the solid curves and can be considered a "gain" curve. (After Koch, *Die reflektorische Selbststeuerung des Kreislaufes.* Leipzig, D. Steinkopff, 1931.)

(If the aortic arch receptors are functioning, change in frequency of impulses from them tends to correct any change in systemic pressure caused by alteration of the carotid sinus pressure.) Effects of the vagus nerve on the heart are thus also eliminated.

As the pressure in the receptor area is raised, systemic pressure at first falls slightly because of vasodilation. The response to a similar increase in pressure is much greater at moderate pressures. At very high pressures the response again decreases. The curve showing the relationship of carotid pressure to systemic pressure is sigmoidal in shape (Fig. 10–16). The ratio of change in systemic pressure to change in carotid pressure can be considered an "amplification factor." The amplification is not constant but is peaked at some input pressure level. Koch[52] found the greatest sensitivity of the reflex response near the animal's "normal" pressure. Scher and Young[73] found a maximal amplification factor of 10 to 15 in some cats.

If the receptor is subjected to steps of pressure, the systemic pressure changes in a direction opposite to the imposed change, often overshoots and then returns to a new level. If the pressure at the receptor changes sinusoidally and the systemic pressure response is monitored, systemic pressure change is in phase with the carotid pressure only at low frequencies (0.01 imp/sec); at higher frequencies (0.4 imp/sec) the systemic pressure lags the carotid pressure. The systemic pressure response is larger at the lower frequencies (Fig. 10–17).

A third characteristic of the response can be seen when phasic pressures are imposed on the receptor, while the mean pressure at the receptor does not change (Fig. 10–18). In this case, systemic blood pressure reflexly decreases to an extent which depends on the frequency and amplitude of the phasic inputs to the receptor (*i.e.*, the greater the frequency and amplitude, the greater the decrease in blood pressure). This feature can be loosely referred to as "rectification"; *i.e.*, a varying pressure at the receptor causes a change in *mean* systemic pressure.

So far three reflex properties have been described: (i) a varying effectiveness or amplification with mean pressure, (ii) a slow response with overshoot to steps and a lagging response to sine waves and (iii) "rectification." These properties are described for a preparation which has mainly vasomotor responses. They can lead to a family of equations describing this portion of the reflex.[56, 73]

The first characteristic, the change in sensitivity with mean input pressure level, appears to result from the varying thresholds and sensitivities of the many receptors

Figure 10–17 Bottom curve shows a sinusoidal pressure change imposed on the carotid sinus receptors. At the left of the figure, the frequency of this sine wave is 1/50 imp/sec, and at the right, 1/100 imp/sec. The upper curve shows the systemic pressure response to this carotid change. Note that the response is larger at the slower frequency. Note also that the mean systemic pressure is higher when the input frequency is lower. (After Scher and Young, *Circulat. Res.*, 1963, *12*, 152–162.)

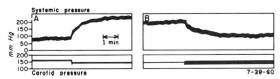

Figure 10-18 Lower curve shows pressure change imposed on the carotid sinus receptors. At left, the pressure at the receptors fell 15 mm Hg, and the systemic pressure rose about 75 mm Hg. On the right, pressure oscillations at 10 imp/sec are imposed on the receptor without changing the mean pressure. The systemic pressure falls.

in the carotid sinus. The overshoot, when steps of pressure are applied at the receptor, appears to result from the sensitivity of the receptor to rate of change of pressure. The reflex response of the blood pressure to steps and sine waves of pressure at the carotid sinus appears to be due to the slow response of the neuroeffector junction or of the smooth muscle itself or both. The "rectification" in the reflex response appears to be a reflection of the receptor properties, *i.e.*, their threshold and their response to rate of change of pressure (see earlier description of receptor).

Sympathetic and Vagal Responses in the Intact Animal. The foregoing description concerns the anesthetized animal. The use of anesthesia alters responses, tending to make experimental animals utilize sympathetic rather than vagal control. Two questions remain unanswered: (i) What type of quantitative description connects pressure and heart rate responses? (ii) What is the relative importance of heart rate, cardiac contractility change and peripheral vasoconstriction in the intact animal and human? To answer these questions some discussion of cardiac effects is necessary, although it overlaps another chapter (Chap. 11, Vol. II). Glick and Braunwald,[33] Robinson *et al.*,[69] Katona *et al.*[49] and Scher and Young[74] have studied heart rate responses to changes in pressure at the baroreceptors. It appears that with pressure increased above normal in the unanesthetized or lightly anesthetized *dog*, the heart rate responses are predominantly due to changes in vagal activity, whereas with decreased pressure the responses are at least partly sympathetic.

Concerning heart rate responses, Glick and Braunwald,[33] using drugs which block sympathetic or vagal responses, have provided evidence that in the intact human the heart rate responses to increased pressure at the receptor are predominantly vagal, whereas responses to decreased pressure are predominantly sympathetic. This conclusion is qualified by Robinson *et al.*,[69] who believe that responses are determined by the antecedent level of sympathetic or parasympathetic tone.

In the studies on intact, awake dogs by Scher and Young,[74] responses to increased pressure often were solely vagal heart rate responses. There was little or no vasodilation. When pressure at the receptor decreased, there was a slight vasoconstriction, but it was far smaller than the heart rate changes. Where heart rate changes are very rapid, vagal control must be operative.[74, 87]

Generalized Equation for the Carotid Sinus Reflex.[73] As an approximation to the static reflex response, the curve of "gain" vs. input pressure can be considered to have the form of a power function, symmetrical about the point of maximal gain. The resulting equation is:

$$\Delta P_0 = -[G_M + K_7(P_I - P_M)^n](\Delta P_I) \qquad (1)$$

where ΔP_0 and ΔP_I are changes in output and input pressure respectively, G_M is maximal gain, P_M is the input pressure at which gain is maximal and n is an even number.

TRANSIENT RESPONSES. Equation (1) must be supplemented when dP_I/dt is not zero as indicated below.

SQUARE AND SINE WAVES. When the input pressure change is a square wave, 2 to 4.5 sec elapse before any response; the output pressure then changes slowly, taking up to 100 sec to reach a new level. In many experiments, the time constant is as long, but the output pressure shows an early overshoot and then slowly attains the new level.

The fact that the output pressure slowly reaches a new level requires that the output side of the equation contain a term one derivative higher than any term on the input side. If there are no derivative terms on the input side (no overshoot), the output is a function of the input and the derivative of the output. With overshoot the output is further responding to the rate of change of input. A linear approximation is:

$$K_2(dP_0/dt) + K_3(d^2P_0/dt^2) = K_1(dP_I/dt)(t - t_0) \qquad (2)$$

where dP_0/dt is rate of change of output pressure, d^2P_0/dt^2 is second derivative of output pressure, dP_I/dt is rate of change of input pressure and $(t - t_0)$ shows a time delay. The sine and square wave data are consistent with equation (2).

Equation (2) accounts only for the transient or phasic responses to transient or phasic inputs and must be used with equations (1) and (3) (below); (1) and (3) must make equation (2) inaccurate to some degree.

EFFECTS OF PHASIC INPUT ON MEAN PRESSURE. As phasic inputs are imposed, with the mean input pressure held constant, the mean pressure falls. This "rectification" requires an additional equation:

$$\overline{\Delta P_0} = \frac{-G_R(P_1 + dP_1/dt - K)[Sgn(P_1 + dP_1/dt - K) + 1]}{2}$$

$$(3)$$

where $\overline{\Delta P_0}$ is the change in mean output pressure due to phasic inputs, G_R is an amplification factor for rectification and Sgn (signum) is a function having a value of $+1$ when positive and -1 when negative.

"GAIN" OR SENSITIVITY. The effectiveness of any reflex control depends in major part on the sensitivity or "gain," the ratio of a change in output to a change in input. In the equation above, gain is specified for an "open-loop" system, i.e., a system in which the receptor can be separately controlled, and the reflex effect of the control can be monitored. The effectiveness of control in a normally functioning system, referred to as a "closed-loop" system, is related to this open-loop gain by the equation:

$$G_C = G_0/(1 + G_0) \qquad (4)$$

where G_C is a closed-loop gain and G_0 an open-loop gain. It is assumed in this equation that the receptor or measuring device is exposed to the entire change made by the control system. If a system has an open-loop gain of 1.0, the closed-loop gain is 0.5. In such a system, any disturbance of the measured variable will be only one-half corrected.

In the experiments of Scher and Young,[73] some cats had a high open-loop gain, up to 15, and dogs a gain of up to 8. Although this is far below what an engineer would build into a control system, the gain of 15 indicates that any external factor which tends to alter the blood pressure will be more than 90 per cent corrected, and the reflex thus appears to be quite powerful.

CAROTID AND AORTIC RECEPTORS COMPARED. Several studies have compared reflex responses to pressure changes in the carotid and aortic regions.[4, 5] The general responses appear similar, although the response to pulsatile pressure changes appears smaller for the aortic receptors.

CHEMORECEPTOR REFLEXES

Although chemoreceptor reflexes have their major effect on respiration and are extensively discussed later (Chap. 23, Vol. II), important effects on the cardiovascular system will be considered here. There are both peripheral and central chemoreceptors.

Peripheral Chemoreceptors. The peripheral chemoreceptors—the carotid and aortic bodies—are found near the carotid and aortic pressoreceptors, but they should not be confused with the pressoreceptors. The carotid body, sometimes visible as a small reddish ball of tissue, lies between the external and internal carotid arteries a few millimeters ventral to the carotid sinus. It is supplied with blood from the occipital and ascending pharyngeal arteries. Venous drainage is via the internal jugular vein. Fibers from the carotid body join the carotid sinus nerve, which merges with the glossopharyngeal trunk. The aortic bodies are found in two main sites: (i) scattered around the aortic arch, particularly between the root of the aorta and the origin of the left subclavian artery on the curvature of the aortic arch, and (ii) at the root of the right subclavian artery. They are not easily visible, but lie close to the large vessels. Carotid and aortic chemoreceptors are similar histologically and probably physiologically (see Figs. 10–1, 10–2 and 10–5). They consist of epithelioid cells surrounded by a large network of sinusoidal blood vessels. The nerves from the chemoreceptors travel with the corresponding pressoreceptor nerves, i.e., with the carotid sinus nerve from the carotid body and with the vagus from the aortic body. The blood supply of the receptors appears to be extremely large per gram of tissue, although the receptors are very small. The blood flow of the carotid body (which weighs about 2 mg) has been estimated at 2000 ml per min per 100 g of tissue. This should be compared with a flow for the kidney of 400 ml per min per 100 g; the kidney is the most favored among the larger organs of the systemic circulation with respect to blood flow per gram of tissue.

The physiology of the chemoreceptors is studied in much the same fashion as that of the pressoreceptors.[44] The carotid body is commonly studied because of the ease of separating the important sensory nerve fibers and of controlling the vascular supply to it. In such studies the vessels of the carotid region are isolated for perfusion with fluids of various composition, and the firing of the nerve fibers is recorded (often both pressoreceptor and chemoreceptor firings are seen at the same time). Chemoreceptor fibers are between 2 and 5 μ in diameter.[61] Conduction velocities are 7 to 12 m per sec. In general, recording is more difficult from chemoreceptor than from pressoreceptor fibers, probably because of the smaller size of the former. Von Euler et al.[82] found that chemoreceptor fibers in cats, anesthetized with chloralose but breathing normal room air, had slight tonic activity. The chemoreceptors responded to a 4 per cent fall in blood oxygen saturation— i.e., from 100 to 96 per cent (Fig. 10–19). Changes in carbon dioxide concentrations caused the chemoreceptors to fire even when the alveolar CO_2 tension was below normal (30 mm Hg). The firing was greatly increased by anoxia or by increased carbon

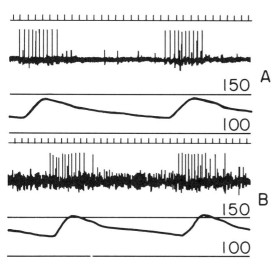

Figure 10–19 Upper record, time marks at 50 imp/sec; middle record, firing of carotid sinus nerve; lower record, arterial blood pressure. A, Animal breathing air. A burst of impulses from a pressoreceptor fiber is seen during each systolic rise in arterial pressure, and a few nerve impulses are seen between these bursts. B, Animal breathing 10 per cent oxygen in nitrogen. Note the greatly increased chemoreceptor activity seen as an increase in baseline firing between the bursts of activity of the pressoreceptor fibers. (After Heymans and Neil, *Reflexogenic areas of the cardiovascular system.* London, J. & A. Churchill, Ltd., 1958.)

dioxide or decreased pH (for gases the partial pressure of the dissolved gas, rather than the gas content, is important). There is some question as to the importance of regulation by the chemoreceptors under normal conditions, because the arterial concentrations of gases at the carotid and aortic sites do not appear to change significantly, except in fairly severe anoxia, hemorrhage, hypercapnia and the like.[44, 77, 82] Sinusoidal changes in inspired CO_2 sufficient to change respiration markedly did not significantly alter heart rate in the studies of Stoll.[77]

Whether or not the chemoreceptors are constantly responding over the normal range of blood gas concentration, they certainly do function in severe stress. *Increased* firing of the chemoreceptors has cardiovascular effects similar in some respects to the effects of *decreased* arterial pressure; i.e., increased chemoreceptor firing increases peripheral resistance. When the oxygen in arterial blood is decreased or the carbon dioxide is increased, there is ordinarily a reflex tachycardia. Although this response seems appropriate, it apparently is *not* due to increased firing of the peripheral chemoreceptors. Bernthal et al.[11] have shown that hypoxia or hypercapnia of the receptors alone—i.e., and not of the medullary regions—produces a bradycardia. The tachycardia, resulting from generalized hypoxia or hypercapnia, is probably due to an effect of altered blood gases on the medullary centers (*vide infra*). Further, as postulated by Daly and Scott,[23, 24] the tachycardia may be secondary to hyperpnea produced by the gases.

Central Chemoreceptors. As far as we know, the nerve fibers from peripheral chemoreceptors run to the same medullary regions as the fibers from pressoreceptors. The medullary centers are sensitive to changes in the tension of O_2 and CO_2 in the perfusing blood. The effects are most prominent in regulation of respiration and are treated in detail elsewhere in this text (Chap. 23, Vol. II), but cardiovascular effects are also noted. The effects of systemic anoxia or hypercapnia are similar to those from decreased stimulation of the baroreceptors. Gernandt et al.[32] believe that the sensitivity is primarily to CO_2, because effects on splanchnic nerve discharge of increasing or decreasing oxygen in cerebral perfusion fluid (but not effects of CO_2) can be eliminated by buffer nerve section. Anoxia alone appears to depress the vasomotor center; *i.e.*, with a fall in O_2 or an increase in CO_2, there are increases in peripheral resistance, heart rate and possibly strength of cardiac contraction. Of course, effects of breathing abnormal concentrations of O_2 or CO_2 could be due to either peripheral or medullary receptors.

OTHER REFLEXES

Spinal Vasomotor Effects. Sectioning the spinal cord is followed initially by a period of spinal shock in which the blood pressure is extremely low; after this period, some vasomotor tone returns in blood vessels below the section. In addition, there is some evidence of residual reflex activity,[1, 26] including limited cutaneous vasoconstriction in man from stimulation of pain fibers. Some reciprocal vasomotor activity also crosses the cord; *i.e.*, warming one extremity in monkeys (see Chap. 12, Vol. I) below a spinal section leads to a vasoconstriction of the opposite extremity, whereas a vasodilation occurs in the warmed extremity. Stimulating the central end of a cut sensory or motor nerve to an extremity may lead to vasomotor changes in that extremity.

The Axon Reflex. The axon reflex apparently does not require participation of spinal vasomotor neurons or of higher centers. For instance, stimulation of an afferent fiber running from the skin can result in a dilatation of nearby cutaneous blood vessels. This reflex is considered to arise because afferent fibers from cutaneous mechanoreceptors may run to ganglia which give off collateral nerves to blood vessels. The axon reflex may be important in the response to inflammation, and the spinal reflexes may operate in response to pain or in the regulation of cutaneous blood flow, but the contribution of these reflexes to cardiovascular regulation is not yet quantitated; indeed, they have been referred to as pseudoreflexes.[45]

Low Pressure Baroreceptors. The arterial baroreceptors and the arterial and medullary chemoreceptors are generally accepted as the major receptors controlling arterial pressure and blood gas composition. Additional receptors which discharge at cardiac frequencies can be electrophysiologically identified in both atria and in the pulmonary artery, and are referred to as low pressure receptors. The impulses from these travel centrally in the vagus nerve. Low pressure receptors regulate the composition and tonicity of the blood (Chap. 26, Vol. II). Atrial receptors show a discharge pattern that corresponds to the changes in atrial pressure and are thus clearly different from aortic receptors. Receptors are also found in the pulmonary artery. When pressure changes are imposed on the pulmonary artery, at times with different changes in the arterial system, the discharge pattern, in sensory fibers from these, follows the pulmonary arterial pressure changes.

There is a substantial literature, extending over 50 years, pertaining to the reflex effect of these receptors. Some writers consider the entire low pressure area as a source of sensory signals which influence cardiovascular performance. In some studies, pressure in a single part of the low pressure area is controlled and the reflex effects are studied. As indicated (Chap. 26, Vol. II), atrial receptors have a clearly established role in regulation of the volume and composition of body fluids.[41] Our concern here is with their effect on cardiac output and peripheral resistance.

The Bainbridge Reflex; the McDowall Reflex. In 1915 Bainbridge found that cardiac acceleration was induced by a rise in venous pressure, produced by intravenous infusion of saline or blood.[47] Section of the sympathetic fibers to the heart reduced this response, as did atropinization (which causes a chemical motor vagotomy). Bilateral vagotomy, however, prevented the response. Bainbridge concluded that the increased heart rate was due to stimulation of receptors in the great veins and possibly in the right atrium. The history of this reflex since Bainbridge's day is confusing. Indeed, several investigators have since been unable to reproduce it, despite intense effort. In experiments of Aviado *et al.*[7] increased pressure in the isolated right atrium caused a reflex bradycardia and hypotension. Kinnison *et al.*[51] have provided evidence supporting the existence of the reflex. Possibly related to this proposed reflex are the passive effects on heart rate of stretching the atrium and great veins.[47, 63] Many pharmacologic and physiologic studies have shown that strips of excised atrial muscle must be stretched slightly to make them beat spontaneously. This might account for the acceleration of heart rate with increased venous (and atrial) pressure. The apparent loss of the supposed reflex response[8] after vagotomy might have resulted from the fact that vagotomy leads to an increase in heart rate. McDowall[59] in 1924 proposed a reflex quite opposite in its general effects to the Bainbridge reflex. After sectioning the vagi in cats which had been subjected to severe hemorrhage, McDowall observed a decrease in systemic blood pressure. He proposed that, prior to vagotomy, atrial receptors stimulated by a fall of atrial pressure had prevented a fall of systemic blood pressure. He believed that the atrial receptors induced a vasoconstriction which maintained arterial pressure. Vagotomy, he believed, abolished the afferent impulses from the atria. Although the effector organs are different, the effect of this reflex (on blood pressure) would be opposite to that proposed in the Bainbridge reflex.

Ledsome and Linden[55] found a cardiac acceleration after distension of the left atrium. Edis *et al.*,[28] however, reported that stimulation of left atrial receptors produces a vasodilation accompanied by bradycardia if the initial heart rate is over 140 to 150 and by tachycardia when the heart rate is below this figure. Pelletier *et al.*[65] studied reflex responses to hemorrhage in dogs after procedures designed to denervate the arterial receptors. They found a vasoconstriction, tachycardia and splanchnic venoconstriction which they attributed to stimulation of low pressure baroreceptors.

Ventricular Receptors. Receptors within the ventricular walls (strangely) can be stimulated to increase their firing rate by administration of drugs from the veratrine family. There is some evidence that increased pressure in the left ventricle leads to reflex bradycardia and vasodilation.[7, 22] The place of ventricular receptors in circulation regulation remains obscure.

Motor Effects Involving the Veins. Like other blood vessels, many veins have a smooth muscle layer in their wall and in some vascular beds they can reflexly change their caliber. The general view of venous regulation assumes that, if the veins did constrict and change their capacity, the filling pressure of the ventricles would be increased, resulting in greater cardiac pumping through the Starling mechanism and a *rise* in systemic arterial pressure. It has additionally been theorized that, since the veins contain such a large portion of the total blood volume (two-thirds if the lungs are included), they can easily change the arterial pressure if slight venoconstriction moves blood into the arterial side of the circulation. Support for this view comes from Alexander[3] (Fig. 10–20), who observed that changes in carotid pressure led to changes in the pressure-volume relationship of the veins. Recently, Bevegård and Shepherd[12] have found evidence for cutaneous *veno*constriction in exercise, although they do not consider venous responses important in ordinary pressure regulation.[13] Greenway *et al.*[37, 38] have found substantial release of blood by the feline spleen and liver in response to sympathetic nerve stimulation, and Brooksby and Donald[18] have found similar results. Pelletier *et al.*[65] found a variety of responses,

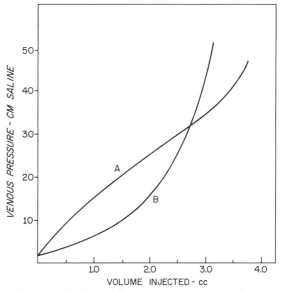

Figure 10–20 Volume-pressure relationship of veins in an isolated intestinal loop. Curve *A* was taken when the mean arterial pressure was raised to 310 mm Hg by vagal stimulation. Curve *B* is a control taken at mean arterial pressure of 82 mm Hg. Curve *A* is considered to show venoconstriction. (After Alexander, *Circulat. Res.*, 1954, 2, 405–409.)

including splanchnic venoconstriction, after hemorrhage in the dog. The decrease of blood volume in the vascularly isolated canine splanchnic circulation occurred in response to sympathetic activation through carotid occlusion, sympathetic stimulation and hemorrhage. Note that the canine spleen differs from the human spleen in having a contractile muscular coat innervated by the sympathetic nervous system. Active participation of the veins in circulatory control is difficult to prove because of the lack of measurable changes in venous pressure. There has been some conflict in the literature regarding venous reflexes,[3, 15] although venous constriction is now generally accepted.[75] Venous reactions will be discussed when separate vascular beds are discussed.

Sympathetic Vasodilator Fibers. Blood vessels of skeletal muscle are innervated by adrenergic sympathetic fibers which respond in an expected fashion to baroreceptor activity. In addition, a cholinergic sympathetic vasodilator innervation of muscle blood vessels has been described, particularly by Barcroft and Swan.[9] In their experiments, sympathetic vasodilator fibers were shown to be active when fainting was caused by passive tilting in human subjects (Fig. 10–21). Many other attempts have been made to find the physiologic importance of these reflexes. In certain studies, sympathetic vasodilator fibers appeared active during severe fright.[14] However, the writer shares the opinion of Uvnäs regarding sympathetic vasodilation: "We know virtually nothing of its functional significance."[80]

Respiratory-Circulatory Interrelationship. In 1847 de Cyon and Ludwig[21] described changes in heart rate associated with respiration. This condition, referred to as sinus arrhythmia, is most common in children and young adults. Angelone and Coulter[6] have shown that the phase relationship between respiration and heart rate changes with respiratory frequency, although at normal respiratory rates there is usually an increased heart rate during inspiration. Several mechanisms might be responsible for the arrhythmia: (i) There may be a direct effect of the respiratory center on the cardiovascular centers in the medulla. (ii) Discharge of lung afferent nerves, which are concerned with respiratory regulation, may alter the activity of the respiratory center

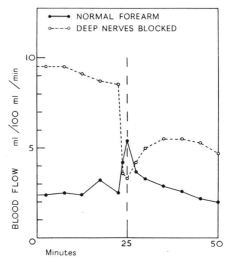

Figure 10–21 Changes in blood flow during induced posthemorrhagic fainting. The solid curve shows the increase in blood flow in normal forearms. The dotted curve shows the decrease in flow in "chemically sympathectomized" forearms. The fact that the normal forearms show greater flow than the blocked forearms is considered to show active sympathetic vasodilation. (From Barcroft and Swan, *Sympathetic control of human blood vessels.* London, Edward Arnold, 1952.)

and therefore contribute to the aforementioned effect, both of these leading to an increase in heart rate during inspiration. Possible indirect effects are as follows: (i) The increase in movement of blood into the thorax during inspiration may increase the discharge of some cardiovascular receptors and reflexly alter the heart rate. (ii) Increased ventricular filling, according to the Starling mechanism, may result in an increase in stroke volume, raising the pressure, activating the arterial presso-receptors and reflexly changing rate.

Reactions During Hemorrhage; Shock. The effects of hemorrhage on the heart can be likened to depriving a pump of the fluid to be pumped. Obviously, with complete loss of circulating blood volume the heart cannot function as a pump. Insufficiency of fluid for normal pumping is apparent fairly early in hemorrhage (stroke volume falls). Slow hemorrhage (15 per cent of blood volume) may not decrease blood pressure.[70] However, as stated, the stroke volume is decreased, and the heart rate does not increase sufficiently to compensate for the decreased stroke volume; *i.e.*, the cardiac output is decreased. If the pressure is maintained, peripheral resistance must increase. Concerning the carotid and aortic reflexes, we have seen that the receptors are sensitive not only to mean pressure but also to the rate of change of pressure. A decrease in the pressure pulse, as occurs in hemorrhage, is thus interpreted as a lower mean pressure; i.e., the receptors send fewer impulses to the central nervous system. In this condition we would expect a reflex vasoconstriction and possibly a venoconstriction. As indicated earlier, some investigators believe that low pressure receptors play a role in adjustment to hemorrhage.[65] The term "venoconstriction" can be extended to include not only constriction of the veins, but also a release of blood from the liver, spleen, lungs and any other hypothetical blood depots. Most of these effects have been noted.[60, 65]

With severe hemorrhage, blood flow to the brain may be insufficient. Here, the chemoreceptor reflexes, involving both the sinoaortic receptors and the medullary centers, may be activated by changes in the tissue concentrations of O_2, CO_2 or H^+. These reflexes are so potent that they can often raise the pressure after the sinoaortic receptors have exerted their maximal effect.[71] Also, with a fall in capillary pressure caused by vasoconstriction, the balance of fluid transfer across the capillary is so changed that fluid moves from the extracellular space back into the blood vessels, replacing some of the blood loss.

Neurogenic and Renal Hypertension. If afferent nerves from both aortic and carotid receptors are sectioned in experimental animals, the arterial blood pressure rises, sometimes to a mean level of 180 to 300 mm Hg. This procedure is not easy to accomplish, because, as stated above, vagotomy leads to death from noncirculatory causes. However, the rise in pressure from denervation has been clearly demonstrated in several studies.[29, 52] This maneuver produces a vasoconstriction similar to that expected if the pressoreceptors are exposed to pressures near zero. Sometimes it produces an increased cardiac output. High blood pressure in humans does not often appear to be caused by decreased sensitivity of the carotid sinus receptors. The most important form of high blood pressure in humans, referred to as "essential hypertension," is of unknown origin. This condition is operationally defined as a blood pressure greater than 135/90 (120/80 is "normal"), although other levels are used at times. Since "normal" blood pressure changes with age, the clinical presence of disease is diagnosed at different pressures for different age groups. In essential hypertension, the systolic, diastolic and mean blood pressures are all elevated. In the early stages this condition is classified as benign. Here blood pressure fluctuates widely without much evidence of arteriosclerosis or other abnormalities. After a time the pressure remains high, does not return to normal and cannot be controlled by sedatives. Because of the high blood pressure, the heart hypertrophies, and blood vessel walls become thickened, decreasing the size of the vessel lumina.

Malignant hypertension is far more serious, with blood pressures as high as 260/150. Changes are much more severe and may even include necrosis of the blood vessels (commonly seen in the retina as papilledema) and, at times, renal failure. This condition, if prolonged sufficiently, leads to death from heart failure, hemorrhage, vascular thrombosis or renal failure.

These conditions appear quite different from neurogenic hypertension, in which the heart rate is high and there is little sign of reflex response to carotid sinus pressure changes. In essential hypertension, at least in the benign stage, there is a reflex response to pressure changes in the carotid artery.

Renal Hypertension. In 1934 Goldblatt[34] demonstrated that chronic partial occlusion of one renal artery can produce a maintained high blood pressure. Hypertension occurs more surely and rapidly and is more severe if the second kidney is removed. There are several techniques for producing hypertension by manipulation of the kidney or its blood supply. This type of hypertension is not neurogenic, because it can be caused by altering the circulation to a kidney transplanted into the neck or to a denervated kidney. Present theory relates this hypertension to a series of chemical changes. The ischemic kidney produces a substance known as renin, which acts on a pseudoglobulin of the blood to produce a pressure substance known as angiotonin. Angiotonin, also called hypertensin, appears to be a potent vasoconstrictor. Hypertension in the human is at times clearly of renal origin and can be cured by removal of a diseased kidney or of a kidney with inadequate blood supply. Whether or not essential hypertension is of renal origin has been a subject of continued controversy.[34, 78]

Edematous Hypertension. If total body fluid is accidentally allowed to increase, through dialysis (artificial kidney) or other means, hypertension ensues. This finding, initially important in the discussion of the origin of renal hypertension, is now especially important when patients with chronic renal failure are treated by periodic dialysis. Although no direct evidence exists, it appears that increased peripheral resistance in this condition may be due to (i) edema of the arterioles, causing the lumina to become smaller and the resistance to increase; or (ii) edema of the pressoreceptors, which lose their ability to respond to the blood pressure, particularly to the pulsatile component of the normal blood pressure. Suggestions relating these effects to hypertension have been made by Peterson[66] and Tobian.[78]

The role of the carotid and aortic pressoreceptors in human hypertension remains to be elucidated. The questions may be stated as follows: (i) Are the ordinary pressoreceptor reflexes inactive during hypertension—*i.e.*, is the level of pressure regulation reset? (ii) If that question is answered negatively, how can hypertension develop if the pressoreceptor reflexes are active? Studies by McCubbin *et al.*[58] of chronic experimental renal hypertension in the dog indicate that sensitivity of the pressoreceptors may be decreased.

Changes During Exercise. During exercise the arterial pressure and pulse rate are generally higher than at rest, whereas, as indicated in Chapter 11, Volume II, stroke volume appears to be relatively unaffected. Probably because of the increase in locally produced metabolites, there is extensive vasodilation in exercising muscle. This appears to be compensated for by the increase in heart rate with a maintained stroke volume and by a vasoconstriction in other vascular beds. Even here the picture is confusing. Van Citters and Franklin[81] find no renal vasoconstriction in Alaskan sled dogs during maintained heroic exercise. In the human, evidence for renal and splanchnic vasoconstriction is more convincing.[16, 83] Detailed discussion appears in Chapters 12 to 14, Volume II, where changes in muscle, skin and viscera are dis-

cussed. The cardiovascular and respiratory responses to exercise cannot be satisfactorily explained at present. There is no known "error signal" for the base response —*i.e.*, no known variable (pressure, blood gases) which is sensed by physiological receptors is displaced from its normal value so that it might induce observed changes in exercise.

Important nervous influences probably descend from higher centers to the hypothalamus and thence to the medullary regions, causing an increased sympathetic discharge to the periphery during exercise.[75a] Certainly, emotional cardiovascular responses would appear to be similar. The fact that blood pressure at times rises slightly above normal during exercise should not be taken as an indication that baroreceptor reflexes are not functioning. In fact, they might be functioning maximally during the exercise response; or if discharge from some higher center is changed, thereby effectively changing the "level setting" of the baroreceptors, they might be keeping pressure at a new elevated level.

SUMMARY OF VASOMOTOR REGULATION

A summary of the regulation discussed in this chapter is relatively simple. Indeed, if one adds to it the major components in respiratory regulation and ignores the details of regulation from higher centers (plus several smaller reflex control systems), a single diagram which is not too complicated can summarize all regulation (Fig. 10–22). Regulation arises from two kinds of receptors: chemoreceptors and baroreceptors. When arterial pressure falls, activity in the baroreceptor fibers is decreased, resulting in decreased vagal activity and increased sympathetic activity. This will increase heart rate, stroke volume and peripheral resistance. These in turn will alter cardiac output and the blood pressure will be restored, thus removing the fall in blood pressure which triggered the reflex.

Similarly, if there is a rise in CO_2 or a fall in O_2 at the chemoreceptors, it will also decrease vagal and increase sympathetic activity and have similar effects on heart rate, stroke volume and peripheral resist-

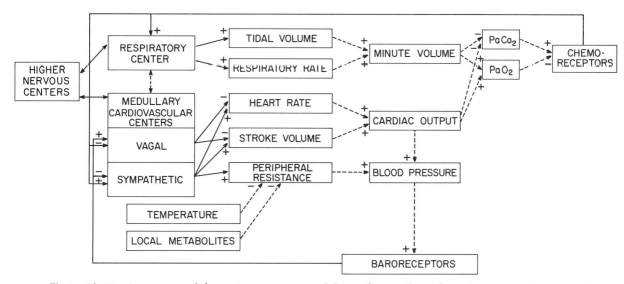

Figure 10–22 A summary of the major components of the cardiovascular and respiratory control systems. A plus sign indicates that an increase in the input into one of the components increases its output. Thus, when the pressure rises at the baroreceptor the sensory nerves will increase firing, and there will be a decreased firing in sympathetic fibers with an increased firing in the vagal fibers. This will produce a vasodilation, decreased heart rate, decreased stroke volume, and a fall in peripheral resistance. The changes in heart rate and stroke volume will decrease the cardiac output which, combined with the fall in resistance, will decrease the blood pressure which, in turn, will decrease the pressure at the baroreceptors.

Increase in CO_2 or decrease in O_2 at the chemoreceptors will increase the sensory discharge, causing a decreased vagal and increased sympathetic activity. This will result in increased heart rate, increased stroke volume and increased peripheral resistance. The heart rate and stroke volume changes produce an increase in cardiac output which, combined with the increased peripheral resistance, increases the blood pressure. In addition, there will be respiratory effects increasing the respiratory minute volume, and these two factors should result in increased carriage of O_2 to the tissues.

ance, with consequent increase in flow which hopefully will remove the change in blood gases. In addition, these changes in gas concentrations at the chemoreceptors increase activity of the respiratory center which increases respiration, so that a larger amount of gas exchange takes place in the lungs. This maintains the blood gas concentration in the face of increased cardiac output.

REFERENCES

1. Alexander, R. S. The effects of blood flow and anoxia on spinal cardiovascular centers. *Amer. J. Physiol.*, 1945, *143*, 698–708.
2. Alexander, R. S. Tonic and reflex functions of medullary sympathetic cardiovascular centers. *J. Neurophysiol.*, 1946, 9, 205–217.
3. Alexander, R. S. The participation of the venomotor system in pressor reflexes. *Circulat. Res.*, 1954, 2, 405–409.
4. Allison, J. L., Sagawa, K., and Kumada, M. An open-loop analysis of the aortic arch barostatic reflex. *Amer. J. Physiol.*, 1969, *217*, 1576–1584.

5. Angell James, J. E., and Daly, M. de B. Effects of graded pulsatile pressure on the reflex vasomotor responses elicited by changes of mean pressure in the perfused carotid sinus-aortic arch regions of the dog. *J. Physiol. (Lond.)*, 1971, *214*, 51–64.
6. Angelone, A., and Coulter, N. A., Jr. Respiratory sinus arrhythmia: a frequency dependent phenomenon. *J. appl. Physiol.*, 1964, *19*, 479–482.
7. Aviado, D. M., Jr., Li, T. H., Kalow, W., Schmidt, C. F., Turnbull, G. L., Peskin, G. W., Hess, M. E., and Weiss, A. J. Respiratory and circulatory reflexes from the perfused heart and pulmonary circulation of the dog. *Amer. J. Physiol.*, 1951, *165*, 261–277.
8. Bainbridge, F. A. The influence of venous filling upon the rate of the heart. *J. Physiol. (Lond.)*, 1915, *50*, 65–84.
9. Barcroft, H., and Swan, H. J. C. *Sympathetic control of human blood vessels.* Baltimore, Williams and Wilkins, 1953.
10. Bernard, C. *De la physiologie générale.* Paris, Hachette et Cie., 1872.
11. Bernthal, T., Greene, W., Jr., and Revzin, A. M. Role of carotid chemoreceptors in hypoxic cardiac acceleration. *Proc. Soc. exp. Biol. (N. Y.)*, 1951, *76*, 121–124.
12. Bevegård, B. S., and Shepherd, J. T. Changes in tone of limb veins during supine exercise. *J. appl. Physiol.*, 1965, *20*, 1–8.

13. Bevegård, B. S., and Shepherd, J. T. Circulatory effects of stimulating the carotid arterial stretch receptors in man at rest and during exercise. *J. clin. Invest.*, 1966, *45*, 132–142.

14. Blair, D. A., Glover, W. E., Greenfield, A. D. M., and Roddie, I. C. Excitation of cholinergic vasodilator nerves to human skeletal muscles during emotional stress. *J. Physiol. (Lond.)*, 1959, *148*, 633–647.

15. Blair, D. A., Glover, W. E., Greenfield, A. D. M., and Roddie, I. C. The increase in tone in forearm resistance blood vessels exposed to increased transmural pressure. *J. Physiol. (Lond.)*, 1959, *149*, 614–625.

16. Bradley, S. E. Variations in hepatic blood flow in man during health and disease. *New Engl. J. Med.*, 1949, *240*, 456–461.

17. Bronk, D. W., and Stella, G. The response to steady pressures of single end organs in the isolated carotid sinus. *Amer. J. Physiol.*, 1935, *110*, 708–714.

18. Brooksby, G. A., and Donald, D. E. Dynamic changes in splanchnic blood flow and blood volume in dogs during activation of sympathetic nerves. *Circulat. Res.*, 1971, *29*, 227–238.

19. Cannon, W. B. *The wisdom of the body.* London, Kegan Paul, 1932.

20. de Castro, F. Sur la structure de la synapse dans les chemocepteurs: leur mécanisme d'excitation et rôle dans la circulation sanguine locale. *Acta physiol. scand.*, 1951, *22*, 14–43.

21. de Cyon, E., and Ludwig, C. F. Die Reflexe eines der sensiblen Nerven des Herzens auf die Motorischen der Blutgefässe. *Arb. Physiol. Anst. Leipz.*, 1866, *1*, 128–149.

22. Daly, I. de B., and Verney, E. B. The localisation of receptors involved in the reflex regulation of the heart rate. *J. Physiol. (Lond.)*, 1927, *62*, 330–340.

23. Daly, M. de B., and Scott, M. J. The effects of stimulation of the carotid body chemoreceptors on heart rate in the dog. *J. Physiol. (Lond.)*, 1958, *144*, 148–166.

24. Daly, M. de B., and Scott, M. J. An analysis of the primary cardiovascular reflex effects of stimulation of the carotid body chemoreceptors in the dog. *J. Physiol. (Lond.)*, 1962, *162*, 555–573.

25. DeGeest, H., Levy, M. N., Zieske, H., and Lipman, R. I. Depression of ventricular contractility by stimulation of the vagus nerves. *Circulat. Res.*, 1965, *17*, 222–235.

26. Downman, C. B. B., and McSwiney, B. A. Reflexes elicited by visceral stimulation in the acute spinal animal. *J. Physiol. (Lond.)*, 1946, *105*, 80–94.

27. Ead, H. W., Green, J. H., and Neil, E. A comparison of the effects of pulsatile and non-pulsatile blood flow through the carotid sinus on the reflexogenic activity of the sinus baroceptors in the cat. *J. Physiol. (Lond.)*, 1952, *118*, 509–519.

28. Edis, A. J., Donald, D. E., and Shepherd, J. T. Cardiovascular reflexes from stretch of pulmonary vein–atrial junctions in the dog. *Circulat. Res.*, 1970, *27*, 1091–1100.

29. Ferrario, C. M., McCubbin, J. W., and Page, I. H. Hemodynamic characteristics of chronic experimental neurogenic hypertension in unanesthetized dogs. *Circulat. Res.*, 1969, *24*, 911–922.

30. Franz, G. N. Nonlinear rate sensitivity of the carotid sinus reflex as a consequence of static and dynamic nonlinearities in baroreceptor behavior. *Ann. N.Y. Acad. Sci.*, 1969, *156*, 811–824.

31. Franz, G. N., Scher, A. M., and Ito, C. S. Small signal characteristics of carotid sinus baroreceptors of rabbits. *J. appl. Physiol.*, 1971, *30*, 527–535.

32. Gernandt, B., Liljestrand, G., and Zotterman, Y. Efferent impulses in the splanchnic nerve. *Acta physiol. scand.*, 1946, *11*, 230–247.

33. Glick, G., and Braunwald, E. Relative roles of the sympathetic and parasympathetic nervous systems in the reflex control of heart rate. *Circulat. Res.*, 1965, *16*, 363–375.

34. Goldblatt, H. *The renal origin of hypertension.* Springfield, Ill., Charles C Thomas, 1948.

35. Gootman, P. M., and Cohen, M. I. Efferent splanchnic activity and systemic arterial pressure. *Amer. J. Physiol.*, 1970, *219*, 897–903.

36. Green, J. H. Physiology of baroreceptor functions: mechanism of receptor stimulation. In: *Baroreceptors and hypertension*, P. Kezdi, ed. Oxford, Pergamon Press, 1967.

37. Greenway, C. V., Lawson, A. E., and Stark, R. D. Vascular responses of the spleen to nerve stimulation during normal and reduced blood flow. *J. Physiol. (Lond.)*, 1968, *194*, 421–433.

38. Greenway, C. V., Stark, R. D., and Lautt, W. W. Capacitance responses and fluid exchange in the cat liver during stimulation of the hepatic nerves. *Circulat. Res.*, 1969, *25*, 277–284.

39. Grodins, F. S. *Control theory and biological systems.* New York, Columbia University Press, 1963.

40. Hellner, K., and von Baumgarten, R. Über ein Endigungsgebiet afferenter, kardiovasculärer Fasern des Nervus vagus im Rautenhirn der Katze. *Pflügers Arch. ges. Physiol.*, 1961, *273*, 223–234.

41. Henry, J. P., Gauer, O. H., and Reeves, J. L. Evidence of the atrial location of receptors influencing urine flow. *Circulat. Res.*, 1956, *4*, 85–90.

42. Hering H. E. *Die Karotissinus Reflexe auf Herz und Gefässe.* Leipzig, D. Steinkopff, 1927.

43. Heymans, C. *Le sinus carotidien.* London, H. K. Lewis, 1929.

44. Heymans, C., and Neil, E. *Reflexogenic areas of the cardiovascular system.* Boston, Little, Brown, 1958.

45. Hillarp, N. A. Peripheral autonomic mechanisms, *Handb. Physiol.*, 1960, sec. 1, vol. II, 979–1006.

46. Humphrey, D. R. Neuronal activity in the medulla oblongata of the cat evoked by stimulation of the carotid sinus nerve. In: *Baroreceptors and hypertension*, P. Kezdi, ed. Oxford, Pergamon Press, 1967.

47. Jensen, D. *Intrinsic cardiac rate regulation.* New York, Appleton–Century–Crofts, 1971.

48. Katona, P. G., Poitras, J. W., Barnett, G. O., and Terry, B. S. Cardiac vagal efferent activity and heart period in the carotid sinus reflex. *Amer. J. Physiol.*, 1970, *218*, 1030–1037.

49. Katona, P. G., Barnett, G. O., and Jackson, W. D. Computer stimulation of the blood pressure control of the heart period. In: *Baroreceptors and hypertension*, P. Kezdi, ed. Oxford, Pergamon Press, 1967.

50. Kezdi, P., and Geller, E. Baroreceptor control of postganglionic sympathetic nerve discharge. *Amer. J. Physiol.*, 1968, *214*, 427–435.

51. Kinnison, G. L., Breeden, C. J., Carmack, R. M., Ballard, B. M., Mel, P. J., and Hemingway, A. Reflex changes in heart rate and ventilation induced by central blood pressure changes. *Amer. J. Physiol.*, 1965, *208*, 1222–1230.

52. Koch, E. *Die reflektorische Selbststeuerung des Kreislaufes.* Dresden, Steinkopff, 1931.

53. Korner, P. I. Integrative neural cardiovascular control. *Physiol. Rev.*, 1971, *51*, 312–367.

54. Landgren, S. On the excitation mechanism of the carotid baroceptors. *Acta physiol. scand.*, 1952, *26*, 1–34.

55. Ledsome, J. R., and Linden, R. J. The effect of distending a pouch of the left atrium on the heart rate. *J. Physiol. (Lond.)*, 1967, *193*, 121–129.

56. Levison, W. H., Barnett, G. O., and Jackson, W. D. Nonlinear analysis of the baroreceptor reflex system. *Circulat. Res.*, 1966, *18*, 673–682.

57. Lindgren, P., and Uvnäs, B. Postulated vasodilator center in the medulla oblongata. *Amer. J. Physiol.*, 1954, *176*, 68–76.

58. McCubbin, J. W., Green, J. H., and Page, I. H. Baroceptor function in chronic renal hypertension. *Circulat. Res.*, 1956, *4*, 205–210.

59. McDowall, R. J. S. A vago-pressor reflex. *J. Physiol. (Lond.)*, 1924, *59*, 41–47.

60. Oberg, B., and White, S. Role of vagal cardiac nerves and arterial baroreceptors in the circulatory adjustments to hemorrhage in the cat. *Acta physiol. scand.*, 1970, *80*, 395–403.

61. Paintal, A. S. The response of pulmonary and cardiovascular vagal receptors to certain drugs. *J. Physiol. (Lond.)*, 1953, *121*, 182–190.

62. Parry, C. H. *An inquiry into the symptoms and causes of the syncope commonly called angina pectoris.* Bath, England, Cruttwell, 1799.

63. Pathak, C. L. The fallacy of the Bainbridge reflex. *Amer. Heart J.*, 1966, *72*, 577–581.

64. Peiss, C. N. Concepts of cardiovascular regulation: past, present and future. In: *Nervous control of the heart*, W. C. Randall, ed. Baltimore, Williams and Wilkins, 1965.

65. Pelletier, C. L., Edis, A. J., and Shepherd, J. T. Circulatory reflex from vagal afferents in response to hemorrhage in the dog. *Circulat. Res.*, 1971, *29*, 626–634.

66. Peterson, L. H. Systems behavior, feed-back loops, and high blood pressure research. *Circulat. Res.*, 1963, *12*, 585–594.

67. Pitts, R. F., Larrabee, M. G., and Bronk, D. W. An analysis of hypothalamic cardiovascular control. *Amer. J. Physiol.*, 1941, *134*, 359–383.

68. Ranson, S. W., and Billingsley, P. R. Vasomotor reactions from stimulation of the floor of the fourth ventricle. Studies in vasomotor reflex arcs. III. *Amer. J. Physiol.*, 1916, *41*, 85–90.

69. Robinson, B. F., Epstein, S. E., Beiser, G. D., and Braunwald, E. Control of heart rate by the autonomic nervous system. Studies in man on the interrelation between baroreceptor mechanisms and exercise. *Circulat. Res.*, 1966, *19*, 400–411.

70. Rushmer, R. F., Van Citters, R. L., and Franklin, D. L. Shock: a semantic enigma. *Circulation*, 1962, *26*, 445–459.

71. Sagawa, K., Taylor, A. E., and Guyton, A. C. Dynamic performance and stability of cerebral ischemic pressor response. *Amer. J. Physiol.*, 1961, *201*, 1164–1172.

72. Salmoiraghi, G. C. "Cardiovascular" neurones in brain stem of cat. *J. Neurophysiol.*, 1962, *25*, 182–197.

73. Scher, A. M., and Young, A. C. Servoanalysis of carotid sinus reflex effects on peripheral resistance. *Circulat. Res.*, 1963, *12*, 152–162.

74. Scher, A. M., and Young, A. C. Reflex control of heart rate in the unanesthetized dog. *Amer. J. Physiol.*, 1970, *218*, 780–789.

75. Shepherd, J. T. Role of the veins in the circulation. *Circulation*, 1966, *33*, 484–491.

75a. Smith, O. A., Jr. Cardiovascular integration by the central nervous system. In: *Physiology and biophysics*, 19th ed., T. C. Ruch and H. D. Patton, eds. Philadelphia, W. B. Saunders Co., 1965.

76. Spickler, J. W., and Kezdi, P. Dynamic response characteristics of carotid sinus baroreceptors. *Amer. J. Physiol.*, 1967, *212*, 472–476.

77. Stoll, P. J. Respiratory system analysis based on sinusoidal variations of CO_2 in inspired air. *J. appl. Physiol.*, 1969, *27*, 389–399.

78. Tobian, L. Interrelationship of electrolytes, juxtaglomerular cells and hypertension. *Physiol. Rev.*, 1960, *40*, 280–312.

79. Tuckman, J., Slater, S. R., and Mendlowitz, M. The carotid sinus reflexes. *Amer. Heart J.*, 1965, *70*, 119–135.

80. Uvnäs, B. Central cardiovascular control. *Handb. Physiol.*, 1960, sec. 1, vol. II, 1131–1162.

81. Van Citters, R. L., and Franklin, D. L. Cardiovascular performance of Alaska sled dogs during exercise. *Arctic Res.*, 1969, *24*, 35–42.

82. Von Euler, U. S., Liljestrand, G., and Zotterman, Y. The excitation mechanism of the chemoreceptors of the carotid body. *Skand. Arch. Physiol.*, 1939, *83*, 132–152.

83. Wade, O. L., and Bishop, J. M. *Cardiac output and regional blood flow.* Oxford, Blackwell, 1962.

84. Waller, A. Experimental researches on the functions of the vagus and the cervical sympathetic nerves in man. *Proc. roy. Soc. (Lond.)*, 1862, *11*, 302–307.

85. Wang, S. C., and Borison, H. L. An analysis of the carotid sinus cardiovascular reflex mechanism. *Amer. J. Physiol.*, 1947, *150*, 712–728.

86. Wang, S. C., and Ranson, S. W. Descending pathways from the hypothalamus to the medulla and spinal cord. Observations on blood pressure and bladder responses. *J. comp. Neurol.*, 1939, *71*, 457–472.

87. Warner, H. R., and Cox, A. A mathematical model of heart rate control by sympathetic and vagus efferent information. *J. appl. Physiol.*, 1962, *17*, 349–355.

CHAPTER 11 CONTROL OF CARDIAC OUTPUT

by ALLEN M. SCHER

CARDIAC OUTPUT, HEART RATE, STROKE VOLUME

Adequate cardiovascular performance is necessary for all functions of the living organism. However, alterations in cardiovascular performance are brought about by a limited number of motor responses. These motor responses include changes in the caliber of resistance and capacitance vessels, discussed in Chapters 3 and 10, Volume II, and changes in the output of the heart. These changes in cardiac output are the subject of this chapter.

The regulation of the output of the heart is accomplished by alteration of the volume ejected during each heart beat, the stroke volume, and by changes in heart rate. Each ventricle is a hollow chamber with muscular walls. During diastole, as the heart fills, these chambers reach a maximal size. During systolic contraction, as the heart empties, the muscle fibers shorten and the

chambers become smaller as blood flows into the aorta and pulmonary artery. The stroke volume is the difference between end diastolic and end systolic volumes. When an individual is in a constant state of rest or activity, the stroke volume is ordinarily the same for the right and left ventricles. The product of stroke volume and heart rate is cardiac output ($CO = HR \times SV$). The cardiac output is, by definition, the minute output of one ventricle.

CONTROL OF STROKE VOLUME

Stroke volume is altered by mechanical factors and by factors which alter the contractility of the heart. The mechanical factors include the degree of ventricular diastolic filling before ejection begins, sometimes called preload, and the opposition to ventricular ejection presented by the aortic pressure, sometimes called after-

170

load. The state or contractility of the cardiac muscle may change because of several influences. In essence these alterations in contractility (inotropic changes as contrasted to chronotropic changes which involve heart rate) are chemical and involve the molecular reactions underlying muscle contraction. They are brought about by changes in heart rate, by the autonomic nervous system and by anoxia among other factors.

The mechanical factors and those concerned with the contractile state of the heart are closely interwoven in consideration of changes in stroke volume. Characteristics of cardiac muscle and its contraction, presented earlier in Chapter 4, Volume II, serve as a basis for consideration of the heart as a pumping organ.

Mechanical Factors Affecting Stroke Volume. STARLING'S LAW OF THE HEART: THE LENGTH-TENSION RELATIONSHIP. Both skeletal and cardiac muscle have a characteristic length-tension relationship. As a muscle is stretched the resting tension increases, and the maximal active tension

recorded during contraction increases to a maximum and then declines. For a cardiac chamber, resting tension would probably be proportional to maximal diastolic volume, and active tension might be approximated by the maximal pressure generated or the volume ejected during contraction.

In 1914, Starling and co-workers[33] devised a heart-lung preparation (Fig. 11–1) which made it possible to determine the equivalent of the length-tension relationship for the isolated heart. In their preparation, ventricular pressure and right atrial pressure were measured with mercury manometers. The combined volume of the two ventricles was measured by a cardiometer which enclosed both chambers and was attached to the heart at the atrioventricular groove. The right atrial pressure (considered to equal left ventricular filling pressure) could be altered by raising an infusion bottle. The (downstream) pressure against which the left ventricle ejected could be varied by altering an artificial resistance placed in series with the aorta. A pressure reservoir in the aortic circuit tended to

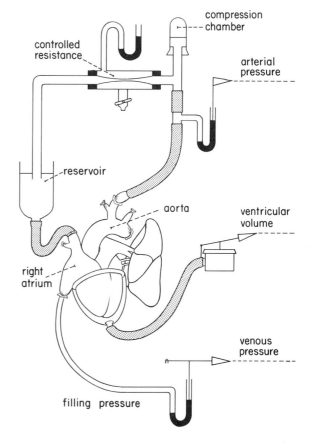

Figure 11–1 Apparatus used by Starling to study the performance of the isolated heart. The filling pressure of the right atrium is controlled by the reservoir at the center left. Pressure in the right atrium is shown by the manometer at the bottom. This pressure is considered the filling pressure of the left ventricle. The two ventricles are in a closed container called a cardiometer, which is sealed at the AV groove and which communicates with a second closed container covered with a flexible rubber membrane. Movement of that membrane moves the lever and is an indication of the combined ventricular volume. Blood flows out of the aorta through a separate cannula inserted into one of the aortic branches, then into a compression chamber which depulsates the flow, and then into a variable resistance which compresses the tubing between the aorta and the inflow reservoir. Arterial pressure is measured with a manometer as shown. The heart can thus be made to eject blood against an altered resistance, or the filling pressure of the heart can be raised by elevating the reservoir. (After Rushmer, *Cardiovascular dynamics*, 3rd ed. Philadelphia, W. B. Saunders Co., 1970.)

maintain and depulsate the aortic pressure. Blood flowing through the resistance returned to an inflow reservoir connected to the right atrium to be pumped through the circuit again. Starling and co-workers were primarily interested in the effects of cardiac muscle diastolic fiber length (filling pressure) on cardiac performance. Ventricular performance was estimated from cardiac output; ventricular pressure or diameter was also measured. The output of the heart could be measured by diverting flow into a calibrated container. Stroke volume could also be estimated as the difference between systolic and diastolic volumes measured by the cardiometer.

In one experiment performed with this preparation, the inflow reservoir was raised (Fig. 11–2). The right atrial pressure and the diastolic dimensions of the heart increased rapidly over several beats, and this led to a marked and parallel increase in *arterial* pressure. The difference between diastolic and systolic ventricular size also increased, showing an increase in the stroke output of the heart. The increased diastolic volume resulting from the increased filling ("venous return") caused the heart to eject more at each beat.

Starling's major conclusions are presented in the form of a classic diagram relating maximal diastolic filling volume or end diastolic volume (fiber length) to systolic pressure or stroke volume. This relation-

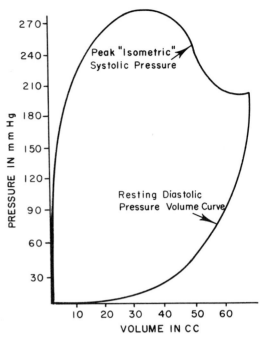

Figure 11–3 Diastolic volume changes and their effect on diastolic and peak systolic pressures on a number of heart beats. This figure is traditionally presented as Starling's "law of the heart." It is also known as a representation of the "Frank-Starling mechanism." It was taken by Starling from Frank's measurements of the diastolic pressure and volume and of the peak systolic pressure in the isometrically contracting frog ventricle. Starling added figures he thought were more reasonable for the canine heart. (After Frank, Z. Biol., 1895, 32, 370–447; Amer. Heart J., 1959, 58, 282–317, and Patterson, et al., J. Physiol. (Lond.), 1914, 48, 465–513.)

ship is known as *Starling's law of the heart* (Fig. 11–3). This curve was adapted by Starling from a curve produced by Frank[15] which described the results of diastolic volume changes on the peak *isovolumetric* pressure developed in the frog ventricle. Starling modified the units to correspond to proper values for the dog. "Starling's law" is sometimes referred to as the "Frank-Starling mechanism."

Starling felt that the surface area of the cardiac fibers at diastole determined the velocity of some (unknown) chemical reactions, which in turn determined the vigor of contraction. The effect of filling pressure on cardiac performance has been attributed to the finding that cardiac muscle and skeletal muscle contract optimally when stretched so that the sarcomeres (contractile units) have a diastolic length of 2.2 μ.[17] It has been proposed that at this sarcomere length the overlap of the sliding filaments, which develop tension in the Huxley model of muscle contraction, is optimal for con-

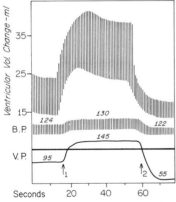

Figure 11–2 An experiment performed by Starling with the apparatus shown in Figure 11–1. The venous pressure is increased; the ventricular diastolic and systolic volumes increase; the distance between them, possibly a measure of the stroke volume, increases; and the arterial pressure rises slightly. The heart has responded to an increase in its filling pressure by increasing the end diastolic ventricular size (and probably the stroke volume). V.P. is atrial pressure in mm H$_2$O, B.P. in mm Hg.

tractile strength and that sarcomeres are, at this length, at the peak of the Starling curve.[17]

At the time when Starling described this relationship and for many years thereafter, it was the only known mechanism by which stroke volume could be altered. For that reason, almost all regulation of stroke volume (and much regulation of cardiac output) was considered to result from changes in diastolic filling.[8] During exercise it was theorized that the heart moved up the ascending limb of the Starling curve, getting larger and ejecting more blood with increased exertion.[4, 20] The failing heart was considered to be on the descending limb of the Starling curve, in which further increases in diastolic volume would decrease rather than increase stroke volume.[4] Changes in stroke volume accompanying changes in heart rate were considered to reflect changes in ventricular filling, because the time available for filling changes with heart rate. Starling's law is still widely invoked to explain stroke volume changes — even when the evidence is quite indirect.

At a later date, Sarnoff and Mitchell[38] utilized flowmeters to measure cardiac output and also measured pressure in the atria and aorta. They altered the ventricular filling pressure in several ways and plotted ventricular "function curves" in which the ventricular end diastolic pressure, or mean left atrial pressure (these were considered indicators of maximal ventricular diastolic volume), was plotted against the left ventricular stroke *work* (Fig. 11–4). Work was computed as the product of mean arterial pressure (actually the mean *driving* pressure, arterial minus atrial pressure, was used) and mean flow (cardiac output). Work increased along a smooth curve as filling pressure increased. Work was used as the "output" variable because of a feeling that metabolism, which should parallel work, was important in the relationship between diastolic volume and systolic pressure. Unfortunately, when work is plotted in this fashion we may not know the pressure and flow, which may be as important or interesting variables as work.* A variety of factors

---

*As indicated in Chapter 4, Volume II, external work is zero when pressure is zero and when pressure is high enough to prevent ejection from the ventricle (flow). Further, as indicated by Levy et al.,[25] work is influenced by arterial pressure when filling pressure is constant. Arterial pressure was not constant in many of Sarnoff's experiments.

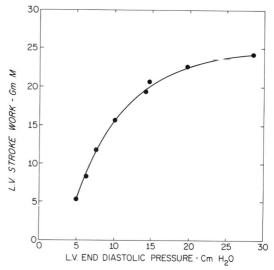

Figure 11–4 If left ventricular end diastolic pressure is raised from 5 cm to approximately 29 cm of water, the left ventricular stroke work increases from 5 to almost 25 gm M. Work is computed as the product of mean pressure and mean flow. (From Sarnoff and Mitchell, *Handb. Physiol.*, 1962, sec. 2, vol. I, 489–532.)

which change the vigor of the contraction of the heart, such as the infusion of sympathomimetic drugs or stimulation of the sympathetic nerves to the heart, produce a family of such curves in which work changes *at a given diastolic pressure*. These changes are discussed below.

Rushmer[36] developed a number of new instruments to assess the performance of the heart in intact unanesthetized animals. These included ventricular diameter gauges (differential transformers, ultrasonic diameter gauges), implantable pressure transducers and ultrasonic blood flowmeters. In animals studied without anesthesia, during the onset of exercise and in other circumstances, there was not a clear relationship between the diastolic filling pressure and the systolic pressure or ventricular dimension changes (indicators of stroke volume) on a beat. Indeed, the relationship sometimes seemed to contradict the Starling hypothesis. This was particularly true of exercise when, contrary to a widely accepted hypothesis (*vide supra*), the heart was found to be smaller in diastole than at rest. Other factors seemed more important than filling pressure in determining the stroke volume (Fig. 11–5).

In addition, technical advances made it possible to study the performance of the human heart. The technical advances included (i) techniques for high-speed radio-

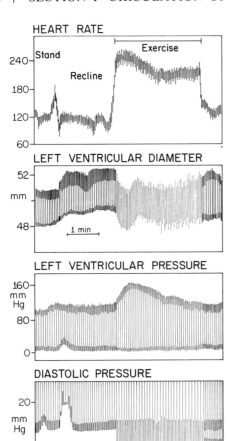

Figure 11–5 The effects of changes in position and of exercise on heart rate, left ventricular diameter, left ventricular pressure and left ventricular diastolic pressure in the dog. As the animal lay down, the left ventricular diameter became larger during both systole and diastole, ventricular pressure did not change appreciably, and ventricular diastolic pressure changed only transiently. Heart rate did not change appreciably. With exercise, the heart rate increased, the left ventricle became smaller in both systole and diastole, the left ventricular systolic pressure rose and the left ventricular diastolic (filling) pressure, as indicated by the lowest pressure recorded in the ventricle, fell. (After Rushmer, *Handb. Physiol.*, 1962, sec. 2, vol. I, 533–550.)

logical viewing of the heart;[26] (ii) the determination of cardiac output by the indirect and direct Fick procedures, particularly the direct Fick technique with cardiac catheterization;[44] and (iii) the recording of intracardiac pressures. The x-ray procedures[26] indicated that during exercise the heart became smaller rather than larger, and as indicated this finding was substantiated by direct measurements on dogs.[36] Further, during a number of procedures that altered the return of blood to the heart[44]

(including hemorrhage, occlusion of cuffs around the extremities and rapid cardiac filling), there was not a clear relationship between ventricular filling pressure and cardiac output.[45] (Heart rate changes were often overlooked in such studies, which indicates the strength of the assumption that the effect of filling pressure on stroke volume was all important.) A number of papers and reviews can be consulted for further discussion of the effects of ventricular filling on stroke volume.[8, 18, 36, 38]

THE EFFECT OF ARTERIAL PRESSURE ON VENTRICULAR CONTRACTION. In addition to the filling pressure, a second mechanical factor of importance in determining the stroke volume is the aortic pressure or "afterload," or downstream pressure. Briefly stated, the heart ejects less against a higher aortic pressure (other factors remaining constant). This factor has been studied most recently by Levy and co-workers[25] (Fig. 11–6). In carefully controlled experiments, increased aortic pressure (when filling pressure was held constant) decreased the stroke volume. The stroke work and power were also *decreased*.

THE ANREP EFFECT (HOMEOMETRIC AUTOREGULATION). One exception to the above has been noted in experiments of Anrep,[1] Sarnoff and Mitchell[38] and Sagawa[37] and is referred to by Sarnoff as "homeometric autoregulation." In Sarnoff's experiments an increased arterial pressure caused an initial decrease in the stroke work, but after this initial decrease and even when filling pressure was held constant, the work done by the ventricle increased and the stroke volume did not appear depressed by the increased afterload. Homeometric autoregulation has not been found in studies of intact animals.[39, 47] Recent evidence[2, 31] indicates that the depression of work immediately following the pressure rise is due to temporary ischemia of the endocardial layers of the ventricular wall, caused by a sudden pressure rise. Work probably rises as the ischemia is overcome (possibly owing to metabolic vasodilation, *vide infra*).

Variables Which Alter the Contractile State of the Heart: Heart Rate and Autonomic Discharge. The two variables discussed previously—filling pressure and arterial pressure—are *mechanical*, and if the contractile *state* of the heart is constant they will determine its performance. Other variables, however, can alter the cardiac state or "contractility." Three such variables seem most important in normal cardiac

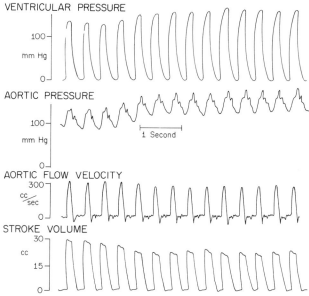

VENTRICULAR PRESSURE

AORTIC PRESSURE

1 Second

AORTIC FLOW VELOCITY

STROKE VOLUME

Figure 11–6 As the ventricular and aortic pressures increased, there was a fall in aortic flow velocity and in stroke volume. These experiments were conducted so that the left atrial pressure and heart rate were constant (pressure was increased by occluding a balloon on the aorta while maintaining filling pressure by occluding the vena cava). Experiments were conducted in an unanesthetized dog.

function: (i) heart rate, (ii) the sympathetic nervous fibers which innervate the heart, and (iii) the vagal fibers which innervate the heart. Of these three, one — heart rate — is (at least during physiological experiments when heart rate is altered by electrical stimulation) an intrinsic factor, *i.e.*, one which originates within the heart. The other two are extrinsic to the heart in that they originate in the central nervous system, even though they involve effects of hormones secreted by the nerve terminations within the heart. Of these three, the sympathetic discharge to the heart seems most important.

HEART RATE AND STRENGTH OF CONTRACTION — THE INTERVAL-STRENGTH RELATIONSHIP. In 1871, Bowditch[6] found that after a period of inactivity the frog ventricle showed a progressive increase in strength of contraction over several beats when stimulated at a constant frequency. He referred to this as *Treppe* (steps) (Chap. 4, Vol. II). The effect of changing stimulating frequency has been extensively investigated recently by several groups.[5, 23, 28] Clear-cut experiments can be conducted, utilizing muscle strips or an isovolumic heart. These preparations allow control of filling pressure or diastolic muscle length and aortic pressure or systolic load. Note that heart rate has an independent effect of its own, but that physiological rate

changes most often occur owing to changes in autonomic discharge to the heart.

The effects of heart rate are complicated, but can be summarized in two statements: (i) A plot of peak systolic pressure vs. interval for a number of maintained intervals shows an increased pressure and presumably increased strength of contraction as the interval increases or decreases from a given level (Chap. 4, Vol. II) (Fig. 11–7). (ii) With transient changes in interval, the strength of contraction increases for longer intervals and decreases with shorter intervals (Chap. 4, Vol. II).

Effects of changes in heart rate on contractility seem associated with movements of calcium ion and particularly with its effects on the coupling of electrical excitation to mechanical contraction. A detailed analysis of these phenomena in terms of two independent inotropic effects has been put forward by Blinks and Koch-Weser.[5] A more detailed discussion is found in Chapter 4, Volume II.

PAIRED PULSE AND COUPLED STIMULATION. The effect of changes in heart rate on contractility can be seen when stimuli at short intervals are delivered to the heart immediately following a normal beat. The extrasystolic beats may be "coupled," that is, a stimulus is delivered at a short interval after a number of beats at a normal interval, or the heart rate may be controlled by a succession of stimuli, some of which are de-

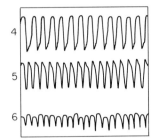

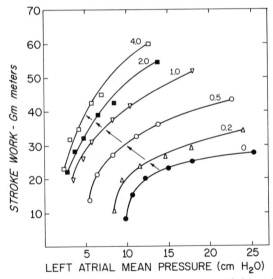

Figure 11-7 Effect of changes in heart rate on peak pressure (Y axis) of an isolate, isovolumic, perfused rat heart stimulated at frequencies of 1 Hz, 3 Hz, 5 Hz, 6 Hz, 8 Hz and 10 Hz respectively in panels 1 through 6. Note that the strength of contraction is greatest at 6 Hz (panel 4), and is less than maximal both above and below this frequency. (After Meijler, *Amer. J. Physiol.*, 1962, *202*, 636–640.)

livered at short intervals. When successive stimuli are delivered at long and short intervals, this latter procedure is called "paired pulse stimulation." If the extrasystole or interpolated beat is sufficiently close in time to the normal beat, there may be no obvious contraction following the extrasystole, but the beat *following* the extrasystole will have a larger-than-normal force of contraction. This phenomenon is referred to as postextrasystolic potentiation. It is seen when a contraction follows the interpolated stimulus and also when there is no contraction. The phenomenon had been noted well before the turn of the last century,[10] but was neglected until recently.[9, 10, 11, 22, 29] The output of the heart may be greatly increased by these types of stimulation. Susceptibility to a standard type of experimental cardiac failure was decreased or eliminated by paired pulse stimulation.[21] Unfortunately, the metabolic cost of paired pulse stimulation is excessive. Studies of Meijler and Durrer[29] indicate that the oxygen requirements of the heart are elevated even more than the contractility is increased. Despite some early hopes, it does not seem profitable to utilize either coupled or paired stimulation to aid the patient with cardiac problems.

THE SYMPATHETIC NERVOUS SYSTEM AND CARDIAC CONTRACTILITY. The sympathetic fibers running to the heart originate in the intermedio-lateral columns of the spinal thoracic segment. They synapse with postganglionic neurons in the paravertebral ganglia and in the cervical ganglia. The postganglionic fibers travel to the heart, either directly from the thoracic ganglia or from the superior cervical ganglia, as the cardiac nerves. As indicated elsewhere (Chap. 10, Vol. II), the sympathetic fibers innervate the atrial and ventricular musculature, the sinus node (pacemaker) and the atrioventricular node. The terminations at the heart are of the β-receptor type, and the transmitter at these junctions is therefore norepinephrine (the effects can be blocked by propranolol) (Chap. 8, Vol. II).

Utilizing procedures which increase the discharge of the sympathetic nerve fibers to the myocardium or which mimic such discharge, Sarnoff produced "families" of function curves, each of which related filling pressure to work. When sympathetic activity was increased, the work was greater at a given filling pressure (Fig. 11–8). Rushmer also stressed the importance of sympathetic discharge to the heart and felt that this was a major factor in stroke volume control.[36]

If other factors which can alter the heart are constant, increased sympathetic discharge to the heart alters the performance of the heart in several ways, all of which characterize a state of increased "contractility." (i) The heart empties more

Figure 11-8 Stroke work as a function of left atrial mean pressure during stimulation of the stellate ganglion at a variety of frequencies. The heart rate was constant. The numbers indicate stimulating frequency in Hz. Note that as the frequency of stimulation is increased the work for a given filling pressure increases, but that one can plot a very smooth curve for work against atrial pressure at each stimulation frequency. (After Sarnoff and Mitchell, *Handb. Physiol.*, 1962, sec. 2, vol. I, 489–532.)

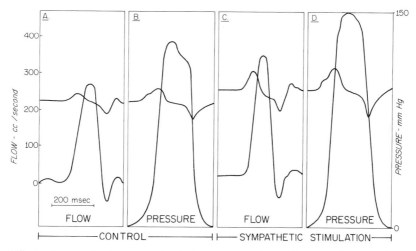

Figure 11-9 Effects of sympathetic nerve stimulation on aortic flow velocity and ventricular pressure, and on the rates of change (time derivatives) of pressure and flow. Small curves near the center of each curve are these derivatives. Sympathetic stimulation increases peak flow and pressure and the derivatives of these. Time required to eject the stroke volume decreases slightly.

completely from a given diastolic volume; stroke volume is thus increased. (ii) The contraction of the muscle fibers is more rapid so that rates of change of pressure and flow in time — dp/dt and df/dt — are greater than in control beats, and peak flow velocities are also increased. (iii) The duration of contraction, or of ejection, is decreased. (iv) Contraction will produce a higher ventricular and aortic systolic pressure. Some of these changes are shown in Figure 11–9.

Surprisingly, the diastolic compliance of the heart (*i.e.*, the diastolic pressure-volume relationship) is not affected by increased sympathetic discharge to the heart[38] (Chap. 4, Vol. II) (Fig. 11–10).

Usually, increased sympathetic discharge to the heart also causes an increase in heart rate, discussed below. The inotropic effects (effects related to strength of contraction, as contrasted to chronotropic effects which concern heart rate) of the sympathetic nerves on the heart are due to direct effects of sympathetic neurohumors on the contractile muscle fibers.

Much controversy over the control of the stroke volume (or stroke work) has concerned the relative importance of control by the diastolic fiber length, or diastolic volume, vs. control by sympathetically induced changes in contractility. Starling and others[8] referred to the importance of sympathetic discharge, as did many individuals who criticized Starling.[26, 36, 42] It is not yet clear how these factors relate in the intact human, because it is difficult to accurately measure ventricular filling and even more difficult to know how much activity there is at any instant in the sympathetic fibers running to the heart.

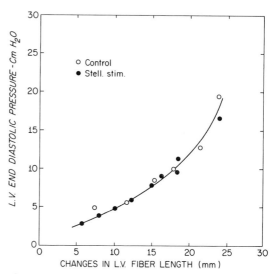

Figure 11–10 The diastolic pressure-volume relationship of the ventricle is shown in control conditions and during stimulation of the stellate ganglion. This stimulation of the stellate ganglion, as indicated previously, increases the force of ventricular contraction but, as indicated in this figure, does not alter the diastolic pressure-volume relationship of the ventricle. (After Sarnoff and Mitchell, *Handb. Physiol.*, 1962, sec. 2, vol. I, 489–532.)

PARASYMPATHETIC EFFECTS ON CARDIAC CONTRACTILITY. Parasympathetic fibers originate in the motor nuclei of the vagus nerve and in the nucleus solitarius, and run with the vagus nerve to the thorax. The fibers run directly to the heart, and the postganglionic cells are located in the cardiac walls. Most of these are near the sinus and atrioventricular nodes.

Vagal effects on strength of contraction of the heart have not been considered to be important. However, in recent years careful studies by Levy[25] have indicated that the isovolumetric left ventricle preparation of the dog displays a decrease in contractile strength when the parasympathetic nerves are stimulated (Fig. 11–11). This effect varies with the background level of sympathetic activity. A change of this sort can be reflexly produced in the anesthetized animal,[25] but it is not known if it is important in the unanesthetized animal.

OTHER FACTORS WHICH ALTER CARDIAC CONTRACTION — ANOXIA, CATECHOLAMINES, HORMONES. A variety of other factors can alter cardiac contractility. These include available oxygen which can be decreased by respiratory changes, by lowered arterial pressure or by obstructions to coronary perfusion. When oxygen delivery is sufficiently decreased, by generalized anoxia or decreased coronary flow, strength of contraction decreases.

Circulating catecholamines, released by the adrenal gland or elsewhere, have similar effects to sympathetic stimulation of the heart, increasing contractility. Contractility seems to also decrease if the levels of thyroid hormones or adrenal cortical hormones are not adequate, but these effects may reflect the general trophic activity of these hormones rather than a specific cardiac effect.

Multivariable Models of Stroke Volume Control. It is implicit in all of the above that no single factor determines the stroke volume.[19] The stroke volume during any heart beat is the sum of a number of instantaneous flows, each reflecting the state of the heart at that instant, the volume within the heart at that instant, the time since the beginning of contraction (active state) and the pressure against which the heart is ejecting. Instantaneous flow might appear a proper variable to consider, but the aortic flow "quantum" in mammalian cardiac function is the stroke volume. Sagawa[37] approached this problem in the anesthetized dog and produced a three-dimensional plot showing effects of both filling pressure and aortic pressure on cardiac output at a constant heart rate (Fig. 11–12). Another recent study of stroke volume control[39] used statistical techniques to set up a model of stroke volume in the awake *resting* dog, which showed spontaneous changes in interval, filling pressure (preload) and aortic pressure (afterload). Analysis of a large number of heart beats in each of many animals indicated that in the resting dog the stroke volume can be predicted from the combined effects of the filling pressure, the interval between beats and the aortic pressure. When the first two of these increase, the stroke volume is increased (note that interval changes are of the transient sort in this case). When aortic pressure increases, stroke volume decreases. The study cited above includes only three of the four major variables which we have considered to be of importance in changing the stroke volume and neglects autonomic

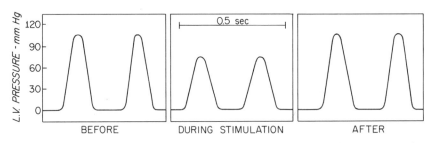

Figure 11–11 Stimulation of the vagus nerve, while other factors were held constant, produced decreased strength of left ventricular contraction. (From Levy *et al.*, *Circulat. Res.*, 1963, *12*, 107–117.)

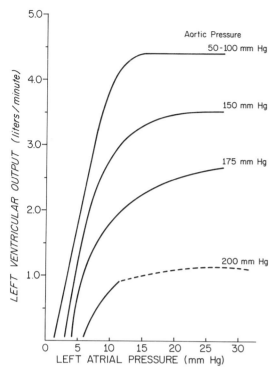

Figure 11-12 Left ventricular output as a function of left atrial pressure, plotted at several fixed aortic pressures. Note that as arterial pressure is increased the left ventricular output decreases at each filling pressure. The major contributor to the changes in cardiac output shown was a change in stroke volume, since the heart rate was virtually constant owing to prior vagotomy. This figure indicates the combined *mechanical* control of stroke volume through filling pressure and arterial pressure. (After Sagawa, in *Physical bases of circulatory transport: regulation and exchange,* E. B. Reeve and A. C. Guyton, eds. Philadelphia, W. B. Saunders Co., 1967.)

(sympathetic) discharge, which is considered by some to be most important in adaptations to exercise and other stress. Hopefully, later studies will incorporate the latter factor.

CONTROL OF HEART RATE

As indicated previously (Chaps. 5 and 6, Vol. II), the rate of the heart is normally controlled by the rate of impulse generation in the SA node (pacemaker of the heart). The heart rate reflects the rate of diastolic depolarization of the pacemaker cells in the sinus node. The pacemaker is innervated by sympathetic and parasympathetic fibers. The anatomy of these fibers has been considered previously in this chapter.

Sympathetic and Vagal Effects on Heart Rate. Increased sympathetic activity increases the rate of diastolic depolarization of the pacemaker cells and thereby increases heart rate. Increased vagal activity causes a slower diastolic depolarization and a decreased heart rate (Chap. 5, Vol. II).

Heart Rate and the Baroreceptor Reflexes. The major control of heart rate is exerted reflexly by the baroreceptors of the carotid sinus and aortic arch. When the arterial pressure falls, there is a reflex by decreased vagal activity and increased sympathetic activity. Conversely, when the pressure rises, vagal activity increases, tending to slow the heart. This is accompanied by a decreased sympathetic activity. As indicated elsewhere (Chap. 10, Vol. II), this picture of a reciprocal vagal and sympathetic activity originates largely from experiments on anesthetized animals. It does not hold exactly for the unanesthetized animal or for the normal human, and there are probably species differences in the relative importance of the two branches of the autonomic system.[40, 41]

HEART RATE AND STROKE VOLUME IN THE INTACT HEART

As has been indicated, changes in heart rate imposed under controlled conditions change contractility, and it is also apparent that changes in rate may also, if imposed without control, change diastolic filling by decreasing filling time. Whether *cardiac output* increases or decreases with heart rate depends on the relative magnitude of changes in rate and stroke volume. An increase in heart rate is usually accompanied by an increase in cardiac output up to a rate of about 120, at which time the increase in rate encroaches on the time available for ventricular filling. Cardiac output in a resting animal in the laboratory may not increase as heart rate increases beyond this figure. When animals are under strong sympathetic influence, however, the increase in heart rate is accompanied by an increase in contractility, and the cardiac output will increase up to and beyond a heart rate of 180.

CARDIAC OUTPUT, HEART RATE AND STROKE VOLUME IN EXERCISE

We have discussed a number of factors that may influence the heart rate and the stroke volume, and thereby change the cardiac output. If stroke volume changes, or even if it remains constant, it is not always clear which variables are affecting stroke volume in a given physiological situation. Measurements in humans indicate that in exercise when cardiac output rises greatly the heart rate changes far more than does the stroke volume (Figs. 11–13 and 11–14). Indeed, the stroke volume, which is high in the supine position, falls when the individual assumes the upright position, probably owing to a decreased ventricular filling pressure as blood fills the dependent veins. Stroke volume rises again to about the value seen at supine rest during minimal leg exercise. The veins are compressed by the movement of the extremities, which moves blood into the heart, probably restoring the filling pressure. There may also be increased sympathetic discharge, increasing ventricular contractility. Stroke volume rises slightly with further exercise, and during very severe exercise it may fall slightly.[3]

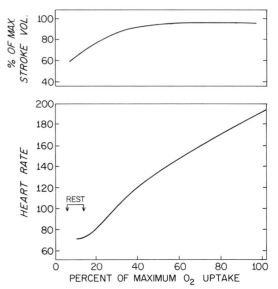

Figure 11–14 During exercise, as shown in Figure 11–13, cardiac output increases close to six-fold; stroke volume rises early in exercise and continues at about the same level and then falls slightly near maximal exercise (top curve). Heart rate, however, rises from a resting value of about 70 beats per min to approximately 195 beats per min. Change in heart rate is thus far greater. Note that the resting stroke volume in a recumbent individual would be close to that achieved during exercise. (After Åstrand *et al.*, *J. appl. Physiol.*, 1964, *19*, 268–274.)

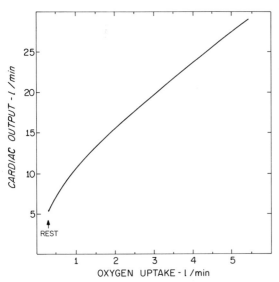

Figure 11–13 Change in cardiac output as a function of oxygen uptake during exercise. As oxygen consumption rises from its normal value of about 250 cc per min to close to 5 liters per min with increasing exertion, cardiac output increases from approximately 5 to almost 30 liters per min. (After Åstrand *et al.*, *J. appl. Physiol.*, 1964, *19*, 268–274.)

During exercise, all or many of the factors indicated above are working to change the cardiac output. Heart rate has increased, the ventricular filling pressure falls, the arterial pressure usually increases and of course there is a change in sympathetic discharge to the heart. If one considers these factors individually, one would expect that owing to the effects of increased rate, of decreased filling pressure and of increased arterial pressure the stroke volume would fall. However, probably because of the increased sympathetic discharge to the heart, the stroke volume is maintained. A complicated interaction of factors thus maintains stroke volume, and the increase in heart rate leads to increased cardiac output. The heart rate can triple in many normal individuals, whereas the range of change in stroke volume is far smaller. Dogs with denervated hearts who had lost much of the normal ability to change the heart rate (Fig. 11–15) managed to increase the cardiac output

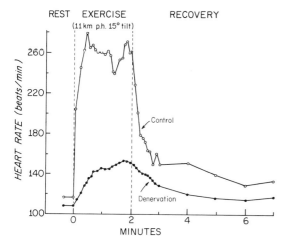

REST | EXERCISE | RECOVERY
(11 km p.h. 15° tilt)

Figure 11–15 A dog with a denervated heart shows a greatly reduced ability to increase heart rate during exercise. The exercise which caused the heart rate change in the top (control) curve was of the same magnitude as that in the bottom curve recorded after cardiac denervation. (After Donald and Shepherd, *Amer. J. Physiol.*, 1963, *205*, 393–400.)

during exercise (Fig. 11–16).[14] Humans accustomed to continuous strenuous exercise develop large hearts with a large resting stroke volume.

In such conditions as hemorrhage, heart rate again will provide a major control of cardiac output, but a number of factors will tend to keep the stroke volume high despite a fall in filling pressure. In hemorrhage the pressure against which the heart ejects may be decreased, which will tend to increase the stroke volume, as will the increased sympathetic discharge to the heart. The major compensatory responses are those which one would expect from traditional baroreceptor responses. These include vasomotor changes, which also compensate for the fall in arterial pressure and cause a redistribution of blood flow (Chap. 10, Vol. II).

EVALUATION OF THE STATE OF THE HEART

The contractile state of the heart (contractility) can, as indicated, be influenced by heart rate, autonomic discharge, adrenal and thyroid hormones, anoxia and other factors. The depressed contractility of the heart in shock or severe hemorrhage, although associated with anoxia, appears to possibly be influenced by a "myocardial depressant factor."[24] It would be most useful to be able to assess the contractile (inotropic) state of the human heart for diagnostic purposes, particularly in an individual who is being considered for cardiac or other surgery.

A large number of suggestions have been made concerning techniques for such assessments. Most of these suggestions are based on techniques thought to provide indices of contractility (discussed in Chap. 4, Vol. II, and in this chapter). A detailed discussion of methods used in the attempt to assess ventricular contractility, using data gathered by catheterization, can be found in Yang *et al.*[48] It appears that (Starling or Sarnoff) curves relating ventricular filling pressure to systolic pressure or to stroke work or to stroke volume might be useful (Figs. 11–3 and 11–4). Indeed, papillary muscles from cats with experimentally induced heart failure do show decreased contractile force at a given length when compared with papillary muscles from normal cats.[43] There are difficulties in applying this to the human case – ventricular filling pressure is not easily measured in the human. Note that measurement of ventricular filling pressure (or distending pressure) ideally requires a measurement of ventricular end diastolic pressure and of intrathoracic pressure. Stroke volume or stroke work is not only a function of filling pressure, but each of these is also influenced by arterial pressure, heart rate and so forth. In many experimental studies,

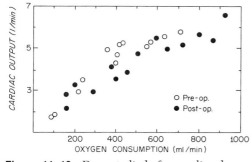

Figure 11–16 Dogs studied after cardiac denervation are able to achieve virtually the same cardiac output during exercise as before denervation. This, taken together with Figure 11–15, indicates that these animals undoubtedly show large changes in stroke volume during exercise. (After Donald and Shepherd, *Amer. J. Physiol.*, 1963, *205*, 393–400.)

a given heart is its own experimental control, because it may be studied at a variety of diastolic volumes. We do not, however, know if the hearts of different individuals can be compared, assuming that a heart can somehow be taken through the range of filling pressures required to produce a Starling or Sarnoff curve; nor do we know how such curves are affected by heart size, physical training and the like.

It has been suggested that the rate of change of ventricular pressure (dp/dt) might be an indicator of cardiac contractility,[16, 30] because it is altered by autonomic discharge.[35, 36] Other suggestions based on "rate of change" effects include the use of the various intervals of the cardiac cycle (duration of isometric contraction, time to reach peak ventricular pressure) which are altered by autonomic activity.[46] Since many of the aforementioned variables are also influenced by some of the variables listed earlier (such as heart rate, filling pressure and aortic pressure), the usefulness and reproducibility of these indices seem doubtful at present,[48] although each index has its advocates.

The force-velocity curves originally used to describe skeletal muscle have been considered by some to also characterize the contractile state of cardiac (papillary) muscle (Chap. 4, Vol. II). It has further been considered that the maximal shortening velocity of the contractile element of cardiac muscle (V_{max}) is independent of muscle fiber length and is an indication of the inotropic or contractile state.[42] Unfortunately, not all experimenters find the same results in the studies of force-velocity.[7] Nevertheless, it has been suggested that the equivalent of a force-velocity curve can be produced for the human heart through calculations performed on ventricular pressure recordings.[27, 42] Attempts to do this have been criticized.[32, 34, 48]

High-speed, biplane cineangiocardiography (rapid x-ray photography using a contrast material to outline the left ventricular cavity) is often performed in man,[12, 13] and it is possible to get an accurate estimation of ventricular volumes during the cardiac cycle. Other measurements, such as intraventricular pressure, can be combined with this determination. These measurements have been used in attempts to

characterize cardiac muscle.[13] Approximations of Starling curves, Sarnoff curves or force-velocity curves can be made. Interestingly, the ratio of stroke volume to end diastolic volume, which is known as the systolic ejection fraction, appears to be depressed in numerous conditions in which the myocardial contractility is depressed. The ejection fraction remains within normal limits in many conditions in which the heart dilates or hypertrophies, unless there is associated myocardial disease.[13] However, this index may be abnormal in the absence of decreased contractility.[48] It is not clear that any of the aforementioned techniques can independently demonstrate loss of myocardial contractile force when such loss is not readily apparent from other observations. Again, the various procedures described above may be useful to detect changes in a single heart, but may be far less useful when one heart is compared with others.

REFERENCES

1. Anrep, G. von. On the part played by the suprarenals in the normal vascular reactions of the body. *J. Physiol. (Lond.)*, 1912, 45, 307–317.
2. Arnold, G., Morgenstern, C., and Lochner, W. The autoregulation of the heart work by the coronary perfusion pressure. *Pflügers Arch. ges. Physiol.*, 1970, 321, 34–55.
3. Åstrand, P.-O., Cuddy, T. E., Saltin, B., and Stenberg, J. Cardiac output during submaximal and maximal work. *J. appl. Physiol.*, 1964, 19, 268–274.
4. Bainbridge, F. A. *The physiology of muscular exercise.* London, Longmans, Green, 1919.
5. Blinks, J. R., and Koch-Weser, J. Analysis of the effects of changes in rate and rhythm upon myocardial contractility. *J. Pharmacol. exp. Ther.*, 1961, 134, 373–389.
6. Bowditch, H. P. Über die Eigenthümlichkeiten der Reizbarkeit, welche die Muskelfasern des Herzens zeigen. *Ber. sächs. ges. (Akad.) Wiss.*, 1871, 23, 652–689.
7. Brady, A. J. Time and displacement dependence of cardiac contractility: Problems in defining the active state and force-velocity relations. *Fed. Proc.*, 1965, 24, 1410–1420.
8. Chapman, C. B., and Mitchell, J. H. *Starling on the heart.* London, Dawsons of Pall Mall, 1965.
9. Chardack, W. M., Gage, A. A., and Dean, D. C. Paired and coupled electrical stimulation of the heart. *Bull. N. Y. Acad. Med.*, 1965, 41, 462–480.
10. Cranefield, P. F. The force of contraction of ex-

trasystoles and the potentiation of force of the postextrasystolic contraction: A historical review. *Bull. N. Y. Acad. Med.,* 1965, *41,* 419–427.

11. Cranefield, P. F. (ed.). Conference on paired pulse stimulation and postextrasystolic potentiation in the heart: Part II. *Bull. N. Y. Acad. Med.,* 1965, *41,* 575–747.

12. Dodge, H. T., Hay, R. E., and Sandler, H. An angiocardiographic method for directly determining left ventricular stroke volume in man. *Circulat. Res.,* 1962, *11,* 739–745.

13. Dodge, H. T., and Hefner, L. L. Physical and mechanical aspects of muscle contraction. In: *Cardiovascular diseases.* Philadelphia, Lea & Febiger, 1971.

14. Donald, D. E., and Shepherd, J. T. Response to exercise in dogs with cardiac denervation. *Amer. J. Physiol.,* 1963, *205,* 393–400.

15. Frank, O. Zur Dynamik des Herzmuskels. *Z. Biol.,* 1895, *32,* 370–447. (Translated by Chapman, C. B., and Wasserman, E. On the dynamics of cardiac muscle. *Amer. Heart J.,* 1959, *58,* 282–317.)

16. Gleason, W. L., and Braunwald, E. Studies on first derivative of the ventricular pressure pulse in man. *J. clin. Invest.,* 1962, *41,* 80–91.

17. Gordon, A. M., Huxley, A. F., and Julian, F. J. The variation in isometric tension with sarcomere length in vertebrate muscle fibres. *J. Physiol. (Lond.),* 1966, *184,* 170–192.

18. Hamilton, W. F. Role of the Starling concept in regulation of the normal circulation. *Physiol. Rev.,* 1955, *35,* 161–168.

19. Hamilton, W. F., and Remington, J. W. Some factors in the regulation of the stroke volume. *Amer. J. Physiol.,* 1948, *153,* 287–297.

20. Harrison, T. R. *Failure of the circulation.* Baltimore, Williams and Wilkins Co., 1939.

21. Hoffman, B. F., Bartelstone, H. J., Scherlag, B. J., and Cranefield, P. F. Effects of postextrasystolic potentiation on normal and failing hearts. *Bull. N. Y. Acad. Med.,* 1965, *41,* 498–534.

22. Katz, L. N. Effects of artificially induced paired and coupled beats. *Bull. N. Y. Acad. Med.,* 1965, *41,* 428–461.

23. Kruta, V., and Braveny, P. Restitution de la contractilité du myocarde entre les contractions et les phénomènes de potentiation. *Arch. int. Physiol. Biochem.,* 1961, *69,* 645–667.

24. Lefer, A. M., and Martin, J. Relationship of plasma peptides to the myocardial depressant factor in hemorrhagic shock in cats. *Circulat. Res.,* 1970, *26,* 59–69.

25. Levy, M. N., Imperial, E. S., and Zieske, H., Jr. Ventricular response to increased outflow resistance in absence of elevated intraventricular end-diastolic pressure. *Circulat. Res.,* 1963, *12,* 107–117.

26. Liljestrand, G., Lysholm, E., and Nylin, G. The immediate effects of muscular work on the stroke and heart volume in man. *Skand. Arch. Physiol.,* 1938, *80,* 265–282.

27. Mason, D. T., Spann, J. F., Jr., and Zelis, R. Quantification of the contractile state of the intact human heart. Maximal velocity of contractile element shortening determined by the instantaneous relation between the rate of pressure rise and pressure in the left ventricle during isovolumic systole. *Amer. J. Cardiol.,* 1070, *26,* 248 257.

28. Meijler, F. L. Staircase, rest contractions, and potentiation in the isolated rat heart. *Amer. J. Physiol.,* 1962, *202,* 636–640.

29. Meijler, F. L., and Durrer, D. Physiological and clinical aspects of paired stimulation. *Bull. N. Y. Acad. Med.,* 1965, *41,* 575–591.

30. Miller, G. A. H., Kirklin, J. W., and Swan, H. J. C. Myocardial function and left ventricular volumes in acquired valvular insufficiency. *Circulation,* 1965, *31,* 374–384.

31. Monroe, R. G., LaFarge, C. G., Kumar, A. E., Plenge, R., Phornphutkul, C., and Davis, M. The Anrep effect reconsidered. *Fed. Proc.,* 1972, *31,* 315.

32. Noble, M. I. M., Bowen, T. E., and Hefner, L. L. Force-velocity relationship of cat cardiac muscle, studied by isotonic and quick-release techniques. *Circulat. Res.,* 1969, *24,* 821–833.

33. Patterson, S. W., Piper, H., and Starling, E. H. The regulation of the heart beat. *J. Physiol. (Lond.),* 1914, *48,* 465–513.

34. Pollack, G. H. Maximum velocity as an index of contractility in cardiac muscle. A critical evaluation. *Circulat. Res.,* 1970, *26,* 111–127.

35. Randall, W. C., and Kelso, A. F. Dynamic basis for sympathetic cardiac augmentation. *Amer. J. Physiol.,* 1960, *198,* 971–974.

36. Rushmer, R. F. Effects of nerve stimulation and hormones on the heart; the role of the heart in general circulatory regulation. *Handb. Physiol.,* 1962, sec. 2, vol. I, 533–550.

37. Sagawa, K. Analysis of the ventricular pumping capacity as a function of input and output pressure loads. In: *Physical bases of circulatory transport: Regulation and exchange,* E. B. Reeve and A. C. Guyton, eds. Philadelphia, W. B. Saunders Co., 1967.

38. Sarnoff, S. J., and Mitchell, J. H. The control of the function of the heart. *Handb. Physiol.,* 1962, sec. 2, vol. I, 489–532.

39. Scher, A. M., Young, A. C., and Kehl, T. H. The regulation of stroke volume in the resting, unanesthetized dog. *Computers & Biomed. Res.,* 1968, *1,* 315–336.

40. Scher, A. M., and Young, A. C. Reflex control of heart rate in the unanesthetized dog. *Amer. J. Physiol.,* 1970, *218,* 780–789.

41. Scher, A. M., Ohm, W. W., Bumgarner, K., Boynton, R., and Young, A. C. Sympathetic and parasympathetic control of heart rate in the dog, baboon and man. *Fed. Proc.,* 1972, *31,* 1219–1225.

42. Sonnenblick, E. H. Force-velocity relations in mammalian heart muscle. *Amer. J. Physiol.,* 1962, *202,* 931–939.

43. Spann, J. F., Jr., Buccino, R. A., Sonnenblick, E. H., and Braunwald, E. Contractile state of cardiac muscle obtained from cats with experimentally produced ventricular hypertrophy and heart failure. *Circulat. Res.,* 1967, *21,* 341–354.

44. Stead, E. A., Jr., and Warren, J. V. Cardiac output

in man: An analysis of the mechanisms varying the cardiac output based on recent clinical
studies. *Arch. int. Med.,* 1947, *80,* 237–248.

45. Warren, J. V., Brannon, E. S., Stead, E. A., Jr., and
Merrill, A. J. The effect of venesection and
the pooling of blood in the extremities on the
atrial pressure and cardiac output in normal
subjects with observations on acute circulatory collapse in three instances. *J. clin. Invest.,*
1945, *24,* 337–344.

46. Weissler, A. M., Harris, W. S., and Schoenfield,
C. D. Bedside technics for the evaluation of
ventricular function in man. *Amer. J. Cardiol.,*
1969, *23,* 577–583.

47. Wilcken, D. E. L., Charlier, A. A., Hoffman, J. I. E.,
and Guz, A. Effects of alterations in aortic
impedance on the performance of the ventricles. *Circulat. Res.,* 1964, *14,* 283–293.

48. Yang, S. S., Bentivoglio, L. G., Maranhao, V., and
Goldberg, H. *From cardiac catheterization
data to hemodynamic parameters.* Philadelphia, F. A. Davis Co., 1972.

CHAPTER 12

THE CUTANEOUS
CIRCULATION

by LORING B. ROWELL

INTRODUCTION

Despite the unique accessibility of cutaneous blood vessels, quantitative measures of skin blood flow are lacking.[2] Detailed architecture of the cutaneous vascular bed is exceedingly complex and poorly understood. Human skin has a rich vascular supply with a dense system of capillary loops in the papillae of the corium, which in turn empties into a capacious, subpapillary venous plexus. The venous plexus contains a major fraction of the cutaneous blood volume. Because of low linear flow velocity within these veins and their proximity to the skin surface, the plexus allows rapid dissipation of heat. In effect, this plexus increases the vascular surface area from which heat can be conducted through the dermal layers. Also, the rich capillary network around the sweat glands is functionally important in sweating.

Also important in heat transfer are the arteriovenous (A-V) anastomoses found in apical structures—fingers, toes, ears and nose and palmar surfaces of hands and feet.[29, 34] These richly innervated small muscular channels directly connect dermal arterioles and venules and can effect large changes in local skin blood flow. They aid in body temperature regulation by utilizing the potentially high rates of heat loss from exposed hands and feet, structures which have large surface to mass ratios.

Skin is the thermal insulator of the body, the effectiveness of which changes inversely with skin blood flow. Skin blood flow determines not only the rate of transfer of heat between body and environment but also the rate of heat transfer from deep tissues to the periphery; because body tissues are poor thermal conductors, deep heat is mainly conducted to the body surface by blood (see Chap. 5, Vol. III).

METHODS OF SKIN BLOOD FLOW MEASUREMENT

Many methods commonly used for measuring regional blood flow are not applicable to skin. For example, the Fick principle cannot be used because there is no known substance which is removed from blood exclusively by skin—and because there is no site for sampling mixed venous blood from any identifiable region of the vascular bed. The complex vascular architecture and small distance between the separate vascular networks preclude localized injection for indicator washout determinations and placement of flow transducers. Measurement of surface temperature is an unreliable means of assessing changes in cutaneous blood flow.

A variety of methods have been used to estimate changes in cutaneous blood flow.[2, 30, 64] Changes in superficial (cutaneous) venous oxygen saturation, thermal conductivity, rates of heat loss by calorimetry, rates of radioactive isotope clearance and absorption of light by transilluminated skin have been used. Problems associated with these various techniques have been reviewed.[2, 30, 61, 64]

The method of choice for quantitative assessment of changes in cutaneous blood flow is venous occlusion plethysmography (see Chap. 3, Vol. II) and most of our knowledge of skin blood flow and its control is derived from this technique when applied under conditions wherein blood flow to underlying muscle is not changing.

GENERAL DESCRIPTION OF SKIN BLOOD FLOW

Range of Blood Flow. Skin, like the kidney (but in contrast to muscle, brain and myocardium) normally has a blood flow which far exceeds its nutritional needs. Because of the methodological problems already cited, no accurate direct measurement of total skin blood flow exists and most observations are derived from accessible regions on the extremities. Extrapolation of these measurements to the total body surface is probably invalid because of the regional variations in normal skin blood flow.[34] Different regions show different thresholds as well as different mechanisms of response to a given thermal stimulus.[24, 29, 34, 61]

The estimates most commonly given for cutaneous blood flow in a man weighing approximately 70 kg, with a surface area of 1.73 m², nude and at rest in a normal indoor environment, range from 200 to 500 ml per min or 116 to 289 ml per m² per min.[29] Estimates for maximal cutaneous blood flow range from 1.2 to 2.0 liters per m² per min or 2.1 to 3.5 liters per min in the 70 kg man.[2, 29, 34, 64] These estimates are derived by a variety of methods and under a wide range of thermal conditions. Prolonged heating and/or intense direct heating can produce much higher estimates of maximal cutaneous flow.[39, 54] For example, prolonged *direct* heating of the whole body to the limits of subjective thermal tolerance (with skin temperature held at 41° C) produces an average increase in cardiac output of 6.6 liters per min in normal young men and maximal increases are as high as 10 liters per min. Figure 12–1 shows that the increase in blood flow under these conditions is directed primarily to skin. The other major vascular beds display decreased blood flow caused almost entirely by reflex vasoconstriction.[51, 55] Taken together, the increase in cardiac output and the redistribution of blood flow away from visceral organs (and to a small extent from skeletal muscle) provide to skin approximately 7 to 8 liters per min or 3.5 to 4 liters of blood per m² body surface per min.[20] Thus, maximally vasodilated skin has a total flow second only to that for maximally vasodilated skeletal muscle.

When skin is cooled to 14° C, its blood flow falls so low that measurement is difficult—*i.e.*, from the normal 3 to 5 ml 100 ml⁻¹ min⁻¹ down to 0.3 ml 100 ml⁻¹ min⁻¹.[29, 34] When skin temperature is lowered further, vasodilation progressively develops and the cutaneous vasomotor state fluctuates between marked dilation and constriction. This phenomenon, called "cold vasodilation," will be discussed later.

Regional Variation in Cutaneous Blood Flow. Qualitative estimates of regional

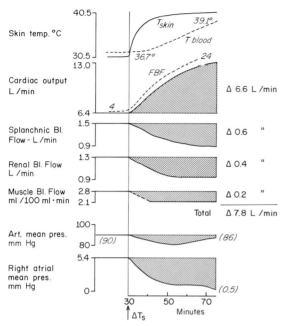

Figure 12–1 Overall cardiovascular response to direct heating of skin over the whole body. Between 30 and 75 min skin temperature was raised to 40.5° C, while right atrial blood temperature (T_blood) rose to 39.1° C. The rise in forearm blood flow (FBF in ml 100 ml⁻¹ min⁻¹) parallels the 6.6 liter per min rise in cardiac output, while blood flow to splanchnic region and kidneys and often to skeletal muscle is reduced by reflex vasoconstriction. Regional vasoconstriction redistributes an extra 1 to 1.2 liters per min of blood to skin, making the average skin blood flow approximately 7.6 to 7.8 liters per min. This represents a maximal response to combined reflex and direct local effects of heat. Arterial mean pressure is well-maintained but right atrial pressure falls markedly.

skin blood flow indicate that considerable variation exists over different portions of the body surface at any given body and skin temperature.[33] Apical structures such as fingers and toes can have extremely high cutaneous blood flow, partly because of the opening of A-V anastomoses. For example, in the hand, which is mainly skin, flow can increase from 3 to 10 ml up to 70 to 100 ml per min per 100 g of tissue when local skin temperature is raised to 44° C. Correcting this value for the proportion of other tissues in the hand yields an estimated maximal flow rate for skin alone of about 180 ml per min per 100 g of skin.[34]

In a comfortable resting subject nearly one-half of the total cutaneous blood flow is directed to the palms, soles of the feet and head, according to estimates taken from Hertzman's photoelectric plethysmograph.[33]

Cutaneous Blood Volume. The *warm* skin may comprise one of the major blood "depots" or "reservoirs" from which redistribution of blood volume into the central circulation can occur under certain conditions.[62, 73] As will be shown, the volume of blood in skin depends in part upon the tone in the resistance and capacity vessels, and tone in turn depends largely upon ambient and body temperatures. Under normothermic conditions 80 per cent of the blood volume in a limb is in skin and muscle veins.[74] Sympathectomy of the cat's paw, for example, adds 40 per cent to its blood volume.[44] In the cat or dog maximal stimulation of appropriate sympathetic vasoconstrictor nerves causes a 35 to 50 per cent reduction in limb volume, most of which comes from the skin.[10]

Maximal dilation of cutaneous resistance vessels causes not only a large fall in peripheral vascular resistance but also, because of the high distensibility and capacity of cutaneous venous plexuses, a major shift in blood volume from central to peripheral vasculature. The magnitude of volume increase with heating is not known, but the accompanying changes in the central circulation (to be discussed) suggest that the increase is large. As cutaneous blood volume increases at any given rate of flow, linear flow velocity decreases proportionally. A large volume of blood passing at reduced linear velocity just below the skin surface increases the rate of heat transfer across the skin. However, combined with a large fall in peripheral vascular resistance, a shift of blood volume into skin poses a significant challenge to regulation of blood pressure and blood flow to other organs.

EXTRINSIC NEURAL CONTROL OF CUTANEOUS RESISTANCE VESSELS

Sympathetic Vasoconstrictor and Vasodilator Control. Cutaneous blood flow is determined mainly by direct local effects and reflex effects of central and peripheral heating. Two types of heating and the different responses they cause must be distinguished. In *indirect* heating, warming one portion of the body causes reflex cutaneous vasodilation in other portions. In *direct* heating, heat is applied locally and the local vasomotor state is dominated by the physical effects of heat. When the whole body is heated, the neural and physical effects of heat are summated. The following discussions of neural and reflex control of skin vessels apply only to conditions in which direct vascular effects of local temperature changes are absent.

The small nutritional needs of skin never seriously affect its great range of local and neural controls. Not all regions of skin are controlled by the same mechanisms. Nonuniformities in regional perfusion are related to anatomy as well as to extrinsic regulatory factors.

Cutaneous resistance vessels are significantly regulated by the sympathetic nervous system. In 1852 Claude Bernard[5] showed that severing the cervical sympathetic chain caused marked vasodilation in the rabbit's ear, demonstrating the tonic vasoconstrictor influence on these cutaneous vessels.[43] One factor that regulates cutaneous vascular resistance is the frequency of impulses over these sympathetic vasoconstrictor nerve fibers. Skin of the cat's paw, for example, is extremely sensitive to variations of the rate of stimulation of sympathetic nerves over the range of 0 to 10 per sec;[44] measurable decreases in blood flow occur at stimulus rates of only 0.25 per sec, with a 100-fold decrease at 10 per sec. This control normally subserves temperature regulation but under some circumstances may serve blood pressure and volume regulation as well (see below).

At normal ambient temperatures, cutaneous resistance vessels exhibit considerable basal tone—i.e., vascular resistance which is independent of innervation[44, 45] (see Chap. 8, Vol. II). Cutaneous A-V anastomoses have little basal tone and are regulated by tonic neural discharge,[44] the elimination of which produces maximal passive dilation.

A recurrent question is whether cutaneous vascular resistance is predominantly controlled only by variation of sympathetic vasoconstrictor tone (passive vasodilation) or whether vasodilation is also neurogenic. Experiments have been designed to test whether maximal reflex vasodilation can occur after blocking all sympathetic neural pathways to skin. The results indicate that different mechanisms of control operate in hands and feet as opposed to the forearm.[4]

In the former, sympathetic blockade causes maximal reflex vasodilation (higher flows are possible only if direct effects of local heating are superimposed) (see Fig. 12–2). Although forearm skin also receives tonic sympathetic vasoconstrictor outflow, local sympathectomy increases flow only if core and skin temperatures are relatively low.[48] Increases in flow are small compared to those in the hand, and indirect body heating does not increase it further. The dramatic increase in forearm skin blood flow during heating requires sympathetic nerve fibers which somehow actively cause vasodilation (see Fig. 12–2).

Other cutaneous regions have been less exhaustively studied. The foot, lips, nose and ear appear similar to the hand, i.e., vasoconstrictor control dominates. In the upper arm, thigh and calf, like the forearm, a weak vasoconstrictor tone is maintained in a cool subject, but active vasodilation dominates when the subject is heated.[29]

Sweat glands, but apparently not cutaneous vessels, are supplied by cholinergic sympathetic nerve endings. Sweat glands are embryologically related to the salivary glands in which vasodilation is partially due to bradykinin formation accompanying cholinergic, neural stimulation.[27, 28] Fox and Hilton[25] postulated a similar mechanism to explain vasodilation in sweat glands and in skin to which bradykinin might spread. Their postulate is supported by two observations: (i) onsets of cutaneous (forearm) vasodilation and local sweating frequently coincide; (ii) a bradykinin-forming enzyme appears in sweat and bradykinin itself appears in subcutaneous perfusates during body heating. (Further details of the bradykinin mechanism are discussed below.)

Reflex Control of Cutaneous Vascular Resistance. THERMORECEPTORS. The separate but interacting roles of central and cutaneous thermoreceptors are now well documented.[36] Heating or cooling of either local or remote cutaneous sites induces vasomotor changes in skin of the extremities. These changes are partly attributable to the return of heated or cooled blood to central areas such as the hypothalamus which may mediate the vasomotor responses.[29, 63] In man, central thermoreceptors are apparently very sensitive; a rise in oral temperature of only 0.15° C is accompanied by reflex vasodilation in the hand.[29]

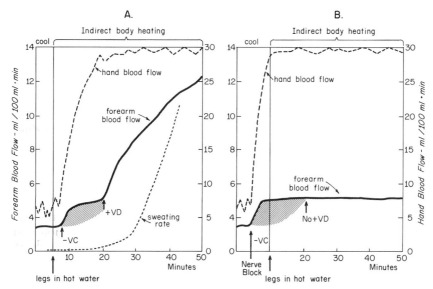

Figure 12–2 Reflex effects on *indirect* heating upon skin blood flow in the normal hand and forearm *(A)* and the same regions with sympathetic nerves blocked *(B)*. The figure reveals the basic differences between control of skin blood flow in the hand or other acral regions (feet, ears, nose) and the rest of the skin. Vasodilation in the hand is mediated only by release of tonic vasoconstrictor tone (–VC), as evidenced by the equal effects of heat *(A)* and nerve block *(B)*. Vasodilation in the forearm results initially from a release of tonic vasoconstrictor tone if subjects are cool at the start of heating. After 10 to 15 min of heating, a second phase of vasodilation coincides with the onset of sweating (+VD) *(A)* and is active as evidenced by its prevention through nerve block *(B)*. Note the absence of sweating in the nerve-blocked forearm *(B)*. (After Shepherd, *Physiology of the circulation in human limbs in health and disease.* Philadelphia, W. B. Saunders Co., 1963.)

The role of cutaneous thermoreceptors is demonstrated by heating the legs during total occlusion to prevent returning blood from raising central temperature. Such heating causes sudden vasodilation in the hands.[36] The short latency indicates that the response is mediated via afferent nerve impulses from thermoreceptors in the heated area. To be effective, relatively large areas of skin must be stimulated. Also, central temperature must be above 36.5 to 36.8° C[15, 17] —*i.e.*, a low central temperature suppresses the response. Since the response is absent in patients after lumbar sympathectomy, the afferent arm of the reflex appears to travel for part of its route, at least, with the sympathetic chain.[16]

Cutaneous vascular responses to cooling also depend on both central and cutaneous thermoreceptors.[29] In addition, local factors govern responses to cooling (see below).

BARORECEPTORS. Formerly, baroreceptors were thought to have no sustained effect on cutaneous resistance vessels.[61] A variety of maneuvers which stimulate systemic arterial baroreceptors[47] — or baroreceptors in the cardiopulmonary region (see Chap. 13, Vol. II) — exert no appreciable effect on cutaneous vascular resistance in the hand.[49] However, application of negative pressure to the lower body, which draws several hundred milliliters of blood into the legs, causes sustained vasoconstriction of skin,[4a] even when cutaneous vessels were previously dilated by moderate body warming.[19] This response is blocked by sympathectomy.[3] In man, stimulation of carotid sinus hypertension by suction over the neck causes reflex cutaneous vasodilation.[4a] In cats, reflex changes in cutaneous vascular resistance are induced by variations in baroreceptor activity, but these responses are generally small relative to the thermoregulatory responses described above.[45] In man and dogs, severe hemorrhage produces intense cutaneous vasoconstriction; however, these responses are not solely ascribable to baroreceptor reflexes. In the dog, adrenal release of catecholamines is mainly responsible, for neither sympathectomy nor nephrectomy (to eliminate participation of the reninangiotensin system) prevents the vasoconstrictor response to hemorrhage.

The response is, however, largely prevented by phenoxybenzamine which blocks alpha-receptors upon which the catecholamines act.[7]

CHEMORECEPTORS. Unlike skeletal muscle and splanchnic vascular beds which constrict to hypoxic stimulation of chemoreceptors, the rabbit's skin shows little or no reflex response to chemoreceptor drive.[38]

EXERCISE. Exercise increases sympathetic vasomotor outflow to the resistance vessels in nonexercising regions so that resistance rises in proportion to the severity of exercise (see Chap. 13 and Fig. 14–6, Chap. 14, Vol. II). Cutaneous resistance vessels and veins constrict at the onset of exercise but subsequently relax as central body temperature rises.[6, 53] The maximal increase of skin blood flow during exercise is unknown, but maximal values observed at rest undoubtedly cannot be reached during exercise because of competing demands for high muscle blood flow. Under these conditions skin blood flow is the result of opposing drives for vasoconstriction (exercise) and vasodilation (heat).

Changes in skin blood flow during exercise are deduced from increments in cardiac output and/or from decrements in blood flow to other regions. Normally, renal and splanchnic blood flows are reduced in proportion to the severity of exercise.[50] Thermal stress increases sympathetic vasomotor outflow so that a greater reduction in blood flow to these regions occurs when heat stress and exercise are combined[50] (see Chap. 14, Vol. II). The redistribution of splanchnic and renal blood flows could supply an additional 600 to 800 ml per min of blood flow to skin. In addition, cardiac output may also increase by 2 to 3 liters per min in a hot environment, further increasing skin blood flow by this amount. This will occur if skin temperature is very high or when the severity of exercise is well below maximal exercise capacity.[57] Endogenous heat stress (pyrogen-induced fever) also raises cardiac output 2 to 3 liters per min at relatively low rates of work.[31] Thus, skin blood flow could not exceed 3 to 4 liters per min during moderately heavy exercise in a hot environment. However, when metabolic demands of exercise exceed 50 to 60 per cent of the subject's maximum capacity for oxygen consumption (see Chap. 13, Vol. II), cardiac output fails to respond to heat stress. Most of the cardiac output goes to working muscle.[31, 56] Therefore, any further increments in skin blood flow must be entirely met by redistribution of blood flow from visceral organs and possibly from working and nonworking skeletal muscle as well.[56] This indicates that skin must be relatively vasoconstricted despite elevated body temperature. At maximal rates of work in the heat, cardiac output, oxygen consumption and total body A-VO$_2$ difference can reach for short periods the same maximal values seen in cool environments. This indicates that the degree of cutaneous vasoconstriction is similar in both environments, for an increase in flow to skin, which extracts little oxygen, would lower the A-VO$_2$ difference. Thus, thermoregulation is temporarily sacrificed to metabolic demands of working muscle, but work capacity is drastically reduced because of accelerated heat storage.

When exogenous and endogenous heat stresses during exercise become sufficiently severe, the cutaneous vessels dilate. Usually the rise in cardiac output is not quite adequate to maintain arterial blood pressure. Pressure falls only slightly at first, but during prolonged stresses it declines to levels causing collapse, particularly when further augmentation of cardiac output is prevented by the decline in stroke volume and the attainment of maximal heart rate.[56, 57]

OTHER REFLEXES. The lability of finger and hand blood flow is well known. A wide variety of emotional, sensory and physical stimuli contribute to this lability.

One such mechanism is the so-called *axon reflex*, misnamed because it is not a reflex at all. Some dorsal root afferent fibers terminating in skin bifurcate, with one branch ending in receptors and the rest in terminals on or near blood vessels. These latter fibers cause intense vasodilation. A stimulus applied to the receptor generates action potentials which are conducted orthodromically to the central nervous system over the dorsal roots but also antidromically down the vasomotor branches, causing regional dilation of the cutaneous vessels. The axon reflex is responsible for the flare of reddening that appears on the margins of a scratch on the skin.

INTRINSIC LOCAL CONTROL OF CUTANEOUS RESISTANCE VESSELS

Physical Stimuli. HEAT. Skin experiences a greater range of temperature than any other organ. To a limited extent the sudomotor system exerts a local feedback regulation of blood flow; the secretion and evaporation of sweat, by lowering skin temperature, tend to reduce its blood flow. However, the primary response of skin to elevated skin temperature is a 4- to 5-fold increase in flow which occurs even in denervated skin. This local reaction to direct heat on cutaneous resistance vessels adds to those neural effects described above.

Local effects of heat on cutaneous vascular resistance depend upon the general thermal state of the body (Fig. 12–3). In general, when vasoconstrictor tone is reflexly diminished by indirect body heating, direct local effects of heating are augmented. Similarly, local effects of cooling are augmented by lowering body temperature. On the other hand, local effects of cooling are counteracted by central warming. For example, if one hand is kept in a cool bath while the body is warmed, the hand vessels dilate. However, because of the direct local effect of cold on the vessels, the increase in blood flow to the cool hand is less than that of a heated one no matter how much the body is heated.[46] Qualitatively, the foot responds similarly but the flows are somewhat lower. Raising cutaneous

vasoconstrictor tone of the whole body by general cooling produces the opposite effects; direct local effects of heating are depressed so that hand blood flow increases less during the same intensity of local heating.[46]

Just how direct local heating raises skin blood flow is unknown. Effects of heat on vascular smooth muscle, axon reflexes, indirect effects on production of vasodilator metabolites and reduced blood viscosity may contribute.[61] Crockford et al.[18] found that vasodilation spread a distance of 12 cm away from a site of restricted heat application on one forearm. Since the spread was blocked by an anesthetic barrier between the heated and unheated distal areas, an axon reflex mechanism was implicated. However, the distal spread was not affected by cervical sympathectomy or by lesions disrupting the brachial plexus; thus, somatic nerves and the sympathetic neurons with cell stations in the paravertebral ganglia are not involved.[61] Injecting a ring of epinephrine into the forearm completely vasoconstricts the underlying skin vessels and also blocks the spread of vasodilation beyond them. One idea is that thermal vasodilation is somehow conducted along vascular smooth muscle.

Vasodilation of heated skin adds heat to the body because heat is continually gained rather than lost when surface to deep body temperature gradients are reversed. Nevertheless, the high skin blood flow serves

Figure 12–3 Changes in hand blood flow in two experiments during local and indirect heating. Right-hand (○) plethysmograph filled with water at 32° C throughout the experiment. Left-hand (●) temperature of plethysmograph raised from 32 to 44° C at time 0 and kept at this temperature for the remainder of the experiment. At 25 min, in order to release sympathetic vasomotor tone, the legs were immersed to knee level in water at 44° C (left), or the body was immersed in a bath at 42° C (right). (From Roddie and Shepherd, J. Physiol. (Lond.), 1956, 131, 657–664.)

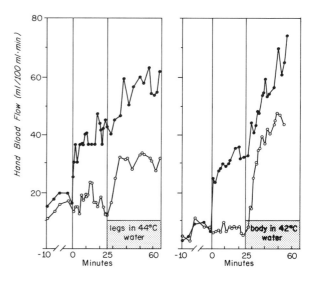

local needs by conducting heat away from underlying tissues and thus prevents burning.

COLD. Cold directly constricts cutaneous blood vessels. The local response depends upon the general thermal state of the subject's body. Responses to local cold are greatest in cold subjects and least in warm ones.[29, 61] Blood flow to the hand reaches a minimum value at a local skin temperature of about 15° C in a cold subject. If local temperature is lowered below 10° C, intense vasoconstriction in fingers or toes is interrupted by periods of vasodilation, a phenomenon which Lewis[41] called the "hunting reaction." Although most marked in fingers and toes, some "hunting" is also noted in other regions normally exposed to cold (ears, nose, hands) and which are rich in A-V anastomoses. Cold vasodilation has also been reported in contact regions such as elbows, knees, soles of feet, palms and buttocks where pressure on a cold surface produces marked cooling.[32] At temperatures below 20° C, a progressive increase in forearm blood flow is observed beginning after 5 to 10 min of local immersion at 1° C with a continuing increase for 20 min thereafter.[14]

The mechanism of cold-induced vasodilation is obscure. It is independent of sympathetic nerve supply (it can occur in nerve-blocked fingers), but an intact somatic nerve supply does appear to be necessary for full development of cold vasodilation — presumably by an axon reflex.[32] Attempts to accelerate the onset of cold vasodilation by intra-arterial injection of histamine or acetylcholine, both potent vasodilators of skin, were without effect.[2] Nor do antagonists of these substances, such as antihistamine and atropine, affect cold vasodilation. Chilled (6.5° C), isolated arteries are unresponsive to constrictor drugs.[35] Cold vasodilation can be elicited in fingers which have been fully constricted by iontophoresis[*35] with epinephrine or norepinephrine. When adrenergic stimulation is increased by body cooling, persistent cold

vasodilation is partially attributable to local loss of vascular response to constrictor hormones. When the body is warm and adrenergic stimulation is minimal, finger vessels relax, presumably because the direct local constrictor effect of cold is abolished by paralysis of vascular smooth muscle. The vasodilation, which restores flow and local temperature, restores responsiveness to cold and vasoconstriction recurs.[35]

MECHANICAL STIMULATION. Local changes in cutaneous vasomotor state can be induced by mildly stroking the skin with a blunt object. The *white reaction* which follows appears to be due to mechanical expulsion of blood from venous plexuses which determine skin color.[29, 42] Some active constriction of venules may also attend this stimulation.

A *triple response* results from stroking the skin more forcefully. A red line is probably caused by dilation of vessels from mechanical stimulation. This red line is followed by a *flare* or blush which is dependent on intact sensory nerves and is probably an axon "reflex." If the stimulus is very strong, a raised *wheal* develops after a few minutes and is caused by increased capillary permeability with loss of fluid and protein. The red line and the wheal are independent of nerves. Lewis[42] postulated that wheal formation depends on release from cells of an "H-substance" — possibly histamine; local intradermal injection of histamine produces a similar response.

Hormones. EPINEPHRINE AND NOREPINEPHRINE. These two hormones both act on alpha-receptors and are potent cutaneous vasoconstrictors. Unlike muscle, skin has only "alpha-receptors" (see Chap. 8, Vol. II). In fact, cutaneous A-V anastomoses are so reactive to these hormones that their lack of basal tone is indirect evidence that circulating catecholamines — in regions where this tone is high — cannot be mediators of basal tone. If so, skin A-V anastomoses would always be constricted.[45] Under normal conditions, cutaneous vascular resistance is controlled by sympathetic vasoconstrictor nerves.[12, 44] Stimulation of sympathetic nerves over a range of frequencies (up to 10 per sec) produces a vasoconstriction in skin which was four to five times greater than that caused by catecholamines administered in *physiological* amounts by

---

*By iontophoresis, epinephrine, which at pH 4.5 has a positive charge, is forced by electrical current into the skin by a positive electrode. The current is taken from a distal region of skin via a negative electrode.

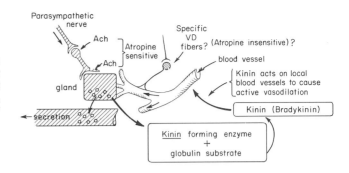

Figure 12–4 Schematic representation of the proposed bradykinin mechanism for dilating blood vessels to and around sweat (and other) glands. (After Folkow and Neil, *Circulation.* London, Oxford University Press, 1971.)

infusion into blood or by graded stimulation of adrenal medullary nerves.[44, 45]

BRADYKININ. This potent cutaneous vasodilator holds a key position, as mentioned previously, in schemes designed to explain active, neurogenic vasodilation in specific regions of skin such as the forearm. However, the entire question of how vasodilation occurs in glandular tissues is still intensely debated.[40, 59] The general scheme for the bradykinin system is illustrated in Figure 12–4. For skin, the supporting findings are that substances with bradykinin-like activity appear both in sweat and in subdermal perfusates when human sweat glands are stimulated by heating, and the observation that abrupt cutaneous vasodilation often coincides with the onset of sweating, as shown in Figure 12–2. Note that the hand, in which vasodilation is caused by reduced sympathetic vasomotor outflow, responds more promptly to heating and that a cool forearm skin displays an initial small vasodilation prior to sweating. This is due to release of tonic sympathetic vasoconstriction which was initially augmented by the cool environment. In accord with Fox and Hilton's analogy between salivary glands and sweat glands, vasodilation in both forearm skin and salivary glands is atropine resistant.[25] Both salivary secretion and sweating are blocked by atropine, as is acetylcholine-induced vasodilation, but kinin release in salivary glands (and presumably in sweat glands) is not. Although the vasodilation is delayed and/or reduced somewhat by atropine (Fig. 12–5), forearm cutaneous vasodilation appears not to be dominated by atropine-sensitive cholinergic nerve fibers to skin vessels.

Critics of the bradykinin hypothesis point out that vasodilation occurs in stimulated glands which are either desensitized to bradykinin or have had their local kinin formation blocked.[59] Apparently vasodilation by a kinin mechanism is not unique to all glands, as originally postulated. Recent experiments indicate that both kinin formation and dilator nerve fibers participate in salivary gland vasodilation, the latter initiating the response and kinin release sustaining it.[27, 28]

Numerous contradictions indicate that the bradykinin hypothesis does not warrant general acceptance. The following observa-

Figure 12–5 Forearm skin blood flow before and during indirect body heating. Note the delay in active vasodilation caused by blockade of cholinergic nerves to sweat glands by atropine (given at times marked ▨) in one arm (●——●). Blockade was checked by repeated administration of acetylcholine (▬). (From Roddie *et al., J. Physiol. (Lond.),* 1957, *136,* 489–497.)

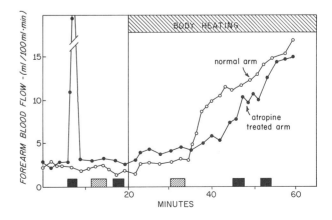

tions are contradictory to the theory: (i) Bradykinin but not the bradykinin-forming enzyme is found in subdermal perfusates; both must be there for the theory to be correct. (ii) Bradykinin appears in the sweat of hands which is thought to be controlled exclusively by adrenergic nerves. (iii) Onsets of sweating and skin vasodilation do not coincide during very gradual body heating; during such heating, cooling a small local area of skin suppresses vasodilation but not sweating.[33, 60] (iv) Sweating rate and intensity of skin vasodilation often change in opposite directions during acclimatization to heat.[23, 76] (v) In shock and insulin-induced hypoglycemia, sweating coincides with cutaneous vasoconstriction. However, a strong vasomotor outflow may override local vasodilatory effects in these cases.

Cutaneous vasodilation is clearly the result of a complex interaction of reduced sympathetic nervous tone (in cool subjects), direct local effects of heat, and some active mechanism requiring sympathetic innervation. The active mechanism may indeed require a kinin, a possibility yet to be established or disproved.

NEURAL CONTROL OF CUTANEOUS VEINS

Sympathetic Venoconstrictor Control. In contrast to resistance vessels, the important relationship in veins is between *pressure and volume* rather than between pressure and flow. The potentially great capacity of the skin venous plexus places particular importance upon active control of its pressure-volume or compliance* characteristics. Assignment of the role of "blood reservoir" to skin implies active regulation of cutaneous venous volume. Sudden variations in this volume can, as will be shown, have major effects upon the central circulation.

Unlike veins of skeletal muscle,[1, 21, 71, 77] cutaneous veins are richly innervated with sympathetic nerve fibers, as evidenced by their very high endogenous concentration of norepinephrine[5, 43] and vigorous response

to low frequency sympathetic nerve stimulation.[10, 44, 69-72] At maximum venoconstriction, 35 to 50 per cent of the total blood content of the cat hindlimb is expelled. Cutaneous veins appear to be more responsive to sympathetic nerve stimulation than are resistance vessels.[44] The maximum venoconstrictor response in terms of the decrease in total limb volume is achieved at stimulation frequencies of 4 to 6 per sec; in contrast, resistance vessels are not maximally constricted until frequencies reach 16 per sec. The difference between veins and arterioles may be mainly due to their different structure rather than to different humoral sensitivity or to numbers of constrictor fibers. Veins have thin walls and large lumens in contrast to the thick walls and relatively small lumens of resistance vessels or arterioles. Given a fixed change in circumference in a vein and an artery of the same total diameter, the percentage change in volume of a vein is greater than the percentage change in resistance in an arteriole. Some investigators have suggested that the venomotor response is overestimated by measuring a volume change of the entire limb.[10] Part of the decreased venous volume is the passive result of lowering venous pressure from constriction of the resistance vessels. The concepts of active vs. passive changes in venous volume are expanded in Chapters 13 and 14, Volume II.

Veins, like cutaneous resistance vessels, are normally dominated by constrictor motor outflow. In physiological amounts, *epinephrine* and *norepinephrine* generally have only 20 to 25 per cent of the effect produced by graded stimulation of sympathetic vasoconstrictor fibers.[44] Cutaneous veins appear insensitive to bradykinin.

LOCAL CONTROL OF CUTANEOUS VEINS

Temperature. Local skin cooling decreases venous compliance; the volume at a given distending pressure decreases. In man, cutaneous veins are almost maximally compliant at normal skin temperatures[37, 61, 75] (Fig. 12–6). Increasing local venous temperature above normal has little *direct* effect on compliance.[26, 64, 75]

Local temperature changes alter the responsiveness of veins to adrenergic stim-

---

*Compliance refers to the change in venous volume (ΔV) for a given change in venous pressure (ΔP)— *i.e.*, $\Delta V/\Delta P$.

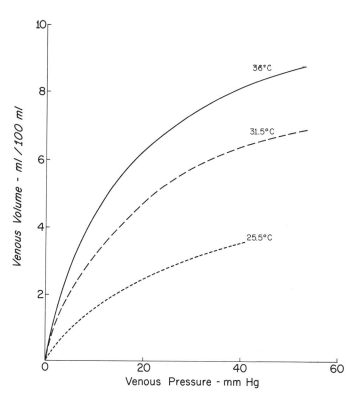

Figure 12–6 Venous volume at different distending pressures in the human hand at different skin temperatures. Note the large decrease in venous compliance ($\triangle V/\triangle P$) when skin temperature is reduced from 31.5° C (normal temperature) to 25.5° C and the relatively small increase when temperature is raised to 36° C. (After Thauer, *Handb. Physiol.*, 1965, sec. 2, vol. III, 1921–1966.)

uli. Local cooling potentiates, and local heating attenuates, venomotor response to reflexly induced sympathetic discharge in the intact limb.[69, 70, 72] Heating also attenuates the response of isolated veins to direct adrenergic stimulation.[67]

The mechanism by which local cooling potentiates the response of cutaneous veins to adrenergic stimuli is unknown. It is not due to (i) inhibition of neuronal uptake of norepinephrine,[68] (ii) interference with enzymatic inactivation of catecholamines[66] or (iii) direct effects of temperature on the contractile process of venous smooth muscle which is depressed rather than enhanced by cooling.

Whatever the cause, the physiological advantage of this control is obvious. Active constriction of superficial veins during cooling reduces heat loss by reducing venous surface area and by increasing linear velocity of blood flow directly beneath the skin. Since deep veins lack these constrictor mechanisms[1, 21, 43, 71, 77] and some even relax with cooling and constrict with heating,[67] blood is diverted from superficial cutaneous veins and returned via deeper veins. The rerouting contributes to heat

conservation in the limb (see Chap. 5, Vol. III).

Epinephrine, Norepinephrine. Both these hormones have a strong constrictor effect on cutaneous veins, but normally cutaneous venomotor tone is dominated by neural control.

REFLEX CONTROL OF CUTANEOUS VEINS

Temperature—Interaction Between Neural and Local Control. Cutaneous veins are reflexly reactive to a wide variety of central and peripheral stimuli,[22, 34, 61] of which the most important is temperature. Since the rate of heat gain and loss from the body largely depends upon the quantity of blood in cutaneous venous plexuses, these thermal reflexes are important in overall temperature regulation. Venous compliance is reflexly altered partly by changes in central body temperature; sympathetic nerve fibers are the efferent limb.[69, 70]

In normothermic resting man, sympathetic venomotor outflow is minimal. A rise

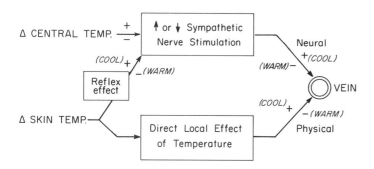

Figure 12–7 Interaction between reflex effects of changing central temperature and the reflex plus direct local effects of changing skin temperature on cutaneous venomotor tone. Lowering central body or skin temperature increases (↑) sympathetic venomotor outflow (+), while lowering local skin temperature at the vein increases (+) its sensitivity to neural stimulation by a direct local effect of temperature on venous smooth muscle.

in central temperature reflexly dilates cutaneous resistance vessels and passively increases cutaneous venous volume. This increases venous pressure and, if compliance remains constant, venous volume increases until a new level of wall tension is reached.

Central cooling reduces cutaneous venous volume by active, reflex venoconstriction and also by *passive* emptying (caused by reduced venous pressure accompanying reflex constriction of resistance vessels) (see Figs. 14–9 and 14–10, Chap. 14, Vol. II). The magnitude of constriction is least in veins which are locally warmed and greatest in those locally cooled, *i.e.*, sensitivity of cutaneous veins to sympathetic stimulation is modified by their local temperature.[69]

In addition to these direct local and central thermal reflex effects, venous compliance in man is also *reflexly* altered by changes in skin temperature at remote sites.[53] This is analogous to the response of resistance vessels.[36] The reflex is apparently mediated via cutaneous thermoreceptors which are effective only when large areas of skin are stimulated. How these receptors interact is shown in Figure 12–7.

Exercise. A potentially important cutaneous venous reaction already referred to is the response to exercise. If body and skin temperatures are cool, cutaneous veins actively constrict in proportion to the severity of exercise.[6] Some view this response as a means of reducing any transient lag in venous return at the onset of exercise. Subsequently, venous tone reflexly declines as exercise continues owing to the rise in central body temperature. For reasons already mentioned, the shift in blood volume to skin increases the rate of heat loss; it also drops central venous pressure, cardiac

filling pressure, intrathoracic blood volume and stroke volume.[56] Cardiac output must be maintained by higher heart rate while increased blood *flow* to skin must be met, in part, by reduced blood flow to visceral organs.[50, 56] Under these conditions work capacity is limited but is dramatically restored by suddenly lowering skin and central body temperatures; cutaneous veins then constrict and the resulting redistribution of blood flow centrally abruptly restores cardiac filling pressures, central blood volume and stroke volume.[57]

Reflex dilation of cutaneous veins during exercise also occurs when skin temperature is raised so that in hot environments combined drives from skin and central thermoreceptors accelerate effects described above. Raising skin temperature, even before central temperature rises, abolishes the reflex venoconstrictor response to exercise in two ways (see Fig. 12–7). First, local effects of heat on veins make them insensitive to the exercise-induced increase in sympathetic discharge. Second, increased body skin temperature reflexly suppresses any sympathetic discharge to cutaneous veins.[53] These effects can be dramatically reversed by lowering skin temperature which increases sympathetic discharge and restores venous sensitivity to it.[53, 70, 72]

Baroreceptor Reflexes. Because of their potentially large volume in warm skin, cutaneous veins could serve as an important blood "reservoir," able to send a sudden "autotransfusion" of blood back into the central venous system. However, a role of veins in subserving pressure regulation is unsupported by the data.[8, 9, 24, 26, 58] In man, maneuvers which alter arterial pressure, such as upright tilting, lower body negative pressure and increased transmural pressure on the carotid sinus, produce at best only transient venoconstriction.[24, 26, 58] Although this may transiently diminish cutaneous venous capacity, total venous capacity of the limb thereafter is regulated passively by changes in vascular resistance. Because increased cutaneous venous tone is lacking, venous volume gradually increases until gravitational force and passive changes in venous wall tension balance out. This accumulation will be rapid in situations such as hyperthermia which suppress cutaneous vasoconstriction, and sudden fainting can result. If sufficient blood volume *gradually* accumulates in limb veins, cutaneous venomotor reflexes may eventually be elicited. Experimentally, hemorrhage causes marked changes in both resistance and capacity of the skin,[13] but variations in species and severity of hemorrhage make generalizations difficult. Other variables such as hypoxia and epinephrine release also contribute to venoconstriction in hemorrhage.[7, 13] A notable point is that excessively warming patients in hemorrhagic shock (or after cardiac surgery) can further deplete central vascular volume by sequestering increased blood volume in cutaneous veins dilated by local or central warming. How much warming can overpower the strong sympathetic vasoconstrictor output attending severe hemorrhage is not known. Burton[11] cites clinical evidence suggesting that warming may sometimes decrease chances for survival.

Other Reflexes. Cutaneous veins are *transiently* reactive to a variety of stimuli, many of which appear valueless to the organism. For example, changes in intrathoracic pressure caused by deep inspiration, hyperventilation or Valsalva's maneuver and emotional stimuli (startle reactions, apprehension, discomfort, etc.) all cause marked but transient venoconstriction in skin.[61] Such extraneous sources of variation complicate the study of veins in man. They must be carefully controlled when studying mechanisms of venomotor reactions to temperature or exercise.

Cutaneous veins fit conveniently into many schemes of peripheral and central cardiovascular regulation. Surprisingly, many active venous responses appear to be transient. Fortunately, responses to cooling, when heat conservation may be vital for survival, are more persistent.

REFERENCES

1. Abboud, F. M., and Eckstein, J. W. Comparative changes in segmental vascular resistance in response to nerve stimulation and to norepinephrine. *Circulat. Res.*, 1966, *18*, 263–277.
2. Abramson, D. I. *Circulation in the extremities.* New York, Academic Press, 1967.
3. Ardill, B. L., Bannister, R. G., Fentem, P. H., and Greenfield, A. D. M. Circulatory responses of supine subjects to the exposure of parts of the body below the xiphisternum to subatmospheric pressure. *J. Physiol. (Lond.),* 1967, *193*, 57–72.
4. Barcroft, H. Sympathetic control of vessels in the hand and forearm skin. *Physiol. Rev.*, 1960, *40*(suppl. 4), 81–91.
4a. Beiser, G. D., Zelis, R., Epstein, S. E., Mason, D. J., and Braunwald, E. The role of skin and muscle resistance vessels in reflexes mediated by the baroreceptor system. *J. clin. Invest.*, 1970, *49*, 225–231.
5. Bernard, C. Influence du grand sympathique sur la sensibilité et sur la calorification. *C. R. Soc. Biol. (Paris),* 1851, *3*, 163–164; quoted by H. H. Hoff and R. Guillermin, C. Bernard and the vasomotor system. In: *Claude Bernard and experimental medicine,* F. Grande and M. B. Visscher, eds. Cambridge, Schenkman, 1967.
6. Bevegård, B. S., and Shepherd, J. T. Changes in tone of limb veins during supine exercise. *J. appl. Physiol.*, 1965, *20*, 1–8.
7. Bond, R. F., Lackey, G. F., Taxis, J. A., and Green, H. D. Factors governing cutaneous vasoconstriction during hemorrhage. *Amer. J. Physiol.*, 1970, *219*, 1210–1215.
8. Brender, D., and Webb-Peploe, M. M. Influence of carotid baroreceptors on different components of the vascular system. *J. Physiol. (Lond.),* 1969, *205*, 257–274.
9. Browse, N. L., Donald, D. E., and Shepherd, J. T. Role of the veins in the carotid sinus reflex. *Amer. J. Physiol.*, 1966, *210*, 1424–1434.
10. Browse, N. L., Lorenz, R. R., and Shepherd, J. T. Response of capacity and resistance vessels of dog's limb to sympathetic nerve stimulation. *Amer. J. Physiol.*, 1966, *210*, 95–102.
11. Burton, A. C. *Physiology and biophysics of the circulation.* Chicago, Year Book Medical Publ., 1965.
12. Celander, O., and Folkow, B. A comparison of the sympathetic vasomotor fibre control of the vessels within the skin and the muscles. *Acta physiol. scand.*, 1953, *29*, 241–250.
13. Chien, S. Role of the sympathetic nervous system in hemorrhage. *Physiol. Rev.*, 1967, *47*, 214–288.
14. Clarke, R. S. J., Hellon, R. F., and Lind, A. R. Vascular reactions of the forearm to cold. *Clin. Sci.*, 1958, *17*, 165–179.
15. Cooper, K. E., Johnson, R. H., and Spalding, J. M. K. The effects of central body and trunk skin temperatures on reflex vasodilatation in the hand. *J. Physiol. (Lond.),* 1964, *174*, 46–54.

16. Cooper, K. E., and Kerslake, D. McK. Abolition of nervous reflex vasodilatation by sympathectomy of the heated area. *J. Physiol. (Lond.)*, 1953, *119*, 18–29.

17. Crockford, G. W., and Hellon, R. F. Vascular responses of human skin to infra-red radiation. *J. Physiol. (Lond.)*, 1959, *149*, 424–432.

18. Crockford, G. W., Hellon, R. F., and Parkhouse, J. Thermal vasomotor responses in human skin mediated by local mechanisms. *J. Physiol. (Lond.)*, 1962, *161*, 10–20.

19. Crossley, R. J., Greenfield, A. D. M., Plassaras, G. C., and Stephens, D. The interrelation of thermoregulatory and baroreceptor reflexes in the control of the blood vessels in the human forearm. *J. Physiol. (Lond.)*, 1966, *183*, 628–639.

20. Detry, J.-M. R., Brengelmann, G. L., Rowell, L. B., and Wyss, C. Skin and muscle components of forearm blood flow in directly heated resting man. *J. appl. Physiol.*, 1972, *32*, 506–511.

21. Donegan, J. F. The physiology of the veins. *J. Physiol. (Lond.)*, 1921, *55*, 226–245.

22. Duggan, J. J., Love, V. L., and Lyons, R. H. A study of reflex venomotor reactions in man. *Circulation*, 1953, *7*, 869–873.

23. Eichna, L. W., Park, C. R., Nelson, N., Horvath, S. M., and Palmes, E. D. Thermal regulation during acclimatization in a hot, dry (desert type) environment. *Amer. J. Physiol.*, 1950, *163*, 585–597.

24. Epstein, S. E., Beiser, G. D., Stampher, M., and Braunwald, E. Role of the venous system in baroreceptor-mediated reflexes in man. *J. clin. Invest.*, 1968, *47*, 139–152.

25. Fox, R. H., and Hilton, S. M. Bradykinin formation in human skin as a factor in heat vasodilatation. *J. Physiol. (Lond.)*, 1958, *142*, 219–232.

26. Gauer, O. H., and Thorn, H. L. Properties of veins in vivo: Integrated effects of their smooth muscle. *Physiol. Rev.*, 1962, *42*(suppl. 5), 283–308.

27. Gautvik, K. Studies on kinin formation in functional vasodilatation of the submandibular salivary gland in cats. *Acta physiol. scand.*, 1970, *79*, 174–187.

28. Gautvik, K. The interaction of two different vasodilator mechanisms in the chorda-tympani activated submandibular salivary gland. *Acta physiol. scand.*, 1970, *79*, 188–203.

29. Greenfield, A. D. M. The circulation through the skin. *Handb. Physiol.*, 1963, sec. 2, vol. 2, 1325–1351.

30. Greenfield, A. D. M., Whitney, R. J., and Mowbray, J. F. Methods for the investigation of peripheral blood flow. *Brit. med. Bull.*, 1963, *19*, 101–109.

31. Grimby, G., and Nilsson, N. J. Cardiac output during exercise in pyrogen-induced fever. *Scand. J. clin. Lab. Invest.*, 1963, *15*(suppl. 69), 44–61.

32. Hellon, R. F. Local effects of temperature. *Brit. med. Bull.*, 1963, *19*, 141–144.

33. Hertzman, A. B. Regulation of cutaneous circulation during body heating. In: *Temperature, its measurement and control in science and industry*, J. M. Herzfeld, ed., vol. III, part 3. New York, Reinhold Publishing Corp., 1963.

34. Hertzman, A. B. Vasomotor regulation of the cutaneous circulation. *Physiol. Rev.*, 1959, *39*, 280–306.

35. Keatinge, W. R. Direct effects of temperature on blood vessels: Their role in cold vasodilatation. In: *Physiological and behavioral temperature regulation*, J. D. Hardy, A. P. Gagge and J. A. J. Stolwijk, eds. Springfield, Ill., Charles C Thomas Co., 1970.

36. Kerslake, D. McK., and Cooper, K. E. Vasodilatation in the hand in response to heating the skin elsewhere. *Clin. Sci.*, 1950, *9*, 31–47.

37. Kidd, B. S. L., and Lyons, S. M. The distensibility of the blood vessels of the human calf determined by graded venous congestion. *J. Physiol. (Lond.)*, 1958, *140*, 122–128.

38. Korner, P. I., Chalmers, J. P., and White, S. W. Some mechanisms of reflex control of the circulation by the sympatho-adrenal system. *Circulat. Res.*, 1961, *20–21* (suppl. III), 157–172.

39. Koroxenidis, G. T., Shepherd, J. T., and Marshall, R. J. Cardiovascular response to acute heat stress. *J. appl. Physiol.*, 1961, *16*, 869–872.

40. Lewis, G. P. Active polypeptides derived from plasma proteins. *Physiol. Rev.*, 1960, *40*, 647–676.

41. Lewis, T. Observations upon the reactions of the vessels of the human skin to cold. *Heart*, 1930, *15*, 177–208.

42. Lewis, T. *The blood vessels of the human skin and their responses.* London, Shaw, 1927.

43. Mayer, H. E., Abboud, F. M., Ballard, D. R., and Eckstein, J. W.: Catecholamines in arteries and veins of the foreleg of the dog. *Circulat. Res.*, 1968, *23*, 653–661.

44. Mellander, S. Comparative studies on the adrenergic neuro-hormonal control of resistance and capacitance blood vessels in the cat. *Acta physiol. scand.*, 1960, *50* (suppl. 176), 1–86.

45. Mellander, S., and Johansson, B. Control of resistance, exchange and capacitance functions in the peripheral circulation. *Pharmacol. Rev.*, 1968, *20*(3), 117–196.

46. Roddie, I. C., and Shepherd, J. T. The blood flow through the hand during local heating, release of sympathetic vasomotor tone by indirect heating, and a combination of both. *J. Physiol. (Lond.)*, 1956, *131*, 657–664.

47. Roddie, I. C., and Shepherd, J. T. The effects of carotid artery compression in man with special reference to changes in vascular resistance in the limbs. *J. Physiol. (Lond.)*, 1957, *139*, 377–384.

48. Roddie, I. C., Shepherd, J. T., and Whelan, R. F. The contribution of constrictor and dilator nerves to the skin vasodilatation during body heating. *J. Physiol. (Lond.)*, 1957, *136*, 489–497.

49. Roddie, I. C., Shepherd, J. T., and Whelan, R. F. Reflex changes in human skeletal muscle blood flow associated with intrathoracic pressure changes. *Circulat. Res.*, 1958, *6*, 232–238.

50. Rowell, L. B., Blackmon, J. R., Martin, R. H., Mazzarella, J. A., and Bruce, R. A. Hepatic clearance of indocyanine green in man under thermal and exercise stresses. *J. appl. Physiol.*, 1965, *20*, 384–394.

51. Rowell, L. B., Brengelmann, G. L., Blackmon, J. R., and Murray, J. A. Redistribution of blood flow during sustained high skin tempera-

ture in resting man. *J. appl. Physiol.*, 1970, 28, 415–420.

52. Rowell, L. B., Brengelmann, G. L., Blackmon, J. R., Twiss, R. D., and Kusumi, F. Splanchnic blood flow and metabolism in heat-stressed man. *J. appl. Physiol.*, 1968, 24, 475–484.

53. Rowell, L. B., Brengelmann, G. L., Detry, J.-M. R., and Wyss, C. Venomotor responses to local and remote thermal stimuli to skin in exercising man. *J. appl. Physiol.*, 1971, 30, 72–77.

54. Rowell, L. B., Brengelmann, G. L., and Murray, J. A. Cardiovascular responses to sustained high skin temperature in resting man. *J. appl. Physiol.*, 1969, 27, 673–680.

55. Rowell, L. B., Detry, J,-M. R., Profant, G. R., and Wyss, C. Splanchnic vasoconstriction in hyperthermic man—role of falling blood pressure. *J. appl. Physiol.*, 1971, 31, 864–869.

56. Rowell, L. B., Marx, H. J., Bruce, R. A., Conn, R. D., and Kusumi, F. Reductions in cardiac output, central blood volume, and stroke volume with thermal stress in normal men during exercise. *J. clin. Invest.*, 1966, 45, 1801–1816.

57. Rowell, L. B., Murray, J. A., Brengelmann, G. L., and Kraning, K. K., II. Human cardiovascular adjustments to rapid changes in skin temperature during exercise. *Circulat. Res.*, 1969, 24, 711–724.

58. Samueloff, S. L., Browse, N. L., and Shepherd, J. T. Response of capacity vessels in human limbs to head-up tilt and suction on lower body. *J. appl. Physiol.*, 1966, 21, 47–54.

59. Schachter, M. Kallikreins and kinins. *Physiol. Rev.*, 1969, 49, 509–547.

60. Senay, L. C., Prokop, L. D., Conau, L., and Hertzman, A. B. Relation of local skin temperature and local sweating to cutaneous blood flow. *J. appl. Physiol.*, 1963, 18, 781–785.

61. Shepherd, J. T. *Physiology of the circulation in human limbs in health and disease.* Philadelphia, W. B. Saunders Co., 1963.

62. Shepherd, J. T. Role of the veins in the circulation. *Circulation*, 1966, 33, 484–491.

63. Strom, G. Central nervous regulation of body temperature. *Handb. Physiol.*, 1960, sec. 1, vol. II, 1173–1196.

64. Thauer, R. Circulatory adjustments to climatic requirements. *Handb. Physiol.*, 1962, sec. 2, vol. III, 1921–1966.

65. Vanhoutte, P. M., and Lorenz, R. R. Effect of temperature on reactivity of saphenous,

mesenteric and femoral veins of the dog. *Amer. J. Physiol.*, 1970, 218, 1746–1750.

66. Vanhoutte, P. M., and Shepherd, J. T. Activity and thermosensitivity of canine cutaneous veins after inhibition of monoamine oxidase and cathechol-o-methyl transferase. *Circulat. Res.*, 1969, 25, 607–616.

67. Vanhoutte, P. M., and Shepherd, J. T. Effect of temperature on reactivity of isolated cutaneous veins of the dog. *Amer. J. Physiol.*, 1970, 218, 187–190.

68. Webb-Peploe, M. M. Cutaneous venoconstrictor response to local cooling in the dog: Unexplained by inhibition of neuronal re-uptake of norepinephrine. *Circulat. Res.*, 1969, 24, 607–615.

69. Webb-Peploe, M. M., and Shepherd, J. T. Peripheral mechanism involved in response of dogs' cutaneous veins to local temperature change. *Circulat. Res.*, 1968, 23, 701–708.

70. Webb-Peploe, M. M., and Shepherd, J. T. Response of dogs' cutaneous veins to local and central temperature changes. *Circulat. Res.*, 1968, 23, 693–699.

71. Webb-Peploe, M. M., and Shepherd, J. T. Response of large hindlimb veins of the dog to sympathetic nerve stimulation. *Amer. J. Physiol.*, 1968, 215, 299–307.

72. Webb-Peploe, M. M., and Shepherd, J. T. Responses of the superficial limb veins of the dog to changes in temperature. *Circulat. Res.*, 1968, 22, 737–746.

73. Webb-Peploe, M. M., and Shepherd, J. T. Veins and their control. *New Engl. J. Med.*, 1968, 278, 317–322.

74. Wood, J. E. *The veins, normal and abnormal function.* Boston, Little, Brown & Co., 1965.

75. Wood, J. E., and Eckstein, J. W. A tandem forearm plethysmograph for study of acute responses of the peripheral veins of man: The effect of environmental and local temperature change, and the effect of pooling blood in the extremities. *J. clin. Invest.*, 1958, 37, 41–50.

76. Wyndham, C. H. Effect of acclimatization on circulatory responses to high environmental temperatures. *J. appl. Physiol.*, 1951, 4, 383–395.

77. Zelis, R., and Mason, D. T. Comparison of the reflex reactivity of skin and muscle veins in the human forearm. *J. clin. Invest.*, 1969, 48, 1870–1877.

CHAPTER 13 CIRCULATION TO SKELETAL MUSCLE

by LORING B. ROWELL

INTRODUCTION

Blood flow to skeletal muscle, in contrast to that of skin, subserves mainly metabolic needs which can vary more than 100-fold. Meeting such high needs often places maximal demands upon the cardiovascular system for large changes in blood flow. Muscle beds exhibit *dual control* to a greater extent than other vascular beds; that is, regulation of muscle blood flow can be predominantly *extrinsic* or *neural*, whereas at other times *intrinsic, local factors* dominate. The latter mode of control is attributed to vasoactive metabolites which are produced in proportion to the metabolic activity of the tissue. Because of its total circulatory dimensions, muscle is a major target for vasomotor reflexes concerned with overall cardiovascular regulation.

ANATOMY

The total mass of skeletal muscle exceeds that of any other organ, comprising 40 to 50 per cent of total body mass in normal young men. Its vascular supply is structurally uniform, but the capacities for maximal blood flow in red and white muscles are different.[34, 59] There is no anatomical evidence of true A-V shunts in skeletal muscle,[5, 10] but some experimental findings indirectly suggest their existence. For example, when muscle vessels are dilated by stimulation of sympathetic cholinergic vasodilator nerve fibers, changes in muscle oxygen uptake and venous oxygen saturation are in a direction suggesting that the increased blood flow has by-passed the capillaries and thus escaped tissue exchange.[63, 67] Also, the constancy of I[131] clearance during neurogenic vasodilation, in contrast to the

200

increase during metabolic vasodilation, supports the idea that increased blood flow does not pass through capillaries during neurogenic vasodilation.[46] Alternatively, two entirely separate circulations in muscle with different flows, one to muscle fibers and the other to connective tissue, may be independently affected by these two types of vasodilation.[10, 81] Contradicting the notion of shunts is the more recent finding in *intact* dog gastrocnemius that isotope clearance from muscle is increased in relation to blood flow (*i.e.*, increased capillary exchange) in both metabolic and neurogenic vasodilation, but the clearance to blood flow ratio is two to three times greater in the former. Apparently metabolic vasodilation opens precapillary sphincters (see Chap. 9, Vol. II), causing increased capillary surface area, whereas neurogenic vasodilation does not increase this area but acts only on precapillary resistance vessels,[18] causing more blood to pass through already opened capillaries.

MEASUREMENT

Muscle blood flow is difficult to measure. The most reliable technique is venous occlusion plethysmography (see Chap. 3, Vol. II); accordingly, most of our knowledge of muscle blood flow is confined to limb musculature. For the same reasons skin blood flow is often included in the total flow measurement unless the overlying skin circulation is arrested by topically infiltrating it with epinephrine (epinephrine iontophoresis).

Another method used to determine muscle blood flow is measuring the clearance rate of radioactive indicators injected into muscle[4, 42] (see Chap. 3, Vol. II). Major problems with this technique are differential solubility of labeled ions in different structural components of muscle, dissociation of labeled moieties from diffusible compounds which as free ions have diffusion-limited clearance, and mechanical damage to small vessels by injection. Results using thermal washout techniques are influenced by the proximity of thermal detectors to blood vessels. Much of our knowledge of control of muscle blood flow comes from animal experiments in which flow is directly measured after surgical isolation of a muscle and its vasculature.

GENERAL DESCRIPTION OF MUSCLE BLOOD FLOW

Range of Flow. In resting normal man somewhere between 15 and 20 per cent of cardiac output is apportioned to skeletal muscle, or about 750 to 1000 ml per min in a 70 kg man. Estimated blood flow per 100 g of muscle is approximately 2 to 4 ml 100 ml^{-1} min^{-1}.[5, 71] Because skeletal muscle has a greater variation in metabolic rate than most other tissues and also because of its great mass, the range of muscle blood flow is huge. During exercise at the maximal metabolic rate (*e.g.*, at the highest cardiac output and systemic A-VO$_2$ difference attainable), 80 to 85 per cent of the cardiac output is distributed to working skeletal muscle.[15, 80] In a normal man having a maximal oxygen consumption by the muscle of 3.5 liters per min and maximal cardiac output of 22 liters per min, maximal muscle blood flow must be about 18.6 liters per min. The minimum muscle blood flow required for this oxygen uptake, assuming 100 per cent extraction of oxygen by the muscle, would be 17.5 liters per min. Oxygen consumption in endurance athletes can reach as high as 6.3 liters per min while cardiac output is 42 liters per min.[31] Since the maximal oxygen extraction efficiency is 85 per cent in these athletes, no less than 36 liters of blood perfuses working muscle each minute. Assuming that two-thirds of the total muscle is utilized during running, maximal muscle blood flow would be at least 140 ml min^{-1} 100 g^{-1}. Flows this high are not seen in isolated muscle.

Data from cat and dog muscle indicate that both basal and maximal blood flow rates differ in white (fast) muscle and red (slow, tonic) muscle.[34, 59] Values given for white muscle at rest and during "maximal" vasodilation are 2 to 5 and 40 to 60 ml min^{-1} 100 g^{-1} muscle, respectively. In contrast, red (sometimes called aerobic) muscle blood flow ranges from 20 to 30 up to 115 ml min^{-1} 100 g^{-1} at rest and at "maximal" vasodilation, respectively.

Blood Volume. Muscle blood volume is relatively low at 2 to 3 ml blood per 100 g muscle.[51] If these values from cat and dog muscle are extrapolated to a 70 kg man, the total volume of blood in human muscle must be in the order of 700 to 1000 ml. Approximately 70 to 80 per cent of this volume is in the veins.[84]

Basal Tone. Because of their enormous capacity for blood flow when dilated, muscle resistance vessels must normally be maintained in a state of tonic vasoconstriction during rest. Part of this tone is maintained neurally,[5, 59] but a major factor is the high degree of basal tone in muscle arterioles.[55, 59] Abolishment of this tone by arterial injection of potent vasodilators such as ATP or acetylcholine decreases vascular resistance by 80 to 85 per cent. Basal tone is ascribed to inherent myogenic activity of vascular smooth muscle, resulting in maintained rhythmic, phasic contraction. Basal tone forms a background state of relative vasoconstriction in muscle around which all active and passive vasodilation and vasoconstriction can occur—that is, a setting from which these vasomotor effects are added or subtracted. This concept is illustrated in Figure 8–8, Chapter 8, Volume II.

NEURAL CONTROL OF SKELETAL MUSCLE RESISTANCE VESSELS

Active Neurogenic Vasoconstriction. In addition to basal tone, resistance vessels receive tonic sympathetic vasoconstrictor nervous outflow. Interruptions of these sympathetic nerves in resting muscle cause a 2- to 3-fold increase in blood flow, a minor increment in comparison with that resulting from maximal metabolic vasodilation (Fig. 8–9, Chap. 8, Vol. II).[7, 65] Stimulation of these nerves at a frequency of 10 imp/sec or more produces a maximum (8- to 10-fold) increase in vascular resistance of white muscle (in the cat) by their release of norepinephrine which stimulates alpha-receptors[58] (see Chap. 8, Vol. II). The same stimulation apparently produces only a 2.0- to 2.5-fold increase in resistance above the normal resting value in red muscle.[34, 58] Control of muscle vascular resistance over a wide range is possible by variation of discharge rates over sympathetic vasoconstrictor fibers.[25, 58, 59] This forms the basis for most neurogenic or reflex control of muscle vascular resistance, discussed in more detail later.

Active Neurogenic Vasodilation. Two different neural mechanisms, secondary in importance to non-neurogenic mechanisms discussed later, produce vasodilation in muscle. The first neural mechanism, described above, is *passive* and is caused by inhibition of tonic sympathetic vasoconstrictor nerve outflow. The second is *active*, requiring activation of sympathetic vasodilator fibers, of which there may be at least two types—cholinergic, which release acetylcholine, and histaminergic, which release histamine.

Sympathetic Cholinergic Vasodilator System. Stimulation of sympathetic axons supplying skeletal muscle in which alpha-receptors (vasoconstrictor) have been pharmacologically blocked produces active vasodilation. Since the response can be blocked by atropine, a cholinergic mediator is proposed.[18, 41, 59] Also, acetylcholinesterase has been identified in or around nerve terminals within small arterial vessels in dog and cat skeletal muscle but was not found in human or primate muscle vessels.[19] These cholinergic fibers appear to originate in the motor cortex and to have relay stations in the hypothalamus.[76]

At stimulation frequencies of 10 imp/sec, muscle vascular resistance decreases five times, which is about 70 per cent of the maximal vasodilation induced by the intra-arterial injections of acetylcholine. The vasodilator response is transient; despite continued stimulation, vascular resistance returns to normal within less than a minute.[18]

Stimulation of certain hypothalamic regions, however, may cause a pure vasodilator response which is observable without previous adrenergic blockade. Some investigators find that hypothalamic stimulation will not cause vasodilation if there is a high background intensity of tonic vasoconstrictor activity.[36] However, others find reproducible vasodilation under these conditions.[68] This discrepancy is unexplained.

The functional significance of the vasodilator system is unknown. Active vasodilation is thought to be part of a centrally integrated alerting and defense reaction. One postulate is that such vasodilation may transiently prepare the muscle for a sudden burst of activity by circumventing the time lag of normal local vasodilator mechanisms; that is, high muscle blood flow is momentarily held in readiness until the local buildup of vasoactive metabolites exert their effects. Although active dilator systems are established in cats and dogs, they have not been demonstrated in primates.

Histaminergic Vasodilator Nerves. Some pharmacologic evidence suggests the existence of a histaminergic vasodilator system of nerve fibers.[13, 22] The evidence is as follows: (i) histamine and its synthesizing enzymes are present in peripheral nerve tissue; (ii) injected histamine produces responses similar to those of the neurogenic transmitter; (iii) the neurogenic response is blocked by antihistamines; and (iv) histamine is released into venous blood during nerve stimulation.[13] However, histaminergic nerve fibers have not been specifically identified, nor does direct stimulation of lumbar sympathetic nerves produce active vasodilation unless the adrenergic nerve fibers are blocked.

The functional significance of a histaminergic vasodilator system is also difficult to understand, because its function is manifest only when normal adrenergic function is abolished and its vasodilator effects last only for a few seconds.[22]

The picture is further complicated by a recent suggestion of still another neurogenic vasodilator system which elicits sustained vasodilation in the dog limb. Its unknown chemical mediator is apparently neither cholinergic nor adrenergic.[13] Since histamine markedly increases capillary permeability, a sustained release of this substance into a large mass of muscle should cause considerable loss of fluid into the extravascular space. Thus, a long-acting histaminergic system appears highly unlikely.

REFLEX CONTROL OF MUSCLE VASCULAR RESISTANCE

Baroreceptors. Peripheral vascular effects of baroreceptor stimulation have been extensively reviewed by Heymans and Neil[44] (see Chap. 10, Vol. II). Extensive studies on anesthetized animals yield a consistent picture. Stimulation of the aortic and carotid sinus baroreceptors by increased wall-stretch at the receptor site causes vasodilation in skeletal muscle. Vasoconstriction occurs when stretch on the baroreceptors is diminished by a fall in

blood pressure. Either response can be abolished by cutting the sympathetic nerves or by pharmacological blockade of norepinephrine action at the receptor site which releases all neural tone in these vessels. Thus, both responses appear to depend on changes in the rate of tonic discharge over sympathetic vasoconstrictor nerves.[23] No sympathetic vasodilator nerves are known to participate in these reflexes.[76]

Skeletal muscle is generally considered to be a major target organ for regulation of blood pressure because of its high total blood flow. But the apparently clear picture of the role of skeletal muscle in these reflexes obtained from anesthetized animals has been difficult to achieve in man.[64, 65, 71] A major question is whether muscle vascular resistance is reflexly altered by arterial baroreceptors, low pressure baroreceptors (see below) or both.

Because of its ready accessibility, forearm muscle has been most frequently studied in man. Despite earlier conclusions that muscle vascular resistance is not modified by reflexes originating from arterial baroreceptors,[64, 71] muscle is now considered an important target for these reflexes in man. For example, when carotid sinus transmural pressure is increased by applying suction to the neck, blood pressure and forearm vascular resistance are reflexly reduced.[13a, 14] Direct electrical stimulation of the carotid sinus nerve in man[32] and dogs[79] also causes muscle vasodilation. This dilation is mainly responsible for the resultant fall in peripheral vascular resistance. In dogs the decrement in muscle vascular resistance is more than two times that seen in renal and splanchnic vasculature.[79] However, this mode of stimulation includes the undetermined effects of stimulating afferent chemoreceptor nerves contained in the sinus nerve. Equivalent data are not available in humans.

Another problem in interpretation is caused by experimental stimulation of one arterial baroreceptor exclusive of the other; this leaves the unstimulated receptor free to counteract that stimulation. When, for example, the carotid sinus sees *reduced* pressure, the aortic receptors will see a *rise* in blood pressure, in heart rate and in the rate of rise of the pressure pulse, all of which may inhibit sympathetic nervous outflow (see Chap. 10, Vol. II). The question is how much the cardiovascular response to altered carotid sinus stretch will be modified by counteracting reflex drive from the aortic baroreceptors. Data from anesthetized dogs indicate that pressure changes over a given range at the carotid sinus elicit greater vasomotor effects than those elicited from aortic receptors by the same stimuli. Further, a carotid sinus induced pressor response obscures a simultaneous aortic induced depressor response of equal magnitude.[29a]

Evidence indicates that baroreceptors located in cardiac chambers and major pulmonary vessels ("low pressure baroreceptors") have important influence on muscle vascular resistance in man.[64, 66] For example, raising central venous pressure and thoracic volume by raising the legs of a supine man sometimes elicits vasodilation of forearm skeletal muscle by reflex inhibition of tonic sympathetic vasoconstrictor tone. Similar changes in muscle blood flow could be elicited by head-down tilting, coughing and negative pressure breathing— all of which increase thoracic venous transmural pressure[71] (see Fig. 13–1). In contrast, maneuvers which decrease the stretch on these regions, such as Valsalva's maneuver, positive pressure breathing, a head-up tilt and lower body negative pressure, cause reflex vasoconstriction of skeletal muscle.[71] During these maneuvers measurable changes in arterial blood pressure are either absent or very small; vasoconstriction coincides with the fall in central venous pressure[86] (see Fig. 13–2). In general, the findings suggest that reflexes from both low and high pressure baroreceptors are important in regulating muscle vascular resistance. Situations such as moderate to severe hemorrhage probably stimulate both receptors, leaving unanswered in man the question of which receptor plays the dominant role. In anesthetized dogs and cats, in apparent contrast to man, low pressure baroreceptors exert their major effects on splanchnic and renal vasculature, with minor effects on muscle vascular resistance.[61a, 61b] In man, neck suction experiments reveal that the reflex can be generated exclusively by the carotid sinus receptors.

Chemoreceptors. Stimulation of aortic and carotid body chemoreceptors by hypoxia produce reflex vasoconstriction in skeletal muscle.[50, 72] This effect is most pronounced when the competing local vasodilator effects of hypoxia are eliminated by separately perfusing muscle with normoxic blood. Depending upon the degree of central hypoxic stimulation, local vasodilator mechanisms can override central, reflex vasoconstrictor mechanisms, or the latter can override the local effects when systemic hypoxia is severe.[72] In addition to local muscle oxygen tension, other factors such as hyperpnea and accelerated epinephrine release may counteract reflex vasoconstriction.[50]

Emotion. Vascular resistance of human skeletal muscle can be drastically reduced by emotional stress.[5, 16, 33, 45, 71] Total fore-

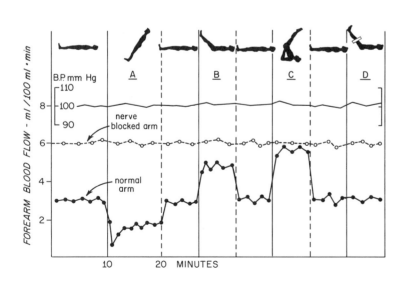

Figure 13–1 Schematic representation of changes in blood flow in a normal (●—●) and sympathetic nerve-blocked (○—○) forearm. Data from different experiments are combined into a single series of maneuvers which at *A* reduce intrathoracic blood volume. This volume at *B* is increased by raising legs and in *C*, raising legs plus lower torso. In *D* occlusion of the legs prevents the response. Constant blood flow in the nerve-blocked arm indicates that changes result from changes in tonic sympathetic vasoconstrictor outflow. Arterial mean pressure remains constant, but arterial pulse pressure may change. (After Shepherd, *Physiology of the circulation in human limbs in health and disease.* Philadelphia, W. B. Saunders Co., 1963.)

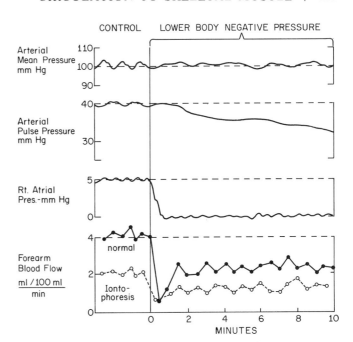

Figure 13–2 Changes in forearm blood flow in a normal arm (●) and one in which skin circulation is arrested by topical infiltration of epinephrine by iontophoresis (○). At the onset of lower body suction, which draws blood into the legs and simulates hemorrhage, central venous pressure and forearm muscle and skin blood flow decrease suddenly. Arterial mean pressure stays constant. Arterial pulse pressure may decline simultaneous with or slightly after the beginning of suction. The latter suggests the presence of low pressure (right atrial) baroreceptors. Responses of the two arms indicate that about one-half of the total forearm vasoconstriction occurs in muscle while the other half is in skin.

arm blood flow may increase as much as eight to ten times under such conditions[33] (Fig. 13–3), and also during a faint.[9, 71] Muscle vasculature is generally considered

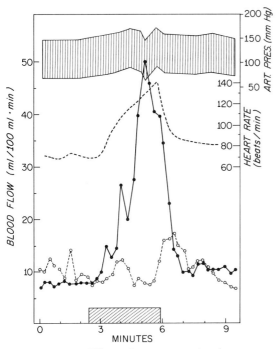

Figure 13–3 Effect of severe emotional stress on arterial pressure. (— — —) = Heart rate; (●) = Forearm blood flow; (○) = Hand blood flow. During the time represented by the rectangle, it was suggested to the subject that he was suffering from severe blood loss. (From Blair *et al.*, *J. Physiol. (Lond.)*, 1959, 148, 633.)

to be the site of this dilation, which appears to be neurogenic because blockade of nerves to vessels reduces the response.[8, 16] Because this vasodilation is not affected by adrenergic blockade[17] and is reduced (but not abolished) by cholinergic blockade with atropine, it appears partly dependent upon a cholinergic mechanism.[16] But neither is this vasodilation completely abolished by total sympathectomy of the arm.[45] Since adrenalectomy reduces vasodilation, release of epinephrine from the adrenal medulla may play a role in the response. Finally, suppression of the skin circulation in an arm can also reduce the vasodilator response in that arm; thus, skin may dilate some (possibly passively). Although the consensus is that emotional stress causes vasodilation in skeletal muscle via the cholinergic vasodilator system,[41] responses appear to be variable both in the site of action and in the mechanism by which the action is mediated.

Exercise. In man, muscular activity is accompanied by increased sympathetic vasoconstrictor nerve activity which causes vasoconstriction in nonexercising muscles and other nonactive vascular beds.[15, 80] This reflex activity increases in proportion to the intensity of exercise and ultimately produces intense vasoconstriction in these inactive regions during severe work (Fig. 13–4). This response serves to diminish

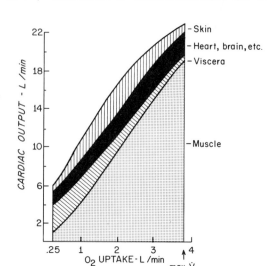

Figure 13–4 Approximate distribution of cardiac output at rest and at different levels of exercise up to the maximal oxygen consumption (max $\dot{V}_{O_2}$) in a normal young man. Section labeled Viscera reveals a progressive reduction in absolute blood flow and percentage of cardiac output distributed to splanchnic region and kidneys to augment muscle blood flow. Even skin is constricted during brief periods of exercise at high oxygen consumption.

disparities between cardiac output and local oxygen demand by redistributing increasingly large percentages of cardiac output to working muscle as the severity of work is augmented.[15, 80]

LOCAL CONTROL OF SKELETAL MUSCLE RESISTANCE VESSELS

Neurally mediated changes in muscle blood flow are rather small in contrast to the changes mediated by locally released metabolites. At rest neural regulation predominates; during muscular contraction, local metabolic regulation supervenes as muscle oxygen consumption reaches very high rates. The question is how this local control adjusts muscle blood flow so that it is closely related to its oxygen consumption. To answer this it is necessary to ascertain whether the blood flow at any given intensity of work is the result of a *balance* between the dilating action of local metabolites and the tonic vasoconstrictor action of sympathetic nerves (which increases in proportion to the severity of exercise, as described above). Some find in the anesthetized dog that a "functional sympatholysis" occurs in working muscle so that

normal rates of sympathetic nerve stimulation are without effect.[62]

Sympathectomy of one hindlimb of an unanesthetized dog causes no change in blood flow to that limb in comparison to the normal contralateral limb during mild to severe, graded exercise.[30] Thus, vascular resistance of working skeletal muscle in dogs can be closely regulated in proportion to metabolic demands by local factors alone and without the influence of sympathetic vasoconstrictor nerve activity. Furthermore, working muscles are still responsive to sympathetic nerve activity when it is increased in unanesthetized dogs by stimulating electrodes implanted on the lumbar sympathetic trunk.[30] Even weak (0.5 imp/sec) stimulation reduces blood flow to these muscles at all intensities of exercise. The magnitude of the effect depends on the frequency of stimulation (Fig. 13–5), and at any frequency the effect is smallest at the highest rate of work (Fig. 13–6). The results suggest that sympathetic vasoconstrictor outflow to muscle in the exercising dog is normally so weak that local, humoral factors dominate in the regulation of blood flow. The lack of vasomotor activity is further indicated by the unaltered mesenteric and renal artery blood flow in dogs during mild[70] to severe exercise,[77] in contrast to man who shows striking vasoconstriction in these regions.[69] In man, further increasing the intensity of sympathetic vasoconstrictor outflow during exercise by stimulating baroreceptors (applying negative pressure to the legs during work with the arms) reduces blood flow to exercising forearm muscle.[75] Although still unproved, the possibility remains that, in man, local metabolic factors cancel or reduce effects of increased sympathetic nerve activity so that muscle blood flow is the resultant of the two opposing functions. Since hypoxia, increased osmolality and increased potassium concentration in muscle can counteract reflex sympathetic vasoconstriction (induced by systemic hypoxia), these factors may be of major importance in local, vascular control.[72, 73]

What are the local agents which produce a functional hyperemia so closely related to metabolic requirements? The list of possible chemical factors is impressive.[43, 59] Gaskell in 1877[40] was apparently the first to postulate that regional vascular tone was inhibited by local accumulation of meta-

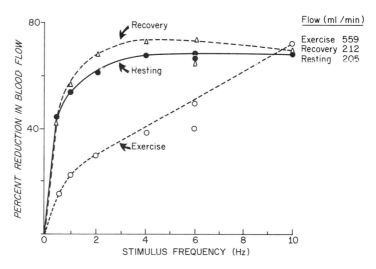

Figure 13–5 Effect on limb blood flow of electric stimulation of lumbar sympathetic chain at various frequencies during exercise and rest. Stimulus voltage and duration were constant. Studies were made with dogs resting quietly (●), during mild exercise (○) and again resting quietly after exercise (△). Stimulation was maintained for 30 sec, and response was plotted as the percentage of reduction from the blood flow recorded immediately before stimulation. Exercise lasted 20 min. (From Donald *et al., Circulat. Res.,* 1970, *26,* 185–199.)

bolic by-products. Such a feedback system could regulate vascular resistance as a close function of metabolic rate, because the latter determines production rate of metabolic end-products. Another statement of the hypothesis is that the blood flow to metabolism ratio sets the tissue concentration of oxygen and vasodilator metabolites and thus determines the diameter of resistance vessels. This concept predicts that any change in the ratio, whether by altering metabolism or blood flow, will produce compensatory vasomotion in a direction to restore the ratio. Supporting this, studies on dog limbs have demonstrated the converse of vasodilation caused by increased metabolism; namely, vasoconstriction was

caused by increased blood flow (at a given mean pressure).[43]

The most convincing evidence of a chemical vasodilator comes from bioassay experiments wherein venous blood draining either an active or a resting limb is diverted into arterial blood supplying another limb; vasodilation is elicited in the latter,[43] but dilatory effects in this assay muscle often do not match those of the donor muscle. This is probably due to dilution of the active substances in the arterial blood supplying the assay organ. In general, venous blood is vasoactive with respect to arterial blood. Unfortunately, these bioassay studies have shed little light upon the chemical nature of any specific vasoactive mediator. Many

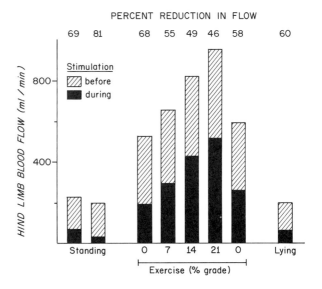

Figure 13–6 Effect of electric stimulation of lumbar sympathetic chain at a fixed frequency (6 imp/sec) for 30 sec on limb blood flow during different intensities of exercise. Combined cross-hatched and solid areas equal blood flow before stimulation, and solid areas alone equal the flow during the last 10 sec of stimulation. Each level of exercise was run separately. (From Donald *et al., Circulat. Res.,* 1970, *26,* 185–199.)

substances such as bradykinin, acetylcholine and histamine have been suggested without experimental support. Increments in blood flow to active muscle are paralleled by decreases in venous oxygen and increases in venous concentration of hydrogen ion, potassium ion, adenosine, adenine nucleotides (ATP, ADP, AMP) and intermediate metabolites of the tricarboxylic acid (Krebs) cycle. Individually, however, their effects on blood flow over a maximum range of physiological values is far too small to account for anything but modest hyperemia. Of this list adenosine and adenine nucleotides are on a mole for mole basis the most potent vasodilators. ATP has been found in human forearm venous blood in concentrations which roughly parallel the pattern of hyperemic forearm blood flow after isometric exercise.[37] In the dog, the quantities of ATP in venous blood after 5 min of ischemic (insufficient oxygen) muscle contraction are too small to account for much vasodilation.[29] However, adenosine, which may play an important role in vasodilating cardiac muscle (see Chap. 4, Vol. II), appears in venous blood draining dog muscle along with its deaminated byproduct inosine at concentrations sufficient to reduce local vascular resistance.[29]

No substance by itself meets all the criteria. The evidence is basically circumstantial. Concentrations of naturally occurring substances or of test substances added to arterial blood are followed in venous blood rather than in the tissue fluid around the vessel where their effects are manifest. Presently methods for determining tissue concentration of many of these substances and their temporal relationship to local vascular events are not available.

A refreshing change from the traditional search for "the dilator substance" is the attempt to identify how various substances might interact and react locally. For example, reduced local oxygen tension potentiates the vasodilator response to potassium, and these effects are further potentiated by increases in local osmolality.[73] The old observation that an increase in muscle osmolality by infusion of hypertonic solutions can dilate arterioles has received renewed attention.[56, 59, 60, 74] Breakdown of larger substrate molecules to smaller ones in muscle cells during accelerated metabolism increases local osmolality. In addition, all substances discussed above may also have entirely nonspecific functions in metabolic vasodilation by serving collectively to alter muscle osmolality. But to some extent this effect must be subsequently counterbalanced by the shift in water from intravascular to intracellular space. Since this fluid shift is complete in a few minutes of exercise, it seems unlikely that increased muscle osmolality could by itself maintain vasodilation over prolonged periods. Rather direct action of local metabolites, potassium, reduced oxygen tension and the like probably sustain the vasodilation. In short, chemical factors are clearly important in muscle vasodilation, but undoubtedly they exist in a myriad of combinations wherein any given substance may produce additive, synergistic and/or potentiating effects upon any other factor which by itself may be unreactive.

Nonchemical Mechanisms. In 1902 Bayliss[12] postulated that changes in transmural pressure may alter vascular tone so that blood vessels of muscle and other regions can "autoregulate" their own blood flow. This scheme implies that increased perfusion pressure will raise vascular transmural pressure, causing vessels to constrict so as to maintain constant blood flow. Alternatively, a fall in perfusion pressure would reduce vascular wall tension and produce vascular relaxation and dilation. This is the *myogenic hypothesis* of autoregulation of blood flow (see Chap. 8, Vol. II). Skeletal muscle under appropriate experimental conditions exhibits this characteristic. The concept fits what is known of the electrical behavior of vascular smooth muscle cells. The frequency of action potentials is increased in response to passive stretch.[59]

Studies of autoregulation in skeletal muscle also indicate the presence of a metabolic component which depends upon a low muscle oxygen tension.[48] Interaction between local chemical and myogenic reactions may occur so that accumulation of metabolites serves as a brake on myogenic mechanisms. This would act to further smooth out reactions to stretch.[47]

The functional significance of autoregulation in intact muscle is unknown. Its existence as a laboratory phenomenon in muscle which is denervated or exposed to adrenergic blockage is generally accepted.

However, blood flow to normally innervated muscle appears subservient to reflex neural control at rest and certainly to local metabolic control in exercise.

Reactive Hyperemia—Metabolic, Myogenic or Both? Temporary arrest or reduction of blood flow to muscle is followed upon release by a marked overshoot in flow called reactive hyperemia. This is commonly attributed to the local accumulation of vasodilator metabolites. Accordingly, one might expect the magnitude and duration of hyperemia to be related to the duration of occlusion in a debt-repayment relationship. The debt is an estimate of the total amount of blood which did not reach the muscle during occlusion (normal resting blood flow times occlusion time), and repayment is the area beneath the recovery curve which is in excess of the preocclusion resting level. But the problem is complex, because often no such simple relationship is found.[71] For example, in the human forearm, hyperemia following occlusion or exercise can be abolished by mechanically reducing (by compression) brachial arterial inflow for 5 min after total occlusion is released.[5] Also, metabolic recovery from exercise is not noticeably affected in man when blood flow to recovering muscle is experimentally reduced.[75] In dogs, the hyperemic response to limb occlusion can be prevented by stimulation of the lumbar sympathetic chain just before and for several minutes after releasing occlusion. Reduction of muscle blood flow by the same means during exercise does not cause any overshoot or "repayment" after stimulation is stopped.[30] Thus, no cause and effect relationship between blood flow and metabolic events is apparent in hyperemic reactions.

Bayliss originally suggested that reduced stretch upon arterial vessels, distal to the point of occlusion, caused a sustained myogenic release in vascular tone which caused hyperemia after release of occlusion. However, Lewis[52] later showed that circulatory arrest by venous occlusion, which *increased* stretch on the arterial vessels, also produced reactive hyperemia. How myogenic and metabolic factors might interact to produce reactive hyperemia is still obscure.

Mechanical Factors. During contractions, skeletal muscle mechanically reduces its own blood flow. Isotonic contractions, *i.e.*, those in which muscle moves at constant tension, produce only minor variations in blood flow at low tensions; but if this tension is great or if muscles contract isometrically (*i.e.*, a static contraction), then blood supply becomes the resultant of two opposing forces: locally produced vasodilator substances and mechanical compression of blood vessels by surrounding tissue. Blood flow in isometrically contracted human forearm increases until about 30 per cent of the subject's capacity for maximum voluntary contraction is reached. Above this level of contraction, increased intramuscular pressure reduces blood flow. Above 70 per cent of maximal voluntary contraction, blood flow is practically nil despite a very marked rise in the systemic blood pressure.[53] This rise in blood pressure is due to a reflex from isometrically contracted muscle. [38, 54]

Temperature. As stated in Chapter 12, Volume II, neither indirect nor even direct whole body heating produces vasodilation of muscle. If anything, a small reflex vasoconstriction occurs.[6, 28] Under the conditions which have been studied, muscle temperatures were not very high (below 39° C) because of quantities of heat carried away from the directly heated regions by skin blood flow. In those cases in which the skin circulation is locally arrested by epinephrine iontophoresis, muscle warms very gradually because of the low thermal conductivity of the surrounding tissues. Little is known about the direct effect of heat on muscle resistance vessels, but some evidence suggests that intense local heating by penetrating radiation (*e.g.*, short wave diathermy) may directly dilate muscle vessels.[4]

Central cooling, without shivering, appears to have no effect on muscle vascular resistance. Intense local cooling of a limb in cold water markedly reduces total forearm blood flow, but the role played by muscle vs. skin is undefined. In water at 0° C vasodilation occurs in a forearm where skin circulation is arrested by epinephrine iontophoresis, indicating a vasodilation either of muscle[5] or of skin owing to cold-induced insensitivity to local epinephrine.[49]

Humoral Agents. In addition to vasoactive metabolites discussed above, catecholamines may under certain conditions (*e.g.*, hemorrhage) cause changes in muscle vascular resistance. However, since ratios of epinephrine to norepinephrine released from the adrenal medulla vary considerably in different species,[59] comparisons between man and other species are difficult. Quantitative studies have been confined to animals other than man.[25]

Norepinephrine. Norepinephrine stimulates alpha-receptors in skeletal muscle. The direct local effect of a close arterial injection of the hormone is vasoconstriction. This should not be confused with the biphasic effects of an intravenous injection. The systemic effects of the latter are first to produce a generalized vasoconstriction in many regions, including muscle, which raises systemic arterial blood pressure. This in turn activates arterial baroreceptors which reflexly inhibit sympathetic vasoconstrictor tone, causing a second phase, reflexly induced passive vasodilation.[83]

Norepinephrine also appears to have a small effect on beta-receptors (see Chap. 8, Vol. II) in muscle. Since pharmacological blockade of these "receptors" increases the vasoconstrictor action of the hormone, direct stimulation of beta-receptors may oppose somewhat its potent vasoconstrictor action.

Normally, however, vascular resistance of quiescent muscle is dominated by variations in adrenergic sympathetic vasomotor outflow (see above). As with skin, muscle resistance vessels are at least four or five times more responsive to sympathetic vasoconstrictor nerve stimulation than to comparable effects of norepinephrine either infused in graded, physiological amounts or released from the adrenal medulla during graded stimulation of its secretory nerves at low frequencies.[25] However, an important physiological role for the adrenomedullary hormone system is not precluded. Under prolonged stresses such as severe hemorrhage, release of high catecholamine concentrations may exert an important sustaining effect on elevated vascular resistance in muscle, skin and possibly the splanchnic region.[26]

Epinephrine. Despite considerable effort to understand the local action of epinephrine upon muscle, no completely clear picture of its physiological or pharmacological actions has emerged.[83] Injected intravenously, epinephrine causes in man a transient 5- to 6-fold increase in muscle blood flow which within 2 to 3 min returns to two times the resting level despite continued infusion. Since this pattern of response is not affected by sympathectomy, it is not neurogenic.[5, 83] Intra-arterial infusion of epinephrine also produces an initial vasodilation but, in contrast to intravenous infusion, blood flow rapidly returns back to resting levels or even lower. Vascular effects of epinephrine seem to represent a balance between its constrictor and dilator actions. When the local constrictor action of epinephrine is blocked, a marked and sustained vasodilation is obtained.

The mechanism of these responses has been explained around the concept of alpha- and beta-receptors (see Chap. 8, Vol. II). The transient increase in blood flow is attributed to stimulation of beta-receptors which are postulated to have a low threshold for epinephrine. At higher epinephrine concentrations, the "high-threshold-alpha-receptors" are stimulated, causing an opposing vasoconstrictor effect. The result is a balance between the two effects. The difference between intravenous and intra-arterial infusions is thought to result from the concentration-time course of epinephrine in blood and its differential stimulation of alpha- and beta-receptors as it reaches muscle at different degrees of dilution.

The vasodilator action of epinephrine has also been attributed to its effect on muscle metabolism. However, vasodilator actions of lactate and other metabolites released during epinephrine infusion are far too weak to account for this effect. Also, the ability of epinephrine to vasodilate muscle which contains no phosphorylase and, thus, cannot metabolize glycogen indicates that the dilator effect is not attributable to accelerated carbohydrate metabolism.[83]

NEURAL AND LOCAL CONTROL OF SKELETAL MUSCLE VEINS

It was once held that the peripheral venous bed serves the entire cardiovascular system as a unit which is under reflex neurogenic control.[11, 20, 44] This control allegedly serves the needs of the central circulation by active redistribution of blood volume from peripheral to central regions. Such a function is described for skin and splanchnic circulations under special conditions in Chapters 12 and 14, Volume II.

Despite its small volume per unit of mass, the muscle venous bed has been postulated to serve a role in reflex control of blood volume distribution. This concept rested upon the changes in total blood volume of the cat or dog limb during maximum stimulation of its vasoconstrictor nerves.[58, 59] Results have been confused by the heterogeneity of reacting vasculature, e.g., skin, muscle, bone vessels and the different segments of the vascular tree as well. Another important source of confusion is the role played by constriction of resistance vessels which reduces venous distending pressure, causing passive emptying by elastic recoil of venous walls. The importance of any active or neurogenic vs. passive changes depends upon the geometry of the vein, which in turn depends upon venous distending pressure and compliance.[61] This is explained diagrammatically in Figure 14–9, Chapter 14, Volume II. The important point here is that when distending pressures are reduced to the point that veins will change their geometry with any further small decrement in distending pressure, large passive decreases in venous volume will occur. Volume is released as a result of collapse of rounded veins to elliptical or flattened shapes. Over this range of low transmural pressures, active venoconstriction, if present, could play little role and in the more flattened configurations could even increase volume. The concept of active vs. passive changes in venous volume and their significance is presented in more detail in Chapter 14, Volume II.

Maximum sympathetic nerve stimulation of the dog or cat hindlimb expels approximately 30 to 35 per cent of the total regional vascular volume.[58] The maximum decrease in the muscle blood volume (normally 2.5 ml per 100 g) is 10 to 20 per cent or 0.25 to 0.5 ml of blood per 100 g of muscle in the dog.[51] Scaling these data to man with approximately 35 kg of muscle, it appears that at most only 100 to 200 ml of blood could be mobilized from muscle by vaso- and venoconstriction.

Most evidence now indicates that changes in muscle blood volume must be passive (active changes in the limb are apparently confined to skin). In 1921 Donegan[30a] first showed that deep (muscle) veins

are nonreactive to neural stimulation as well as to norepinephrine. Muscle veins, in contrast to the richly innervated cutaneous and splanchnic veins, are virtually devoid of any adrenergic nerve endings, as shown by absence of norepinephrine in venous walls[39] (norepinephrine content of the venous wall is closely correlated with reactivity of veins to adrenergic stimulation[57]). Direct electrical stimulation of the lumbar sympathetic chain in dogs over a range of 0 to 10 imp/sec produces no constriction in veins draining muscle.[82] Thus, these veins do not redistribute blood volume by any direct reflex mechanism.[11, 44] As with other regions, muscle veins receive no sympathetic cholinergic innervation.[82]

Reflexes. In support of the aforementioned findings, active changes in venous wall tension are not present in muscle during carotid sinus hyper- or hypotension, even over extreme ranges.[21, 24] A variety of stimuli known to produce reflex venoconstriction in human skin (e.g., leg exercise, deep breaths, cold pressure tests) are without effect on veins draining human forearm muscle.[85]

Hormones. In contrast to skin, veins in muscle show extremely small reaction to epinephrine, norepinephrine and a variety of constrictor drugs. So far, vasopressin is the only substance found to cause venoconstriction in muscle.[2] Muscle veins are also unresponsive to local vasodilator metabolites produced during exercise.[59]

Temperature. Although muscle veins lack reactivity to adrenergic stimuli, their smooth muscle responds directly to local temperature changes. Local cooling of isolated femoral veins in the dog causes them to dilate by direct local action of cooling.[78] This creates a thermal short circuit in which heat is conserved by redistribution of venous blood away from reflexly constricted cutaneous veins into directly relaxed deep veins of muscle.[3] The opposite response occurs with heating, with heat directly constricting femoral veins and reflexly releasing constrictor tone in cutaneous veins.[78]

REFERENCES

1. Abboud, F. M., and Eckstein, J. W. Comparative changes in segmental vascular resistance in response to nerve stimulation and to

norepinephrine. *Circulat. Res.*, 1966, *18*, 263–277.

2. Abdel-Sayed, W. A., Abboud, F. M., and Ballard, D. R. Contribution of venous resistance to total vascular resistance in skeletal muscle. *Amer. J. Physiol.*, 1970, *218*, 1291–1295.

3. Abdel-Sayed, W. A., Abboud, F. M., and Calvelo, M. G. Effect of local cooling on responsiveness of muscular and cutaneous arteries and veins. *Amer. J. Physiol.*, 1970, *219*, 1772–1778.

4. Abramson, D. I. *Circulation in the extremities.* New York, Academic Press, 1967.

5. Barcroft, H. Circulation in skeletal muscle. *Handb. Physiol.*, 1963, sec. 2, vol. II, 1353–1385.

6. Barcroft, H., Bock, K. D., Hensel, H., and Kitchin, A. H. Die Muskeldurchblutung des Menschen bei indirekter Erwärmung und Abkühlung. *Pflügers Arch. ges. Physiol.*, 1955, *261*, 199–210.

7. Barcroft, H., Bonnar, W. McK., Edholm, O. G., and Effron, A. S. On sympathetic vasoconstrictor tone in human skeletal muscle. *J. Physiol. (Lond.)*, 1943, *102*, 21–31.

8. Barcroft, H., Brod, J., Hezl, Z., Hirsjaroi, E. A., and Kitchin, A. H. The mechanism of the vasodilatation in the forearm during stress (mental arithmetic). *Clin. Sci.*, 1960, *19*, 577–586.

9. Barcroft, H., and Edholm, O. G. On the vasodilatation in human skeletal muscle during post-haemorrhagic fainting. *J. Physiol. (Lond.)*, 1945, *104*, 161–175.

10. Barlow, T. E., Haigh, A. L., and Walder, D. N. Evidence for two vascular pathways in skeletal muscle. *Clin. Sci.*, 1961, *20*, 367–385.

11. Bartelstone, H. J. Roles of the veins in venous return. *Circulat. Res.*, 1960, *8*, 1059–1076.

12. Bayliss, W. M. On the local reactions of arterial wall to changes of internal pressure. *J. Physiol. (Lond.)*, 1902, *28*, 220–231.

13. Beck, L., Pollard, A. A., Kayaalp, S. O., and Weiner, L. M. Sustained dilatation elicited by sympathetic nerve stimulation. *Fed. Proc.*, 1966, *25* (I), 1596–1606.

13a. Beiser, G. D., Zelis, R., Epstein, S. E., Mason, D. J., and Braunwald, E. The role of skin and muscle resistance vessels in reflexes mediated by the baroreceptor system. *J. clin. Invest.*, 1970, *49*, 225–231.

14. Bevegård, B. S., and Shepherd, J. T. Circulatory effects of stimulating the carotid arterial stretch receptors in man at rest and during exercise. *J. clin. Invest.*, 1966, *45*, 132–142.

15. Bevegård, B. S., and Shepherd, J. T. Regulation of the circulation during exercise in man. *Physiol. Rev.*, 1967, *47*, 178–213.

16. Blair, D. A., Glover, W. E., Greenfield, A. D. M., and Roddie, I. C. Excitation of cholinergic vasodilator nerves to human skeletal muscles during emotional stress. *J. Physiol. (Lond.)*, 1959, *148*, 633–647.

17. Blair, D. A., Glover, W. E., Kidd, B. S. L., and Roddie, I. C. Peripheral vascular effects of bretylium tosylate in man. *Brit. J. Pharmacol.*, 1960, *15*, 466–475.

18. Bolme, P., and Edwall, L. The disappearance of Xe^{133} and I^{125} from skeletal muscle of the anesthetized dog during sympathetic cholinergic vasodilatation. *Acta physiol. scand.*, 1970, *78*, 28–38.

19. Bolme, P., and Fuxe, K. Adrenergic and cholinergic nerve terminals in skeletal muscle vessels. *Acta physiol. scand.*, 1970, *78*, 52–59.

20. Braunwald, E., Ross, J., Jr., Kahler, R. L., Gaffney, T. E., Goldblatt, A., and Mason, D. T. Reflex control of the systemic venous bed. Effect on venous tone of vasoactive drugs, and of baroreceptor and chemoreceptor stimulation. *Circulat. Res.*, 1963, *12*, 539–552.

21. Brender, D., and Webb-Peploe, M. M. Influence of carotid baroreceptors on different components of the vascular system. *J. Physiol. (Lond.)*, 1969, *205*, 257–274.

22. Brody, M. J. Neurohumoral mediation of active reflex vasodilation. *Fed. Proc.*, 1966, *25* (I), 1583–1592.

23. Bronk, D. W., Pitts, R. F., and Larrabee, M. G. Role of hypothalamus in cardiovascular regulation. *Ass. Res. nerv. Dis. Proc.*, 1940, *20*, 323–341.

24. Browse, N. L., Donald, D. E., and Shepherd, J. T. Role of the veins in the carotid sinus reflex. *Amer. J. Physiol.*, 1966, *210*, 1424–1434.

25. Celander, O. The range of control exercised by the "sympathico-adrenal system." *Acta physiol. scand.*, 1954, *32* (suppl. 116), 1–132.

26. Chien, S. Role of the sympathetic nervous system in hemorrhage. *Physiol. Rev.*, 1967, *47*, 214–288.

27. Costin, J. C., and Skinner, N. S., Jr. Competition between vasoconstrictor and vasodilator mechanisms in skeletal muscle. *Amer. J. Physiol.*, 1971, *220*, 462–466.

28. Detry, J.-M. R., Brengelmann, G. L., Rowell, L. B., and Wyss, C. Skin and muscle components of forearm blood flow in directly heated resting man. *J. appl. Physiol.*, 1972, *32*, 506–511.

29. Dobson, J. G., Jr., Rubio, R., and Berne, R. M. Role of adenine nucleotides, adenosine, and inorganic phosphate in the regulation of skeletal muscle blood flow. *Circulat. Res.*, 1971, *29*, 375–384.

29a. Donald, D. E., and Edis, A. J. Comparison of aortic and carotid baroreflexes in the dog. *J. Physiol. (Lond.)*, 1971, *215*, 521–538.

30. Donald, D. E., Rowlands, D. J., and Ferguson, D. A. Similarity of blood flow in the normal and the sympathectomized dog hind limb during graded exercise. *Circulat. Res.*, 1970, *26*, 185–199.

30a. Donegan, J. F. The physiology of the veins. *J. Physiol. (Lond.)*, 1921, *55*, 226–245.

31. Ekblom, B., and Hermansen, L. Cardiac output in athletes. *J. appl. Physiol.*, 1968, *25*, 619–625.

32. Epstein, S. E., Beiser, G. D., Goldstein, R. E., Stampfer, M., Wechsler, A. S., Glick, G., and Braunwald, E. Circulatory effects of electrical stimulation of the carotid sinus nerves in man. *Circulation*, 1969, *40*, 269–276.

33. Fencl, V., Hejl, Z., Jirka, J., Madlafousek, J., and Brod, J., Changes of blood flow in forearm muscle and skin during an acute emotional stress (mental arithmetic). *Clin. Sci.*, 1959, *18*, 491–498.

34. Folkow, B., and Halicka, H. D. A comparison between "red" and "white" muscle with respect

to blood supply, capillary surface area and oxygen uptake during rest and exercise. *Microvascular Res.*, 1968, *1*, 1–14.

35. Folkow, B., and Mellander, S. Veins and venous tone. *Amer. Heart J*, 1964, *68*, 397–408.

36. Folkow, B., Öberg, B., and Rubinstein, E. H. A proposed differentiated neuro-effector organization in muscle resistance vessels. *Angiologica*, 1964, *1*, 197–208.

37. Forrester, T., and Lind, A. R. Identification of adenosine triphosphate in human plasma and the concentration in the venous effluent of forearm muscles before, during and after sustained contractions. *J. Physiol. (Lond.)*, 1969, *204*, 347–364.

38. Freychuss, U. Cardiovascular adjustment to somatomotor activation. *Acta physiol. scand.*, 1970, *79* (suppl. 342), 1–63.

39. Fuxe, K., and Sedvall, G. The distribution of adrenergic nerve fibers to the blood vessels in skeletal muscle. *Acta physiol. scand.*, 1965, *64*, 75–86.

40. Gaskell, W. H. On the changes of the bloodstream in muscles through stimulation of their nerves. *J. Anat. (Lond.)*, 1877, *11*, 360–402.

41. Greenfield, A. D. M. Survey of the evidence for active neurogenic vasodilatation in man. *Fed. Proc.*, 1966, *25* (I), 1607–1610.

42. Greenfield, A. D. M., Whitney, R. J., and Mowbray, J. F. Methods for the investigation of peripheral blood flow. *Brit. med. Bull.*, 1963, *19*, 101–109.

43. Haddy, F. J., and Scott, J. B. Metabolically linked vasoactive chemicals in local regulation of blood flow. *Physiol. Rev.*, 1968, *48*, 688–707.

44. Heymans, C., and Neil, E. *Reflexogenic areas of the cardiovascular system.* London, Churchill, 1958.

45. Holling, H. E. Effect of embarrassment on blood flow to skeletal muscle. *Trans. Amer. clin. climat. Ass.*, 1964, *76*, 49–59.

46. Hyman, C., Rosell, S., Rosén, A., Sonnenschein, R. R., and Uvnäs, B. Effects of alterations of total muscular blood flow on local tissue clearance of radio-iodide in the cat. *Acta physiol. scand.*, 1959, *46*, 358–374.

47. Johnson, P. C., ed. Autoregulation of blood flow. *Circulat. Res.*, 1964, *15* (suppl. 1), 1–291.

48. Jones, R. D., and Berne, R. M. Evidence for a metabolic mechanism in autoregulation of blood flow in skeletal muscle. *Circulat. Res.*, 1965, *17*, 540–554.

49. Keatinge, W. R. Direct effects of temperature on blood vessels: Their role in cold vasodilatation. In: *Physiological and behavioral temperature regulation,* J. D. Hardy, A. P. Gagge and J. A. J. Stolwijk, eds. Springfield, Ill., Charles C Thomas, 1970.

50. Korner, P. I., Chalmers, J. P., and White, S. W. Some mechanisms of reflex control of the circulation by the sympatho-adrenal system. *Circulat. Res.*, 1967, *21* (suppl. III), 157–172.

51. Lesh, T. A., and Rothe, C. F. Sympathetic and hemodynamic effects of capacitance vessels in dog skeletal muscle. *Amer. J. Physiol.*, 1969, *217*, 819–827.

52. Lewis, T. *The blood vessels of the human skin and their responses.* London, Shaw, 1927.

53. Lind, A. R., Taylor, S. H., Humphreys, P. W., Kennelly, B. M., and Donald, K. W. The circulatory effects of sustained voluntary muscle contraction. *Clin. Sci.*, 1964, *27*, 229–244.

54. Lind, A. R., McNicol, G. W., Bruce, R. A., MacDonald, H. R., and Donald, K. W. The cardiovascular responses to sustained contractions of a patient with unilateral syringomyelia. *Clin. Sci.*, 1968, *35*, 45–53.

55. Löfving, B., and Mellander, S. Some aspects of the basal tone of the blood vessels. *Acta physiol. scand.*, 1956, *37*, 134–141.

56. Lundvall, J., Mellander, S., and White, T. Hyperosmolality and vasodilatation in human skeletal muscle. *Acta physiol. scand.*, 1969, *77*, 224–233.

57. Mayer, H. E., Abboud, F. M., Ballard, D. R., and Eckstein, J. W. Catecholamines in arteries and veins of the foreleg of the dog. *Circulat. Res.*, 1968, *23*, 653–661.

58. Mellander, S. Comparative studies on the adrenergic neuro-hormonal control of resistance and capacitance blood vessels in the cat. *Acta physiol. scand.*, 1960, *50* (suppl. 176), 1–86.

59. Mellander, S., and Johansson, B. Control of resistance, exchange and capacitance functions in the peripheral circulation. *Pharmacol. Rev.*, 1968, *20*(3), 117–196.

60. Mellander, S., and Lundvall, J. Role of tissue hyperosmolality in exercise hyperemia. *Circulat. Res.*, 1971, *28* (suppl. 1), (I–39)–(I–45).

61. Öberg, B. The relationship between active constriction and passive recoil of the veins at various distending pressures. *Acta physiol. scand.*, 1967, *71*, 233–247.

61a. Öberg, B., and White, S. Role of vagal cardiac nerves and arterial baroreceptors in the circulatory adjustments to hemorrhage in the cat. *Acta physiol. scand.*, 1970, *80*, 395–403.

61b. Pelletier, C. L., Edis, A. J., and Shepherd, J. T. Circulatory reflex from vagal afferents in response to hemorrhage in the dog. *Circulat. Res.*, 1971, *29*, 626–634.

62. Remensnyder, J. P., Mitchell, J. H., and Sarnoff, S. J. Functional sympatholysis during muscular activity. Observations on influence of carotid sinus on oxygen uptake. *Circulat. Res.*, 1962, *11*, 370–380.

63. Renkin, E. M., and Rosell, S. Effects of different types of vasodilator mechanisms on vascular tonus and on transcapillary exchange of diffusible material in skeletal muscle. *Acta physiol. scand.*, 1962, *54*, 241–251.

64. Roddie, I. C., and Shepherd, J. T. The effects of carotid artery compression in man with special reference to changes in vascular resistance in the limbs. *J. Physiol. (Lond.)*, 1957, *139*, 377–384.

65. Roddie, I. C., and Shepherd, J. T. Nervous control of the circulation in skeletal muscle. *Brit. med. Bull.*, 1963, *19*, 115–119.

66. Roddie, I. C., Shepherd, J. T., and Whelan, R. F. Reflex changes in human skeletal muscle blood flow associated with intrathoracic pressure changes. *Circulat. Res.*, 1958, *6*, 232–238.

67. Rosell, S., and Uvnäs, B. Vasomotor nerve activ-

ity and oxygen uptake in skeletal muscle of the anesthetized cat. *Acta physiol. scand.*, 1962, *54*, 209–222.

68. Rosell, S., Bolme, P., and Ngai, S. H. Interaction between cholinergic vasodilator and adrenergic vasoconstrictor nerves on canine skeletal muscle blood vessels. In: *Circulation in skeletal muscle*, O. Hudlicka, ed. Oxford, Pergamon Press, 1968.

69. Rowell, L. B., Blackmon, J. R., and Bruce, R. A. Indocyanine green clearance and estimated hepatic blood flow during mild to maximal exercise in upright man. *J. clin. Invest.*, 1964, *43*, 1677–1690.

70. Rushmer, R. F., Franklin, D. L., Van Citters, R. L., and Smith, O. A. Changes in peripheral blood flow distribution in healthy dogs. *Circulat. Res.*, 1961, 9, 675–687.

71. Shepherd, J. T. *Physiology of the circulation in human limbs in health and disease.* Philadelphia, W. B. Saunders Co., 1963.

72. Skinner, N. S., Jr., and Costin, J. C. Role of O_2 and K^+ in abolition of sympathetic vasoconstriction in dog skeletal muscle. *Amer. J. Physiol.*, 1969, *217*, 438–444.

73. Skinner, N. S., Jr., and Costin, J. C. Interactions of vasoactive substances in exercise hyperemia: O_2, K^+ and osmolality. *Amer. J. Physiol.*, 1970, *219*, 1386–1392.

74. Stainsby, W. N., and Barclay, J. K. Effect of infusions of osmotically active substances on muscle blood flow and systemic blood pressure. *Circulat. Res.*, 1971, 28 (suppl. 1), (I-33)–(I-38).

75. Strandell, T., and Shepherd, J. T. The effect in humans of increased sympathetic activity on the blood flow to active muscles. *Acta med. scand.*, 1967, (suppl. 472), 146–167.

76. Uvnäs, B. Cholinergic vasodilator nerves. *Fed. Proc.*, 1966, *25* (I), 1618–1622.

77. Van Citters, R. L., and Franklin, D. L. Cardiovascular performance of Alaska sled dog during exercise. *Circulat. Res.*, 1969, *24*, 33–42.

78. Vanhoutte, P. M., and Lorenz, R. R. Effect of temperature on reactivity of saphenous, mesenteric, and femoral veins of the dog. *Amer. J. Physiol.*, 1970, *218*, 1746–1750.

79. Vatner, S. F., Franklin, D., Van Citters, R. L., and Braunwald, E. Effects of carotid sinus nerve stimulation on blood-flow distribution in conscious dogs at rest and during exercise. *Circulat. Res.*, 1970, 27, 495–503.

80. Wade, O. L., and Bishop, J. M. *Cardiac output and regional blood flow.* Oxford, Blackwell, 1962.

81. Walder, D. N. Vascular pathways in skeletal muscle. In: *Circulation in skeletal muscle*, O. Hudlicka, ed. Oxford, Pergamon Press, 1968.

82. Webb-Peploe, M. M., and Shepherd, J. T. Response of large hindlimb veins of the dog to sympathetic nerve stimulation. *Amer. J. Physiol.*, 1968, *215*, 299–307.

83. Whelan, R. F. *Control of the peripheral circulation in man.* Springfield, Ill., Charles C Thomas, 1967.

84. Wood, J. E. *The veins, normal and abnormal function.* Boston, Little, Brown & Co., 1965.

85. Zelis, R., and Mason, D. T. Comparison of the reflex reactivity of skin and muscle veins in the human forearm. *J. clin. Invest.*, 1969, *48*, 1870–1877.

86. Zoller, R. P., Mark, A. L., Abboud, F. M., Schmid, P. G., and Heistad, D. D. The role of low pressure baroreceptors in reflex vasoconstrictor responses in man. *J. clin. Invest.*, 1972, *51*, 2967–2972.

CHAPTER 14   THE SPLANCHNIC CIRCULATION

by LORING B. ROWELL

INTRODUCTION

The splanchnic circulation supplies an extremely complex series-parallel arrangement of organs which includes the gastrointestinal tract (stomach, small and large intestines), spleen, pancreas and liver. Adequate treatment of control of blood flow to individual splanchnic organs would far exceed the scope of this chapter. Individual organs are discussed only when necessary to illustrate some unique features. For the most part, these organs will be lumped together and treated as a single hypothetical organ. The anatomical features which permit this treatment are described below. A major objective of this chapter is to show how reflexly and/or locally induced changes in total blood flow, vascular resistance and venous capacity of the entire lumped system aid in overall regulation of blood pressure, distribution of blood volume, body temperature and regional oxygen de-livery. This simplified approach is intended to reveal the central importance of the total splanchnic organ system in overall cardiovascular regulation. Detailed treatments of blood flow and its control in the different organs of this region are available in recent books[3, 15, 19] and reviews.[6, 17]

ANATOMY

Splanchnic organs receive their blood supply from a number of arteries with interconnected branches. Superior and inferior mesenteric arteries and the celiac artery, which divides into three branches, supply the stomach, pancreas, duodenum, gallbladder, large and small intestines and the pancreas. The hepatic artery supplies the liver; some of its branches supply other splanchnic organs. A great deal of inter- and intraspecies variation exists in the arrangement of arterial supply and in venous drain-

215

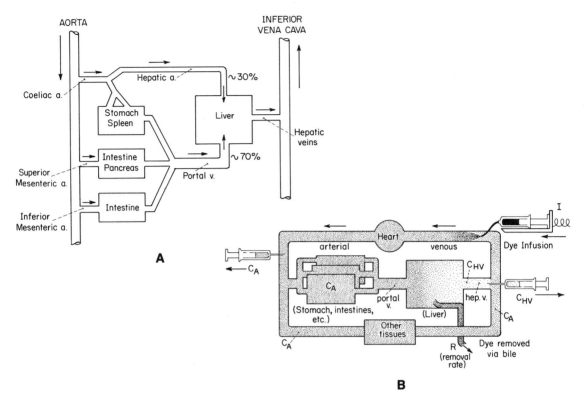

Figure 14–1 *A,* Simplified schematic representation of the splanchnic vascular bed. The major points are (i) the parallel arrangement of various arteries supplying stomach, portions of the intestine, etc; (ii) the series arrangement of all these segments with the liver to which they are connected by the portal vein; and (iii) the dual blood supply to the liver via the portal vein which carries approximately 70 per cent of total splanchnic blood flow and the hepatic artery which supplies the remaining 30 per cent.

B shows the uniform distribution of dye at arterial concentration (C_A) over the entire vasculature, except in the hepatic vein (HV) and liver where dye is removed (R) from the circulation via bile secretion. Sampling (C_A and C_{HV}) and infusion (I) sites are shown. See text.

age. The simplified schematic diagram in Figure 14–1 reveals that the gastrointestinal and splenic vascular beds are in a parallel configuration, with all drainage going into the portal vein which supplies the liver with about 70 to 75 per cent of its blood flow. The liver receives the remaining 25 to 30 per cent of its blood supply from the hepatic artery. Blood flow from all splanchnic organs drains from the liver into the hepatic veins, and from them into the inferior vena cava. The common venous drainage of this system allows all the splanchnic organs to be grouped together as a single hypothetical organ. It should be clear from Figure 14–1 that any given change in blood flow to one splanchnic organ will produce the same magnitude of change in total splanchnic blood flow, if not compensated by an equivalent and opposite change in another parallel organ. Because the methods commonly applied to measure total splanchnic blood flow in man and intact animals usually reflect only changes in total blood flow and cannot discriminate where changes have occurred, it is frequently necessary to treat the splanchnic system as a single organ.

MEASUREMENT OF SPLANCHNIC BLOOD FLOW

Many methods have been applied to study total and regional splanchnic blood flow.[3, 15, 19] There are two basic approaches. One, useful in animals, is to place one of a variety of cannulating or noncannulating flow transducers (see Chap. 3, Vol. II) in or around one of the major arteries supplying some portions of the splanchnic region—usually a mesenteric artery.[47, 48] Changes in total splanchnic blood flow are estimated from flow changes in a single vessel. The assumptions are that other arteries supplying different portions of the splanchnic region respond identically and that the flow transducer has not damaged underlying nerves supplying downstream resistance vessels. Total splanchnic blood flow can be measured only if these transducers are placed on the hepatic artery and portal vein. Obviously the technique is not applicable to humans.

The second approach, applicable to both man and animals, is an application of the Fick principle.[4] Nontoxic dyes (e.g., indocyanine green or sulfobromophthalein sodium) are infused into the blood stream at a constant and precisely known rate (see Fig. 14–1). These dyes are removed from the blood by the liver. When arterial dye concentration (C_A) becomes constant, dye infusion rate (I) equals hepatic dye removal rate (R). Thus, knowing hepatic removal rate, arterial and hepatic venous dye concentration (C_{HV}), hepatic or splanchnic blood flow (SBF) can be calculated as indicated in the following equation.

$$SBF = \frac{I}{C_A - C_{HV}} = \frac{R}{C_A - C_{HV}}$$

The quantity $C_A - C_{HV}$ represents the extraction of dye by the liver. This is called the constant infusion technique, and it is usually applied under conditions of constant splanchnic blood flow. Variations of this technique can be applied to conditions of rapidly changing splanchnic blood flow by making appropriate calculations for the rates of change of C_A, C_{HV} and the difference between I and R in the aforementioned equation. Details of the latter method and also one which employs a single bolus injection of dye are in the literature.[3, 4, 39]

With the dye technique three intravascular catheters must be inserted; one catheter is needed for dye infusion, one for measuring C_A and a third for measuring C_{HV}. To measure C_{HV} the catheter must be positioned into a large hepatic vein under fluoroscopic guidance.

Measurements employing dyes yield what is commonly called "estimated" hepatic (or splanchnic) blood flow because C_{HV} measured from one hepatic vein is actually an estimate of the true average C_{HV} for all hepatic veins.* That is, since several veins drain the liver, using only one could produce an atypical C_{HV} value and thus an erroneous flow measurement. However, the evidence indicates that this is not an important source of error in the measurement of splanchnic blood flow. For example, simultaneous sampling of different dyes from different hepatic veins, simultaneous comparison of flow measured with dye and with flowmeters[11] and finally comparison of flows derived from indicators which are almost completely extracted by the liver in one pass—all yield very similar values for splanchnic blood flow. For more detailed review of the methodological concepts and problems of these techniques, see references.[3, 23, 39]

GENERAL DESCRIPTION OF SPLANCHNIC BLOOD FLOW

Range of Splanchnic Blood Flow. The aforementioned methods all show average splanchnic blood flow in average-sized man to be close to 1500 ml per min. This constitutes approximately 25 per cent of the normal resting cardiac output and is a high perfusion rate for the liver alone which weighs approximately 1.5 kg (100 ml per 100 gm) or about 2 per cent of total body weight. Liver blood flow has approximately the same relationship to liver mass in dogs and cats.[15] Since splanchnic organs extract only 15 to 20 per cent (3.0 to 4.0 ml per 100 ml) of the oxygen delivered by each 100 ml of blood, large decrements in splanchnic blood flow can occur without compromising local oxygen consumption. Accordingly, the splanchnic system is a major source from which blood flow can be redistributed to other organs. Also large changes in peripheral vascular resistance elsewhere can be partially compensated by splanchnic vasomotor changes. The splanchnic region, in man at least, plays a major role in correcting disparities between cardiac output on the one hand and regional blood supply or

---

*Total splanchnic blood flow is greater than hepatic blood flow when significant shunts exist between the portal vein and inferior vena cava. In such cases the dye techniques validly estimate only hepatic blood flow, but such shunts are quantitatively insignificant in normal man.

peripheral vascular resistance on the other. This will be discussed more fully below (see section on exercise).

Splanchnic Blood Volume. The best estimates of total splanchnic blood volume in man and dogs, derived from the local dilution of I^{131} labeled albumin, indicate that it comprises approximately 20 to 25 per cent of total blood volume.[5, 15, 34, 52] This high proportion of total blood volume plus the rich innervation of splanchnic veins with sympathetic nerves[31, 37a] lends support to an old postulate that the splanchnic vascular bed is the "*venosector* and blood giver of the circulation."[26] In short, the concept is that the splanchnic venous bed is under reflex control which can actively constrict these veins and translocate blood volume. In hemorrhage, for example, this function is viewed as providing an "autotransfusion" of blood into the central circulation where it is needed to maintain cardiac output and arterial blood pressure. This will be discussed more fully in the section on control of splanchnic veins.

NEURAL CONTROL OF SPLANCHNIC RESISTANCE VESSELS

Sympathetic Vasoconstrictor Nerves. The splanchnic region is richly innervated by sympathetic vasoconstrictor nerves, originating primarily from the splanchnic nerve. The tonic sympathetic vasoconstrictor discharge to splanchnic organs is apparent from the vasodilation often noted after sectioning the splanchnic nerve.[15, 28] However, blockade of sympathetic nerves in some preparations is without effect upon splanchnic blood flow.[15] Basal tone (see Chap. 8, Vol. II) of intestinal vascular smooth muscle seems to be very high—at least in the cat. For example, maximal drug-induced relaxation of intestinal resistance vessels produces flows which, if extrapolated to human intestine, would reach approximately 5.5 liters per min (from 0.7 liter per min).[17, 31]

Direct electrical stimulation of splanchnic nerves in anesthetized dogs or cats causes vasoconstriction, particularly in the intestine. If such stimulation is continued for several minutes, blood flow gradually returns toward prestimulus values in what is termed "autoregulatory escape."[16, 17, 31]

Interpretation is complicated, however, because vessels in the outer mucosal region of the intestine may remain constricted throughout nerve stimulation, whereas during the so-called "autoregulatory escape" phase, submucosa vessels may dilate so that total intestinal blood flow is restored to normal. Similar "escape" has been observed for hepatic arterial blood flow, indicating either some intrahepatic redistribution of blood flow as in the intestine, or perhaps a local action of vasodilator metabolites upon the blood vessels which override the neural vasoconstrictor drive.[17] The mechanism for this phenomenon remains obscure.

Some find that the splanchnic region in dogs will remain vasoconstricted throughout 10 min of continuous, direct stimulation of splanchnic nerves.[7] The decrease in blood flow is rapid (within 12 sec) and its magnitude is directly proportional to the stimulation frequency over the range of 1.5 to 15 imp/sec, as shown in Figure 14–2. In these studies inflow and outflow of blood from the *total* splanchnic region were continuously monitored; thus, *net* effects of sympathetic nerve stimulation on blood flow to the whole organ system were demonstrated.[7] At stimulation frequencies of 10 to 15 imp/sec blood flow through the total splanchnic region is reduced 80 per cent. This corresponds to the largest reduction in splanchnic blood flow seen in man during

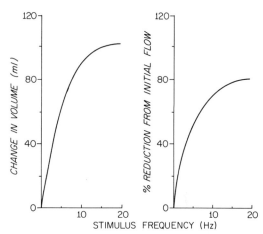

Figure 14–2 Change in splanchnic blood volume and flow in dogs during 18-sec electrical stimulation of left thoracic splanchnic nerve at different frequencies. (From Brooksby and Donald, *Circulat. Res.,* 1971, 29,227–238. By permission of The American Heart Association, Inc.)

maximal exercise. In man, vasoconstriction of the splanchnic bed can be sustained for long periods during severe stress without any evidence of net "escape."[41–45]

Sympathetic Vasodilator Nerves. There appears to be no cholinergic, sympathetic vasodilator innervation of the splanchnic circulation. There is no vasodilator effect from either efferent nerve stimulation[15, 31] or central nervous stimulation. The latter, which sometimes produces active cholinergic vasodilation in skeletal muscle (see Chap. 13, Vol. II), produces vasoconstriction in the intestinal circulation.[15]

REFLEX CONTROL OF SPLANCHNIC VASCULAR RESISTANCE

Baroreceptors. The splanchnic region is a major target in reflex regulation of blood pressure by the carotid sinus and aortic baroreceptors.[9, 15, 28] Reflex splanchnic vasoconstriction is important in restoring blood pressure under conditions which produce hypotension, such as hypovolemia, massive vasodilation of other vascular beds, or orthostasis. Experimental evidence both from man and animals tends to support this concept, but exceptions exist.[3, 9, 15, 17] Alternatively, a rise in pressure at the carotid and aortic sinuses causes splanchnic vascular resistance to fall because of a reduction in impulse frequency over sympathetic

nerves.[7, 22] It is not known whether the latter response occurs in man.

In the cat, carotid sinus occlusion causes hepatic arterial vasoconstriction.[16] However, this vasoconstriction is transient if the animal is adrenalectomized, suggesting that release of catecholamines is required for sustained vasoconstrictor responses. The transient vasoconstriction was shown not to be due to exhaustion of the neural transmitters from sympathetic nerve endings.[16] It is doubtful that the adrenal medulla plays a similar synergistic role in maintaining elevated splanchnic vascular resistance in man because epinephrine in physiological concentrations raises splanchnic blood flow (see below).

In the anesthetized dog, reducing carotid sinus pressure below its normal level causes rapid (complete within 12 sec) splanchnic vasoconstriction, whereas raising carotid sinus pressure causes vasodilation by reflex inhibition of vasoconstrictor tone (Fig. 14–3).[7] Most of the net change in total splanchnic blood flow and resistance occurs within ±30 mm Hg of the normal sinus pressure. For example, splanchnic blood flow at constant perfusion pressure is raised or reduced approximately 30 to 40 per cent above or below its normal level when sinus pressure is raised or lowered by this amount. Reflex vasoconstriction is a stable response, whereas reflex vasodilation is attended by some biphasic or oscil-

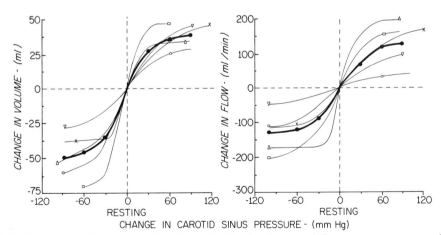

Figure 14–3 Increases and decreases in splanchnic blood volume and flow in response to graded changes in carotid sinus pressure in five dogs. Each line indicates response to increase or decrease, in carotid sinus pressure, from constant systemic arterial pressure. Heavy lines indicate the averages. Change in volume is total change with each alteration in carotid sinus pressure; change in flow is that change observed when responses reach a steady state after adjusting sinus pressure. (From Brooksby and Donald, *Circulat. Res.*, 1971, 29,227–238. By permission of The American Heart Association, Inc.)

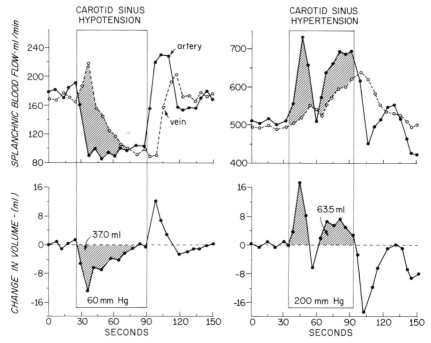

Figure 14–4 Temporal changes in blood flow into splanchnic arteries and out of splanchnic veins accompanying step changes in pressure within vascularly isolated carotid sinus in two dogs. Volume changes were derived from differences between inflow and outflow of blood. *Left,* Intrasinus pressure was decreased from control of 105 mm Hg to 60 mm Hg. *Right,* Intrasinus pressure was increased from control of 145 mm Hg to 200 mm Hg. Total change in splanchnic blood volume is indicated in each case by the shaded area. (From Brooksby and Donald, *Circulat. Res.,* 1971, 29, 227–238. By permission of The American Heart Association, Inc.)

latory patterns of blood flow (Fig. 14–4). The decrements in splanchnic blood flow are far less than the 80 per cent decrements seen during maximally effective nerve stimulation frequencies (10 to 15 imp/sec)— or during maximal exercise in man (see below). This may reflect the depressant effects of anesthesia on reflex sympathetic nerve activity, or it could indicate a limited participation of the splanchnic bed in arterial baroreceptor reflexes in this species. Some[29] find substantial differences in regional vascular reactions to alteration in arterial baroreceptor activity. For example, cat and dog skeletal muscle seems most affected by a change in carotid sinus pressure; the same stimulus produces a less marked response in the intestines and spleen.[7, 29, 33a]

In man, reduction in arterial mean or pulse pressure induced either by assumption of upright posture (head-up tilting) or by simulated hemorrhage (induced by

lower body negative pressure[*]) causes splanchnic vasoconstriction. Interruption of the splanchnic nerves prevents the splanchnic vasoconstrictor response to head-up tilting.[56] Recent evidence from man indicates that the relative magnitude of splanchnic vasoconstriction is comparable to that for skin and muscle beds.[44] During sustained (up to 20 min) lower body negative pressure of −50 mm Hg, splanchnic blood flow falls 32 per cent (resistance rises 30 per cent), whereas forearm skin and muscle blood flow fall 35 to 45 per cent. The cardiovascular responses are illustrated in Figure 14–5, which shows that the sudden withdrawal of blood from the central circu-

---

[*]In lower body negative pressure the legs are enclosed in a box sealed at the iliac crests and suction is applied to the box. Depending upon the force of suction, a considerable quantity of blood (possibly up to 1 liter) can be drawn into the legs, primarily into veins, and sequestered there.

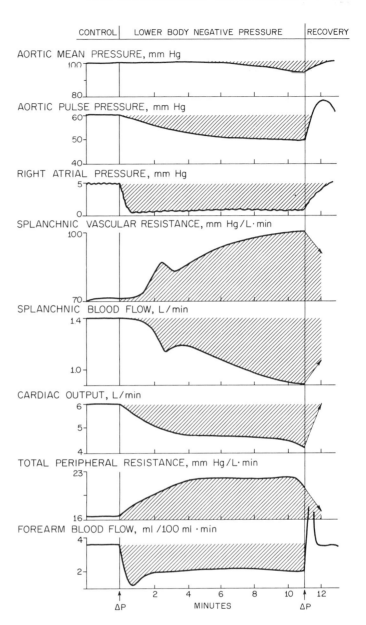

Figure 14-5 Circulatory changes in man during lower body negative pressure (simulated hemorrhage). From top to bottom are the time courses for aortic mean pressure, aortic pulse pressure, right atrial pressure and so on, as labeled. Splanchnic vasoconstriction, as indicated by the increase in splanchnic vascular resistance and fall in splanchnic blood flow, accounted for one-third of the compensatory vascular adjustment, which maintained mean pressure up to 21 min despite a marked fall in cardiac output. Note also the sudden fall in forearm blood flow. (After Rowell *et al.*, *J. appl. Physiol.*, 1972, 32, 213–220.)

lation causes an immediate vasoconstriction in the forearm coincident with the fall in right atrial pressure. Splanchnic blood flow tends to fall later, along with the decline in arterial pulse pressure. The contribution of splanchnic vasoconstriction to the maintenance of arterial mean pressure can be calculated by computing the change in total and regional conductances (conductance is the reciprocal of resistance [see Chap. 2, Vol. II]). Under the conditions obtained in the experiment illustrated in Figure 14–5, the decrease in splanchnic vascular conductance accounted for 33 per cent of the total compensatory, peripheral vascular adjustment. Presumably renal and combined cutaneous and muscle vascular beds make similar contributions.

Low pressure baroreceptors in the cardiopulmonary region may exert their main effects upon splanchnic (and renal) resistance vessels. In anesthetized dogs with carotid sinus pressure held constant and with aortic baroreceptors presumably denervated, lowering cardiopulmonary pressures causes a marked mesenteric arterial constriction, which is abolished by vagotomy. In contrast, hindlimb resistance vessels are much less sensitive to lowering cardiopulmonary pressures, but are much more sensitive to changes in carotid sinus pressure.[33a]

In summary, most recent evidence from intact man and anesthetized animals indicates that sympathetic vasoconstrictor outflow to splanchnic resistance vessels is reflexly increased by a fall in arterial pressure. In man the splanchnic bed, along with skeletal muscle, plays a major and sustained role in maintaining arterial pressure. In some other species skeletal muscle vessels appear more reactive to arterial baroreceptor reflexes than splanchnic vessels, but both regions are major targets for pressure regulation. It is not known yet whether a rise in arterial pressure causes splanchnic vasodilation in man as in dogs. In dogs, low pressure baroreceptors appear to exert effects which equal those of carotid sinus stimulation on splanchnic vasomotor tone. Similar data are not yet available from man.

Chemoreceptors. Distinguishing the reflex effects of chemoreceptor stimulation from local effects of hypoxia on splanchnic blood flow and resistance is difficult. The primary reflex effect of hypoxia on the splanchnic circulation is, as in muscle, sustained vasoconstriction.[27] In hemorrhage, stimulation of chemoreceptors by hypoxia is, along with the baroreceptor input, an important drive to splanchnic vasoconstrictor nerves.

Hemorrhage. Reflex responses to hemorrhage originate from chemoreceptors and arterial and low pressure baroreceptors. Ischemia and hypoxia attending severe hemorrhage excite spinal sympathetic neurons.[9] The relative importance of high vs. low pressure baroreceptors has not been clearly defined. Denervation of arterial baroreceptors reveals their importance in regulating blood pressure during hemorrhage; without them blood pressure in anesthetized dogs falls more markedly.[8, 9, 28] In severe hemorrhage, firing rates of splanchnic nerves can be increased to maximal levels.[9]

Minor blood loss mainly activates low pressure baroreceptors of the cardiopulmonary region. Experiments described above in which anesthetized dogs were bled while carotid sinus pressure was held constant (aortic arch denervated) indicate a major role of these receptors in eliciting splanchnic (and renal) vasoconstriction during hemorrhage.[33a] No equivalent data are available from man, but simulation of minor hemorrhage by lower body negative pressure (LBNP) will cause marked vasoconstriction in skin and muscle.[57]

General vasoactive responses to severe hemorrhage appear to occur in two steps.[9] During the first few minutes after sudden hemorrhage, total splanchnic vascular resistance is increased. This sympathetic activation is transient, but then a second phase of sustained sympathetic activity develops, as documented by increased frequency of action potentials from splanchnic sympathetic nerves. This second phase probably results from reduced afferent input from baroreceptors plus an increase in chemoreceptor impulses caused by hypoxia.[8, 9, 27] Also, the concentration of circulating catecholamines is increased markedly. A reduced cardiac output is therefore drastically redistributed, and circulation to the heart and brain is preserved. Eventually, accumulation of metabolites in tissues where blood flow is too low to supply sufficient oxygen overwhelms the neurogenic adjustments and vasodilation occurs. This frequently leads to an unstable and poorly understood condition called "shock."

Reproducible or comparable responses to hemorrhage, even within a given species, have been difficult to obtain. In dogs, anesthesia appears to shift responses from primarily cardiac output changes to predominantly vasomotor changes.[9] Different techniques of bleeding cause different responses; for example, reduction of blood volume to a constant low pressure may require repeated bleedings which ultimately exhaust vasoconstrictor reserve, whereas removal of an estimated percentage of total blood volume may alter pressure very little. These problems and the different origins of reflex changes complicate our understanding of reflex adjustments to hemorrhage.

Exercise. In man, exercise in either supine[2, 52] or upright[1a, 38, 40, 41] positions reduces splanchnic blood flow and raises splanchnic vascular resistance despite a large increase in cardiac output and a moderate rise in mean arterial pressure. This response is minor in dogs,[47-50] even during severe exercise,[48] wherein mesenteric arterial resistance rises only 20 to 50 per cent; values are ten times greater in man. However, interventions which reduce the cardiac output response to exercise, such as experimentally induced heart block, result in a very marked splanchnic vasoconstrictor response to exercise in dogs.[50]

Returning to man, during exercise, splanchnic blood flow has an inverse linear relationship to oxygen uptake (see Fig. 14-6A). But this relationship in any individual depends upon the capacity of the circulation to deliver oxygen to working muscle. A measure of this capacity is the maximal oxygen consumption, which by definition equals the product of maximal cardiac output times maximal systemic arteriovenous oxygen difference. Figure 14-6A illustrates the reduction in splanchnic blood flow in response to graded exercise in three groups, patients with "pure" mitral stenosis (no congestive heart failure), sedentary young men and endurance athletes. Each regression line ends at the average maximal oxygen uptake of the designated group (corrected for difference in body weight). Two major points are notable. First, the lower the circulatory capacity for oxygen delivery, the greater

is splanchnic vasoconstriction at any given level of oxygen uptake during exercise. Second, Figure 14-6B shows that splanchnic blood flow in all groups is reduced in the same proportion to the *relative* severity of exercise, or relative oxygen uptake expressed as the percentage of maximal oxygen uptake required. Thus, scaling splanchnic blood flow responses with respect to *relative* demands upon the circulation reveals an underlying similarity between individuals. Thus, splanchnic vascular resistance is regulated during exercise in proportion to the individual's circulatory capacity for oxygen delivery. This is also true for renal blood flow during exercise.[20]

The cause of splanchnic vasoconstriction during exercise is unknown. It is not triggered by a fall in arterial pressure at the onset of exercise, for the pressure usually rises immediately. Recent findings, reviewed elsewhere,[38] suggest that this response may be reflexly mediated by "receptors" in working muscle. The receptors (possibly unmyelinated fibers) increase their firing rate in response to local chemical changes in working muscle. Chemical stimulation of muscle nerves by capsaicin causes reflex constriction of both splanchnic and renal vasculature.[55a] The response of receptors activated by muscle metabolites should be proportional to the adequacy of muscle perfusion at any given meta-

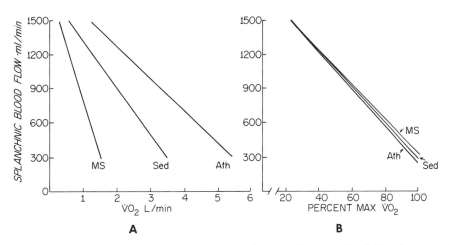

A **B**

Figure 14-6 Schematic representation of progressive changes in splanchnic blood flow during graded upright exercise and increasing oxygen uptake (A)—up to maximal oxygen uptake (max $\dot{V}o_2$)—in three groups: (i) patients with pure mitral stenosis (MS), (ii) normal sedentary young men (Sed), and (iii) endurance athletes (Ath). (From Rowell *et al.*, *J. clin. Invest.*, 1964, *43*, 1677-1690; Blackmon *et al.*, *Circulation*, 1967, *36*, 497-510; and Rowell, *Physiol. Rev.*, in press.) Each line in A ends at the average maximal oxygen uptake of the group. In B, the regression lines become almost superimposed when oxygen uptake is expressed as percentage of maximal for each group.

bolic rate. This places a possible link between neural and humoral events in working muscle and circulatory delivery of oxygen to muscle. The close relationship between splanchnic (and renal) vasomotor responses, heart rate and the relative circulatory demands of exercise might eventually be explained by such a linkage.

The major physiological significance of splanchnic vasoconstriction during exercise is the redistribution of blood flow and oxygen to working muscle. This can be accomplished without reduction of splanchnic oxygen uptake, because arteriovenous oxygen difference across this region is only 3 to 4 ml per 100 ml. The following expressions

$$\dot{V}O_{2_{spl}} = \dot{Q}_{spl} \times A\text{-}VO_{2_{spl}}$$

rest $\quad\quad 60 = 1500 \times 4$ ml/100 ml

maximal exercise $60 = 350 \times 17$ ml/100 ml

where $\dot{V}O_{2_{spl}}$, $\dot{Q}_{spl}$ and $A\text{-}VO_{2_{spl}}$ are splanchnic oxygen uptake, flow and arteriovenous oxygen difference, respectively, illustrate this point. As splanchnic blood flow falls from 1500 to 350 ml per min, the arteriovenous oxygen difference increases to 17 ml per 100 ml; thus, 85 per cent of available oxygen is extracted while the 1150 ml per min difference in splanchnic blood flow during rest and exercise makes 230 ml of oxygen available to working muscle each minute (i.e., 1150 ml per min × arterial oxygen content [20 ml per 100 ml] = 230 ml oxygen per min). Combining this with a similar redistribution of blood flow away from the kidneys, which receive approximately 1250 ml per min and extract only 1.3 ml of oxygen from each 100 ml of blood, then a total of 400 ml or more of additional oxygen is made available from both regions. Nonworking skeletal muscle also vasoconstricts under these conditions, but accurate estimates of the amount of blood thus redistributed are not available. As a conservative estimate, redistribution of cardiac output during severe exercise makes a total of approximately 600 ml of additional oxygen available to working muscle each minute. The splanchnic organs make a major contribution to this reallocation of oxygen.

The quantitative importance of redistribution of blood flow during exercise depends upon maximal oxygen consumption and thus the maximal cardiac output of the individual. Going from the highest to the lowest values of maximal oxygen consumption shown in Figure 14–6, the redistribution of cardiac output can account for approximately 11 per cent, 17 per cent and 38 per cent of maximal oxygen consumption, respectively. That is, the increase in oxygen supply to muscle is the same in the three groups, but the adjustment provides a greater fraction of total oxygen transport capacity in those with the lowest cardiac output. Thus, as the capacity of the cardiovascular system is reduced by disease or inactivity, the quantitative significance of blood flow redistribution increases proportionally. The importance of redistribution of splanchnic blood flow is most dramatically illustrated in cardiac patients who are unable to raise their cardiac outputs in response to exercise so that nearly all the increase in oxygen transport must be met by redistribution of blood flow. In severe cardiac failure, adequate oxygen supply to vital tissues, even at rest, is met in part by sustained vasoconstriction of splanchnic and renal beds. Wade and Bishop have written an excellent review of this subject.[51]

Regional vasoconstriction also contributes, along with the rise in cardiac output, to the maintenance of arterial blood pressure during exercise. Normally, the fall in muscle vascular resistance is compensated for mainly by the rise in cardiac output so that in a normal man failure of regional vasoconstriction would lower blood pressure 15 to 20 mm Hg. In the other two examples cited, reductions in blood pressure would be only 10 to 12 mm Hg in athletes but as great as 40 mm Hg in mitral stenosis patients. The lower the capacity to increase cardiac output, the more blood pressure regulation, like oxygen transport, is served by this regional vasoconstriction.

Thermoregulatory Reflexes. Direct whole body heating of resting man causes a progressive reduction in splanchnic blood flow and a rise in splanchnic vascular resistance which parallels the rise in central body temperature (the kidneys behave similarly and skeletal muscle flow may also decrease).[10, 42] The overall cardiovascular response to whole body heating at rest is shown in Figure 21–1, Chapter 12, Volume II. The rise in splanchnic vascular resistance during heating appears to be of thermoregulatory reflex origin. In dogs, rabbits or cats, selective heating of either the spinal cord[53] or the hypothalamus[25] produces opposite changes in blood flow and sympathetic nerve activity to skin and intestine. Increased activity of splanchnic nerves reduces intestinal blood flow, while at the same time reduced activity to cutaneous sympathetic nerves causes passive vasodilation of skin. Cooling of these two thermosensitive regions is attended by similar opposing changes but of the opposite sign. In the human and animal experiments just cited, splanchnic vasoconstriction during heat stress is not due to reflex activation of arterial baroreceptors by falling arterial mean pressure or pulse pressure. The response occurs in man independently of whether arterial pressure declines slightly, remains constant or increases.[45]

The major significance of elevated splanchnic vascular resistance during heating is the potentially important role in redistributing blood volume to a dilated, capacious cutaneous venous bed (see Chap. 12, Vol. II). Constriction of splanchnic resistance vessels causes a passive reduction in splanchnic venous volume by lowering venous distending pressure (see section on splanchnic veins and Fig. 14–9). This effect combined with the large fall in right atrial pressure during heating will further augment passive collapse of splanchnic veins. This displaces blood volume toward the heart and compensates for the expansion of cutaneous venous volume. In contrast to hemorrhage or exercise, for example, the 40 per cent decrease in splanchnic blood flow with heating could make only minor contributions to blood pressure regulation or to increased skin blood flow, because cardiac output increases markedly, and this increase is directed to skin (see Fig. 12–1, Chap. 12, Vol. II).

Combined Exercise and Thermal Stress. When *exercise* and *thermal stress* are combined, two competitive demands are placed upon the circulation: demands for oxygen transport to working muscle and demands for heat transport from deep tissues to the skin (see Chap. 5, Vol. III, and Chap. 12, Vol. II). The demand for increased skin blood flow under these conditions is met in part by further redistribution of blood flow away from splanchnic organs and also

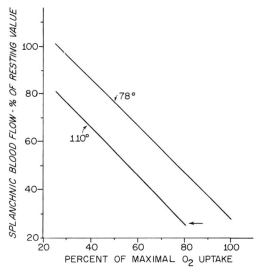

Figure 14–7 The progressive reduction in splanchnic blood flow with increasing severity of exercise (expressed as per cent of maximal oxygen uptake) is further increased by addition of heat stress; a greater percentage of cardiac output is distributed to skin during exercise. The slope of the line relating splanchnic blood flow to the percentage of maximal oxygen uptake required during exercise is shifted to the left so that at any given intensity of exercise, splanchnic blood flow is 20 per cent lower in hot (110°F) than in cool (78°F) environments. (After Rowell *et al., J. appl. Physiol.*, 1965, 20:384–394.)

the kidneys.[35, 41, 43, 46] The effects of the combined stresses are then additive so that at any *relative* oxygen consumption the slope of the line relating splanchnic blood flow to the percentage of maximal oxygen consumption required during exercise is shifted to the left, as shown by the 110° F line in Figure 14–7. The blood flow diverted from the splanchnic region can contribute additional blood flow to skin. As with thermal stress during rest, the response does not appear to be mediated by arterial baroreceptors, because blood pressure is well maintained (if heating is not too severe) despite the added cutaneous vasodilation.

In summary, a variety of stresses have been described which in man produce splanchnic vasoconstriction. In some situations (hemorrhage, orthostasis or lower body negative pressure), the response aids in maintaining arterial blood pressure. In others the responses allow a greater fraction of left ventricular output to perfuse regions where demand for oxygen or heat transfer is very high. In instances in which cardiac output cannot fully meet peripheral demands for oxygen, this redistribution plays a vital role in maintaining adequate supply of oxygen to these peripheral tissues. When circulatory demands for heat transport are superimposed upon demands for oxygen transport during exercise, additional splanchnic (and renal) vasoconstriction allows a greater fraction of cardiac output to be distributed to skin. The origin(s) of these reflex responses is unknown.

Figure 14–8 reveals an underlying feature in the control of human splanchnic blood flow and splanchnic vascular resistance during some of the stresses discussed above.[38] Slopes of the lines relating decrements in splanchnic blood flow to heart rate are essentially the same under all these stresses. Renal blood flow follows nearly the same slope for the one stress studied, upright exercise. The rightward shift in intercepts of the regression lines determined for exercise probably reflects the initial withdrawal of vagal outflow to the heart which precedes increased sympathetic nerve stimulation at the lower intensities of exercise.[37] Otherwise, the similarities suggest a close correspondence between sympathetic accelerator outflow to the heart and sympathetic vasoconstrictor outflow to the splanchnic and renal vessels.

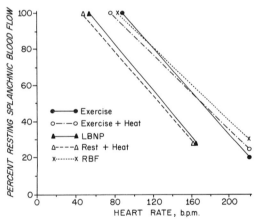

Figure 14–8 In man the sympathetic's outflow increasing heart rate is closely paralleled by sympathetic vasomotor outflow to the splanchnic region and kidneys under a variety of stresses. Stresses listed in the figure all reduce splanchnic blood flow by the same amount for each heart rate increase. The difference in curves for supine rest and upright exercise is significant and probably indicates a dominance in vagal outflow reduction vs. increased sympathetic outflow in response to the lowest levels of exercise. LBNP, Lower body negative pressure (see Fig. 14–5). RBF, renal blood flow during exercise. (From Rowell, *Physiol. Rev.*, 1974, 54, 75–159.)

The same relationship holds during exercise for those patients with mitral stenosis discussed above.

LOCAL CONTROL OF SPLANCHNIC RESISTANCE VESSELS

Splanchnic vascular resistance is influenced by a number of circulating hormones.[23] Secretin increases splanchnic blood flow either by direct action on vascular smooth muscle or by an indirect action dependent upon the release of other vasoactive substances such as kinins (similar to bradykinin—see Chap. 12, Vol. II). Glandular structures in the gastrointestinal mucosa may release kinins which cause functional hyperemia. However, this vasodilation may also be controlled by specific vasodilator nerve fibers to resistance vessels supplying glandular structures, so that kinin and neural mechanisms may operate synergistically[12, 13] (see Chap. 12, Vol. II).

The mechanism which raises splanchnic blood flow after protein but not after carbohydrate meals is not understood. It may be partly attributable to secretin stimulation. Insulin, glucagon and glucose have no direct effect upon splanchnic blood flow despite their effects upon hepatic metabolism.[17, 23] Their effects are considered secondary to those of altered local metabolism. In most cases the increase in blood flow appears to be primarily in the intestine. In some conditions, such as after infusion of amino acids, splanchnic oxygen consumption rises markedly without an increase in blood flow.[17] Thus, in this region blood flow and metabolic rate can vary independently. In fasting man during heavy exercise, splanchnic blood flow is strikingly reduced while at the same time splanchnic uptake of oxygen, free fatty acids and lactate is maintained or increased and production of glucose is accelerated to the point at which 20 per cent of the total metabolic costs can be met by its oxidation in muscle.[43] Most of these metabolic changes are confined to the liver. Thus, in contrast to skeletal muscle, variations in local hepatic metabolism appear to have little effect on its blood flow under conditions in which sympathetic adrenergic outflow to the organ is augmented.

Both *epinephrine* and *norepinephrine*

have marked effects upon splanchnic blood flow.[3, 15, 17] In man, the splanchnic region contains both *alpha-* and *beta*-receptors Intravenous infusion of epinephrine at physiological concentration causes, in man, marked vasodilation of the splanchnic bed.[3] Epinephrine also causes marked metabolic effects in the liver such as increased glycogenolysis and release of potassium. Studies in animals have produced conflicting findings; in the dog, epinephrine appears to cause splanchnic vasoconstriction. However, others find that intravenous administration of low concentrations causes vasodilation.[17]

Norepinephrine stimulates splanchnic alpha-receptors and is, in many species, a strong vasoconstrictor. As with cutaneous and muscle resistance vessels, hepatic and mesenteric resistance vessels are less responsive to physiological concentrations of norepinephrine than to sympathetic nerve stimulation.[27, 31]

NEURAL CONTROL OF SPLANCHNIC VEINS

Like cutaneous veins, but unlike muscle veins, splanchnic capacity vessels are richly innervated with sympathetic adrenergic fibers. Considerable evidence indicates that large quantities of blood can be rapidly mobilized from this region.[26] Direct electrical stimulation of splanchnic nerves at relatively low frequencies (4 to 6 imp/sec) produces maximal decrements in splanchnic blood volume. For example, in the cat, 50 per cent of the liver blood volume is expelled; this constitutes 7 per cent of the total blood volume. The total quantity of blood expelled from liver, intestines and spleen constitutes approximately 20 to 25 per cent of the total blood volume in cats; approximately 10 to 15 per cent is expelled by constriction of the splenic capsule and veins.[18] In contrast to hepatic and mesenteric resistance vessels in cats, venoconstriction appears to be sustained.[17, 18, 31] Findings from dogs are similar; direct sympathetic nerve stimulation at only 2 imp/sec expels 48 per cent of the total splanchnic blood volume, whereas at higher frequencies (up to 15 imp/sec) 66 per cent and up to 90 per cent of splanchnic blood volume is expelled (see Fig. 14–2). The response is rapid, being 50 per cent complete in 18 sec.[7]

Recent efforts to differentiate between *active* and *passive* changes in splanchnic venous volume indicate that 65 per cent of the total splanchnic blood volume released during splanchnic nerve stimulation is released passively when splanchnic and central venous pressures are normal.[7a] Passive, elastic recoil of the venous wall occurs if transmural or distending pressure is suddenly reduced by vasoconstriction of the series-coupled resistance vessels. This passive component becomes progressively less important as central and splanchnic venous pressures are elevated above normal. Because of their importance in splanchnic volume regulation, some general mechanical features of veins are reviewed here. It is obvious from a "characteristic" venous pressure-volume curve (Fig. 14–9) that the magnitude of passive volume changes will vary greatly with venous distending pressure. For example, in Figure 14–9 a small change in pressure in the range from 0 to 10 mm Hg causes a large change in venous volume—that is, venous "distensibility" is very high in the low pressure range. The curve in Figure 14–9 is from human forearm veins. The relationship between venous pressure and volume will vary among different venous beds and depends upon venous size and wall structure. For example, equivalent volume changes would occur at much higher pressures in human leg veins where distending pressures are high and the walls have a thick muscular layer. Major volume shifts would be expected over a lower pressure range in splanchnic veins which, being closer to the heart, are exposed to lower hydrostatic pressures. The major point in Figure 14–9 is that for any given venous bed there is some distending pressure below which *passive* geometrical changes rather than *active* constriction dominate volume changes. That is, as distending pressure falls below the level necessary to hold the vein in a rounded shape, the vein collapses from elastic recoil of the wall, causing a major fraction of its volume to be *passively* dispelled.[33] Thus, the steep portion of the pressure-volume curve is not an index of true venous distensibility but rather reflects the changes in venous cross-sectional geometry. At low distending pressures active venoconstric-

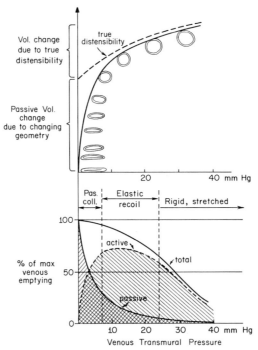

Figure 14–9 The upper portion of the figure shows hypothetical relationships between pressure and volume in splanchnic veins; actual relationships are unknown. The illustration, from a single large vein, shows the relative importance of *active* vs. *passive* changes in venous volume and their dependence upon (i) venous distending pressure and (ii) venous geometry at the time of constriction. Over the volume range of true venous distensibility (dashed line on top), *active* volume changes (active changes are those caused by venoconstriction and are explained in Figure 14–10) predominate as the diameter of *round* veins is altered. But as pressure in this hypothetical vein falls below 7 mm Hg, its contents are displaced by passive collapse as its cross-sectional geometry changes. Below 7 mm Hg active changes have little effect on volume and could even increase volume as veins become elliptical or flat (*i.e.*, venoconstriction could make them rounder).

The lower section of the figure shows the percentage of maximal venous emptying which would be expected to occur over the venous pressure range inscribed by the upper curve. The three curves show total emptying (active plus passive) and the separate contributions of active and passive emptying. (After Oberg, *Acta physiol. scand.*, 1967, *71*, 233–247.)

tion has little or no effect on venous volume as shown in the lower portion of Figure 14–9. In contrast, once veins attain a circular cross section, further increments in transmural pressure cause relatively small changes in volume, and large volume changes must be accomplished predominantly by venoconstriction. This would be the case when central and splanchnic venous pressures are elevated. The distensibility of human splanchnic veins is unknown, but in supine man splanchnic venous pressures would be expected on hydrostatic principles to be low. Consequently, passive emptying would be expected to dominate in situations in which splanchnic resistance vessels constrict, and when central venous pressure also falls. This produces a pressure gradient favoring a passive shift of splanchnic blood volume toward the heart. The combination of reduced arterial pressure and intensive vasoconstriction in severe hemorrhage, for example, would be expected to cause a *passive* depletion of splanchnic blood volume. In these situations active venoconstriction would be expected to play a minor role, if any, in the redistribution of blood volume from the splanchnic veins. The situation in upright man may be quite different owing to the hydrostatic increase in splanchnic venous pressure – that is, distending pressures in these veins may be increased to the point that any expulsion of splanchnic venous volume would require active venoconstriction.

REFLEX CONTROL OF SPLANCHNIC VEINS

Baroreceptors. Reductions in carotid sinus pressure reflexly decrease the compliance of mesenteric veins;[1, 24, 55] carotid sinus hypertension increases venous compliance.[1] Thus, at a given distending pressure, venous volume decreases from point A to point B in Figure 14–10, for example, with a fall in arterial pressure. If venous pressure also fell from P_1 to P_2 owing to constriction of resistance vessels, then the volume expelled would increase from ΔV_1 to ΔV_2. Mesenteric veins in dogs, kept partially distended by raised venous pressure, can sustain a predominantly active venoconstriction mediated via baroreceptor reflex.[21] No "autoregulatory escape" is observed in veins. Reducing carotid sinus pressure in dogs down to 40 mm Hg causes a sudden (within 24 sec) release of approximately 16 per cent of total splanchnic blood volume (Figs. 14–3 and 14–4). Increasing carotid sinus pressure to 200 mm Hg causes an increase in splanchnic blood volume of similar magnitude. Changes in splanchnic blood volume parallel the changes in splanchnic blood flow (Fig. 14–3). Presumably about half of these volume changes

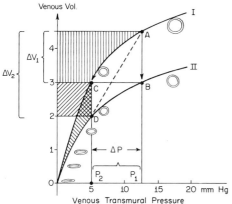

Figure 14–10 Two volume-pressure curves for the same vein reveal how *active* changes in venous volume result from a change in venous compliance ($\triangle V/\triangle P$). The active volume change ($\triangle V_1$) caused by venoconstriction is illustrated by points A and B where veins constrict at constant pressure (P_1). The same volume change ($\triangle V_1$) would be passively caused by reducing venous pressure from P_1 to P_2 or from points A to C on curve I. Note the smaller volume change between P_1 and P_2 on curve II, the pressure volume curve for constricted (lower compliance) veins. Maximum volume change ($\triangle V_2$) over the pressure range (P_1 to P_2) would accompany the combined effects of reducing both pressure and compliance—points A to D.

are due to passive changes discussed above. In general, available evidence indicates that splanchnic venous capacity is under reflex control mediated by arterial baroreceptors.

Hemorrhage. In an extensive review, Chien[9] concludes that in the dog and cat, hemorrhage reflexly reduces splanchnic blood volume. Since hemorrhage causes an immediate decrease in mesenteric volume which exceeds that predicted from total blood loss,[36] reflex constriction of the capacity vessels is suggested. However, part of the decrease is undoubtedly due to passive reduction in venous capacity through reduced arterial and central venous pressure. Splanchnic volume reaches a minimum during the first minute after hemorrhage and is well maintained even with prolonged hypotension.[9] This response is reduced in intestines by carotid sinus denervation; thus baroreceptors appear to be involved.

When 90 to 130 ml (7.2 ml per kg) of blood is removed from dogs over a 2 min period, 17 per cent of total splanchnic blood volume or 54 per cent of the total volume bled is gradually mobilized from the splanchnic circulation while splanchnic blood flow declines and vascular resistance rises (Fig. 14–11).[7] After splanchnic denervation the fraction of splanchnic blood volume passively expelled is 68 per cent of that expelled in the innervated state.[7a]

In man, removal of 17 per cent of total blood volume causes an estimated 40 per cent fall in total splanchnic blood volume. Thus, 50 per cent of the total volume removed was expelled from the splanchnic region.[34] Since arterial mean pressure, splanchnic blood flow and splanchnic vascular resistance appeared not to change, these results suggest an active mobilization of splanchnic blood volume. However, the extent to which a fall in central venous pressure might contribute to passive changes is unknown.

In general, the conditions usually attending hemorrhage would lead one to expect a predominance of *passive* over *active* changes in splanchnic venous capacity. Under some conditions active volume changes appear to occur, but their significance is unknown.

Exercise. The increase in sympathetic nerve activity in man during exercise (illustrated in Figs. 14–6, 14–7 and 14–8) might

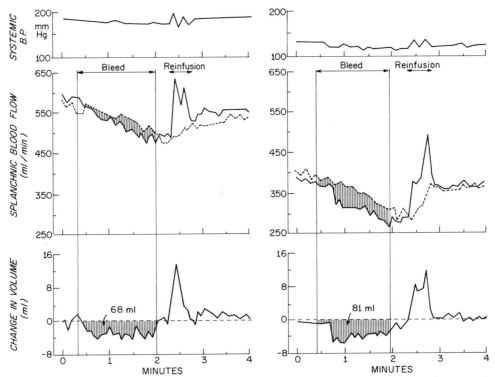

Figure 14–11 Changes in mean systemic arterial blood pressure, splanchnic arterial and venous flow and volume during hemorrhage of 100 ml in two dogs. Total volume of blood mobilized from splanchnic circulation from start of hemorrhage until reinfusion was begun is shown for each dog. Note the gradual decline in splanchnic inflow (——) and outflow (– – –) during hemorrhage, which appears, particularly in the left-hand figure, to result primarily from a rise in resistance (blood pressure falls only slightly). In contrast, the decrement in volume appears to happen suddenly and is sustained throughout the bleeding. This could be due to active venoconstriction. (From Brooksby and Donald, *Circulat. Res.*, 1971, 29, 227–238. By permission of The American Heart Association, Inc.)

be expected to constrict splanchnic veins as well as resistance vessels. Splanchnic blood volume in man decreases 34 per cent during *mild supine* exercise.[52] Three factors suggest that the response is predominantly active rather than passive: arterial blood pressure is maintained or increases slightly; splanchnic blood flow decreases only slightly; and central venous pressure would not be expected to fall (it actually may rise).

Mobilization of splanchnic blood volume at the onset of exercise was once regarded as necessary to reduce any lag in venous return which was assumed to attend sudden filling of dilated muscle vessels and veins, but muscle volume increases very little.[38] There is no real evidence that such volume shifts are necessary at the onset of exercise. However, if exercise is performed in hot environments, a large volume of blood is displaced into cutaneous veins (see Chap. 12, Vol. II). It is not known if splanchnic

venoconstriction occurs in this situation, but the fall in central venous pressure and the large increase in splanchnic vascular resistance should passively mobilize splanchnic blood volume into the central circulation. This partially compensates for the volume displaced to skin.[38] Since dogs show little or no increase in sympathetic nerve activity to splanchnic resistance vessels during exercise (see above), it is probable that reflex venomotor changes are also absent.

Thermoregulatory Reflexes. The large increase in cutaneous venous volume attending heat stress clearly must be compensated by a shift of blood volume from veins elsewhere. In 1905 Muller[32] used a "partial weighing" technique wherein the horizontally supported body of warm and cool human subjects was weighed in segments. His findings suggested that heating produces a shift of blood volume away from abdominal-visceral segments, whereas cooling produces the opposite effect. Also, X-rays of heated human subjects reveal a decrease in liver dimensions.[14] Reductions in splanchnic volume during heating could be due to the elastic recoil of veins accompanying two major effects of the stress: (i) the marked decline in central venous pressure produces a pressure gradient which should shift splanchnic blood volume centrally (see Fig. 12-1, Chap. 12, Vol. II); (ii) the progressive rise in splanchnic vascular resistance accompanying rising body temperature would reduce splanchnic venous pressure and further augment passive venous emptying. With cooling, cutaneous veins constrict and blood volume shifts centrally, causing central venous pressure to rise while the splanchnic resistance vessels dilate;[45] now, splanchnic blood volume must increase. There is no direct evidence suggesting the occurrence of reflex splanchnic venoconstriction during heat stress. However, since sympathetic nerve activity to the region is augmented by heating in man (see Fig. 14-8), venoconstriction might be expected. Raising central body temperature in dogs does not cause a reflex venomotor response in the isolated spleen or loop of ileum.[55] But as in exercise, dogs may also rely very little on vasomotor (or venomotor) adjustments during heat stress.

LOCAL CONTROL OF SPLANCHNIC VEINS

As with muscle and cutaneous veins, splanchnic veins appear to be influenced to a relatively minor extent by local metabolic changes and circulating catecholamines.[9, 30, 31]

REFERENCES

1. Alexander, R. S. The participation of the venomotor system in pressor reflexes. *Circulat. Res.*, 1954, 2, 405–409.
1a. Blackmon, J. R., Rowell, L. B., Kennedy, J. W., Twiss, R. D., and Conn, R. D. Physiological significance of maximal oxygen intake in pure mitral stenosis. *Circulation*, 1967, 36, 497–510.
2. Bradley, S. E. Variations in hepatic blood flow in man during health and disease. *New Engl. J. Med.*, 1949, 240, 456–461.

3. Bradley, S. E. The hepatic circulation. *Handb. Physiol.*, 1963, sec. 2, vol. II, 1387–1438.
4. Bradley, S. E., Ingelfinger, F. J., Bradley, G. P., and Curry, J. J. The estimation of hepatic blood flow in man. *J. clin. Invest.*, 1945, 24, 890–897.
5. Bradley, S. E., Marks, P. A., Reynell, P. C., and Meltzer, J. The circulating splanchnic blood volume in dog and man. *Trans. Ass. Amer. Phycns.*, 1953, 66, 294–302.
6. Brauer, R. W. Liver circulation and function. *Physiol. Rev.*, 1963, 43, 115–213.
7. Brooksby, G. A., and Donald, D. E. Dynamic changes in splanchnic blood flow and blood volume in dogs during activation of sympathetic nerves. *Circulat. Res.*, 1971, 29, 227–238.
7a. Brooksby, G. A., and Donald, D. E. Release of blood from the splanchnic circulation in dogs. *Circulat. Res.*, 1972, 31, 105–118.
8. Chalmers, J. P., Korner, P. I., and White, S. W. Effects of haemorrhage on the distribution of peripheral blood flow in the rabbit. *J. Physiol. (Lond.)*, 1967, 192, 561–574.
9. Chien, S. Role of the sympathetic nervous system in hemorrhage. *Physiol. Rev.*, 1967, 47, 214–288.
10. Detry, J-M. R., Brengelmann, G. L., Rowell, L. B., and Wyss, C. Skin and muscle components of forearm blood flow in directly heated resting man. *J. appl. Physiol.*, 1972, 32, 506–511.
11. Drapanas, T., Kluge, D. N., and Schenk, W. G., Jr. Measurement of hepatic blood flow by Bromsulphalein and by the electromagnetic flowmeter. *Surgery*, 1960, 48, 1017–1021.
12. Gautvik, K. Studies on kinin formation in functional vasodilatation of the submandibular salivary gland in cats. *Acta physiol. scand.*, 1970, 79, 174–187.
13. Gautvik, K. The interaction of two different vasodilator mechanisms in the chorda-tympani activated submandibular salivary gland. *Acta physiol. scand.*, 1970, 79, 188–203.
14. Glaser, E. M., Berridge, F. R., and Prior, K. M. Effects of heat and cold on the distribution of blood within the human body. *Clin. Sci.*, 1950, 9, 181–187.
15. Grayson, J., and Mendel, D. *Physiology of the splanchnic circulation.* Baltimore, Williams & Wilkins Co., 1965.
16. Greenway, C. V., Lawson, A. E., and Mellander, S. The effects of stimulation of the hepatic nerves, infusions of noradrenaline and occlusion of the carotid arteries on liver blood flow in the anesthetized cat. *J. Physiol. (Lond.)*, 1967, 192, 21–41.
17. Greenway, C. V., and Stark, R. D. Hepatic vascular bed. *Physiol. Rev.*, 1971, 51, 23–65.
18. Greenway, C. V., Stark, R. D., and Lautt, W. W. Capacitance responses and fluid exchange in the cat liver during stimulation of the hepatic nerves. *Circulat. Res.*, 1969, 25, 277–284.
19. Grim, E. The flow of blood in the mesenteric vessels. *Handb. Physiol.*, 1963, sec. 2, vol. II, 1439–1456.
20. Grimby, G. Renal clearances during prolonged supine exercise at different loads. *J. appl. Physiol.*, 1965, 20, 1294–1298.
21. Hadjiminas, J., and Öberg, B. Effects of carotid

baroreceptor reflexes on venous tone in skeletal muscle and intestine of the cat. *Acta physiol. scand.*, 1968, *72*, 518–532.

22. Heymans, C., and Neil, E. *Reflexogenic areas of the cardiovascular system.* London, Churchill, 1958.

23. Hultman, E. Blood circulation in the liver under physiological and pathological conditions. *Scand. J. clin. Lab. Invest.*, 1966, *18* (suppl. 92), 27–41.

24. Iizuka, T., Mark, A. L., Wendling, M. G., Schmid, P. G., and Eckstein, J. W. Differences in responses of saphenous and mesenteric veins to reflex stimuli. *Amer. J. Physiol.*, 1970, *219*, 1066–1070.

25. Iriki, M., Riedel, W., and Simon, E. Regional differentiation of sympathetic activity during hypothalamic heating and cooling in anesthetized rabbits. *Pflügers Arch. ges. Physiol.*, 1971, *328*, 320–331.

26. Katz, L. N., and Rodbard, S. The integration of the vasomotor responses in the liver with those in other systemic vessels. *J. Pharmacol. exp. Ther.*, 1939, *67*, 407–422.

27. Korner, P. I., Chalmers, J. P., and White, S. W. Some mechanisms of reflex control of the circulation by the sympatho-adrenal system. *Circulat. Res.*, 1967, *20* and *21* (suppl. III), 157–172.

28. Kremer, M., and Wright, S. The effects on blood-pressure of section of the splanchnic nerves. *Quart. J. exp. Physiol.*, 1932, *21*, 319–335.

29. Löfving, B. Differentiated vascular adjustments reflexly induced by changes in the carotid baro- and chemoreceptor activity and by asphyxia. *Med. exp. (Basel)*, 1961, *4*, 307–312.

30. Mellander, S. Comparative studies on the adrenergic neuro-hormonal control of resistance and capacitance blood vessels in the cat. *Acta physiol. scand.*, 1960, *50* (suppl. 176), 1–86.

31. Mellander, S., and Johansson, B. Control of resistance, exchange and capacitance functions in the peripheral circulation. *Pharmacol. Rev.*, 1968, *20*(3), 117–196.

32. Müller, O. Über die Blutverteilung im menschlichen Körper unter dem Einfluss thermischer Reize. *Dtsch. Arch. klin. Med.*, 1905, *82*, 547–585.

33. Öberg, B. The relationship between active constriction and passive recoil of the veins at various distending pressures. *Acta physiol. scand.*, 1967, *71*, 233–247.

33a. Pelletier, C. L., Edis, A. J., and Shepherd, J. T. Circulatory reflex from vagal afferents in response to hemorrhage in the dog. *Circulat. Res.*, 1971, *29*, 626–634.

34. Price, H. L., Deutsch, S., Marshall, B. E., Stephen, G. W., Behar, M. G., and Neufeld, G. R. Hemodynamic and metabolic effects of hemorrhage in man, with particular reference to splanchnic circulation. *Circulat. Res.*, 1966, *18*, 469–474.

35. Radigan, L. R., and Robinson, S. Effects of environmental heat stress and exercise on renal blood flow and filtration rate. *J. appl. Physiol.*, 1949, *2*, 185–191.

36. Reynell, P. C., Marks, P. A., Chidsey, C., and

Bradley, S. E. Changes in splanchnic blood volume and splanchnic blood flow in dogs after haemorrhage. *Clin. Sci.*, 1955, *14*, 407–419.

37. Robinson, B. F., Epstein, S. E., Beiser, G. D., and Braunwald, E. Control of heart rate by the autonomic nervous system. *Circulat. Res.*, 1966, *19*, 400–411.

37a. Ross, G. The regional circulation. *Ann. Rev. Physiol.*, 1971, *33*, 445–478.

38. Rowell, L. B. Human cardiovascular adjustments to exercise and thermal stress. *Physiol. Rev.*, 1974, *54*, 75–159.

39. Rowell, L. B. Dye technique for estimating hepatic-splanchnic blood flow in man. In: *Dye curves: The theory and practice of non-diffusible indicator dilution*, D. A. Bloomfield, ed. University Park, Pennsylvania, University Park Press, 1973.

40. Rowell, L. B., Blackmon, J. R., and Bruce, R. A. Indocyanine green clearance and estimated hepatic blood flow during mild to maximal exercise in upright man. *J. clin. Invest.*, 1964, *43*, 1677–1690.

41. Rowell, L. B., Blackmon, J. R., Martin, R. H., Mazzarella, J. A., and Bruce, R. A. Hepatic clearance of indocyanine green in man under thermal and exercise stresses. *J. appl. Physiol.*, 1965, *20*, 384–394.

42. Rowell, L. B., Brengelmann, G. L., Blackmon, J. R., and Murray, J. A. Redistribution of blood flow during sustained high skin temperature in resting man. *J. appl. Physiol.*, 1970, *28*, 415–420.

43. Rowell, L. B., Brengelmann, G. L., Blackmon, J. R., Twiss, R. D., and Kusumi, F. Splanchnic blood flow and metabolism in heat-stressed man. *J. appl. Physiol.*, 1968, *24*, 475–484.

44. Rowell, L. B., Detry, J-M. R., Blackmon, J. R., and Wyss, C. Importance of the splanchnic vascular bed in human blood pressure regulation. *J. appl. Physiol.*, 1972, *32*, 213–220.

45. Rowell, L. B., Detry, J-M. R., Profant, G. R., and Wyss, C. Splanchnic vasoconstriction in hyperthermic man—role of falling blood pressure. *J. appl. Physiol.*, 1971, *31*, 864–869.

46. Rowell, L. B., Marx, H. J., Bruce, R. A., Conn, R. D., and Kusumi, F. Reductions in cardiac output, central blood volume, and stroke volume with thermal stress in normal men during exercise. *J. clin. Invest.*, 1966, *45*, 1801–1816.

47. Rushmer, R. F., Franklin, D. L., Van Citters, R. L., and Smith, O. A. Changes in peripheral blood flow distribution in healthy dogs. *Circulat. Res.*, 1961, *9*, 675–687.

48. Van Citters, R. L., and Franklin, D. L. Cardiovascular performance of Alaska sled dog during exercise. *Circulat. Res.*, 1969, *24*, 33–42.

49. Vatner, S. F., Franklin, D., Van Citters, R. L., and Braunwald, E. Effects of carotid sinus nerve stimulation on blood-flow distribution in conscious dogs at rest and during exercise. *Circulat. Res.*, 1970, *27*, 495–503.

50. Vatner, S. F., Higgens, C. B., White, S., Patrick, T., and Franklin, D. The peripheral vascular response to severe exercise in untethered

dogs before and after complete heart block. *J. clin. Invest.*, 1971, *50*, 1950–1960.

51. Wade, O. L., and Bishop, J. M. *Cardiac output and regional blood flow.* Oxford, Blackwell, 1962.

52. Wade, O. L., Combes, B., Childs, A. W., Wheeler, H. O., Cournand, A., and Bradley, S. E. The effect of exercise on the splanchnic blood flow and splanchnic blood volume in normal man. *Clin. Sci.*, 1956, *15*, 457–463.

53. Walther, O.-E., Iriki, M., and Simon, E. Antagonistic changes of blood flow and sympathetic activity in different vascular beds following central thermal stimulation. *Pflügers Arch. ges. Physiol.*, 1970, *319*, 162–184.

54. Webb-Peploe, M. M. The isovolumetric spleen: Index of reflex changes in splanchnic vascular capacity. *Amer. J. Physiol.*, 1969, *216*, 407–413.

55. Webb-Peploe, M. M. Effect of changes in central body temperature on capacity elements of limb and spleen. *Amer. J. Physiol.*, 1969, *216*, 643–646.

55a. Webb-Peploe, M. M., Brender, D., and Shepherd, J. T. Vascular responses to stimulation of receptors in muscle by capsaicin. *Amer. J. Physiol.*, 1972, *222*, 189–195.

56. Wilkins, R. W., Culbertson, J. W., and Ingelfinger, F. J. The effect of splanchnic sympathectomy in hypertensive patients upon estimated hepatic blood flow in the upright as contrasted with the horizontal position. *J. clin. Invest.*, 1951, *30*, 312–317.

57. Zoller, R. P., Mark, A. L., Abboud, F. M., Schmid, P. G., and Heistad, D. D. The role of low pressure baroreceptors in reflex vasoconstrictor responses in man. *J. clin. Invest.*, 1972, *51*, 2967–2972.

CHAPTER 15 THE RENAL CIRCULATION

by ALLEN M. SCHER, ROBERT B. STEPHENSON, *and* MARSHAL H. CALEY*

INTRODUCTION

The kidneys maintain and regulate the osmolarity and solute concentration of the blood plasma, holding both of these within narrow limits. Blood plasma becomes the interstitial fluid in all vascular beds. Thus, by maintaining the plasma composition, the kidneys regulate the fluid environment in which all the cells of the body function. Regulation of plasma composition by the kidneys is accomplished by processes whose net effect is to extract material from plasma and produce urine.

There is a close anatomical and functional relationship between the renal vasculature and the renal urinary tubules (nephrons) which facilitates exchange between the two systems. Blood enters the kidney via the renal artery. In the production of urine this arterial blood becomes two separate streams, one confined to the vascular system, the other within the renal tubular system. These two streams become, respectively, the renal venous blood and the urine. The formation of urine, discussed in detail in another chapter (Chap. 25, Vol. II), involves a sequence of passive filtration of fluid from the vascular system into the urinary tubules, followed by passive and active

transport of fluid and solute between the tubules and the vascular system.

RENAL ANATOMY (Fig. 15–1)

Major Renal Vessels. The renal artery on each side originates from the abdominal aorta. The major renal arterial branches, which radiate from the hilum toward the periphery between the renal pyramids (lobes), are called the interlobar arteries. At the junction between the cortex and the medulla of the kidney, these divide to form the arcuate arteries which course parallel to the surface of the kidney, demarcating the boundary between the renal cortex and medulla. The arcuate arteries give rise to interlobular arteries, which run radially outward through the cortex toward the surface of the kidney. Interlobular arteries give rise to smaller branches, which again parallel the surface of the kidney and which are, or give rise to, the renal afferent arterioles. The afferent arterioles supply the urine-producing units, the nephrons. The major arteries—renal, interlobar, arcuate and interlobular—are paralleled by an

---

*Deceased.

234

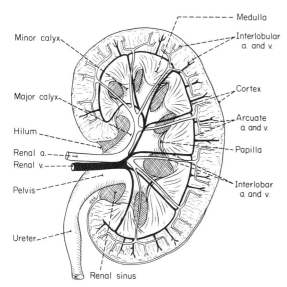

Figure 15–1 Anterior view of human kidney, frontal section, showing blood vessels and major anatomical features. Arteries (a) shown in white; veins (v) in black. Hilum is point of entry of artery, vein and ureter. Pelvis (expanded end of ureter) divides and subdivides to form major and minor calyces. Tip (papilla) of each medullary lobe projects into minor calyx. Sinus is a fat-filled space.

identically named set of venous vessels. Circulation to the nephron is described below.

The Nephron and Its Circulation. The regulatory functions of the kidney are performed by approximately two million nephrons (one million in each kidney) which make up the renal parenchyma. The nephron consists of Bowman's capsule, the proximal tubule, the loop of Henle and the distal tubule which leads into the collecting duct. Figure 15–2 diagrams this functional unit. In Figure 15–3 a schematic diagram of the renal circulation is superimposed upon the shadows of two nephrons, showing the close functional relationship of the renal vasculature to the nephron. Virtually all renal arterial blood flows through afferent arterioles to glomerular capillaries. Here, high hydrostatic pressure favors filtration of a portion of the blood plasma across the capillary walls into Bowman's capsule, the initial portion of the tubular nephron (Chap. 9, Vol. II). As indicated below, proteins do not pass into Bowman's capsule. Glomerular filtration thus removes an ultrafiltrate of plasma from the blood and increases the hematocrit and the protein concentration of the remaining blood.

Each glomerular capillary tuft empties into an efferent arteriole, and from these arterioles the blood flows into a network of peritubular capillaries. The peritubular capillary network surrounds proximal and distal tubules and collecting ducts. Low hydrostatic pressure and increased protein osmotic pressure in these capillaries favor reabsorption of fluid from the tubules. Through both active and passive reabsorption, most of the filtered water and solutes are returned from the nephrons to the blood of the peritubular capillaries. The blood vessels which exit from a glomerulus continue their close association with the tubule which leaves the same glomerulus.

Figure 15–2 Schematic diagram of the two types of mammalian nephrons. Bowman's capsule gives rise to the proximal tubule, followed by Henle's loop (thick and thin segments), distal tubule and collecting duct. Outer cortical nephrons have shorter loops of Henle than juxtamedullary nephrons. Collecting duct opens into minor calyx at the papilla.

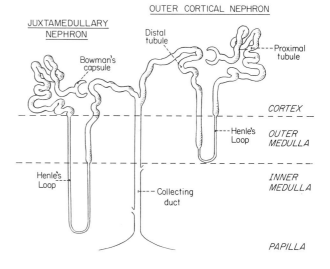

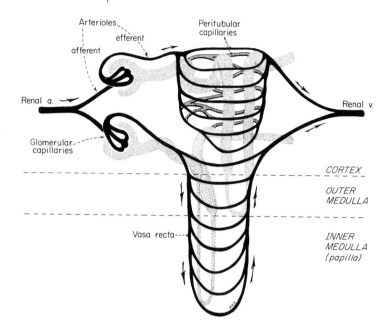

Figure 15–3 Schematic of renal circulation (black) superimposed on two nephrons (gray). Bold arrows show direction of blood flow. The several arteries (a) and renal veins (v) are shown. Vasa recta descend into the medulla.

Renal circulation is unusual in that there are two sets of arteriolar vessels and two capillary beds arranged in series. The sequence is: afferent arteriole, glomerular capillary, efferent arteriole, peritubular capillary. The latter exit into the venules and veins. There are important anatomical differences between the outer cortex and the juxtamedullary cortex of the kidney. Nephrons in the outer cortex have relatively short loops of Henle, and the associated peritubular capillaries also lie mostly in the cortex. By contrast, juxtamedullary nephrons have long loops of Henle, which extend deep into the medulla. The peritubular capillaries which arise from juxtamedullary efferent arterioles perfuse both the juxtamedullary cortex and the outer medulla. Juxtamedullary efferent arterioles also give rise to the vasa recta, long, straight arterioles which course deeply into the medulla, even extending to the tips of the renal papillae (Fig. 15–3). Vasa recta give rise to capillary networks which surround the descending and ascending limbs of Henle's loops and the collecting ducts in the medulla. Medullary capillaries anastomose freely. The blood perfusing medullary capillaries is returned to juxtamedullary veins. The medullary blood supply thus follows a looplike pattern in which vessels originate and terminate in the juxtamedullary region of the cortex.

Juxtaglomerular Apparatus. Another specialized feature of the renal vasculature is found at the junction of afferent arteriole and glomerulus, particularly in the outer renal cortex. On histological section, the cells in the medial layer of the wall of the afferent arteriole are swollen, epithelioid and fibrillar in appearance, and they contain granules which vary in number with changes in blood pressure or sodium excretion. These specialized cells of the afferent arteriole (called the juxtaglomerular apparatus) are also in contact with the distal convoluted tubule of the same nephron (Fig. 15–4). Where the afferent arteriole is in close apposition to the distal tubule, some of the cells of the distal tubule are tall and columnar, a specialization called the macula densa.

The granules in the juxtaglomerular cells of the afferent arteriole are thought to contain renin. Renin is a proteolytic enzyme which acts on a plasma protein called angiotensinogen to produce angiotensin I. Angiotensin I is converted in the lungs to angiotensin II (both are polypeptides). Angiotensin II is a potent vasoconstrictor; and, in addition to this cardiovascular effect, it also stimulates the secretion by the

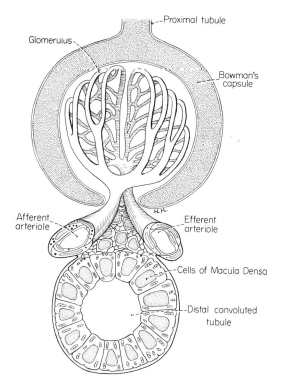

Figure 15-4 The juxtaglomerular region. Bowman's capsule and the distal tubule shown in cross section. Afferent arteriole, glomerulus and efferent arteriole drawn in perspective. Cells between afferent and efferent arterioles are supporting cells. Specialized afferent arteriolar cells (juxtaglomerular apparatus) are granulated.

[Figure labels: Proximal tubule; Glomerulus; Bowman's capsule; Afferent arteriole; Efferent arteriole; Cells of Macula Densa; Distal convoluted tubule]

adrenal gland of aldosterone, a hormone which acts on the kidney to promote reabsorption of sodium ions. Thus, renin, through angiotensin II, has profound effects on the entire cardiovascular system and on ionic balance (Chap. 10, Vol. III). Renin production appears to be increased by a decrease in renal arterial pressure, by a decrease in extracellular fluid volume, by increased sympathetic discharge to the kidney, or by a change in the sodium load delivered to the distal tubule* [44] (see also Chap. 10, Vol. III).

---

*The early observation that partial constriction of the renal artery could produce hypertension (Goldblatt hypertension),[14] coupled with the subsequent observation[37] that renin was released by such constriction, led to the conclusion that renin is the causative agent in chronic renal hypertension. Later studies have indicated that renin elevation following renal arterial narrowing is not maintained, although hypertension persists.[2] Also, rabbits immunized against angiotensin II still develop renal hypertension.[26]

INTRARENAL VASCULAR, INTERSTITIAL AND TUBULAR PRESSURES

The peculiarities of the renal circulation, particularly the presence of two capillary beds and two arteriolar vessels in series, would lead us to expect something unusual about the changes in pressure along the renal vasculature. Pressure measurements are most commonly made in the rat kidney. Unlike most other animals, the rat has a transparent renal capsule, and the tubules and peritubular capillaries are visible under the microscope so that micropipettes can be introduced into these structures with visual control.[5] Unfortunately, not all segments of the vasculature are accessible. In dogs, measurements have been made of venous pressure above and below the level of the arcuate veins. In addition, interstitial fluid pressure measurements have been made by inserting needles into the renal parenchyma, a technique which has been somewhat controversial. Interstitial pressure has also been approximated by measuring the pressure in a cannula inserted retrograde into the venous system to or beyond the level of the arcuate veins.

With a mean arterial pressure of 100 mm Hg (as in the micropuncture studies of Brenner *et al.*),[6] the pressure in the glomerular capillaries of the rat is about 45 mm Hg. Pressure in the efferent arteriole is about 11 mm Hg. Pressure in the peritubular capillaries is about 6 mm Hg. A further small pressure drop occurs in the venous system down to the level of the arcuate veins. In contrast, Hinshaw[19] and Swann[40] give values of 25 mm Hg or more for venous pressure in the dog (probably at the level of arcuate veins), although others find lower venous pressures. Hinshaw and Swann find interstitial pressures of about 25 mm Hg, whereas others give values as low as 3 to 4 mm Hg. Whatever the interstitial pressure, venous pressure must be higher or the thin-walled veins would collapse. In studies by Brenner *et al.* in the rat,[5] tubular pressures (16 mm Hg) were found to exceed pressures in adjacent peritubular capillaries (6 mm Hg). This seems to indicate that such renal structures as tubules and capillaries may be rigid enough to be hydraulically isolated from one another. These pressures will be discussed further with reference to regulation of renal blood flow.

RENAL CAPILLARIES AND CAPILLARY FILTRATION

Renal circulation is once more unusual in that it is the only circulation in which a very large volume of fluid passes out of the blood vessels into the interstitial space. Most of this fluid is reabsorbed during its passage along the tubules. According to the calculations of Landis and Pappenheimer,[24] in the rest of the body, a total of 20 liters of fluid per day is filtered out of the plasma into the interstitial space in man. Of this, 16 to 18 liters is reabsorbed directly into the blood and 2 to 4 liters appears as lymph.[24] By contrast, glomerular filtration is about 120 cc per min, or 7.2 liters per hour, or 173 liters per day. Since glomerular filtration is so much greater than capillary filtration in the rest of the body, we might consider the kidney to have the most permeable capillaries of any organ. The glomerular epithelium has wide pores, about one-tenth of a micron in diameter. These pores are so large that plasma proteins could pass through them. No proteins appear in urine, however, because the basement membrane and the visceral cell layer of Bowman's capsule, particularly the latter, form narrow slits which restrain plasma proteins from passing through into the inside of the tubule.

RENAL BLOOD FLOW, OXYGEN EXTRACTION, ARTERIOVENOUS OXYGEN

Renal tubular reabsorption and secretion involve transport of certain solutes across the tubular epithelial cells against electrochemical potential gradients. Metabolic energy is utilized to perform this work, and the oxygen utilization per gram of tissue is high. In mammals, the kidneys constitute roughly 0.5 per cent of body weight and account for approximately 8 per cent of resting oxygen consumption.

Heart and brain also extract relatively large *amounts* of oxygen from the blood flowing through them, and in these organs there is a large difference in oxygen *content* between arterial and venous blood. Because of the extremely high rates of renal blood flow (1200 cc per min in the human, 20 to 25 per cent of the cardiac output at rest), the high renal oxygen consumption is not observable as a large arteriovenous oxygen (A-VO$_2$) difference. The renal A-VO$_2$ difference is lower than that of most other tissues in the body—between 1 and 2 ml O$_2$ per 100 ml renal blood vs. an overall systemic A-VO$_2$ difference of 5 ml O$_2$ per 100 ml blood.

The oxygen utilization of the kidney seems to be determined by two separate needs. The first of these is a constant basal oxygen uptake, probably concerned with normal cellular metabolism. This basal oxygen uptake is found in the kidney which is not filtering or reabsorbing because of very marked hypotension. Superimposed on this, there is a variable oxygen uptake which in normally functioning kidneys is correlated very closely with the quantity of sodium reabsorbed. This leads to the hypothesis that sodium reabsorption represents the main renal work and the main determinant of renal oxygen uptake.[25]

INTRINSIC REGULATION OF RENAL BLOOD FLOW

The kidney shows important blood flow regulation by both intrinsic and extrinsic mechanisms. Evidence for intrinsic regulation of the blood flow is seen in the relationship between pressure and flow in the isolated, blood-perfused kidney. This preparation lacks extrinsic nerve supply, and yet the renal blood flow tends to remain constant when arterial pressure is varied over the range from 90 to 200 mm Hg. This constancy of flow in an isolated organ, called "autoregulation," results from a pressure-dependent increase in resistance to blood flow (Fig. 15–5). Autoregulatory mechanisms are also active in the kidney *in situ*, although they may at times be hard to separate from extrinsic (autonomic) regulation. It further seems reasonable to assume that autoregulatory mechanisms are active in the kidney of the unanesthetized animal, although their effects might again be suppressed or masked by extrinsic regulation. As indicated elsewhere (Chap. 8, Vol. II),

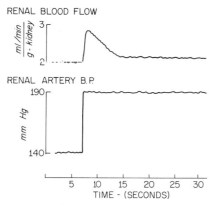
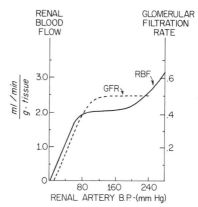

Figure 15–5 Autoregulation of renal blood flow. *Left:* Response of denervated, blood-perfused kidney to a step change in perfusion pressure. Flow rises and then returns nearly to control level despite increased pressure. (Redrawn from Semple and de Wardener, *Circulat. Res.,* 1959, 7, 643–648.) *Right:* Steady state levels of renal blood flow (RBF) and glomerular filtration rate (GFR) as a function of perfusion pressure in denervated, blood-perfused kidney. Note the relative independence of pressure exhibited by flow and GFR. (Redrawn from Shipley and Study, *Amer. J. Physiol.,* 1951, *167,* 676–688.)

there are several hypotheses which attempt to explain autoregulation. The increase in resistance as pressure rises demands that some blood vessels in the kidney actually *decrease* their caliber (so as to *increase* the resistance to blood flow) as the pressure rises. If the caliber of all vessels remained constant, the resistance would be constant and flow would be proportional to the pressure. Glomerular filtration rate (the volume of plasma filtered per minute at the glomeruli) shows autoregulation which parallels the autoregulation of renal plasma flow. Although other factors might account for this (tubular pressure changes, for instance), a constriction of the afferent arteriole seems likely to decrease both flow and filtration. Brenner *et al.*[6] have produced evidence that the autoregulation of glomerular filtration is a passive result of autoregulation of plasma flow.

Myogenic Hypothesis.[42, 45, 46]   Many years ago, Bayliss[4] presented some evidence for a "myogenic response" of smooth muscle. Following rapid stretch, a strip of smooth muscle was observed to contract, *i.e.,* stretch depolarized the muscle. The most popular theory of renal blood flow autoregulation invokes this mechanism and is known as the myogenic theory. Specifically, some renal vessel is considered to constrict as the pressure within it increases. The proponents of this theory consider the afferent arteriole to be the site of

the pressure-sensitive resistance change. Unfortunately, neither Bayliss' original description of this phenomenon nor the experimental observations in the succeeding years has demonstrated any instance in which a length of smooth muscle subjected to increased tension actually becomes *shorter* than before. Such a shortening of muscle fibers in a vessel wall is required to increase resistance. Such a mechanism might be an irreversible response, an example of positive feedback. In an extreme model of such a process, increased pressure within a vessel leads to constriction of that vessel, which further increases pressure, leading to further constriction, and so on. Indirect support for the myogenic theory comes from the observation that autoregulation is abolished by drugs which block contraction of smooth muscle (in some cases, the drugs are used in massive doses). When exposed to these drugs, the kidney shows a constant low resistance. Such drug action might not be consistent with the tissue pressure theory of autoregulation (see below). The proponents of the myogenic response have little direct evidence to present for this hypothesis, but feel that other hypotheses are inadequate. It is possible that a myogenic type of response accounts for a part of renal autoregulation, but it would seem that such a response could, at best, keep resistance constant. It is possible that there is a myogenic response in which increased pressure leads to an oscillatory contraction and relaxation of a blood vessel around a smaller diameter, but this type of oscillation has not been demonstrated in mammalian arterioles.

Metabolic Theories. In this theory, suggested by Selkurt,[35] a chemical vasodilator which is a normal product of renal metabolism is washed out of the kidney by the renal circulation. Increased blood flow would wash out more of this vasodilator, always tending to keep the flow at a fixed value. To date, no vasodilator has been found that would fit the requirements of this theory. Also, the kidney does not show decreased resistance after prolonged arterial occlusion,[39] as many organs do. Finally, whereas the metabolic theory would make flow dependent on metabolic rate, the metabolism of the kidney seems to be determined by the blood flow, which changes the rate of sodium delivery to the tubules. Modifications of the metabolic theory have been proposed in which an increase in the total solute load, the osmotic load or the sodium load delivered to the distal tubule (these conceivably might increase with increased arterial pressure) activates an intrarenal reflex, causing the afferent arteriole to constrict. Renin release from the juxtaglomerular apparatus (described earlier) has been considered by some to be involved in such a vasoconstrictor response to increased distal tubular sodium concentration.[18] These complicated modifications of the general metabolic theory are contrary to the following: (i) The concentration of sodium at the distal tubule does not rise with increased pressure; (ii) there is controversy concerning the neural or other mechanism whereby information about osmolarity or solute load can be transmitted to the juxtaglomerular apparatus;[44] (iii) the "standard" and traditional stimulus for renin production is a *fall* in renal arterial pressure,[14, 37] whereas the "renin" version of the metabolic theory of autoregulation requires that renin production increase secondary to a rise in pressure. Although they may not account for renal autoregulation, there undoubtedly are important links between the excretory and hormonal functions of the kidney and its vasomotor control.

Tissue Pressure Hypothesis. The kidney, perfused at higher and higher pressures, grows turgid and increases in weight and resistance. Further, if the kidney is perfused with fluid of low osmotic pressure, the resistance will increase, at times sufficiently to prevent flow.[18] These observations have led to the theory that increased pressure in the glomerulus or perhaps in the peritubular capillaries causes an extravasation (by filtration) of fluid into the extravascular spaces, and the resulting increased interstitial pressure compresses some venous vessels and thus increases the resistance to flow.[20, 34] This mechanism of passive regulation of renal resistance is similar to a mechanism which regulates flow through the lungs,[31] but the tissue pressure theory has not been popular among renal vascular physiologists. Evidence supporting this theory was provided by Caley and Scher,[7] who found that as the arterial pressure of blood perfusing the isolated kidney was increased, the renal *vascular* volume actually decreased, although the total kidney volume increased markedly. Other evidence was provided by Hinshaw *et al.*,[20] who found that tissue pressure measured by a needle inserted into the renal parenchyma increased with arterial pressure in a fashion which would tend to increase the renal venous resistance. Pressure measured by a catheter inserted retrograde into the renal vein and advanced into the arcuate veins also rose in similar fashion with increased arterial pressure. Increased renal venous pressure decreases flow far more than would be predicted by the decreased arteriovenous pressure drop (turgor and weight also increase). The tissue pressure theory is subject to the following criticisms: (i) increasing ureteral pressure does not increase vascular resistance[36] (strangely, it appears to cause dilatation of the afferent arterioles), and (ii) pressures measured in the glomerulus, efferent arteriole and renal tubules do not change with arterial pressure as the tissue pressure theory would seem to require.[17]

Elucidation of the mechanism of renal autoregulation requires, as a beginning, measurements of pressure and flow at several sites in the renal microcirculation. Brenner has furnished some of these measurements.[5, 6, 33] Even if the mechanism is elucidated, it remains to be shown if autoregulation is a major regulatory process or if it is overridden in the intact animal by the extrinsic controls discussed below.

EXTRINSIC REGULATION

Extrinsic regulation of the renal blood flow follows the conventional pattern for visceral organs. The renal arterioles are extensively innervated by the sympathetic nervous system through the renal branches of the splanchnic nerves, which originate in spinal segments T4 to L4 (but mainly from T10 to T12). There is anatomical evidence for sympathetic innervation of the large arteries and veins, afferent arterioles, and descending vasa recta. In contrast, efferent arterioles show sparse or negligible sympathetic innervation, especially in the outer cortex.[27, 30] Action potentials have been recorded in the splanchnic nerves in many conditions,[15, 22] and stimulation of these nerves increases renal resistance markedly.[9] The renal resistance is also increased by increased levels of circulating catecholamines or angiotensin II. Drugs which block alpha adrenergic receptors (alpha antagonists) (see Chap. 8, Vol. II) inhibit the renal vascular response to renal nerve stimulation or to circulating catecholamines. There is some evidence for cholinergic innervation of the kidney,

but its physiological significance is not clear.[13, 16]

The sympathetic control of the renal vasculature can be reflexly activated to increase resistance when pressure falls at the carotid sinus region, as it might in hemorrhage. Resistance also increases when arterial oxygen falls. This latter response originates in the arterial chemoreceptors.[23] In both cases, renal vasoconstriction can (teleologically) be considered to divert blood flow to other, more essential circulations (e.g., coronary or cerebral). Renal vasoconstriction can be induced in the cat, dog and primate by stimulation of the hypothalamus,[11] a structure which is important in the integration of cardiovascular responses to behavioral situations (Fig. 15–6). Renal vasoconstriction occurs in monkeys in response to stressful environmental stimuli.[12] Renal resistance has also been reported to rise in humans during exercise or emotional stress.[10, 38, 47] In dogs, severe exercise does not lead to a marked rise in renal vascular resistance or to much change in renal blood flow, which seems to show a major species difference between dog and man.[43]

RENAL CLEARANCE AND RENAL BLOOD FLOW

Renal excretory function and renal blood flow can be measured by studying the rela-

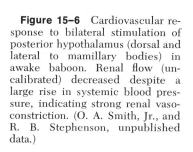

Figure 15–6 Cardiovascular response to bilateral stimulation of posterior hypothalamus (dorsal and lateral to mamillary bodies) in awake baboon. Renal flow (uncalibrated) decreased despite a large rise in systemic blood pressure, indicating strong renal vasoconstriction. (O. A. Smith, Jr., and R. B. Stephenson, unpublished data.)

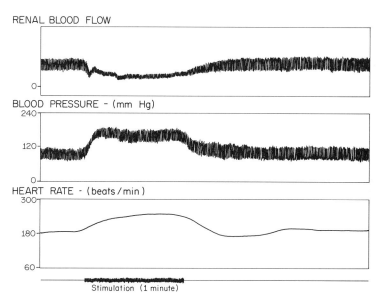

tionship between plasma concentration and renal excretion of various substances. These studies involve the concept and calculation of "renal clearance." The clearance of a substance is the volume of arterial plasma which would furnish the quantity of the substance excreted in the urine in 1 min (if all the substance were removed from that plasma). Mathematically, clearance is:

$$\text{Clearance} = \frac{U \cdot V}{P_a} \left(\frac{mg/ml \cdot ml/min}{mg/ml} = \frac{ml}{min} \right)$$

where U is the concentration of the substance in urine (mg/ml) and V is the rate of urine formation (ml/min). The numerator, U·V, is the rate at which the substance is excreted in the urine (mg/min). P_a is the concentration of the substance in the arterial plasma (mg/ml). As indicated, clearance has the units of ml/min. When arterial plasma samples are not available, P_a is often approximated by assaying a sample of venous blood. Accurate determination of excretion (U·V) requires collection of urine via a catheter inserted into the bladder. Renal clearance is discussed in detail elsewhere (Chap. 25, Vol. II) and briefly below.

The rate at which fluid is filtered across the glomerular capillaries can be measured as the clearance of inulin. Inulin is a polysaccharide which is filtered across the glomerular capillaries and into Bowman's capsule, but is not secreted or reabsorbed as it passes through the tubule. That is, when inulin is given intravenously, it appears in the urine solely through glomerular filtration, and all the inulin filtered shows up in the urine. The clearance of inulin is thus a measure of the amount of plasma which is filtered per minute at the glomeruli. This value is called the glomerular filtration rate (GFR).

Whereas the clearance of inulin measures glomerular filtration, the clearance of certain other compounds can be used to measure or approximate renal plasma flow. These substances are completely or virtually completely removed from the plasma in one passage through the kidney. One such substance is para-aminohippuric acid (PAH). When PAH is administered intravenously in a small, continuous dose, its renal clearance is a close approximation to renal plasma flow. That this is so can be seen by

recalling the Fick technique for measuring blood flow (Chap. 3, Vol. II). In the Fick calculation:

$$\text{Flow} = \frac{dQ/dt}{P_a - P_v}$$

where dQ/dt is the excretion or extraction per minute of an indicator substance, and P_a and P_v are the arterial and venous concentrations of the substance. If we consider PAH excretion by the kidney, the rate at which PAH is extracted from the plasma equals the rate of PAH addition to the urine. That is, dQ/dt in the Fick equation is equal to U·V. Since PAH is virtually completely removed from the plasma (90 per cent) in one passage through the kidney, P_v may be assumed to be zero. In this case, the Fick equation for PAH excretion becomes identical to the equation for PAH clearance, and PAH clearance is a measure of renal plasma flow. (To correct for the fact that the extraction of PAH from plasma is only 90 per cent complete in one circuit through the kidney, the total renal plasma flow is sometimes estimated to be about 10 per cent higher than the measured PAH clearance.)

The ratio of glomerular filtration rate (inulin clearance) to renal plasma flow (PAH clearance) indicates what fraction of the plasma entering the kidney is filtered at the glomeruli to form the tubular ultrafiltrate.

$$\text{Filtration fraction} = \frac{\text{Glomerular filtration rate}}{\text{Renal plasma flow}}$$

Renal plasma flow can be used to calculate renal blood flow by taking into account the hematocrit (Hct) (per cent of red cells):

$$\text{Renal blood flow} = \frac{\text{Renal plasma flow}}{(1 - \text{Hct})}$$

Nominal values for these quantities are indicated in Table 15–1.

INTRARENAL DISTRIBUTION OF BLOOD FLOW

Intrarenal blood flow is not evenly distributed to all parts of the kidney. Almost all arterial blood entering the kidney first passes through glomerular capillaries in the cortex. This is shown by studies on dogs and rats in which radioactive plastic spheres

Table 15–1 *Normal Renal Hemodynamic Data*

Renal plasma flow (PAH clearance)	660	ml/min
Glomerular filtration rate (GFR) (inulin clearance)	130	ml/min
Filtration fraction	0.20	
Hematocrit (Hct)	0.45	
Renal blood flow	1200	ml/min

Nominal values normalized for 1.73 m² body surface area for adult males.

(15 to 19 μ in diameter) are injected into the renal artery. Spheres of this size lodge in the first capillaries they encounter, and in the kidney virtually all the injected spheres lodge in cortical glomeruli. About 1 per cent of the injected spheres lodge in medullary capillaries, indicating that only this small fraction of incoming arterial blood reaches the medulla without first passing through glomerular capillaries.[28] Microsphere studies also show that the outer cortical glomeruli receive a larger fraction of the incoming arterial blood than the inner cortical (juxtamedullary) glomeruli. Two factors contribute to this uneven cortical blood flow: (i) there are more glomeruli per gram of tissue in the outer cortex than in the juxtamedullary cortex, and (ii) the flow rate in each glomerulus is somewhat higher in the outer cortex.[21]

As mentioned earlier, blood leaving the renal glomeruli travels via efferent arterioles or vasa recta to the peritubular capillaries of the cortex or medulla. The complexity of the renal vasculature makes it very difficult to measure the distribution of peritubular capillary flow. Current ideas about this distribution are inferred from a number of procedures, including measuring the intrarenal circulation of red cells, measuring tissue flow with thermal flowmeters in various parts of the kidney and monitoring the washout of diffusible dyes or radioactive gases (^{85}Kr or ^{133}Xe). As an example of the washout procedure, one can measure the radioactivity of the kidney following the rapid injection into the renal artery of saline containing dissolved ^{85}Kr. The radioactivity rapidly reaches a peak as the ^{85}Kr enters the kidney and diffuses into the parenchyma; then the radioactivity falls more slowly as the ^{85}Kr is washed out of the kidney, to be eliminated by the lungs. A typical washout curve of radioactivity (Fig. 15–7) can be resolved into several exponential curves. These are considered to indicate that several flow channels with different flow rates are present. The anatomical identity of these flow channels is determined by freezing kidneys at various stages of the ^{85}Kr washout and determining the distribution of the remaining radioactivity by autoradiography. Unfortunately, the ^{85}Kr washout from a particular region of the kidney is not solely regulated by the peritubular capillary flow through that region. Other factors which might affect the washout rate include exchange across glomerular capillaries, transport of ^{85}Kr into or out of the region by the tubular fluid, and "trapping" of ^{85}Kr in the region by diffusion of ^{85}Kr from blood leaving the region to blood entering it. Thus, the blood flow rates computed from washout studies are identified

Figure 15–7 A typical washout curve of radioactivity following injection of ^{85}Kr into the renal artery can be resolved into four constituent curves. The nonlinear washout curve (bold line) is approximated by the sum of four exponential curves (which appear as straight lines in this log plot). The slopes of the four exponential constituents are used to calculate blood flow through four "flow channels" within the kidney. (Redrawn from Thorburn *et al.*, *Circulat. Res.*, 1963, *13*, 290–307.)

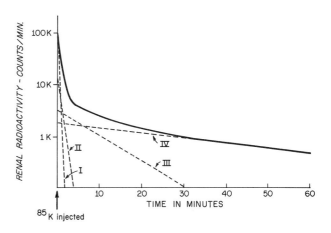

as "indicated flows" rather than true peritubular capillary flows.*

For the resting, unanesthetized dog, the "^{85}Kr-indicated" flow to the outer cortex is 4.7 ml per min per gram of tissue. Indicated flow in the juxtamedullary cortex and outer medulla is lower (1.3 ml per min per gram), and indicated flow to the renal papilla is much less (0.2 ml per min per gram).[41] These flow rates (taken together with the relative volumes of cortex, medulla and papilla) imply that 94 per cent of the blood entering the kidney traverses the outer cortical glomeruli and peritubular capillaries. About 5 per cent of the entering blood traverses the juxtamedullary cortex and outer medulla, and about 1 per cent of the blood passes through the inner medulla (papilla). All these routes then lead into the venous system. Figure 15–8 diagrams these relationships. The ^{85}Kr-indicated flows probably underestimate the true peritubular capillary flows in the outer and inner medulla.[41]

Variations from the normal distribution of intrarenal blood flow are considered to have important consequences for renal function. For example, increased medullary and papillary flows are accompanied by increased water excretion. This is thought to

*Some authors use the term "indicated nutrient flow" or simply "nutrient flow."[41]

be because high medullary flow rates decrease the interstitial concentration gradient normally associated with the production of concentrated urine (Chap. 25, Vol. II). For a second example, severe hemorrhage, which decreases total renal blood flow, also appears to cause a redistribution of the remaining blood flow. Juxtamedullary glomeruli receive a larger-than-normal fraction of the incoming flow, at the expense of outer cortical glomeruli. Outer cortical ^{85}Kr indicated flow drops sharply.[8, 32] It has been suggested that the redistribution caused by hemorrhage is functionally significant because the salt-and-water-retaining capability of the long-loop juxtamedullary nephrons is presumably greater than that of the short-loop outer cortical nephrons.[3] Finally, redistribution of renal blood flow may be important in autoregulation of blood flow when perfusion pressure changes. If renal perfusion pressure is decreased in a dog by constricting the renal artery, a redistribution of blood flow from outer cortex to juxtamedullary cortex occurs, but total renal flow is unchanged.[1, 29] The mechanisms linking changes in perfusion pressure with changes in distribution of blood flow have not been elucidated.

Present ideas about the functional significance of changes in flow distribution are speculative, and the various methods for studying flow distribution often yield

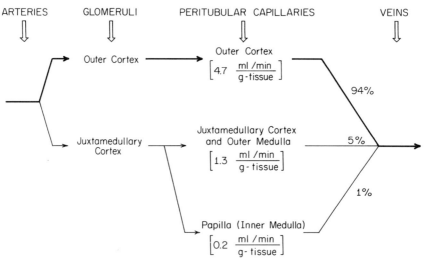

Figure 15–8 Normal intrarenal distribution of blood flow. The predominant blood flow pathway traverses the outer cortical glomeruli. The small fraction of arterial blood not going to glomeruli is ignored. Numerical values are computed from ^{85}Kr-indicated flows. (Data of Thorburn *et al.*, *Circulat. Res.*, 1963, *13*, 290–307.)

conflicting results. Hopefully, a comprehensive picture of renal function will soon emerge which unites the regulation of renal blood flow with the production of urine and also explains the mechanism of autoregulation.

REFERENCES

1. Abe, Y. Intrarenal blood flow distribution and autoregulation of renal blood flow and glomerular filtration rate. *Jap. Circulat. J.*, 1971, *35*, 1163–1173.
2. Ayers, C. R., Harris, R. H., Jr., and Lefer, L. G. Control of renin release in experimental hypertension. *Circulat. Res.*, 1969, *24*, Suppl. 1, 103–112.
3. Barger, A. C., and Herd, J. A. The renal circulation. *New Engl. J. Med.*, 1971, *284*, 482–490.
4. Bayliss, W. M. On the local reactions of the arterial wall to changes of internal pressure. *J. Physiol. (Lond.)*, 1902, *28*, 220–231.
5. Brenner, B. M., Troy, J. L., and Daugharty, T. M. Pressures in cortical structures of the rat kidney. *Amer. J. Physiol.*, 1972, *222*, 246–251.
6. Brenner, B. M., Troy, J. L., Daugharty, T. M., Deen, W. M., and Robertson, C. R. Dynamics of glomerular ultrafiltration in the rat. II. Plasma-flow dependence of GFR. *Amer. J. Physiol.*, 1972, *223*, 1184–1190.
7. Caley, M. H., and Scher, A. M. Effect of arterial pressure on renal vascular and extravascular volumes. *Physiologist*, 1969, *12*, 192.
8. Carriere, S., Thorburn, G. D., O'Morchoe, C. C. C., and Barger, A. C. Intrarenal distribution of blood flow in dogs during hemorrhagic hypotension. *Circulat. Res.*, 1966, *19*, 167–179.
9. Celander, O. The range of control exercised by the 'sympathico-adrenal system.' *Acta physiol. scand.*, 1954, *32* (Suppl. 116), 1–132.
10. Chapman, C. B., Henschel, A., Minckler, J., Forsgren, A., and Keys, A. The effect of exercise on renal plasma flow in normal male subjects. *J. clin. Invest.*, 1948, *27*, 639–644.
11. Feigl, E. O. Vasoconstriction resulting from diencephalic stimulation. *Acta physiol. scand.*, 1964, *60*, 372–380.
12. Forsyth, R. P., and Harris, R. E. Circulatory changes during stressful stimuli in Rhesus monkeys. *Circulat. Res.*, 1970, *27* (Suppl. 1), 13–20.
13. Fourman, J., and Moffat, D. B. *The blood vessels of the kidney.* Oxford & Edinburgh, Blackwell Scientific Publications, 1971.
14. Goldblatt, H., Lynch, J., Hanzal, R. F., and Summerville, W. W. Studies on experimental hypertension. I. The production of persistent elevation of systolic blood pressure by means of renal ischemia. *J. exp. Med.*, 1934, *59*, 347–379.
15. Gootman, P. M., and Cohen, M. I. Efferent splanchnic activity and systemic arterial pressure. *Amer. J. Physiol.*, 1970, *219*, 897–903.
16. Gosling, J. A. Observations on the distribution of intrarenal nervous tissue. *Anat. Rec.*, 1969, *163*, 81–88.
17. Gottschalk, C. W., and Mylle, M. Micropuncture study of pressures in proximal tubules and peritubular capillaries of the rat kidney and their relation to ureteral and renal venous pressures. *Amer. J. Physiol.*, 1956, *185*, 430–439.
18. Guyton, A. C., Langston, J. B., and Navar, G. Theory for renal autoregulation by feedback at the juxtaglomerular apparatus. *Circulat. Res.*, 1964, *15* (Suppl. 1), 187–196.
19. Hinshaw, L. B. Mechanism of renal autoregulation: Role of tissue pressure and description of a multifactor hypothesis. *Circulat. Res.*, 1964, *15* (Suppl. 1), 120–129.
20. Hinshaw, L. B., Day, S. B., and Carlson, C. H. Tissue pressure as a causal factor in the autoregulation of blood flow in the isolated perfused kidney. *Amer. J. Physiol.*, 1959, *197*, 309–312.
21. Katz, M. A., Blantz, R. C., Rector, F. C., Jr., and Seldin, D. W. Measurement of intrarenal blood flow. I. Analysis of the microsphere method. *Amer. J. Physiol.*, 1971, *220*, 1903–1913.
22. Kezdi, P., and Geller, E. Baroreceptor control of postganglionic sympathetic nerve discharge. *Amer. J. Physiol.*, 1968, *214*, 427–435.
23. Korner, P. I. Effects of low oxygen and of carbon monoxide on the renal circulation in unanesthetized rabbits. *Circulat. Res.*, 1963, *12*, 361–374.
24. Landis, E. M., and Pappenheimer, J. R. Exchange of substances through the capillary walls. *Handb. Physiol.*, 1963, sec. 2, vol. II, 961–1034.
25. Lassen, N. A., Munck, O., and Thaysen, J. H. Oxygen consumption and sodium reabsorption in the kidney. *Acta physiol. scand.*, 1961, *51*, 371–384.
26. Macdonald, G. J., Louis, W. J., Renzini, V., Boyd, G. W., and Peart, W. S. Renal-clip hypertension in rabbits immunized against angiotensin II. *Circulat. Res.*, 1970, *27*, 197–211.
27. McKenna, O. C., and Angelakos, E. T. Adrenergic innervation of the canine kidney. *Circulat. Res.*, 1968, *22*, 345–354.
28. McNay, J. L., and Abe, Y. Pressure-dependent heterogeneity of renal cortical blood flow in dogs. *Circulat. Res.*, 1970, *27*, 571–587.
29. McNay, J. L., and Abe, Y. Redistribution of cortical blood flow during renal vasodilatation in dogs. *Circulat. Res.*, 1970, *27*, 1023–1032.
30. Norvell, J. E. A histochemical study of the adrenergic and cholinergic innervation of the mammalian kidney. *Anat. Rec.*, 1969, *163*, 236.
31. Permutt, S. Effects of interstitial pressure of the lung on pulmonary circulation. *Med. Thorac.*, 1965, *22*, 118–131.
32. Rector, J. B., Stein, J. H., Bay, W. H., Osgood, R. W., and Ferris, T. F. Effect of hemorrhage and vasopressor agents on distribution of renal blood flow. *Amer. J. Physiol.*, 1972, *222*, 1125–1131.
33. Robertson, C. R., Deen, W. M., Troy, J. L., and Brenner, B. M. Dynamics of glomerular

ultrafiltration in the rat. III. Hemodynamics and autoregulation. *Amer. J. Physiol.*, 1972, *223*, 1191–1200.

34. Scher, A. M. Mechanism of autoregulation of renal blood flow. *Nature (Lond.)*, 1959, *184* (Suppl. 17), 1322.

35. Selkurt, E. E. Der Nierenkreislauf. *Klin. Wochschr.*, 1955, *33* (15), 359–362.

36. Selkurt, E. E., Brandfonbrener, M., and Geller, H. M. Effects of ureteral pressure increase on renal hemodynamics and the handling of electrolytes and water. *Amer. J. Physiol.*, 1952, *170*, 61–71.

37. Skinner, S. L., McCubbin, J. W., and Page, I. H. Renal baroreceptor control of renin secretion. *Science*, 1963, *141*, 814–816.

38. Smith, H. W. Physiology of the renal circulation. *Harvey Lect.*, 1940, *35*, 166–222.

39. Smith, H. W. *The kidney: Structure and function in health and disease.* New York, Oxford University Press, 1951.

40. Swann, H. G. Some aspects of renal blood flow and tissue pressure. *Circulat. Res.*, 1964, *15* (Suppl. 1), 115–119.

41. Thorburn, G. D., Kopald, H. H., Herd, J. A., Hollenberg, M., O'Morchoe, C. C. C., and Barger, A. C. Intrarenal distribution of nutrient blood flow determined with krypton[85] in the unanesthetized dog. *Circulat. Res.*, 1963, *13*, 290–307.

42. Thurau, K., and Kramer, K. Weitere Untersuchungen zur myogenen Natur der Autoregulation des Nierenkreislaufes. *Pflügers Arch. ges. Physiol.*, 1959, *269*, 77–93.

43. Van Citters, R. L., and Franklin, D. L. Cardiovascular performance of Alaska sled dogs during exercise. *Circulat. Res.*, 1969, *24*, 33–42.

44. Vander, A. J. Control of renin release. *Physiol. Rev.*, 1967, *47*, 359–382.

45. Waugh, W. H., and Shanks, R. G. Cause of genuine autoregulation of the renal circulation. *Circulat. Res.*, 1960, *8*, 871–888.

46. Winton, F. R. Present concepts of the renal circulation. *A. M. A. Arch. intern. Med.*, 1959, *103*, 495–502.

47. Wolf, S., Pfeiffer, J. B., Ripley, H. S., Winter, O. S., and Wolff, H. G. Hypertension as a reaction pattern to stress; summary of experimental data on variations in blood pressure and renal blood flow. *Ann. intern. Med.*, 1948, *29*, 1056–1076.

CHAPTER 16 THE CORONARY CIRCULATION

by ERIC O. FEIGL

ANATOMY

ARTERIES. The human heart is supplied by two branches of the aorta—the right and left coronary arteries, which course over the epicardial surface of the heart before penetrating the myocardium (Fig. 16–1). The arteries continue branching within the myo-cardium, the finer branches becoming arterioles and capillaries which drain into venules and small veins. The small veins drain outward from the myocardium and form epicardial veins which empty into the right atrium via the coronary sinus. In addition, venous channels drain directly into the cardiac chambers. An outline of human

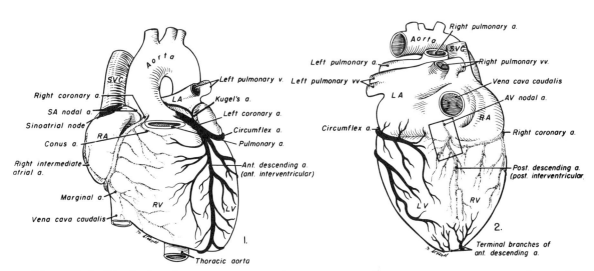

Figure 16–1 Distribution of coronary arteries as they most commonly occur in man. *1,* Anterior surface of the heart. *2,* Posterior surface of the heart. (From Truex, in *Coronary heart disease,* Likoff and Moyer, eds. New York, Grune & Stratton, 1963.)

cardiac vascular anatomy is given below; however, the anatomy is variable and other sources should be consulted for details.[3,18,30]

Left Coronary Artery. The left coronary artery originates in the sinus of Valsalva behind the left cusp of the aortic valve. The ostium lies approximately at the level of the free margin of the aortic valve cusp. After coursing between the pulmonary artery and the left atrial appendage, the left coronary artery divides into the left anterior descending branch and the left circumflex branch. The *left anterior descending branch* forms a fairly direct extension of the main left coronary artery as it descends in the anterior interventricular sulcus to the apex of the heart. Several *anterior septal branches* perforate the interventricular septum from the posterior aspect of the anterior descending branch.

The *left circumflex branch* of the left coronary artery curves away from the main left coronary artery in the groove between left atrium and ventricle. Small branches are given off to the left atrium and larger ones to the left ventricle. The *left marginal branch* arises from the left circumflex and descends toward the apex, left of the posterior interventricular sulcus. *Left diagonal branches* arise from the anterior descending and/or the left circumflex. In some cases the main left coronary artery may form three branches: the anterior descending, left diagonal and circumflex branches. Diagonal branches course across and down over the left ventricle.

Right Coronary Artery. The right coronary artery originates in the sinus of Valsalva behind the right aortic valve cusp and passes behind the pulmonary artery and into the right atrio-ventricular groove, giving off branches to the right atrium and ventricle. The *conus branch*, the first major branch of the right coronary artery, supplies the conus of the right ventricle. The branching occurs immediately at the ostium in about one-third of human hearts. In approximately one-sixth of human hearts, the conus artery is a vessel with its own separate ostium in the right sinus of Valsalva.[3] A *right marginal branch* descends over the right ventricle toward the apex. The terminal portion of the right coronary artery forms the *posterior descending branch* which follows the posterior sulcus between right and left ventricles. *Posterior septal branches* of the posterior descending branch penetrate the interventricular septum.

CAPILLARIES. Normal human hearts have a capillary density of 3300 per mm²,[68] whereas skeletal muscle contains only 400 capillaries per mm². Both cardiac and skeletal muscles have approximately one capillary for each muscle fiber, the difference in capillary densities being due to the cross-sectional diameter differences between cardiac (17 to 20 μm) and skeletal (ca 50 μm) fibers. Taking the capillary surface area and relative capillary permeability into account, myocardial capillaries are almost 15 times more effective than skeletal muscle capillaries in exchanging small molecules.[56]

When cardiac muscle hypertrophies, the ratio of capillaries to fibers remains unchanged at 1:1. Hypertrophy increases both the distance over which materials must diffuse to reach all parts of the myocardium and the mass of tissue which must be nourished by cardiac capillaries.

Measurements in rat hearts indicate that approximately 50 per cent of cardiac capillaries are open under normal conditions and that more capillaries open (recruitment) when arterial oxygen tension is decreased.[41] This is the microcirculatory aspect of metabolic control of coronary blood flow discussed below.

VEINS. The left ventricle is drained by a venous system which begins as the anterior interventricular vein which courses parallel to the left anterior descending artery. At approximately the level of the main left coronary artery bifurcation, the anterior interventricular vein turns to run parallel to the left circumflex artery in the atrio-ventricular groove. At this point the vein, enlarged by several tributaries, is called the great cardiac vein. Somewhat further, a fold of endothelium forms the incompetent valve of Vieussens which marks the beginning of the coronary sinus. Embryologically, the valve of Vieussens marks the portion of the left superior vena cava which forms the coronary sinus. After receiving additional tributaries, the coronary sinus empties via its ostium into the right atrium.

Much of the right ventricle is drained by two to four epicardial *anterior cardiac veins* which run toward the base of the heart and drain directly into the right atrium.

VESSELS CONNECTING DIRECTLY WITH

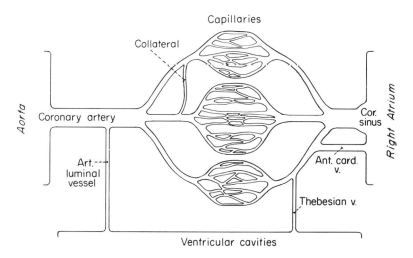

Figure 16-2 The blood vessels of the heart are shown schematically. In addition to the usual arrangement of arteries-capillaries-veins, there are arterial and venous channels which drain directly into the ventricular chambers. There are collateral channels on both the arterial and venous sides.

THE CARDIAC CHAMBERS. In addition to the more usual arteries and veins supplying the myocardium, there are small vessels which drain directly into the heart chambers without passing through an epicardial collecting vein.

Thebesian Veins. Thebesius demonstrated in 1708 that material injected retrograde into the coronary sinus entered the atria and ventricles via multiple small ostia. The relative magnitude of Thebesian flow depends on the pressure gradients between the venous system and the cardiac chamber. Thebesian drainage in the left ventricle is slight. The Thebesian drainage in the right ventricle is greater than in the left ventricle because of a more favorable pressure gradient. Eighty per cent of canine septal artery flow has been reported to drain directly into the right ventricle.[47]

Arterioluminal Vessels. Wearn[68] has reported, and others have confirmed, that arterial injection of plastic material which does not pass through capillaries appears in the ventricles through arterial shunts. These arterioluminal shunts are from 80 to 200 μm in diameter. Usually less than six arterioluminal vessels are found in the left ventricle; a similar number are found in 50 per cent of human right ventricles.[3] The function of arterioluminal vessels is unknown.

COLLATERAL VESSELS. The normal myocardium contains numerous collateral arteries between the major branches of a coronary artery (homocoronary) and anastomoses between right and left coronary arteries (intercoronary). The internal diameter of collaterals is normally about 40 μm. In man, arterial collaterals are most frequent near the endocardium, in dogs, near the epicardium. In no sense are coronary vessels "end arteries" without collaterals. With coronary artery stenosis, the collateral vessels enlarge greatly and often show a characteristic tortuous shape.[64]

The cardiac venous system has abundant collateral channels between cardiac veins, the coronary sinus, anterior cardiac veins, and Thebesian veins.

CARDIAC LYMPHATICS. The myocardium has a rich lymphatic system, consisting of lymphatic capillaries and vessels with valves. Subepicardial and subendocardial lymphatic plexuses communicate with the myocardial plexus in the ventricular wall.[31, 53] The mitral and tricuspid valves are also supplied with lymphatics.[31, 44] Cardiac lymph flow is about 3 ml per hour in the dog.[12, 43] The total protein in cardiac lymph is approximately three-fourths of that in plasma, whereas sodium, potassium and chloride concentrations are essentially the same as in plasma.[43]

NORMAL CORONARY BLOOD FLOW

Normal human coronary blood flow during resting conditions ranges from 60 to 80 ml/min per 100 g of tissue.[6, 7, 42, 54] With a normal heart weight of 300 g and a normal resting cardiac output of 5.5 liters per min, coronary flow preempts approximately 4 per cent of the resting cardiac output.

Resting coronary blood flow in unanesthetized dogs ranges from 40 to 60 ml per min per 100 g of tissue.[23, 28] Blood flow per unit mass of tissue to the right ventricle is approximately two-thirds of the flow to the left ventricle.[11, 54] A probable reason is that the right ventricle develops less tension and consumes less oxygen. Similarly, atrial blood flow per unit weight of atria is about one-half that to the left ventricle.[11]

CONTROL OF CORONARY BLOOD FLOW

Myocardial blood flow is determined by four major factors: (i) coronary perfusion pressure, (ii) myocardial compression, (iii) myocardial metabolism and (iv) neural control. These factors are easily understood one at a time, but their complex interactions sometimes provide surprises. The reason is that the heart is responsible for providing its own perfusion pressure. Thus an alteration in the myocardium can trigger a complex chain of events which in turn affects coronary blood flow. For example, a drop in aortic pressure caused by sudden hemorrhage decreases coronary perfusion pressure. This decrease may compromise the heart and alter myocardial compression and metabolism; these changes, in turn, alter coronary blood flow and the perfusion pressure generated by the heart. On all of this are superimposed reflex changes in the heart and coronary vessels triggered from the carotid sinus and other reflexogenic areas.

Coronary Perfusion Pressure. The pressure gradient between the root of the aorta and the right atrium is the effective driving pressure for coronary blood flow. As in other vascular beds, the simple hemodynamic effects of perfusion pressure also control coronary blood flow; the greater the perfusion pressure, the greater the flow of viscous blood. Aortic perfusion pressure is also an important determinant of myocardial metabolism which alters coronary blood flow by a metabolic rather than hemodynamic mechanism as discussed below. Provided other variables are constant, the coronary blood flow changes little when perfusion pressure is varied between 60 and 180 mm Hg. Outside this range, coronary blood flow changes more steeply with changes in perfusion pressure. The flat pressure-flow relation has been termed "autoregulation." See Chapter 8, Volume II, for a discussion of autoregulation.

Myocardial Systolic Compression. With each heart beat the myocardium squeezes its own vasculature. In the left ventricle systolic myocardial compression nearly stops arterial inflow. The right ventricular systolic compression is less as reflected by the lower right ventricular pressure. The extravascular compressive force can be estimated by measuring intramyocardial pressure by either of two methods. In the first, a small liquid-filled space is created in the myocardium and connected to a pressure transducer. The space may be created by inserting a vein segment or a small balloon, or injecting a pocket of saline into the myocardial wall. Since creating the space distorts the myocardium which produces a pressure artifact, it is desirable to keep the artificial space as small as possible. The second method is to measure flow through a needle tract or small vein segment threaded through the heart wall and perfused with saline from a pressure bottle. The perfusion pressure where flow is just stopped during systole or diastole is taken as the intramyocardial pressure at the corresponding times of the cardiac cycle. Both methods distort the myocardium and are subject to error. Both give similar results.

Intramyocardial pressure is a few mm Hg during diastole and increases rapidly with the onset of cardiac contraction (Fig. 16–3). It is controversial whether left ventricular intramyocardial pressure exceeds ventricular intracavitary blood pressure during systole. Most observers agree that intramyocardial pressure approaches or exceeds intraventricular cavity pressure during systole, particularly in the inner layers of myocardium. A flow record from an artery supplying the left ventricle demonstrates the throttling effect of myocardial systolic compression on coronary blood flow (Figs. 16–8, 16–9 and 16–10).

Theoretical analysis predicts, and measurement confirms, that systolic intramyocardial pressure is greater in the inner layers of the myocardium adjacent to the endocardium than in the outer layers adjacent to the epicardium.[34]

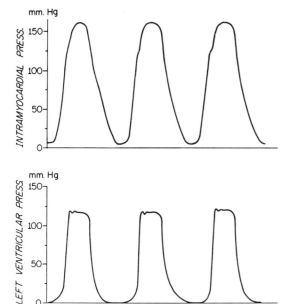

Figure 16–3 Intramyocardial pressure and intraventricular pressure records from the left ventricle of a dog. Pressure was measured with a needle inserted into a small liquid-filled space within the myocardium. (After Kreuzer *et al.*, *Pflügers Arch. Ges. Physiol.*, 1963, *278*, 181–198.)

is much lower in an area of myocardial infarction than in normal myocardium, despite the fact that the infarcted tissue is bulging and is subjected to the same transmural pressure.[2, 37]

Subendocardial myocardial infarctions are more frequent than subepicardial infarctions. Also an infarction which extends through the wall is characteristically larger near the ventricular lumen than near the epicardium. The larger compressive forces in the subendocardium probably contribute to these pathological findings. Another contributing factor is that coronary arteries perforate from the epicardium so that the vessels of the deeper layers come from more distal branches.

REGIONAL MYOCARDIAL BLOOD FLOW. Whether the relatively greater systolic compression of the inner layers of the myocardium significantly restricts subendocardial blood flow is controversial. Investigations employing extravascular methods to estimate blood flow (clearance of needle deposits of diffusible isotopes or determinations of tissue oxygen tension with inter-

The simplest geometrical model of the heart is a thick-walled cylinder. The compressive stress (σ_r) in the radial direction for a thick-walled cylinder subjected to an internal pressure and no external pressure is given by the equation:[65]

$$\sigma_r = \frac{a^2 P_i}{b^2 - a^2}\left(1 - \frac{b^2}{r^2}\right)$$

where a is the internal radius, b the external radius, P_i the internal pressure and σ_r the compressive stress at radius r between a and b (see Fig. 16–4). The equation predicts a greater compressive stress near the inner wall than near the outer wall. This equation or more complicated ones describing spheres and ellipsoids subjected to pressure gradients are derived for homogeneous, isotropic noncontractile materials, such as steel. Although the forces caused by pressure and radius of curvature must be involved in the heart wall, it is important to realize that the forces in the heart wall are due to myocardial contraction, not to a transmural pressure gradient being imposed on the ventricle. This is emphasized by the finding that intramyocardial pressure

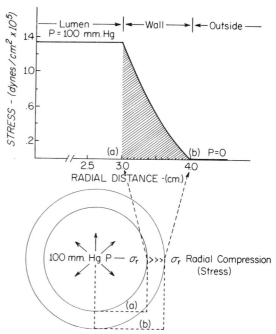

Figure 16–4 The calculated compressive stress in the radial direction of a cylinder with an internal radius of 3 cm and an external radius of 4 cm. The pressure inside the cylinder is 100 mm Hg and zero outside. Note that the compressive stress is greatest at the inner layers of the wall.

stitial oxygen electrodes) indicate relative underperfusion of the endocardium. In contrast, studies utilizing intravascular techniques (tissue uptake of intravascularly administered diffusible isotopes or the trapping of injected radioactive microspheres) indicate equal perfusion of the inner and outer layers of the left ventricle.[45]

A more pertinent question is how well coronary blood flow supports cardiac metabolism. Normal myocardium has a net uptake of lactate and pyruvate (arterial concentration exceeds coronary sinus concentration). Underperfused hypoxic myocardium manifests net lactate production (coronary sinus concentration exceeds arterial concentration) but a continued pyruvate uptake. This is a consequence of a shift from aerobic to anaerobic metabolism in the myocardium. Thus the ratio of lactate/pyruvate concentration in coronary sinus blood or myocardial tissue may be used as a biochemical indicator of the state of myocardial oxygenation. Under normal conditions and during catecholamine stimulation of the heart tissue, lactate/pyruvate ratios are the same in inner and outer layers of the left venticle.[25, 26] However, subendocardial lactate/pyruvate ratios are elevated with coronary constriction or under the combined influence of aortic stenosis and catecholamine stimulation.[24, 25] This indicates that subendocardial blood flow is adequate during normal conditions but relatively vulnerable in pathological circumstances.

Metabolic Control. The coronary circulation is particularly sensitive to myocardial oxygen tension and less so to carbon dioxide tension and lactic acid levels. Myocardial hypoxia results in prompt sustained coronary vasodilation. Coronary vasodilation during hypoxia may be a direct effect of interstitial oxygen tension on vascular smooth muscle or may involve an intermediate link such as adenosine, as discussed below.

Under basal conditions, coronary venous oxygen tension is the lowest found in the body, approximately 20 mm Hg. However, the concept of a basal resting state is ambiguous for the heart, which never "rests." Intracellular myocardial oxygen tension, as determined with micro pO_2 electrodes, is less than 10 mm Hg.[69] Such a low intracellular value preserves an oxygen diffusion gradient from capillary to tissue.

The mammalian heart has a very limited capacity for anaerobic metabolism. For practical purposes, the metabolism of the heart may be equated with its oxygen consumption.[51] Because the oxygen content of venous blood leaving the heart is low during resting conditions, the arterial-venous oxygen difference can be only slightly widened during exertion. Thus increased myocardial oxygen demand is primarily met by an increase in coronary blood flow. Since alterations in myocardial metabolism produce secondary changes in coronary blood flow, the factors which determine myocardial oxygen consumption are important.

MYOCARDIAL OXYGEN CONSUMPTION. Normal human cardiac oxygen consumption at rest is approximately 8 cc per min per 100 g of tissue.[6, 42] Cardiac oxygen consumption in resting unanesthetized dogs is approximately 6 cc per min per 100 g of tissue.[23] The oxygen consumption of the inactive heart has been measured in preparations in which the heart was arrested by perfusion with high potassium concentration blood. Under these conditions, cardiac oxygen consumption was approximately 2 ml O_2 per min per 100 g of heart or 27 per cent of the oxygen consumption rate of the beating heart.[39] Thus, approximately one-fourth of a normal heart's metabolism is related to the basic "housekeeping" of maintaining cardiac cells, and is unrelated to the blood pressure generated by the heart. The metabolic cost of electrical depolarization and repolarization that is associated with each heart beat has been estimated by studying hearts electromechanically uncoupled by perfusion with solutions deficient in calcium. Under these circumstances, electrical depolarization can be elicited by stimulation, but there is no mechanical response. The energy cost of electrical activation was found to be less than 1 per cent of normal cardiac oxygen consumption.[35]

Cardiac oxygen consumption is more closely related to the development of ventricular pressure than to external cardiac work.[13, 49, 58, 62, 63] The external work of the left ventricle is calculated as the mean arterial pressure during systole minus left atrial pressure times stroke volume ejected, and is usually expressed as the work performed per beat (stroke work), or the work per minute (stroke work times the heart rate). Since work is the product of pressure and flow, equal amounts of work may be obtained at high pressure and low flow ("high pressure work") or at low pressure and high flow ("low pressure work"). However, cardiac oxygen consumption differs for these two conditions at equal levels of external work, the high pressure work being more costly than low pressure work. Therefore myocardial oxygen consumption is estimated by an index, taking into account the pressure generated.

The summed areas under the systolic portions of a ventricular pressure recording for 1 min have been called the "tension-time index."[62] The index is the pressure-time integral of ventricular pressure during systole times the heart rate for a 1 min period (Fig. 16–5). A more appropriate name would be the pressure-time index, because pressure rather than ventricular wall tension is measured. An increase in heart rate augments the systolic pressure-time index by more frequent systoles in the 1 min period. An increase in the systolic pressure generated by the heart ("high pressure work") increases the area under the systolic portion of an intraventricular pressure recording. Thus the pressure-time index predicts changes in cardiac oxygen consumption associated with alterations in blood pressure and heart rate. As discussed below, it predicts oxygen consumption less well when heart size and contractility are altered. The development of ventricular pressure during the first half of systole accounts for most of the myocardial oxygen consumption, as demonstrated in isolated heart preparations with a sophisticated valve arrangement which permitted "decompressing" the ventricle to resting pressure at any moment during systole. When sudden decompressions were made during the ascending phase of ventricular systolic pressure (greatly reducing the pressure-time index) there was a good correlation between the pressure-time integral of these abbreviated beats and myocardial oxygen consumption (Fig. 16–5). The maintenance of pressure after peak values were reached and during relaxation required little additional oxygen. The increment of oxygen consumption associated with an increase in the pressure-time integral during relaxation had a slope of one-sixth that observed during the ascending phase of ventricular pressure (Fig. 16–5).[48]

A variable even better correlated with myocardial oxygen consumption is the development of calculated wall tension.[38, 57] Tension in the wall of the ventricle is proportional to the transmural pressure (ventricular pressure minus pericardial pressure), but is also related to the size of the ventricle; the larger the ventricular radius, the greater the myocardial wall tension for a given transmural pressure, *i.e.*, wall tension is greater in a dilated heart

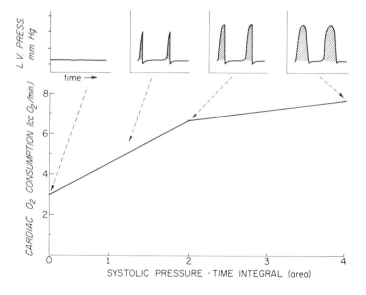

Figure 16–5 The pressure-time integral (shaded area) under the systolic portion of left ventricular pressure is compared with cardiac oxygen consumption in a preparation where ventricular pressure can suddenly be decompressed with a valve. Oxygen consumption correlates well with the pressure-time integral up to the development of peak tension but not beyond. This indicates that the development of pressure requires more oxidative energy than pressure maintenance. (After Monroe, *Circulat. Res.*, 1964, *14*, 294–300.)

than in an undistended heart at the same intraventricular pressure (Laplace's law). This relation can be illustrated using a thick-walled cylinder as a simple approximate geometrical model of the ventricle. The tangential stress (tension) in a thick-walled isotropic cylinder subjected to an internal pressure is given by:

$$\sigma_t = \frac{a^2 P_i}{b^2 - a^2}\left(1 + \frac{b^2}{r^2}\right)$$

where P_i is the internal pressure, a is the internal radius, b the external radius and r the radius at the point where the stress is being calculated.[65] As shown in Figure 16–6, enlargement of the ventricle at constant transmural pressure increases wall tension both because the radius of curvature increases and because the wall thins (constant wall volume). The total force is thus distributed over a smaller cross-section.

Although myocardial oxygen consumption is poorly correlated with external work alone, shortening incurs a metabolic cost. A muscle twitch is not like releasing a stretched spring, where a predetermined amount of energy is released by activation. Rather, the mechanical loading of the muscle during contraction determines the extent of energy conversion. This is known as the Fenn effect, and it was first demonstrated by the observation that the total energy converted to heat was greater in isotonic twitches than in isometric contractions.[17] That is, more energy is utilized when a muscle is allowed to shorten and perform external work than under isometric conditions. This effect holds for cardiac muscle[10] and is a factor (in addition to myocardial tension development) which determines cardiac oxygen requirements.

In addition to the mechanical factors of

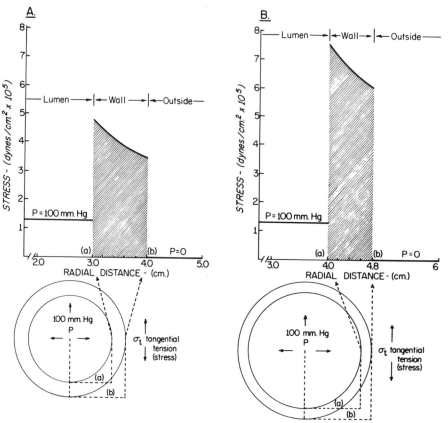

Figure 16–6 The calculated tangential stress (hoop stress) in the wall of a cylinder. The internal pressure is 100 mm Hg and the external pressure is zero in both cases. The cylinder is shown with an internal radius of 3 cm and an external radius of 4 cm on the left. The same cylinder is shown "dilated" to an internal radius of 4 cm on the right. The external radius is now 4.8 cm (assuming a constant wall volume). Note that the larger cylinder manifests a larger tangential tension at the same transmural pressure (100 mm Hg) because of the greater radius of curvature.

tension and shortening mentioned above, the inotropic state of the heart also influences cardiac oxygen consumption. Catecholamines administered to the potassium-arrested heart produce a small increase in cardiac oxygen consumption independent of changes in contractility and tension development observed in beating hearts.[36] The augmented oxygen consumption associated with increased contractility is greater in beating hearts.[21] Myocardial tension development is more costly in oxygen consumption when myocardial contractility is elevated. The oxygen wasting effect of increased contractility is observed with catecholamines, digitalis and paired electrical pacing.[9]

In summary, the basal oxygen consumption of the potassium-arrested heart is approximately one-fourth that of the beating heart under resting conditions. The major correlates of oxygen consumption are heart rate, myocardial tension development, shortening and increased contractility. The factors associated with oxygen wasting are (i) cardiac dilation, which increases wall tension by the Laplace relation, (ii) elevated aortic pressure, which results in increased myocardial tension, and (iii) augmented contractility. In general, coronary blood flow follows changes in myocardial oxygen consumption by an intrinsic vasodilator mechanism.

ADENOSINE HYPOTHESIS. The observation that adenosine is released from hearts subjected to coronary occlusion and myocardial hypoxia[5] has led to the hypothesis that adenosine is the mediator of coronary vasodilation in response to the oxygen needs of the heart. Adenosine triphosphate (ATP), adenosine diphosphate (ADP), adenosine monophosphate (AMP) and adenosine are all potent smooth muscle relaxants and coronary vasodilators. However, the nucleotides (ATP, ADP and AMP) are normally only found intracellularly and do not cross cell membranes, whereas the nucleosides (including adenosine) do cross cell membranes. The biochemical energy stores of cardiac muscle are in the form of high energy phosphates, mostly as ATP and creatine phosphate (CP). Energy utilization is accompanied by a degradation of ATP to ADP, AMP and adenosine; and ATP is replenished by oxidative phosphorylation of these adenosine compounds. The

adenosine hypothesis suggests that in hypoxia ATP is degraded to adenosine, which diffuses out of the myocardial cell to reach and dilate coronary vascular smooth muscle. Vasodilation in turn increases perfusion of the myocardium and oxygen delivery, permitting oxidative phosphorylation and reduction of adenosine levels. Thus adenosine is considered the link between myocardial metabolism and coronary vasomotion. The critical question is whether enough adenosine is released from myocardial cells to modulate coronary resistance in physiological conditions.

The adenosine hypothesis is outlined in Figure 16–7. Adenosine has a unique position among the compounds which are formed in the degradation of ATP, in that it both is a vasodilator and can penetrate cell membranes. The amino group on adenosine is apparently necessary for the vasodilating action. Since cell membranes are impermeable to ATP, ADP and AMP, these compounds are unlikely mediators of coronary vasomotion. The breakdown products of adenosine — inosine and hypoxanthine — permeate cell membranes but have only weak vasodilator action. Histochemical evidence indicates that 5′ nucleotidase, the enzyme responsible for dephosphorylation of AMP, is bound to membranes in cardiac muscle.[60] Dephosphorylation of AMP at the inner surface of the cell wall permits the prompt outward diffusion of adenosine before being deaminated and inactivated by adenosine deaminase, which is found in myocardial cytoplasm. Adenosine in the interstitial space is free from degradation by intracellular enzymes. A potential transmitter such as adenosine must also be inactivated if it is to have a dynamic role in blood flow control. Adenosine buildup in the interstitial space results in its diffusion into cells which contain adenosine deaminase. These include capillary endothelial cells, vascular pericytes, red blood cells and myocardial cells (which adenosine may re-enter).

The adenosine hypothesis is supported by the observation that adenosine and its breakdown products appear in hypoxic hearts,[4, 19, 32, 52, 61] and that adenosine is the result of cardiac catabolism of ATP.[20, 29] The hypothesis is challenged by the recent finding that vascular pericytes rather than myocardial cells are the sites of 5′ nucleo-

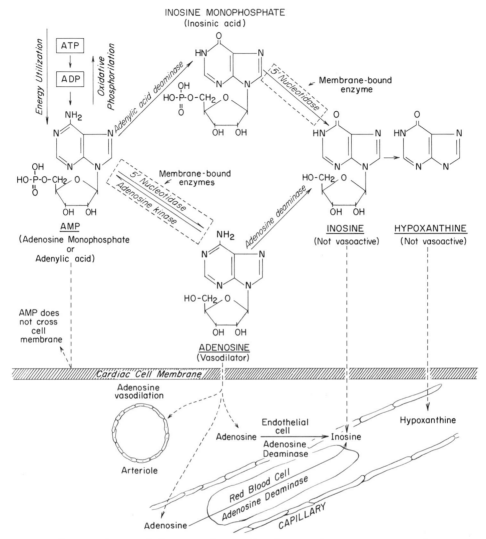

Figure 16–7 A diagram of myocardial tissue indicating the adenosine hypothesis. Myocardial energy utilization leads to the breakdown of ATP to AMP, which is further degraded to adenosine, which diffuses out of the myocardial cell to produce coronary vasodilation. The coronary vasodilation results in greater oxygen delivery which permits oxidative phosphorylation and reduction of adenosine levels. Thus, adenosine is considered the link between myocardial metabolism and coronary vasomotion. (After Rubio and Berne, *Circulat. Res.*, 1969, 25, 407–415.)

tidase.[8] Also, agents which block the action of injected adenosine fail to blunt reactive hyperemia in the coronary bed.[1] Finally, even large amounts of injected adenosine fail to reproduce the intense coronary vasodilation observed with hypoxia.[46]

In summary, the adenosine hypothesis proposes that adenosine is the link between myocardial oxygen metabolism and coronary vasomotion. The evidence that adenosine is the "transmitter" between myocardial cells and coronary smooth muscle is at present suggestive, but not conclusive.

Neural Control. The coronary vessels are innervated by both the sympathetic and parasympathetic divisions of the autonomic nervous system. The nerve fibers characteristically travel with the coronary arteries, often in the outer adventitial layers. The perivascular fibers are both afferent and efferent.

Studies of direct neural control of coronary blood flow are severely complicated by the fact that the pacemaker, atrial and ventricular myocardium are also innervated. Activation of autonomic fibers to the heart alters heart rate and contractility, which in turn change cardiac metabolism and, by an intrinsic mechanism, produce secondary changes in coronary blood flow. The experimental problem is to separate direct neural effects on the coronary vessels from those secondary to metabolic changes.

PARASYMPATHETIC. Stimulation of the cardiac end of the cut vagus nerve results in pronounced bradycardia, systemic hypotension, and a decreased coronary blood flow. This observation led to the erroneous conclusion that vagal fibers are vasoconstrictors to coronary vessels. Actually, the decreased coronary blood flow is secondary to the bradycardia and hypotension. Both factors decrease the systolic pressure-time integral which initiates both a marked fall in myocardial metabolism and a decrease in coronary blood flow. If the heart is electrically paced at a constant rate so that bradycardia is prevented during vagal stimulation, the coronaries dilate.[16] This is illustrated in Figure 16–8.

SYMPATHETIC. Stimulation of the stellate ganglion causes tachycardia, increased aortic pressure and augmented myocardial contractility, accompanied by an increased coronary blood flow. The cardiac acceleration, increased contractility and elevated aortic pressure all contribute to increased myocardial oxygen consumption which results in a secondary increase in coronary blood flow. The question is whether there is also a direct effect of sympathetic stimulation on the coronary vessels.

The positive inotropic and chronotropic effects of sympathetic stimulation can be prevented by beta-receptor blocking agents, such as propranolol. When cardiac sympathetic nerves are stimulated after administration of beta-receptor blocking agents, coronary vasoconstriction is consistently observed,[14, 40, 59] indicating that the direct effect of sympathetic activation on coronary vessels is vasoconstriction. This vasoconstriction is mediated by alpha-receptors, because the effect is prevented by alpha-receptor blocking agents, such as phentolamine.[14] In awake animals without beta-receptor blockade, cardiac sympathetic stimulation causes transient coronary vasoconstriction, which is later masked by vasodilation secondary to the increased heart rate and contractility resulting from neural stimulation[22] (Fig. 16–9). These findings are consistent in indicating direct sympathetic coronary vasoconstriction mediated by alpha-receptors.

Coronary beta-receptor vasodilation has been demonstrated pharmacologically with isoproterenol in arrested hearts[36] and iso-

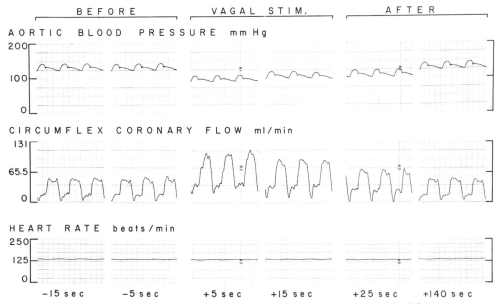

Figure 16–8 Effects of a 20-sec vagal stimulation (time 0 to +20 sec) on coronary blood flow. Coronary vasodilation is evidenced by increased blood flow during both systole and diastole despite a fall in aortic pressure. Heart rate was constantly paced. (From Feigl, *Circulat. Res.*, 1969, 25, 509–519. By permission of the American Heart Association, Inc.)

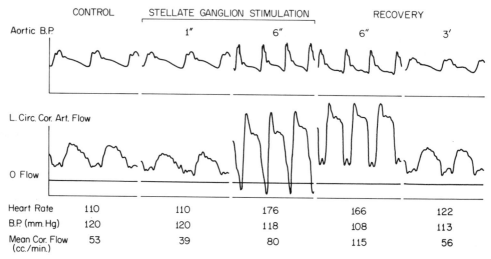

Figure 16–9 The effects of stimulating the left stellate ganglion on coronary blood flow in a conscious dog. The initial decrease in coronary blood flow is interpreted as direct coronary vasoconstriction which is subsequently masked by increased flow secondary to increased metabolism, due in turn to the chronotropic and inotropic effects of sympathetic stimulation. (After Granata *et al.*, *Circulat. Res.*, 1965, *16*, 114–120.)

lated coronary vessels.[70] Cardiac sympathetic stimulation after cardioselective beta-receptor blockade with practolol failed to reveal sympathetic innervation of coronary beta-receptors in one investigation,[40] but vasodilation was observed in another study.[50] However, the interpretation presumes that practolol selectively blocks cardiac inotropic and chronotropic effects while sparing coronary beta-receptor effects. Since some investigators find that pharmacological coronary beta-receptor vasodilation is blocked by practolol, the conclusion that these receptors are sympathetically innervated is not firmly established.

REFLEX CONTROL OF THE CORONARY CIRCULATION. Studies of coronary blood flow during cardiovascular reflexes are complicated by the interaction between direct neural control of the coronary vessels and changes secondary to altered myocardial metabolism. In general, a cardiovascular adjustment which includes tachycardia, hypertension, and increased cardiac contractility will be accompanied by an increase in coronary blood flow secondary to increased cardiac metabolism. Within such a response there may be a relative coronary vasoconstriction or vasodilation mediated by autonomic nerves to the coronary vessels. Lowering carotid sinus distending pressure results in reflex peripheral vascular vasoconstriction, tachycardia, increased cardiac contractility and increased

coronary blood flow. However, when the same reflex is studied in animals treated with beta-receptor blocking agents, reflex coronary vasoconstriction is unmasked. The vasoconstriction is prevented by cardiac sympathectomy.[15] Carotid sinus nerve stimulation, which mimics carotid sinus hypertension, causes coronary vasodilation by the mechanism of inhibiting tonic sympathetic constriction[66] and by vagal parasympathetic coronary vasodilation.[27]

Stimulation of carotid chemoreceptors with nicotine or cyanide produces reflex vagal coronary vasodilation.[27] This observation suggests that the parasympathetic coronary vasodilation observed during carotid sinus nerve stimulation may be due to activation of chemoreceptor fibers rather than baroreceptor fibers.

In summary, active parasympathetic coronary vasodilation independent of cardiac effects has been demonstrated. Parasympathetic coronary vasodilator fibers are reflexly activated by peripheral chemoreceptors and perhaps also by carotid baroreceptors. Sympathetic coronary alpha-receptor vasoconstriction is readily unmasked with beta-receptor blockade. Beta-receptor coronary vasodilation has been demonstrated pharmacologically, but there is no clear evidence at present that direct beta-receptor coronary vasodilation is activated by sympathetic nerves. Reflex sympathetic alpha-receptor coronary vasoconstric-

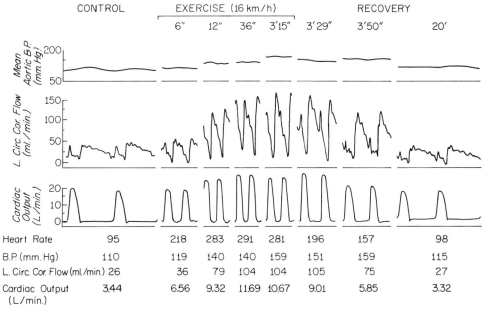

Figure 16-10 The effect of treadmill exercise (3.25 min) on coronary flow in a dog. Note the large increase in coronary blood flow which accompanies increased heart rate and cardiac output. (After Khouri *et al.*, *Circulat. Res.*, 1965, *17*, 427-437.)

tion is elicited by carotid sinus hypotension, and sympathetic vasoconstrictor tone is inhibited by afferent carotid sinus nerve stimulation.

CORONARY FLOW DURING STRESS

The elegant electromagnetic flowmeter implantation techniques of Gregg and co-workers have permitted the measurement of coronary flow in intact unanesthetized animals during exercise and excitement. The cardiac and coronary response during exercise resembles that induced by sympathetic stimulation with cardiac acceleration, an increase in coronary blood flow and a modest decrease in coronary venous oxygen content[33, 67] (Fig. 16-10). The response during excitement is similar to that during exercise, but is less intense and shorter lived.[55]

REFERENCES

1. Afonso, S., Ansfield, T. J., Berndt, T. B., and Rowe, G. G. Coronary vasodilator responses to hypoxia before and after aminophylline. *J. Physiol. (Lond.)*, 1972, *221*, 589-599.
2. Baird, R. J., and Ameli, F. M. The changes in intramyocardial pressure produced by acute ischemia. *J. thorac. cardiovasc. Surg.*, 1971, *62*, 87-94.
3. Baroldi, G., and Scomazzoni, G. *Coronary circulation in the normal and the pathologic heart.* Washington, D. C., Government Printing Office, 1967.
4. Berne, R. M. Cardiac nucleotides in hypoxia: Possible role in regulation of coronary blood flow. *Amer. J. Physiol.*, 1963, *204*, 317-322.
5. Berne, R. M. Regulation of coronary blood flow. *Physiol. Rev.*, 1964, *44*, 1-29.
6. Binak, K., Harmanci, N., Sirmaci, N., Ataman, N., and Ogan, H. Oxygen extraction rate of the myocardium at rest and on exercise in various conditions. *Brit. Heart J.*, 1967, *29*, 422-427.
7. Bing, R. J., Hammond, M. M., Handelsman, J. C., Powers, S. R., Spencer, F. C., Eckenhoff, J. E., Goodale, W. T., Hafkenschiel, J. H., and Kety, S. S. The measurement of coronary blood flow, oxygen consumption, and efficiency of the left ventricle in man. *Amer. Heart J.*, 1949, *38*, 1-24.
8. Borgers, M., Schaper, J., and Schaper, W. Adenosine-producing sites in the mammalian heart: A cytochemical study. *J. mol. cell. Cardiol.*, 1971, *3*, 287-296.
9. Braunwald, E. Control of myocardial oxygen consumption. Physiologic and clinical considerations. *Amer. J. Cardiol.*, 1971, *27*, 416-432.
10. Coleman, H. N., Sonnenblick, E. H., and Braunwald, E. Myocardial oxygen consumption associated with external work: The Fenn effect. *Amer. J. Physiol.*, 1969, *217*, 291-296.
11. Domenech, R. J., Hoffman, J. I. E., Noble, M. I. M., Saunders, K. B., Henson, J. R., and Subijanto, S. Total and regional coronary blood flow measured by radioactive microspheres in conscious and anesthetized dogs. *Circulat. Res.*, 1969, *25*, 581-596.

12. Downey, H. F., and Kirk, E. S. Coronary lymph: Specific activities in interstitial fluid uptake of ^{42}K. *Amer. J. Physiol.*, 1968, *215*, 1177–1182.

13. Evans, C. L., and Matsuoka, Y. The effect of various mechanical conditions on the gaseous metabolism and efficiency of the mammalian heart. *J. Physiol. (Lond.)*, 1915, *49*, 378–405.

14. Feigl, E. O. Sympathetic control of coronary circulation. *Circulat. Res.*, 1967, *20*, 262–271.

15. Feigl, E. O. Carotid sinus reflex control of coronary blood flow. *Circulat. Res.*, 1968, *23*, 223–237.

16. Feigl, E. O. Parasympathetic control of coronary blood flow in dogs. *Circulat. Res.*, 1969, *25*, 509–519.

17. Fenn, W. O. A quantitative comparison between the energy liberated and the work performed by the isolated sartorius muscle of the frog. *J. Physiol. (Lond.)*, 1923, *58*, 175–203.

18. Fulton, W. F. M. *The coronary arteries: Arteriography, microanatomy, and pathogenesis of obliterative coronary artery disease.* Springfield, Ill., Charles C Thomas, 1965.

19. Gerlach, E., and Deuticke, B. Bildung und Bedeutung von Adenosin in dem durch Sauerstoffmangel geschädigten Herzmuskel unter dem Einfluss von 2,6-Bis(diaethanolamino)-4,8-dipiperidinopyrimido(5,4-d)pyrimidin. *Arzneimittel-Forsch.*, 1963, *13*, 48–50.

20. Gerlach, E., Deuticke, B., and Dreisbach, R. Der Nucleotid-Abbau im Herzmuskel bei Sauerstoffmangel und seine mögliche Bedeutung für die Coronardurchblutung. *Naturwissenschaften*, 1963, *50*, 228–229.

21. Graham, T. P., Jr., Covell, J. W., Sonnenblick, E. H., Ross, J., Jr., and Braunwald, E. Control of myocardial oxygen consumption: Relative influence of contractile state and tension development. *J. clin. Invest.*, 1968, *47*, 375–385.

22. Granata, L., Olsson, R. A., Huvos, A., and Gregg, D. E. Coronary inflow and oxygen usage following cardiac sympathetic nerve stimulation in unanesthetized dogs. *Circulat. Res.*, 1965, *16*, 114–120.

23. Gregg, D. E., Khouri, E. M., and Rayford, C. R. Systemic and coronary energetics in the resting unanesthetized dog. *Circulat. Res.*, 1965, *16*, 102–113.

24. Griggs, D. M., Jr., Chen, C. C., and Tchokoev, V. V. Subendocardial anaerobic metabolism in experimental aortic stenosis. *Amer. J. Physiol.*, 1973, *224*, 607–612.

25. Griggs, D. M., Jr., Tchokoev, V. V., and Chen, C. C. Transmural differences in ventricular tissue substrate levels due to coronary constriction. *Amer. J. Physiol.*, 1972, *222*, 705–709.

26. Griggs, D. M., Jr., Tchokoev, V. V., and DeClue, J. W. Effect of beta-adrenergic receptor stimulation on regional myocardial metabolism: Importance of coronary vessel potency. *Amer. Heart J.*, 1971, *82*, 492–502.

27. Hackett, J. G., Abboud, F. M., Mark, A. L., Schmid, P. G., and Heistad, D. D. Coronary vascular responses to stimulation of chemoreceptors and baroreceptors. Evidence for reflex activation of vagal cholinergic innervation. *Circulat. Res.*, 1972, *31*, 8–17.

28. Herd, J. A., Hollenberg, M., Thorburn, G. D.,

Kopald, H. H., and Barger, A. C. Myocardial blood flow determined with krypton 85 in unanesthetized dogs. *Amer. J. Physiol.*, 1962, *203*, 122–124.

29. Imai, S., Riley, A. L., and Berne, R. M. Effect of ischemia on adenine nucleotides in cardiac and skeletal muscle. *Circulat. Res.*, 1964, *15*, 443–450.

30. James, T. N. *Anatomy of the coronary arteries.* New York, Paul B. Hoeber, 1961.

31. Johnson, R. A., and Blake, T. M. Lymphatics of the heart. *Circulation*, 1966, *33*, 137–142.

32. Katori, M., and Berne, R. M. Release of adenosine from anoxic hearts. Relationship to coronary flow. *Circulat. Res.*, 1966, *19*, 420–425.

33. Khouri, E. M., Gregg, D. E., and Rayford, C. R. Effect of exercise on cardiac output, left coronary flow and myocardial metabolism in the unanesthetized dog. *Circulat. Res.*, 1965, *17*, 427–437.

34. Kirk, E. S., and Honig, C. R. Nonuniform distribution of blood flow and gradients of oxygen tension within the heart. *Amer. J. Physiol.*, 1964, *207*, 661–668.

35. Klocke, F. J., Braunwald, E., and Ross, J., Jr. Oxygen cost of electrical activation of the heart. *Circulat. Res.*, 1966, *18*, 357–365.

36. Klocke, F. J., Kaiser, G. A., Ross, J., Jr., and Braunwald, E. Mechanism of increase of myocardial oxygen uptake produced by catecholamines. *Amer. J. Physiol.*, 1965, *209*, 913–918.

37. Kreuzer, H., and Schoeppe, W. Der Myokarddruck bei veränderter Coronardurchblutung und bei Ischämie. *Pflügers Arch. ges. Physiol.*, 1963, *278*, 209–220.

38. McDonald, R. H., Jr., Taylor, R. R., and Cingolani, H. E. Measurement of myocardial developed tension and its relation to oxygen consumption. *Amer. J. Physiol.*, 1966, *211*, 667–673.

39. McKeever, W. P., Gregg, D. E., and Canney, P. C. Oxygen uptake of the nonworking left ventricle. *Circulat. Res.*, 1958, *6*, 612–623.

40. McRaven, D. R., Mark, A. L., Abboud, F. M., and Mayer, H. E. Responses of coronary vessels to adrenergic stimuli. *J. clin. Invest.*, 1971, *50*, 773–778.

41. Martini, J., and Honig, C. R. Direct measurement of intercapillary distance in beating rat heart *in situ* under various conditions of O_2 supply. *Microvasc. Res.*, 1969, *1*, 244–256.

42. Messer, J. V., Wagman, R. J., Levine, H. J., Neill, W. A., Krasnow, N., and Gorlin, R. Patterns of human myocardial oxygen extraction during rest and exercise. *J. clin. Invest.*, 1962, *41*, 725–742.

43. Miller, A. J., Ellis, A., and Katz, L. N. Cardiac lymph: Flow rates and composition in dogs. *Amer. J. Physiol.*, 1964, *206*, 63–66.

44. Miller, A. J., Pick, R., and Katz, L. N. Lymphatics of the mitral valve of the dog: Demonstration and discussion of the possible significance. *Circulat. Res.*, 1961, *9*, 1005–1009.

45. Moir, T. W. Subendocardial distribution of coronary blood flow and the effect of antianginal drugs. *Circulat. Res.*, 1972, *30*, 621–627.

46. Moir, T. W., and Downs, T. D. Myocardial reactive hyperemia: Comparative effects of adenosine, ATP, ADP, and AMP. *Amer. J. Physiol.*, 1972, *222*, 1386–1390.

47. Moir, T. W., Eckstein, R. W., and Driscol, T. E. Thebesian drainage of the septal artery. *Circulat. Res.*, 1963, *12*, 212–219.

48. Monroe, R. G. Myocardial oxygen consumption during ventricular contraction and relaxation. *Circulat. Res.*, 1964, *14*, 294–300.

49. Monroe, R. G., and French, G. N. Left ventricular pressure-volume relationships and myocardial oxygen consumption in the isolated heart. *Circulat. Res.*, 1961, *9*, 362–374.

50. Nayler, W. G., and Carson, V. Effect of stellate ganglion stimulation on myocardial blood flow, oxygen consumption, and cardiac efficiency during beta-adrenoceptor blockade. *Cardiovasc. Res.*, 1973, *7*, 22–29.

51. Neill, W. A., Krasnow, N., Levine, H. J., and Gorlin, R. Myocardial anaerobic metabolism in intact dogs. *Amer. J. Physiol.*, 1963, *204*, 427–432.

52. Olsson, R. A. Changes in content of purine nucleoside in canine myocardium during coronary occlusion. *Circulat. Res.*, 1970, *26*, 301–306.

53. Patek, P. R. The morphology of the lymphatics of the mammalian heart. *Amer. J. Anat.*, 1939, *64*, 203–249.

54. Pitt, A., Friesinger, G. C., and Ross, R. S. Measurement of blood flow in the right and left coronary artery beds in humans and dogs using [133]xenon technique. *Cardiovasc. Res.*, 1969, *3*, 100–106.

55. Rayford, C. R., Khouri, E. M., and Gregg, D. E. Effect of excitement on coronary and systemic energetics in unanesthetized dogs. *Amer. J. Physiol.*, 1965, *209*, 680–688.

56. Renkin, E. M. Blood flow and transcapillary exchange in skeletal and cardiac muscle. In: *International symposium on the coronary circulation and energetics of the myocardium*, G. Marchetti and B. Taccardi, eds. New York, S. Karger, 1967.

57. Rodbard, S., Williams, C. B., Rodbard, D., and Berglund, E. Myocardial tension and oxygen uptake. *Circulat. Res.*, 1964, *14*, 139–149.

58. Rohde, E. Über den Einfluss der mechanischen Bedingungen auf die Tätigkeit und den Sauerstoffverbrauch des Warmblüterherzens. *Naunyn-Schmiedeberg's Arch. exp. Path. Pharmak.*, 1912, *68*, 401–434.

59. Ross, G., and Mulder, D. G. Effects of right and left cardiosympathetic nerve stimulation on blood flow in the major coronary arteries of the anesthetized dog. *Cardiovasc. Res.*, 1969, *3*, 22–29.

60. Rostgaard, J., and Behnke, O. Fine structural localization of adenine nucleoside phosphatase activity in the sarcoplasmic reticulum and the T system of rat myocardium. *J. ultrastruct. Res.*, 1965, *12*, 579–591.

61. Rubio, R., and Berne, R. M. Release of adenosine by the normal myocardium in dogs and its relationship to the regulation of coronary resistance. *Circulat. Res.*, 1969, *25*, 407–415.

62. Sarnoff, S. J., Braunwald, E., Welch, G. H., Jr., Case, R. B., Stainsby, W. N., and Macruz, R. Hemodynamic determinants of oxygen consumption of the heart with special reference to the tension-time index. *Amer. J. Physiol.*, 1958, *192*, 148–156.

63. Sarnoff, S. J., Case, R. B., Welch, G. H., Jr., Braunwald, E., and Stainsby, W. N. Performance characteristics and oxygen debt in a nonfailing, metabolically supported, isolated heart preparation. *Amer. J. Physiol.*, 1958, *192*, 141–147.

64. Schaper, W. *The collateral circulation of the heart.* New York, American Elsevier, 1971.

65. Timoshenko, S. *Strength of materials, part II.* 3rd ed. Princeton, N. J., D. Van Nostrand Company, 1956.

66. Vatner, S. F., Franklin, D., Van Citters, R. L., and Braunwald, E. Effects of carotid sinus nerve stimulation on the coronary circulation of the conscious dog. *Circulat. Res.*, 1970, *27*, 11–21.

67. Vatner, S. F., Higgins, C. B., and Franklin, D. Regional circulatory adjustments to moderate and severe chronic anemia in conscious dogs at rest and during exercise. *Circulat. Res.*, 1972, *30*, 731–740.

68. Wearn, J. T. Morphological and functional alterations of the coronary circulation. *Harvey Lectures*, 1940, *35*, 243–270.

69. Whalen, W. J. Intracellular PO_2 in heart and skeletal muscle. *Physiologist*, 1971, *14*, 69–82.

70. Zuberbuhler, R. C., and Bohr, D. F. Responses of coronary smooth muscle to catecholamines. *Circulat. Res.*, 1965, *16*, 431–440.

CHAPTER 17 THE CEREBRAL CIRCULATION

by LOUIS G. D'ALECY

ANATOMY

The brain and spinal cord are contained in rigid bony chambers, formed by the skull and vertebral column. Just within these bony structures there is a dense, fibrous connective tissue layer, the dura mater. The brain lies within the dura mater, supported and surrounded by the cerebral spinal fluid. The spinal fluid compartment is continuous from the cerebral ventricles to the cranial cavity and down the spinal fluid space of the vertebral column.

The existence of this rigid bony cavity led early observers of the cerebral circulation to believe that blood flow to the brain was constant—the Monro-Kellie doctrine. Monro's theory (1783) was that the rigid case of bone allowed essentially no change in brain volume. Kellie (1824)[26] tested the theory by measuring the postmortem volume of blood in dog brains after a variety of slow and fast hemorrhagic deaths. In animals with intact skulls the vessels of the brain were always full, whereas the vessels of other organs showed variations in fullness, depending on the form of exsanguination. By failing to distinguish between brain blood volume and brain blood flow, Monro and Kellie reasoned that the rigid skull allowed no change in blood flow. We now know that blood flow to the brain is not constant, but can be changed by alterations in arterial carbon dioxide tension as well as by nervous control.

The volume within the skull is only relatively fixed, and some variation in volume is possible by expansion of the dural outpocketings around the spinal and cranial nerve roots. The cerebral spinal fluid volume may also vary with changes in the volume of the veins and arteries on the surface of the brain and spinal cord. Normal spinal fluid pressure is about 10 cm H_2O. The pressure in the spinal fluid may reflect arterial pulsations, venous pulsations and slower oscillations coincident with respiration.[1] These pressure oscillations are transmitted to the spinal fluid by the arteries and veins of the brain and spinal cord.

ARTERIAL SUPPLY TO THE BRAIN. The basilar artery and the circle of Willis are the primary vessels from which blood is distributed to the brain stem, cerebellum and cerebrum. Blood reaches the basilar artery via the right and left vertebral arteries. The circle of Willis is supplied principally by the internal carotid arteries and the basilar artery (Fig. 17–1). Species and individual anatomic variations deter-

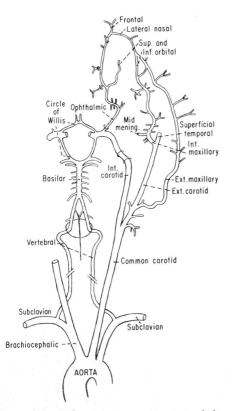

Figure 17–1 Schematic representation of the arteries supplying the brain in man. The distal branches of the circle of Willis and most extracranial arteries have been omitted. Blood can reach the circle of Willis by the vertebral arteries, the internal carotid arteries and the internal and external maxillary branches of the external carotid system.

age in case of occlusion of a single artery. Comparable redundancy is not present in the arteries beyond the circle of Willis to the various regions of the brain. Functional communications between anterior, middle and posterior cerebral arteries are not clearly established.

Intra-arterial cushions are found in pial arteries and larger arterioles.[22, 23, 54] These cushions are analogous to precapillary sphincters, but their function is not yet defined. True precapillary sphincters have not been observed in cerebral arterioles.

CAPILLARIES. Cerebral capillaries have a continuous lining of endothelial cells which are joined by "tight" junctions. There is no anatomic evidence for pores or fenestrations which occur in capillaries of other organs. Surrounding the capillary endothelium is a basement membrane which is continuous with the astrocytes. Astrocytes and oligodendrocytes make up the neuroglia of the central nervous system and separate the vascular system from the neuronal elements of the central nervous system. The neuroglia makes up 50 per cent of the volume of the brain and is arranged to eliminate direct contact between capillaries and neurons.

The anatomy of the endothelium, basement membrane and neuroglia is a possible explanation for the diffusional barrier between the blood and the brain, the "blood-brain barrier." This barrier allows the passage of most ions and glucose but limits the movement of large molecular weight dyes, antibiotics, fatty acids and proteins.[11]

VENOUS DRAINAGE OF THE BRAIN. The major portion of the brain is drained by the superior sagittal sinus and the straight sinus. The straight sinus originates from the inferior sagittal sinus and the great vein of Galen. The superior sagittal sinus and straight sinus merge in the midline to form the right and left transverse sinuses. The transverse sinuses travel laterally and inferiorly to become the sigmoid sinuses and leave the skull as the internal jugular veins. Before leaving the skull, the sigmoid sinuses receive blood from the inferior and superior petrosal sinuses. The petrosal sinuses communicate with the cavernous sinuses, pterygoid plexus, orbital plexus and facial veins. The cavernous sinus communicates via the basilar plexus with numerous vertebral veins and sinuses (Fig.

mine which vessel or vessels provide the major blood supply to the brain. In man, the internal carotid blood is normally distributed to the ipsilateral cerebral hemispheres. However, there is free communication via the circle of Willis to the contralateral side, as well as communication via the ophthalmic arteries to the orbit and face. Blood is distributed to the spinal cord by the right and left vertebral arteries. As the vertebral arteries enter the skull, they join to form the basilar artery which supplies the contents of the posterior cranial fossa and which joins the circle of Willis. Because of communications between the vertebral and carotid systems and between the circle of Willis and the extracranial arteries, occlusion of one or more of these vessels can be tolerated in some individuals.[2] The communications produce a redundant arterial supply which protects the brain from dam-

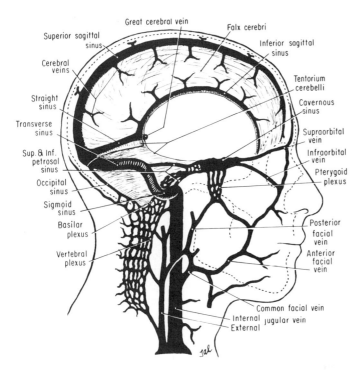

Figure 17-2 Lateral view of the human cranial venous drainage, showing the communication between the internal jugular drainage system and the vertebral plexus and cavernous sinus. The intracranial sinuses and veins lack valves and the direction of flow can vary, depending upon pressure gradients. For simplicity only the right sigmoid sinus and the right jugular vein are shown. In sampling blood from the internal jugular vein, care must be taken to avoid drawing blood from extracerebral tissues.

17–2). The significance of these communications is demonstrated by the observation that both internal jugular veins can be ligated in man and cerebral drainage continues via the vertebral and external jugular system through the spinal venous plexus and through the orbit to the face.[2] Veins on the brain surface form anastomoses interconnecting different regions of the brain and also connecting via diploic veins with veins of the cranium and the scalp.

BLOOD FLOW MEASUREMENT

Cerebral blood flow can be measured by either direct or indirect methods. The direct methods involve a determination of blood volume entering or leaving the brain per unit time, utilizing flowmeters. Indirect methods involve the measurement of changes over time in concentration or quantity of some indicator (isotope, inert gas, heat, etc.) that is dependent on blood flow.

Inasmuch as there is no single artery or vein that supplies or drains the brain, direct measurement of cerebral blood flow requires occlusion of parallel collateral flow channels or multiple flow measurements.

This difficulty of direct flow measurement is one reason indirect techniques have been developed.

DIRECT TECHNIQUES. The application of a blood flow transducer to the arterial side of the circulation poses two particular problems. First, some modification of the multiple arterial supplies is required to assure that the flow measured in a vessel is the sole supply to the brain and there is no flow to extracerebral tissues. Second, dissection, ligation, cannulation or manipulation of the carotid or vertebral arteries for flow transducer placement may damage the autonomic nerves on these vessels. These two problems severely limit the usefulness of direct arterial cerebral blood flow measurements.

In experimental animals in which extensive surgical modification of the venous drainage is tolerable, collateral venous drainage channels can be virtually eliminated. In the dog, cannulation of the confluence of the sagittal and straight sinuses and occlusion of the right and left transverse sinuses allows cerebral blood flow to be measured by an extracorporeal tubing circuit which then returns the blood to the external jugular vein.[50] Another venous outflow technique for use in dogs measures flow with an electromagnetic flow transducer on the retroglenoid vein.[10] The retroglenoid vein is the extracranial continuation of the transverse sinus.

INDIRECT TECHNIQUES. The greatest single advance in the study of the cerebral circulation was the development of the Kety-Schmidt nitrous oxide technique for cerebral blood flow in man.[33] Through this special application of the Fick principle cerebral blood flow can be measured from brain arterial-venous nitrous oxide concentration differences. The principle of the method is that the brain uptake of an indicator is determined during the transition from one steady state equilibrium concentration of indicator to another. The gaseous indicator is taken into the arterial blood via the lungs. During the transition period when the brain takes up indicator from the blood, cerebral venous concentration is less than arterial concentration. The amount taken up by the brain in the transition period may be calculated from arterial-venous difference and used in the Fick principle for determining cerebral blood flow. (See Chap. 3, Vol. II, for a presentation of the Fick principle.)

The procedure is as follows. At time, t_0, the individual starts to breathe a fixed low concentration of nitrous oxide (usually 15 per cent). Serial arterial and jugular bulb venous blood samples are taken and analyzed for nitrous oxide concentration. At some time, t_1 (usually 10 min later) the arterial and venous concentrations become approximately equal. Using the following symbols:

C_a = arterial concentration
C_b = brain concentration
C_v = venous concentration
F = flow
Q = quantity (of indicator)
S = λ/ρ
t = time
V_b = brain volume
λ = brain/blood partition coefficient
ρ = brain density

the Fick principle from Chapter 3, Volume II, may be written:

$$F = \frac{dQ/dt}{C_a - C_v} \quad (1)$$

The arterial and cerebral venous concentrations will be measured directly, but dQ/dt must be calculated.

During equilibrium at concentrations below saturation, an indicator will partition between two phases according to the ratio of the solubility coefficients in the two phases. Thus there will be a constant ratio between the concentration of indicator in the two phases. A partition coefficient for brain and cerebral venous blood may be defined:

$$\lambda = \frac{C_b}{C_v} \quad (2)$$

This permits a calculation of the brain concentration from the venous concentration if the partition coefficient is determined in a separate experiment.

$$C_b = \lambda C_v \quad (3)$$

The quantity of indicator in the brain is the brain concentration times the brain volume:

$$Q = C_b V_b = \lambda C_v V_b \quad (4)$$

The quantity of indicator taken up by the brain per time in the interval t_0 to t_1 may be written:

$$dQ/dt = \frac{\Delta Q}{t_1 - t_0} = \frac{(\lambda C_v V_b)_{t_1} - (\lambda C_v V_b)_{t_0}}{t_1 - t_0} \quad (5)$$

The initial concentration at time t_0 is usually zero, so we may write:

$$\frac{\Delta Q}{t_1 - t_0} = \frac{(\lambda C_v V_b)_{t_1}}{t_1 - t_0} \quad (6)$$

The average arterial-venous indicator concentration difference in the interval t_0 to t_1 is equal to the integral of the arterial-venous difference divided by the time interval:

$$C_a - C_v = \frac{\int_{t_0}^{t_1} (C_a - C_v)dt}{t_1 - t_0} \quad (7)$$

Substituting equations (6) and (7) into equation (1), we have:

$$F = \frac{\dfrac{(\lambda C_v V_b)_{t_1}}{t_1 - t_0}}{\dfrac{\int_{t_0}^{t_1} (C_a - C_v)dt}{t_1 - t_0}} \quad (8)$$

If flow is stated as ml/min per volume of brain, equation (8) may be rewritten:

$$\frac{F}{V_b} = \frac{(\lambda C_v)_{t_1}}{\int_{t_0}^{t_1} (C_a - C_v)dt} \quad (9)$$

To express flow as ml/min per unit mass, the brain density is used.

$$\text{density} = \frac{\text{mass}}{\text{volume}} = \rho$$

$$\text{or mass} = \rho \cdot V$$

Multiplying equation (9) by $1/\rho$ we obtain:

$$\frac{F}{\rho V_b} = \frac{F}{\text{mass}} = \frac{(\lambda C_v)_{t_1}}{\int_{t_0}^{t_1} (C_a - C_v)dt} \cdot \frac{1}{\rho} \quad (10)$$

Kety combines the constants λ and ρ to form a new constant S = λ/ρ. By substitution we have the Kety equation.

$$F_{(ml/min/g)} = \frac{(C_v S)_{t_1}}{\int_{t_0}^{t_1} (C_a - C_v)dt} \quad (11)$$

The numerator is the measured jugular bulb cerebral venous concentration at t_1 multiplied by the constant S. The integral in the denominator may be determined either graphically (Fig. 17–3) or by calculation from a series of arterial and venous concentration measurements.

Numerous modifications of the nitrous oxide technique have been developed principally because of the difficulty in analyzing nitrous oxide in blood. Various radioactive isotopes have been substituted for nitrous oxide as an indicator ([85]Kr, ^{133}Xe), as have other inert gases (H_2).[3] The assumptions for all the arterial-venous difference techniques are the same. The principal assumptions are as follows:

(i) Jugular bulb venous blood is repre-sentative of cerebral venous outflow concentration of indicator. Care must be taken here in considering the anatomic communication of the jugular bulb with vertebral venous plexuses, cavernous sinus and facial veins (Fig. 17–2).

(ii) Brain blood flow is constant throughout the 10 min measurement period.

(iii) Stationarity of flow in various regions of the brain is constant throughout the measurement period. That is, not only total brain flow but the flow to each area of the brain must be constant during the measurement period.

(iv) Equilibration of indicator into tissue is flow limited rather than diffusion limited.

(v) The partition coefficient is a true constant, and the concentrations of indicator in blood and tissue are below saturation.

(vi) The brain capacity for indicator is constant. The indicator is not destroyed, secreted or produced by the tissue, and the indicator is not a drug which alters cerebral blood flow.

Other variations of the Fick principle have been applied to the problem of measuring cerebral blood flow. Indicator washout techniques with isotopes have proved useful in both animal and human determinations of cerebral blood flow. Details of these techniques are given in Chapter 3, Volume II. Heat clearance and hydrogen clearance techniques have yielded useful information on regional variations in cerebral blood flow. All the flow measurements based on the Fick principle are subject to similar assumptions, and all yield average flow determinations over the time required for measurement.

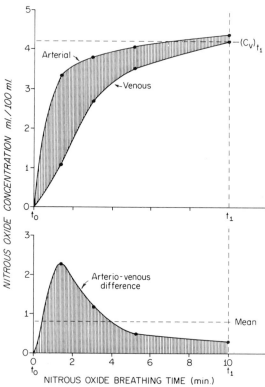

Figure 17–3 Plots of the arterial and venous nitrous oxide and A-V nitrous oxide difference concentrations during a 10 min period of breathing of 15 per cent nitrous oxide air mixture. Graphs can be used to calculate cerebral blood flow in ml/min/gram using equation (11).

FACTORS CONTROLLING CEREBRAL BLOOD FLOW

Carbon Dioxide. The most potent factor which controls cerebral blood flow is the tension of carbon dioxide in arterial blood. Lennox and Gibbs,[39] using arterio-venous oxygen differences across the brain to measure cerebral blood flow in man (assuming constant brain metabolism), demonstrated an increase in flow in response to increased arterial carbon dioxide tension. Hyperventilation produced a decrease in arterial carbon dioxide tension and a decrease in cerebral blood flow. The occur-

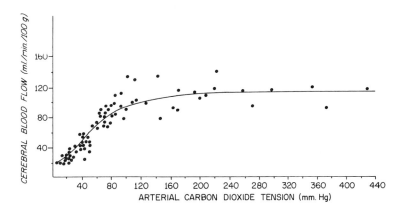

Figure 17-4 The response of cerebral blood flow to alterations in arterial carbon dioxide tension. The curve was computer generated to fit measured data points. Through a range bracketing normal CO_2 tension (40 mm Hg) by $\pm$ 20 mm Hg flow and tension are approximately linearly related. (After Reivich, *Amer. J. Physiol.*, 1964, *206*[1], 25–35.)

rence of cerebral vasodilation in response to increased arterial carbon dioxide tension, and of cerebral vasoconstriction in response to decreased arterial carbon dioxide tension, has been confirmed by several investigators.[32, 34, 46, 47, 51] The data presented by Reivich[51] cover a wide range of arterial carbon dioxide tensions (Fig. 17–4). At extremely high and low arterial carbon dioxide tensions, the curve flattens to a characteristic sigmoid shape. Although the response of cerebral blood flow to arterial carbon dioxide tension can be consistently demonstrated, the exact mechanism of the response is not clear. The cerebral vasodilation in response to increased arterial carbon dioxide tension is attenuated by sympathetic stimulation and accentuated by sympathectomy (Fig. 17–5).[25] Sympathetic modulation of the carbon dioxide response suggests an interaction between reflex and direct effects of carbon dioxide on cerebral blood flow.

Severinghaus and Lassen[59] postulate that change in the pH of the extracellular fluid around the arterioles and not the change in the carbon dioxide tension itself is the stimulus through which arterial carbon dioxide controls cerebral blood flow. The basis of this idea was their observation that cerebral blood flow changed more rapidly than brain tissue carbon dioxide tension (as indicated by venous pCO_2) when a rapid change in arterial carbon dioxide tension was imposed. This implies that tissue carbon dioxide tension was not the stimulus to cerebral vasodilation. In contrast to the slow venous carbon dioxide tension change, spinal fluid pH changes rapidly after an alteration of arterial carbon dioxide tension. This is because the spinal fluid has

far less buffering capacity than blood. Thus the hypothesis is that extracellular pH rather than carbon dioxide tension is the stimulus to cerebral vasomotion.

Acid-Base Balance. Acute arterial acidosis results in cerebral vasodilation, and acute arterial alkalosis results in cerebral vasoconstriction when changes in carbon dioxide tension secondary to the infusion of acid or alkali are not controlled.[56, 57] If the arterial carbon dioxide tension is controlled, then the intravenous infusion of

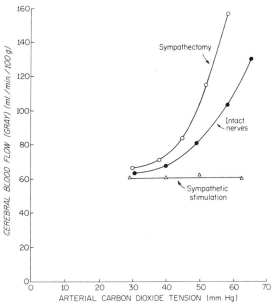

Figure 17-5 The effect of sympathetic neural activity on the responsiveness of cerebral blood flow to arterial CO_2 tension. Sympathetic stimulation greatly attenuates and sympathectomy accentuates the vasodilator effect of increased arterial carbon dioxide tension. (After James *et al.*, *Circulat. Res.*, 1969, *25*, 77–93.)

acid or alkali does not change cerebral blood flow acutely.[20, 37] This observation appears to conflict with the Severinghaus-Lassen hypothesis that extracellular pH is the dominant factor in controlling cerebral blood flow. However, it is difficult to know what the pH is at the critical extracellular site where the arterioles lie during acute changes.

Chronic acidosis results in an increase in cerebral blood flow, and chronic alkalosis results in a decrease in flow. Patients with acidosis from diabetic coma have high cerebral blood flows.[31] Chronic metabolic acidosis or alkalosis can be experimentally induced by feeding ammonium chloride or sodium bicarbonate respectively. Under these conditions cerebral blood flow is higher in acidosis and lower in alkalosis at a given arterial carbon dioxide tension than under normal acid-base balance, indicating a shift from the normal arterial carbon dioxide vs. cerebral blood flow relation (Fig. 17–6). However, these data during normal, alkalotic and acidotic conditions group along a single line when cerebral blood flow is compared to calculated spinal fluid pH (Fig. 17–7).[14] This supports the hypothesis that extracellular pH is the dominant factor controlling cerebral blood flow.

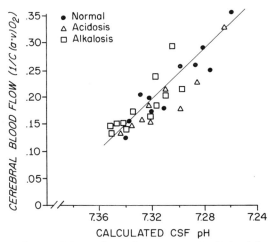

Figure 17–7 Relation between cerebral blood flow in normal conditions, chronic acidosis and chronic alkalosis as a function of calculated CSF pH. Data are from the experiments shown in Figure 17–6. (After Fencl *et al.*, *J. appl. Physiol.*, 1969, 27, 67–76.)

Ions. Increasing perivascular potassium ion from zero to 10 mEq per liter causes dilation of pial vessels on the surface of the brain.[35] At potassium ion concentrations above 20 mEq per liter pial vessels begin to constrict. Increasing bicarbonate concentration produces vasoconstriction. The vasoconstriction produced by increasing perivascular bicarbonate is reduced by elevating potassium concentration. The role of these ions and their interaction with hydrogen ions in the control of vascular diameters and cerebral blood flow are not clearly understood.

Oxygen. Arterial oxygen tension has an opposite effect to that of arterial carbon dioxide tension on cerebral blood flow. The maximum vasodilation in response to anoxia is greater than the maximum vasoconstriction in response to hyperoxia.[16] Breathing hyperbaric oxygen at 3.5 atmospheres produces a 25 per cent decrease in cerebral blood flow.[36] The vascular responses to changes in arterial oxygen tension are less pronounced than the responses caused by changes in arterial carbon dioxide tension.[34] In the normal range of arterial oxygen tension cerebral blood flow is constant. However, if arterial oxygen tension goes below 50 to 60 mm Hg, there is a sharp increase in cerebral blood flow (Fig. 17–8).

The effect of arterial carbon dioxide ten-

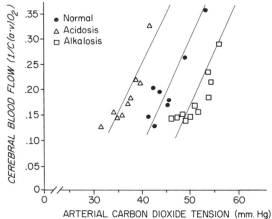

Figure 17–6 The effect of chronic acidosis and alkalosis on the response of cerebral blood flow to arterial carbon dioxide tension. The curve is shifted to the right in alkalosis, to the left in acidosis, *i.e.*, cerebral blood flow is higher in chronic acidosis and lower in chronic alkalosis. Blood flow was determined by arterio-venous oxygen differences assuming a constant cerebral oxygen metabolism. (After Fencl *et al.*, *J. appl. Physiol.*, 1969, 27, 67–76.)

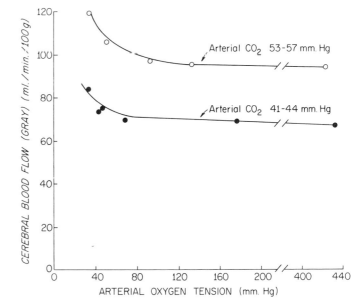

Figure 17–8 The response of cerebral blood flow to alterations in arterial oxygen tension. There is little change in cerebral blood flow until the arterial oxygen tension falls below 50 to 60 mm Hg, when flow increases. The changes are smaller than those resulting from alteration of arterial CO_2 tension. (After James et al., Circulat. Res., 1969, 25, 77–93.)

sion, oxygen tension and pH may be summarized as follows: Increased arterial carbon dioxide tension has a profound vasodilator effect on cerebral blood flow; oxygen has, to a lesser degree, the opposite effect. Chronic acidosis increases cerebral blood flow, and chronic alkalosis decreases cerebral blood flow. Changes in spinal fluid or tissue pH are probably important in mediating the effects of carbon dioxide on arterioles of the brain.

Autonomic Control. The cerebral vessels are innervated by both the sympathetic and parasympathetic divisions of the autonomic nervous system. Sympathetic fibers to cerebral vessels originate in the thoracic segments of the spinal cord and relay through the stellate, inferior, middle and superior cervical ganglia. Fibers from the stellate ganglion also travel along with the vertebral arteries up the neck to the basilar artery. The autonomic nerve fibers travel adjacent to or in the adventitia of the vessels they supply. Adrenergic fibers have been traced throughout the cerebral vascular tree to the point at which the continuous layer of vascular smooth muscle is lost from arterioles.[43] The light microscope,[5, 6, 17, 21, 41, 48, 62] fluorescent histochemistry[12, 13, 44] and the electron microscope[24, 42, 43, 54, 55] have all been used to demonstrate an abundant innervation of the

cerebral vessels. Electron microscopic studies show dense core synaptic vesicles which are characteristic of adrenergic terminals, and also empty synaptic vesicles characteristic of cholinergic terminals.[24] The greater superficial petrosal nerve originating with the seventh (facial) cranial nerve carries parasympathetic fibers to the internal carotid artery and the circle of Willis. The cerebral vessels therefore have demonstrable autonomic innervation.

SYMPATHETIC CONTROL. Sympathetic nerve fibers exert little or no tonic influence on cerebral blood flow; in resting recumbent man, cerebral blood flow is unchanged after sympathectomy or stellate ganglion blockade.[19, 27, 29, 30, 33, 40, 60] In animals, electrical stimulation of the stellate ganglion produces a marked vasoconstriction and decrease in cerebral blood flow (Fig. 17–9). The magnitude of this vasoconstriction increases, with increased stimulation frequency reaching a maximum at 15 to 20 stimuli per sec (Fig. 17–10).[10] This sympathetic vasoconstrictor mechanism competes with the vasodilator effects of arterial carbon dioxide (Fig. 17–5). Sympathetic activation attenuates the vasodilator response to elevated arterial carbon dioxide. Conversely, sympathectomy accentuates the carbon dioxide vasodilator response.[25] The vasoconstriction resulting from electrical

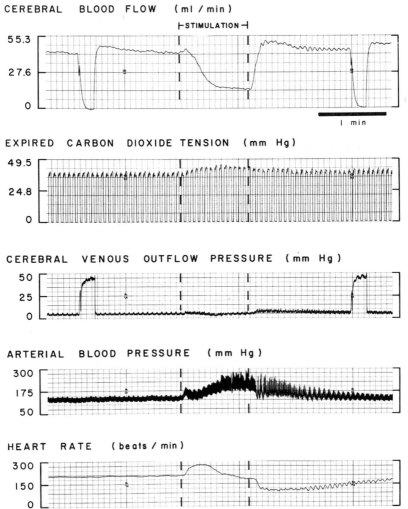

Figure 17–9 Effect of electrical stimulation (15 per sec) of stellate ganglion on cerebral blood flow in dog as measured by a flowmeter on the retroglenoid vein. Sympathetic stimulation decreased flow despite elevation of end-expiratory CO_2 tension, arterial blood pressure and heart rate. Deflections at beginning and end of flow record were caused by occluding the venous outflow to check flowmeter stability. (From D'Alecy, *Stroke*, 1973, *4*, 30–37. By permission of the American Heart Association, Inc.)

stimulation of the stellate ganglion can be eliminated by adrenergic alpha-receptor blockade.[9] The possible existence of adrenergic beta-receptors is less clearly defined.

Circulating catecholamines have little effect on cerebral blood flow. Intra-arterial injections of norepinephrine barely change cerebral blood flow. Nevertheless, when tested in tissue baths, isolated cerebral vessels constrict to norepinephrine, epinephrine, 5-hydroxytryptamine, isoproterenol, acetylcholine and histamine (Fig. 17–11).[45]

The doses required for these constrictions are higher than needed for vessel preparations from other vascular beds.

The reason for the discrepancy between *in vivo* and *in vitro* experiments is not known. The blood-brain barrier, which is a specialized diffusion barrier between the blood and brain tissue, may limit the movement of drugs from blood to vascular receptors. The heavy basement membrane which invests cerebral endothelium is in an appropriate location to do this. It has been suggested that the abundant sympa-

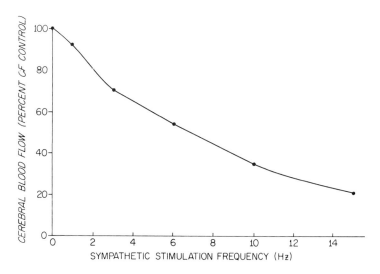

Figure 17–10 Graph showing relation between cerebral blood flow and frequency of electrical stimulation to the stellate ganglion in dogs. Flow measurements made as in Figure 17–9. Maximal effects are reached when stimulation frequencies approach 15 per sec. (After D'Alecy and Feigl, *Circulat. Res.*, 1972, *31*, 267–283.)

thetic nerve terminals on cerebral vessels rapidly take up circulating catecholamines and thus "protect" the vessels from blood-borne catecholamines.[52]

PARASYMPATHETIC CONTROL. The effect of parasympathetic activation was examined by early workers using a pial window technique. Pial vessels dilated when the greater superficial petrosal nerve was electrically stimulated.[4] It would be speculative to conclude that this dilation necessarily caused an increase in cerebral

blood flow, without knowledge of the pressure gradients involved. Stimulation of the distal end of the sectioned seventh cranial nerve elicited an increase in cerebral blood flow as indicated by ^{14}C-antipyrine autoradiography.[53] In monkeys, with sympathetic effects eliminated by cervical cordotomy, brain stem stimulation produced a 40 per cent increase in cerebral blood flow with no change in blood pressure.[38] Topical application of cholinomimetic drugs to surface vessels of the brain caused a local

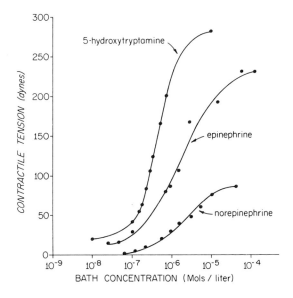

Figure 17–11 Log dose-response curves showing contractile responses of lengths of isolated middle cerebral artery from three different cats after fractionated application of 5-hydroxytryptamine, epinephrine and norepinephrine. (After Nielsen and Owman, *Brain Res.*, 1971, *27*, 33–42.)

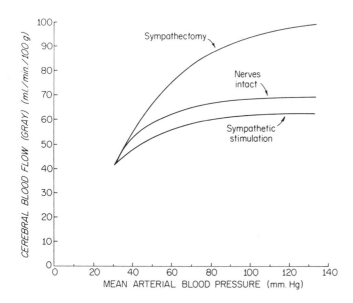

Figure 17–12 The response of cerebral blood flow to alterations in arterial blood pressure under conditions of sympathectomy, intact nerves and sympathetic stimulation. The normal curve shows relative constant flow from 60 to 140 mm Hg arterial pressure. This constancy of flow is modified by the sympathetic nervous influences. (After James *et al., Circulat. Res.*, 1969, 25, 77–93.)

increase in blood flow as indicated by hydrogen clearance techniques.[58]

In summary, the demonstration of a sympathetic vasoconstrictor mechanism does not explain its role in the overall regulation of the cerebral circulation. Whether the autonomic innervation of the cerebral vessels participates in general cardiovascular reflexes is not yet known. The adjustment of cerebral blood flow to postural changes has been postulated to be reflex in mechanism,[15, 61] but baroreceptor reflexes do not seem to be necessary for the "tone" of the cerebral vasculature.[50] Present evidence suggests that the autonomic nervous system can significantly alter cerebral blood flow, but the conditions under which these mechanisms are active has yet to be clarified.

Perfusion Pressure. The term "cerebral autoregulation" is often used to refer to the fact that cerebral blood flow remains constant despite changes in perfusion pressure (Fig. 17–12). This phenomenon probably results from interaction between intrinsic and extrinsic control mechanisms. Rigorously defined, autoregulation is restricted to control by intrinsic mechanisms; that is, to mechanisms remaining in a denervated organ without autonomic control. The literature on "cerebral autoregulation" developed prior to demonstration of the capability of the autonomic

nervous control of the cerebral circulation, and observations of constant cerebral blood flow during changes in perfusion pressure were made with cerebral vessel autonomic innervation intact.[49] Autoregulation, carefully defined, does not appropriately describe the results.

CEREBRAL SPINAL FLUID PRESSURE. Cerebral spinal fluid pressure is usually low (1 to 10 mm Hg) and normally has little effect on cerebral blood flow. However, if cerebral spinal fluid pressure is elevated close to arterial pressure, cerebral blood flow decreases provided arterial pressure is held constant (Fig. 17–13).[18] If arterial pressure is allowed to vary, the Cushing reflex is elicited and increased intracranial pressure causes a dramatic increase in arterial blood pressure (Fig. 17–14).[7, 8] The postulated mechanism for the Cushing reflex is that the increase in cerebral spinal fluid pressure causes generalized compression of the cerebral vessels. This, in turn, activates the medullary vasomotor center by anoxic, hypercapnic or acidotic stimulation, resulting in increased sympathetic outflow, splanchnic vasoconstriction and increased arterial blood pressure. The effect of the Cushing reflex is to elevate arterial blood pressure above cerebral spinal fluid pressure and to maintain cerebral blood flow despite high levels of cerebral spinal fluid pressure.

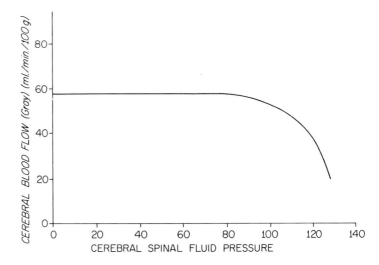

Figure 17–13 Graphic representation of the response of cerebral blood flow to increased cerebral spinal fluid pressure. The "Cushing reflex" was eliminated by holding arterial pressure constant. When cerebral spinal fluid pressure approached arterial pressure, cerebral blood flow began to fall. (After Häggendal *et al.*, *Acta physiol. scand.*, 1970, 79, 262–271.)

CLINICAL OBSERVATIONS

Blood flow to the brain is remarkably constant. Studies on patients using the nitrous oxide technique indicate that cerebral blood flow is normal in hypertension, epilepsy, schizophrenia and multiple sclerosis. In contrast, cerebral blood flow is decreased in cerebral arteriosclerosis and senile psychosis.[28]

Local alterations of cerebral flow occur in ischemic disease states in both the primary area of ischemia and in surrounding areas.

Lassen[38a] has described a "luxury perfusion syndrome," in which brain areas adjacent to diseased brain have a high flow, presumably in response to acid metabolites that have diffused from the diseased region. Ischemic areas distal to a cerebral artery narrowing will also be maximally dilated owing to the build-up of metabolites. The vessels in these areas of ischemia and luxury perfusion are unresponsive to arterial carbon dioxide tension and blood pressure changes, and hence are characterized by flow that is directly proportional to perfusion pressure. These ischemic and sur-

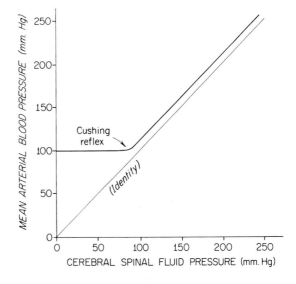

Figure 17–14 Graph illustrating the "Cushing reflex" (change in blood pressure due to increased CSF pressure). As CSF approaches arterial blood pressure, the latter rises and maintains a differential permitting uninterrupted perfusion of cerebral vessels.

rounding areas may show a paradoxical decrease in blood flow when the arterial carbon dioxide tension is elevated. Elevation of arterial carbon dioxide tension dilates the vessels in the normal brain, increasing blood flow to the normal brain and possibly decreasing the pressure in vessels supplying the area of ischemia or luxury perfusion.[2a] This redistribution of flow from diseased areas to normal brain, in the face of a generalized vasodilation, has been termed "intracerebral steal." Conversely, hyperventilation (decrease in arterial carbon dioxide tension) results in a fall in normal brain blood flow and an increase in the flow to the areas of ischemia or luxury perfusion. Lassen has termed this the "Robin Hood effect" (taking from the rich to give to the poor).[38b]

Anesthetics can also affect the cerebral blood flow. Most of the inhalation anesthetics increase cerebral blood flow, whereas the barbiturates decrease it. Virtually all general anesthetics decrease the cerebral metabolic rate.[61a]

REFERENCES

1. Adolph, R. J., Fukusumi, H., and Fowler, N. O. Origin of cerebrospinal fluid pulsations. *Amer. J. Physiol.*, 1967, *212*, 840–846.
2. Batson, O. V. Anatomical problems concerned in the study of cerebral blood flow. *Fed. Proc.*, 1944, 3(3), 139–144.
2a. Brawley, B. W., Strandness, D. E., and Kelly, W. A. The physiologic response to therapy in experimental cerebral ischemia. *Arch. Neurol.*, 1967, *17*, 180–187.
3. Brock, M., Fieschi, C., Ingvar, D. H., Lassen, N. A., and Schürmann, K., eds. *Cerebral blood flow.* Berlin, Springer-Verlag, 1969.
4. Chorobski, J., and Penfield, W. Cerebral vasodilator nerves and their pathway from the medulla oblongata. With observations on the pial and intracerebral vascular plexus. *Arch. Neurol. Psychiat. (Chic.)*, 1932, *28*, 1257–1289.
5. Clark, S. L. Innervation of the blood vessels of the medulla and spinal cord. *J. comp. Neurol.*, 1929, *48*, 247–265.
6. Clark, S. L. Innervation of chorioid plexuses and the blood vessels within the central nervous system. *J. comp. Neurol.*, 1934, *60*, 21–35.
7. Cushing, H. Concerning a definite regulatory mechanism of the vaso-motor center which controls blood pressure during cerebral compression. *Bull. Johns Hopkins Hosp.*, 1901, *12*, 290–292.
8. Cushing, H. Some experimental and clinical observations concerning states of increased intracranial tension. *Amer. J. med. Sci.*, 1902, *124*, 375–400.
9. D'Alecy, L. G. Sympathetic cerebral vasoconstriction blocked by adrenergic alpha receptor antagonists. *Stroke*, 1973, *4*, 30–37.
10. D'Alecy, L. G., and Feigl, E. O. Sympathetic control of cerebral blood flow in dogs. *Circulat. Res.*, 1972, *31*, 267–283.
11. Dunn, J. S., and Wyburn, G. M. The anatomy of the blood brain barrier: A review. *Scot. med. J.*, 1972, *17*, 21–36.
12. Falck, B., Mchedlishvili, G. I., and Owman, C. Histochemical demonstration of adrenergic nerves in cortex-pia of rabbit. *Acta Pharmacol. Toxicol.*, 1965, *23*, 133–142.
13. Fang, H. C. H. Cerebral arterial innervations in man. *Arch. Neurol. (Chic.)*, 1961, *4*, 651–656.
14. Fencl, V., Vale, J. R., and Broch, J. A. Respiration and cerebral blood flow in metabolic acidosis and alkalosis in humans. *J. appl. Physiol.*, 1969, *27*, 67–76.
15. Galindo, A., Savolainen, V. P., and Suutarinen, T. Craniotomy and internal carotid blood flow. *Ann. Surg.*, 1964, *159*, 437–444.
16. Gibbs, F. A., Gibbs, E. L., and Lennox, W. G. Changes in human cerebral blood flow consequent on alterations in blood gases. *Amer. J. Physiol.*, 1935, *111*, 557–563.
17. Gulland, G. L. The occurrence of nerves on intracranial blood vessels. *Brit. med. J.*, 1898, *2*, 781–782.
18. Häggendal, E., Löfgren, J., Nilsson, N. J., and Zwetnow, N. N. Effects of varied cerebrospinal fluid pressure on cerebral blood flow in dogs. *Acta physiol. scand.*, 1970, *79*, 262–271.
19. Harmel, M. H., Hafkenschiel, J. H., Austin, G. M., Crumpton, C. W., and Kety, S. S. The effect of bilateral stellate ganglion block on the cerebral circulation in normotensive and hypertensive patients. *J. clin. Invest.*, 1949, *28*, 415–418.
20. Harper, A. M., and Bell, R. A. The effect of metabolic acidosis and alkalosis on the blood flow through the cerebral cortex. *J. Neurol. Neurosurg. Psychiat.*, 1963, *26*, 341–344.
21. Harper, A. M. Physiology of cerebral bloodflow. *Brit. J. Anaesth.*, 1965, *37*, 225–235.
22. Hassler, O. A systematic investigation of the physiological intima cushions associated with the arteries in five human brains. *Acta Soc. Med. upsalien.*, 1962, *67*, 35–41.
23. Hassler, O. Physiological intima cushions in the large cerebral arteries of young individuals. I. Morphological structure and possible significance for the circulation. *Acta path. microbiol. scand.*, 1962, *55*, 19–27.
24. Iwayama, T., Furness, J. B., and Burnstock, G. Dual adrenergic and cholinergic innervation of the cerebral arteries of the rat. An ultrastructural study. *Circulat. Res.*, 1970, *26*, 635–646.
25. James, I. M., Millar, R. A., and Purves, M. J. Observations on the extrinsic neural control

of cerebral blood flow in the baboon. *Circulat. Res.*, 1969, *25*, 77–93.

26. Kellie, G. An account of the appearances observed in the dissection of two of three individuals presumed to have perished in the storm of the 3rd and whose bodies were discovered in the vicinity of Leith on the morning of the 4th November 1821: With some reflections on the pathology of the brain. *Trans. Edinb. med. chirurg. Soc.*, 1824, *1*, 84–169.

27. Kety, S. S. Stellate ganglion blockade and the cerebral circulation (editorial). *Anesthesiology*, 1959, *20*, 697.

28. Kety, S. S. The cerebral circulation. *Handb. Physiol.*, 1960, sec. 1, vol. III, 1751–1760.

29. Kety, S. S. Circulation and energy metabolism of the brain. *Clin. Neurosurg.*, 1963, *9*, 56–66.

30. Kety, S. S. The control of cerebral circulation (editorial). *Circulation*, 1964, *30*, 481–483.

31. Kety, S. S., Polis, B. D., Nadler, C. S., and Schmidt, C. F. The blood flow and oxygen consumption of the human brain in diabetic acidosis and coma. *J. clin. Invest.*, 1948, *27*, 500–510.

32. Kety, S. S., and Schmidt, C. F. The effects of active and passive hyperventilation on cerebral blood flow, cerebral oxygen consumption, cardiac output, and blood pressure of normal young men. *J. clin. Invest.*, 1946, *25*, 107–119.

33. Kety, S. S., and Schmidt, C. F. The nitrous oxide method for the quantitative determination of cerebral blood flow in man: Theory, procedure and normal values. *J. clin. Invest.*, 1948, *27*, 476–483.

34. Kety, S. S., and Schmidt, C. F. The effects of altered tensions of carbon dioxide and oxygen on cerebral blood flow and cerebral oxygen consumption of normal young men. *J. clin. Invest.*, 1948, *27*, 484–492.

35. Kuschinsky, W., Wahl, M., Bosse, O., and Thurau, K. Perivascular potassium and *p*H as determinants of local pial arterial diameter in cats. *Circulat. Res.*, 1972, *31*, 240–247.

36. Lambertsen, C. J., Kough, R. H., Cooper, D. Y., Emmel, G. L., Loeschcke, H. H., and Schmidt, C. F. Oxygen toxicity. Effects in man of oxygen inhalation at 1 and 3.5 atmospheres upon blood gas transport, cerebral circulation and cerebral metabolism. *J. appl. Physiol.*, 1953, *5*, 471–486.

37. Lambertsen, C. J., Semple, S. J. G., Smyth, M. G., and Gelfand, R. H$^+$ and pCO_2 as chemical factors in respiratory and cerebral circulatory control. *J. appl. Physiol.*, 1961, *16*(3), 473–484.

38. Langfitt, T. W., and Kassell, N. F. Cerebral vasodilatation produced by brain-stem stimulation: Neurogenic control vs. autoregulation. *Amer. J. Physiol.*, 1968, *215*, 90–97.

38a. Lassen, N. A. The luxury-perfusion syndrome and its possible relation to acute metabolic acidosis localized within the brain. *Lancet*, 1966, *2*, 1113–1115.

38b. Lassen, N. A., and Pálvölgyi, R. Cerebral steal during hypercapnia and the reverse reaction during hypocapnia observed by the ^{133}Xenon technique in man. *Scand. J. clin. lab. Invest.*, 1968 (Suppl. 102), XIII:D.

39. Lennox, W. G., and Gibbs, E. L. The blood flow in the brain and the leg of man, and the changes induced by alteration of blood gases. *J. clin. Invest.*, 1932, *11*, 1155–1177.

40. McHenry, L. C., Jr. Cerebral blood flow. *New Engl. J. Med.*, 1966, *274*, 82–91.

41. McNaughton, F. L. The innervation of the intracranial blood vessels and dural sinuses. *Assoc. Res. nerv. ment. Dis.*, 1938, *18*, 178–200.

42. Nelson, E., and Rennels, M. Neuromuscular contacts in intracranial arteries of the cat. *Science*, 1969, *167*, 301–302.

43. Nelson, E., and Rennels, M. Innervation of intracranial arteries. *Brain*, 1970, *93*, 475–490.

44. Nielson, K. C., and Owman, C. Adrenergic innervation of pial arteries related to the circle of Willis in the cat. *Brain Res.*, 1967, *6*, 773–776.

45. Nielson, K. C., and Owman, C. Contractile response and amine receptor mechanisms in isolated middle cerebral artery of the cat. *Brain Res.*, 1971, *27*, 33–42.

46. Norcross, N. C. Intracerebral blood flow. An experimental study. *Arch. Neurol. Psychiat., (Chic.)*, 1938, *40*, 291–299.

47. Patterson, J. L., Jr., Heyman, A., Battey, L. L., and Ferguson, R. W. Threshold of response of the cerebral vessels of man to increase in blood carbon dioxide. *J. clin. Invest.*, 1955, *34*, 1857–1864.

48. Penfield, W. Intracerebral vascular nerves. *Arch. Neurol. Psychiat. (Chic.)*, 1932, *27*, 30–44.

49. Purves, M. J. *The physiology of the cerebral circulation.* Cambridge, University Press, 1972.

50. Rapela, C. E., Green, H. D., and Denison, A. B., Jr. Baroreceptor reflexes and autoregulation of cerebral blood flow in the dog. *Circulat. Res.*, 1967, *21*, 559–568.

51. Reivich, M. Arterial pCO_2 and cerebral hemodynamics. *Amer. J. Physiol.*, 1964, *206*(1), 25–35.

52. Rosenblum, W. I. Increased binding of norepinephrine by nerves to cerebral blood vessels: Evidence from the effects of reserpine on nerves to cerebral and extracerebral blood vessels. *Stroke*, 1973, *4*, 42–45.

53. Salanga, V. D., and Waltz, A. G. Regional cerebral blood flow during stimulation of seventh cranial nerve. *Stroke*, 1973, *4*, 213–217.

54. Samarasinghe, D. D. The innervation of the cerebral arteries in the rat: An electron microscope study. *J. Anat. (Lond.)*, 1965, *99*, 815–828.

55. Sato, S. An electron microscopic study on the innervation of the intracranial artery of the rat. *Amer. J. Anat.*, 1966, *118*, 873–889.

56. Schieve, J. F., and Wilson, W. P. The changes in cerebral vascular resistance of man in experimental alkalosis and acidosis. *J. clin. Invest.*, 1952, *32*, 33–38.

57. Schmidt, C. F. The influence of cerebral blood-flow on respiration. I. The respiratory responses to changes in cerebral blood-flow. *Amer. J. Physiol.*, 1928, *84*, 202–222.

58. Scremin, O. U., Rovere, A. A., Raynald, A. C., and Giardini, A. Cholinergic control of blood flow in the cerebral cortex of the rat. *Stroke*, 1973, *4*, 232–239.

59. Severinghaus, J. W., and Lassen, N. Step hypocapnia to separate arterial from tissue pCO_2 in

the regulation of cerebral blood flow. *Circulat. Res.*, 1967, *20*, 272–278.

60. Shenkin, H. A., Groff, R. A., Spitz, E. B., Scheuerman, W. G., and Cabieses, F. Effects of bilateral stellectomy on the cerebral circulation of man. *Arch. Neurol. Psychiat. (Chic.)*, 1950, *64*, 289–291.

61. Shenkin, H. A., Scheuerman, W. G., Spitz, E. B., and Groff, R. A. Effect of change of position upon the cerebral circulation of man. *J. appl.*

Physiol., 1949, *2*, 317–326.

61a. Smith, A. L., and Wollman, H. Cerebral blood flow and metabolism: Effects of anesthetic drugs and techniques. *Anesthesiology*, 1972, *36*, 378–400.

62. Stohr, P. Nerves of the blood vessels, heart, meninges, digestive tract and urinary bladder. In: *Cytology and cellular pathology of the nervous system*, vol. 1, W. Penfield, ed. New York, Paul B. Hoeber, Inc., 1932.

CHAPTER 18 PULMONARY CIRCULATION

by CLAUDE LENFANT

CHARACTERISTICS OF THE PULMONARY CIRCULATION

The primary function of the pulmonary circulation is to expose the blood to alveolar air so that gas can be exchanged between them (external respiration). The pulmonary circulation is in series with the systemic circulation. Although the systemic circulation acts similarly by conveying blood to capillaries where gas exchange also occurs (internal respiration), the pulmonary circulation exhibits some unique features.

Anatomical Characteristics. ARTERIES. The pulmonary artery divides repeatedly into smaller arteries which become arterioles (less than 100 μ caliber), finally reaching a diameter of approximately 50 μ at the level of the alveolar ducts. No vessel in the human pulmonary circulation has a thick muscular media similar to the arterioles of the systemic circulation. Even the small pulmonary arteries of 1000 to 100 μ in diameter contain only a thin layer of smooth muscle. Smooth muscle is a prerequisite for vasomotoricity, and is responsible for change in resistance. Thus, such low density of muscle precludes intensive vasomotor activity.

CAPILLARIES. The terminal arterioles precede the capillaries which subdivide into a thin network lying between the alveolar walls. The capillaries are 10 to 14 μ in length and 7 to 9 μ in diameter. Not all the capillaries are open at any one time. However, a larger number probably open in the event of increasing lung volume or to accommodate an increase in blood flow, as in exercise.

VEINS. The capillaries join to form the pulmonary veins which also contain little or no muscle. Although this suggests a lack of significant vasomotor activity in the pulmonary veins, there is now evidence of venoconstriction in response to some pharmacological agents.[11]

These anatomical characteristics largely explain the *extensibility* and *collapsibility* of the pulmonary vessels. The large arteries and the small veins are more extensible than the small arteries and the large veins. The venous portion is the more compliant part of the vascular bed,[5] and contains more than half of the pulmonary blood volume. The collapsibility of the thin-walled pulmonary vessels has important functional implications which are discussed extensively below and under blood flow distribution (Chap. 21, Vol. II). Briefly, this property bears on the distribution of blood flow in the lung, for if the surrounding alveolar pressure exceeds the intraluminal vascular pressure, the vessels collapse.

Functional Characteristics. Functionally the pulmonary circulation contrasts strikingly with the systemic circulation.

Low Pressure. The pulmonary circulation is a low pressure system. Pulmonary arterial pressure oscillates from about 10 mm Hg in diastole to about 25 mm Hg during systole. Mean pressure is approximately 18 mm Hg, or one-sixth that of the systemic circulation. Mean pressure in the left atrium is generally 5 to 6 mm Hg; in the large pulmonary veins, outside the lung tissues, venous pressure is similar and, as in the atrium, oscillates during the cardiac cycle from 3 to 10 mm Hg (Fig. 18–1). The low pressure prevailing in the pulmonary artery prevents filtration of fluid from blood into the alveoli. Indeed, in normal subjects the mean arterial pressure is lower than the colloidal osmotic pressure of the plasma (25 to 30 mm Hg), which is the driving force for fluid transudation. However, when the pulmonary vascular pressures increase, as in mitral stenosis, the total pressure inside the vessels exceeds osmotic pressure, resulting in pulmonary edema.

High Blood Flow. The pulmonary circulation is a high blood flow system. The entire cardiac output perfuses the pulmonary circulation because of its position in series with the systemic circulation. No other single organ circulation receives as high a blood flow. Although the total flow in the systemic circulation is just as great, it is distributed to many different organs in accordance with their needs.

Low Resistance. The pulmonary circulation is a low resistance system. The total resistance to blood flow in the lung is given in terms of the ratio of driving pressure to blood flow. The mean total driving pressure is $(18 - 5) = 13$ mm Hg, one-seventh or one-eighth times that of the systemic circulation. Thus, since total blood flow is the same in both circulations, pulmonary resistance is seven or eight times less than systemic resistance. The difference in resistance is due partly to the structural properties of the pulmonary vessels which allow for a remarkable extensibility. This, plus the opening of a larger number of capillaries, explains why, in exercise, the blood flow can increase many times without a proportionate increase in pulmonary artery pressure.

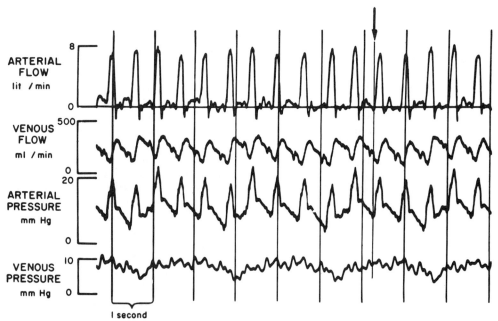

Figure 18–1 Typical pressure recording from the pulmonary artery and the left atrium. Note effect of respiration, *i.e.*, a decrease in absolute value during inspiration. Flow velocity is shown in two upper rows. (From Wiener *et al.*, *Circulat. Res.*, 1966, *19*, 834.)

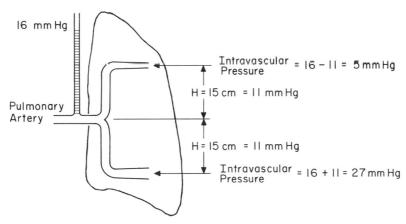

Figure 18–2 Effect of hydrostatic pressure in the lung. In the main pulmonary artery the mean pressure generated by the right ventricle is 16 mm Hg. At the apex of the lung the *absolute* pressure is equal to the main pulmonary artery pressure minus that exerted by the hydrostatic gradient or 5 mm Hg; at the base the absolute pressure is (16 + 11) = 27 mm Hg. The total hydrostatic pressure is given by the total height of the blood "columns," or 22 mm Hg in this example.

EFFECT OF HYDROSTATIC PRESSURE. The pulmonary circulation is greatly affected by the hydrostatic pressure. Because the pulmonary capillary bed is surrounded by air spaces (alveoli) where the pressure is nearly atmospheric, hydrostatic pressure differences between the apex and the base of the lung influence the distribution of the blood circulation. At the bottom of the lung, hydrostatic pressure is maximum (Fig. 18–2) and increases *transmural pressure*. (Transmural pressure is the outwardly acting difference in pressure between the inside and the outside of the vessel.) Thus, the vessels are distended and offer the least resistance to blood flow. The pressure generated by the right ventricle is hardly sufficient to raise the blood to the upper parts of the lung where hydrostatic pressure is at a minimum. The vessels have small diameters and exert enough resistance to considerably decrease blood flow to the upper lobe. In the systemic circulation, the weight of the tissues is equal to or greater than that of the blood and opposes the hydrostatic pressure of the blood at any vertical level of the body, with the result that the vessels are not distended.

BLOOD RESERVOIR. The pulmonary vascular bed is a distensible blood reservoir interposed between the left and right heart. It normally contains 10 to 12 per cent of the total blood volume,[5] but this amount can change greatly under various conditions.[8]

An increase occurs (i) with increasing blood flow as in exercise, (ii) with reduction of left ventricular output as in left heart failure, (iii) with assumption of the supine position (which increases the cardiac output), (iv) with immersion of the body in water (except the head) and (v) with systemic vasoconstriction. A decrease in pulmonary blood volume is noted (i) on standing up, (ii) with a decrease in cardiac output, (iii) with systemic vasodilation as in heat exposure and (iv) with positive pressure breathing. Because of the large amount of blood normally present in the pulmonary vasculature, more than half of which is in the pulmonary veins, the left heart can transiently increase its output over that of the right heart. The pulmonary capillary blood volume is relatively small, 75 to 100 ml. This volume is spread out in numerous capillaries which occupy an area of approximately 70 m². The transmit time for blood in the capillaries is about 0.75 sec.[9]

INFLUENCE OF RESPIRATION ON THE PULMONARY CIRCULATION

The diameter of the pulmonary vessels, like that of most vessels, is determined by the *transmural pressure* which depends partly on the external pressure, *i.e.*, the location of the vessels in relation to the lung itself. The larger vessels are in the intrapleural space and thus are influenced by the negative pressure prevailing in that space, but the smaller ones traverse the diaphanous alveolar walls and are therefore exposed to atmospheric pressure.

Influence of the Respiratory Cycle. Since the pleural and alveolar pressures vary during the respiratory cycle, the change in transmural pressure produces significant hemodynamic and geometric alterations (Fig. 18–3A and B).

During inspiration the lungs enlarge and the pleural pressure decreases (becomes more negative). Hence the outward transmural force exerted on the heart and extrapulmonary vessels increases, resulting in the following effects: (i) widening and lengthening of the extrapulmonary vessels and (ii) lowering of the intravascular pressure relative to atmospheric pressure,

immediately followed by (iii) flow and volume increase caused by increased systemic venous return. The net result is significant distension of the larger extra-alveolar pulmonary vessels, which implies a decrease of their resistance to flow.

At the same time, the decreased intrapleural pressure during inspiration is transmitted in part to the alveoli and causes air to enter the airways. The change in alveolar pressure is much less than that in intrapleural pressure, so the effect on the alveolar vessels is as if they were being compressed by a positive pressure. Indeed the alveolar pressure always becomes positive relative to the pressure around the heart, as transpulmonary pressure increases. The geometric changes associated with the lung inflation also have an opposite effect; *i.e.*, they exert an elongating force on the alveolar vessels, stretching them and tending to

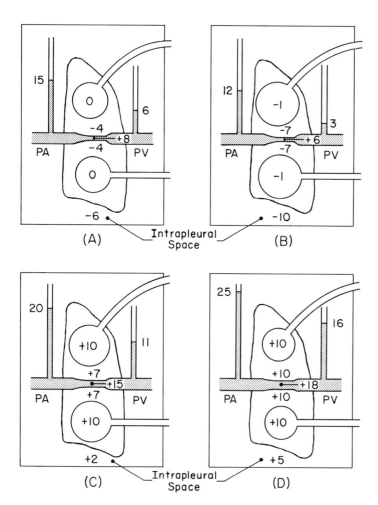

(A)

Intrapleural Space

(B)

(C)

Intrapleural Space

(D)

Figure 18–3 Effect of lung inflation on the pulmonary circulation (all pressures in mm Hg). A, The lung at end of normal expiration. Alveolar pressure is zero and pleural pressure is −6 mm Hg; pericapillary pressure is slightly more positive or −4 mm Hg. B, During inspiration the pleural pressure becomes more negative, decreasing the pressure in the extrapulmonary vessels. Because of the inflation the diameter of the capillaries decreases. C, Positive lung inflation causes an increase of the pressures in the pleural space and in all the pulmonary vessels. Again lung expansion compresses the capillaries. D, If the lung cannot expand (as in a strapped chest), the volume is of course unchanged, and this, the differential pressures and resistance are exactly the same as normal. The only effect then stems from the decrease in venous return, because the pressures in the chest are higher than those in the extrathoracic veins.

reduce their diameter. These two effects increase the resistance to flow in the alveolar vessels.

Overall, these two opposite changes in resistance produce little significant change of total resistance in the range of normal breathing.[14] However, when the functional residual capacity is increased, either voluntarily[3] or in clinical conditions resulting from chronic obstructive airway disease, total vascular resistance increases.

During normal expiration the changes described for inspiration are reversed. Pulmonary blood flow and blood volume reach a lower value because of the increase in intrathoracic pressure which opposes the filling of the right heart. During forceful expiration or in clinical conditions associated with active expiration, the pleural pressure may transiently exceed the pressure in the extrathoracic veins and completely interrupt venous return.

Influence of Pressure Breathing. Positive pressure breathing exists when the pressure applied at the mouth exceeds that around the thorax. The term *intermittent* positive pressure breathing is used if the difference exists only during inspiration, and *continuous* positive pressure breathing is used if the positive pressure is maintained during both inspiration and expiration. Positive pressure breathing affects the pulmonary circulation by increasing lung size and causing pleural pressure to become more positive. The increase in pleural pressure around the cardiac chambers reduces the venous return and diminishes the diastolic filling, and thus the cardiac output. It also reduces the volume of the extrapulmonary vessels. This, plus the high lung volume influencing the size of the capillaries, increases the overall resistance to blood flow (Fig. 18–3C and D).[15, 16] All these hemodynamic changes are immediately reversible upon release of the positive pressure.

Negative pressure breathing is characterized by a lower pressure in the airways than around the body (as for example, during immersion in water up to the neck or during snorkel breathing). Blood is displaced into the lungs and the pulmonary vessels become distended, yielding hemodynamic changes opposite to those caused by positive pressure breathing.

PARTITIONING OF PULMONARY VASCULAR RESISTANCE

Because the pulmonary vessels are collapsible and distensible, and because of the interaction between the intravascular and the perivascular pressures, the intravascular pressure fall is not continuous between the main pulmonary artery and the large pulmonary veins. On the contrary, regions exist where flow and upstream pressure are independent of the downstream pressure. The mechanism causing this situation is comparable to that of a "sluice" or a "waterfall."[12, 13]

The resistance exerted by a collapsible tube surrounded by a pressure is determined by the transmural pressure difference between the inside and outside pressures: (i) if the external pressure is smaller than both inflow pressure and outflow pressure, the tube is wide open and flow is proportional to the difference between inflow and outflow pressures; (ii) if the external pressure is greater than both inflow and outflow pressures, the tube collapses and no flow occurs; (iii) if the external pressure is smaller than the inflow pressure but greater than the outflow pressure, flow and diameter of the tube are proportional to the difference between inflow and external pressures and independent of the outflow pressure.

The difference between the external and the outflow pressures corresponds to the vertical height of the waterfall and does not influence the flow over the edge of the fall. The collapsible tube may develop an intermittent constriction at its distal end where the difference between inside pressure and external pressure is minimum and, at times, nil. In this case, the tube tends to collapse and flow is reduced. Upstream pressure then builds up and forces the tube open. With flow re-established, distal pressure inside the tube drops until it equals the external pressure, at which time the tube collapses again. This cycle repeats and is comparable to the vibration obtained in a flutter valve.

Two waterfall mechanisms have been recognized in the lungs.

The Alveolar Waterfall.[2, 10, 13] This waterfall exists in the pulmonary capillaries where the alveolar pressure (P_A) is smaller

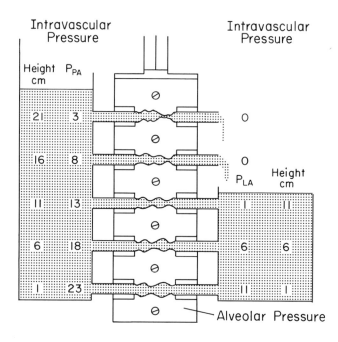

Figure 18-4 The alveolar waterfall. A waterfall effect exists in all the capillaries above the venous reservoir (right hand side), the height of which represents the venous pressure. The absolute pressure (P) in any channel is equal to the pressure generated by the right ventricle (24 mm Hg) in the arterial side and 12 mm Hg in the venous side) minus the height of the column of blood (as in Fig. 18-2). On the venous side the pressure in the veins becomes zero for all the channels at a height greater than that corresponding to the pressure generated by the heart (12 mm Hg). Since alveolar pressure equals 2 cm H_2O, the condition Pa>PA>Pv is reached in all these channels and the waterfall phenomenon is present. (After Caro, *Advances in respiratory physiology.* London, Edward Arnold, 1966.)

than the pressure in the small arteries (Pa) but greater than that in the small intrapulmonary veins (Pv), or (Pa>PA>Pv). (Recall that a vertical hydrostatic gradient exists in the pulmonary bed.) Since the pressure in the alveoli is uniformly atmospheric (no gradient), the condition Pa>PA>Pv pertains between the level where the absolute pressure in the veins equals the atmospheric pressure and the apex of the lung. However, this condition is not present between this level and the bottom of the lung, as illustrated in Figure 18-4, which also shows that the absolute pressure in a vein is equal to the driving pressure minus the height of the column of blood. This explains how hydrostatic gradients play a major role in distribution of blood flow in the lung (Chap. 2, Vol. II).

The Venous Waterfall.[17] The intrapulmonary veins gradually coalesce and emerge into the pleural space before converging into the left atrium. If the pressure fall were continuous from the smaller veins to the left atrium, a progressive rise of the left atrial pressure would be accompanied by an equal increase of the intrapulmonary venous pressures. Figure 18-5, however, reveals a discontinuity. First, raising the left atrial pressure from −5 to +7 cm H_2O does not change the intrapulmonary venous pressure; then a further increase causes an almost equal rise in venous pressure. The

value of +7 cm H_2O can be compared to the height of the waterfall as the veins leave the lung; if the downstream level rises above it, the upstream level increases by the same amount. Figure 18-5 also shows that the arterial pressure does not increase until the left atrial pressure exceeds a value close to the alveolar pressure. This discontinuity corresponds to the alveolar waterfall discussed earlier.

The venous waterfall is located within a few millimeters of the pleural surface. It presumably results from the effect on the veins of the change from the intrapulmonary to the less positive extrapulmonary pressures. Apparently the perivascular pressure is higher just within the edge of the lung than in the pleural space and acts on the vein in a manner analogous to that of high alveolar pressure on the capillaries. The venous waterfall plays a role in limiting drainage from the intrapulmonary vascular bed when the left atrial pressure is lowered during inspiration, *i.e.*, it helps to regulate left atrial filling.

VASOMOTOR ACTIVITY

Since the pulmonary vessels are invested with smooth muscles, they have the ability to contract actively and to dilate. A challenging and still unresolved aspect is

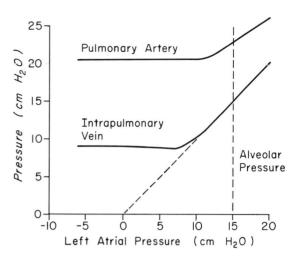

Figure 18-5 The venous waterfall. An increase in the left atrial pressure is not immediately accompanied by an increase in pressure in the intrapulmonary veins. Also note that pressure in the pulmonary vein must be within a few centimeters of water of the alveolar pressure before pulmonary artery pressure begins to rise. This is due to the alveolar waterfall. In all waterfalls the pressure upstream is independent of the pressure downstream. (From Takahashi *et al.*, *J. appl. Physiol.*, 1969, *26*, 578–584.)

whether this mechanism is normally used to control the pulmonary blood flow. Various agents stimulate the vasomotor activity.

Constrictors. HYPOXIA. A reduction below 12 per cent in O_2 concentration of inspired air elicits an increase in pulmonary arterial pressure. When the hypoxia is acute, this response exceeds greatly that attributable to the concomitant small increase in cardiac output. Pulmonary hypertension also exists in chronic arterial hypoxia (arterial oxygen saturation lower than 80 per cent), as in residents at high altitude. At an altitude of 15,000 feet the increase in pulmonary arterial pressure is moderate at rest (about 28 mm Hg, mean pressure), but it becomes very marked during exercise.

The counterpart of hypoxia, hyperoxia, does not reduce pressure in subjects with a normal pulmonary pressure. However, such a decrease does occur in subjects with chronically elevated pressure because of hypoxia.

ACIDOSIS. Pulmonary hypertension is also elicited by acidosis quite independently from the effect of hypoxia, which often appears in concert with it. The effects of metabolic or respiratory acidosis are similar, and the degree of vasoconstriction is related to the degree of acidosis. Acidosis and hypoxia appear to exert their effect directly on the vessel rather than by way of extrapulmonary reflexes.

PHARMACOLOGICAL AGENTS. The following substances have been shown to act as vasoconstrictors in the pulmonary circula-

tion: serotonin, histamine, adenosine triphosphate, bacterial endotoxin, alloxan and hypertonic saline. The site of vasomotor response—artery or vein—varies with the agent.[11]

DILATORS. These agents are grouped as follows: (i) musculotropics such as aminophylline; (ii) parasympathomimetics such as acetylcholine: this substance is especially active and is capable of reversing the constricting effect of hypoxia;[9] (iii) sympathomimetics such as isoproterenol; (iv) adrenergic blockers such as tolazoline; and (v) ganglionic blockers such as hexamethonium.

REFERENCES

1. Aviado, D. M. The pharmacology of the pulmonary circulation. *Pharmacol. Rev.*, 1960, *12*, 159–239.
2. Banister, J., and Torrance, R. W. The effects of the tracheal pressure upon flow: Pressure relations in the vascular bed of isolated lungs. *Quart. J. exp. Physiol.*, 1960, *45*, 352–367.
3. Butler, J., and Paley, H. W. Lung volume and pulmonary circulation. *Med. Thorac.*, 1962, *19*, 69–75.
4. Caro, C. G. Mechanics of the pulmonary circulation. In: *Advances in respiratory physiology*, C. G. Caro, ed. London, Edward Arnold, 1966.
5. Caro, C. G., and Saffman, P. G. Extensibility of blood vessels in isolated rabbit lungs. *J. Physiol. (Lond.)*, 1965, *178*, 193–210.
6. Dock, D. S., Kraus, W. L., McGuire, L. B., Hyland, J. W., Haynes, F. W., and Dexter, L. The pulmonary blood volume in man. *J. clin. Invest.*, 1961, *40*, 317–328.
7. Fishman, A. P. Respiratory gases in the regulation of the pulmonary circulation. *Physiol. Rev.*, 1961, *41*, 214–280.

8. Fishman, A. P. Dynamics of the pulmonary circulation. *Handb. Physiol.*, 1963, sec. 2, vol. 11, 1667–1743.

9. Forster, R. E. Exchange of gases between alveolar air and pulmonary capillary blood: Pulmonary diffusing capacity. *Physiol. Rev.*, 1957, *37*, 391–452.

10. Fritts, H. W., Jr., Harris, P., Clauss, R. H., Odell, J. E., and Cournand, A. The effect of acetylcholine on the human pulmonary circulation under normal and hypoxic conditions. *J. clin. Invest.*, 1958, *37*, 99–110.

11. Glazier, J. B., and Murray, J. F. Sites of pulmonary vasomotor reactivity in the dog during alveolar hypoxia and serotonin and histamine infusion. *J. clin. Invest.*, 1971, *50*, 2550–2558.

12. Howell, J. B. L., Permutt, S., Proctor, D. F., and Riley, R. L. Effect of inflation of the lung on different parts of pulmonary vascular bed. *J. appl. Physiol.*, 1961, *16*, 71–76.

13. Permutt, S., Bromberger-Barnea, B., and Bone, H. N. Alveolar pressure, pulmonary venous pressure and the vascular waterfall. *Med. Thorac.*, 1962, *19*, 47–68.

14. Roos, A., Thomas, L. J., Jr., Nagel, E. L., and Prommas, D. C. Pulmonary vascular resistance as determined by lung inflation and vascular pressures. *J. appl. Physiol.*, 1961, *16*, 77–84.

15. Thomas, L. J., Jr., Griffo, Z. J., and Roos, A. Effect of negative-pressure inflation of the lung on pulmonary vascular resistance. *J. appl. Physiol.*, 1961, *16*, 451–456.

16. Thomas, L. J., Jr., Roos, A., and Griffo, Z. J. Relation between alveolar surface tension and pulmonary vascular resistance. *J. appl. Physiol.*, 1961, *16*, 457–462.

17. Takahashi, S., and Butler, J. A vascular waterfall in extra-alveolar vessels of the excised dog lung. *J. appl. Physiol.*, 1969, *26*, 578–584.

18. Wagenvoort, C. A., Heath, D., and Edwards, J. E. *The pathology of the pulmonary vasculature.* Springfield, Ill., Charles C Thomas, 1964.

CHAPTER 19 COAGULATION OF BLOOD

by ROSEMARY BIGGS

Blood coagulation results from the solidification in the blood of a particular plasma protein, fibrinogen. This alteration in fibrinogen is the only directly observable change that occurs in the blood during clotting, but it is certainly preceded by a long chain of preliminary reactions which can be studied only by their effect on the final stage of clotting. Information about important early stages is therefore not direct but inferred, and is often a matter more of interpretation and opinion than of fact. Thus conflicting views are inevitable. In this presentation much of the conflict is eliminated by giving only one point of view. This method is unsatisfactory, because it does not prepare the reader for the equally dogmatic opposing views encountered in the literature; but in reviewing so complicated a problem both briefly and clearly a personal interpretation is unavoidable. The reader must be warned that if the subject is pursued more deeply the conflict of opinion will very soon be encountered. Another difficulty arises from the fact that much information about the normal process of coagulation is derived from study of patients with abnormal clotting or from study of plasma fractions isolated from blood. The phenomena observed in artificial experiments, or in the blood of patients, may have no simple interpretation in terms of the normal process.

The study of blood coagulation has been simplified by the introduction of the Roman numeral system of nomenclature. In this system each recognized coagulation factor has a Roman number and each activated

Table 19–1

FACTOR	COMMON IMPORTANT SYNONYMS	PROPERTIES
I	Fibrinogen	Converted by thrombin to fibrin with the release of fibrinopeptides
II	Prothrombin	The precursor of thrombin adsorbed by $Al(OH)_3$, $BaSO_4$ and $Ca_3(PO_4)_2$; not present in serum
III	Thromboplastin tissue factor	Present in saline extracts of most tissues, notable activity being present in lung and brain
IV	Calcium	Required for the interaction of all clotting factors except XI + XII + IIa
V	Labile factor Accelerator globulin	Not adsorbed by $Al(OH)_3$, $BaSO_4$, $Ca_3(PO_4)_2$; not present in serum; labile on storage; required by extrinsic and intrinsic clotting
VI	This term is not now used	
VII	Stable factor Serum accelerator globulin Autoprothrombin I Proconvertin	Adsorbed by $Al(OH)_3$, $BaSO_4$ and $Ca_3(PO_4)_2$; present in serum; stable on storage; not required by RVV; not involved in intrinsic clotting
VIII	Antihemophilic factor A Antihemophilic globulin	Not adsorbed by $Al(OH)_3$, $BaSO_4$, $Ca_3(PO_4)_2$, etc.; absent from serum
IX	Autoprothrombin II Anithemophilic factor B Christmas factor Plasma thromboplastin component	Adsorbed by $Al(OH)_3$, $BaSO_4$, $Ca_3(PO_4)_2$, etc.; present in serum
X	Stuart-Prower factor Autoprothrombin III	Adsorbed by $Al(OH)_3$, $BaSO_4$, $Ca_3(PO_4)_2$; present in serum
XI	Plasma thromboplastin antecedent (PTA) factor	Adsorbed by glass and celite sodium stearate, etc.; not adsorbed by $Al(OH)_3$, $BaSO_4$; present in serum
XII	Hageman factor	" " "
XIII	Fibrin stabilizing factor Fibrinase	Required for the formation of cross linkages in fibrin during the conversion of fibrinogen to fibrin by thrombin

derivative is indicated by the lower case letter "a" (Table 19–1).

FIBRINOGEN (FACTOR I), THROMBIN (FACTOR IIa), THE THROMBIN-FIBRINOGEN REACTION AND FIBRIN (FACTOR Ia)

Fibrinogen. Fibrinogen, a globulin, is the plasma protein which is coagulated by a specific enzyme, thrombin. Its molecular weight is about 400,000 to 500,000. It is three to four times as large as other plasma proteins and is needle shaped. Fibrinogen is destroyed by heating to 47°C. It is precipitated from human plasma by one-fourth to one-third saturation with ammonium sulfate, by molar phosphate buffer, by 11 volumes per cent of ether at 0°C and by 6 volumes per cent of ethyl alcohol at −2°C and pH 6.8. A good deal is now known about the chemistry of fibrinogen. There are three fibrinopeptide chains in the molecule, designated α(A) β(B) γ(C).[3]

In normal blood fibrinogen is present at a concentration of from 190 to 330 mg per 100 ml. In advanced liver disease the level may be reduced; experiments on chloroform poisoning, which causes both liver damage and fibrinogen reduction, suggest that fibrinogen may be formed in the liver. On the other hand, acute liver disease is often associated with excessive destruction of fibrinogen by fibrinolytic enzymes.

Thrombin (Factor IIa). Thrombin, an

esterase which splits arginyl-glycine bonds, is an active coagulant of fibrinogen which appears in blood during clotting. It is thus not a normal blood constituent, but rather is derived from a precursor, prothrombin (Factor II). The complications of blood clotting theory derive from attempts to understand the mechanism by which prothrombin is normally activated. Thrombin is a stable substance which can be stored in the freeze-dried state.

Thrombin-Fibrinogen Reaction. The most important property of thrombin is its ability to clot fibrinogen. This reaction is usually studied by obtaining relatively pure solutions of fibrinogen and thrombin and observing the speed at which coagulation occurs on mixing the two. By this method it is found that the clotting time is inversely proportional to the concentration of thrombin. This relationship is important in the interpretation of methods for the measurement of prothrombin. Many factors, such as the concentration of fibrinogen, pH, salt concentration and colloid osmotic pressure, influence the clotting time.

The strength of a thrombin solution can be measured by the clotting time of a standard fibrinogen solution under closely defined conditions of pH, salt concentration and temperature. In this way units of thrombin have been variously defined. In practice thrombin "units" are not usually referred to an international standard but are arbitrarily defined for particular experimental situations.

During the thrombin-fibrinogen reaction two fibrinopeptides, A and B, derived from the α(A) and β(B) chains respectively, are split off from the fibrinogen molecule. The initial stages of the reaction involve splitting off the fibrinopeptides to produce residual fibrin-monomer residues which subsequently polymerize.

Fibrin (Factor Ia). When thrombin is added to fibrinogen, stages in the formation of fibrin can be observed with the electron microscope. It is found that the fibrinogen molecules unite end-to-end to form long fibrils; later these fibrils associate into bundles which have regular cross striations. Fibrinogen and fibrin have much the same gross chemical constituents, though the N-terminal amino acids of fibrinogen and fibrin are different. The transformation of fibrinogen into fibrin is thought to be a process of polymerization, the fibrin monomer molecules being held together by both primary and secondary linkages.

Fibrin-Stabilizing Factor (Factor XIII). When purified fibrinogen reacts with thrombin the resulting clot is soluble in weak acids and in 30 per cent urea. Clots formed in plasma are insoluble in these solvents because of the presence of Factor XIII.

Naturally Occurring Deficiency of Factors I and XIII. Occasionally fibrinogen may be reduced or absent from the blood as a constitutional defect. Patients with this rare defect bleed excessively after injury. A more common anomaly is the reduction of fibrinogen in previously normal people, usually caused by intravascular coagulation.

A naturally occurring deficiency of Factor XIII has also been described, and it too is associated with abnormal bleeding.

PROTHROMBIN (FACTOR II) AND THE CONVERSION OF PROTHROMBIN TO THROMBIN (FACTOR IIa)

Prothrombin and the Classical Theory. Prothrombin is the plasma precursor of thrombin. Its existence was orginally postulated because it was observed that whereas an active coagulant of fibrinogen could be isolated from fresh serum, no such coagulant could be obtained from fresh, unclotted whole blood. The existence of prothrombin was confirmed by numerous observations. Mellanby[9] isolated from plasma a substance which, although not itself a coagulant, could be converted into a coagulant on suitable treatment. The isolation from plasma of prothrombin, more or less freed from other constituents, has now become a frequent step in blood coagulation research.

In citrated or oxalated plasma, prothrombin is readily converted to thrombin. On the addition of $CaCl_2$ this conversion may take 2 to 3 minutes. If a tissue extract, such as that from saline extracted brain, is present in addition to $CaCl_2$, the conversion may be achieved in a few seconds. Thus tissue extracts, usually referred to as thromboplastin, accelerate the conversion of prothrombin to thrombin, and the coagulation process may be represented diagrammatically as in Scheme 1.

$$\text{Prothrombin} \xrightarrow[\text{CaCl}_2]{\text{Thromboplastin}} \text{Thrombin}$$

$$\text{Fibrinogen} \xrightarrow{\text{Thrombin}} \text{Fibrin}$$

Scheme 1

This scheme to explain the reactions of coagulation is often called the classical theory of blood coagulation; it was proposed by Schmidt[16] and supported by Morawitz,[10] as well as by much experimental work in the first 40 years of this century.

In 1935 Quick developed the one-stage prothrombin time test which was based on this theory. It consisted of mixing equal parts of citrated or oxalated plasma, thromboplastin and $CaCl_2$ and recording the clotting time of the mixture. If fibrinogen were present in excess, then the clotting time could, according to the theory, only be influenced by the amount of prothrombin present.

Factors V, VII and X. The introduction of the one-stage prothrombin time test led to a wide survey of patients who bled excessively after injury. Among these, many were found whose plasma had a long one-stage prothrombin time. Careful study of these patients disclosed that a few lacked prothrombin, but many patients lacked other factors which affected the speed of conversion of prothrombin to thrombin but did not influence the amount of thrombin formed. In the end, three of these accelerators of prothrombin conversion were defined and named Factors V, VII and X.

It was shown that purified prothrombin alone is converted to thrombin very slowly or not at all in the presence of tissue extract. The reactions which precede thrombin formation are possibly of the type shown in Scheme 2.

In fact, the one-stage prothrombin time test seems to involve a complex chain reaction. It is an empirical test useful in laboratory medicine but it does not measure any specific factor unless modified to do so.

Naturally Occurring Deficiency of Factors II, V, VII and X. There are patients whose blood lacks one of the four Factors II, V, VII or X. These single deficiencies are very rare and the conditions are inherited through an autosomal recessive mechanism. Patients with these defects all bleed excessively after injury. The type of bleeding in some Factor VII deficient patients is peculiar in that bleeding into tissues after trivial bumps may occur but surgical operation sites may heal normally.

Complex deficiencies involving the simultaneous depletion of several factors will be referred to later.

The Intrinsic and Extrinsic Clotting Systems. The test system of the one-stage prothrombin time involves the addition of tissue extract, usually a potent saline extract of brain. Clotting in the presence of tissue extract is often referred to as the *extrinsic clotting system*, because the test involves the addition of a tissue component not normally present in the blood. But whole blood carefully collected without tissue damage will clot solidly in 4 to 10 minutes, and thus all the essential reagents must be present in the blood. The clotting system which does not involve the addition of tissue extract is often referred to as the *intrinsic system*.

Study of this mode of clotting was prompted by the fact that the blood of patients with the most common type of hereditary clotting defect, hemophilia, contained normal amounts of Factors I, II, V, VII and X. A great deal of experimental work has disclosed a complex chain reaction occurring

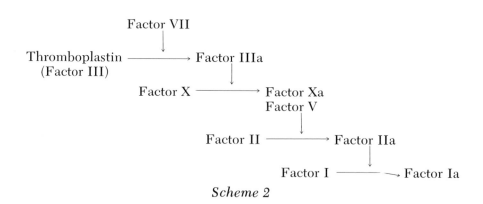

Scheme 2

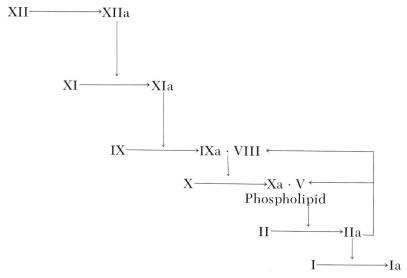

Scheme 3

in blood in the absence of tissue extract. In laboratory experiments this series of reactions is triggered by contact of the blood with a foreign surface, usually glass. Glass contact activates Factor XII, which in turn activates Factor XI and thereafter Factor IX, which in the presence of Factor VIII activates Factor X; thereafter prothrombin conversion occurs as in the extrinsic system. This scheme is shown above.

The difference between the extrinsic and intrinsic systems consists in the mode of activation of Factor X. In the extrinsic system this activation is brought about by tissue extract; in the intrinsic system Factors VIII and IX are of primary importance. In the extrinsic system phospholipid is provided by the tissue extract. In whole blood it is derived from the platelets. The system has an autocatalytic pathway in that thrombin (Factor IIa) activates Factors V and VIII in some way. In the complete absence of thrombin Factor VIII seems to be inactive.[2]

The main disadvantage of this theory is that in life the significance of Factors XI and XII is doubtful. Complete absence of Factor XII produces no abnormality in patients; they undergo surgical operations with no bleeding and can suffer from intravascular clotting. Deficiency of Factor XI is associated with a bleeding tendency, but the degree of laboratory abnormality gives little idea of the severity of bleeding to be expected clinically. The clotting time

of normal blood in glass tubes varies from 4 to 10 minutes. In the vessels the clotting time is almost certainly very much longer, a matter of hours rather than minutes. Thus laboratory testing in glass tubes produces an artifact and we have no certainty that Factors XI and XII are usually concerned in the normal control of bleeding after injury.

Naturally Occurring Defects in the Intrinsic Clotting Mechanism. HEMOPHILIA. Hemophilia is caused by deficiency or total lack of Factor VIII (antihemophilic factor) in the blood. It is inherited as a sex-linked recessive character. A typical family tree is shown in Figure 19–1. Patients suffer from hemorrhagic episodes from early infancy; by adolescence they were, in the old days, usually crippled by the damaging effect of repeated hemorrhages into joints. The coagulation abnormality in hemophilia can be detected by various laboratory tests, including a lengthened whole-blood clotting time, and by other tests of intrinsic clotting function which will be referred to later. Treatment consists of replacing the missing coagulation factor in the blood by transfusion of plasma or concentrates of Factor VIII prepared from plasma.

CHRISTMAS DISEASE. This condition, which is inherited in the same way as hemophilia and has the same clinical features, is caused by lack of Factor IX. Both hemophilia and Christmas disease are rare con-

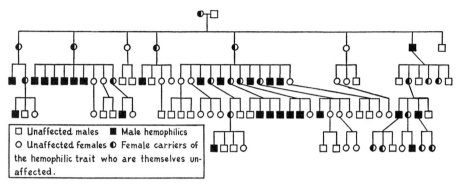

Figure 19–1 Inheritance of hemophilia.

ditions, hemophilia occurring at a rate of two to three cases per 100,000 of the population and Christmas disease at two to three per million. The severity of the bleeding tendency in both conditions is correlated with the results of laboratory tests.

FACTOR XI DEFICIENCY. Factor XI deficiency or plasma thromboplastin antecedent deficiency is rare in many populations, but relatively common in those of Jewish descent. The inheritance is probably autosomal with partial dominance. The bleeding tendency is not well correlated with laboratory test results.

FACTOR XII DEFICIENCY. Factor XII deficiency or Hageman defect is a rare anomaly. The clotting time test in glass tubes is prolonged but the patients seem normal in every way.

Complex Deficiency States. The most common type of complex deficiency state is that in which Factors II, VII, IX and X are all reduced in amount.

VITAMIN K DEFICIENCY. Before Quick's one-stage technique was widely used, a hemorrhagic disease of chickens fed purified diets was investigated. It was found that the condition was cured by giving green vegetables and that the curative substance was not vitamin C, because it was removed by extraction of the diet with fat solvents. A brilliant series of experiments led to the isolation of a fat-soluble substance called vitamin K. It was found that a number of synthetic substances with a structure similar to 2-methyl-3-hydroxy-1,4-naphthoquinone had ability to counteract the disease in chickens, although they were somewhat less effective than the substance extracted from its natural sources.

In human beings deficiency of vitamin K may arise in patients who have a reduced ability to absorb fats and in patients with jaundice. In these patients the one-stage prothrombin time is prolonged and a hemorrhagic diathesis may develop. The hemorrhagic tendency is corrected and the one-stage prothrombin time reduced to normal by the administration of vitamin K. An analysis of the clotting defect in these patients has shown reduction of the four Factors II, VII, IX and X.

HEMORRHAGIC DISEASE OF THE NEWBORN. Before the discovery of vitamin K, a hemorrhagic tendency was observed in about one in 400 newborn infants. The bleeding occurred at the third or fourth day of life. The one-stage prothrombin time of newborn infants is often normal at birth, but the clotting time by this technique lengthens during the first week of life. It is thought that vitamin K is normally made in the intestine as a result of the action of bacterial enzymes on the food ingested and that in newborn infants this ability to synthesize vitamin K is deficient. At birth Factors II, VII, IX and X are slightly reduced. The levels of these substances fall during the first week unless vitamin K is given, when the fall is prevented.

DICOUMARIN. A hemorrhagic disease of Canadian cattle was traced to feeding on spoiled sweet clover hay and found to be attributable to the development in the hay of a certain derivative of coumarin. This substance was identified as 3,3′-methylenebis (3-hydroxycoumarin), also called dicoumarin. It can be synthesized, and if given by mouth to animals it causes bleeding. The bleeding tendency is associated with a long one-stage prothrombin time which, in turn, is caused by marked reduction of the same four Factors II, VII, IX and X.

This substance or modifications such as Tromexan (bis-3,3'-[4-hydroxycoumarinyl] ethylacetate) or substances with a similar physiological action such as Dindevan (phenylindanedione) are often given to patients who suffer from intravascular clotting to reduce the coagulability of the blood. In patients with clots in the cardiac arteries and in patients immobilized after fractures or orthopedic operations the treatment is valuable, for it reduces the tendency to excessive clotting in the patient's veins. For therapeutic purposes, the dose is controlled by measuring the level of the one-stage prothrombin test and maintaining this level between defined limits by varying the dose of the drug. This control is essential for safe treatment, because the amount necessary varies greatly from one patient to another and an overdose may give rise to serious bleeding.

Since the administration of vitamin K preparations to patients who have received an overdose of coumarin type drugs rapidly corrects the clotting defect, it must be assumed that they interfere in some way with the action or metabolism of vitamin K.

The association of vitamin K with simultaneous reduction of four coagulation factors arouses speculation. One group of workers[17] have always believed that the separation of these four factors is artifactual and that they have at least a common precursor. The effect of vitamin K deficiency supports this view; on the other hand, the four factors can be separated from normal whole plasma and the appropriate purified factors will correct the coagulation anomaly in patients with deficiencies of single factors.

THE PLATELETS AND BLOOD COAGULATION. When platelets are reduced in number (thrombocytopenia) the blood usually clots in the normal time but much prothrombin (Factor II) remains in the serum when clotting is complete. When normal blood clots, most of the prothrombin is converted to thrombin and this thrombin is neutralized by an inhibitor, antithrombin. Thus normal serum contains very little prothrombin. The reason for this coagulation defect in thrombocytopenia is that platelets provide a phospholipid (platelet factor 3) which acts at the stage of prothrombin conversion (Scheme 3).

There are some patients whose platelets are present in normal numbers but lack the normal amount of platelet factor 3, or are unable to release this factor in the normal way. These patients are said to have thrombasthenia.

CALCIUM AND BLOOD COAGULATION

Calcium is required for at least three of the reactions which precede thrombin formation. This fact accounts for the well-known anticoagulant efficiency of decalcifying agents, such as citrates or oxalates. It is thought that calcium is adsorbed by various plasma proteins and may act by maintaining an optimum surface charge for their interaction.

TESTS OF CLOTTING FUNCTION

Many of the tests of clotting function are empirical; they do not record the effects of single factors or stages of clotting, but rather merely record a defect.

WHOLE BLOOD CLOTTING TIME. In this test a measured amount of blood is placed in a tube which is tilted at intervals, and the time which elapses before the tube can be inverted without spilling the contents is recorded. Clearly this test will record abnormality at any stage of clotting. It is unfortunately rather insensitive to minor coagulation defects, and normality of this test does not exclude many clinically important diseases.

PROTHROMBIN CONSUMPTION TEST. In this test, serum from the whole blood clotting and the relative amount of prothrombin in it is recorded, comparison being made with normal serum. This test is also quite nonspecific but is somewhat more sensitive in detecting mild coagulation defects than the whole blood clotting time. It will also record abnormality in thrombocytopenia and thrombasthenia, whereas the clotting time is usually normal in these conditions.

THE THROMBOPLASTIN GENERATION TEST. This is a test of intrinsic clotting function. It consists of mixing reagents that contain all the factors necessary for intrinsic activity (Factors XII, XI, IX, VIII, X and V; see Scheme 3) and recording the amount of

Table 19–2 An Example of the Use of the Thromboplastin Generation Test in the Detection of Hemophilia and Christmas Disease

SOURCE OF REAGENTS	FACTORS PROVIDED BY REAGENT								INCUBATION TIME (MINUTES)				
	Al(OH)$_3$ Treated Plasma				Serum				1	2	3	4	6
Normal	V	VIII	XI	XII	IX	X	XI	XII	31	17	12	12	12
Hemophilia (Factor VIII deficiency)	V	XI	XII		IX	X	XI	XII	70	62	65	61	48
Christmas disease (Factor IX deficiency)	V	VIII	XI	XII	X	XI	XII		65	45	36	39	36

activity developed. When plasma is treated with inorganic precipitates such as Al(OH)$_3$, BaSO$_4$ and the like, Factors II, VII, IX and X are removed by adsorption and Factors XII, XI, V and VIII remain in the plasma. Serum on the other hand lacks Factors II, V and VIII but contains Factors XII, XI, VII, IX and X. The required factors with the exception of phospholipid are supplied by mixing Al(OH)$_3$ treated plasma and serum.

The test can be made diagnostic, because if the normal Al(OH)$_3$ treated plasma is replaced by Al(OH)$_3$ treated plasma from a hemophilic patient (Factor VIII deficiency), the results will be abnormal. If the normal serum is replaced by serum from a patient with Christmas disease (Factor IX deficiency), the results will be abnormal. An example of the use of this test is given in Table 19–2. The test can be modified to give a quantitative measure of Factors VIII and IX. These tests are useful in the treatment of patients and in the preparation of concentrates of these factors for clinical administration.[1] In fact, the quantitative modifications of the thromboplastin generation test are now more widely used than the original test.

It should be noted that the name of this test dates from a time when it was thought that something resembling tissue extract (thromboplastin) was built up during intrinsic clotting. The test is not so useful for recording defects of Factors V, X, XII and XI as for deficiency of Factors VIII and IX.

TESTS OF CONTACT PHASE ACTIVITY. Factors XII and XI which affect this phase are adsorbed from plasma by celite, and in plasma these factors are activated by kaolin. With the use of these two reagents quantitative tests for Factors XII and XI can be devised.

THE ONE-STAGE PROTHROMBIN TIME. This test measures extrinsic clotting activity and records defects in Factors II, V, VII and X. Plasma samples from patients with naturally occurring defects can be modified to give quantitative assays of the separate factors. The name of this test dates from the time when Factors V, VII and X were not recognized and the test was thought to give a specific measure of prothrombin (Factor II).

NATURAL INHIBITORS OF COAGULATION

The blood contains a considerable excess of many factors which promote clotting. The prothrombin in 10 ml of plasma, if converted to thrombin, would be sufficient to clot all the fibrinogen in the body. The Factor Xa from 10 ml of platelet-containing plasma is also theoretically capable of converting all the prothrombin in the body to thrombin. This enormous potential coagulant ability must naturally be controlled in the body if life is to proceed. The controlling mechanism probably consists of a series of inhibitory systems capable of neutralizing active coagulants or preventing their formation.

Antithrombin. A powerful mechanism for neutralizing thrombin exists in normal blood. Within 15 to 30 minutes after all the prothrombin in a sample of blood is converted to thrombin, no thrombin can be detected in the serum. This disappearance of thrombin is due in part to adsorption of thrombin by fibrin formed during clotting and in part to the neutralization of thrombin by a substance in the α_2 globulin fraction called antithrombin 3. It appears that this antithrombin reacts quantitatively with thrombin to form a neutralized substance which has been called metathrombin.

The presence of antithrombin means that the formation of thrombin and its neutralization in blood progress simultaneously. Thus the amount of thrombin to be detected in blood is always less than the amount formed. Moreover, the amount to be detected will depend greatly on the speed of thrombin formation; if thrombin formation is slow, the thrombin will be exposed to antithrombin for a longer period of time during its formation and the amount of thrombin to be demonstrated at any one time will be low. An example to show the effect of speed of thrombin formation to be detected at any time is shown in Figure 19–2. In the experiment three different concentrations of tissue extract were added to hemophilic plasma (Factor VIII deficient) and the amounts of thrombin present at different times after recalcification were recorded. With no tissue extract added the amounts of thrombin were too small to record. It will be seen that with the tissue extract added the thrombin concentrations achieved depend on the potency of the tissue extract present. The efficiency of thrombin formation in blood is therefore closely related to the speed of its formation, and reduced speed of activation of Factor X

(as in hemophilia) will greatly reduce the clotting efficiency of the blood.

Theoretically, abnormal clotting could be caused by an increase or decrease of antithrombin in the blood. One or two patients with tendency to abnormal intravascular clotting have been found to have reduced plasma concentrations of antithrombin.

Heparin. In experiments designed to purify tissue extract, McLean,[8] working in Howell's laboratory, found that an inhibitor of blood coagulation could be obtained from liver. Subsequent work showed that this material, heparin, as extracted from beef liver, is a mixture of mucoitin polysulfuric acids. The action of heparin is complex. It inhibits the thrombin-fibrinogen reaction and prevents thrombin formation. If thrombin and fibrinogen isolated from blood react together, heparin has little inhibitory effect. On the other hand, heparin strongly inhibits the reaction of thrombin with fibrinogen in whole plasma. It appears that heparin requires a co-factor from plasma for its action to be effective. Much evidence suggests that the co-factor for heparin in plasma is antithrombin. Heparin therefore probably works by enhancing the normal antithrombin. In addition, heparin inhibits thrombin formation by promoting the inactivation of Factor Xa.

Heparin is often used to depress the coagulability of blood in patients with thrombosis. It is nontoxic, but its action is short-lived and it is effective only when given intravenously. Heparin does not occur in measurable quantity in the blood of normal people. It appears in the blood after anaphylactic shock, and very occasionally the

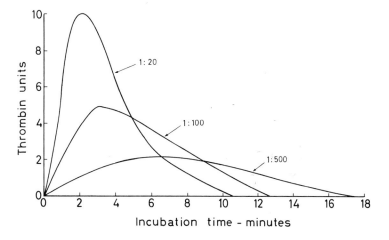

Figure 19–2 Thrombin generation test in hemophilic plasma to which various (stated) concentrations of tissue extract were added.

occurrence of heparin in the blood, for which there is no known cause, may give rise to a hemorrhagic state in patients.

In addition to inhibitors which neutralize thrombin, there are probably naturally occurring inhibitors to all the active coagulants which appear during clotting. Niemetz and Nossel[12] have demonstrated an inhibitor to Factor XIa, and Biggs *et al.* have demonstrated an inhibitor to Factor Xa. These inhibitors undoubtedly play a very important part in limiting normal coagulation and preventing abnormal intravascular clotting (thrombosis) but they have not been studied extensively.

Hemorrhagic Disease Due to Anticoagulants. A number of patients have been described in whom an abnormal circulating anticoagulant has developed. Some of these patients have hemophilia, and in some normal persons the condition has followed pregnancy. These anticoagulants neutralize Factor VIII or inhibit the activation of the intrinsic clotting system. In the hemophilic patients the inhibitor follows repeated transfusion, and probably Factor VIII behaves as a foreign protein to these patients and thus provokes the development of specific antibodies which neutralize Factor VIII. The etiology in the other cases is more obscure; it is possible that, in the cases which follow pregnancy, thromboplastin components from the placenta may enter the circulation at delivery and antibodies to some component may develop. These antigens presumably have antigenic similarity to Factor VIII.

SNAKE VENOMS AND CLOTTING

By a curious twist of evolution the venoms of snakes have been related to mammalian coagulation. Some mimic thrombin and clot fibrinogen, some convert prothrombin to thrombin (Nahas *et al.*)[11] and some activate factor X, *e.g.*, Russell's Viper venom. These properties doubtless account in part for the toxic properties of the venoms. To the research worker the venoms have proved invaluable tools. Some venoms that clot fibrinogen, for example, do not destroy Factor VIII and can be used to free Factor VIII preparations from fibrinogen, with which they are normally associated. Those venoms which are direct activators of

prothrombin can be used to measure prothrombin, and the activators of Factor X can be used to measure Factor X.

THE HOMEOSTATIC BALANCE, HEMOSTASIS AND THROMBOSIS

The blood coagulation factors are maintained in the circulation at a fairly constant level, although the concentrations of some vary a good deal from time to time. From observations of patients deficient in single factors to whom infusions of the missing factor have been given, the half-lives of the various clotting factors have been determined. These, expressed in hours, are approximately as follows:

I	II	V	VII	X	VIII	IX	XI	XII
120	72	30	5	30	12	24	30	30

The remarkable feature in all cases is the very short duration of these factors in the circulation. It is possible that some coagulation goes on all the time and that the clotting factors are continuously used up.

In pyrexia, pregnancy and the postoperative period fibrinogen (Factor I) is increased. Factor VIII is also a variable; it is increased in pregnancy and in the postoperative period. It is also increased after injections of adrenalin.[5] It has been found that this effect is mediated through the β-adrenergic receptors, because the effect can be blocked by the use of propranolol, a β-receptor blocker.[6] Factor VIII also increases after strenuous exercise.[15]

The remarkable chain reactions illustrated in Schemes 2 and 3 suggest a great positive coagulant potential. In life this is balanced by two mechanisms, (i) the inhibitory factors (see Fig. 19–2) and (ii) the presence in the blood of a powerful enzyme system which dissolves fibrin. This fibrinolytic system is complex, but it seems that in many instances the circumstances which may promote excessive clotting also activate the fibrinolytic system. For example, if tissue extract is injected intravenously in animals, activation of coagulation is not usually accompanied by the presence of large clots in the blood vessels. More commonly, all the fibrinogen apparently disappears, and products of its breakdown by lysis are found in the blood. In patients a similar complete disappearance of fibrinogen, known as the defibrination syndrome, may occur. For some years there was doubt about the causes of this syndrome, but it is now usually thought of as caused by widespread activation of the clotting mechanism and the subsequent but rapidly following removal of fibrin or fibrinogen monomer by lysis. Conditions associated with this type of excessive clotting are premature separation of the placenta, diffuse carcinoma in some patients, major surgical operations involving the lungs, leukemia and mismatched transfusions.

These are examples of generalized excessive coagulation, but localized clots are more common. In these the platelets play an important part. The platelets stick to a site of injury, fuse together and release a coagulant phospholipid. A blood clot becomes attached to the platelet mass. This localized thrombosis extending beyond a site of injury is a serious complication of many diseases. Unfortunately, examination of the blood has not yet provided an infallible test by which predisposition to thrombosis may be diagnosed.

In recent years treatment has involved the exposure of the patient's blood to foreign surfaces. For example, plastic valves are used to replace defective heart valves. The blood is passed through machines for dialysis in patients with kidney disease or through heart-lung machines for those undergoing heart surgery. This exposure of the blood to foreign surfaces undoubtedly promotes excessive coagulation, probably through the activation of Factors XII and XI.

Hemostasis. The prevention of the escape of blood from injured vessels is complex and is controlled by two mechanisms. When a vessel is injured, platelets adhere to the site of injury and may completely block the lumen of the vessel. If these platelets are exposed to thrombin (Factor IIa), they fuse together to form a firm mass. This mechanism alone cannot suffice to block all injured vessels, because in most coagulation defects this platelet mechanism is normal and yet the patients bleed excessively after injury. Probably the platelets and blood coagulation cooperate to stem the flow of blood. The platelets first adhere and release platelet factor 3 which encourages clotting and the formation of thrombin and fibrin. Thrombin promotes the solidification of the platelet mass, and the fibrin gives additional firmness to the plug.

Blood coagulation is an extremely complex process. Because of the pertinacity and originality of coagulation specialists, much is now known about the single reactions which complete the process. This knowledge was accumulated through a mixture of the clinical study of patients and ingenuity in observing clots. Despite all this work, very little is yet known about the chemistry of the individual factors.[4] Because there are so many undefined factors, there has been little study of the biochemistry of coagulation. Biochemical experiments are usually conducted to study the reaction of a complex enzyme of unknown structure with a simpler reagent of known structure. This approach has undoubtedly provided much laboratory information, but in the complexities of actual life, it is more probable that complex enzymes react with complex substrates as in blood coagulation. Blood clotting may in fact prove to be the first chain reaction studied in detail in which all the reactants are complex proteins. When more knowledge is accumulated, it may prove that most other biological processes are equally complex.

REFERENCES

1. Biggs, R., and Macfarlane, R. G. *Treatment of haemophilia and other coagulation defects.* Oxford, Blackwell Scientific Publications, 1966.
2. Biggs, R., Macfarlane, R. G., Denson, K. W. E., and Ash, B. J. Thrombin and the interaction of Factors VIII and IX. *Brit. J. Haemat.*, 1965, *11*, 276–295.
3. Blomback, B. Report of the Subcommittee on Nomenclature to the International Committee on Haemostasis and Thrombosis, 1967.
4. Esnouf, M. P., and Macfarlane, R. G. Enzymology and the blood clotting mechanism. *Advanc. Enzymol.*, 1968, *30*, 255–315.
5. Ingram, G. I. C. Increase in antihaemophilic globulin activity following infusion of adrenalin. *J. Physiol. (Lond.)*, 1961, *156*, 217–224.
6. Ingram, G. I. C., and Jones, V. The rise in clotting Factor VIII induced in man by adrenaline: effect of α- and β-blockers. *J. Physiol. (Lond.)*, 1966, *187*, 447–454.

7. Macfarlane, R. G., and Ash, B. J. The activation and consumption of Factor X in recalcified plasma: the effect of added Factor VIII and Russel's viper venom. *Brit. J. Haemat.*, 1964, *10*, 217–224.

8. McLean, J. The thromboplastic action of cephalin. *Amer. J. Physiol.*, 1916, *41*, 250–257.

9. Mellanby, J. J. The coagulation of the blood. *J. Physiol. (Lond.)*, 1909, *38*, 28–112.

10. Morawitz, P. Die Chemie der Blutgerinnung. *Ergebn. Physiol.*, 1905, *4*, 307–422.

11. Nahas, L., Denson, K. W. E., and Macfarlane, R. G. A study of the coagulant action of eight snake venoms. *Thromb. Diathesis haemorrh. (Stuttg.)*, 1965, *12*, 355–367.

12. Niemetz, J., and Nossel, H. L. Method of purification and properties of anti-Xla (inhibitor of the contact product). *Thromb. Diathesis haemorrh. (Stuttg.)*, 1967, *17*, 335–348.

13. Biggs, R., Denson, K. W. E., Akman, N., Borrett, R., and Haddon, M. E. Antithrombin III, antifactor Xa and heparin. *Brit. J. Haemat.*, 1970, *19*, 283–306.

14. Quick, A. J. The prothrombin in hemophilia and in obstructive jaundice. *J. biol. Chem.*, 1935, *109*, lxxiii.

15. Rizza, C. R. The effect of exercise on the level of antihaemophilic globulin in human blood. *J. Physiol. (Lond.)*, 1961, *156*, 128–135.

16. Schmidt, A. *Zer Blutlehre.* Leipzig, F. C. W. Vogel, 1892.

17. Seegers, W. H. *Prothrombin.* Cambridge, Mass., Harvard University Press, 1962.

CHAPTER 20   ANATOMY AND PHYSICS OF RESPIRATION

by J. HILDEBRANDT

With the exception of certain microorganisms and a few parasites, plants and animals generally require free access to molecular oxygen to maintain their metabolic processes. In lower animals this oxygen is supplied in various ways: (i) direct diffusion (protozoa, bacteria), (ii) exchange through the skin to capillary blood (annelids, hibernating frogs), (iii) exchange via gills to blood (lower chordates and other aquatic phyla), and (iv) air tubes, or tracheae, which con-duct gas directly to the tissues (insects). However, in mammals almost all the O_2 is collected by blood during its passage through the capillaries of air-ventilated lungs.

In humans, abnormalities in the mechanical properties of the respiratory system are characteristic of a number of diseases. The leading symptom is breathlessness, which is present to varying degrees in clinical states with localized or diffuse airway obstruction, cardiac disease, pulmonary fibrosis and hyaline membrane disease. Accordingly, the aims of this chapter are to describe the normal physical properties of the system, to account for the changes that occur in certain diseases of the lungs and chest, and to provide guidelines for rational therapy. For purposes of the present discussion the respiratory system may be treated as three main components: (i) the thoracic wall, including rib cage, diaphragm and abdomen; (ii) the lung tissue and surface lining; and (iii) the air and airways. After a brief review of structure, detailed consideration will be given to the mechanics of each component.

ANATOMIC RELATIONSHIPS

Movements of the Thoracic Cage. The lungs may be considered as elastic multi-chambered bags suspended in the pleural cavities and connected with the exterior through the airways (Fig. 20–1). Air is drawn through these tubes and into the gas-exchanging area as a result of the contraction of the diaphragm and certain respiratory muscles which act to enlarge the thoracic cavity. Thus descent of the diaphragm increases the vertical dimension,

displacing the abdominal contents downward, while simultaneously providing a reactive lifting force to the rib cage. Thoracic inspiratory muscles further increase the anteroposterior diameter of the thorax by moving the sternum up and forward ("pump-handle" movement). They also increase the transverse diameter of the chest by pulling the arched ribs up and lateral ("bucket-handle" movement). The consequence of these movements is an increase in the volume of the thorax and lungs. As a result of the expansion, the pressure of the intrapulmonary gas is reduced and there follows an influx of air from a region of higher to one of lower pressure. In quiet respiration (roughly 12 breaths per minute, 600 ml per breath), expiration results passively from the elastic recoil of the lungs after the inspiratory muscles relax. With more vigorous breathing, as in exercise, expiration is accelerated by contraction of expiratory muscles in the chest and abdomen.

Descent of the diaphragm appears to account for the larger part of the inspired volume. Even in maximal respiration, only about half of the ventilation is due to lifting and expansion of the rib cage. Persons with bilateral diaphragmatic paralysis have about one-half the normal vital capacity in the erect position and about one-third

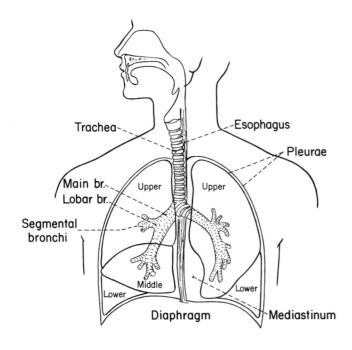

Figure 20–1 The thoracic cavity is divided by the mediastinum into two major chambers that contain the right and left lungs. The right lung has three lobes (upper, middle and lower) separated into ten segments, whereas the left lung is divided into two lobes (upper and lower) comprising nine segments. Air is inhaled when the thoracic cavity is enlarged, both by diaphragmatic contraction and descent, and by anterior and lateral movements of the rib cage.

Table 20–1 *Respiratory Muscles Active During Inspiration and Expiration at Various Minute Volumes (MV)*

	QUIET BREATHING OR MILD ACTIVITY (MV < 50 L/MIN)	MODERATE TO SEVERE EXERCISE (MV > 50 L/MIN)	VITAL CAPACITY (MAXIMAL BREATH)
Inspiration	Diaphragm (always) Internal intercostals of the parasternal region Scaleni (sometimes)	Diaphragm, external intercostals, scaleni, sternomastoids, vertebral extensors	All inspiratory muscles, plus anterolateral abdominals, adductors of the larynx
Expiration	Completely passive, except during early part of expiration, when some inspiratory contraction persists	Transverse and oblique abdominals; internal intercostals	Mainly abdominals, scaleni, sacrospinalis, internal intercostals Transversus thoracis

normal in the supine position.[30] After extensive paralysis (*e.g.*, in poliomyelitis) adequate resting ventilation can sometimes be maintained with the neck and shoulder muscles alone.

The role of various muscles attached to the thorax has been revealed by electromyography. Simultaneous recordings of air movement and of a muscle's electrical activity enable the investigator to estimate the relative strength of the contraction and its timing with respect to the respiratory cycle. In Table 20–1 are summarized data collected from normal subjects.[6, 48]

Airways and the Anatomic Dead Space. The trachea, supported by C-shaped cartilaginous rings, bifurcates into two main bronchi that further divide within the lungs into lobar and segmental bronchi. These branch repeatedly into smaller cartilaginous tubes, or bronchioles. In all, there are about 16 generations of conducting airways. The last few are noncartilaginous and are called terminal bronchioles.

A comprehensive study of pulmonary morphometrics has been published by Weibel.[52] Taking the trachea as the zero generation with a cross-sectional area of 2.54 cm², he finds that the total cross-sectional area of the airways initially decreases to 2.0 cm² at the third generation and then increases steadily by a factor of approximately 7/5 per generation to 180 cm² at the sixteenth. The number of airways at this stage is about 2^{16} or 65,000. It happens that when the factor is 7/5, total flow resistance per unit distance down the bronchial tree (calculated from Poiseuille's equation for laminar flow) is nearly constant,

implying a nearly uniform pressure gradient along the airway at least to within 6 to 8 mm of the terminal unit.

Up to and including the terminal bronchioles, the air passages are lined with cuboidal or columnar epithelium. No appreciable gas exchange occurs in this portion of the respiratory tree and, accordingly, its volume is functionally termed the *anatomic dead space*. The size (in ml) of the anatomic dead space of resting normals is roughly equal numerically to the ideal body weight (in lbs). The dead space increases on inspiration approximately in proportion to the total volume of air in the lungs.

Besides transporting air to and from gas exchanging areas, the dead space is designed to serve a number of additional functions. (i) Air is quickly warmed to body temperature and humidified to saturation (47 mm Hg) as it passes through the nose, pharynx and upper airways. (ii) Dust is effectively removed from inspired air either by straining through nasal hairs or by impingement on moist mucosal surfaces. (iii) The dead space is lined by ciliated and mucus-secreting cells (the "mucociliary escalator") whose role in clearing the airways of fine particulate matter and other impurities is enormously important. (iv) During a cough, the flexible dorsal membrane of the trachea invaginates deep into the lumen, thus increasing air velocity and facilitating the clearance of particles and mucus. Luminal narrowing and effectiveness of expectoration can be further increased by bronchial smooth muscle constriction.[33]

The airways have certain weaknesses which account for a large proportion of human disease and disability. (i) They are exceedingly susceptible to infection, the common cold being the major reason for absenteeism. (ii) Throughout the tracheobronchial tree are found smooth muscle cells innervated by vagal and sympathetic fibers which respond to certain drugs and irritants, altered gas concentrations and psychological factors. In reversible obstructive syndromes (*e.g.*, asthma) bronchospasm, together with mucosal congestion and sputum, increases air flow resistance. (iii) In irreversible obstructive syndromes (*e.g.*, emphysema) the airways are subject to collapse or narrowing, thus providing an additional mechanism for difficult breathing. (iv) Larger foreign bodies lodged in airways are not easily cleared.

The Respiratory Unit and the Air-Blood Barrier. The terminal bronchioles bifurcate or trifurcate into about three generations of respiratory bronchioles, which, like other gas-exchanging surfaces in the lung, are lined with thin squamous epithelium. The last respiratory bronchioles widen and divide into elongated thin-walled chambers, the *alveolar ducts*, which subdivide into a number of pouches or *alveolar sacs*. Ducts and sacs comprise the last four generations of the airway system. Finally, each sac is partitioned by numerous interalveolar septa into about 20 cavities called *alveoli*. To achieve optimal contact between gas and blood, the interalveolar septa are invested with a rich capillary network. All structures peripheral to a cartilaginous bronchiole form a lobule, whereas structures peripheral to a terminal bronchiole form an *acinus* (Fig. 20–2).

At birth a child's lungs contain about 30 million alveoli, which proliferate rapidly during the first few postnatal years. The alveoli reach their final number of 250 to 350 million by the age of eight.[18] Thereafter, lung growth proceeds solely by enlargement of existing spaces. The area of the alveolar surface in the adult human at rest totals about 75 m², and increases further on inflation. Alveolar diameters range from 100 to 300 μ. During the breathing cycle the volumes of the alveolar ducts, sacs and alveoli apparently increase and decrease in almost equal proportions.[47] The total capillary surface area (tissue-blood interface) is estimated to be only slightly less than that of the air-tissue interface,[52] im-

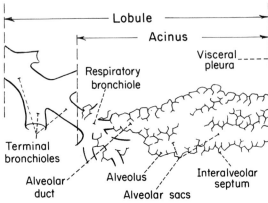

Figure 20–2 The mammalian respiratory unit (diagrammatic). The terminal bronchiole gives rise to several generations of respiratory bronchioles, each of which branches into several alveolar ducts and many hundreds of alveoli. The alveolar duct often has several major divisions or alveolar sacs. Interalveolar septa are shown extending into the sacs; the irregular small spaces or alveoli each have about ten flatsided walls.

plying that the capillary mesh is exceedingly dense.

During a normal breath, fresh air is moved by mass flow only as far as the first alveolar ducts. Movement of O_2 through the remaining generations of ducts and sacs must occur within roughly one second by gaseous diffusion. Simultaneously, CO_2 must diffuse in the other direction. Transport across the alveolar-capillary barrier is likewise by diffusion, after solution in tissue water. This exquisitely thin barrier (Fig. 20–3)—normally having a harmonic (effective) mean thickness of only 0.6 μ as revealed by electron microscopy[53]—consists of a layer of flattened alveolar epithelial cells and a capillary endothelial layer separated by basement membranes. The air-exposed surface of the alveolar membrane appears to be covered with a 100 Å noncellular film (surface lining, surfactant) floating on a fine fluid or gel layer presumably produced as a transudate by alveolar epithelial cells.[29, 54] The thickness of this extracellular layer and that of the interstitial space increase in pulmonary edema and in hyaline membrane disease.[10]

Although gaseous diffusion through liquids is generally about ten thousand times slower than diffusion through another gas, the thin tissue barrier presents no

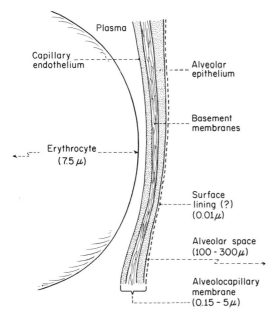

Figure 20-3 The fine structure of the alveolocapillary membrane. From the relative dimensions it is apparent that the principal diffusion barrier is not the membrane, but rather the plasma and red cell itself.

appreciable hindrance to alveolocapillary gas exchange. Even in disturbances such as chronic pulmonary congestion in which membranes are several times thicker than normal, incomplete oxygenation caused by diffusion abnormalities alone is rarely observed. Parallel lung regions are apparently not completely separated by air-tight partitions. Considerable movement of gas can occur into a region through channels other than its conducting airways.[28] This collateral ventilation may be especially significant in the presence of obstructive airway disease. The precise anatomic pathways have yet to be clarified.

PHYSICAL RELATIONSHIPS

Pressure and Volume. Because of the elasticity of lung, its properties are most easily compared with those of a balloon. Thus a plot of lung volume against distending pressure should provide an important description of lung mechanical characteristics. However, the measurement of lung properties on excised lungs or in the open chest results in some oversimplification when extrapolated to lungs in the closed chest. The problem may be illustrated in the following way. Suppose a certain balloon has the PV (pressure-volume) relationship shown by the middle curve of Figure 20-4. At each point on the surface the transmural pressure ΔP obeys the Young-Laplace relation for curved surfaces: $\Delta P = \dfrac{Tx}{Rx} + \dfrac{Ty}{Ry}$, where Rx and Ry are the two principal (greatest and least) radii of curvature and Tx and Ty are the corresponding tensions. The same ΔP exists at every point over the surface.

However, if this balloon is inflated inside another of a different shape, the presence of nonuniform external surface pressures ensures that the same ΔP is no longer present at every point. Consequently, a single PV curve does not suffice to describe the properties of either balloon. For example, at points where the two are in contact the inner balloon is flattened, increasing the radii of curvature and decreasing ΔP (lower curve, Fig. 20-4). On the other hand, in the vicinity of a space, where the radii of curvature are diminished, the transmural pressure is increased. Thus, a range of pressures could exist in the pleural space simply because of the shape discrepancy between lung and chest. Slippage between lobes

Figure 20-4 Model illustrating effect of enclosing the lung in the chest. The PV curve of the inner balloon by itself is given by the middle curve. At point 1 the transmural pressure is below that present in a free balloon, whereas at point 2 it is above. A single PV curve can no longer describe the properties of an enclosed elastic bag.

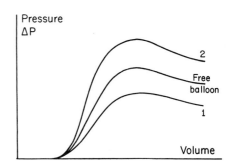

probably helps to relieve these stress non-uniformities.

Additional factors can contribute to unequal pleural pressures. The effect of gravity on the thoracic contents tends to create more negative pressures in uppermost regions (see also Distribution). Furthermore, the lungs, consisting of spongy plastic viscoelastic materials, do not act like a fluid where pressure applied at one point is transmitted equally to all other points. Thus, increases in lung volume near the diaphragm could be greater than those near the apex, contributing to intrapleural pressure differences. Nevertheless, in order to abbreviate discussion, one commonly speaks of a single mean pleural pressure to be associated with a particular lung volume.

Definition of Lung Volumes and Capacities. For convenience, a description of lung-chest mechanics should first include a definition of certain volumetric parameters in common usage; details of measurement will be treated later. Four *lung volumes* and the four *lung capacities* have been named as shown in Figure 20–5. Note that the volumes do not overlap, whereas each capacity includes more than one volume.

The volume of air inhaled and exhaled during any respiratory cycle is termed the *tidal volume* (TV). The additional volume of air that could be inspired by maximal effort from any end-inspiratory level is the *inspiratory reserve volume* (IRV), and the volume that could be forcibly expelled from any end-expiratory level is the *expiratory reserve volume* (ERV). Tidal volume is therefore variable and during exercise may increase by encroaching on both IRV and ERV. The volume of air remaining in the lungs after maximal voluntary expiration is called the *residual volume* (RV).

The *functional residual capacity* (FRC) is the volume of gas remaining in the chest at the resting end-expiratory level. The maximal volume of gas which can be inspired from this resting level is the *inspiratory capacity* (IC). *Vital capacity* (VC) is the maximal volume that can be voluntarily expelled from the lungs after a maximal inspiration (IRV + TV + ERV). Finally, the sum of all four volumes is the *total lung capacity* (TLC).

Physical Properties of the Lung-Chest System. The volume-pressure relationship of the entire system can be investigated by either of two methods. (i) The subject inspires air from a spirometer to a

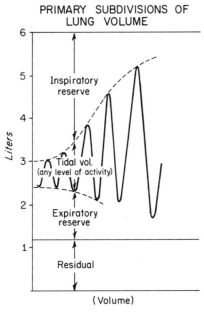
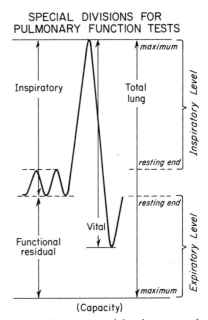

Figure 20–5 Lung volumes and capacities. In normal subjects, the resting tidal volume is only about one-tenth of the total lung capacity. (After *Fed. Proc.*, 1950, 9, 602–605.)

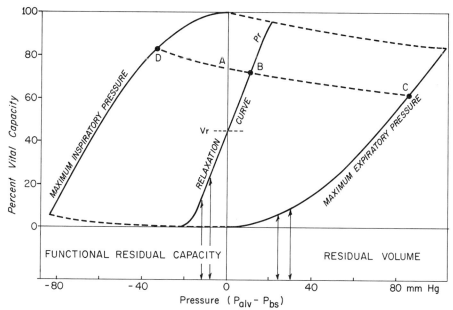

Figure 20–6 Properties of the lung-chest system. The relaxation curve represents the passive inflation curve of the lung and chest together. The pressure axis refers to the pressure in the alveoli (Palv) relative to the pressure on the body surface (Pbs) which is usually atmospheric. Thus, while the muscles are relaxed, Palv − Pbs = Pr. The maximum inspiratory and maximum expiratory pressure curves are derived at a series of lung volumes during maximal muscular effort against a manometer at the mouth. The glottis must be open. The dashed lines represent sections of hyperbolas from the gas equation (PB − PH$_2$O) VL = nRT, where n represents the total number of moles of dry air. Notice that whereas FRC is normally equal to Vr, it could equal any point on the relaxation curve when pressure is applied at the mouth. RV can be similarly interpreted. (After Rahn *et al.*, *Amer. J. Physiol.*, 1946, *146*, 161–178.)

certain lung volume (*e.g.*, point A, Fig. 20–6). He then relaxes with open glottis against a manometer placed in his mouth. As he relaxes, the alveolar pressure increases to point B, representing the pressure required to hold the lungs and chest at that volume. The maneuver is repeated at several lung volumes in the vital capacity range to obtain the entire *relaxation curve*. The relaxation pressure becomes negative at volumes less than Vr (the usual FRC) because the tendency for the chest wall to spring out at lower volumes is greater than the tendency for the lung to collapse. Many untrained subjects are unable to relax satisfactorily at lung volumes far above or below FRC. (ii) As a second but less practical method, the subject can be curarized and his mouth connected to a pressure source. Volumes are measured at each of a number of pressures and a relaxation curve is thus generated. (The same results are obtained if *negative* pressures are applied by a tank enclosing the body from the neck down, *e.g.*, the Drinker respirator.)

The relaxation curve is useful in giving a measure of the pressures required for respiration at any lung volume and in assessing the "stiffness" of the system as a whole.

The two additional curves in Figure 20–6 represent maximum expiratory pressures and maximum inspiratory pressures. These may be determined when, for example, the subject after relaxing to reach point B forcibly tries to exhale against a manometer. With maximum expiratory effort, point C is reached (note the gas compression) and conversely with maximum inspiratory effort, point D. The horizontal difference between the relaxation curve and the maximum expiratory curve represents the pressure developed by the expiratory muscles at that volume. This pressure can therefore be used as a test for respiratory muscle weakness. Healthy subjects can attain pressures of up to 250 cm H$_2$O.[13] However, this is only about 4 psi, and explains why one cannot inflate a flat tire by mouth.

The air pressure in the alveoli (Palv) at both end-inspiration and end-expiration

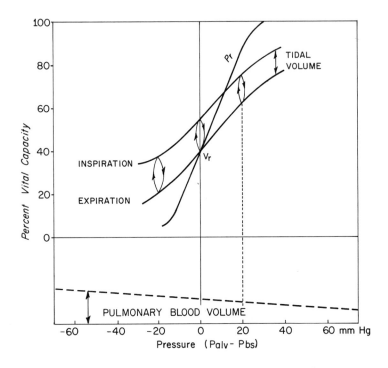

Figure 20–7 Effect of mouth pressure on breathing. Tidal breathing at atmospheric pressure (Pbs = 0) is shown by the center loop. Vr is the lung volume when pressure applied at the mouth is zero and the muscles are relaxed. Whenever the mouth pressure is very positive, *e.g.*, 20 cm H_2O, complete relaxation of the expiratory muscles may never occur at any phase of the respiratory cycle (right hand loop). When a large negative pressure is applied at the mouth, the inspiratory muscles are continuously active (left loop). Reflex muscular activity accounts for the difference between the actual end-expiratory volumes and the volumes predicted by the relaxation curve. (After Rahn *et al.*, *Amer. J. Physiol.*, 1946, *146*, 161–178.)

is equal to zero (taking atmospheric pressure as zero reference). However, during the period that gas is flowing into the lungs, Palv is slightly negative and during expiration slightly positive. The middle loop of Figure 20–7 illustrates cyclic changes in Palv. Since expiration is largely passive in quiet breathing, the lung-chest volume at end-expiration comes close to the point Vr, where the relaxation curve crosses the zero pressure line.

Pressure breathing has become a topic of increasing practical importance. In Figure 20–7, pressure breathing curves have been superimposed on the relaxation curve. When a subject breathes from a tank which contains air under constant positive pressure (+20 cm H_2O, for example), his lung volume may alternate between the ends of the loop shown on the right. Since neither of these points is on the relaxation curve, muscular activity must continue throughout the respiratory cycle. The vertical difference between the end of this loop and the relaxation curve represents the volume change caused by the reflex contraction of the expiratory muscles at end-inspiration. At this particular pressure, breathing would be accomplished entirely by the expiratory muscles (since the loop is completely below the relaxation curve). With a snorkel, the

mouth pressure is negative relative to body surface pressure (left loop) and the inspiratory muscles may fail to relax completely. Apparently, the stretch reflexes of the thorax tend to bring the FRC back toward the accustomed level. The extent to which the pressure breathing curve deviates from the relaxation curve is variable.[27] In many subjects the curves are nearly coincident for moderate pressures (up to ± 15 cm H_2O), but a substantial deviation is present at higher positive and negative pressures. With scuba apparatus, mouth pressure is automatically adjusted to equal ambient, so that almost no additional respiratory effort is required.

Physical Properties of Isolated Lung. Although sometimes inexact for reasons already outlined, analysis of the passive properties of the two constituent components of the lung-chest system is desirable. One may start by studying the mechanical properties of lung by itself, that is, either excised or in the open chest. In the next section methods for obtaining estimates of lung properties within the closed chest will be described.

AIR-FILLED LUNG. A functional relationship between pressure and volume of two excised human lungs was first published by Hutchinson in 1849.[26] From measurements of volume at four increasing pressures, Hutchinson obtained values from which approximately linear inflation pressure-volume (PV) curves could be obtained. Later, in 1900, van der Brugh published several deflation curves which were quite nonlinear.[49] The apparent contradiction was subsequently resolved as arising from hysteresis.[43] The failure to obtain identical paths upon inflation and deflation of a lung is only one example of hysteresis. The phenomenon is in fact present in varying degrees during stretch and release in all tissues and other materials. The area of a hysteresis loop represents the amount of energy degraded into heat by various frictional processes during each cyclic deformation of the material.

Because of hysteresis, the pressure developed by the lung at any volume depends strongly on the prior inflation history. Some PV patterns obtained from healthy excised monkey lungs are shown in Figure 20–8A. If one begins with a lung which is completely atelectatic (all alveoli airless and collapsed), the first inflation requires a high pressure (15 to 20 cm H_2O) to open the cohering bronchioles, ducts and alveoli. Isolated lobules then successively pop open, gradually merging as full inflation is reached at about 20 to 25 cm H_2O.

Much of the PV work of the first breath is expended in establishing the large surface-to-air interface of the lung. This work, Ws, may be estimated from the integral of surface tension, γ, times area, A: $Ws = \oint \gamma dA = \overline{\gamma} Amax$. Taking the average tension $\overline{\gamma}$ equal to 50 dynes/cm, and Amax as 100 m^2 (human), we find $Ws = 5 \times 10^7$ ergs. The total work of inflation, WT, is $WT = \oint pdV = \overline{p}Vmax$. Taking the average pressure $\overline{p}$ as 15 cm H_2O, and Vmax as 6L (human), we find $WT = 9 \times 10^7$ ergs. Thus more than half the work done on the first breath may be surface work, part of which is recoverable on deflation. About 20 per cent of WT appears to be elastic work done on tissue elements (Fig. 20–8B), and the remainder is apparently lost as heat owing to gas flows.

On deflation from TLC one always notices an immediate rapid reduction in pressure, followed by a gradual nearly linear fall. Volume does not return to zero when the inflation pressure is completely removed, indicating some air trapping in the alveoli. Accordingly, succeeding inflations require less work, since part of the surface is already established. A second factor reducing the work of re-inflation is the low surface tension present in the lung at end-deflation, as will be discussed presently.

The re-inflation curve of excised lung

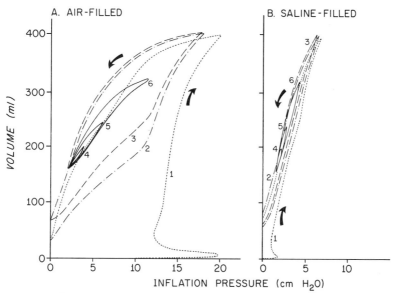

Figure 20–8 PV curves of excised monkey lungs, filled first with air *(A)* and subsequently with saline *(B)*. Successive cycles numbered 4, 5 and 6 represent tidal volumes of 40, 80 and 160 ml respectively. (Courtesy of H. Bachofen.)

exhibits a characteristic "knee" or inflection point at 10 to 12 cm H_2O, possibly representing the reopening of finer alveoli and airways which may have collapsed near minimal volume. If the lungs are re-expanded from a volume more closely approaching the FRC of the living animal, and if TV is of normal magnitude, one notices that the amount of the hysteresis becomes very small. This remaining loop area is sometimes ascribed to "tissue resistance." Like the viscous fluid flow resistance occurring in blood vessels and airways, tissue resistance has the characteristic that at a given cycling frequency the area of the hysteresis loop varies roughly as $(TV)^2$. However, unlike viscous resistance, the loop area is virtually independent of respiratory rate.[24] Behavior such as this is best described in terms of the viscoelasticity and plasticity thought to be present both in the surface lining and in the tissue parenchyma of the lung.

A widely used measure of the mechanical status of the lung is the compliance, or slope of the PV curve, dv/dp. Since there are many such slopes for any one lung (Fig. 20–8A), it becomes necessary to specify the conditions of measurement. One convenient set of conditions is this: the lung is expanded to TLC several times (deep breaths), returned to FRC, and normal breathing continued. The slope of the tidal inspiratory curves is then taken to be the compliance of the lung. Since compliance obviously must depend on lung size, the value should be normalized to unit volume by dividing by TLC or FRC, yielding specific compliance (see Pulmonary Function Tests).

LIQUID-FILLED LUNG. As already implied, events occurring in an air-filled lung are strongly dependent on surface phenomena. One means of assessing the role of the tissue component separately, as first performed by von Neergaard,[51] is to suppress surface effects by filling the lung with liquid. In this way no air interface is allowed to form. The saline PV curves shown in Figure 20–8B were obtained immediately subsequent to the air-filling illustrated by Figure 20–8A. Inflation pressures (and thus total work of expanding the lung tissue) are only a fraction of the magnitude required for the first air inflation. The deflation curve follows the inflation curve fairly closely, such that the hysteresis is not striking. During tidal volume maneuvers near FRC, the area of each loop is less than one-fourth the corresponding area obtained during air breathing.

By comparing fluid and air PV curves, a number of important deductions can therefore be made. First, roughly three-quarters of the lung's hysteresis in physiological ranges is attributable to surface effects and the remainder to tissue properties. Second, the total retractive pressure of the lungs can be resolved into two components: the tissue retractive pressure plus the surface retractive pressure. At each volume the sum of these two equals the total pressure. Toward TLC, the surface provides most of the recoil; near RV the tissue may dominate. Finally, an important inference concerning the lung's surface tension characteristics can be drawn from the convergence of air and liquid deflation curves near minimal volume (Fig. 20–8). If, for the moment, one considers each alveolus to be a spherical bubble with a constant surface tension, γ, one would expect from the Laplace relation $P = 2\gamma/R$ that the surface component of the retractive pressure would increase as $2\gamma/R$ decreased. In other words, the two curves should diverge as volume is decreased. Since precisely the opposite is true, it is most likely that the surface tension is not a constant, but rather diminishes toward zero as deflation progresses. Confirmation of this supposition is provided by direct measurement of the surface tension of lung washes and extracts, as will be described shortly.

AN ALVEOLAR MODEL. An exact analysis of the role of surface tension in lung mechanics would have to be developed around a realistic model of alveolar geometry, such as that illustrated in Figure 20–2. Instead, a highly idealized model of the lung will first be presented which permits simple calculations, while nevertheless yielding provocative suggestions concerning the shape of PV curves and the tendency of certain lungs toward atelectasis (instability).

In the model, interdependent polygonal alveoli sharing septa are replaced by discrete spherical bubbles situated at the ends of tubes representing the terminal bronchial tree.[11, 32] Before proceeding further, it is best to digress briefly into bubble mechanics.

If a soap film covering the end of a tube is

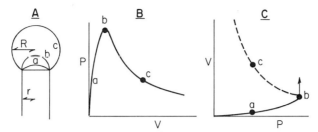

Figure 20–9   The single bubble. A, The radius of curvature of a bubble blown on the end of a tube decreases rapidly to a minimum at b, then increases slowly.

B, Bubble pressure as a function of volume can be calculated from tube radius r and Laplace's law, $P = 2\gamma/R$. As bubble volume is increased beyond b, pressure decreases asymptomatically toward zero. No instability occurs.

C, When pressure is the controlled (independent) variable, the bubble is stable only up to point b. A slight increase in pressure above this point bursts the bubble.

blown to form a bubble (Fig. 20–9A), the film's radius of curvature first decreases until the bubble is hemispherical in shape and then increases. Consequently, the pressure across the bubble wall first rises to a maximum when the radius of curvature is a minimum and then falls once more (Fig. 20–9B). Here volume is the independent variable and the system is always stable, *i.e.*, the equation $P = f(V)$ represents a single-valued function (for each V only a single P exists).

However, if one varies the pressure, as in Figure 20–9C, the bubble is stable only to point *b*; beyond this point it is unstable, *i.e.*, there is no V corresponding to a greater P, and so the bubble bursts. Even if one drops the pressure after passing point *b*, *e.g.*, to *c*, a moment's reflection will verify that since dv/dp is negative, the slightest perturbation again leads to bursting.

The situation represented by Figure 20–9C is encountered whenever two or more

bubbles are connected in parallel; here one cannot adjust the individual volumes independently. For example, if two bubbles are connected together (Fig. 20–10), the bubble having the smaller radius of curvature (and therefore the higher pressure) empties into the larger. The parallel system comes to rest at either of two stable equilibrium points – *d* or *f* of Figure 20–10 – moving away from the unstable equilibrium point *e*. The problem of maintaining "bubbles" in parallel is critical in the lungs where millions of alveoli exist side by side. One might suspect that the lung ought to be an unstable structure with only the large alveoli open. The fact that this is not true implies that other factors are involved.

TISSUE ELASTICITY.   In addition to the retractive forces arising from the surface tension in the alveolar lining, the tissue elastic recoil of the alveolar wall must also be considered. This component of the total lung retractive force has been illustrated

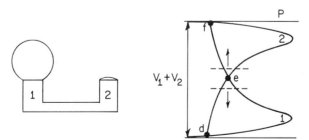

Figure 20–10   The double bubble. Two bubbles, 1 and 2, are connected in parallel. The smaller promptly empties into the larger until the radii of curvature of the two are equal – *i.e.*, until the pressures are equal. In the 2-bubble system, the total volume is represented by the constant, $V_1 + V_2$. Curves 1 and 2 represent the PV curves for the respective bubbles. If the pressure in bubble 1 is initially greater than in 2, the system will come to rest at point *d* – *i.e.*, most of the volume is in 2, and 1 is almost collapsed. If the greater pressure is initially in bubble 2, the system will come to rest at point *f*.

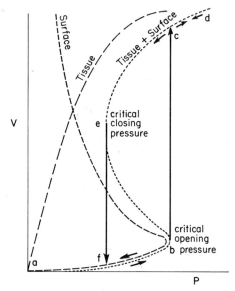

Figure 20-11   A tissue component is added to the surface component of the bubble illustrated in Figure 20-10. The net PV curve is now given by the dotted line *afbecd*. After reaching *b*, such a bubble "pops" open to *c*; when deflated, it suddenly collapses from *e* to *f*. It is prevented from bursting by the presence of the limiting tissue sheath.

in Figure 20–8B. One might, therefore, enclose each bubble of the model in a tissue net, as shown by Figure 20–11, where a tissue elastic curve representing that obtained from fluid-filled lungs is plotted along with the PV curve of a bubble. The pathway *afbecd* (dotted line) represents the sum of the pressures of these two curves. The stable portions of this curve are those where dv/dp is positive, but finite. Thus as the pressure in the bubble is increased, volume follows this summed curve to *b* where, as before, the bubble pops open. However, it is prevented from bursting by the tissue, and thus volume comes to rest at point *c*. Point *b* marks the critical opening pressure. As already mentioned, "popping" units are easily seen on the surface of lungs inflated from the collapsed state.

Further increase of pressure carries the volume to point *d*. As pressure is diminished, volume decreases along *dce*. At *e*— the *critical closing pressure*—a corresponding abrupt decrease in volume to point *f* occurs. This theoretical curve displays some hysteresis reminiscent of that obtained from the first inflation curves of the lung. However, the model differs from normal lung by closing at a relatively high pressure.

Thus it resembles more closely the curves obtained either from the unstable lungs of infants with the idiopathic respiratory distress syndrome, or from lungs washed with detergents (see Fig. 20–13B). The deflation curve of unstable atelectatic lungs is shifted far toward the inflation curve, indicating that the lung units collapse at high pressures. Since the liquid PV curves of these unstable lungs are similar to those of normal lungs, some disturbance relating to surface tension must have taken place to bring about the instability.

Thus, the simple tissue-enclosed bubble model accounts fairly well for the properties of the abnormal lung, but fails to explain why the alveoli of normal lungs remain open at exceedingly low pressures. This felicitous property of normal lung appears to be, at least in part, a consequence of the lining of alveoli—the lung "surfactant."

THE LUNG LINING LAYER.   The pressure inside a closed spherical bubble is greater than that outside by the amount given by Laplace's equation. As a result, the contained gas tends to diffuse out through the wall and the bubble soon disappears. Bubbles of lung edema fluid, on the other hand, are extremely durable, lasting many hours. Pattle, in 1955, deduced from the stability of the edema bubble that its internal pressure must be exceedingly low and therefore its surface tension proportionately small.[41] His calculations in fact indicated a value for the surface tension of lung edema bubbles of only about 0.05 dynes/cm. This could happen only if the bubbles from the lung were lined by a very stable, poorly soluble, compressible film.

In an alternative approach to the study of the lining film, Clements[9] used a surface balance (Fig. 20–12A) designed to measure the surface tension of a film as a function of its area. In this technique the material to be examined is spread on the surface of a fluid-filled Teflon tray where a movable barrier compresses or expands the film, thereby mimicking the changes in alveolar surface area produced during deep breathing. A hanging plate attached to a force transducer measures γ.

Be spreading small quantities of saline washes or minced extracts from healthy lungs on the fluid filling the tray, Clements found that as the film area was decreased to one-fifth of its initial value, γ fell to very low

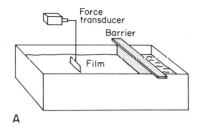

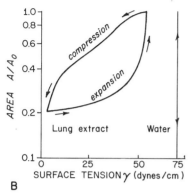

Figure 20–12 Surface tension as a function of film area.

A, The surface balance consists of a tray partially filled with fluid on which materials to be studied can be spread and compressed by a moving barrier. The hanging plate continuously records the downward force, proportional to surface tension.

B, Lung extracts from healthy mammals show extensive hysteresis and a minimum tension of 1 to 10 dynes per cm. For comparison, the surface tension of water is shown to be constant at about 70 dynes per cm.

$\frac{V}{Vo} = \left(\frac{A}{Ao}\right)^{3/2} = \left(\frac{1}{2}\right)^{3/2} \simeq \frac{1}{3}.$ This means that if the film in the lung behaved like the film on the tray, the volume of the lung could be reduced to about one-third without large volume-pressure hysteresis. Experimental data are fairly consistent with this deduction (Fig. 20–8A).

ALVEOLAR STABILITY. Surface tension in the lung model has thus far been assumed to be constant. The finding that surface tension of alveolar lining material depends on area may now be taken into account. When the film is stretched, γ rises rapidly to about 50 dynes/cm and remains fairly constant on further expansion (Fig. 20–12B). Consequently, a bubble lined with surfactant would pop open much as described earlier. But during deflation from point 1 (Fig. 20–13A), the surface tension falls and the pathway must deviate from that of the original model. For example, when γ drops to 30 dynes/cm, the PV point must lie on a new curve (point 2). As the bubble continues to deflate, the curve eventually proceeds through point 3 where γ equals 10 dynes/cm and so on. If, for this example, γ is assumed to fall only to 10, deflation follows the solid curve shown in Figure 20–13A. The significant new feature is the greatly extended range of stability; the bubble does not pop shut until very low values of pressure and volume are reached. It can be verified, by the graphic method demonstrated in Figure 20–10, that two or more such bubbles in parallel can stably deflate to low volumes without collapsing. In this way, a variable surface tension would help to account for (i) the striking fall in pressure seen in normal deflation curves, (ii) the instability of lungs where γ cannot assume low values, (iii) the large hysteresis loop of the collapsed lung, and (iv) the presence of hysteresis even where no collapse of lung units occurs.

Composition and Origin of the Lining Layer. Pattle[41] presented evidence that the material lining the lung was an insoluble lipoprotein layer. On the other hand, analyses of lung washes and extracts have found principally phospholipids, lecithin being the most prevalent. When extract film is compressed to 0.2 Ao and held at this area, the surface tension rises toward 24 dynes/cm with a time constant at room temperature of about an hour. When ex-

values, e.g., 5 to 10 dynes per cm (Fig. 20–12B). On expansion, a different curve was generated where γ rapidly increased toward that of plasma (near 50 dynes per cm) but, upon reaching this value, remained fairly constant.

When the area A of the film was decreased from its initial area, Ao, to only half Ao, the hysteresis loop was not as evident.[5] Within this range the film behaved more elastically. The change in lung volume corresponding to halving the surface area can be estimated as follows. If the shape of the lung, including all its subdivisions, remains fixed as it contracts, the alveolar surface area would depend on volume according to the relationship: $A = KV^{2/3}$. K is a constant called the shape factor which can be calculated for each geometrical configuration. Thus for a sphere, K = 4.85. When A/Ao = 1/2, the ratio of volumes would be:

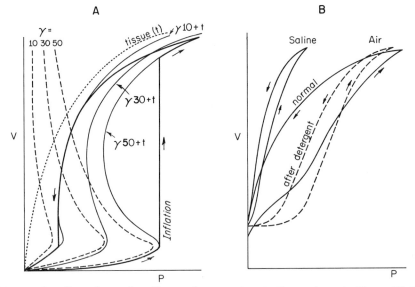

Figure 20–13 *A,* The effect of a surface film, with properties similar to those in Figure 20–12*B,* on the PV curve of the model alveolus. The inflation curve is not significantly different from the one shown in Figure 20–11. If, however, as deflation progresses the surface tension drops rapidly from 50 to 30 to 10 dynes per cm (or often less), the path on the PV plane must join the (γ_{10} + tissue) line after partial deflation. The net result is that pressure in the alveolus falls rapidly to low values, and all alveoli may be deflated to small volumes before any collapse.

B, Air, detergent and saline filling. The curve of the saline-filled lung represents the elastic recoil of the lung tissue alone. In the curve of normal air-filled lung (solid curve), both tissue and surface forces are acting. However, when a lung is flushed with a detergent solution and then refilled with air, the surface lining material is modified. Instead of the area-dependent surface tension characteristic of lung surfactant, the γ of a detergent such as Tween 20 is nearly constant at about 30 dynes per cm. As a result, alveoli collapse at high pressures and the PV curve more closely resembles that of the hyaline lung. Third inflation-deflation cycles of cat lung are shown.

panded to Ao and held, γ falls toward 24 dynes/cm. This time-dependent behavior is much less marked in edema bubbles. The several physical and chemical discrepancies between results from bubbles and extracts have not yet been resolved.[46]

The ability to form a normal alveolar film appears suddenly during fetal development—*e.g.,* on the seventeenth to eighteenth day in the mouse, which has a gestation period of 19 days, and around the seventh month in the human fetus. Coincident with appearance of osmiophilic inclusion bodies in the large Type II alveolar cells, phospholipid synthesis increases and, shortly afterward, surfactant activity can be demonstrated. Although not a proof, this sequence suggests a link between the inclusions and the origin of the lining.[37, 46]

If stored at body temperature, lungs soon lose their high surface activity, suggesting that the lining material must be continuously replenished in the alveoli of live animals. If lungs are refrigerated, the activity is unaffected for many weeks.

Interdependence. The subject of alveolar stability has thus far been treated in the context of a lung model having independent alveoli (discrete bubbles). In reality, as already depicted in Figure 20–2, each alveolus forms a unit of an interconnected mesh. Therefore small units which might have a tendency to collapse are supported and stabilized by surrounding parenchyma. The net effect of the "interdependence"[34] of air-space distension is to render alveoli considerably more stable and resistant to spontaneous atelectasis than has been calculated on the basis of bubble concepts. Collapse of individual units is in fact rarely observed in normal lungs, even near zero transpulmonary pressure. During inflation interdependence has the converse effect, in that a localized collapsed region in an expanding lung opens more readily when the tethers from neighboring regions begin to exert significant stresses.

Despite the mechanically stabilizing effect imposed by interdependence, regional lung collapse does occur when surface properties are significantly altered, as for example after saline or detergent washing, or in states of respiratory distress. Typically whole groups of alveoli empty simultaneously. Thus one must conclude that although the bubble stability theory probably has a continuing place in lung mechanics, it is necessary to introduce the notion of interdependence to achieve a realistic picture of alveolar and lobular mechanics.

Pleural and Esophageal Pressure: Lungs in Situ. A description of the normal lung as it functions *in situ* is desirable in terms of PV curves similar to those obtained from excised lungs. In the closed chest system, the lung distending or transpulmonary pressure, Ptp, is the difference between the alveolar pressure, Palv, and the pleural pressure, Ppl, the latter being the pressure within the pleural space: Ptp = Palv − Ppl. When the airways are open and no flow is taking place, Palv = 0, and so Ptp = −Ppl. Thus it is possible to obtain the recoil pressure by measuring the pleural pressure. Under static conditions Ppl is found to be negative because the lungs and the chest wall pull in opposite directions, creating suction or "negative pressure."

It should be emphasized that in mechanics all pressures, unless otherwise specified, are taken relative to the pressure on the outside of the chest (usually atmospheric). For example, the statement that Palv = +10 cm H_2O means that alveolar pressure is 10 cm H_2O greater than the pressure outside the chest.

Direct measurement of Ppl is inconvenient because it entails puncturing the chest wall with a large blunt needle. Instead, it has been found that under certain conditions esophageal pressure closely approximates pleural pressure. Where it passes through the thoracic cavity the esophagus is subjected to the distending force created by lung recoil. A small, flaccid, tubular balloon containing a small quantity of air and placed in the midthoracic esophagus will therefore reflect surrounding pressure changes. Some limitations and sources of error should be recognized. For example, the esophageal wall is muscular and therefore may not transmit pressures faithfully; indeed, peristaltic waves often generate spurious pressures. The heartbeat produces a regular artifact. Furthermore, the pressures measured depend on the location of the balloon in the esophagus, and on the body position of the subject. The gas in the balloon tends to accumulate at the point where the surrounding pressure is least, such that one probably records the most negative pressure appearing along the length of the balloon. When tidal volumes are small, changes in Ppl measured directly compare well with changes in esophageal pressure (ΔPes), except at the apex and base of the lung. For larger tidal volumes, esophageal pressures swing less than pleural pressures (ΔPes < ΔPpl) throughout the lung.[7]

Physical Properties of the Chest Wall. Elastic properties of the chest wall are described by the relationship between chest volume and the pressure across the wall of the relaxed chest (Pw). Since the chest wall contains contractile elements, measurements of passive elastic chest compliance can be made only when these muscles are completely inactive. Although pleural pressure is equal to trans-chest pressure at all times, it is clear that Pw = Ppl ≈ Pes only when the pressure component caused by muscular activity is zero.

The chest compliance curve (Fig. 20–14A) can thus be derived from esophageal pressure while the subject relaxes against a closed airway at a series of lung volumes. It is worth emphasizing that Pes (or Ppl) can represent two very different quantities, depending on the experimental conditions. When the muscles are relaxed and the airway is closed, Pes = Pw, and data can be obtained for the PV curve of the chest. However, when the muscles hold the chest at any given volume and the airway is open, then Pes = −Ptp, and data can be obtained for the PV curve of the lung. Obviously, since Ptp + Pw = Pr (the relaxation pressure), the chest PV curve may also be obtained indirectly by subtracting the lung PV curve from the total relaxation curve.

Mechanically, the lung-chest system at rest can be represented by two springs with different resting lengths (Fig. 20–14B). In the intact state the lungs, by elastic recoil, are continually "sucking in" the chest wall and diaphragm such that the combined resting length of the two springs is intermediate between the individual resting lengths. For this reason pleural pressure is negative.

The chest wall, like all other tissues, has

A

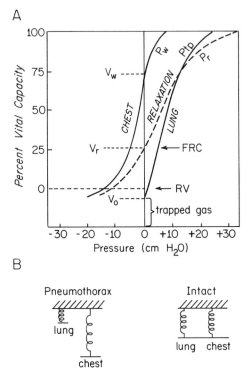

B

Figure 20-14 The elastic components of the relaxation curve (Pr).

A, At any given volume, the sum of the transpulmonary pressure (Ptp) and the passive chest-wall recoil pressure (Pw) is equal to the relaxation pressure, *i.e.*, the total pressure of the respiratory system (Pr). At a volume corresponding to FRC the tendency of the chest wall to expand is exactly balanced by the lung retractive force, so that Pr = 0. If air is admitted to the pleural space, the chest expands from FRC to its own resting volume (Vw), and the lung collapses to Vo.

B, The lungs and the chest may be represented crudely by springs of different lengths. When combined, their length is intermediate between the original lengths.

some hysteretic properties, but for most purposes these can be neglected.

ABSORPTION OF TRAPPED GAS AND FLUID FROM BODY SPACES

Since there is normally a negative pressure in the pleural space, one might ask why this "space" does not fill with gas or fluid. Actually the opposite occurs; an induced pneumothorax or hydrothorax disappears over a period of hours or days.

Gas absorption from any closed space in the body can be accounted for as follows. The total pressure on gases trapped in the

pleural space is about 5 mm Hg below atmospheric, or about 755 mm Hg at sea level. The sum of the partial pressures of O_2, N_2, CO_2 and H_2O in the space and in blood might be as shown in Table 20-2. As O_2 is extracted from arterial blood by the tissues, the blood PO_2 drops sharply because extraction occurs over the flat part of the sigmoid O_2-Hb dissociation curve (cf. next chapter). However, PCO_2 does not rise an equivalent amount because addition of CO_2 occurs on a steep CO_2 dissociation curve. This is the key to the process which enables gas absorption. PN_2 remains unchanged between arterial and venous blood, and PH_2O is, of course, constant. As a result the total pressure of the gases in venous blood is only about 706 mm Hg. In the trapped gas, PO_2 and PCO_2 almost equilibrate with surrounding tissues and venous blood. PH_2O is fixed, whereas PN_2 makes up the remainder. Consequently, a diffusion gradient is set up for molecular nitrogen. As N_2 diffuses into the venous blood and is carried to the lungs, PN_2 in the space tends to fall. Simultaneously, since the sum of all partial pressures in the space is 755 mm Hg, PO_2 and PCO_2 rise. This rise in PO_2 and PCO_2 in turn sets up a diffusion gradient for CO_2 and O_2. As these gases leave, PN_2 rises and so forth. The process continues until all the gases disappear.

In the case of fluid resorption, it will be remembered that effective plasma protein osmotic pressure, or plasma oncotic pressure, is 25 to 30 mm Hg. As long as Ptp is less than 20 to 25 mm Hg the net driving force on the fluid is directed toward the blood capillary. Abnormal stiffness of the lung (which leads to more negative Ppl),

Table 20-2 *Partial Pressures of Body Gases**

	ARTERIAL	VENOUS	TRAPPED GAS
N_2	573	573	620
O_2	90	40	38
CO_2	40	46	50
H_2O	47	47	47
Total	750	706	755

*Partial pressures at sea level in arterial blood, in venous blood and in gas in a pneumothorax some time after trapping. The large arteriovenous drop in PO_2, without a corresponding increase in PCO_2, lowers the total gas tension of venous blood. Trapped gases (notably N_2) then move into venous blood in the direction of their diffusion gradients.

plus increased permeability of the capillary walls and pleura to plasma proteins, would tend to produce pleural effusion. A similar argument applies to the etiology of pulmonary edema. Whenever the hydrostatic pressure in intrapulmonary blood vessels, plus perivascular traction caused by parenchymal stress, exceeds the osmotic drive in the reverse direction, accumulation of fluid in the interstitial and perivascular spaces may be expected. The hydrostatic drive is heightened by increased pulmonary vascular pressure, and traction is increased by higher alveolar surface tension, whereas the osmotic drive is affected by protein leakage through the capillary endothelium. Blocked lymphatic drainage would enhance the edematous effects.

AIRWAY RESISTANCE

Normally we are not conscious of the forces which move air in and out of the lungs; air movement becomes a problem only during severe exercise or illness. Disorders such as asthma, emphysema, bronchitis, laryngitis and pneumonia are characterized by moderate to severe increases in resistance caused by partial obstruction of the airways.

The resistance to gas flow within the airways (Raw) is defined as the total pressure drop along the airway (Paw) divided by the volume flow: $Raw = Paw/\dot{V}$. The Poiseuille-Hagan equation states that for laminar flow in straight tubes, Raw is a simple function of the tube dimensions and of the viscosity of the gas. Flow in a branching distensible system where turbulence may exist does not obey convenient theoretical relations. As an approximation, Rohrer[45] proposed that the airway pressure drop be described by two terms of a parabola: $Paw = K_1\dot{V} + K_2(\dot{V})^2$. The first term was considered to be due to laminar flow, and the second to turbulent flow at higher velocities. The factors K_1 and K_2 were proportionality constants obtained by fitting the equation to the experimental data of each subject. One sees that Raw is not a constant, but rather a function of $\dot{V}$: $Raw = Paw/\dot{V} = K_1 + K_2\dot{V}$. However, since K_2 is small, the second term is insignificant during normal breathing. Consequently, K_1 can be considered to be

the airway resistance. Although the parabolic fit is adequate, its interpretation in terms of laminar and turbulent components is hazardous. Other empirical formulas, such as $Paw = K(\dot{V})^n$, with n between one and two, might serve as well as Rohrer's.[1] It has in fact been shown [42] that if one takes into account nonparabolic flow profiles in branching tubes, a theoretical expression can be derived in which n = 3/2, and K depends on both viscosity and density. The simple Poiseuille equation underestimates Paw by a factor of around 3 or more.

Alveolar Pressure. Determining the airway resistance is not as simple as the equation $Raw = Paw/\dot{V}$ might suggest. Although flow can be satisfactorily measured by a variety of modern flowmeters (pneumotachographs), a difficulty arises in evaluating Paw. Doing so requires that the pressure be measured at the inner end of the airway (in the alveoli) as well as at the mouth. Several means for estimating the alveolar pressure indirectly have been devised. Although the original constant-volume whole-body box of Dubois[17] is widely used, the volume-displacement modification of Mead[31] will be described. Two measurements must be made in quick succession (Fig. 20–15). First, the subject rebreathes from a bag within the box. Whenever he exhales, the spirometer deflects downward by a certain amount (ΔV) owing to compression of alveolar air, and vice versa on inspiration. Alveolar pressure is found using the Boyle-Mariotte Law: $\overset{*}{P}_B V_L = (\overset{*}{P}_B + Palv)(V_L - \Delta V)$, where $\overset{*}{P}_B$ is the dry gas pressure, $P_B - P_{H_2O}$. Solving for Palv, one obtains: $Palv = \dfrac{\overset{*}{P}_B \Delta V}{V_L - \Delta V} \approx \dfrac{\overset{*}{P}_B \Delta V}{V_L}$. Second, a shutter valve stops air flow for a few moments while the subject continues to make respiratory efforts. The same form of equation pertains: $\text{Pálv} = \dfrac{\overset{*}{P}_B \Delta V'}{V_L}$, where Pálv is alveolar pressure with no flow and $\Delta V'$ is the gas compression at no flow. Now, eliminating $\overset{*}{P}_B/V_L$ from the two equations gives: $Palv = \left(\dfrac{Palv}{\Delta V'}\right)\Delta V$. The ratio of the simultaneously measured values of Pálv and $\dot{V}$ gives the total resistance, from which ap-

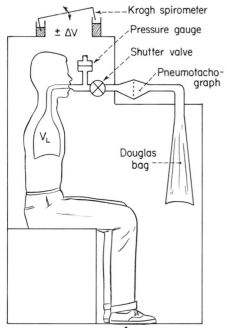

Krogh spirometer
± ΔV
Pressure gauge
Shutter valve
Pneumotacho-
graph
V_L
Douglas
bag

Figure 20–15 Volume-displacement whole body plethysmograph. The subject is seated in a closed chamber where he rebreathes from a bag. Since this maneuver merely moves warm, moist gas from one part of the system to another, the total box volume should be constant except for the slight expansion or compression of gas in V_L. The extremely sensitive Krogh spirometer is used to measure this volume change under two conditions: (i) airway open, breathing through the pneumotachograph; and (ii) respiratory effort against a closed airway. Total resistance, calculated as described in the text, includes the resistance of the instrumental tubing. (After Bachofen, *Die mechanishen Eigenschaften der Lunge.* Bern, Hans Huber, 1969.)

paratus resistance is subtracted to yield Raw.

Partitioning Resistance. The total airway resistance has been partitioned into a series of smaller resistances, each representing a section of the air-conducting system.[20] Measured by mouth breathing in the plethysmograph, a typical value for Raw would be 1.4 cm H_2O/l/sec. Near FRC the upper airways (larynx and above) account for more than 40 per cent of the total, whereas at 80 per cent VC they account for more than 60 per cent.[50] Even though each small respiratory bronchiole has a high resistance to flow, the number of fine airways in parallel is so large that their total resistance per unit length is actually substantially lower than that of the larger airways.

Turbulence which occurs in the trachea and larynx at even moderate velocities never occurs in the bronchioles. (Turbulence arises in long, straight, round tubes whenever the dimensionless quantity, Reynold's number (Re), is greater than about 2000: $Re = d\bar{v}\rho/\eta$ [d = diameter of the airway, $\bar{v}$ = mean velocity, ρ = fluid density, η = viscosity]. When the tubular geometry is irregular, turbulence begins at much lower values of Re, *e.g.*, 500.)

Dependence on Flow and Volume. Implicit in Rohrer's pressure-flow equation is the assumption that airway dimensions are constant. We know now that airway size depends on the degree of lung inflation and, as Einthoven predicted,[19] on the flow-dependent pressure difference across the bronchial walls. Thus it would seem that considerations of resistance should involve at least three variables: pressure, volume and flow, or P, V, V̇. A model of the lung and airways[21] which may be helpful in visualizing the relations of these three variables is shown in Figure 20–16. One may view the lung parenchyma as an elastic mesh transmitting stress from the visceral pleura to the airways (interdependent). During inspiration, while the lung volume is increasing, the augmented tension in the lung tissue (the springs in the model) enlarges the intrapulmonary airways. Meanwhile, as Ppl becomes more negative, the extrapulmonary airways (the lower trachea and primary bronchi) become distended. In addition, the bronchial transmural difference in gas pressure acts to enlarge the intrapulmonary airways. Consequently, the rate of inspiration is limited mainly by rate of muscle contraction; the faster the piston moves, the larger the airways become and the higher the flow rates.

During expiration, as the piston is forced upward, Ppl becomes more positive. This pressure is applied directly to the lower trachea and primary bronchi and compresses them. Within the lung, Palv is elevated narrowing the intrapulmonary airways. Bernouilli effects accentuate compression in larger airways. Furthermore, as the lung volume diminishes, so does the tension in parenchyma, allowing airway dimensions to be reduced even further. As a consequence of these mechanisms, the rate of expiration is self-limiting. The most striking example of this "check-valve

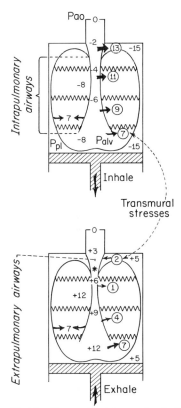

Figure 20–16 Shown in this mechanical model of the respiratory system are factors that influence tracheobronchial dimensions. The airways are represented by a semiflaccid tube leading into an elastic lung. The lung parenchyma is stretched between pleura and airways.

A, Inhalation. Suppose at some instant Ppl is −15 cm H_2O, Palv is −8 cm H_2O and Ptp is 7 cm H_2O. Pressure along the airway falls steadily toward the alveoli from 0 to −8, whereas outside the airway the pressure is everywhere −8. At the point in the airway where pressure is −4, the net transmural stress would be −4 − (−8) + 7 = 11. Similarly, the other stresses along the airway are seen to distend the lumen.

B, Exhalation. On reversing the flow, let Palv = +12, Ppl = +5, and Ptp again be 7. The intrabronchial pressures are positive, dropping toward 0 at the mouth. At the point in the airway where pressure is +6, the net transmural stress is +6 − (+12) + 7 = +1. The distending forces on the upper airways are thus very low, even becoming compressive in the extrapulmonary segment. Consequently, airway caliber decreases when breathing out. Above the point ° (the equal pressure point), the net force is compressive and pronounced collapse of a flaccid airway could occur.

mechanism"[15] is seen in clinical emphysema: the harder the patient blows, the more his airways collapse, with the result that flow rate is almost independent of effort. The added effort is merely wasted. His inspiratory flow rates may be nearer

normal. Patients with obstructive syndromes find partial relief through several mechanisms: (i) unconsciously maintaining a high FRC to distend the airways, (ii) inspiring more rapidly to allow time for slower expiration and (iii) pursing the lips or narrowing the glottis to limit the flow rate and thus reduce airway collapse. In asthma, the heightened bronchomotor tone may constrict airways throughout the respiratory cycle, such that both inhalation and exhalation are labored. However, constricted airways are more rigid and therefore resist dynamic collapse to a larger extent.

Figure 20–17 illustrates graphically how the variables $\dot{V} = f(Ppl, VL)$ are related. It is apparent that, in a lung with distensible airways, as volume increases Raw should decrease. An approximate relation for normals has been found[4] to be: $\frac{1}{Raw} \approx 0.24 VL$. For example, at a lung volume of 3 l, Raw ≈ 1.4 cm H_2O/l/sec.

Interestingly, the exercising normal subject never exerts pressures greater than those on the *Max. flow* line in Figure 20–17.[38] He seems automatically to adjust his effort in such a way that energy will not be wasted.

DISTRIBUTION OF INSPIRED GAS IN THE LUNG

The discussion thus far has treated the properties of the whole lung without regard to the properties of its anatomic subdivisions. There is good reason to believe that in a complex structure such as the lung, and especially in the diseased lung, not all units have identical properties. One result of this nonuniformity is unequal distribution of gas on inspiration, and unequal emptying of lung units on expiration. Equal distribution occurs whenever ΔV/Vo is the same for every region (ΔV is the volume of fresh air inspired into the region and Vo its initial volume). In other words, with equal distribution the dilution in all regions must be equal throughout inspiration. Nonuniform distribution may arise from any of the following conditions: (i) unequal specific compliances leading to unequal volume increments in the various lung regions, (ii) unequal airway resistance to various parts of the lung (important only in

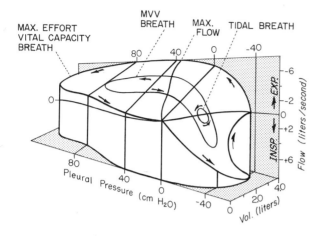

Figure 20–17 Surface showing dependence of V̇ on Ppl and on VL. If V̇ is plotted on the Z-axis, a surface is obtained which describes the flow characteristics of a lung. Shown here are normal data. Expiration is limited by airway collapse as indicated by the fact that the surface becomes level for large positive pleural pressures. The intersection of the surface with the PV plane is the static PV curve of the lung. (After Fry and Hyatt, *Amer. J. Med.*, 1960, 29, 672–689.)

states of greatly increased resistance), (iii) unequal forces either on the pleural surface or within the lung, or (iv) the presence of regions anatomically in series in which longitudinal diffusion is incomplete.

Unequal compliances might occur most obviously when the tissue and surface properties vary from place to place. Non-uniform compliances may also be induced by gravitational effects.[35] For example, insofar as the lung is an elastic mesh having mass, it must sag under its own weight. Respiratory units near the apices of an upright man are enlarged and therefore situated on the less compliant part of the PV curve (Fig. 20–8), whereas units toward the base are less distended and hence highly compliant. Consequently, when a subject inhales from FRC, relatively more air enters the lower lobes.

The better ventilation of lower lobes is further enhanced by wider swings in pleural pressure at the base of the lung. These probably result from the fact that the thoracic cavity enlarges most at the base.

Nonuniform ratios of $\Delta V/Vo$ are particularly prevalent in obstructive airway disease. Uncompensated, they result in ventilation-perfusion abnormalities which are the most common cause of hypoxemia in clinical medicine.

TESTS OF PULMONARY MECHANICAL FUNCTION

Tests of pulmonary function may be grouped into four classes: measurements of flow, volume, pressure and nonuniform distribution of inspired air. The first three are mechanical measurements, whereas the last employs methods involving alveolar gas analysis and will be discussed in the following chapter. Compliance is derived secondarily from pleural pressure and volume; resistance from flow and alveolar pressure.

Flow. *Maximum Voluntary Ventilation (MVV).* The subject is asked to breathe from a spirometer as rapidly and deeply as he can for 12 to 15 sec. The spirometer has a summing device which adds each inspired tidal volume to the previous ones so that the total ventilation may be read directly. Observed values are compared with predicted values obtained from equations relating MVV to age, body surface area and sex. A typical figure for a young male would be 150 l/min.

MVV measures airway resistance as well as physical strength and motivation in the patient.

Forced Expiratory Volume (FEVt) or Timed Vital Capacity. The subject inspires to TLC, then exhales as rapidly as possible to RV while a spirometer records both the volume expired over a preset timed interval and the vital capacity. A normal person can exhale about 68 per cent of his VC in the first 0.5 sec, 77 per cent in 0.75 sec, 84 per cent in 1.0 sec, 94 per cent in 2.0 sec, and 97 per cent in 3.0 sec.[36] If observed percentages are below those predicted, obstructive changes may be presumed. This is a commonly used test in pulmonary function laboratories.

Maximum Expiratory Flow (MEF). A value close to the highest flow rate reached during a maximal forced expiration can be recorded with a Wright Peak Flow Meter.[56] Healthy subjects reach peaks of 6 to 12 liters per sec. The Puffmeter,[29] a more damped flow meter, measures the flow that can be maintained over a short period of time, rather than the instantaneous peak. Both tests correlate well with MVV and can be made with relatively inexpensive equipment. They are therefore extremely useful in office tests.

Volume. Three of the four lung volumes can be measured with any simple spirometer. The residual volume, however, must be determined indirectly (usually by first measuring FRC) by methods such as plethysmography, gas dilution or N_2 washout.

Plethysmography has the virtue of detecting gas trapped in nonventilated regions as well as air in the ventilated regions. The technique is illustrated in Figure 20–15.

Dilution methods are based on the simple principle that the number of mols of a gas is unchanged when the gas occupies a larger space. By rebreathing from a bag, a relatively insoluble gas such as helium (He) is mixed with the air already in the lung. The computation proceeds as follows: $CiVi$

$$= CfVf = Cf(Vi + V_L), \text{ or } V_L = Vi\left(\frac{Ci - Cf}{Cf}\right).$$

Here Ci refers to initial He concentration and Vi to the initial volume of gas in the bag; Cf and Vf are the final concentration and volume in the bag-lung system. V_L, the desired volume, is equal to FRC if the rebreathing is begun at resting end-tidal expiration. The same principle may be applied to the N_2 washout technique where a subject takes several deep breaths of O_2, and the mixed exhaled air is analyzed for N_2.[55]

Compliance. STATIC AND DYNAMIC. *"Static" Lung Compliance (Cl).* Measurement of the slope of the lung PV curve in the tidal volume range has been described earlier. Pleural pressure is measured at a number of volumes a few seconds after flow is stopped. Because of viscoelasticity and the resultant stress relaxation, the pleural pressure depends somewhat on the length of the stop, but is generally fairly stable after 3 sec. In men over age 60, Cl is about 25 per cent greater than in younger men; very little change with age appears to occur in women.[12] Normal compliances in adults range from 0.18 to 0.27 l/cm H_2O.

Static Chest Wall Compliance (Cw). As described earlier, Cw is the slope of the chest PV curve, obtained while the respiratory muscles are relaxed. It usually has about the same value as Cl.

Specific Compliance. To compare individuals of different sizes, the values of Cl and Cw must be standardized. As an example, the parenchyma of mouse lungs having a compliance of 0.0001 l/cm H_2O is not 2000 times as stiff as that of human lungs. In practice, therefore, the static compliance is divided by some characteristic lung volume (*e.g.*, FRC, VC or TLC) to yield a compliance per unit volume or "specific" compliance.

Dynamic Compliance (Cdyn). The "static" compliance is ideally determined at zero frequency, meaning, in effect, that Pes has equilibrated before a reading is taken. Thus, if Pes has become only approximately constant when the pressure is read, a "quasi-static" compliance is obtained. However, if the subject is breathing regularly at a fixed frequency, the measurement must be called the "dynamic" compliance at that frequency.

Although the data are easily obtained, the interpretation of Cdyn requires some comment. Measurement is conveniently made by plotting the pressure drop from airway opening (mouth) to esophagus (Pao-Pes) on the X-axis of an oscilloscope, versus V on the Y-axis. Figure 20–18A shows three such dynamic PV loops made at the same tidal volume, but at three different frequencies. Loop *a* was obtained from a very slowly executed respiratory cycle (this curve is like the smaller loops of Figure 20–8). To generate loop *b* the subject breathed at a moderately rapid rate, and then still more rapidly for loop *c*. Care should be taken to distinguish these apparent hysteresis loops from the lung hysteresis already described. Here the dynamic loops arise almost entirely from gas flow-resistance. In other words, the chest must exert an additional pressure, Paw, to move the gas. Paw is zero at those two instants of the respiratory cycle when $\dot{V}$ is zero (the maxima and minima of V), as shown in Figure 20–18B. The slope of the line joining these two points is the dynamic compliance of the lung at that particular frequency. In elderly men and in emphysematous patients, Cdyn clearly decreases as breathing frequency is increased, whereas in younger, healthy

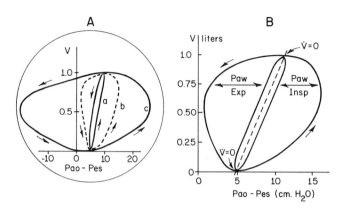

Figure 20–18 Dynamic PV loops.

A, These oscilloscopic figures represent simultaneously recorded tidal volume (vertical axis) and esophageal pressure (horizontal axis) at three breathing rates: (a) very slow, where flow resistive pressure is negligible, (b) moderately rapid and (c) fast. Pao-Pes is the differential pressure between esophagus and airway opening (mouth).

B, The slope of the line joining the two points at which $\dot{V} = 0$ gives the dynamic compliance of the lung. Paw is the added pressure necessary to propel gas through the conducting airways.

individuals it may remain more nearly constant. A fall in compliance as respiratory rate rises is due primarily to unequal parallel time constants (RC) in different lung regions.[40] Compliance may become frequency dependent before other tests of lung function become abnormal,[28] suggesting that its measurement could be useful in detection of early lung disease.

Resistance. Abnormal resistance may be inferred if low values are obtained from any flow test. However, these tests depend to some extent on the patient's cooperation, his effort and his physical condition. An objective measure of airway resistance is obtainable from the body plethysmograph as already described.

WORK OF BREATHING

In one sense (the physicist's), it is correct to say simply that the work done by a muscle as it lifts a weight is the external force (F) times the distance moved (x), or $\int F dx$. Yet, from the amount of work performed on the surroundings, we have no information about the dissipation of metabolic energy within the muscle while it is holding a fixed length, or lowering the weight or acting against antagonists. The total energy used in breathing can best be related to the metabolic cost, or O_2 consumption, associated with respiratory movements.

Two dissipative mechanisms can be identified in the respiratory system: hysteresis in lungs and chest wall, and flow resistance in the airways. Although both are non-linear, they may be lumped for purposes of

discussion into a dashpot, R (Fig. 20–19A). Two nondissipative elements in the mechanical analog are the lung-chest compliance, C, and inertance, I (related to the weight of tissues, bones and blood). Inertance is normally neglected since it is significant only at frequencies in excess of two breaths per second. The equation describing the motion of the system is therefore $P = (1/C)V + R\dot{V}$.

Except in moderate or severe exercise, expiration is almost entirely passive. Metabolic work is therefore expended mainly during inspiration by extending the spring and dashpot. The frictional energy loss appears as heat, while the energy put into the spring is temporarily stored. Since this stored energy is not returnable to the muscles, it is used to bring the system back to its resting volume; in other words, it is dissipated in the dashpot and in the muscle itself during expiration. When an active muscle is stretched by an external force, it cannot convert mechanical energy into chemical energy; it merely degrades chemical energy at a lower rate. Work done on an active muscle has been called negative work.

The work (W) of stretching a spring may be computed as follows: $W = \int F dx$, represents the cross-hatched area in Figure 20–19B. A simple development of this equation leads to a method for finding the work of inflating a balloon: $W = \int F dx = \int (F/A) \cdot A dx = \int P dV$, where A is the surface area. This is the cross-hatched region shown under the PV curve in Figure 20–19C. When friction is present the total applied pressure equals the sum of elastic pressure (Pr from the relaxation curve) and resistive

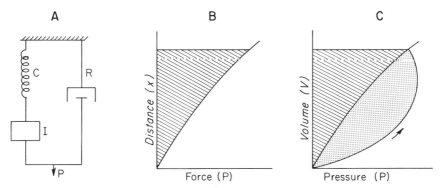

A   B   C

Figure 20–19 *A*, Simplified mechanical analog of the respiratory system. The elastance, resistance and inertance are represented by three pure elements, C, R and I, respectively. Inertance can be neglected except at very high volume accelerations.

B, The work of stretching a spring is the shaded area to the left of the force-elongation curve.

C, Similarly, the work of inflating a balloon is the shaded area to the left of the PV curve. The stippled area represents the work done in overcoming inspiratory flow resistance.

pressure (Paw): $P = Pr + Paw$; thus $W = \int P dV = \int (Pr + Paw) dV = \int Pr dV + \int Paw dV$. The latter term is represented by the stippled area of Figure 20–19*C*; the sum is approximately the external work done for each breath. Whenever expiration is not passive, another term must be computed for this part of the cycle as well.

Otis *et al.*[39] have approximated the flow by a sinuosidal function to permit an analytic instead of a graphic integration. They were able to compute the energy required to maintain a certain alveolar ventilation at various respiratory rates. A graph of computed power versus respiratory rate for a fixed ventilation shows a power minimum at some frequency. Interestingly, it appears that most animals have adjusted their resting respiratory rates to fall somewhere near this minimum.[14]

If a subject ventilates at a series of minute volumes ($\dot{V}E$), from resting levels up to MVV, it may be assumed that the respiratory muscles account for the additional O_2 consumption above the resting rate (Fig. 20–20). From this figure, the plight of the emphysematous patient is quite apparent. By increasing his ventilation on exertion, he may consume more O_2 in the respiratory muscles than can be provided by this additional ventilation, and consequently, the arterial O_2 tension falls (perhaps also as a result of increased maldistribution). Normally, the limiting factor in exercise is likely to be the cardiac output.

A comparison of the external work with the caloric input (O_2 consumption) yields a ratio called the "efficiency." Studies of the respiratory apparatus indicate that its efficiency is equal to that of other skeletal muscles, *i.e.*, in the order of 20 per cent.

In patients, sensations of dyspnea are often associated with increased work of breathing in connection with either elevated airway resistance or decreased compliance. Frequently, psychological factors or "life stresses" seem to correlate well with

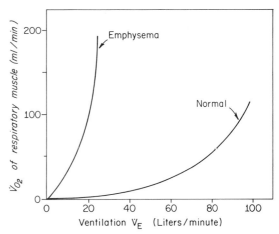

Figure 20–20 These curves were obtained from one normal subject and one patient with emphysema. Ventilation was stimulated by adding dead space. The patient with emphysema had a greater O_2 cost of breathing even at resting levels of ventilation, but most striking was the rate of rise of $\dot{V}_{O_2}$ as ventilation was increased. (After Campbell *et al.*, *J. appl. Physiol.*, 1957, *11*, 303–308.

episodes of dyspnea even if respiratory function appears to be normal. At present it is unclear if emotional stress leads to bronchoconstriction, thence to dyspnea; or if stress evokes the sensation of dyspnea directly, without increased work of breathing; or if one's awareness of increased work of breathing varies with the psychic state.

LUNG MECHANICS AND THE PULMONARY CIRCULATION

It is fairly easy to understand that alterations in the pulmonary vascular resistance would arise from changes in vasomotor tone or from obstruction to flow caused by embolism, thrombosis, chronic obstructive emphysema, fibrosis or metastatic carcinoma. However, the influences of transpulmonary pressure, degree of lung inflation and other mechanical factors on vascular resistance are more subtle.

The heart and its large conducting vessels are suspended in the mediastinum and, like the esophagus, are subject to distending forces caused by the retraction of the lungs. These forces, as Donders recognized in 1853,[16] have a "sucking" effect on all structures opening into the thoracic cavity, thus facilitating venous return and slightly encumbering arterial flow out of the thorax. Significant alterations in intrathoracic pressures occur whenever the chest is opened; this should be kept in mind if physiologic deductions regarding cardiovascular or respiratory function are to be made from open-chest experiments.

Studies of flow through collapsible blood vessels embedded in tissue must take into account both the resistive pressure difference between the ends of the vessel and the transmural stress. The latter determines the luminal size of a given vessel. The concept may be applied to the lung by grouping the pulmonary vessels into two broad categories: large and small.[25] The lung parenchyma is considered to be like an elastic mesh stretched from the visceral pleura to the major blood vessels, as well as to the airways. An increasing force in this mesh (as during inflation of the lungs) tends to distend not only the airways but also the larger blood vessels. However, quite unlike the condition in airways, alveolar pressure *does* act transmurally in blood vessels to cancel almost completely the parenchymal distending forces. For this reason, the size of the intrapulmonary large vessel is not so greatly affected by lung inflation. On the other hand, the capillaries within the interalveolar septa are *not* under radial stress from tissue or surface forces. In fact, they are primarily flattened or "squeezed" by a rise in alveolar pressure.

A simple model which may help to clarify the effect of Palv on small vessels is shown in Figure 20–21A from Riley.[44] The pulmonary arterial pressure, Ppa, is represented by the height of the first reservoir. The thin-walled, collapsible capillaries pass through a chamber whose pressure is Palv, and empty via a "waterfall" into the venous reservoir. Several points are immediately obvious: (i) a rise in Palv constricts the small vessels, acting as a "sluice gate;"[2] (ii) moving the venous reservoir up or down has no effect on flow from the "waterfall," except when Ppv becomes positive (as in mitral stenosis); (iii) either a rise in Ppa or a positive Ppv enlarges partially compressed vessels; (iv) blood flow therefore normally depends on (Ppa − Palv), not on (Ppa − Ppv), except in abnormal cases in which there is "back pressure" from the pulmonary vein. The effects of lung inflation on blood flow are shown in Figure 20–21B. For a fixed Ppa blood flow drops to near zero when Palv = Ppa. Thus to minimize the work load on the right heart and to maintain cardiac output, it is desirable to keep the "sluice" pressure, Palv, as low as possible. This is an important consideration where positive pressure ventilatory-assist devices are used, as in intensive care units.

MECHANICAL PARAMETERS AND LUNG DISEASE

Respiratory problems are becoming increasingly important as smoking and air pollution persist at a high level; roughly 10 per cent of the American population over age 40 present readily detectable evidence of pulmonary dysfunction. In this section a few common disorders have been selected to illustrate changes in lung mechanical properties in disease (Fig. 20–22). A comprehensive text is now available which provides an excellent discussion of respira-

A   B

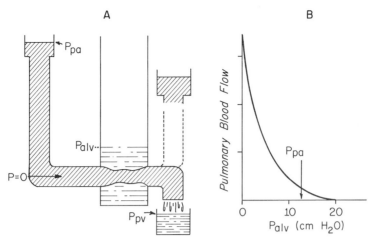

Figure 20–21 *A,* Schematic representation of the "waterfall" hypothesis. The collapsible segment is sometimes called a "Starling resistor." Flow is independent of pulmonary venous pressure, Ppv, except when Ppv > Palv. When either Palv or Ppv is equal to pulmonary arterial pressure (Ppa), flow ceases. (All pressures here are referred to pleural pressure, Ppl.)

B, Plot of flow through the cat's pulmonary vascular system at increasing levels of inflation (Palv) when the arterial pressure is held constant at 12.3 cm above the base of the lung and the venous reservoir is below the base of the lung. (After Riley, *Ciba foundation symposium on pulmonary structure and function,* 1962.)

tory diseases from a functional (physiological) point of view.[3]

Asthma. Bronchospastic respiratory diseases can be differentiated from nonspastic obstructive airway diseases by the former's positive response to bronchodilators. It should be remembered, however, that a considerable proportion of the airway resistance in asthma is due to the presence of edema fluid and mucus in the smaller bronchioles. In common with other obstructive diseases, the higher airway resistance in asthma leads to an elevated mean lung volume, and fortunately this tends to enlarge the airways. Lung compliance is generally not greatly affected by constriction of the airways. Distribution of inspired gas may be normal.

Figure 20–22 Static and dynamic lung PV curves illustrate the effects of certain diseases on lung compliance, lung volumes and airway resistance. The upper horizontal bar on each PV curve represents a typical value for the FRC, and the lower bar the RV. Augmented airway resistance is the predominant feature in asthma and emphysema (obstructive diseases), whereas decreased compliance is most obvious in pulmonary fibrosis and respiratory distress syndrome (restrictive diseases). Cardiac patients show some changes in both characteristics.

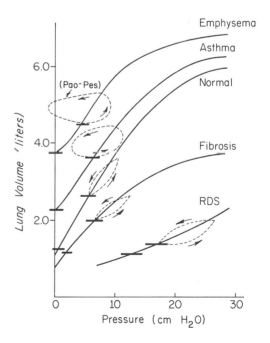

Emphysema. Emphysema has been defined as "a condition of the lung characterized by increase beyond the normal in the size of air spaces distal to the terminal bronchiole either from dilation or from destruction of their walls" (Ciba Foundation Guest Symposium, *Thorax*, 1959, *14*, 236). According to this definition, emphysema can be diagnosed with certainty only by a pathologist on postmortem examination, and occasionally by a radiologist if the spaces or bullae are greatly enlarged. For clinical purposes the irreversible obstructive pulmonary syndrome (clinical emphysema) is described somewhat differently. The static compliance and airway resistance are increased, FRC and TLC are elevated, maldistribution of inspired air occurs and ventilation-to-perfusion ratios are abnormal. The airways are more susceptible to collapse for several reasons: (i) Inflammation may produce mucosal congestion and sputum which, together with bronchospasm, result in increased airway resistance and consequent airway collapse owing to higher bronchial transmural pressures. (ii) Destructive changes in the walls of the bronchioles or loss of lung parenchyma supporting the bronchioles increase airway resistance and lung compliance.

Fibrosis. Thickening and stiffening of the pulmonary membranes by granulomatous or collagenous tissue results in decreased compliance and a decreased diffusing capacity. When severe, the result is dyspnea and hypoxia (cyanosis) in exercise, or even at rest. Increases in airway resistance are not consistent. Distribution of inspired gas may be normal. Often dyspnea is not an outstanding symptom until late in the course of the disease.

Respiratory Distress Syndrome (RDS). RDS accounts for approximately 30,000 deaths annually in the United States. Premature infants are particularly susceptible. Usually within hours after birth such babies develop severe breathing difficulty as evidenced by marked inspiratory effort (very low compliance), sternal retraction, grunting expiration and hypoxia or cyanosis. Blood shunt (blood bypassing open ventilated alveoli) increases from the normal 3 per cent to 10 to 80 per cent. Mortality is highest on the first or second day; thereafter chances for survival improve sharply. Postmortem examination of the lungs reveals a dark liver-like appearance, widespread atelectasis, alveolar instability, and very little gas trapping. After several hours of respiratory difficulty, the infant's lungs show thickened, glassy (hyaline) membranes in the gas-exchanging portions, hence the name "hyaline membrane disease." The surface tension of lung extracts made from atelectatic lungs generally does not fall below 20 dynes/cm as compared to 5 dynes/cm for normal lungs.

It has been suggested[23] that prior to birth an unknown form of stress affects the fetus, in some way reducing the activity of lung surfactant. The deficiency apparently corrects itself after the second day, making the disease self-limited. Meanwhile high surface tension promotes fluid transudation from blood, and fibrin deposition. As an alternative explanation of the etiology,[8] it is possible that a perinatal stress such as hypoxemia, acidemia, hypothermia or hypovolemia may bring about pulmonary vasoconstriction, thereby causing a ductus arteriosus shunt. The surfactant deficiency might then follow secondarily. Rational therapy, therefore, should be directed at correcting the vasoconstriction as well as relieving the initiating stress.

REFERENCES

1. Ainsworth, M., and Eveleigh, J. W. A method of estimating lung resistance in humans. Ministry of Supply, C.D.E.E. Tech. Paper No. 320, Porton, Wiltshire, England, 1952.
2. Banister, J., and Torrance, R. W. The effects of the tracheal pressure upon flow. *Quart. J. exp. Physiol.*, 1960, 45, 352–367.
3. Bates, D. V., Macklem, P. T., and Christie, R. V. *Respiratory function in disease. An introduction to the integrated study of the lung.* Philadelphia, W. B. Saunders Co., 1971.
4. Briscoe, W. A., and DuBois, A. B. Relationship between airway resistance, airway conductance, and lung volume in subjects of different age and body size. *J. clin. Invest.*, 1958, 37, 1279–1285.
5. Brown, E. S., Johnson, R. P., and Clements, J. A. Pulmonary surface tension. *J. appl. Physiol.*, 1959, 14, 717–720.
6. Campbell, E. J. M. *The respiratory muscles and the mechanics of breathing.* Chicago, Year Book Publishers, 1958.
7. Cherniack, R. M., Farhi, L. E., Armstrong, B. W., and Proctor, D. F. A comparison of esophageal and intrapleural pressure in man. *J. appl. Physiol.*, 1955, 8, 203–211.
8. Chu, J., Clements, J. A., Cotton, E., Klaus, M. H., Sweet, A. Y., Thomas, M. A., and Tooley, W. H. The pulmonary hypoperfusion syndrome: preliminary report. *Pediatrics*, 1965, 35, 733–742.
9. Clements, J. A. Surface tension of lung extracts. *Proc. Soc. exp. Biol. (N.Y.)*, 1957, 95, 170–172.

10. Clements, J. A. Pulmonary edema and permeability of alveolar membranes. *Arch. environ. Health*, 1961, *2*, 280–283.

11. Clements, J. A., Hustead, R. F., Johnson, R. P., and Gribetz, I. Pulmonary surface tension and alveolar stability. *J. appl. Physiol.*, 1961, *16*, 444–450.

12. Cohn, J. E., and Donoso, H. D. Mechanical properties of lung in normal man over 60 years old. *J. clin. Invest.*, 1963, *42*, 1406–1410.

13. Cook, C. D., Mead, J., and Orzalesi, M. M. Static volume-pressure characteristics of the respiratory system during maximal efforts. *J. appl. Physiol.*, 1964, *19*, 1016–1022.

14. Crosfill, M. L., and Widdicombe, J. G. Physical characteristics of the chest and lungs and the work of breathing in different mammalian species. *J. Physiol. (Lond.)*, 1961, *158*, 1–14.

15. Dayman, H. Mechanics of airflow in health and emphysema. *J. clin. Invest.*, 1951, *30*, 1175–1190.

16. Donders, F. C. Beiträge zum Mechanismus der Respiration und Circulation im gesunden und kranken Zustande. *Z. rat Med.*, 1853, *3*, 287–319.

17. DuBois, A. B., Botelho, S., and Comroe, J. H., Jr. Resistance to airflow through the tracheobronchial tree as measured by means of a body plethysmograph. *J. clin. Invest.*, 1954, *33*, 929.

18. Dunnill, M. S. Postnatal growth of the lung. *Thorax*, 1962, *17*, 329–333.

19. Einthoven, W. Ueber die Wirkung der Bronchialmuskeln, nach einer neuen Methode untersucht, und über Asthma nervosum. *Pflügers Arch. ges. Physiol.*, 1892, *51*, 367–445.

20. Ferris, B. G., Jr., Mead, J., and Opie, L. H. Partitioning of respiratory flow resistance in man. *J. appl. Physiol.*, 1964, *19*, 653–658.

21. Fry, D. L., and Hyatt, R. E. Pulmonary mechanics; a unified analysis of the relationship between pressure, volume, and gas flow in the lungs of normal and diseased human subjects. *Amer. J. Med.*, 1960, *29*, 672–689.

22. Goldsmith, J. R. A simple test of maximal expiratory flow for detecting ventilatory obstruction. *Amer. Rev. Tuberc.*, 1958, *78*, 180–190.

23. Gruenwald, P. The course of the respiratory distress syndrome of newborn infants, as indicated by poor stability of pulmonary expansion. *Acta ped.*, 1964, *53*, 470–477.

24. Hildebrandt, J. Dynamic properties of excised air-filled cat lung determined by liquid plethysmograph. *J. appl. Physiol.*, 1969, *27*, 246–250.

25. Howell, J. B. L., Permutt, S., Proctor, D. F., and Riley, R. L. Effect of inflation of the lung on different parts of the pulmonary vascular bed. *J. appl. Physiol.*, 1961, *16*, 71–76.

26. Hutchinson, J. Thorax. In: *The cyclopedia of anatomy and physiology*, R. B. Todd, ed., Vol. 4, Part II. London, Longman, Brown, Green, Longmans & Roberts, 1835–1859.

27. Johnson, L. F., Jr., and Mead, J. Volume-pressure relationships during pressure breathing and voluntary relaxation *J. appl. Physiol.*, 1963, *18*, 505–508.

28. Macklem, P. T. Airway obstruction and collateral ventilation. *Physiol. Rev.*, 1971, *51*, 368–436.

29. Macklin, C. C. Pulmonary alveolar mucoid film and the pneumocytes. *Lancet*, 1954, *266*, 1099–1104.

30. McCredie, M., Lovejoy, F. W., Jr., and Kaltreider, N. L.: Pulmonary function in diaphragmatic paralysis. *Thorax*, 1962, *17*, 213–217.

31. Mead, J Volume displacement body plethysmograph for respiratory measurements in human subjects. *J. appl. Physiol.*, 1960, *15*, 736–740.

32. Mead, J. Mechanical properties of lungs. *Physiol. Rev.*, 1961, *41*, 281–330.

33. Mead, J. The distribution of gas flow in lungs. In: *Circulatory and respiratory mass transport*, Ciba Foundation Symposium. Boston, Little, Brown and Co., 1969.

34. Mead, J., Takishima, T., and Leith, D. Stress distribution in lungs: A model of pulmonary elasticity. *J. appl. Physiol.*, 1970, *28*, 596–608.

35. Milic-Emili, J., Henderson, J. A. M., and Kaneko, K.: Distribution of ventilation as investigated with radio-active gases. *J. biol. nucl. Med.*, 1967, *11*, 63–68.

36. Miller, W. F., Johnson, R. L., Jr., and Wu, N. Relationships between fast vital capacity and various timed expiratory capacities. *J. appl. Physiol.*, 1959, *14*, 157–163.

37. Morgan, T. E. Pulmonary surfactant. *N. Engl. J. Med.*, 1971, *284*, 1185–1193.

38. Olafsson, S., and Hyatt, R. E. Ventilatory mechanics and expiratory flow limitation during exercise in normal subjects. *J. clin. Invest.*, 1969, *48*, 564–573.

39. Otis, A. B., Fenn, W. O., and Rahn, H. Mechanics of breathing in man. *J. appl. Physiol.*, 1950, *2*, 592–607.

40. Otis, A. B., McKerrow, C. B., Bartlett, R. A., Mead, J., McIlroy, M. B., Selverstone, N. J., and Radford, E. P., Jr. Mechanical factors in distribution of pulmonary ventilation. *J. appl. Physiol.*, 1956, *8*, 427–443.

41. Pattle, R. E. Properties, function and origin of the alveolar lining layer. *Nature (Lond.)*, 1955, *175*, 1125–1126.

42. Pedley, T. J., Schroter, R. C., and Sudlow, M. F. The prediction of pressure drop and variation of resistance within the human bronchial airways. *Resp. Physiol.*, 1970, *9*, 387–405.

43. Radford, E. P., Jr. Recent studies of mechanical properties of mammalian lungs. In: *Tissue elasticity*, J. W. Remington, ed. Washington, D.C., Amer. Physiol. Soc., 1957.

44. Riley, R. L. Effect of lung inflation upon the pulmonary vascular bed. In: *Ciba foundation symposium: pulmonary structure and function*, A. V. S. DeReuck, and M. O'Connor, eds. Boston, Little, Brown and Co., 1962.

45. Rohrer, F. Der Strömungswiderstand in den menschlichen Atemwegen und der Einfluss der unregelmassigen Verzweigung des Bronchialsystems auf den Atmungsverlauf in versshiedenen Lungenbezirken. *Pflügers Arch. ges. Physiol.*, 1915, *162*, 225–300.

46. Scarpelli, E. M. *The surfactant system of the lung*. Philadelphia, Lea & Febiger, 1968.

47. Storey, W. F., and Staub, N. C. Ventilation of terminal air units. *J. appl. Physiol.*, 1962, *17*, 391–397.

48. Taylor, A. The contribution of the intercostal muscles to the effort of respiration in man. *J. Physiol. (Lond.)*, 1960, *151*, 390–402.

49. Van der Brugh, J. P. Über eine Methode zur

Messung des interpleuralen Druckes. *Pflügers Arch. ges. Physiol.*, 1900, *82*, 591–602.

50. Vincent, N. J., Knudson, R., Leith, D. E., Macklem, P. T., and Mead, J. Factors influencing pulmonary resistance. *J. appl. Physiol.*, 1970, *29*, 236–243.

51. von Neergaard, K. Neue Auffassungen über einen Grundbegriff der Atemmechanik. Die Retraktionkraft der Lunge, abhängig von der Oberflächenspannung in den Alveolen. *Z. ges. exp. Med.*, 1929, *66*, 373–394.

52. Weibel, E. R. *Morphometry of the human lung.* Berlin, Springer, 1963.

53. Weibel, E. R., and Knight, B. W. A morphometric study on the thickness of the pulmonary air-blood barrier. *J. Cell. Biol.*, 1964, *21*, 367–384.

54. Weibel, E. R., and Gil, J. Electron microscopic demonstration of an extracellular duplex lining layer of alveoli. *Resp. Physiol.*, 1968, *4*, 42–57.

55. Wilmore, J. H. A simplified method for determination of residual lung volumes. *J. appl. Physiol.*, 1969, *27*, 96–100.

56. Wright, B. M., and McKerrow, C. B. Maximum forced expiratory flow rate as a measure of ventilatory capacity. *Brit. med. J.*, 1959, *2*, 1041–1047.

CHAPTER 21 GAS TRANSPORT AND GAS EXCHANGE

by CLAUDE LENFANT

GAS TRANSPORT

The process of respiration entails the transport of gas from a point of high partial pressure to a point of low partial pressure.* Thus O_2 is transported from the environment to the mitochondria where it is con-

Partial pressure. In respiratory physiology, it is convenient to speak in terms of gas partial pressure. This mode of expression permits better comparison between certain situations because it takes into account both concentration and total pressure which may vary independently of each other: for instance, the concentration of O_2 in ambient air is the same at sea

(Footnote continues on following page.)

325

sumed, and inversely CO_2 goes from the cells where it is produced to the ambient air. The overall transport system comprises four steps: (i) ventilation, (ii) gas exchange in the lung, (iii) circulation and (iv) gas exchange in the tissues. Each step is marked by a fall in partial pressure (Fig. 21–1), which can be modified by the influence of various factors.

Although each step is described separately, for convenience they are actually closely interrelated. When any step is altered, one or more of the others compensates in part or totally for the resulting deficiency.

Step One: Ventilation. Man cannot store a significant amount of gas; rather, he depends upon a continuous exchange with the air surrounding him. Gas exchange is made possible by *ventilation*, the active process by which gas molecules are transported between the environment and the alveoli. The effect of ventilation on gas exchange is discussed later in this chapter (page 343), and its regulation is described in Chapter 23, Volume II.

Two factors independent of the rate of gas exchange influence the transport of gas by ventilation, ambient conditions and dead space.

EFFECT OF TEMPERATURE AND HUMIDITY. In quiet breathing, inspired gases assume body temperature and full water vapor saturation in the lower part of the trachea. Since ambient air is much cooler than the body, it undergoes changes with inspiration which are governed by simple physical laws.

Water vapor obeys Dalton's law, thus

Footnote *(Continued)*
level and at altitude, but the partial pressures are different. Also, diffusion of a gas is governed by the gradient of partial pressure, not concentration. This is particularly important in blood where concentration and partial pressure may not be linearly related. *Dalton's law* states that each gas in a mixture of gases behaves as if it alone occupied the total volume and exerts its partial pressure independently of the other gases present. The sum of the partial pressures of the individual gases is equal to the total pressure. The partial pressure of a gas in a mixture is easily calculated from the composition of the mixture; it is equal to the product of the mole fraction (concentration) and the total pressure. Therefore in dry air at 1 atmosphere total pressure (760 mm Hg), the partial pressures of O_2, N_2 and CO_2 are $O_2 = 0.21 \times 760 = 160$ mm Hg; $N_2 = 0.79 \times 760 = 600$ mm Hg; $CO_2 = 0.0004 \times 760 = 0.30$ mm Hg.

Symbols Used in Respiratory Physiology

General variables:
V volume
V̇ a gas volume per unit time
Q volume flow of blood per unit time
P a gas pressure, or partial pressure
F fractional concentration in dry gas phase
C concentration of gas in blood
S O_2 saturation of blood
R exchange ratio (CO_2 output divided by O_2 uptake)
D diffusing capacity
f frequency of breathing

Symbols used to modify the above:
I inspired gas
E expired gas
A alveolar gas
D dead space gas
S shunt blood
M membrane
B barometric
L lung
a arterial blood
c capillary blood
v venous blood

The chemical symbols, O_2, CO_2, N_2, H_2O, etc., are used as modifying subscripts.

A horizontal line over a symbol indicates a mean value.

---

exerting a pressure independent of other gases present. Gases in contact with water receive water molecules by evaporation until the number leaving the liquid phase is equal to the number of water molecules returning from the gas phase. Since the number of molecules leaving the liquid phase is proportional to the temperature of the liquid, the partial pressure of water vapor in the gas phase is also proportional to the temperature. The temperature of air in the lungs is 37° C. If air is in equilibrium with water, the partial pressure of water in the lungs is 47 mm Hg, or about 6.2 per cent of the total gas mixture at sea level. When inhaled gas is saturated with water vapor at 37° C, the oxygen and nitrogen in the inspired air are diluted (Fig. 21–2). The dilution factor is equal to the fractional change in pressure exerted by O_2 plus N_2. If ambient air is dry, this factor is ($P_B - P_{H_2O_{37}}$)/P_B or (760–47)/760 when the barometric pressure (P_B) is 760 mm Hg. Thus:

$$F_{O_2} \text{ or } F_{N_2} \text{ (wet)} = F_{O_2} \text{ or } F_{N_2} \text{ (dry)} \times \frac{P_B - P_{H_2O_{(37)}}}{P_B} \quad (1)$$

If the ambient air is saturated with water vapor at the ambient temperature, the de-

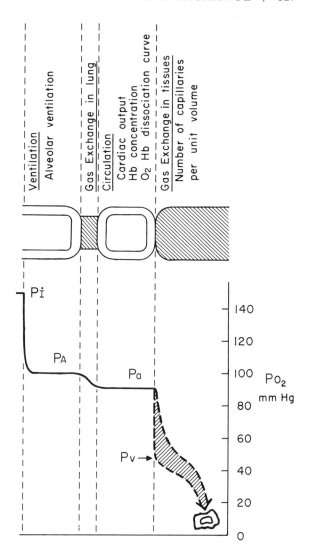

Figure 21–1 Schematic representation of the stepwise fall in PO_2 from environment to tissues. The PCO_2 pattern is reversed. Values apply to normal man at sea level.

nominator PB must be replaced by $PB–PH_2O_{ambient}$.

The volume of gas respired is also affected by temperature and humidity. *Boyle's law* states that the pressure of a gas is inversely proportional to its volume if the temperature remains constant, and *Charles' law* states that the pressure of a gas is directly proportional to its absolute temperature at constant volume. *Avogadro's principle*

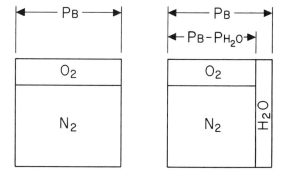

Figure 21–2 Diluting effect of water vapor. Note that the ratio of O_2 to N_2 remains constant.

states that different gases having the same volume at the same temperature and pressure contain an equal number of molecules. This principle, together with the laws of Boyle and Charles, can be combined in a simple mathematical expression, the *ideal gas law:*

$$PV = nRT \qquad (2)$$

In this expression, P is the pressure exerted by the gas; V is the volume of the gas; n is the number of moles of the gas; T is the absolute temperature ($0° C = 273° A$); R is a constant whose value depends upon the units in which the variables are expressed. When pressure is expressed in atmospheres, volume in liters and temperature in degrees absolute, R is equal to 0.082 liter-atmospheres per mole per degree.

Writing equation (2) as:

$$nR = PV/T \qquad (3)$$

shows that a certain number of moles of gas occupies a volume which depends upon the pressure it exerts and its temperature. Thus when these conditions change, the volume also changes according to the expression:

$$nR = P_1V_1/T_1 = P_2V_2/T_2 \qquad (4)$$

This can be rearranged as follows:

$$V_1 = V_2 \frac{P_2}{P_1} \times \frac{T_1}{T_2} \qquad (5)$$

When a number of molecules changes from Ambient Temperature ($273 + t_{ambient}$), Pressure ($P_B - P_{H_2O\,ambient}$) and water vapor Saturation to Body Temperature ($273 + t_{body}$), Pressure ($P_B - P_{H_2O\,ambient}$) and water vapor Saturation, equation (5) becomes:

$$V_{BTPS} = V_{ATPS} \times \frac{P_B - P_{H_2O\,ambient}}{P_B - 47} \times \frac{273 + 37}{273 + t_{ambient}} \qquad (6)$$

Equation (6) shows that, because of increased temperature and water vapor in the lung, the volume of air entering the lung expands, and fewer moles of "fresh air" enter the lung than if temperature and pressure did not change.

Although respiratory volumes such as lung volumes and ventilation are measured under ambient conditions, they are always corrected to body conditions. However, to standardize the metabolic activity between species, the amounts of oxygen consumed

($\dot{V}_{O_2}$) and of carbon dioxide produced ($\dot{V}_{CO_2}$) are expressed in standard condition of temperature ($0° C$), pressure (760 mm Hg), dry (or simply STPD).

EFFECT OF DEAD SPACE. Not all the gas entering the lungs with each breath reaches those regions where gas is actually exchanged, *i.e.,* the alveoli (Fig. 21–3A). The ineffective fraction constitutes the *anatomical dead space* (V_D) and corresponds to the volume of the airways (from buccal and nasal cavities to terminal bronchioles). The volume of the anatomical dead space varies with sex and lung volume. It constitutes a fraction of each tidal volume (V_T):

$$V_D \simeq 0.33\ V_T \qquad (7)$$

To relate the effective ventilation ($\dot{V}_A =$ alveolar ventilation per min) to the partial pressure of gases in the lungs, the dead space fraction of each breath must be deducted from the inhaled or tidal volume (V_T):

$$V_A = V_T - V_D \qquad (8)$$

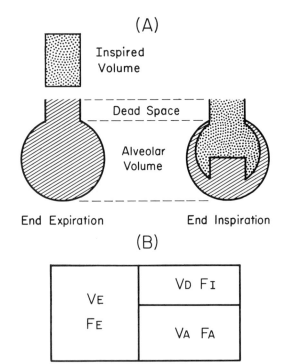

Figure 21–3 *A,* During inspiration, only part of the "fresh air" enters the alveolar space. The gas contained in the dead space at end expiration goes into alveoli at next inspiration. *B,* Mixed expired air is composed of air coming from alveolar space plus air from dead space.

or per minute:

$$\dot{V}_A = \dot{V}_E - (V_D \times f)$$

in which f is the respiratory rate. These expressions show that the smaller the dead space, the more efficient the ventilation.

Figure 21–3B shows the relationship between dead space volume and the respired volume and shows how the total exhaled gas differs in composition from the alveolar gas only as a result of the dead space. Let us say V_x represents the volume of O_2 or CO_2 exhaled in one breath:

$$V_x = V_T \, FE_x = V_D \, FI_x + V_A \, FA_x \quad (9)$$

Eliminating V_x and solving for V_D after replacing V_A by $(V_T - V_D)$, we may express this relationship as:

$$V_D = V_T \frac{FE_x - FA_x}{FI_x - FA_x} \quad (10)$$

Table 21–1 shows the changes in gas concentration between inspired, expired and alveolar air.

Step Two: Gas Exchange in the Lungs. In the lung, inspired air loses O_2 and gains CO_2; conversely, blood absorbs O_2 and loses CO_2 according to the body's needs. The amount of O_2 absorbed is greater than the amount of CO_2 released. This is explained by the fact that O_2 oxidizes both the carbon and the hydrogen of ingested food; consequently, although most of the O_2 is eliminated in exhaled air as CO_2, some is excreted as H_2O. The ratio of the amount of CO_2 exhaled to the amount of O_2 absorbed is called the *respiratory gas exchange ratio* (R), which is not to be confused with respiratory quotient (RQ) (Chap. 4, Vol. III). The RQ reflects food consumed, whereas R is an index of the respective exchange of O_2 and CO_2 in the lung. Only during steady state conditions is R equal to RQ.

Gas exchange in the lungs is passive in that it depends only upon the permeability of the alveolar membrane to gas molecules. The factors which can impair gas exchange are discussed later in this chapter.

Step Three: Circulation. Supply of O_2 to, and removal of CO_2 from the tissues depend upon circulation of the gases between lung and tissues. Blood flow by heart action is obviously essential to this phase of gas transport. However, without its respiratory properties[4] blood could transport only an insignificant amount of gas.

SOLUBILITY AND PARTIAL PRESSURES OF GASES IN LIQUIDS. The quantity of a gas in physical solution at constant temperature is directly proportional to its partial pressure in the gas phase (*Henry's law* of solubility of gases). At equilibrium the number of gas molecules leaving the liquid per unit time equals the number entering, and any change in partial pressure of the gas produces a corresponding change in equilibrium. Gas in the liquid phase also has a partial pressure, and at equilibrium the partial pressures of the gas and liquid phases are equal. The amount of gas in physical solution must be carefully distinguished from its pressure in solution. At partial pressures equivalent to those in the alveoli, 100 ml of blood contains 0.30 ml of O_2, 2.69 ml of CO_2 and 1.14 ml of N_2 in physical solution. The amounts of O_2 and CO_2 in circulating blood are much greater than the amounts of physically dissolved gases, because the blood carries O_2 and CO_2 largely in chemical combination. Chemically combined gas does not contribute to partial pressure of physically dissolved gas.

INTERACTION OF GAS AND BLOOD (RESPIRATORY PROPERTIES OF BLOOD). Hemoglobin (HHb[*]) combines reversibility with both O_2 and CO_2. If blood contained no HHb, a circulating blood volume 75 times larger than normal would be needed to satisfy the O_2 requirements. The reversible combination of HHb with O_2 is described by the equation:

$$HHb_4 + 4O_2 \longleftrightarrow HHb_4 \, (O_2)4 \quad (11)$$

The symbol HHb_4 indicates that the molecule of hemoglobin contains four atoms of iron and that oxygen combines with

Table 21–1 *Composition of Dry Inspired, Expired and Alveolar Air in Man at Rest at Sea Level*

	N_2 MOLES %	O_2 MOLES %	CO_2 MOLES %
Inspired air	79.02	20.94	0.04
Expired air	79.2	16.3	4.5
Alveolar air	80.4	14.0	5.6

[*]*HHb* indicates un-ionized hemoglobin and also indicates that hemoglobin acts as an acid.

hemoglobin in the ratio of one molecule of O_2 to one atom of iron. The molecular weight of one equivalent containing one iron atom is 16,400 g.

Blood contains about 15 g of HHb per 100 ml. Since one HHb molecule can combine with four O_2 molecules, one g of HHb can combine with $\dfrac{4 \times 22,400}{4 \times 16,400} = 1.36$ ml of O_2 so that fully oxygenated blood contains 20 ml of O_2 per 100 ml of blood (the O_2 capacity of the blood). Saturation is not complete until Po_2 reaches 150 mm Hg. An increase in Po_2 beyond this value only adds dissolved oxygen to the blood. Saturation (S%) in oxygen is given by the relationship:

$$S\% = \frac{O_2 \text{ Content}}{\text{Capacity}} \times 100 \qquad (12)$$

In the lungs, at a partial pressure of 95 mm Hg, 97 per cent of the HHb is converted to $HHbO_2$ at equilibrium. Twenty-five per cent of the O_2 in the blood can be released into the tissues when the O_2 pressure falls by 50 mm Hg (Table 21–2).

OXYHEMOGLOBIN. When whole blood is equilibrated with air containing O_2 at various partial pressures, the amount of O_2 taken up by the blood is not directly proportional to the partial pressures. A plot of the observed O_2 content of each sample against the corresponding partial pressures of O_2 is distinctly S-shaped (Fig. 21–4).

This shape of the O_2-Hb dissociation curve of blood has definite physiologic significance. The curve is flat above a pressure of 80 mm Hg, thus ensuring practically constant composition of arterial blood despite wide variations in alveolar O_2 pressure. The steep portion between 20 and 60 mm Hg permits delivery of a large amount of the blood O_2 to the tissues with a limited fall in pressure (Fig. 21–5). The partial pressure of O_2 necessary to reach a determined saturation depends on the affinity of hemoglobin for O_2. The affinity

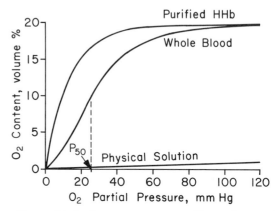

Figure 21–4 Oxygen content in a dilute solution of purified Hb and in whole blood, compared to O_2 in plasma (physical solution) at various partial pressures of O_2. Note difference in shape of absorption curves of purified Hb and arterial blood and small amount of O_2 carried in physical solution.

is high if a low O_2 partial pressure gives a high saturation, and inversely it is low if a high O_2 partial pressure is needed to reach a high saturation. Affinity is usually assessed by the partial pressure at which 50 per cent of the hemoglobin is saturated with oxygen (P_{50}). Increased P_{50} is affected by increased acidity, temperature and various red cell components. In normal blood at $pH = 7.4$ and temperature $= 37°$ C, P_{50} is 27.0 mm Hg.

Effect of CO_2 and pH (Bohr Effect). An increase in CO_2 pressure or acidity of blood increases its P_{50}, *i.e.*, its affinity for O_2 is decreased (Fig. 21–6A and B). This effect is particularly important physiologically because the taking up of CO_2 by the blood perfusing the tissues automatically favors

Table 21–2 *Normal Blood Gas Values*

	ARTERIAL	VENOUS
Po_2, mm Hg	95–85	43–40
O_2 saturation, %	97–95	72–70
O_2 content, vol %	19.5	14.5
Pco_2, mm Hg	40	46
CO_2 content, vol %	48–50	52–54

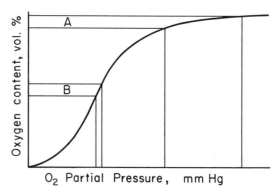

Figure 21–5 Difference in change of partial pressure with an identical change in O_2 content between the flat part of the oxygen dissociation curve (A) and the steep part of the curve (B).

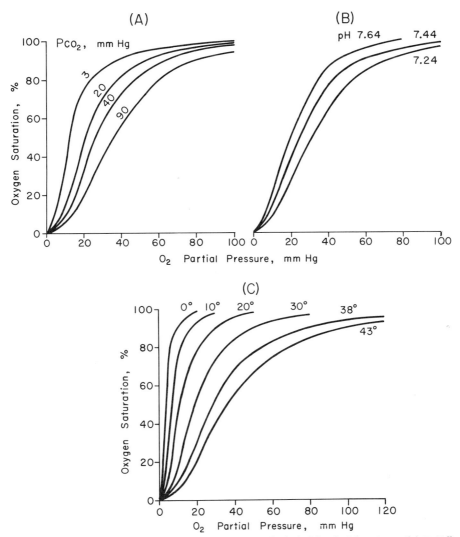

Figure 21-6 *A*, Effect of CO_2 on oxygen dissociation curve of whole blood. (After Barcroft.) *B*, Effect of acidity on oxygen, dissociation curve of blood. (After Barcroft.) *C*, Effect of temperature on oxygen dissociation curve of blood. (From Roughton, In: *Handbook of respiratory physiology*, W. M. Boothby, ed. Randolph Air Force Base, Texas, Air University, USAF School of Aviation Medicine, 1954.)

the release of O_2 from $HHbO_2$. In fact, the amount of O_2 the blood holds at any P_{O_2} is inversely related to the P_{CO_2} in the blood. The action of CO_2 in promoting the O_2 released from the blood is twofold: it increases the acidity of blood (which lowers its *p*H) and it forms carbamino compounds ($HHbCO_2$) with the hemoglobin. $HHbCO_2$ has much less affinity than HHb for O_2. Consequently, the amount of O_2 which the blood holds at a given O_2 pressure decreases and more O_2 is available to the tissues. The *p*H change contributes about 80 per cent of the Bohr effect.[47]

Effect of Temperature. An increase in temperature also shifts the O_2-Hb dissociation curve to the right (Fig. 21-6C). The temperature effect aids in releasing O_2 to the tissues; since the temperature is higher near actively metabolizing cells than near resting tissues, more O_2 is released in these cells. Conversely, some tissues are consistently cooler than central body temperature. For instance, temperature of the hand or foot may be 20° C or lower; thus less O_2 is released into the cooler extremities than to active muscles for an equal O_2 pressure fall.

Effect of 2,3-Diphosphoglycerate. The

most abundant of the organic phosphates found in the human red cell is 2,3-diphosphoglycerate (2,3-DPG). Its concentration is not constant, but appears to be determined by the concentration of reduced hemoglobin and by the plasma pH. When its concentration increases, affinity of hemoglobin for oxygen decreases, *i.e.*, P_{50} increases. Since 2,3-DPG levels do not change quickly, it plays a role only in circumstances involving a relatively long period of time such as in chronic hypoxia[41] or marked acid-base disorders. The transient shift as a result of pH and Pco_2 change during respiration is not affected by the 2,3-DPG effect but by the Bohr effect.

Dilute purified HHb has a much higher affinity for O_2 than whole blood. Studies of purified HHb solutions have shown that the reaction of HHb with O_2 is affected by the CO_2 content, the acidity, the ionic concentration, the 2,3-diphosphoglycerate[6] and the temperature of the medium containing HHb. Also the concentration of HHb itself affects its affinity for O_2; concentrated solutions of purified HHb have less affinity than diluted solutions. The difference between HHb solution and whole blood most likely results from a summation of the various factors enumerated, the more important being the salt composition of blood, the 2,3-DPG concentration and the highly concentrated form of HHb in the interior of the erythrocyte.

TRANSPORT OF CO_2 IN BLOOD. The blood contains relatively little CO_2 in physical solution; the major portion is carried in chemical combination. The forms of combined CO_2 are carbonic acid (H_2CO_3), bicarbonate ion (HCO_3^-), and carbamino hemoglobin ($HHbCO_2$). $HHbCO_2$ is found only in the blood cells, but H_2CO_3 and HCO_3^- are present in both cells and plasma. An enzyme in the red cells, carbonic anhydrase, speeds up the hydration of CO_2 and the dehydration of H_2CO_3. If carbonic anhydrase is present, all forms of CO_2 are in chemical equilibrium with one another. Even at equilibrium the concentration of HCO_3^- in the red cell differs from that in the plasma because the red cell is relatively impermeable to cations.

A CO_2 absorption curve for whole blood can be obtained in the same way as the O_2-Hb dissociation curve: blood is equilibrated with gases containing CO_2 at various partial pressures, and the CO_2 content

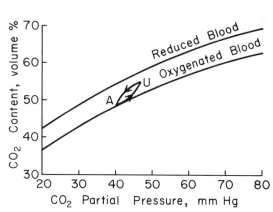

Figure 21-7 CO_2 titration curve of whole blood. Note that oxygenated blood contains less CO_2 at a given pressure of CO_2 than reduced blood. Blood goes through a cycle, as indicated by A (arterial blood) and U (venous blood), in the capillaries of tissues and lungs.

of the equilibrated blood is determined by blood gas analysis. Figure 21-7 shows the form of the CO_2 absorption curves for oxygenated and reduced blood. These curves demonstrate that dissociation of CO_2 is affected by oxygen saturation in a fashion similar to that in which the O_2 dissociation curve is affected by CO_2 pressure. The absorption of O_2 aids in the unloading of CO_2 in the lungs, and the absorption of CO_2 aids in the unloading of O_2 in the tissues (Haldane effect).

Bicarbonates. In a vacuum, a $NaHCO_3$ solution releases only half its CO_2. This release is described by the equation $2NaHCO_3 = Na_2CO_3 + CO_2 + H_2O$. Plasma behaves like a simple bicarbonate solution, except that more of its CO_2 is extracted in a vacuum because of the presence of acid phosphates and other weak acids. Whole blood will release most of its CO_2 in a vacuum. The difference in behavior between whole blood and plasma and bicarbonate solutions is due to the acid properties of hemoglobin. Both reduced and oxygenated hemoglobin can furnish sufficient H^+ to carry the reaction $H^+ + HCO_3^- = H_2CO_3 = H_2O + CO_2$ to completion. In the lungs, the following series of reversible chemical reactions occurs as O_2 enters the blood:

$$O_2 + HHb = HHbO_2 = HbO_2^- + H^+$$
$$H^+ + HCO_3^- = H_2CO_3 = H_2O + CO_2$$

Because $HHbO_2$ is a stronger acid than

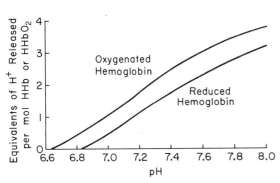

Figure 21-8 Acid-base titration curves of oxygenated and reduced Hb. As O_2 is released from HbO_2, the weaker base Hb^- can combine with H^+ to form HHb. For each mole of O_2 released to tissues, 0.7 mole of H^+ from the ionization of H_2CO_3 can be neutralized by the Hb without change in pH (isohydric cycle).

Other protein molecules in the blood besides the HHb molecules probably carry CO_2 in the same manner. The compound formed by combining CO_2 with HHb is physiologically the more important of the carbamino compounds because it enters into a reversible reaction with O_2:

$$O_2 + HHbCO_2 = HHbO_2 + CO_2$$

This reaction is important in respiratory exchange, because it provides a rapid method for exchange of CO_2 without significant changes in pH. Of the total changes in CO_2 in whole blood during the respiratory cycle, 30 per cent are accounted for by changes in carbamino-bound CO_2. In the red cells, the carbamino compounds contribute 60 to 75 per cent of the total changes.

O_2–CO_2 Interrelationship. The chemical reactions occurring in the blood as O_2 is delivered to the tissues can now be summarized. Figure 21-9 shows the important steps in this sequence of events. The series of reactions is reversed in the lungs. Carbon dioxide is continually produced in the tissue cells where it achieves its highest partial pressures. It then diffuses from the cells, through the interstitial fluid and capillary walls into the plasma. Some of the CO_2 reacts slowly with H_2O in the plasma to form H_2CO_3, which in turn ionizes and liberates H^+. Many of the H^+ immediately combine with the plasma proteins which have a buffering effect. The major portion of the CO_2 diffuses into the red corpuscles, where

HHb, it easily releases H^+ to combine with HCO_3^-. Since the reactions are reversible, an increase of either O_2 or CO_2 in the blood will, in accordance with the law of mass action, drive the reaction in the appropriate direction. Titration curves of $HHbO_2$ and HHb with NaOH are given in Figure 21-8, which shows that the oxygenation of HHb can furnish 0.7 mole of H^+ per mole of O_2 absorbed for combining with HCO_3^-.

Carbamino Compounds ($HHbCO_2$). Approximately one-fifth of the total CO_2 in the blood is carried as $HHbCO_2$, in which the CO_2 is combined directly with amino groups of the HHb molecules:

$$HHbNH_2 + CO_2 = HHbNHCOOH$$

Figure 21-9 The more important chemical reactions by which O_2 is made available to tissues. Forces initiating and controlling exchange of gases are gradients in partial pressure of CO_2 and O_2 between capillary blood and tissue cells.

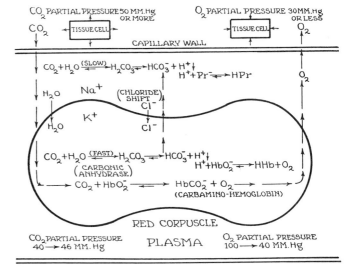

it enters into two reactions. The CO_2 can combine with water exactly as it did in plasma; this reaction is rapid, because it is catalyzed by carbonic anhydrase. The H^+ eventually released is then taken up by HbO_2^- to form HHb and O_2, and the resultant O_2 diffuses out of the cell to supply the largest fraction of the O_2 gained by the tissues from the blood.

This series of chemical reactions in the erythrocyte has been termed the *isohydric cycle* because CO_2 is taken up and O_2 is released without producing excess H^+. The buffering power of HbO_2^- allows a large amount of CO_2 to be absorbed and O_2 to be released without a marked change in acidity. The excess HCO_3^- diffuses out of the cell into the plasma. This diffusion, if uncompensated, would leave an excess of positive ions in the cell. To keep the positive and negative ions in balance, Cl^- move into the cell simultaneously. The balance cannot be restored by K^+ because the red cell membranes are almost impermeable to positive ions. This exchange, called the *chloride shift*, allows a great quantity of HCO_3^- to be carried in the plasma.

Some of the CO_2 combines with the various forms of HHb, the most important reaction being with HbO_2, which releases O_2 without changing *p*H. The interrelationship of O_2 and CO_2 in the red cell is only one example of the mutual dependence of these two gases during the whole process of O_2 utilization.

Step Four: Gas Exchange in the Tissues. Gases are exchanged continuously between the blood and the tissues. The basic process of this exchange is *physical diffusion*, which, just as in the lung, occurs because of a pressure gradient between the capillaries and the active cells.

As the blood progresses from the arterial to the venous end of the capillary, Po_2 decreases and Pco_2 increases as a function of the metabolic rate of the tissues and of the blood flow. Normally, at rest, at any point along the capillaries, Po_2 does not reach a critically low level and the changes in Po_2 as a function of the radial distance from the capillary are independent of capillary Po_2. Variation of Po_2 in the tissues is also related to the metabolic activity and to the distance between capillaries. Should this distance be very large, and/or the metabolic demand increase, a *critical pressure*

may be reached. This pressure is defined as one which does not permit sufficient transfer of O_2 to satisfy the demand, and thus results in a low Po_2 in the cells. Similarly, insufficient delivery may result from insufficient supply, as in hypoxemia.

HYPOXIA. Four general types of hypoxia can be differentiated on the basis of physiologic factors: (i) *Stagnant hypoxia* arises when the flow of blood through a tissue is reduced. (ii) *Hypoxic hypoxia* results from interference with the exchange of O_2 across the lungs or some preceding step in respiration. (iii) *Anemic hypoxia* follows reduction of the O_2 carrying capacity of the blood. (iv) *Histotoxic hypoxia* occurs when the tissue cells cannot properly use the O_2 available.

Figure 21–10 illustrates how one can distinguish the various types of hypoxia from the O_2 content and the percentage saturation of arterial and venous blood. The O_2 content of arterial blood is normal in stagnant and histotoxic hypoxia, and is reduced in hypoxic and anemic hypoxia. Venous blood contains less O_2 than normal in stagnant, hypoxic or anemic hypoxia; in histotoxic hypoxia the O_2 content of venous blood is above normal. In each kind of hypoxia adaptive mechanisms intervene to maintain or restore adequate O_2 delivery. These mechanisms improve gas transport. Also, the number of capillaries in the tissues may increase, either by development of new capillaries or by the opening of existing, but previously closed, capillary beds.[11] In either case, the decrease in distance between capillaries helps to overcome the diffusion limitations resulting from low Po_2.

GAS EXCHANGE IN THE LUNG

The exchange of O_2 and CO_2 between alveolar air and blood is the most important function of the lung. Although the rate of gas exchange is essentially controlled by the body's requirements, four factors in the lungs may limit it and affect the partial pressures of O_2 and CO_2 in alveolar air and in arterial blood. These factors are (i) diffusion limitation, (ii) shunt, (iii) uneven ventilation-to-perfusion ratios and (iv) hypoventilation.

The first three of these factors have an opposite effect on the partial pressure of

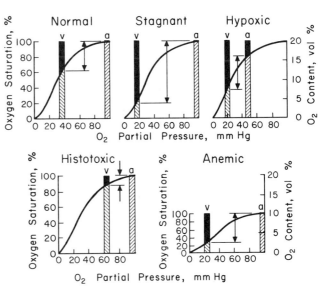

Figure 21-10 Composition of arterial (a) and venous (v) blood in various types of hypoxia and their relation to O_2 absorption curve of hemoglobin. Hatched areas indicate amount of $HHbO_2$ in blood. Amount of O_2 removed from blood as it passes through tissues is expressed as difference in O_2 content of arterial and venous blood.

O_2, CO_2 or N_2 in the alveolar air and in the arterial blood. Thus they cause *gradients*, or alveolar-arterial differences in partial pressure.

Diffusion. Whether or not a diffusion limitation causes an end capillary–to–alveolar gas partial pressure gradient at sea level is a controversial and unanswered question. However, the physical and physiological principles governing the movement of gas between alveolar air and blood are fundamental to understanding the overall gas exchange.

Diffusion, then, is the passive movement of molecules from a point of high concentration to one of low concentration. In the lungs this molecular movement causes a transfer of gas between the alveolar air and the red blood cells across several well identified layers of tissue: the alveolar epithelium, the interstitium, the capillary endothelium, the plasma and, finally, the red cell, where loose chemical reactions with hemoglobin occur (see Fig. 20–3, Chap. 20, Vol. II). The air-plasma distance is approximately 6 μ thick, but the layer of plasma is obviously variable.

At each point along the pulmonary capillary the amount of oxygen transferred across the tissue-blood barrier is given by

$$\dot{V}O_2 = DL\ (PA_{O_2} - Pc_{O_2}) \qquad (13)$$

If the entire capillary is considered, this equation becomes

$$\dot{V}O_2 = DL\ (PA_{O_2} - P\bar{c}_{O_2}) \qquad (14)$$

which indicates that $\dot{V}O_2$ is proportional to a factor DL and to the mean difference in O_2 pressure $(PA_{O_2} - P\bar{c}_{O_2})$ between the alveolar air and the red cell. DL is called diffusing capacity and depends upon the physical characteristics of the gas (coefficient of solubility in tissues and plasma and molecular weight) and the geometry of the alveolar-capillary membrane (area and thickness). The mean PO_2 difference is that which, if it remains unchanged along the entire length of the capillary, permits a continuous flow of O_2 at a constant rate across the tissue-blood barrier. Figure 21–11A shows the changes in PO_2 in the blood along the capillary, as determined by the classic Bohr integration procedure[8] and the derived $PA_{O_2} - P\bar{c}_{O_2}$. When blood enters the capillary, the alveolar-capillary blood PO_2 gradient is large. As Pc_{O_2} increases owing to O_2 transfer, the gradient decreases. Finally, when the blood approaches the end of the capillary, Pc_{O_2} nearly equals PA_{O_2}, as shown in Figure 21–11A.

Since at any instant along the capillary the rate of O_2 transfer is proportional to the pressure difference between alveolar air and capillary blood (equation 13), the area in Figure 21–11 between the curve Pc and the horizontal PA_{O_2} is proportional to the total amount of O_2 transferred. The mean capillary PO_2 can be determined graphically from this area. By definition, the mean PO_2 difference is that necessary for a continuous and constant flow of O_2 at the exist-

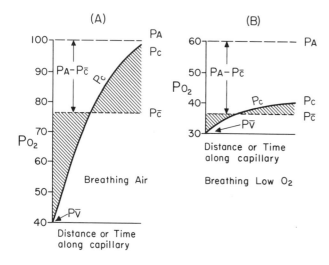

Figure 21–11 Moment-to-moment changes in Po_2 along the alveolar capillary during (A) air breathing at sea level and (B) low O_2 breathing. The mean alveolar-capillary Po_2 difference is the same in both cases. The end-capillary gradient ($PA_{O_2} - P\bar{c}_{O_2}$) is small during air breathing and large during low O_2 breathing.

ing rate. Thus, $P\bar{c}_{O_2}$ corresponds to the horizontal line intersecting Pc so that, as shown in Figure 21–11, the shaded area below the line (rate of O_2 transfer greater than average) equals that above it (rate of O_2 transfer smaller than average).

EFFECT OF THE O_2-HB DISSOCIATION CURVE ON THE END CAPILLARY GRADIENT. Changing the concentration of inspired oxygen has a marked effect on the end capillary gradient[42] because of the shape of the O_2-Hb dissociation curve. When room air (21 per cent O_2) is inhaled, the pressure of O_2 in the venous blood entering the capillary (40 mm Hg) and that in the alveolar air (100 mm Hg) correspond respectively to the curvature of the O_2-Hb dissociation curve where the slope is steep and to the upper part where it is more gradual (Fig. 21–12). This explains the high gradient shown in Figure 21–11A. This large gradi-

ent is associated with a rapid transfer of O_2 at the beginning of the capillary, which in turn rapidly increases the oxygen saturation of the capillary blood. For this reason P_{cO_2} moves on the O_2-Hb dissociation curve from the curved to the upper, more gradual, segment where small changes in O_2 saturation cause large changes in Po_2. Thus, in spite of a decreasing $PA_{O_2} - Pc_{O_2}$ and a decreasing rate of O_2 transfer as the blood progresses along the capillary, Pc_{O_2} rises quite rapidly until it nearly equals PA_{O_2} at the distal end of the capillary.

When an O_2 concentration lower than in room air is inspired (12 per cent, for instance), both the venous Po_2 (30 mm Hg) and the alveolar Po_2 (50 mm Hg) correspond to the steepest part of the O_2-Hb dissociation curve. In these conditions the initial Po_2 gradient is smaller, and consequently O_2 transfer is slower. This causes a slow rise of Pc_{O_2} which at the end of the capillary remains markedly lower than PA_{O_2} (Fig. 21–11B). If the dissociation curve were vertical in the applicable range, the curve of Pc_{O_2} vs. time would be a horizontal line superimposed on $P\bar{c}_{O_2}$, since $PA_{O_2} - Pc_{O_2}$ would be constant throughout. The closer the actual dissociation curve approaches a vertical, the flatter the curve of Pc_{O_2} vs. time and the larger the end capillary–to–alveolar Po_2 difference becomes. Conversely, with high O_2 breathing, the initial Po_2 gradient is very large; thus O_2 transfer is so fast that complete equilibrium between alveolar air and capillary blood is reached long before the blood leaves the capillary.

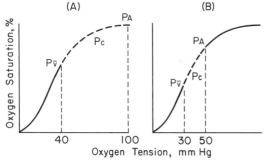

Figure 21–12 Position of $P\bar{v}$ and PA on the oxygen dissociation curve during (A) air breathing and (B) low O_2 breathing. Dashed line represents the path of Pc.

The end capillary–to–alveolar gradient reflects a diffusion limitation. Clearly, at any level of inspired P_{O_2} this gradient decreases if the oxygen requirements $\dot{V}_{O_2}$ are smaller or if D_L increases, for instance, because of an increase of the total area of gas exchange. On the contrary, this gradient might become larger with increased oxygen requirements and/or decreased diffusing capacity because of some pathological process. All investigators agree that in a normal resting man breathing room air, the end capillary gradient is insignificant (<1 mm Hg). At low inspired O_2 concentration (12 per cent) the gradient has been estimated at approximately 8 mm Hg.[42] These calculated values are not to be confused with the alveolar-arterial gradient of which the end capillary gradient is only part (diffusion component).

The capability of changing the size of the end capillary–to–alveolar difference by changing the inspired O_2 concentration has yielded a method of calculating D_L from equation.[15] $\dot{V}_{O_2}$ is measured directly and the mean difference in P_{O_2} can be determined by graphical analysis if the end capillary–to–alveolar gradient is known. Because this gradient is extremely small with room air breathing, but increases with low oxygen breathing (whereas the total alveolar-arterial gradient remains the same in both conditions), the low O_2 condition is chosen to determine it with maximum accuracy.[42] This method gives results which vary considerably between individuals. A value of 15 ml per min per mm Hg or greater is considered normal.

Components of D_L. The previous discussion assumed D_L constant throughout the capillary and an immediate P_{O_2} equilibrium between plasma and red cells during capillary transit. Recent evidence suggests that the speed of reaction of O_2 with reduced hemoglobin limits the rate of O_2 uptake by the red cells and that a P_{O_2} gradient exists between plasma and intracellular hemoglobin.[61–63] Thus, at any instant along the capillary the total P_{O_2} difference between alveolar air and red cell has two components:

$$P_{A_{O_2}} - P_{RBC_{O_2}} = (P_{A_{O_2}} - P_{c_{O_2}}) + (P_{c_{O_2}} - P_{RBC_{O_2}}) \quad (15)$$

in which $P_{c_{O_2}}$ is P_{O_2} in the plasma and P_{RBC} is P_{O_2} inside the erythrocyte.

The corollary of this is that O_2 is transferred in two steps. One depends on diffusion across the tissue-plasma barrier (D_M) and the other on the rate at which O_2 combines with hemoglobin. This rate, expressed by the symbol θ, can be experimentally determined. If θ is multiplied by the capillary blood volume V_c in which the reaction occurs at any instant, we obtain the total volume of O_2 transferred (or taken up) θV_c per time unit and per mm Hg. Since this volume is the same as that having transferred across the tissue-plasma barrier, the following equations hold:

$$\dot{V}_{O_2} = D_L(P_{A_{O_2}} - P_{RBC_{O_2}}) \quad (16)$$

$$\dot{V}_{O_2} = D_M(P_{A_{O_2}} - P_{c_{O_2}}) \quad (17)$$

$$\dot{V}_{O_2} = \theta V_c(P_{c_{O_2}} - P_{RBC_{O_2}}) \quad (18)$$

Each can be rearranged, for example, as:

$$\frac{\dot{V}_{O_2}}{D_L} = P_{A_{O_2}} - P_{RBC_{O_2}}$$

and the three equations may then be combined because of equation (15):

$$\frac{\dot{V}_{O_2}}{D_L} = \frac{\dot{V}_{O_2}}{D_M} + \frac{\dot{V}_{O_2}}{\theta V_c}$$

or:

$$\frac{1}{D_L} = \frac{1}{D_M} + \frac{1}{\theta V_c} \quad (19)$$

In expression (16) D_L can be compared to a conductance (ml/min per mm Hg); then the inverse $1/D_L$ represents total resistance to diffusion of gas between alveolar air and red cells. It is equal to the resistance offered by the membrane $1/D_M$ plus that exerted in the capillary $1/\theta V_c$. The membrane diffusing capacity D_M is determined only by the cross-section area, the thickness of the membrane and the physical characteristics of the gas. Therefore the resistance $1/D_M$ is constant from the venous to the arterial end of the capillary. However, the diffusion resistance in the red cell $1/\theta V_c$ varies along the capillary because θ is not a constant; it is directly related to the overall velocity constant for the uptake of oxygen by the red cell, k'_c, which in turn varies in part with O_2-Hb saturation. Figure 21–13 shows the changes in k'_c and θ as a function of oxygen saturation; θ is relatively constant until S_{O_2} reaches 75 per cent and then drops sharply to approach zero as S_{O_2} approaches 100 per cent.

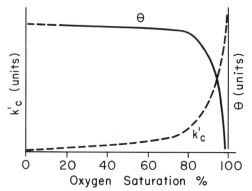

Figure 21-13 Change in k'_c, the velocity constant for O_2 uptake by the red cell, and θ, rate of combination of O_2 with hemoglobin as a function of oxygen saturation.[59]

This indicates that the conductance θVc between plasma and red cells is high at the venous end of the capillary and low at the arterial end, and that DL is not constant during the capillary transit.

The effect of considering the components of DL on the time dependent change in Pc_{O_2} and $PRBC_{O_2}$ along the capillary and on the end capillary gradient is shown in Figure 21-14. The middle curve labeled $PRBC_{O_2}$ is obtained by a Bohr integration procedure which accounts for the change in θVc. When room air is breathed, as in this example, almost all the oxygen absorbed

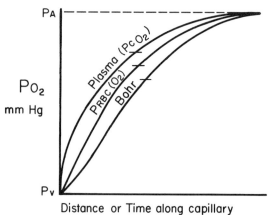

Figure 21-14 Moment-to-moment change in Po_2 in the red blood cell and in the plasma during room air breathing. As soon as blood enters the capillary, Po_2 in plasma increases sharply because of a massive flux of O_2 across the membrane. The Po_2 increase in RBC lags behind because of slow reaction rate. At the end of the capillary, equilibrium between plasma and red cell is almost reached.

enters the red cell where it combines with hemoglobin. Thus at any instant equation (18) can be substituted into equation (16) to obtain the difference in pressure between plasma and the red cell:

$$Pc_{O_2} - PRBC_{O_2} = \frac{DL}{\theta Vc}(PA_{O_2} - PRBC_{O_2})$$

From this equation the curve labeled Pc_{O_2} is obtained.

The respective position of these two curves indicates that Po_2 in the plasma is always higher than in the red cells. However, at the end of the capillary $PRBC$ approaches Pc because of the shape of the O_2-Hb dissociation curve. The change in Po_2 obtained with DL assumed constant (equation 14) is shown by the curve labeled *Bohr* (Fig. 21-14). In this case, Po_2 in the blood is always lower than in the red blood cell, as determined by the method separating DL into its components. The reason for this is that when the rate of O_2 uptake is taken into account, DL is higher early in the capillary, which speeds up the equilibration. The two methods differ in that the Po_2 end capillary–to–alveolar gradient owing to diffusion is overestimated when the method in which DL is assumed constant over the entire capillary is used.

Normal, resting individuals are generally accepted as having no significant diffusion limitation, whether air or a low O_2 mixture is breathed.[61] Heavy exercise at low O_2 can produce an increased Po_2 gradient, however, as shown by a decreasing arterial O_2 saturation.[2] This situation is probably an extreme resulting from the effect of the shape of the O_2-Hb dissociation curve and from the extremely high oxygen uptake. In patients with pulmonary disease, no general opinion prevails in spite of some indirect evidence that other factors (shunt and ventilation-perfusion ratio unevenness) might be primarily responsible for the Po_2 gradient.[23] Actually, since the determination of an O_2 gradient as a result of diffusion limitation requires numerous assumptions (because end capillary blood is not accessible and all capillaries are not similar), the direct measurement of diffusing capacity DL is a more widely accepted clinical test than the attempts to assess the end capillary gradient owing to diffusion limitation.

The diffusing capacity for oxygen DL_{O_2}

can be calculated by the methods described above,[42, 63] which give quite comparable results. However, the difficulties of the procedure and the assumption involved render these calculations impractical. This has led to the development of methods using carbon monoxide, which were in fact successfully used long before the methods for measuring DL_{O_2} had reached maturity.[8, 34]

MEASUREMENT OF LUNG DIFFUSING CAPACITY USING CO. Carbon monoxide has special properties which make it an extremely convenient gas with which to assess alveolar-capillary transfer characteristics. Up to the point at which CO enters the red blood cell, its behavior is much like that of O_2. Its mixing and diffusion in the gas medium are similar to that of O_2 and it is therefore distributed throughout the lung in a comparable manner. It diffuses in the fluid medium of the alveolar wall at a rate about 80 per cent of that of O_2. When CO reaches the red blood cell, however, its behavior differs sharply from that of O_2, because its affinity for hemoglobin is approximately 210 times as great as that of O_2. Thus, in the presence of CO alone, the blood takes up 210 times as much CO as O_2 per unit of partial pressure. Despite this huge difference in affinity for hemoglobin, CO, like O_2, forms a reversible combination with hemoglobin, and, when the PCO scale is expanded by a factor of 210, the shape of the CO dissociation curve is similar to that of the O_2 dissociation curve. On the other hand, when the CO dissociation curve is plotted on the same scale as the O_2 dissociation curve, it is practically vertical until full saturation is reached (Fig. 21–15).

As noted in the discussion of alveolar-capillary O_2 transfer, if the dissociation curve were vertical, the end-capillary gradient and the mean gradient would be the same. In the case of CO this happy circumstance exists, and estimating mean capillary PCO involves none of the difficulties present with O_2. Furthermore, because of the very high affinity of Hb for CO, and since CO is not normally present in the blood in significant quantities (in nonsmokers), the mean $PRBC_{CO}$ can ordinarily be considered zero. Then the mean alveolar-capillary gradient $PA_{CO} - PRBC_{CO}$ becomes simply $PA_{CO} - 0$ and

$$DL_{CO} = \dot{V}CO/PA_{CO} \qquad (20)$$

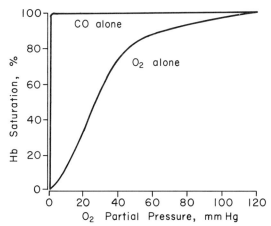

Figure 21–15 Dissociation curves for HbO_2 and HbCO plotted on the same scale. Saturation of hemoglobin with O_2 is not maximal until the PO_2 is greater than 120 mm Hg; with CO, however, maximal saturation is attained with a PCO of less than 1 mm Hg. (From Comroe et al., The lung. Chicago, Year Book Publishers, 1955.)

DL_{CO} can be measured by any of several methods[24] which are all based on equation (20). The major drawback of this determination is that a truly representative sample of alveolar air cannot be obtained, especially in patients in whom inspired gas is evenly distributed. Carbon monoxide, as O_2, does not react instantaneously with the intracellular hemoglobin; this means that the plasma CO is not negligible during CO uptake. Then:

$$(PA_{CO} - 0) = (PA_{CO} - Pc_{CO}) + (Pc_{CO} - 0)$$

If this expression is divided by $\dot{V}CO$ (amount of CO transferred across the tissue-blood barrier in 1 min), it yields, by analogy with equations (16), (17) and (18):

$$\frac{1}{DL_{CO}} = \frac{1}{DM_{CO}} + \frac{1}{\theta_{CO}Vc} \qquad (21)$$

which is similar to equation (19) for oxygen.

In vitro studies have shown that the rate of reaction between CO and hemoglobin is affected by the amount of O_2 present,[57, 58] because CO and O_2 compete for the available hemoglobin. The reaction rate θ is faster when the O_2 concentration is low. *In vivo* studies of the lung diffusing capacity DL_{CO} at high and low levels of O_2 show that DL_{CO} varies inversely with the level of O_2.[25] Since the diffusional characteristics of the alveolar membrane or the volume of blood in the pulmonary capillaries are not likely to be greatly affected by the level of O_2, the observed changes in DL_{CO} are attributed to changes in the rate of reaction between CO and hemoglobin.

By using the *in vivo* data for DL_{CO} and the *in vitro* data for θ, values for DL_{CO} and θ can be introduced into equation (21) for conditions of high and low O_2. This yields two statements of the equation with two unknowns, DM_{CO} and VC, which are assumed to remain constant at the two O_2 levels. A solution is therefore possible, as shown in Figure 21–16. These studies suggest that the reaction between CO and hemoglobin constitutes a significant part of the total resistance to transfer of CO between alveoli and red blood cells, and that the membrane diffusing capacity, DM_{CO}, is considerably larger than the total lung diffusing capacity. In normal subjects at rest and breathing air, a rather large range of DL_{CO} values is obtained between the various methods (20 to 30 ml CO per min per mm Hg). With one method a change in DL can result from a decrease in DM_{CO} such as may occur in certain lung diseases (fibrosis from sarcoidosis), or from alterations in the blood component $\theta COVc$. The latter is usually due to a change in capillary blood volume—for instance, a decrease in cases of pulmonary emboli and in anemia, or an increase as in polycythemia.

DIFFUSION OF CO_2 AND N_2.[24] Carbon dioxide diffusion across the lung tissue is rapidly and greatly enhanced by the presence of carbonic anhydrase. Because of the high effective solubility of CO_2 in the blood, the PCO_2 difference between venous blood and alveolar air is small. This causes the equilibration of CO_2 between plasma and air to be extremely rapid. Equilibration has been calculated to reach 99 per cent in 0.072 sec, which is about 10 per cent of the total transit time (0.79 sec). This compares with approximately 0.3 sec for oxygen.[61] However, this is only part of the problem, because CO_2, like O_2, forms compounds with the blood's buffers and with the hemoglobin. Despite carbonic anhydrase in the red cells, the reaction rate plus the transfer of ions between plasma and cells may be factors influencing the time required to reach PCO_2 equilibrium between red cell and plasma in the capillary. Data are insufficient to obtain quantitative information. Because the reactions between O_2 and Hb continue after blood leaves the capillary, more CO_2 will be freed from compounds. Thus PCO_2 will increase slightly in the plasma, but not enough to cause a measurable gradient between capillary blood and alveolar air.

Unevenness of ventilation causes unevenness of PN_2 in the local alveolar air (see p. 345). Consequently a gradient of PN_2 can exist in various parts of the lung between the incoming venous blood and the alveolar gas. As N_2 is only dissolved, not bound in the blood, the absolute amount exchanged is extremely small, and their reaction rate causes no limitation. Consequently, equilibration is extremely rapid: 99 per cent within approximately 0.01 sec.

Pulmonary Anatomical Shunt. Although the existence of arteriovenous anastomoses in the lungs has been suggested for many years,[56, 59] indisputable anatomical evidence for them has been obtained only recently.[64]

An anatomical pulmonary shunt is a pathway by which venous blood can reach the left pulmonary circulation (pulmonary vein) and mix with the arterial blood without being exposed to any gas. The effect of this venous admixture is to lower the oxy-

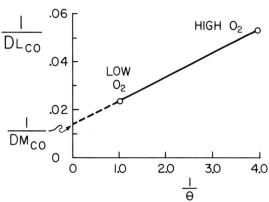

Figure 21–16 Graphic solution of equation (21) for Vc and DM_{CO}. In the form $y = (a) + (b)$, equation (21) becomes: $\frac{1}{DL_{CO}} = \left[\frac{1}{Vc}\right]\frac{1}{\theta} + \left[\frac{1}{DM_{CO}}\right]$. The term $1/Vc = a$ and $1/DM_{CO} = b$; $1/Vc$ is the slope of line connecting values of $1/DL_{CO}$ obtained during high and low O_2 breathing; $1/Vc = \frac{0.054 - 0.014}{4 - 0} = 0.01$; $Vc = 100$ ml; $1/DM_{CO}$ = zero intercept $[1/DL_{CO}$ when $1/\theta = 0]$ 0.014; $DM_{CO} = 71$ ml CO/min per mm Hg difference in PCO across the membrane.

gen content, or oxygen saturation of the postcapillary blood, to an extent related to the percentage of blood shunted (see equation 24). Thus, a difference in partial pressure of oxygen exists between the end capillary blood and the arterial blood.

MORPHOLOGY. There are two kinds of anatomical shunts: *true shunts*, or direct communications from the right to the left side of the pulmonary circulation, and the *collateral circulations* (Fig. 21–17). The true shunts include (i) anastomoses present in the peripheral parts and in the hilus of the lungs, and (ii) some capillaries corresponding to collapsed alveoli (these alveoli have undetermined locations). Both true shunts permit mixed venous blood from the pulmonary artery to mix directly with the end capillary blood.

Collateral circulations originate from the pulmonary veins or from the aorta, and return to either one after perfusing a specific tissue or organ. Several have been identified: bronchial circulation, pleural circulation, thebesian circulation and aortic vasa vasorum. All these collateral circulations have one common characteristic which

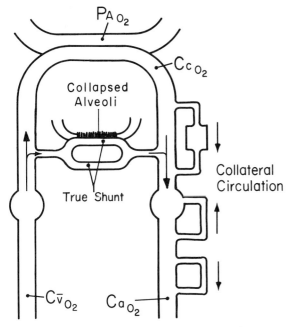

Figure 21–17 Shunt pathways. From top to bottom, true shunts are collapsed alveoli and direct anastomosis; collateral circulations are bronchial or pleural circulation, thebesian circulation and aortic vasa vasorum.

differentiates them from the true shunt, *i.e.*, their venous blood may have different oxygen contents, and each content may differ from that of the mixed systemic venous blood. These venous bloods are quite inaccessible for sampling, and therefore all the channels for venous admixture, whether true shunt or collateral circulation, are grouped together.

TOTAL SHUNT MEASUREMENT USING O_2. This measurement is based upon the statement that the quantity of oxygen in the arterial blood is equal to the quantity of oxygen in the capillary blood, plus that in the shunted blood:

$$\dot{Q}T Ca_{O_2} = \dot{Q}c Cc_{O_2} + \dot{Q}s C\bar{v}_{O_2} \qquad (22)$$

in which the total blood flow $\dot{Q}T$ is equal to the capillary flow $\dot{Q}c$, plus the flow through the shunt $\dot{Q}s$. This expression yields:

$$\frac{\dot{Q}s}{\dot{Q}T} = \frac{Cc_{O_2} - Ca_{O_2}}{Cc_{O_2} - C\bar{v}_{O_2}} = \frac{Sc_{O_2} - Sa_{O_2}}{Sc_{O_2} - S\bar{v}_{O_2}} \qquad (23)$$

This classic and very important equation includes the assumption that the oxygen content in the blood returning from all the collateral circulations is the same as in the mixed venous blood, Cv_{O_2}. As noted above, this assumption is most likely untrue. Fortunately, examination of equation (23) shows the error of overestimation or underestimation of any venous oxygen content. For instance, if certain Cv_{O_2} is lower than that $C\bar{v}_{O_2}$ (O_2 content of the mixed venous blood), the calculated shunt will be too high. Conversely, if it is higher, the shunt will be too low.

Rewriting equation (23):

$$Cc_{O_2} - Ca_{O_2} = \frac{\dot{Q}s}{\dot{Q}T} (Cc_{O_2} - C\bar{v}_{O_2}) \qquad (24)$$

shows that the difference in oxygen content between the end capillary blood and the arterial blood depends on the size of the shunt. Corresponding to this difference in oxygen is a difference in oxygen pressure, the magnitude of which in turn depends on the slope of the oxygen dissociation curve:

$$(Pc_{O_2} - Pa_{O_2}) \, \alpha'_{O_2} = \frac{\dot{Q}s}{\dot{Q}T} (Cc_{O_2} - C\bar{v}_{O_2}) \qquad (25)$$

in which α'_{O_2} is the effective slope of the O_2-Hb dissociation curve. Practically,

Pc_{O_2} can be replaced by PA_{O_2} because no measurable end capillary gradient exists between blood and alveolar gas.

ALVEOLAR-ARTERIAL PO_2 DIFFERENCE DUE TO SHUNT. For any given value of shunt the alveolar arterial gradient is affected by two factors: (i) the shape of the O_2-Hb dissociation curve and (ii) the oxygen content of the mixed arterial blood. If the shunt is measured at an inspired O_2 concentration corresponding to the steep part of the O_2-Hb dissociation curve (α' O_2 is large), the gradient is less than with high O_2 concentration breathing (Fig. 21–18). However, after PA_{O_2} has increased to a level at which both the end capillary blood and the arterial blood are fully saturated, further increases do not increase the gradient because the relationship between oxygen content and PO_2 has become linear.

As seen by equation (25), the alveolar arterial gradient is also related to $C\bar{v}_{O_2}$. Thus any condition which decreases mixed venous blood oxygen content increases the gradient above normal.[53] Such a situation prevails in anemic patients or in patients with low cardiac output caused by shock or cardiac failure. In these conditions, the oxygen transported to the tissues is decreased, which is compensated for by increased extraction of O_2 from the blood.

EFFECT OF SHUNTING ON PCO_2 AND PN_2. As with oxygen, a shunt causes a difference in CO_2 content between the end capillary blood and the arterial blood. Thus by substituting CO_2 for O_2 in equation (23), a shunt could be calculated with carbon dioxide. However, since the slope of the CO_2 absorption curve is much greater than that of the oxygen dissociation curve, the PCO_2 gradient is not measurable and PCO_2 in the arterial blood does not significantly increase.

The partial pressure of nitrogen in the alveolar air and in the capillary blood is determined by the exchange of O_2 and CO_2, or the ratio of ventilation to perfusion in the lungs (see next section: hypoventilation). Thus, in a steady state, N_2 is not exchanged in the lungs. Furthermore, since molecular nitrogen is not exchanged between blood and tissues, PN_2 is the same in the capillary, arterial and venous blood.

QUANTITATIVE ESTIMATION OF THE ANATOMICAL SHUNT. Anatomical shunt is often determined while the subject is breathing an O_2-enriched gas mixture (equation 23). This is done on the assumption that breathing high O_2 concentration eliminates or attenuates the effect of other factors contributing to the alveolar-arterial difference in PO_2 diffusion and ventilation-to-perfusion unevenness. Since measurement of O_2 content is difficult and requires great accuracy, the PO_2 in the capillary blood (or alveolar air) and in the arterial blood is measured instead, and then translated into O_2 content by using the expression:

$$O_2 \text{ content} = (PO_2 \times \alpha_{O_2}) + (Hb \times 1.34)$$

in which α_{O_2} = coefficient of solubility for O_2, Hb = hemoglobin concentration and 1.34 = Hb combining power for O_2. PO_2 can easily be measured in the arterial blood, and PA_{O_2} is derived from the "alveolar air" equation (20):

$$PA_{O_2} = PI_{O_2} + \frac{Pa_{CO_2} \; FI_{O_2} \; (1-R)}{R} - \frac{Pa_{CO_2}}{R}$$

where PC_{O_2} is assumed to be the same in

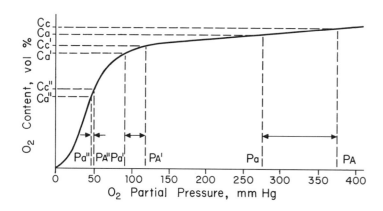

Figure 21–18 Effect of shunt at three levels of O_2 breathing. In all cases the shunt is the same, and therefore $(Cc_{O_2} - Ca_{O_2})$ is also the same. $(PA_{O_2} - Pa_{O_2})$ increases with a rise in PA_{O_2}. It becomes constant when both PA_{O_2} and Pa_{O_2} are higher than the PO_2 corresponding to complete oxygen saturation of the hemoglobin (150 mm Hg).

alveolar air and in arterial blood. The last unknown parameter, $C\bar{v}_{O_2}$, is either measured directly or derived from the assumption that $(Ca_{O_2} - C\bar{v}_{O_2}) = 5$ or 6 ml of oxygen per 100 ml of blood.

Calculated by this method, the shunt in normal subjects varies from 3 to 5 per cent of the cardiac output. This value is affected, of course, by all the assumptions inherent in the equation used for the calculation. The hypothesis that, with high O_2 breathing, the alveolar arterial difference in O_2 is due mainly to the anatomical shunt is most important; this hypothesis is true only if pure O_2 has been breathed long enough to eliminate nitrogen in the lungs. In all other instances, unevenness of ventilation-to-perfusion ratios in the lungs contributes markedly to the P_{O_2} gradient (see p. 345) even at high O_2 breathing, and thus increases the calculated shunt.[38]

Although not really practical for clinical use, the contribution of this factor can be separated from the total alveolar arterial gradient.[37, 43] This shows that the true anatomical shunt amounts to less than 1 per cent of the blood flow. The difference between the true anatomical shunt and the one measured by the O_2 method corresponds to the *physiological shunt*. This shunt is composed of blood leaving perfused but poorly or unventilated alveoli. The O_2 content of this blood is, of course, lower than the blood perfusing the well-ventilated alveoli, and it can therefore be compared to the venous admixture of the anatomical shunt.

In disease in which the anatomical shunt plays only a small role in reducing the oxygen saturation of the arterial blood, the oxygen method is commonly used to evaluate the physiological shunt.

Hypoventilation. At a given metabolic level, which determines the rate of oxygen uptake, $\dot{V}_{O_2}$, and of carbon dioxide elimination, $\dot{V}_{CO_2}$, and for a given pulmonary blood flow, the gas partial pressures in the alveolar air are determined by the alveolar ventilation ($\dot{V}_A$).

The amount of gas lost or gained by alveolar air per unit time is the difference between the amount entering and the amount leaving the respiratory units. Hence:

$$\dot{V}_{O_2} = \dot{V}_{A_I}F_{I_{O_2}} - \dot{V}_{A_E}F_{A_{O_2}} \qquad (26)$$

where the inspired and expired alveolar minute volumes ($\dot{V}_{A_I}$ and $\dot{V}_{A_E}$) are unequal. In equation (26) the difference between the inspired and expired volume can be corrected because no net exchange of nitrogen occurs across the lung:

$$\dot{V}_{A_I}F_{I_{N_2}} = \dot{V}_{A_E}F_{A_{N_2}} \qquad (27)$$

Hence, by eliminating $\dot{V}_{A_I}$

$$\dot{V}_{O_2} = \dot{V}_A\left(\frac{F_{A_{N_2}}}{F_{I_{N_2}}} F_{I_{O_2}} - F_{A_{O_2}}\right) \qquad (28)$$

and similarly:

$$\dot{V}_{CO_2} = \dot{V}_A\left(F_{A_{CO_2}} \frac{F_{A_{N_2}}}{F_{I_{N_2}}} - F_{I_{CO_2}}\right) \qquad (29)$$

In these equations, $\dot{V}_{O_2}$, $\dot{V}_{CO_2}$ and $\dot{V}_A$ are expressed in the same conditions; after adjustments for conditions in which $\dot{V}_{O_2}$, $\dot{V}_{CO_2}$ and $\dot{V}_A$ are usually expressed $\dot{V}_{O_2}$ or $\dot{V}_{CO_2}$ in ml per min STPD and $\dot{V}_A$ in 1 per min BTPS and for changing the fractional concentration in partial pressures, equations (28) and (29) become (letting $F_{I_{CO_2}} = 0$):

$$P_{A_{O_2}} = P_{I_{O_2}} \frac{P_{A_{N_2}}}{P_{I_{N_2}}} - \frac{0.863\dot{V}_{O_2}}{\dot{V}_A} \qquad (30)$$

and:

$$P_{A_{CO_2}} = \frac{0.863\dot{V}_{CO_2}}{\dot{V}_A} \qquad (31)$$

Equations (30) and (31) show that when $\dot{V}_A$ diminishes, $P_{A_{O_2}}$ decreases and $P_{A_{CO_2}}$ increases (Fig. 21–19). Because of this relationship between ventilation and gas pressure, hypoventilation is defined by the elevation of P_{CO_2} above normal (40 mm Hg). Primary alveolar hypoventilation is a clinical phenomenon frequently found in obese individuals; it can also be observed in nonobese individuals with various pathological conditions.[5]

An important feature of hypoventilation is that it does not cause a gas pressure gradient between alveolar air and capillary blood. In fact, any change in gas partial pressure of alveolar air is accompanied by a similar change in capillary blood.

VENTILATION-PERFUSION RELATIONSHIP. Thus far, a constant pulmonary blood flow has been assumed, and hence only the effect of changing ventilation on the partial pressures has been considered. However, the actual level of P_{O_2} and P_{CO_2} in alveolar air and in capillary blood is related to blood flow (perfusion) as well. The amount of gas

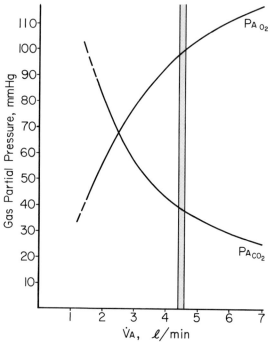

Figure 21–19 Alveolar P_{O_2} and P_{CO_2} as a function of alveolar ventilation. Values were calculated with equations (30) and (31). $\dot{V}_{O_2}$ and $\dot{V}_{CO_2}$ were assumed to be 250 and 200 ml/min. With no diffusion barrier, $P_{A_{O_2}}$ and $P_{A_{CO_2}}$ are equal to $P_{c_{O_2}}$ and $P_{c_{CO_2}}$ respectively. The shaded area indicates the normal range of values at rest.

lost or gained by the blood is given by the following equations:

$$\dot{V}_{O_2} = \dot{Q}Ca_{O_2} - \dot{Q}Cv_{O_2}$$
$$= \dot{Q}(Ca_{O_2} - Cv_{O_2}) \qquad (32)$$

and

$$\dot{V}_{CO_2} = \dot{Q}(Cv_{CO_2} - Ca_{CO_2}) \qquad (33)$$

Since O_2 lost by alveolar air is taken up by the capillary blood, and inversely for CO_2, the following equalities become obvious:

$$\frac{\dot{V}_A}{0.863}\left(P_{I_{O_2}}\frac{F_{A_{N_2}}}{F_{I_{N_2}}} - P_{A_{O_2}}\right) =$$
$$\dot{Q}(Ca_{O_2} - C\bar{v}_{O_2}) \qquad (34)$$

and

$$\frac{\dot{V}_A}{0.863} \times P_{A_{CO_2}} = \dot{Q}(C\bar{v}_{O_2} - Ca_{O_2}) \qquad (35)$$

Rearranging

$$\frac{\dot{V}_A}{\dot{Q}} = \frac{0.863(Ca_{O_2} - C\bar{v}_{O_2})}{P_{I_{O_2}}(F_{A_{N_2}}/F_{I_{N_2}}) - P_{A_{O_2}}} \qquad (36)$$

and

$$\dot{V}_A = \frac{0.863(C\bar{v}_{CO_2} - Ca_{CO_2})}{P_{A_{CO_2}}}$$

Since the gas exchange ratio R is the ratio of CO_2 elimination to O_2 uptake, we have:

$$R = \frac{\dot{V}_{CO_2}}{\dot{V}_{O_2}} = \frac{(Cv_{CO_2} - Ca_{CO_2})}{(Ca_{O_2} - Cv_{O_2})} \qquad (37)$$

Equation (37) can then be rewritten:

$$\frac{\dot{V}_A}{\dot{Q}} = \frac{0.863R(Ca_{O_2} - Cv_{O_2})}{P_{A_{CO_2}}} \qquad (38)$$

Equations (36), (37) and (38) indicate that, for fixed inspired and mixed venous blood gas tensions, the composition of alveolar air, as shown by $P_{A_{O_2}}$ and $P_{A_{CO_2}}$, and of blood leaving the capillary, as shown by Ca_{O_2}, depends upon the ventilation-perfusion relationship. They further indicate that alveolar and capillary gas tensions are independent of metabolism and that an increase in blood flow has the same effect as a decrease in alveolar ventilation.

TEMPORAL CHANGES IN ALVEOLAR GAS PRESSURES. Actually, because of the cyclic nature of ventilation, P_{O_2} and P_{CO_2} constantly vary rhythmically even with constant minute volumes of ventilation and blood flow. The events in one respiratory cycle (one breath) may be visualized with the aid of Figure 21–20. At the end of expiration, the air in the respiratory passages is that expelled last from the alveoli. During the interval after expiration and just before inspiration, O_2 continues to be removed from, and CO_2 to be added to, the alveoli at rates determined by the blood flow and the respective partial pressures. On inspiration, the first gas to enter the alveoli is the end expired alveolar gas from the previous breath remaining in the respiratory passages (dead space), followed by fresh inspired gas. This inspired gas mixes with and dilutes the gas already present in the alveoli.

Unevenness of the Ratios of Ventilation to Perfusion. In the previous discussion, alveolar air was assumed to be uniform in composition throughout the lung. Besides the *temporal* variations in composition described earlier, *regional* differences (*i.e.*, from place to place within the lung) also exist in the alveolar air composition.[28] This implies, then, that the ratio of ventila-

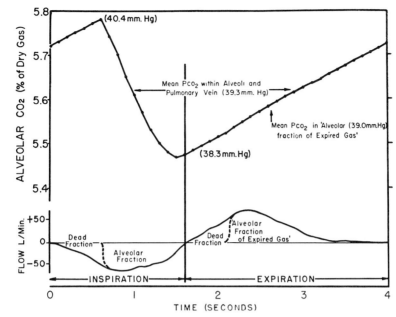

Figure 21-20 Fluctuations in alveolar P_{CO_2} during normal respiratory cycle. A damping effect occurs owing to variation of blood flow and capacity of tissue for CO_2. The best time for an expired sample of mean alveolar gas for CO_2 determination is shortly after midexpiration. If most of the alveolar gas expired leaves early in expiration, the CO_2 therein will have a lower tension than that of mean arterial blood.[16]

tion to perfusion is not even throughout the lung. Remembering that the adult lung is composed of about 300 million alveoli, one realizes that nonuniformities are indeed highly probable. Unevenness of ventilation-to-perfusion ratios ($\dot{V}_A/\dot{Q}$) can result from inequalities of ventilation and/or blood flow. This factor is currently considered the most important among the causes of impaired oxygenation in normal and diseased subjects.

UNEVENNESS OF VENTILATION. Ventilation is uneven when inspired gas is not equally distributed among all the alveoli. Two mechanisms have been proposed for this unevenness: (i) some areas expand more than others; (ii) stratification. In the former case the amount of ventilation in relation to the preinspiratory volume ($\dot{V}_A/V_0$) is not uniform. An enhancing factor is that the dead space is not evenly distributed; if some respiratory units receive more dead-space gas and less fresh air than others, the effect is as if they had in fact received less ventilation since the dead space-gas is poor in O_2, but rich in CO_2. In the latter case, movement of gas by diffusion is not fast enough to attain a homogeneous mixture of gas in the most distal generation of

conducting airways. Convective movement of gas (ventilation) is limited to the larger airways; in the smaller ones, gas is distributed by diffusional movement. With relatively slow movement, *layers* (strata) of gas mixtures of different concentration exist, which in fact are equivalent to uneven distribution of inspired gas.

Historically, these two mechanisms have been alternately favored or opposed. Current opinion represents a compromise: regional differences in expansion are the main cause of gas nonuniformity, but within the respiratory lobule, inhomogeneities may exist as a result of stratification.

EVIDENCE OF UNEVEN GAS DISTRIBUTION. Uneven gas distribution can be detected by several methods which have great clinical value.

Single Breath Technique. After a single deep breath of pure O_2, the nitrogen concentration in the succeeding expiration is continuously recorded. The increase in N_2 concentration between the first 750 ml exhaled and the next 500 ml should not exceed 1.5 per cent.[12] The N_2 concentration increase will be greater in obstructive lung diseases. This test reveals asynchronous emptying of different lung units as well as unequal distribution. In the absence of sequential emptying, a flat N_2 alveolar plateau would be expected because the relative contribution of the different lung regions to the expirate would remain constant.

Multiple Breath Technique. When a lung region is poorly ventilated, elimination of N_2 from this region during pure O_2 breathing is slower than from the better ventilated regions. Thus, the N_2 concentration in mixed expired air diminishes more slowly than if N_2 were rapidly eliminated from all regions of the lung. After breathing O_2 for 7 min, the subject with a normal lung has a N_2 concentration of only 2.5 per cent in forced expired air, whereas the patient with poor mixing may have up to 20 per cent.[13] Also, with this method distribution is considered uniform when a plot of $F\bar{E}_{N_2}$ (mean fraction of N_2 in expired air) against the number of breaths is linear. However, not even normal subjects exhibit such a simple relationship. The plot obtained is curvilinear and is the summation of several linear relationships, each representing a lung compartment grouping all the respiratory units with an identical ventilation-to-volume ratio.[27]

Topographical Methods. These methods permit a functional differentiation of the regions of the lung according to anatomic localization.

With *lobar spirometry or lobar gas sampling,* the *interlobar* sequential emptying was demonstrated. In normal subjects, the upper lobes received most of the early part of the inspired volume, whereas the lower lobes received most of the last part.[33, 44] Yet the lower lobe had a greater ventilation-to-volume ratio than the upper lobe. In obstructive airway disease this distribution is reversed and the upper lobe becomes the better ventilated in relation to its volume. During expiration the poorly ventilated part of the lung contributes to the early part and the well ventilated to the latter part of the expiration. This is exemplified in Figure 21–21, which shows a smaller difference in N_2 concentration between the trachea and the upper lobe at the beginning, and inversely between the curves of the trachea and the lower lobe at the end of the expiration. This pattern of emptying prevails whether the poorly ventilated lobe is the lower, as in patients with airway disease, or the upper, as in normal subjects. This method has also provided evidence of *intralobar* sequential emptying. The rising slope of the N_2 concentration in the gas coming from each lobe indicates that the well ventilated units of each lobe empty first and the poorly ventilated units empty last. In other words, the intralobar sequence of emptying is opposite to the interlobar sequence.[68]

Techniques using radioactive gas also permit topographic studies of gas distribution.[9, 46, 65] Several counters are placed over the chest to obtain local count densities after a single inhalation, or washin from a closed circuit, or

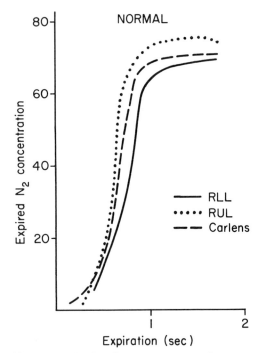

Figure 21–21 Simultaneous tracings of N_2 concentration from right upper lobe, right lower lobe and mixed expired gas from the right lung. Mixed expired N_2 concentration is closer to that in the upper lobe at the beginning and closer to the concentration in the lower lobe at the end of the expiration. Note that N_2 concentration increases in each lobe and in the mixed expired gas.[64]

washout from the alveolar gas after intravenous injection of the radioactive isotope of a gas. Xenon[133] and CO_2 labeled with oxygen[15] have been used. All these methods give identical results; ventilation per unit of lung volume decreases linearly with distance up the lung (Fig. 21–22). This, and observations on sequential filling, confirm the results obtained with the lobar gas sampling technique.

DISPERSION OF VENTILATION–TO–LUNG VOLUME RATIOS. Dispersion of a variable within a system or population can be quantitatively estimated by the range of this variable. Such an estimation of the uneven distribution of ventilation, defined by $\dot{V}_A/V_0$, can be made by most of the methods described above. In normal subjects, the radioactive gas method reveals a small difference in $\dot{V}_A/V_0$ from the upper to the lower region of the lung (about 1:1.6). Better resolution is obtained with the multiple breath technique, because all regions with the same $\dot{V}_A/V_0$ are lumped together regardless of location, whereas the radioactive

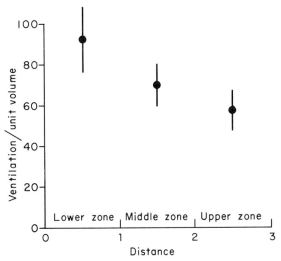

Figure 21-22 Distribution of ventilation-to-volume ratios from apex to bottom of the lung in upright man.[9]

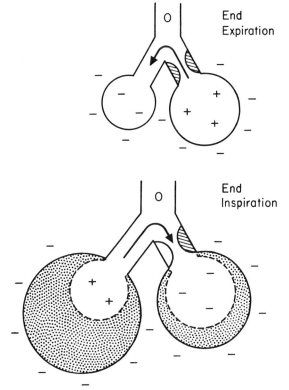

Figure 21-23 Effect of increased airway resistance. Alveolar pressure in the corresponding alveolus is higher (end expiration) or lower (end inspiration) than in the alveolus with low resistance airways. As a result, air passes from one group of alveoli to the other.[47] Higher pressure relative to atmosphere: +, lower: −, equal: 0. Shaded area indicates lung expansion.

gas method gives the average of high and low $\dot{V}_A/V_0$ within one anatomical region. Table 21-3 shows that the calculated dispension is actually related to the number of compartments into which the system is analyzed.

CAUSES OF UNEVEN GAS DISTRIBUTION. Nonuniform distribution may be caused by any of the following mechanisms:

Unequal Airway Resistance. If airway resistance to flow is not uniform throughout the lungs, the airways with the low resistance receive the larger proportion of the volume inspired. This difference in inspired gas distribution is further accentuated when the respiratory rate is increased, because alveolar pressure cannot reach equilibrium during each phase of the respiratory cycle (Fig. 21-23).[50] Experimental evidence indicates that the aforementioned situation exists in diseases with increased airway resistance such as asthma.[7] In normal subjects, however, it plays no significant role, because the distribution is the same under static (airway resistance independent) or dynamic (air-

way resistance dependent) conditions.[9] This indicates that in normal lungs alveolar pressure indeed reaches zero throughout the lungs between phases of the respiratory cycle and that differences in lung tissue properties would more likely cause nonuniform gas distribution.

Unequal Specific Compliance. If tissue of different lung regions has different elastic properties, the more "compliant" respiratory units tend to receive the larger proportion of the volume moved. In seeking the cause of tissue differ-

Table 21-3 Range of $\dot{V}_A/V_0$ in Normal Subjects

NUMBER OF COMPARTMENTS	MIN $\dot{V}_A/V_0$	MAX $\dot{V}_A/V_0$	RATIO	REFERENCE NUMBER
2	0.12	0.37	1 : 4.2	25
6	0.05	0.60	1 : 12.6	65
Infinite	0.05	0.72	1 : 14.4	45

ences, one can postulate that the regions of the lung which have the higher perfusion are "stiffer" or less compliant than the less perfused regions. However, experimentation indicates that the better perfused regions are also the better ventilated. Also, direct measurements of static compliance between the upper and lower lobe have failed to show a difference.[45] Thus, differences in tissue elasticity most likely do not play a significant role in apex-to-base variation in gas distribution.

Unequal Intrapleural Pressure. Differences in intrapleural pressure between the top and bottom of the lung, in upright chest, are experimentally well established: pleural pressure is more negative at the top and progressively decreases 0.25 cm H_2O per cm down to the bottom of the lung. This gradient is probably caused by a similar gradient in transpulmonary pressure which can be accounted for by a difference in lung expansion. The expansion, in turn, is related to the force of gravity operating on the lungs, causing distension at the top and compression at the bottom. In other words, regional functional residual volume is greater at the top than at the bottom (Fig. 21–24).

Lung tissue has a density of approximately 0.25 g/cc. Therefore, a 1 cm² "column" of lung 30 cm high weighs 7.5 g, which actually corresponds to the apex-to-bottom intrapleural pressure difference of 7.5 cm H_2O.

The difference in lung expansion is comparable to the difference in distance between the coils of a spring suspended at one end, as a result of the spring's weight. Because of the nonuniform expansion, tissues at different vertical levels operate on different parts of the pressure

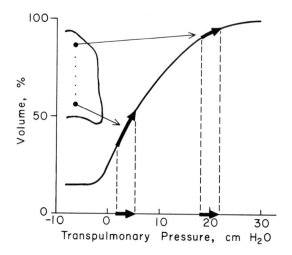

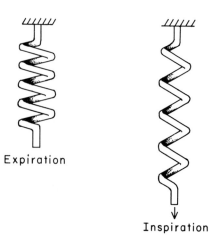

Figure 21–25 Effect of a gradient in intrapleural pressure on the distribution of inspired gas. Because top and bottom of the lung are on a different part of the *same* compliance curve, an equal change in transpulmonary pressure during inspiration gives a lesser expansion. This is comparable to pulling downward on a nonlinear spring suspended from its top.[61]

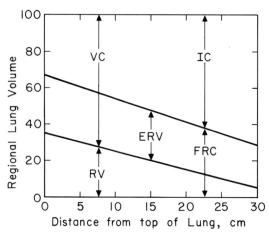

Figure 21–24 Percentage of regional lung volumes (right lung) at full inspiration, from top to bottom of the lung. RV = Residual volume, FRC = functional residual capacity, ERV = expiratory reserve, IC = inspiratory capacity, VC = vital capacity.[43]

volume curve, even though the elastic properties are the same (Fig. 21–25). During inspiration the lower regions which are on the steep part of the PV curve have a greater volume change per unit change of transpulmonary pressure than the apex regions. This is comparable to pulling downward on the spring, which results in an increased total distance and a more even spacing between the coils because the lower coils expand more than the upper.

Gravitational forces are certainly an important mechanism of uneven gas distribution. Such a complex phenomenon cannot, however, be entirely explained by gravity. The distribution within one lobe has been observed to be just as uneven as that within the whole lung. This, of course, cannot be accounted for by gravity but may be explained by randomly distributed unequal tissue properties (as yet poorly understood) or by stratification inhomogeneities.

STRATIFICATION INHOMOGENEITIES. The concept of stratification, originally developed by Rauwerda in 1946,[54] has regained new vigor from experimental data suggesting that the inspirate does not reach the terminal units by convective movement.[1, 15] However, some theoretical analyses of the time course of gaseous diffusion support the notion that the interpretation of the physiological data may be open to question.[16, 35]

When a subject takes one breath of a mixture of two inert gases of different densities, and if diffusion is rapid, the ratio of their concentrations should remain constant through expiration, irrespective of sequential emptying. Experimental results with a mixture of light gas, neon, and a heavy gas, sulfur hexafluoride, do not confirm this pattern (Fig. 21–26). The high ratio of SF_6 to Ne at the beginning of expiration suggests that more of the light gas has left this volume, whereas the low ratio at the end indicates that less of the heavy gas has reached the more remote parts of the lungs.[15] Thus, diffusion equilibrium does not appear to be attained for SF_6. This observation, however, does not answer the question of whether the diffusion process is normally too slow for O_2 and CO_2.

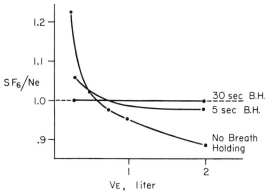

Figure 21–26 Change in the ratio, SF_6/Ne, during expiration after inhalation of a mixture of SF_6 and Ne.[19] If the inhomogeneity was regional rather than diffusion dependent, this ratio should remain uniform throughout expiration, as it does after breath holding.

If stratified inhomogeneity causes the upward slope of the alveolar plateau to be increased as observed with the "single breath technique," holding the breath should reduce the slope, or even abolish it. Such has indeed been reported for expired N_2 concentration after a breath of pure O_2.[32] Similarly, if a breath of the mixture containing SF_6 and Ne is held for 30 sec, their ratio is constant during the following expiration (Fig. 21–26).[15]

After a breath of 20 per cent O_2 in argon, expiration reveals a decreasing A to N_2 ratio as the volume expired increases. This indicates that the early part of the expirate represents, probably, a greater contribution from the better ventilated regions in relation to their volume. However, the ratio becomes more uniform as the breath is held up to 20 sec, thus establishing the time requirement for a better diffusion mixing.[60] Unfortunately, this type of observation does not separate diffusion within respiratory units from that between groups of alveoli with different $\dot{V}_A/V_0$ ratios.

The differentiation between regional and stratified inhomogeneity has been attempted by observing the effect of breath holding on the *cardiogenic oscillation.* Cardiac oscillations are periodic changes in expired gas concentration which are synchronous with the heart beats. They arise from transient differences in emptying between the apex and the base of the lung, or, in other words, between regions with different $\dot{V}_A/V_0$.[26] Since these oscillations persist after breath holding, whereas the slope of the plateau decreases significantly (Fig. 21–27), it appears that regional inhomogeneities are not suppressed by breath holding and that the slope of the alveolar plateau is mainly due to stratification.[19]

These experimental findings have been supported by a mathematical analysis which showed that diffusion mixing was complete during a single respiratory cycle.[14] Some more recent analyses[16, 35] challenge the original observation and shed some doubt on the interpretation of the experimental findings. The discrepancies between the earlier and more recent analysis are due to the type of lung model used for the calculations.

Although the recent work is based on a model which is clearly a closer approximation of the real lung, there is still some need

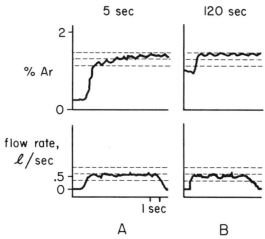

Figure 21-27 Effect of breath holding on cardiogenic oscillations and slope of the alveolar plateau. Only the slope disappears with long breath holding.[17]

to reconcile all the experimental interpretations with the calculation-based conclusions.

Unevenness of Pulmonary Blood Flow. Unevenness of blood flow in the upright lung has been known since 1887.[49] Investigators have unanimously found an increase in blood flow from the apex to the dependent parts of the lung.

EVIDENCE OF UNEVEN PULMONARY BLOOD FLOW DISTRIBUTION. Two methods are used to demonstrate blood flow unevenness. The first method is by bronchospirometry with a triple lumen catheter. Two lumens permit separation of gas going to the upper lobe from that going to the lower lobe of the right lung. The third lumen serves to ventilate the left lung. Measurement of oxygen consumption in each lobe reveals that the upper lobe has 22 per cent of the total O_2 uptake of the right lung. These data demonstrate a lower blood flow in the upper lobe.[44]

The second is by the radioactive gases technique. The principle of this method is the same as for the determination of ventilation unevenness. A very soluble gas, such as oxygen, or a poorly soluble gas, such as xenon[133], can be used. In the first instance, the gas is inhaled and the breath is held: blood flow is proportional to the rate of disappearance of the gas during breath holding. In the second case the gas is injected intravenously and again the breath is held while the gas evolves into the alveolar air

because of its low solubility; buildup in radioactivity in any area is proportional to the local blood flow.[65]

With this technique, distribution of blood flow can easily be determined across the vertical dimension of the chest. In normal man the blood flow per alveolus increases nearly linearly from the apex where it is almost nil toward the dependent part of the lung.[3, 66] At the bottom of the lung blood flow is again reduced. The level at which this reduction appears is inversely related to the lung volume.[31] For example, at maximum inspiration only a small basal region of reduced flow is present; at the end of normal expiration the zone of reduced flow increases markedly and the peak blood flow appears about 10 cm above the bottom of the lung (Fig. 21–28). Further decrease in lung volume down to residual volume progressively raises the level of peak blood flow.

CAUSES OF UNEVEN PULMONARY BLOOD FLOW DISTRIBUTION.[52] The unevenness of pulmonary blood flow distribution results from the combination of two factors. One is gravitational force: if the density of blood is 1.0, the hydrostatic pressure in the arterioles increases at the rate of 1 cm H_2O per cm distance from top to bottom. The other factor is lack of rigidity of the pulmonary vessels: they are distensible but also collapsible. If they were rigid, they

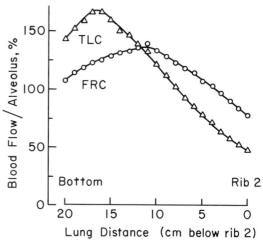

Figure 21-28 Index of blood flow per alveolus as a function of distance from bottom of the lung at two lung volumes: functional residual capacity and total lung capacity.[29]

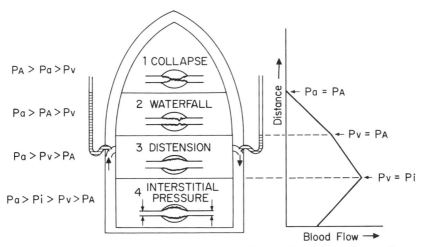

Figure 21–29 Pattern of uneven blood flow distribution in the lung.[29] In zone 1 the capillaries are collapsed because PA exceeds Pa. In zone 2, Pa exceeds PA, allowing blood to flow. The flow is independent of Pv, and thus this zone is comparable to a *waterfall*. In zone 3, the flow becomes dependent upon Pv because it is higher than PA. In zone 4, Pi is higher than Pv; thus the gradient Pa–Pi governs the flow.

would behave like siphons and the blood flow would be uniform throughout the lung.

With the presence of these two factors, the relationship between the arterial, venous and alveolar pressure determines the blood flow distribution. Furthermore, since these relationships vary from the top to the bottom of the lung, three zones can be distinguished.[67] In addition, at the bottom of the lung an interstitial—*i.e.*, extra-alveolar—pressure intervenes. This pressure depends upon a balance of forces and contributes to form a fourth zone (Fig. 21–29).

At the top is *zone 1* where PA is greater than Pa. Thus, the capillaries are collapsed and blood does not flow until the level, $Pa = PA$, is reached at the top of zone 2. This level depends upon the relative arterial and alveolar pressures and can be raised by vasoconstriction or increasing flow.

Zone 2 corresponds to the part of the lung in which arterial exceeds alveolar pressure, but PA exceeds venous pressure. Here the flow depends upon the difference, Pa–PA, since the thin, collapsible wall of the vessel offers no resistance to the collapsing pressure which is reached when $Pa = PA$. As Pa diminishes along the capillary, a constriction develops at the downstream end. This constriction varies on the vertical axis of zone 2 and disappears when $Pv = PA$, at the demarcation between zones 2 and 3.

Since in all zones PA is the same, the linear increase of flow in zone 2 is explainable by the linear increase of Pa owing to the increased hydrostatic pressure.

Zone 3 extends below the level at which venous and alveolar pressures are equal. Here flow is no longer controlled by Pa–PA but by Pa–Pv. Since this difference is constant, the continued flow increase down this zone may at first seem surprising. Distension of the vessel, thus decreasing its resistance, is the cause of this flow increase. The degree of distension depends on the transpulmonary pressure which increases as the outside pressure (PA) stays constant, while rising hydrostatic pressure raises the inside pressure. Except for interstitial pressure, the pattern of zone 3 would extend to the bottom of the lung.

Zone 4 appears at a level which varies with lung expansion. It may result from the action of two opposing forces on the extra-alveolar vessels. One is the tension in the vessel walls which tends to constrict them; the other results from lung expansion which generates a negative pressure around the vessel, increasing its capacity.[30, 51] The algebraic sum of these forces develops an interstitial pressure Pi which affects the flow when Pi is greater than Pv; in other words, flow becomes dependent on Pa–Pi. Since the expansion decreases toward the base,[46] the constrictive force has progressively greater effect in the most dependent part, thus causing a progressive reduction of blood flow.

UNEVENNESS OF VENTILATION-TO-PERFUSION RATIO. Unevenness of ventilation and pulmonary blood flow is important in terms of gas exchange only if their ratio is also uneven. Should ventilation and perfusion be uneven but their ratio constant, gas exchange would not be affected. In the previous section on hypoventilation, this ratio was shown to determine the gas tensions in the alveolar air and in the blood. A variance in this ratio is then accompanied by variances in P_{O_2}, P_{CO_2} and P_{N_2} which in turn cause an alveolar arterial difference for each of these gases.

Direct evidence of $\dot{V}_A/\dot{Q}$ unevenness can be obtained by numerous methods, including local gas analysis by small catheter, lobar spirometry and radioactive gases. The latter technique shows that, from apex to base, perfusion per unit lung volume increases much faster than ventilation; thus, $\dot{V}_A/\dot{Q}$ decreases from top to bottom (Fig. 21–30). The drawback of this technique is its lack of resolution; each counter gives the average for an anatomically defined region. Since no region has a uniform $\dot{V}_A/\dot{Q}$, the range of $\dot{V}_A/\dot{Q}$ in the whole lung is under-

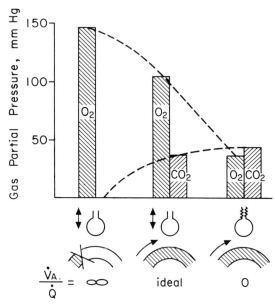

Figure 21–31 Effect of $\dot{V}_A/\dot{Q}$ on the O_2 and CO_2 tensions. The ideal $\dot{V}_A/\dot{Q}$ is equal to the mean $\dot{V}_A/\dot{Q}$. At $\dot{V}_A/\dot{Q} = 0$, P_{O_2} and P_{CO_2} are equal to those in the venous blood; at $\dot{V}_A/\dot{Q} = \infty$ they equal those in the inspired air.[18]

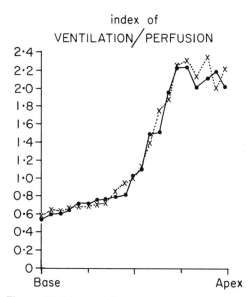

Figure 21–30 Distribution of ventilation to perfusion ratios from top to bottom of the lung. Because blood flow varies more than ventilation, $\dot{V}_A/\dot{Q}$ is four times higher at the apex than at the base.[15]

estimated. A study of the effect of $\dot{V}_A/\dot{Q}$ inhomogeneity permits a better assessment of its dispersion.

Effect of $\dot{V}_A/\dot{Q}$ Distribution. Gas partial pressure may be considered in respiratory units with a low and a high $\dot{V}_A/\dot{Q}$ in comparison to the average $\dot{V}_A/\dot{Q}$ of the lungs as a whole (ideal $\dot{V}_A/\dot{Q}$). The ideal $\dot{V}_A/\dot{Q}$ is defined by the ratio of total $\dot{V}_A$ to total $\dot{Q}$; in normal man it varies between 0.8 and 1. In the low $\dot{V}_A/\dot{Q}$ units, P_{O_2} is low and P_{CO_2} is high, and the opposite is true in high $\dot{V}_A/\dot{Q}$ units. The possible range of P_{O_2} and P_{CO_2} within the lung extends from the partial pressure values in the mixed venous blood entering the lung (approximately $P\bar{v}_{O_2} = 40$ and $P\bar{v}_{CO_2} = 46$ mm Hg) to those in the inspired air ($P_{I_{O_2}} = 150$ and $P_{I_{CO_2}} = 0$ mm Hg). One end of the range corresponds to units perfused but not ventilated, whereas at the other end there is ventilation but no perfusion. These changes in gas tension with $\dot{V}_A/\dot{Q}$ are depicted in Figure 21–31. Since the total gas pressure in all alveoli is equal to the barometric pressure, any difference between O_2 leaving the alveolar air and CO_2 entering it must be compensated for by an equal change in N_2, as shown in Figure 21–32.

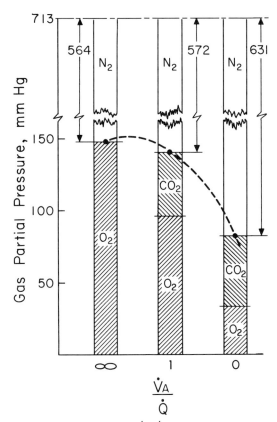

Figure 21–32 Effect of $\dot{V}A/\dot{Q}$ on PO_2, PCO_2 and PN_2. Between ideal $\dot{V}A/\dot{Q}$ and high $\dot{V}A/\dot{Q}$, PN_2 does not vary greatly but PO_2 and PCO_2 change markedly. Between ideal $\dot{V}A/\dot{Q}$ and low $\dot{V}A/\dot{Q}$, PN_2 and PO_2 change by almost the same amount but PCO_2 remains almost the same.[18]

The cause and direction of the alveolar arterial difference can now be easily understood:

(i) Since the units with a high $\dot{V}A/\dot{Q}$ contribute more to total ventilation than those with low $\dot{V}A/\dot{Q}$, and since the units with a low $\dot{V}A/\dot{Q}$ contribute more to the mixed arterial blood than those with a high $\dot{V}A/\dot{Q}$, the gas tensions in the mixed alveolar air differ from those in the mixed arterial blood. That is, the mixed alveolar gas composition tends to reflect $\dot{V}A/\dot{Q}$ in the better ventilated units, but mixed arterial blood tends to reflect the units with low $\dot{V}A/\dot{Q}$.

(ii) Since PO_2 is higher and PCO_2 and PN_2 are lower in the units with high $\dot{V}A/\dot{Q}$, PO_2 is higher in mixed alveolar air than in mixed arterial blood; on the other hand, PA_{CO_2} and PA_{N_2} are lower than Pa_{CO_2} and Pa_{N_2}. This explains why in lung disease

where the number of units with a low $\dot{V}A/\dot{Q}$ increases, PO_2 decreases progressively, and PCO_2 progressively increases in the arterial blood.

(iii) Figure 21–32 shows that PO_2, PCO_2 and PN_2 are not influenced to the same degree with changes in $\dot{V}A/\dot{Q}$. This is because of the respective shapes of the O_2-Hb dissociation and CO_2 absorption curves. Between ideal $\dot{V}A/\dot{Q}$ and very low $\dot{V}A/\dot{Q}$, PO_2 and PCO_2 vary greatly by about the same magnitude but change in opposite directions, while PN_2 stays almost constant. Since the alveolar-arterial PO_2 difference results from the difference in gas tension between the various lung units, the $(A-a)_{O_2}$ quite obviously is affected by both low and high $\dot{V}A/\dot{Q}$ units. Further, $(A-a)_{CO_2}$ is chiefly influenced by high $\dot{V}A/\dot{Q}$ units where PCO_2 is much lower than in other units, whereas $(A-a)_{N_2}$ reflects principally the low $\dot{V}A/\dot{Q}$ units with the highest PN_2.

Dispersion of $\dot{V}A/Q$ in the Lung. Measurement of alveolar to arterial gas tension differences in normal man breathing various concentrations of O_2 reveals a sharp increase in $(A-a)_{O_2}$ and in $(A-a)_{N_2}$ when FI_{O_2} increases from 0.2 to 0.4 while $(A-a)_{CO_2}$ remains constant (Fig. 21–33).[36] The existence of the CO_2 gradient indicates an uneven distribution of $\dot{V}A/\dot{Q}$ because diffusion limitation, shunt or hypoventilation cannot cause it. It also shows that the distribution of $\dot{V}A/\dot{Q}$ includes units with a high $\dot{V}A/\dot{Q}$, because

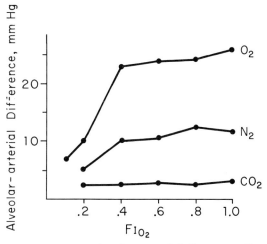

Figure 21–33 Alveolar arterial difference in PO_2, PCO_2 and PN_2 at various concentrations of inspired oxygen.[33] The subjects breathed each mixture for 15 minutes before the measurements were made.

units with a low $\dot{V}A/\dot{Q}$ contribute negligibly to the $(A\text{-}a)_{CO_2}$. The presence of an $(A\text{-}a)_{N_2}$ demonstrates that the distribution also includes alveoli with a low $\dot{V}A/\dot{Q}$. The true range of $\dot{V}A/\dot{Q}$ can be obtained, however, only by measuring of the increase in $(A\text{-}a)_{N_2}$ with higher FI_{O_2} concentrations. Since the size of the gradient depends upon the range of partial pressure, the increase in $(A\text{-}a)_{N_2}$ with increased FI_{O_2} means an increase in PN_2 range, despite a decreased N_2 concentration in the inspired gas. This situation is illustrated in Figure 21–34, which shows that only units with an extremely low $\dot{V}A/\dot{Q}$ cause a PN_2 range increase (see discussion of "air trapping," Chap. 20, Vol. II).

By using the PN_2 alveolar arterial difference, $\dot{V}A/\dot{Q}$ in the hypoventilated lung units can be calculated in normal subjects and is found to be smaller than 0.01. This means that some units receive at least 100 times more perfusion than ventilation.[38] The location of these units is difficult to determine; they may be scattered over the whole lung, because they cannot be located by either lobar or segmental gas analysis, or by radioactive gas technique.

The $\dot{V}A/\dot{Q}$ distribution is currently viewed as a *bimodal distribution*. One mode is located near the overall mean which includes the largest number of lung units, including those with a high $\dot{V}A/\dot{Q}$. The second mode corresponds to a group of units with very low $\dot{V}A/\dot{Q}$: the size of this group and its mean $\dot{V}A/\dot{Q}$ appear to vary

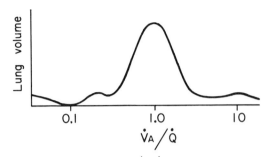

Figure 21–35 Bimodal $\dot{V}A/\dot{Q}$ distribution curve in normal subjects. The ordinate represents the relative proportion of the lung volume with the corresponding $\dot{V}A/\dot{Q}$.[37] The small compartment on the left-hand side was determined from the $(A\text{-}a)_{N_2}$.

constantly as shown by continuous variation in the size of the $(A\text{-}a)_{N_2}$.[38, 39] In lung disease, hypoxia and hypercarbia result from the increase in size of, or an increased perfusion of, the low $\dot{V}A/\dot{Q}$ group or a combination of the two (Fig. 21–35).[40]

$\dot{V}A/\dot{Q}$ *Distribution and Anatomical Shunt.* If nitrogen tension is high in the low $\dot{V}A/\dot{Q}$ units, PO_2 is low and is about the same in mixed venous blood and air. Consequently, from the point of view of O_2 exchange, these units behave just like an anatomical shunt and thus contribute to the rise in $(A\text{-}a)_{O_2}$ with increased FI_{O_2} (see Fig. 21–33).

Since the sum of the partial pressures in each alveolus is equal to the atmospheric pressure, the sum of the partial pressure in the blood of each capillary is also atmospheric. From this, it follows that the PO_2 alveolar arterial difference caused by $\dot{V}A/\dot{Q}$ unevenness would equal the sum of the PCO_2 and PN_2 alveolar arterial difference, were it not for the curvature of the O_2-Hb dissociation curve (Fig. 21–36).[10] However, if the atmospheric pressure is raised to the point at which the blood of all capillaries is fully saturated, the effect of nonlinearity is abolished and the relationship $(A\text{-}a)_{O_2}$ (due to $\dot{V}A/\dot{Q}$) $= (A\text{-}a)_{N_2} + (A\text{-}a)_{CO_2}$ is exactly true.[37] Thus, by subtracting the CO_2 and N_2 differences from the actual $(A\text{-}a)_{O_2}$, the contribution of the anatomical shunt can be accurately determined.

For clinical purposes, this method is obviously not practical. Since the anatomical shunt is constant, or affected by lung disease to a much lesser extent than the $\dot{V}A/\dot{Q}$ distribution, anatomical shunt and the effect of low $\dot{V}A/\dot{Q}$ units can be combined for convenience under the name of *physiological shunt*. Correspondingly, the anatomical

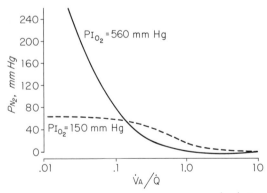

Figure 21–34 Local PN_2 as a function of $\dot{V}A/\dot{Q}$ during air breathing (PI_{O_2} : 150 mm Hg) or high O_2 concentration ($PI_{O_2} = 560$ mm Hg). Between $\dot{V}A/\dot{Q} = 10$ and $\dot{V}A/\dot{Q} = 0.01$, PN_2 increases by only 65 mm Hg when air is breathed but by more than 260 mm Hg when high O_2 is breathed. Note that PN_2 changes most when $\dot{V}A/\dot{Q}$ is lower than 0.1.

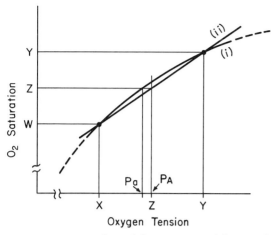

Figure 21-36 Effect of the curvature of the O_2-Hb dissociation curve on the $(A-a)_{O_2}$ due to $\dot{V}A/\dot{Q}$. Assume that the blood and the gas of two separate compartments, X and Y, mix equally (Z). Because the relationship between tension and saturation is curved (i), a difference ($PA - Pa$) occurs; if this relationship were linear (ii), Pa would equal PA.

dead space may be grouped with the high $\dot{V}A/\dot{Q}$ to form the *physiological dead space*. As a result, the lung may be divided into three functional parts: one which is ventilated but not perfused with blood, a second which is evenly ventilated and perfused, and a third which is perfused but not ventilated. All gas exchange takes place in the second part, the only part where blood and gas come into close contact. The second compartment is called the "effective" compartment in contrast to the other two which contribute nothing to air-blood gas exchange. The assumption of complete nonventilation and complete nonperfusion in certain parts of the lung creates a model which, although quantitatively inexact, is equivalent in terms of gas exchange to the real lung.[55]

REFERENCES

1. Altshuler, B., Palmes, E. D., Yarmus, L., and Nelson, N. Intrapulmonary mixing of gases studied with aerosols. *J. appl. Physiol.*, 1959, *14*, 321–327.
2. Asmussen, E., and Nielsen, M. Alveolo-arterial gas exchange at rest and during work at different O_2 tensions. *Acta physiol. scand.*, 1960, *50*, 153–166.
3. Ball, W. C., Stewart, P. B., Newsham, L. G. S., and Bates, D. V. Regional pulmonary function studied with Xenon[133]. *J. clin. Invest.*, 1962, *41*, 519–531.
4. Barcroft, J. Lessons from high altitude. In: *The respiratory function of the blood.* Cambridge, Cambridge University Press, 1925.
5. Bates, D. V., and Christie, R. V. *Respiratory function in disease.* Philadelphia, W. B. Saunders Co., 1964.
6. Benesch, R., Benesch, R. E., and Yu, C. I. Reciprocal bending of oxygen and diphosphoglycerate by human hemoglobin. *Proc. nat. Acad. Sci. (Wash.)*, 1968, *59*, 526–532.
7. Bentivoglio, L. G., Beerel, F., Bryan, A. C., Stewart, P. B., Rose, B., and Bates, D. V. Regional pulmonary function studied with Xenon[133] in patients with bronchial asthma. *J. clin. Invest.*, 1963, *42*, 1193–1200.
8. Bohr, C. Über die spezifische Tätigkeit der Lungen bei der respiratorischen Gasaufnahme und ihr Verhalten zu der durch die Alveslarwand stattfindenden Gasdiffusion. *Skand. Arch. Physiol.*, 1909, *22*, 221–280.
9. Bryan, A. C., Bentivoglio, L. G., Beerel, F., MacLeish, H., Zidulka, A., and Bates, D. V. Factors affecting regional distribution of ventilation and perfusion in the lung. *J. appl. Physiol.*, 1964, *19*, 395–402.
10. Canfield, R. E., and Rahn, H. Arterial-alveolar N_2 gas pressure differences due to ventilation-perfusion variations. *J. appl. Physiol.*, 1957, *10*, 165–172.
11. Cassin, S., Gilbert, R. D., and Johnson, E. M. Capillary development during exposure to chronic hypoxia. *Report USAF School Aviat. Med.* SAM-TR-66-16-1966.
12. Comroe, J. H., and Fowler, W. S. Lung function studies: VI. Detection of uneven alveolar ventilation during a single breath of oxygen. *Amer. J. Med.*, 1951, *10*, 408–413.
13. Cournand, A., Baldwin, E. deF., Darling, R. C., and Richards, D. W. Studies on intrapulmonary mixture of gases. IV. The significance of the pulmonary emptying rate and a simplified open circuit measurement of residual air. *J. clin. Invest.*, 1941, *20*, 681–689.
14. Cumming, G., Crank, J., Horsfield, K., and Parker, I. Gaseous diffusion in the airways of the human lung. *Resp. Physiol.*, 1966, *1*, 58–74.
15. Cumming, G., Horsfield, K., Jones, J. G., and Muir, D. C. F. The influence of gaseous diffusion on the alveolar plateau at different lung volumes. *Resp. Physiol.*, 1967, *2*, 386–398.
16. Cumming, G., Horsfield, K., and Preston, S. B. Diffusion equilibrium in the lungs examined by nodal analysis. *Resp. Physiol.*, 1971, *12*, 329–345.
17. Dollery, C. T., and Gillam, P. M. S. The distribution of blood and gas within the lungs measured by scanning after administration of Xe[133]. *Thorax*, 1963, *18*, 316–325.
18. DuBois, A. B., Britt, A. G., and Fenn, W. O. Alveolar CO_2 during the respiratory cycle. *J. appl. Physiol.*, 1952, *4*, 535–548.
19. Farber, J. P. Study of factors influencing regional emptying of the lung with soluble inert gases. Ph.D. dissertation. Buffalo, State University of New York, 1969.
20. Farhi, L. E. Ventilation-perfusion relationship and its role in alveolar gas exchange. In: *Advances in respiratory physiology*, C. G. Caro, ed. London, Edward Arnold, 1966.
21. Farhi, L. E. Diffusive and convective movement

of gas in the lung. In: *Circulatory and respiratory mass transport, Ciba Foundation Symposium,* G. E. W. Wolstenholme and J. Knight, eds. Boston, Little, Brown and Co., 1969.

22. Fenn, W. O., Rahn, H., and Otis, A. B. A theoretical study of the composition of the alveolar air at altitude. *Amer. J. Physiol.,* 1946, *146,* 637–653.
23. Finley, T. N., Swenson, E. W., and Comroe, J. H. The cause of arterial hypoxemia at rest in patients with "alveolar-capillary block syndrome." *J. clin. Invest.,* 1962, *41,* 618–622.
24. Forster, R. E. Diffusion of gases. *Handb. Physiol.,* 1964, Sec. 3, Vol. 1, 839–872.
25. Forster, R. E., Roughton, F. J. W., Cander, L., Briscoe, W. A., and Kreuzer, F. Apparent pulmonary diffusing capacity for CO at varying alveolar O_2 tensions. *J. appl. Physiol.,* 1957, *11,* 277–289.
26. Fowler, K. T., and Read, J. Cardiac oscillations in expired gas tensions and regional pulmonary blood flow. *J. appl. Physiol.,* 1961, *16,* 863–868.
27. Fowler, W. S., Cornish, E. R., and Kety, S. S. Lung function studies. VIII. Analysis of alveolar ventilation by pulmonary N_2 clearance curves. *J. clin. Invest.,* 1952, *31,* 40–50.
28. Grehant, N. Du renouvellement de l'air dans les poumons de l'homme. *C. R. Acad. Sci. (Paris),* 1862, *55,* 278–280.
29. Hashimoto, T., Young, A. C., and Martin, C. J. Compartmental analysis of the distribution of gas in the lungs. *J. appl. Physiol.,* 1967, *23,* 203–209.
30. Howell, J. B. L., Permutt, S., Proctor, D. F., and Riley, R. L. Effect of inflation of the lung on different parts of pulmonary vascular bed. *J. appl. Physiol.,* 1961, *16,* 71–76.
31. Hughes, J. M. B., Glazier, J. B., Maloney, J. E., and West, J. B. Effect of lung volume on the distribution of pulmonary blood flow in man. *Resp. Physiol.,* 1968, *4,* 58–72.
32. Kjellmer, I., Sandqvist, L., and Berglund, E. "Alveolar plateau" of the single breath nitrogen elimination curve in normal subjects. *J. appl. Physiol.,* 1959, *14,* 105–108.
33. Koler, J. J., Young, A. C., and Martin, C. J. Relative volume changes between lobes of the lung. *J. appl. Physiol.,* 1959, *14,* 345–347.
34. Krogh, A., and Krogh, M. On the rate of diffusion of CO into lungs of man. *Skand. Arch. Physiol.,* 1910, *23,* 236–247.
35. La Force, R. C., and Lewis, B. M. Diffusional transport in the human lung. *J. appl. Physiol.,* 1970, *28,* 291–298.
36. Lenfant, C. Measurement of ventilation/perfusion distribution with alveolar-arterial differences. *J. appl. Physiol.,* 1963, *18,* 1090–1094.
37. Lenfant, C. Measurement of factors impairing gas exchange in man with hyperbaric pressure. *J. appl. Physiol.,* 1964, *19,* 189–194.
38. Lenfant, C. Effect of high $F_{I_{O_2}}$ on measurement of ventilation/perfusion distribution in man at sea level. *Ann. N.Y. Acad. Sci.,* 1965, *121,* 797–808.
39. Lenfant, C. Time dependent variations of pulmonary gas exchange in normal man at rest. *J. appl. Physiol.,* 1967, *22,* 675–684.
40. Lenfant, C., and Okubo, T. Distribution function of pulmonary blood flow and ventilation-perfusion ratio in man. *J. appl. Physiol.,* 1968, *24,* 668–677.
41. Lenfant, C., Torrance, J. D., and Finch, C. A. The regulation of hemoglobin affinity for oxygen in man. *Trans. Ass. Amer. Phycns.,* 1969, *82,* 121–128.
42. Lilienthal, J. L., Riley, R. L., Proemmel, D. D., and Franke, R. E. An experimental analysis in man of the oxygen pressure gradient from alveolar air to arterial blood during rest and exercise at sea level and at altitude. *Amer. J. Physiol.,* 1946, *147,* 199–216.
43. Mallemgaard, K., Lassen, N. A., and Georg, J. Right-to-left shunt in normal man determined by use of tritium and krypton 85. *J. appl. Physiol.,* 1962, *17,* 778–782.
44. Martin, C. J., and Young, A. C. Ventilation-perfusion variations within the lung. *J. appl. Physiol.,* 1957, *11,* 371–376.
45. Martin, C. J., Young, A. C., and Ishikawa, K. Regional lung mechanics in pulmonary disease. *J. clin. Invest.,* 1965, *44,* 906–913.
46. Milic-Emili, J., Henderson, J. A. M., Dolovich, M. B., Trop, D., and Kaneko, K. Regional distribution of inspired gas in the lung. *J. appl. Physiol.,* 1966, *21,* 749–759.
47. Naeraa, N., Petersen, E. S., Boye, E., and Severinghaus, J. W. pH and molecular CO_2 components of the Bohr effect in human blood. *Scand. J. clin. Lab. Invest.,* 1966, *18,* 96–102.
48. Okubo, T., and Lenfant, C. Distribution function of lung volume and ventilation determined by lung N_2 washout. *J. appl. Physiol.,* 1968, *24,* 658–667.
49. Orth, J. *Ätiologisches und Anatomiches über Lungen-Schwindsucht.* Berlin, Hirschwald, 1887.
50. Otis, A. B., McKerrow, C. B., Bartlett, R. A., Mead, J., McIlroy, M. B., Selverstone, N. J., and Radford, E. P. Mechanical factors in distribution of pulmonary ventilation. *J. appl. Physiol.,* 1956, *8,* 427–443.
51. Permutt, S. Effect of interstitial pressure of the lung on pulmonary circulation. *Med. Thorac.,* 1965, *22,* 118–131.
52. Permutt, S., Bromberger-Barnea, B., and Bane, H. N. Alveolar pressure, pulmonary venous pressure and the vascular waterfall. *Med. Thorac.,* 1962, *19,* 239–260.
53. Prys-Roberts, C., Kelman, G. R., and Greenbaum, R. The influence of circulatory factors on arterial oxygenation during anaesthesia in man. *Anaesthesia,* 1967, *22,* 257–275.
54. Rauwerda, P. E. Unequal ventilation of different parts of the lung and the determination of cardiac output, Thesis. Groningen, The Netherlands, State University of Groningen, 1946.
55. Riley, R. L. Gas exchange and transportation. In: *Physiology and Biophysics,* T. C. Ruch and H. D. Patton, eds. 19th ed. Philadelphia, W. B. Saunders Co., 1965.
56. Riley, R. L., and Cournand, A. Ideal alveolar

air and the analysis of ventilation perfusion relationships in the lungs. *J. appl. Physiol.*, 1949, *1*, 825–847.

57. Roughton, F. J. W. The kinetics of the reaction CO + O_2Hb in human blood at body temperature. *Amer. J. Physiol.*, 1945, *143*, 609–620.

58. Roughton, F. J. W., Forster, R. E., and Cander, L. Rate at which carbon monoxide replaces oxygen from combination with human hemoglobin in solution and in the red cell. *J. appl. Physiol.*, 1957, *11*, 269–276.

59. Sackur, I. Weiters zur Lehre vom Pneumothorax. *Virchows Arch. path. Anat.*, 1897, *150*, 151–160.

60. Sikand, R., Cerretelli, P., and Farhi, L. E. Effect of $\dot{V}A$ and $\dot{V}A/\dot{Q}$ distribution and of time on the alveolar plateau. *J. appl. Physiol.*, 1966, *21*, 1331–1337.

61. Staub, N. C. Alveolar-arterial oxygen tension gradient due to diffusion. *J. appl. Physiol.*, 1963, *18*, 673–680.

62. Staub, N. C., Bishop, J. M., and Forster, R. E. Velocity of O_2 uptake by human red blood cells. *J. appl. Physiol.*, 1961, *16*, 511–516.

63. Staub, N. C., Bishop, J. M., and Forster, R. E.

Importance of diffusion and chemical reaction rates in O_2 uptake in the lung. *J. appl. Physiol.*, 1962, *17*, 21–27.

64. Tobin, C. E. Arteriovenous shunts in the peripheral pulmonary circulation in the human lung. *Thorax*, 1966, *21*, 197–204.

65. West, J. B. Regional differences in blood flow and ventilation in the lung. In: *Advances in respiratory physiology*, C. G. Caro, ed. London, Edward Arnold, 1966.

66. West, J. B., and Dollery, C. T. Distribution of blood flow and ventilation-perfusion ratio in the lung, measured with radioactive CO_2. *J. appl. Physiol.*, 1960, *15*, 405–410.

67. West, J. B., Dollery, C. T., and Naimark, A. Distribution of blood flow in isolated lung; relation to vascular and alveolar pressure. *J. appl. Physiol.*, 1964, *19*, 713–724.

68. Young, A. C., and Martin, C. J. The sequence of lobar emptying in man. *Resp. Physiol.*, 1966, *1*, 372–381.

69. Young, A. C., Martin, C. J., and Hashimoto, T. Can the distribution of inspired gas be altered? *J. appl. Physiol.*, 1968, *24*, 129–134.

CHAPTER 22 NEURAL CONTROL OF RESPIRATION

by ALLAN C. YOUNG

The main functions of the respiratory system are to provide O_2, eliminate CO_2 and maintain a constant pH of the blood, but this system participates in many other functions. The chest, lungs and upper respiratory tract provide controlled movement of air for sniffing, coughing, sneezing and vomiting; for such expressions of emotion as laughing and sobbing; and for a variety of highly skilled voluntary movements such as speaking, singing and blowing a wind instrument. It is interesting that, as Campbell[10] has emphasized, both the relative and total extent to which various muscle groups are active may be very different in voluntary and involuntary breathing movements.

PERIPHERAL NEURAL MECHANISMS

Efferent Discharge. The phrenic nerves, originating in C2, C3 and C4, provide the innervation for the diaphragm. The inter-costal muscles receive their innervation from T1-6 via the intercostal nerves; the abdominal muscles receive theirs from L1 and T7-12. The scaleni are innervated from C4-8, and the innervation of the sternomastoids is derived from C2 and the spinal accessory nerve. It is reasonable to speculate that some features of the respiratory act are integrated in the brain stem rather than in the spinal cord, because so many segmental levels are involved in the innervation of respiratory muscle.

The periodic respiratory enlargement and contraction of the thoracic space are superimposed upon an underlying postural tone. Tidal exhalations do not leave the thorax in full expiration, but in a state of partial inspiration, which is not far from the midposition of the thorax. This state is maintained by a tetanus in a smaller or larger proportion of the inspiratory motor units, each unit firing at rather slow rates (5 to 20 imp/sec). The diaphragm as well as the intercostal muscles is involved.[8, 22, 23] Undoubtedly the source of this activity is in general postural,

Figure 22–1 Discharge from single motor unit of external intercostal muscle. Unit fired continuously throughout inspiration and expiration, but rate increased during inspiration. Middle line is pneumograph (inspiration down) and lower line is graph of impulse frequency. (From Bronk and Ferguson, *Amer. J. Physiol.*, 1935, *110*, 700–707.)

and it represents participation of the inspiratory mechanism in the maintenance of the upright posture. There is, however, a definite respiratory component, inasmuch as the degree of inspiratory tone is regulated, via the carotid and aortic bodies, by the O_2 and CO_2 content of the blood.[19]

The expiratory muscles also exhibit a basic tonic activity, since certain of them, like the inspiratory muscles, oppose by contraction the force of gravity and participate therefore in the maintenance of upright posture. Thus the muscles of the abdomen, while functioning in respiration as expiratory muscles, are also important postural muscles to retain the abdominal contents.

The act of inspiration begins against a background of tonic innervation of both inspiratory and expiratory muscles. Simultaneously two events centrally coordinated occur: (i) those units supplying inspiratory muscles which are in tonic contraction increase their rate of firing (Fig. 22–1), and (ii) the tonic firing of expiratory units is reciprocally inhibited. As the size of the thorax increases and the diaphragm descends, the expiratory apparatus gives way in equal degree to accommodate for this movement. In addition, as inspiration proceeds, new units are added, or "recruited," so that the inspiratory act gains force as it proceeds (Fig. 22–2). The firing of individual units accelerates in rate, resulting in a progressive increment in the strength of contraction of each unit. By increase in the number of active units, and by augmentation in the strength of each unit's contraction through its increasing rate of discharge, inspiration grows to a peak determined by the various factors that control the depth of respiration. The whole accelerating tempo is then abruptly terminated. Other inspiratory units cease firing more slowly.[10] Units participating in the main-

tenance of inspiratory tone return to their former slow steady tetanus, which is maintained throughout expiration. The tonic expiratory discharge recommences.

In normal quiet breathing, this is at times the whole of the respiratory act, but often some traces of active expiration develop, reciprocating with inspiration (Fig. 22–3).

Virtually nothing is known of the size and number of the motor units of the respiratory system. Much more is known about the rates of discharge in single phrenic and intercostal motor units, which have been studied repeatedly. Basically the rates are slow, particularly at the onset of inspiration, at the end of expiration and during the tonic phases, and may amount to no more than 5 to 10 discharges per sec. Accelerating as respiration progresses, the rates reach the neighborhood of 30 to 40 imp/sec at the

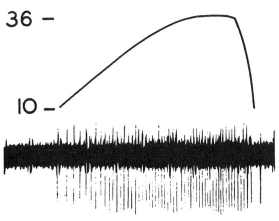

Figure 22–2 Characteristic slowly augmenting inspiratory discharge of single motor fiber of phrenic nerve of dog. This record was obtained during complete motor paralysis produced by intravenous injection of curare. Vagus nerves are sectioned. Frequency of firing is plotted on ordinates above original electrogram. (From Gesell *et al.*, *Amer. J. Physiol.*, 1940, *128*, 629–634.)

Figure 22–3 Simultaneous records of motor nerve impulses to internal intercostal muscle (*upper record*) in expiration, and to external intercostal muscle (*lower record*) in inspiration; vagi and carotid sinus nerves cut; animal completely immobilized with curare. *Bottom line*, time—0.2 sec intervals. (From Bronk and Ferguson, *Amer. J. Physiol.*, 1935, *110*, 700–707.)

end of normal inspiration. Even with an extreme respiratory drive producing a maximum hyperpnea, rates above 100 per sec are rarely seen; this rate is close to the upper limit of motoneuron discharge to skeletal muscle in normal circumstances.

Afferent Pathways. The respiratory system is affected by stimuli, especially noxious stimuli, from many parts of the body; it is also affected to some extent by proprioceptors of the limb and possibly by other muscles. These proprioceptors may be of some importance in the control of breathing (Chap. 23, Vol. II). The afferent nerve fibers which are most effective in controlling and modifying respiration are contained in the glossopharyngeal (ninth) and vagus (tenth) nerves. These nerves carry impulses from the carotid and aortic bodies or glomera (Chap. 11, Vol. I), which contain chemoreceptors responsive to increased pCO_2 and lowered pO_2 in the arterial blood.[13, 14] The tenth nerve carries impulses from many types of receptor, but mainly from (i) chemoreceptors of the aortic glomus, (ii) stretch receptors in the large veins,[3] (iii) nociceptors subserving the cough reflex, and (iv) stretch receptors[20, 21] located in the lungs—the most important type insofar as respiration is concerned.

Receptors which fire when the lungs are inflated were first studied by Adrian.[1] Two types are distinguished on the basis of threshold and rate of adaptation (Fig. 22–4).[28, 29] Slowly adapting receptors fire with relatively slight degrees of lung distension and reflexly decrease activity of phrenic motoneurons. Rapidly adapting receptors respond only to forcible distension of the lungs, *i.e.*, lung volumes exceeding eupneic tidal volumes, and elicit a brief increase in phrenic motoneuron discharge.

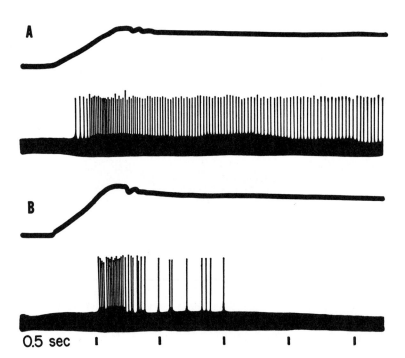

Figure 22–4 Responses of two types of afferent vagal fibers to inflation of lungs (chest wall removed). Upper trace in both records, intratracheal pressure; respiration pump stopped in expiration just before start of both records. Fibers responding as shown in *A* adapt slowly. Fibers responding as shown in *B* adapt rapidly. (After Knowlton and Larrabee, *Amer. J. Physiol.*, 1946, *147*, 100–114.)

Widdicombe[58] localized these receptors in the tracheobronchial tree and believed they are identical with the mechanoreceptors of the cough reflex. Lung stretch receptors are supplied by the larger vagal afferents having conduction velocities ranging from 14 to 59 m per sec. Deflation of the lungs elicits a marked increase in phrenic motoneuron discharge.[29] Knowlton and Larrabee[28] noted that inflation-sensitive receptors often fired also on lung deflation. However, Paintal[39] has identified units which respond to deflation but not to inflation. It is likely that these receptors are responsible for the increased inspiratory discharge when the lungs are deflated.

HERING-BREUER REFLEXES. In 1865, Hering and Breuer[20, 21] discovered that the stretch receptors detect inflation and deflation of the lungs and that the afferent discharge into the brain stem alters the respiratory cycle. Their findings were elaborated by Head in 1880. These investigators found that inflation of the lungs tends to terminate inspiration and that collapse of the lungs tends to initiate it. (The inflation-terminating receptor is the slowly adapting one mentioned above.) The conclusion was that these reflexes provide a self-regulatory mechanism or, in modern language, a regulatory feedback. The inspiratory-terminating reflex is supposedly operative in the range of eupneic breathing. However, the end organs for the inspiration excitatory reflex respond only to extreme deflation (either passive or active) or in deep respiration. Recent studies described below raise the possibility that these afferents do not control respiration cycle by cycle but rather provide a background of respiratory drive.

Cough Reflex.[57, 58] Coughing results from mechanical or chemical irritation of endings in the respiratory passages. Discharge in mechanoreceptors elicited by introducing a tube into the trachea inhibits phrenic motoneuron discharge and causes expiratory efforts and bronchiolar constriction. Mechanoreceptors may also be excited by abrupt volume changes in the isolated tracheobronchial system. The receptive zone most sensitive to mechanical stimuli is the inner surface of the larynx. The tracheal bifurcation and the lower half of the trachea are also sensitive, but the main bronchi are relatively insensitive.

Chemically induced coughing follows inhalation of sulfur dioxide. This gas is effective when introduced through an endobronchial catheter so that the gas comes into contact with only the lungs and smaller bronchi. A weaker cough is produced by perfusing sulfur dioxide through the isolated tracheobronchial system. The individuality of the mechanosensitive and chemosensitive cough reflex afferents has been established by single unit recording. The former adapt rapidly to volume changes and are found in greatest concentration near the larynx and carina. They respond readily to mechanical stimulation or inhalation of powders, but are relatively insensitive to sulfur dioxide. The chemosensitive receptors adapt more slowly to volume changes, are widely distributed in the tracheobronchial system, and respond readily to sulfur dioxide. Both mechanically and chemically induced coughing are decreased by vagotomy, but both vagotomy and sympathectomy are required to abolish the reflex.

CENTRAL NEURAL MECHANISMS

As has been mentioned above, respiration is affected by a wide variety of afferent stimuli, is subject to voluntary control and takes part in a variety of emotional expressions. Respiratory responses follow electrical stimulation of several levels of the central nervous system and several regions of the cerebral cortex.

Cerebral Cortex. Spencer[47] was the first to call attention to respiratory responses to stimulation of the presylvian area. Two general areas within this region have been mapped: (i) An *accelerator area* lies on the anterior sigmoid gyrus and immediately adjacent cortex of the medial surface of the hemisphere in the dog and cat; a comparable area in the monkey lies just rostral to the superior precentral gyrus. Portions of this area are rostral to the motor representation of the face, tongue, glottis, etc.; and stimulation of them gives rise to rhythmic licking, chewing and swallowing movements with salivation. Croaking, grunting and other forms of vocalization occur. (ii) An *inhibitory area* definable in the dog and cat is relatively large, including the gyrus compositus anterior and most of the cortex of the sylvian and ectosylvian gyri; in the cat the gyrus proreus is also included. In the monkey an inhibitory field is located just caudal to the lower end of the inferior precentral sulcus. Mastication is also obtained by stimulating this area. A second area producing acceleration of respiration coincides closely with motor area 2. Respiratory movements appear therefore to be localized in both cortical representations of the body musculature.

More recently, attention has been focused upon the respiratory effects of stimulating the *limbic* area of the cortex, *i.e.*, the region forming the hilus of the hemispheres and embracing the medial and orbital surfaces of the frontal lobe (Chap. 28, Vol. I). Mapping of responsive zones reveals that the cortex of the posterior orbital surface, the cingulate gyrus, the tip of the temporal pole, the anterior temporal operculum and the anterior insula form a continuous strip of cortex, giving rise to respiratory responses. The larger part of the "insular-orbital" area is inhibitory, as is the larger part of the cingulate gyrus. These findings have been confirmed in man by stimulation at the time of operation for prefrontal leukotomy.

Since alteration of autonomic activity may also be elicited by stimulation of the same general regions, they appear to subserve the autonomic and respiratory correlates of certain types of behavior. The close association with the olfactory area naturally recalls that sniffing and breathing are necessary for olfaction, and the nearness of the respiratory area on the convexity of the cortex to Broca's area suggests the integration of speech and breathing at a cortical level. The association of masticatory and swallowing movements with respiratory change on stimulation of area 6 points to a cortical integration of the various components of food-taking. If, as has been suggested, the orbital and other limbic areas are concerned with emotional behavior, it is logical that breathing, which is a component of emotional expression, should be altered by stimulation of these areas.

According to conventional neurology, one would expect to find respiratory representation in cortical area 4, the motor area from which the pyramidal tracts arise, because respiration is subject to voluntary acceleration or inhibition. This does not appear to be so. Although the trunk and thoracic musculature are represented in this area, there is no evidence of coordinated employment of these muscles in acts suggestive of respiration when their cortical representation is stimulated, and pyramidal section in the cat does not influence the respiratory responses to cortical stimulation. Therefore, respiration does not have a pyramidal control and the voluntary regulation of respiration is mediated by extra-pyramidal pathways.

Pons and Medulla. ELECTRICAL STIMULATION. The respiratory responses to stimulation of the pons and medulla are greater and more discrete than are the responses to cortical stimulation. When specific small regions are stimulated, the patterns of breathing change in a manner dependent upon the site of the stimulus. Changes in rate, inspiratory apnea (apneusis), apnea at normal end expiration and expiratory apnea are elicited by stimulation in different pontine and medullary areas.[38] Figure 22–5 shows the more or less discrete localization of the regions from which each of these effects may be produced. Earlier, less detailed work[9, 41] had yielded essentially similar results except for the localization of the regions causing expiratory apnea. It should be noted that stimulation of one side of the pons and medulla affects equally the respiratory muscles on both sides of the body. This finding, together with evidence from lesions,[44, 45] indicates extensive cross connections between the two sides of the brain stem and/or descent of impulses from one-half of the pons and medulla to musculature on both sides of the body.

SECTION AND ABLATION. The results of transection at various levels in the pons and medulla (Fig. 22–6) are quite different when the vagi are intact and when they are sectioned.

Transection with Vagi Intact. In an animal with the standard midcollicular decerebration, breathing is essentially indistinguishable from that in the intact animal. Thus, the influences of the cortical areas on respiration discussed above and of the temperature regulation centers in the hypothalamus (see Chap. 12, Vol. I) are not necessary for normal respiration. Transections in the pons do not markedly affect the pattern of breathing, although the rate is usually slowed. Transection in the rostral medulla,* however, may lead to marked changes in the breathing pattern,[6, 7, 24, 25, 56] which may be eupneic but is usually of a gasping type and may resemble Biot's breathing (brief periods of rapid breathing followed by pauses in expiration). Transection below a plane 2 mm caudal to the ros-

---

*The medullopontine junction is defined here (after Wang) as a plane running from the acoustic stria dorsally to the caudal border of the trapezoid body on the ventral surface.

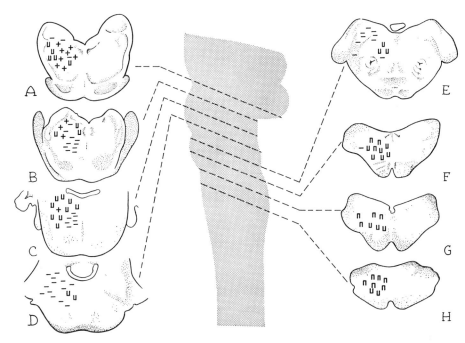

U =Inspiratory spasm   Π =Active expiration   − =Expiratory standstill
+ =Expiratory acceleration

Figure 22–5 Summary diagram showing the types of respiratory responses elicited by stimulating brain stem at various levels between the isthmus of the pons *(A)* and 2 mm caudal to the obex *(H)*. (After Ngai and Wang, *Amer. J. Physiol.*, 1957, *190*, 343–349.)

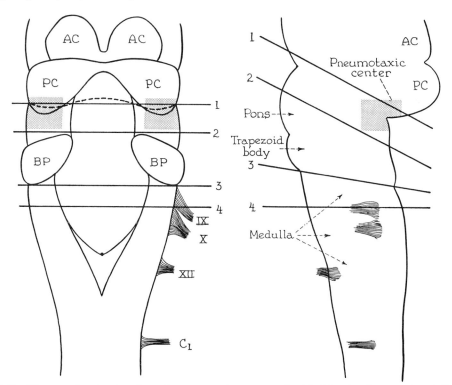

Figure 22–6 Diagram showing location of pneumotaxic center and critical levels of transection. *1*, Lower midbrain section; *2*, high pontine section; *3*, section at medullopontine junction as defined by Wang *et al.; 4*, section 2 mm caudal to rostral border. *AC*, Anterior colliculus; *PC*, posterior colliculus; *BP*, brachium pontis. (After Wang *et al.*, *Amer. J. Physiol.*, 1957, *190*, 333–342.)

tral border of the medulla causes cessation of respiration.

Transection After Section of Vagi. When the aforementioned series of sections is made in an animal with both vagi cut or blocked, the effect of the pontine transections is very different.[6, 7, 24, 25, 30–34, 40–42, 48, 49] Although transection at the rostral border of the pons again produces little effect, transection slightly below this level causes dramatic changes. Initially rhythmic respiration gives way to sustained inspiration, or apneusis, which may last for minutes. This inspiratory spasm is usually followed by *apneustic* breathing. In this type of breathing, the inspiratory effort is strong and sustained, lasting for seconds or even minutes, and expiration consists only of a brief relaxation of the sustained inspiration. More caudal transections in the pons usually decrease the duration of the inspiratory phase and increase the duration of the expiratory phase. The effect on the inspiratory phase is the greater, so the net result is an increase in the mean respiratory rate. Thus the more caudal the lesion, the more nearly normal the rate and the duration of the inspiratory and expiratory phases.

The results of transection through the rostral medulla are not appreciably different when the vagi are cut and when the vagi are intact, even though the tenth cranial nerve enters the medulla below the level of section.

Respiratory Centers. As a result of experiments depending on transection and other surgical techniques, a series of centers (actually subcenters) have been defined. The region in the anterior pons, destruction of which in combination with vagotomy leads to apneusis and apneustic respiration, is called the *pneumotaxic center*. From the transection experiments described above, this center clearly lies in the extreme rostral pons. Tang[52] has shown that bilateral ablation of a few cubic millimeters in the dorsolateral tegmentum of the pons leads to apneusis followed by apneustic breathing (Fig. 22–7). Unilateral lesions are ineffective. Subsequent workers[56] have found that somewhat larger bilateral lesions, placed slightly more medially, will also cause apneusis. All workers agree that midline lesions are ineffective. These differences in localization are very slight, a matter of 1 or 2 mm, and may be only apparent owing to damage beyond the visible sites of the lesions; or it may be necessary to remove only a portion of each pneumotaxic center to produce apneusis.

The *apneustic center* is defined as those regions of the caudal two-thirds of the pons

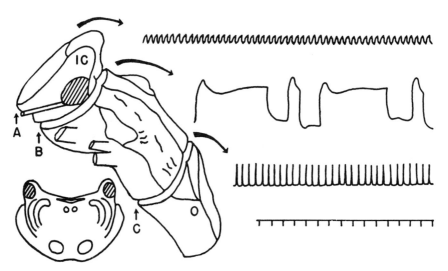

Figure 22–7 Diagram showing respiratory patterns of vagotomized cats after brain stem lesions. *A*, Midcollicular transection results in rhythmic breathing resembling normal pattern. *B*, Transection caudal to inferior colliculi results in hypertonic breathing pattern (apneustic breathing). Same pattern can be obtained by bilateral lesions of hatched area. Lower left diagram shows a cross section through critical hatched area ("pneumotaxic" center or area). *C*, Transection 3 to 5 mm above obex results in atonic breathing pattern (gasping). *IC*, Inferior colliculus; *O*, obex. Time in 10 sec intervals. (From Tang, *Amer. J. Physiol.*, 1953, *172*, 645–652.)

which, subsequent to section of the vagi and ablation of the pneumotaxic center, *support* apneusis and apneustic breathing. This center, as its name implies, provides an inspiratory drive. As more and more of it is eliminated, the inspiratory phase becomes shorter and shorter, or less apneustic. The apneustic center has not been localized to any particular pontine structure but apparently is part of the reticular facilitatory area.

The *medullary respiratory center* is often in error called simply "the respiratory center." Its exact localization and whether it is divided into an inspiratory and an expiratory center are matters of continuing investigation (see below).

EXTRACELLULAR RECORDING. The techniques of stimulation and ablation have certain limitations. With ablation it is difficult to know the exact limits of the destruction of nervous tissue; the question whether the effects are those of destruction of the cell bodies or of interruption of nerve pathways passing through the area is also a problem. With electrical stimulation, it is difficult to know whether the elements being stimulated are a "center" or fibers passing to or issuing from it; with unipolar electrodes, the spread of current is an additional problem. Also, it is not possible to know to what degree the response is abnormal, owing to the nearly synchronous volleys excited by electrical stimulation.

By recording with microelectrodes, one can determine the activity of cells or fibers subjected to only minimal damage (Chap. 1, Vol. I). The respiratory system is particularly suitable for study by this technique, since the firing of at least some cells involved in respiratory movements is periodic at the respiratory rate. However, considerable care is required to minimize the false periodic firing which is related to brain movements induced by breathing and which may cause the electrode to move with respect to the neurons. This source of confusion was not always recognized in early experiments. The criterion most useful in eliminating such an artifact is the constancy of spike amplitude during the respiratory cycle.

Electrical Activity of the Pons. Whether periodically firing cells exist in the pons has been a subject of controversy. However, at least in the decerebrate cat, it seems that there are pontine cells which fire trains of impulses with the same periodicity as respiration, although usually not in phase with either inspiration or expiration.[12, 51] At high CO_2 levels such cells have been recorded and seem to be important in determining the specific nature of the phasic responses of the medullary respiratory cells.[11, 12, 12a, 26, 56] The integrity of the pontine region is also essential for the integration of the Hering-Breuer reflex.[26]

Periodic firing in the pneumotaxic center has been more difficult to see but has been shown when a spike density histogram (averaged near a number of respiratory cycles) was used.[4, 12a]

Electrical Activity of the Medulla. The electrical activity related to respiration recorded from the medulla of decerebrate animals is more striking than that recorded from the pons.[17] Although a few active units may be found in other places, most of the activity is confined to the region near the level of the obex.[1, 18, 35, 37, 45] As shown in Figure 22-8, potentials from cell bodies which fire during inspiration are recorded for the most part slightly rostral to the obex, whereas potentials from cell bodies firing during expiration tend to occur slightly caudal to the obex. These neurons are not afferents of vagal origin, because the activity persists when the vagal inputs are blocked. These cells are mainly in or near *nucleus ambiguus* or *nucleus retroambigualis*. As will be seen by comparing Figures 22-5 and 22-8, the extent and location of the areas yielding electrical activity differ distinctly from those of areas yielding respiratory responses to stimulation. Impulses periodic with respiration which disappear when the vagi are blocked are also found in the medulla. These impulses are recorded mainly in the same general region as the inspiratory neurons, *i.e.*, somewhat rostral to the obex. This activity appears to originate in the region of the *tractus solitarius*. Periodic neuronal activity has also been recorded in the isolated medulla. This activity presumably reflects the inherent rhythmic ability of the medullary respiratory center.

Activity of fibers from the inspiratory and expiratory cells has been traced from the medulla to the level of the second cervical segment.[34] The information from ablation, stimulation and recording can be incor-

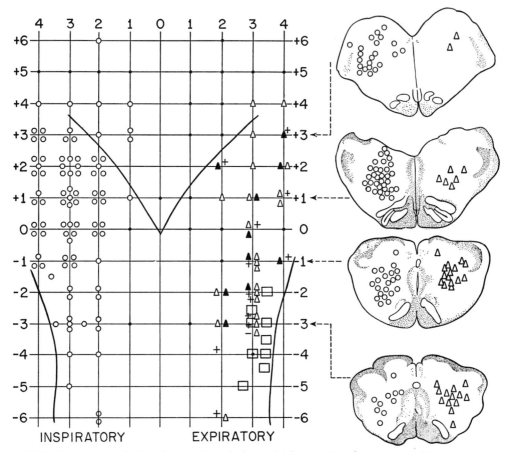

Figure 22–8 Points in medulla and cervical cord of cat which are active during respiration.

Left, Points plotted along stereotaxic coordinates, in millimeters, rostral (+) and caudal (−) to obex; although separated here for clarity, inspiratory and expiratory points are present on both sides of midline. ○, Inspiratory points; ▲, early expiratory points; △, late expiratory; +, expiratory of undefined timing; □, sites of activity appearing to be from fibers. (After Haber *et al., Amer. J. Physiol.,* 1957, *190,* 350–355; and Nelson, *J. Neurophysiol.,* 1959, *22,* 590–598.)

Right, Sections at levels indicated by arrows to stereotaxic grid. ○, Inspiratory points; △, expiratory points. (After Haber *et al., Amer. J. Physiol.,* 1957, *190,* 350–355.)

porated in a schematic organization, as shown in Figure 22–9.

Before continuing into the results of intracellular recordings, we will consider the properties of systems which are necessary for the generation of sustained oscillations.

GENESIS OF THE RESPIRATORY RHYTHM

The experiments described above establish that the medullary portion of the respiratory center is capable of rhythmic activity which roughly resembles normal respiration. Section of the vagus nerve has little effect on this rhythm, which, in fact,

endures in the absence of any neural input whatsoever (isolated medullary center). The initiation and maintenance of a rhythm can therefore be centrogenic.

Just above the medullary level is the apneustic center, which acts in conjunction with the medullary centers but which can exert an inspiratory drive that obscures rhythmic medullary activity in the absence of higher pontine centers and the vagal input. If the vagi are intact, rhythmic breathing is maintained. It may be inferred that the vagi operate through this apneustic center, although they enter the medulla farther down. As we shall see, at least two interpretations of the relation between the vagi and the apneustic center are possible:

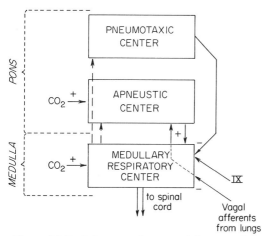

Figure 22-9 Schematic diagram of the central respiratory centers, showing possible interconnections and the points at which CO_2 and impulses from the lungs impinge; + augmentative and − inhibitory in inspiratory discharge.

(i) at each breath the vagal impulses periodic with inspiration inhibit the inspiratory drive of the apneustic center; and (ii) the vagal impulses balance the inspiratory drive of the apneustic center at the medullary respiratory center, and the periodic nature of the vagal input is of secondary importance.

That the effect of vagal impulses is not solely a function of a waxing and waning of their bombardment of the respiratory centers is shown by the following experiment. Apneusis following anterior pontine section may be interrupted and a normal type of rhythmic respiration reinstated by steady stimulation of the vagus.[27, 53] This is done physiologically by holding the lungs distended at a fixed volume,* thus invoking a continuous discharge from the stretch receptors in the lungs. Under this condition the respiratory motor response of the animal is essentially normal in rate and depth. When the steady vagal discharge is terminated by reducing the lung volume, apneusis ensues. Chemical respiratory drives are also involved as a continuous and antagonistic respiratory drive. The amount of vagal discharge necessary to interrupt

apneusis is determined by the pCO_2 of the blood; the higher the pCO_2, the greater the continuous vagal discharge necessary to reinstate rhythmic respiration. The pneumotaxic center, like the vagus, opposes the apneustic center, since, in the absence of vagal input, the pneumotaxic center maintains the normal respiration of the decerebrate preparation. The ability of this center to promote rhythmic respiration is therefore clearly not due to periodic vagal impulses. Lumsden,[30, 31] as his choice of the term "pneumotaxic center" indicates, ascribed the rhythmicity of breathing to this pontine area. Pitts *et al.*[40-42] went further and ascribed respiratory rhythm to a breath-by-breath discharge of the pneumotaxic center. This was visualized as follows. As the "inspiratory center" discharged caudally to the respiratory muscles, it also discharged rostrally to the pneumotaxic center, which in turn discharged caudally to curb the inspiratory center discharge. Even within the reticular substance it is difficult to account for the time lapse in this hypothetical circuit. Moreover, Hoff and Breckenridge[6, 7, 24, 25] showed that the respiratory rhythm could be maintained by the medullary respiratory center separated from both the vagi and the pneumotaxic center.

Increasingly, the tendency is to think that respiratory rhythm is derived from the interplay of descending impulses and from the interaction between neurons. This view can be explained on two levels: (i) by appeal to simple analogy and (ii) by appeal to a knowledge of the factors which make for oscillation in electronic or other control systems.*

Respiration Viewed as an Oscillatory System. The central neural mechanism controlling respiratory movements belongs to

---

*See Young, A. C. In: *Physiology and biophysics,* 19th ed., T. C. Ruch and H. D. Patton, eds. Philadelphia, W. B. Saunders Co., 1965, Chapter 41, Figure 10, page 797.

---

*Analogies to the respiratory system can be found in spinal reflex action, although the neural levels are different.[38] The divided spinal cord tends to yield alternate flexion and extension (stepping), and a favorable condition for bilateral stepping is the concurrent and equal stimulation of homologous nerves on both sides of the body. When the spinal cord is connected with the brain stem, oscillatory phenomena disappear and the system is biased toward extension (decerebrate rigidity). When still higher levels (upper midbrain and hypothalamus) are included in the system, flexor and extensor reflex excitability are again more nearly balanced and oscillating phenomena occur (*e.g.,* effective walking occurs).

a class of systems, including many mechanical and electrical systems, capable of oscillation. To understand the respiratory system, consider rather generally the properties of a system which lead to a rhythmic output, i.e., to oscillation. Since neural systems must be mechanistic, there *must* be a broad similarity between them and non-neural oscillatory systems. Mathematicians and control systems engineers conceive of oscillating systems as falling into the following three classes.[36, 50, 54]

TYPE I. LINEAR SYSTEMS. Consider a system with an input and an output. The system could be of any reasonable type—electrical, mechanical, chemical—or it could be a mixture of types. For example, the input could be mols of O_2 used per minute in an electrochemical reaction, and the output could be amperes of current flowing as a result of the chemical reaction. The system might be one in which the input and the output were of the same type, like the input and output voltage in an amplifier; or, the number of impulses per second in an afferent neural volley could be the input and the number of impulses per second in the resulting efferent discharge the output. If, in any of these systems, the output is directly proportional to the input, that system is called "linear." A large number of physical and chemical systems are linear or nearly so, and their properties have been very thoroughly studied and applied. For our purpose, a few properties of linear systems will be considered.

(i) Sinusoidal inputs always lead to sinusoidal outputs. In general, the outputs are not the same shape as the inputs for other shapes of waves.

(ii) To make an oscillator of a linear system it is necessary to "connect" the output (i.e., feed it back) to the input. The conditions for oscillation are: (a) the phase shift from input to output must be exactly zero or exactly an integral number of cycles and (b) the output must be greater than the input. If a frequency exists at which these conditions can be met, the system will oscillate with the sinusoidal wave at this frequency. Oscillators of this type are in wide general use.

(iii) If the response of a linear system is known for all frequencies of the input, the properties of the system are completely known, and the output for any form of input can be calculated.

TYPE II. NEARLY LINEAR SYSTEMS. The theory of linear systems may be extended to account for the properties of systems which are nearly linear. In these, the output may contain frequencies which are harmonics of the input signal; however, such harmonics become negligible as the input and output amplitudes of the system are reduced.

TYPE III. NONLINEAR SYSTEMS. A third type of system is currently being studied extensively by mathematicians and engineers. It is "nonlinear" because its output is not proportional to its input. Systems of this third type may have properties which cannot be approximated by linear and nearly linear systems. In general, nonlinear systems do not produce sinusoidal waves when used as oscillators, and the oscillations do not become sinusoidal as amplitude is reduced. Nonlinear systems may also produce subharmonics or fractional-order subharmonics; that is, if an input is of a particular frequency, the output may be at a frequency of one-third, one-half, two-thirds, etc., of the input frequency. Further, the properties of nonlinear systems cannot be completely determined from knowledge of their response to any one kind of input. There is no unique general method of studying these systems; each must be approached with a variety of analytic and experimental techniques. Some nonlinear systems do have properties which cannot be obtained in linear or nearly linear systems with the same number of elements. Nonlinear systems are used in counting and scaling circuits, in sweep circuits in oscilloscopes and television sets, and in pulse generators and coincidence counters.

System Analysis of Breathing. In the aforementioned terms, the type of system involved in the neural control of breathing can be determined with experimental techniques already familiar. The volume of the lungs may be controlled in any desired manner in order to control vagal input to the respiratory centers in the brain stem. The output may be measured as pressure produced by the respiratory muscles. Alternatively, nerve impulses in the phrenic nerves can be recorded electrically. With

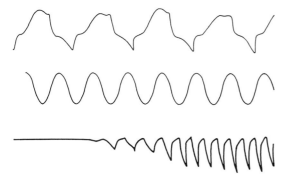

Figure 22-10 Patterns of pressure changes produced by inspiratory activity of respiratory muscles. *Top*, pressure changes produced by inspiratory muscles (*i.e.*, motor output) caused by *center*, sinusoidal variation of lung volume. *Bottom*, pressure changes during recovery from apnea while lung volume was fixed.

the input from the vagus held constant, the chemical input may be controlled by varying the CO_2 level in the blood. Neural drive and chemical effects can thus be measured separately or in combination.

When the animal is rendered apneic by hyperventilation, the output drops to zero, and the system does not oscillate until the CO_2 concentration increases. The pressure changes produced during recovery from apnea are shown in Figure 22-10. Clearly, the output is not sinusoidal even at the lowest amplitudes, and it thus appears that the respiratory system is nonlinear. A further check is obtained when the lung volume is changed sinusoidally and the resulting pressure output is recorded. The output, depending on the input frequency, may be either harmonically or subharmonically related to the input frequency; Figure 22-10 shows an example that is subharmonically related. We may therefore conclude that the respiratory system is nonlinear, having at least two nonoscillating positions, apnea and apneusis, and an oscil-

latory range between these limits in which normal eupneic breathing occurs.

NONLINEAR NEURAL MODEL. A possible neural model of a system which would oscillate is shown in Figure 22-11. This model is much oversimplified in that it shows only two neurons; the actual system would be composed of many connected neurons. A somewhat more realistic model is shown in the Appendix at the end of this chapter. The neurons in Figure 22-11 must have three properties. The first is similar to adaptation in sense organs. Thus, with a continuous bombardment by afferent impulses, the number of impulses from the neurons must at first be high and then gradually decline. Second, the number of efferent impulses per second at first must be actually higher than the number in the afferent volley (gain greater than unity). The third essential property is that at least one of the neurons must have a nonlinear relation between its output frequency and the frequency of the input to it. Both the first and second properties—"adaptation" and a gain greater than one—have been observed in interneurons in the spinal cord. The third condition is commonly encountered in the nervous system. When neurons with these properties are connected as shown in Figure 22-11, the system will give bursts of impulses, the interval of the bursts being determined by the "time constant" of the adaptation-like process and, to some extent, by the nonlinear properties of the cells. This situation can be simulated by an electric analogue and is, in fact, the well known multivibrator.

In such a system, if additional afferents synapse with the cells, a sufficiently strong continuous discharge will stop the oscillation of the system, in which case one of the cells will discharge continuously while the other cell is nearly quiescent. In the analogue, the corresponding effect is observ-

Figure 22-11 Diagram of a simple two-neuron, oscillatory system giving properties described in text.

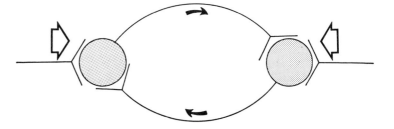

able if the individual tubes in the multi-vibrator are biased with an impressed voltage. In general, systems of this nature have two or more stable positions and an intervening region of continuous oscillation. They also have the general properties listed in the discussion of nonlinear systems. These oscillating pairs of neurons could be in the medullary respiratory area and conform to the cells that fire during inspiration and expiration.

A more complete model would, of course, have to take into account interconnections between individual pairs of cells. Such interconnections are required to lock the oscillator together in order to gain a synchronous discharge like that occurring in breathing. The more complete model would also have to take into account additional interconnections between the neurons via pathways in the pons (apneustic and pneumotaxic center). These centers may act primarily as biases though at times having electrical activity corresponding to respiration (see Appendix at the end of this chapter).[12, 26, 51]

INTRACELLULAR RECORDING—MEDULLA. One critical test for the hypothesis that the periodicity of the respiratory center is due to the interconnection of reciprocally connected groups of neurons should be the presence of EPSP's and IPSP's causing the firing of these cells (Chap. 7, Vol. I). Such intercellular potentials would not, of course, be present in the firing of cells such as those of the sino-atrial node (Chap. 5, Vol. II). Salmoiraghi and von Baumgarten[43] were successful in obtaining such records from *nucleus tractus solitarii*. More extensive recordings have been made by Hildebrandt in or near the region of *nucleus ambiguus* (see above).

The existence of cells with an adapting response is crucial to the generation of periodic activity. An adapting response can be shown in two ways. The first is by observing the firing rate of the neuron while monitoring its normal excitatory or inhibitory input. An example of responses showing adaptation is shown in Figures 22–12 and 22–13. The other cells from the same region do not show adaption. The second

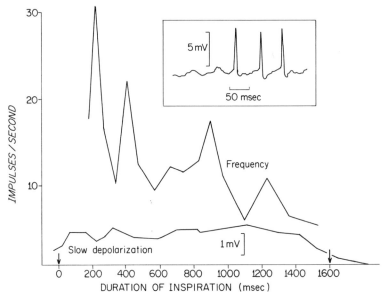

Figure 22–12 Frequency, slow depolarization and action potentials from an inspiratory recurrent laryngeal motoneuron with an adapting discharge pattern.

The small (6 mV) AB spike in this record (inset) was superimposed on a slow depolarization on which there was a great deal of synaptic noise. The irregularities seen in the hyperpolarizing potential following the AB spike may have been due to this large amplitude synaptic noise. Because the AB spike was so small and the synaptic noise relatively large, it seems reasonable to conclude that the electrode was situated in the soma or dendrite, far from the trigger zone of the cell.

This cell was antidromically fired with a latency of 1.5 msec by stimulation of the ipsilateral vagus. Following the antidromic spike, the discharge of the cell was inhibited. (From Hildebrandt, *An analysis of intracellular potentials from respiratory neurons,* Ph.D. Thesis, University of Washington, 1966.)

Figure 22-13 Frequency, slow depolarization and intracellular potentials from an expiratory neuron with an adapting discharge pattern.

It was not possible to determine for certain whether or not the 20 mV action potential of this cell had an inflection on the rising phase. Since no action potentials were evoked after 1 sec even though the slow depolarization remained quite large, the cell was considered to show adaptation.

No attempt was made to antidromically fire this neuron by vagal stimulation. (From Hildebrandt, *An analysis of intracellular potentials from respiratory neurons*, Ph.D. Thesis, University of Washington, 1966.)

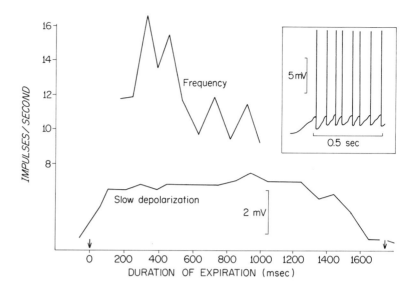

method is to artificially activate or deactivate the cell by pulses of current through the recording electrode and observe the resulting change in firing rate. The response of cells to a depolarizing current is shown in Figure 22–14. The response to removal of a hyperpolarizing current is shown in Figure 22–15. The existence (in the medulla) of adapting cells which fire in synchrony with respiration is definitely proved. The role of such cells in the genesis of the respiratory rhythm is covered above and in the Appendix at the end of this chapter.

In relatively complex systems of this nature it is not usually possible to speak of the oscillation as resulting from connections between specific elements, but it may be a property of the interconnections of the relatively large number of cells. For this reason it is rather dangerous to think of the rhythmicity of such a system as residing in a particular region of the brain stem merely because this part is capable of oscillation when isolated from other parts. Possibly, a large number of other groups of cells would also oscillate if they could be isolated surgically.

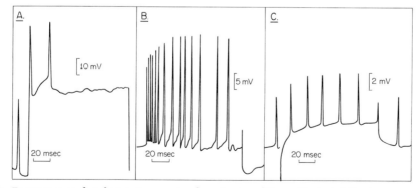

Figure 22–14 Responses to depolarizing current pulses. In *A* and *B* are illustrated the two types of responses to a constant depolarizing current pulse which are classified as adapting. Many cells responded with only one or two spikes as did the inspiratory recurrent laryngeal motoneuron in *A*. Other cells fired for the duration of the pulse, but at a progressively decreasing frequency. Such a response from an inspiratory neuron is illustrated in *B*.

The nonadapting response is illustrated in *C*. In this expiratory cell, the frequency is unchanged during the depolarizing current pulse and there was no rapid initial frequency decline as seen in *A* and *B*. (From Hildebrandt, *An analysis of intracellular potentials from respiratory neurons*, Ph.D. Thesis, University of Washington, 1966.)

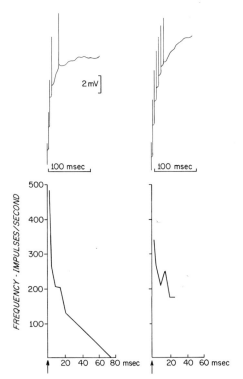

Figure 22–15 Response of an expiratory neuron to off pulses.

A hyperpolarizing current was passed through this expiratory neuron. At the end of the hyperpolarizing pulse, the neuron responded with a short, adapting discharge. In the upper illustration on the left is a tracing of the response occurring during the active phase of the cell (expiration). On the right is the response during inspiration. Below the tracings are the graphs of frequency as a function of time following the end of the hyperpolarizing pulse. The frequency attained during expiration was higher than that reached during inspiration, presumably because the cell was receiving excitatory input during expiration and was thus depolarized (From Hildebrandt, *An analysis of intracellular potentials from respiratory neurons,* Ph.D. Thesis, University of Washington, 1966.)

SPINAL NEURONS

While the medulla and pons are basic to the respiratory activity, the role of the spinal cord neurons in performing the final integration of the respiratory system is not minor. The activity of the thoracic stretch receptors is to modify the discharge rate of the spinal motor neurons.

By recording with intracellular electrodes, Sears showed the periodic nature of the synaptic input to the respiratory motor nuclei and the reciprocal nature of this input to inspiratory and expiratory motor nuclei.[46] It was also found that the gamma efferents to the muscle spindles were ac-

tivated in synchrony with the alpha drive to the corresponding respiratory intercostal muscles. This strong "alpha-gamma" linkage is of importance in maintaining the response of the muscle during the breathing cycle. Cutting the dorsal roots causes a considerable decrease in the respiratory movements.[55]

The respiratory activity resulting from stimulations of fibers in the pyramidal decussation or from circumscribed regions of the contralateral sensorimotor cortex is carried in tracts in the dorsal quadrants of cord, whereas the tracts from the "respiratory center" are in the ventrolateral quadrant. A model for the manner in which polysegmental reflexes are integrated with supraspinal inputs has been proposed by Aminoff and Sears.[2] The monosynaptic excitatory reflexes are not shown in this model (Fig. 22–16).

Intercostal to Phrenic Reflexes. The diaphragm contains relatively few proprio-

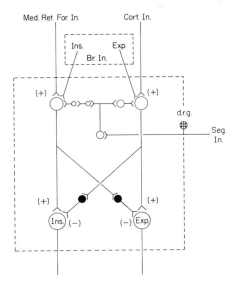

Figure 22–16 Schematic representation of interneuronal pathways transmitting reciprocal excitatory and inhibitory actions to respiratory motoneurons. Excitatory interneurons, open circles; inhibitory interneurons, filled circles. Ins., Inspiratory motoneurons; Exp., expiratory motoneurons; d.r.g., dorsal root ganglion; Med. Ret. For. In., medial reticular formation input; Cort. In., cortical input; Seg. In., segmental input; Br. In., breathing input (Ins., inspiratory; Exp., expiratory). Large square (dashed lines) depicts segmental interneuronal network; small rectangle, source of breathing input. A single excitatory interneuron in the diagram may represent a chain of interneurons. (From Aminoff and Sears, *J. Physiol. (Lond.),* 1971, *215,* 557–575.)

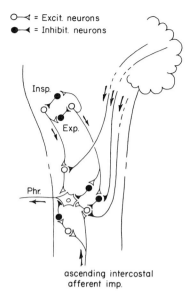

○━◁ = Excit. neurons
●━◁ = Inhibit. neurons

Insp.

Exp.

Phr.

ascending intercostal
afferent imp.

Figure 22–17 A schematic representation of some features of the synaptic arrangement of the phrenic motoneurons which, in an oversimplified way, would explain some of the results from cerebellar stimulation. In this hypothetical diagram inhibitory functions have only been represented by postsynaptic mechanisms. Insp. and Exp. represent the reciprocally occurring inspiratory and expiratory activity descending from the respiratory mechanisms in medulla. The pathways from cerebellum are broken to indicate that they might be interrupted by brain stem relays. (From Decima and von Euler, *Acta physiol. scand.*, 1969, 76, 148–158.)

ceptors and hence is not subject to autogenetic facilitation. However, stimulation of either internal or external intercostal nerves to the lowest intercostal segments gives rise to polysynaptic reflex excitation of phasic motor neurons.[16] The receptive area is limited to the region of insertion of the diaphragm. In contrast to intercostal reflexes, stimulation of cutaneous afferents does not excite the diaphragm.

A model of the interaction of the intercostal-phrenic reflex and cerebellar input to the phrenic is shown in Figure 22–17.

APPENDIX

A possible arrangement of neurons which could lead to sustained oscillations is shown in Figure 22–18. Each neuron in the diagram represents one of a group which are interconnected to other groups to maintain synchronization between groups.

Properties of the Cells. Cells (A_1) and (A_3) are adapting cells. Consider the excita-

tory state, E, of these cells—possibly a function of the membrane potential in the trigger zone. The zero of E is chosen so that when E is positive it is measured by the firing rate of the cell. The magnitude of E negative would be determined by the extra excitatory input needed to just fire the cell. The firing rate J is related to E as shown in Figure 22–19A. The equation of response of these cells is $T\dot{E} + E = TG\dot{I}$ where I is the input to the cells—positive if excitatory, negative if inhibitory and $\dot{E} = \dfrac{dE}{dt}$, etc. The response of such an idealized type "A" cell to square wave pulse inputs positive and negative is shown in Figure 22–19B. These should be compared with Hildebrandt's results as shown in Figures 22–12 through 22–15. Type B cells have an upper limit to their firing rate as shown in Figure 22–19C. The type "C" interneurons are interposed to give an inhibitory output in accord with the general hypothesis that all efferents from a particular cell are either excitatory or inhibitory.

The equations for the system of Figure 22–18 are then, omitting dashed line inputs (I = inputs to cells, J = outputs in firing rates, and E = excitatory state):

$$T_1\dot{E}_1 + E_1 = T_1G_1\dot{I}_1 = T_1\,G_1\,(\dot{J}_2 - \dot{J}_6) \quad (1)$$

$$J_1 = D(E_1) \quad (2)$$

$$J_2 = B\,(I_2) \quad (3)$$

$$T_3\dot{E}_3 + E_3 = T_3\,G_3\,\dot{I}_3 = T_3G_3\,(\dot{J}_4 - \dot{J}_5) \quad (4)$$

$$J_3 = D\,(E_3) \quad (5)$$

$$J_4 = B\,(I_4) \quad (6)$$

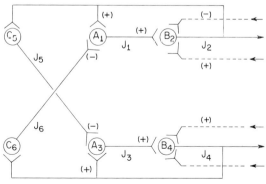

Figure 22–18 A possible arrangement of neurons which could lead to sustained oscillations. Properties of the neurons are given in the text, equations (1–10), and in Figure 22–19.

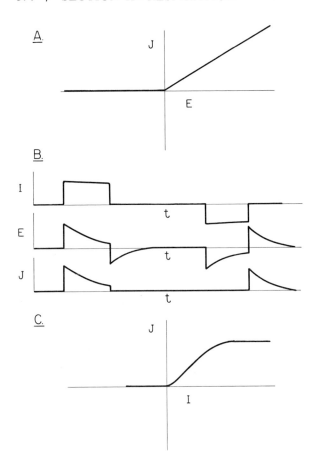

Figure 22–19 *A*, Firing rate of type A cells as a function of their excitatory state. $J = D(E)$. *B*, Response E and J of type A cells to square wave input I. *C*, Output J of B type cells in response to input I. $J = B(I)$.

$$J_6 = I_6 = J_4 \qquad (7)$$

$$I_2 = J_1 \qquad (8)$$

$$J_5 = I_5 = J_2 \qquad (9)$$

$$I_4 = J_3 \qquad (10)$$

For simplicity of solutions we take
$$T_1 = T_3 = T$$
$$G_1 = G_3 = G.$$
Equations (1) and (4) become:

$$T\dot{E}_1 + E_1 = T \, G(\dot{J}_2 - \dot{J}_4) \qquad (11)$$

$$T\dot{E}_3 + E_3 = T \, G(\dot{J}_4 - \dot{J}_2) \qquad (12)$$

Adding (11) and (12):

$$T (\dot{E}_1 + \dot{E}_3) + (E_1 + E_3) = 0 \qquad (13)$$

The solution of this equation is $(E_1 + E_3) = (E_1 + E_3)_0 \; \ell^{-t/T}$, *i.e.*, $(E_1 + E_3)$ approaches zero with time constant T. We may then take $E_1 + E_3 = 0$ or $E_1 = -E_3$ to solve the equation. We now define $X_1 = \int E_1 dt$, *i.e.*, $\dot{X}_1 = E_1$ and $E_3 = -\dot{X}_1$.
On integration of equation (11) above

$$T_1 \, E_1 + E_1 dt = TG(J_2 - J_4) =$$
$$TG[B(I_2) - B(I_4)] =$$
$$TG \, [B(D(E_1)) - B(D(-E_1))]$$
$$= TG \, [B(D(\dot{X}_1)) - B(D(-\dot{X}_1))] \qquad (14)$$

or

$$X_1 = T \, G[B(D(\dot{X}_1)) - B \, (D(-\dot{X}_1))] - T \, \dot{X}_1 \qquad (15)$$

This equation is plotted in Figure 22–20*D*. On a phase diagram such as Figure 22–20*D* it is apparent that for any value of $\dot{X}_1$ there is a unique value of X_1. However, for a specific value of X_1 more than one value of $\dot{X}_1$ may be possible. A determination of a unique value for $\dot{X}_1$ for each X_1 will make possible the determination of X_1 or X as a function of time. First it is apparent that the "direction" of motion along the curve in the diagram is easily determined, because if $\dot{X}$ is positive X must be increasing and if $\dot{X}$ is negative X must be decreasing. There are only three positions on the graph which require special consideration. The first of these at the origin has X=0 and $\dot{X}$=0. This is a position of unstable equilibrium, because any small disturbance will cause $\dot{X}$ to become either positive or negative and the pathway will then be along OA or OB respectively. It is easily seen that no matter what the initial state of the system, the system will first proceed to either position A or B. If the position is A it follows that X_1 has a specific value, but $\dot{X}_1$ is not zero, so the system cannot stay at A. For the specific value X_{1A} the system must be at A or A'. It follows that the system must "jump" from A to A' (note X_1 cannot instantaneously change value, because this would require an infinite value of $\dot{X}_1$, and this in turn would require an infinite input which is impossible in the system where B type cells have a maximum value which is not infinite). The system then transverses the pathway A→A'→B→B'→A→A', *i.e.*, the system oscillates. To plot X_1 as a function of time it is only necessary to start at a particular point and note that $t = \int dt = \int \frac{dt}{dX_1} \cdot dX_1 = \int \frac{dX_1}{\dot{X}_1}$.

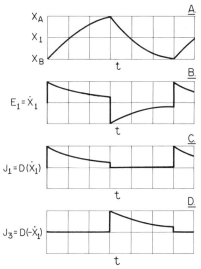

Figure 22–21 A, Plots of X_1, E_1, J_1 and J_3 vs. time. A is obtained by plotting $t = \int \frac{dX_1}{\dot{X}_1}$ where $\frac{1}{\dot{X}_1}$ is determined as a function of X_1 from Figure 22–20D. B, E_1 is obtained directly as a derivative of X_1. C, D, J_1 and J_3 are obtained from B and Figure 22–19A and C.

The final results of J_1 and J_3 as a function of time are shown in Figure 22–21.

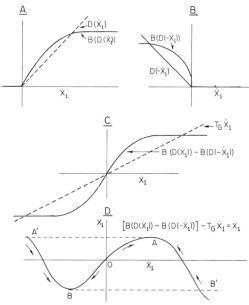

Figure 22–20 A, $D(\dot{X}_1)$ and $B(D(\dot{X}_1))$ as a function of $\dot{X}_1$ from Figure 22–19A and C. B, $D(-\dot{X}_1)$ and $B(D(-\dot{X}_1))$ as a function of $\dot{X}_1$ from Figure 22–19A and C. C, $B(D(\dot{X}_1)) - B(D(-\dot{X}_1))$ from A and B, above. D, Resulting phase diagram of X_1 vs. $\dot{X}_1$ from C, above. This is the same as equation (15).

REFERENCES

1. Adrian, E. D.   Afferent impulses in the vagus and their effect on respiration. *J. Physiol. (Lond.),* 1933, *79,* 332–358.
2. Aminoff, M. J., and Sears, T. A.   Spinal integration of segmental, cortical and breathing inputs to thoracic respiratory motoneurones. *J. Physiol. (Lond.),* 1971, *215,* 557–575.
3. Aviado, D. M., Jr., Li, T. H., Kalow, W., Schmidt, C. F., Turnbull, G. L., Peskin, G. W., Hess, M. E., and Weiss, A. J.   Respiratory and circulatory reflexes from the perfused heart and pulmonary circulation of the dog. *Amer. J. Physiol.,* 1951, *165,* 261–277.
4. Bertrand, F., and Hugelin, A.   Respiratory synchronizing function of nucleus parabrachialis medialis: Pneumotaxic   mechanisms. *J. Neurophysiol.,* 1971, *34,* 189–207.
5. Biscoe, T. J., and Sampson, S. R.   Responses of cells in the brain stem of the cat to stimulation of the sinus, glossopharyngeal, aortic and superior laryngeal nerves. *J. Physiol. (Lond.),* 1970, *209,* 359–373.
6. Breckenridge, C. G., and Hoff, H. E.   Pontine and medullary regulation of respiration in the cat. *Amer. J. Physiol.,* 1950, *160,* 385–394.
7. Breckenridge, C. G., Hoff, H. E., and Smith, H. T.   Effect on respiration in midpontine animal of chemical inhibition of facilitatory system. *Amer. J. Physiol.,* 1950, *162,* 74–79.
8. Bronk, D. W., and Ferguson, L. K.   The nervous

control of intercostal respiration. *Amer. J. Physiol.*, 1935, *110*, 700–707.

9. Brookhart, J. M. The respiratory effects of localized faradic stimulation of the medulla oblongata. *Amer. J. Physiol.*, 1940, *129*, 709–723.

10. Campbell, E. J. M. *The respiratory muscles and the mechanics of breathing.* Chicago, Year Book Medical Publishers, Inc., 1958.

11. Cohen, M. I. Discharge patterns of brain-stem respiratory neurons in relation to carbon dioxide tension. *J. Neurophysiol.*, 1968, *31*, 142–165.

11a. Cohen, M. I. Synchronization of discharge, spontaneous and evoked, between inspiratory neurons. *Acta neurobiol. exp.*, 1973, *33*, 189–218.

12. Cohen, M. I., and Wang, S. C. Respiratory neuronal activity in pons of cat. *J. Neurophysiol.*, 1959, *22*, 33–50.

13. Comroe, J. H. The location and function of the chemoreceptors of the aorta. *Amer. J. Physiol.*, 1939, *127*, 176–191.

14. Comroe, J. H., and Schmidt, C. F. The part played by reflexes from the carotid body in the chemical regulation of respiration in the dog. *Amer. J. Physiol.*, 1938, *121*, 75–97.

15. Decima, E. E., and von Euler, C. Intercostal and cerebellar influences on efferent phrenic activity in the decerebrate cat. *Acta physiol. scand.*, 1969, *76*, 148–158.

16. Decima, E. E., von Euler, C., and Thoden, U. Intercostal-to-phrenic reflexes in the spinal cat. *Acta physiol. scand.*, 1969, *75*, 568–579.

17. Gesell, R., Bricker, J., and Magee, C. Structural and functional organization of the central mechanism controlling breathing. *Amer. J. Physiol.*, 1936, *117*, 423–452.

18. Haber, E., Kohn, K. W., Ngai, S. H., Holaday, D. A., and Wang, S. C. Localization of spontaneous respiratory neuronal activities in the medulla oblongata of the cat: A new location of the expiratory center. *Amer. J. Physiol.*, 1957, *190*, 350–355.

19. Harris, A. S. Inspiratory tonus in anoxia. *Amer. J. Physiol.*, 1945, *143*, 140–147.

20. Hering, E., and Breuer, J. Die Selbststeuerung der Athmung durch den Nervus vagus. *S. B. Akad. Wiss. Wien*, 1868, *57 (2)*, 909–937.

21. Hering, E., and Breurer, J. Die Selbststeuerung der Athmung durch den Nervus vagus. *S. B. Akad. Wiss. Wien*, 1868, *58 (2)*, 909–937.

22. Hess, W. R. Die Rolle des Vagus in der Selbststeuerung der Atmung. *Pflügers Arch. ges. Physiol.*, 1936, *237*, 24–39.

23. Hess, W. R., and Wyss, O. A. M. Die Analyse der physikalischen Atmungsregulierung an Hand der Aktionsstrombilder des Phrenicus. *Pflügers Arch. ges. Physiol.*, 1936, *237*, 761–770.

23a. Hildebrandt, J. An analysis of intracellular potentials from respiratory neurons. Ph.D. Thesis, University of Washington, 1966.

24. Hoff, H. E., and Breckenridge, C. G. The medullary origin of respiratory periodicity in the dog. *Amer. J. Physiol.*, 1949, *158*, 157–172.

25. Hoff, H. E., Breckenridge, C. G., and Cunningham, J. E. Adrenaline apnea in medullary animal. *Amer. J. Physiol.*, 1950, *160*, 485–489.

26. Kahn, N., and Wang, S. C. Electrophysiologic basis for pontine apneustic center and its role in integration of the Hering-Breuer reflex. *J. Neurophysiol.*, 1967, *30*, 301–318.

27. Kerr, D. I. B., Dunlop, C. W., Best, E. D., and Mullner, J. A. Modification of apneusis by afferent vagal stimulation. *Amer. J. Physiol.*, 1954, *176*, 508–512.

28. Knowlton, G. C., and Larrabee, M. C. A unitary analysis of pulmonary volume receptors. *Amer. J. Physiol.*, 1946, *147*, 100–114.

29. Larrabee, M. G., and Knowlton, G. C. Excitation and inhibition of phrenic motoneurones by inflation of the lungs. *Amer. J. Physiol.*, 1946, *147*, 90–99.

30. Lumsden, T. Observations on the respiratory centres in the cat. *J. Physiol. (Lond.)*, 1923, *57*, 153–160.

31. Lumsden, T. Observations on the respiratory centres. *J. Physiol. (Lond.)*, 1923, *57*, 354–367.

32. Marckwald, M. Die Athembewegungen und deren Innervation beim Kaninchen. *Z. Biol.*, 1887, *23*, 149–283.

33. Marckwald, M. *The movements of respiration and their innervation in the rabbit.* Trans. by Haig, T. A. London, Blackie & Son, 1888.

34. Marckwald, M., and Kronecker, H. Ueber die Auslosung der Athembewegungen. *Arch. Anat. Physiol., Physiol. Abt.*, 1880, 441–446.

35. Merrill, E. G. The lateral respiratory neurones of the medulla: Their associations with nucleus ambiguus, nucleus retroambigualis, the spinal accessory nucleus and the spinal cord. *Brain Res.*, 1970, *24*, 11–28.

36. Minorsky, N. *Introduction to non-linear mechanics.* Ann Arbor, Michigan, J. W. Edwards, 1947.

37. Nelson, J. R. Single unit activity in medullary respiratory centers of cat. *J. Neurophysiol.*, 1959, *22*, 590–598.

38. Ngai, S. H., and Wang, S. C. Organization of central respiratory mechanisms in the brain stem of the cat: Localization by stimulation and destruction. *Amer. J. Physiol.*, 1957, *190*, 343–349.

39. Paintal, A. S. The conduction velocities of respiratory and cardiovascular afferent fibres in the vagus nerve. *J. Physiol. (Lond.)*, 1953, *121*, 341–359.

40. Pitts, R. F., Magoun, H. W., and Ranson, S. W. Localization of the medullary respiratory centers in the cat. *Amer. J. Physiol.*, 1939, *126*, 673–688.

41. Pitts, R. F., Magoun, H. W., and Ranson, S. W. Interrelations of the respiratory centers in the cat. *Amer. J. Physiol.*, 1939, *126*, 689–701.

42. Pitts, R. F., Magoun, H. W., and Ranson, S. W. The origin of respiratory rhythmicity. *Amer. J. Physiol.*, 1939, *127*, 654–670.

43. Salmoiraghi, G. C., and von Baumgarten, R. Intracellular potentials from respiratory neurones in brain-stem of cat and mechanism of rhythmic respiration. *J. Neurophysiol.*, 1961, *24*, 203–219.

44. Schiff, J. M. *Lehrbuch der Physiologie des Menschen.* Lahr, Schaufenburg, 1858–59.

45. Schiff, J. M. Moritz Schiff's *Gesammelte Beitrage*

zur Physiologie, Vol. I. Lausanne, B. Benda, 1894.

46. Sears, T. A. The slow potentials of thoracic respiratory motoneurons and their relation to breathing. *J. Physiol. (Lond.)*, 1964, *175*, 404–424.

47. Spencer, W. G. The effect produced upon respiration by faradic excitation of the cerebrum in the monkey, dog, cat, and rabbit. *Phil. Trans.*, 1894, *B185*, 609–659.

48. Stella, G. The dependence of the activity of the "apneustic centre" on the carbon dioxide of the arterial blood. *J. Physiol. (Lond.)*, 1938, *93*, 263–275.

49. Stella, G. The reflex response of the "apneustic" centre to stimulation of the chemoreceptors of the carotid sinus. *J. Physiol. (Lond.)*, 1939, *95*, 365–372.

50. Stoker, J. J. *Nonlinear vibrations in mechanical and electrical systems.* New York, Interscience Publishers, 1950.

51. Takagi, K., and Nakayama, T. Respiratory discharge of the pons. *Science*, 1958, *128*, 1206.

52. Tang, P. C. Localization of the pneumotaxic center in the cat. *J. Physiol. (Lond.)*, 1953, *172*, 645–652.

53. Tang, P. C., and Young, A. C. Interrelations of CO_2, O_2 and vagal influences on respiratory centers. *Fed. Proc.*, 1956, *15*, 184.

54. Truxal, J. G. *Automatic feedback control system synthesis.* New York, McGraw-Hill, 1955.

55. Von Euler, U. S. *Excitatory synaptic mechanisms*, P. Anderson and J. K. S. Jansen, eds. Olso, Universitetsforlaget, 1970.

56. Wang, S. C., Ngai, S. H., and Frumin, M. J. Organization of central respiratory mechanisms in the brain stem of the cat: Genesis of normal respiratory rhythmicity. *Amer. J. Physiol.*, 1957, *190*, 333–342.

57. Widdicombe, J. G. Respiratory reflexes from the trachea and bronchi of the cat. *J. Physiol. (Lond.)*, 1954, *123*, 55–70.

58. Widdicombe, J. G. Receptors in the trachea and bronchi of the cat. *J. Physiol. (Lond.)*, 1954, *123*, 71–104.

59. Woldring, S., and Dirken, M. N. J. Site and extension of bulbar respiratory centre. *J. Neurophysiol.*, 1951, *14*, 227–242.

CHAPTER 23 THE CHEMICAL REGULATION OF VENTILATION

by THOMAS F. HORNBEIN *and* SØREN C. SØRENSEN

Ventilation, the movement of gas into and out of the lungs, serves the primary functions in mammals of gas exchange and heat exchange as well as such secondary functions as coughing, yawning, sighing and talking. In man the major function is gas exchange, delivery of oxygen and removal of carbon dioxide. In furry animals ventilation is also of great importance for heat dissipation. Rapid, shallow breathing (panting) dissipates heat mainly by ventilating the anatomical dead space with minimal influence on homeostasis of alveolar Po$_2$ and Pco$_2$. Regulation and integration of these functions requires a variety of receptors, mechanical and chemical, as well as a center to convert this information into an appropriate ventilatory stimulus. In this chapter we shall describe the way in which the physiological chemical stimuli—hypoxia (decreased Po$_2$), hypercapnia (increased Pco$_2$) and hydrogen ion concentration ([H$^+$])—modify ventilation to maintain adequate gas exchange and a relatively constant [H$^+$] in blood and brain.* Several recent reviews detail these facets of ventilatory control.[5, 44, 65]

Oxygenation is the primary purpose of gas exchange. The secondary importance of CO$_2$ regulation is graphically illustrated by the changes in Pco$_2$ associated with the phylogenetic emergence of life from the sea (Fig. 23–1). Water breathers, inhabiting a medium where O$_2$ is approximately 24 times less soluble than CO$_2$, possess arterial Pco$_2$'s of 1 to 3 mm Hg at the levels of gill ventilation needed to provide an adequate arterial Po$_2$. In contrast the air breather, who exchanges CO$_2$ for O$_2$ on a one-for-one basis, requires only about one-twentieth as much ventilation to maintain oxygenation and, as a result, possesses a Pco$_2$ about 20 times higher than the water breather. In this scheme oxygenation determines ventilation, which then defines Paco$_2$. The kidney serves the secondary function of adjusting [HCO$_3^-$] to regulate the body [H$^+$] (see Chap. 25, Vol. II). The secondary importance of the control of Pco$_2$ to that of oxygenation can also be seen in the adjustments of man to chronic hypoxia, to be discussed later in this chapter. Why Paco$_2$ is 40 mm Hg instead of 20 or 80 mm Hg in man at sea level is not clear; in part this

---

*If necessary, the reader should refer to Chapter 27, Volume II, to refresh his memory concerning the relationship between [H$^+$], Pco$_2$ and [HCO$_3^-$] before treading further into the depths of this dissertation.

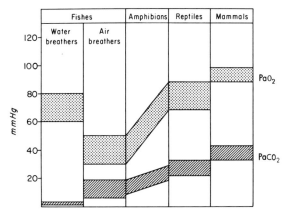

Figure 23–1 Schematic representation of changes in blood gas values during the transition from water breathing to air breathing in vertebrates. (After Lenfant *et al.*, *Fed. Proc.*, 1970, 29, 1124–1129.)

The greater effect of CO_2 than metabolic acids on breathing (Fig. 23–2) might be explained by the observation that gases can diffuse more rapidly than ions through biological membranes. Thus, blood-borne CO_2 could easily penetrate the blood-brain barrier to reach receptors in the brain, altering $[H^+]$ locally, whereas metabolic acids might be confined largely to the blood. The ventilatory response to metabolic H^+ indicates the existence of chemosensitive sites responding to changes in blood, whereas the greater response to CO_2 implies the existence of additional sites less easily accessible to free H^+ in the blood. According to the H^+ theory the ventilatory response to CO_2 or metabolic acids can be characterized by the expression:

$$\Delta \dot{V} = k_1 \, [H^+]_a + k_2 \, [H^+]_x$$

setting may reflect a balance between oxygen needs and the energy cost or effort of breathing.

Although the primary goal of gas exchange is oxygenation, the mammalian organism has developed mechanisms which regulate ventilation not only in response to change in Po_2, but also in response to changes in Pco_2 and $[H^+]$ (Fig. 23–2). The nature of these responses in man was portrayed by the classic telephone-booth studies of J. S. Haldane and Lorraine Smith, reported in 1892.[23] A person was "allowed to remain in a closed chamber until the air became very impure." As carbon dioxide accumulated in the chamber and oxygen content fell, ventilation increased markedly. Repeating the experiment with a carbon dioxide absorber in the chamber, they observed that breathing was unaffected until the oxygen content of the air had dropped from 21 per cent to below 15.5 per cent. They concluded that "excessive hyperpnea" in the first situation was "due almost entirely to excessive carbonic acid and not to deficiency of oxygen," although a sufficient fall in oxygen content "may also cause some hyperpnea." The essence of these simple, significant studies is shown in Figure 23–2, along with the ventilatory response to metabolic pH changes in the blood and to exercise.

In 1911 Hans Winterstein proposed that H^+ is the unique chemical stimulus to ventilation.[67, 68] According to this scheme, carbon dioxide would act by combining with water to form carbonic acid and thus H^+.

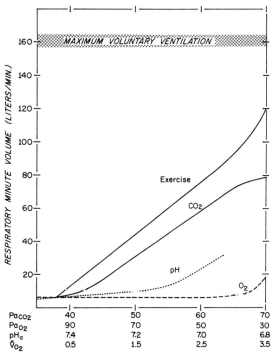

Figure 23–2 Acute ventilatory response to chemical stimuli in a 70 kg male. The abscissae for arterial Pco_2, Po_2, pH and oxygen consumption (Vo_2) are associated respectively with the curves labeled CO_2, O_2, pH and Exercise. Responses are compared to the maximum ventilation which the subject can attain transiently by voluntary effort. The curves represent responses of normal man with intact compensatory readjustments. Thus, the curves for hypoxia and metabolic pH changes show the response obtained as increasing ventilation yields a falling Pco_2.

where a is arterial, x represents chemosensitive areas not accessible to free H^+ from the blood, and k_1 and k_2 are appropriate constants defining the relative magnitude of the contribution from the two sites. In the last few years considerable evidence has accumulated supporting the idea that CO_2 affects ventilation through its role as an acid rather than by specific action of the CO_2 molecule per se.[18, 41, 47] In this chapter we shall analyze ventilatory responses assuming that $[H^+]$, not CO_2, is the primary chemical stimulus to ventilation.

Two sensory systems have been demonstrated. The peripheral or arterial chemoreceptors, the carotid and aortic bodies, respond rapidly to changes in PO_2, PCO_2 and $[H^+]$ of arterial blood. Central chemosensitive areas, found mainly in the brain stem, respond to changes in arterial PCO_2 or to the $[H^+]$ of their localized environment. We shall deal first with current knowledge concerning structure and function of these peripheral and central chemosensitive areas. Then we shall consider how these two systems interact during various acute and chronic changes in arterial PO_2, PCO_2 and $[H^+]$ and in response to exercise.

THE PERIPHERAL CHEMORECEPTORS

The extreme vascularity and rich innervation of the carotid body prompted De Castro[13] in 1926 to suggest that this structure might sense the chemical composition of blood. A year later, J.-F. and C. Heymans[25, 27] observed an increase in ventilation when the isolated aortic arch of a dog was perfused with blood from a donor animal breathing a gas low in oxygen. In 1930, C. Heymans, Bouckaert and Dautrebande[26] unveiled the exact physiologic role of the carotid body as predicted by De Castro 4 years before. For this work, C. Heymans received the Nobel Prize in 1938.

The importance of the carotid and aortic bodies to normal ventilation has been the subject of considerable debate. Some investigators felt that these structures had no significant role in the control of normal breathing at sea level, that they served only an emergency function in response to severe hypoxia.[7] Supporting this view is the fact that transection of the nerves from the peripheral chemoreceptors acutely causes little change in normal ventilation or in the response to inspired carbon dioxide in the dog,[20] whereas lack of oxygen leads to respiratory depression rather than to stimulation.[12] Recent evidence suggests that peripheral chemoreceptors play a definite although unobtrusive role in the regulation of normal sea level ventilation. Studies of structure-function relationships within the peripheral chemoreceptors have enjoyed a renaissance in recent years, as detailed in the excellent but already dated review of Biscoe.[3]

Anatomy. The carotid body is a tiny nodule (about 5 mm long in man) composed of large epithelioid "glomus" or glandlike cells within a network of connecting or sustentacular cells. The body lies just above the carotid sinus, affixed to the ascending pharyngeal or occipital artery, deriving its blood supply from one or two tiny branches of these vessels (Fig. 23–3). Venous blood drains into the internal jugular vein. The carotid nerve (nerve of Hering), containing afferent fibers from the carotid sinus and carotid body, joins the glossopharyngeal nerve. Electron micrographs reveal that nerve fibers terminate in endings making broad contact with the surface of the glomus (Type I) cells.[3] Histologically similar tissue has been found at several sites about the aortic arch and subclavian arteries.[8, 65] Afferent fibers supplying these aortic bodies are carried in the vagus nerve or in a separate thin branch, the depressor nerve, which accompanies the vagus. The anatomic and physiologic characteristics of the aortic bodies have not been so clearly delineated as those of the carotid body. Although the relative participation of the carotid and aortic chemoreceptors in circulatory and ventilatory regulation appears to differ, the entire ventilatory contribution from both sites will be treated in the subsequent discussion as from a single source: the peripheral chemoreceptors.

Blood Flow. With a rather intricate preparation, Daly and co-workers[11] estimated blood flow and metabolism of the cat's carotid body. The flow was found to be 2000 ml per 100 g per min, a value nearly 40 times that of the brain. They could not determine how much of this high flow might have bypassed chemosensitive tissue through arterial-venous shunts. Only when

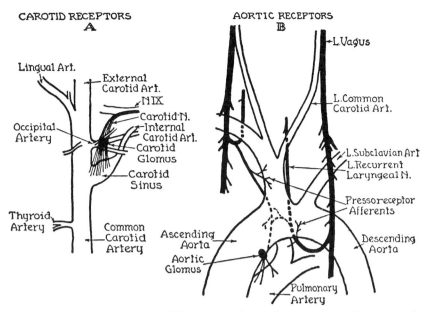

Figure 23–3 Diagrammatic representation of location and neural connections of carotid and aortic chemoreceptors and pressoreceptors (dog). (From Comroe, *Amer. J. Physiol.*, 1939, *127*, 176–191; and Comroe and Schmidt, *Amer. J. Physiol.*, 1938, *121*, 75–97.)

flow was diminished by lowering the blood pressure could an arterial-venous oxygen difference be detected. Under these conditions, oxygen consumption was calculated to be 9 ml per 100 g per min, a very high metabolism compared to other organs, but small in comparison to total blood flow. Studies utilizing a different technique provide a lower value for oxygen consumption, similar to that for the brain,[17] but recent work tends to confirm Daly's original measurements.[3]

Stimuli. The carotid bodies respond to decreases in arterial P_{O_2} and flow and to increases in arterial P_{CO_2}, and arterial [H^+]. Under most normal circumstances the cells appear to respond as if they were sampling arterial blood, *i.e.*, as if there were no significant arterial-venous difference across the carotid body.

P_{O_2}, P_{CO_2}, [H^+]. The carotid nerve of an anesthetized cat with normal arterial P_{O_2} and P_{CO_2} has a low level of tonic activity (Fig. 23–4). As the arterial P_{O_2} falls, or P_{CO_2} increases, the activity increases greatly. Significantly, combined hypoxia and hypercapnia produce more than a simple additive effect on chemoreceptor activity.[28] This interaction may account in part for the increased ventilatory sensitivity to carbon dioxide observed in subjects

acutely exposed to lowered oxygen tension.[46] In the steady state, the response to CO_2 results entirely from changes in arterial [H^+].[22, 29] This is apparent in Figure 23–5, where, for a given arterial [H^+], there is a unique chemoreceptor response in spite

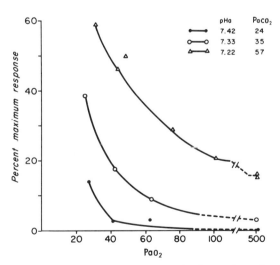

Figure 23–4 Averaged neural discharge from the carotid body of a cat in response to changing PaO_2 at three levels of arterial PCO_2–pH achieved by altering ventilation. (From Hornbein. In: *Arterial Chemoreceptors*, R. W. Torrance, ed. Oxford, Blackwell Scientific Publications, 1968.)

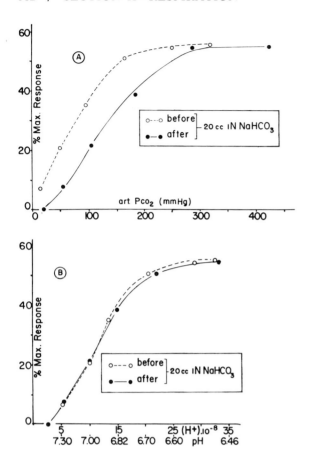

Figure 23-5 Integrated neural discharge (expressed as percentage of maximum asphyxic activity) from the cat's carotid body in response to stepwise alterations in inspired P_{CO_2}. *A*, Response is related to arterial P_{CO_2}. *B*, Response is related to arterial $[H^+]$. Addition of fixed base (20 ml 1 N $NaHCO_3$) alters the steady state response for a given Pa_{CO_2} *(A)*, but not for a given $[H^+]_a$ *(B)*. (From Hornbein and Roos, *J. appl. Physiol.*, 1963, *18*, 580–584.)

of large differences in P_{CO_2} between the curves. In other words, in the steady state the carotid body appears to be H^+-specific in its responses to metabolic or respiratory acidosis. Increased CO_2 evokes a quicker chemoreceptor response than an equivalent addition of fixed acid, probably because CO_2 diffuses more readily into the chemoreceptor cells to produce a faster alteration of the $[H^+]$.[22] As one might anticipate, the interaction between hypoxia and hypercapnia is actually an interaction between hypoxia and $[H^+]$.[29]

FLOW SENSITIVITY. Carotid body blood flow is usually so high that the arterial-venous oxygen difference is immeasurable, suggesting that P_{O_2}, rather than oxygen content of arterial blood, reflects the true stimulus to the carotid body. Indeed, perfusion of the carotid body with Ringer-Locke solution equilibrated at normal gas tensions produces no ventilatory response, although, in the absence of hemoglobin,

oxygen content of the perfusate is low.[7] Similarly, up to 70 per cent of blood hemoglobin can be bound with carbon monoxide without measurably increasing electrical activity in the carotid nerve,[16] an observation which supports the importance of arterial P_{O_2} rather than oxygen content as the chemoreceptor stimulus. However, when perfusion is decreased, arterial gas tensions no longer correlate with carotid body response. Landgren and Neil[39] observed profound chemoreceptor discharge in cats rendered hypotensive by hemorrhage, although arterial P_{O_2}, P_{CO_2} and $[H^+]$ were presumably not greatly changed. They suggested that this activity resulted from a lower P_{O_2}, a greater P_{CO_2} and a higher concentration of acid metabolites within glomus cells rather than in arterial blood because of reduced blood flow to the carotid body.

HYPOXIC TRANSMITTER. Hypoxia is thought to act by causing the release of a chemical transmitter from carotid body cells that would act on nerve endings. Substances suggested for this role include acetylcholine, catecholamines and [H⁺], as well as the ratio of ADP to ATP.[65] Supporting the existence of a transmitter is the observation that the response of the carotid body to a decrease in the oxygen supply produced by a reduction in blood flow is always greater than when the oxygen supply is lowered by decreasing the oxygen-carrying capacity of blood.[16, 65]

CENTRAL CHEMOSENSITIVE AREAS

The greater ventilatory response to hypercapnia than to metabolic acidosis for equivalent changes in arterial [H⁺] indicates an effect of CO_2 separate from its influence or arterial [H⁺].[32, 36, 40] The demonstration that carotid chemoreceptor response to CO_2 correlates solely with [H⁺]$_a$[22, 29] indicates that CO_2 must be acting elsewhere. Denervation of the peripheral chemoreceptors confirms this conclusion: although hypoxia no longer acutely increases ventilation, hypercapnia still does.[20] These observations imply the existence of receptor sites anatomically and functionally distinct from the peripheral chemoreceptors. Assuming that H⁺ is the unique stimulus to these other receptors, these observations also indicate the presence of a diffusion barrier between blood and the receptor that is freely permeable to CO_2 but poorly permeable to H⁺ or HCO_3^-. Early attempts to localize these receptors indicated the presence of chemosensitive areas in the brain. In 1954 Leusen[41] demonstrated that perfusing the ventriculo-cisternal system in dogs with solutions of high [H⁺] stimulated ventilation, whereas solutions with a low [H⁺] caused depression. This study proved the existence of chemosensitive areas in the brain responsive to changes in [H⁺] and indicated that at least some of them are superficially located close to bulk cerebrospinal fluid. Attempts to localize these areas led to the delineation of bilateral sites of chemosensitivity on the ventrolateral surface of the medulla oblongata in the cat.[42] Application of a local anesthetic to these areas in the decerebrate unanesthetized cat with denervated peripheral chemoreceptors reduced the ventilatory

response to CO_2,[10] indicating the presence of other chemosensitive sites as well. Recently, additional superficial brain stem areas sensitive to [H⁺] changes have been described.[54] By applying a ventriculo-cisternal perfusion technique to the awake goat, Pappenheimer et al.[47] attempted to quantify the effect of changes in [H⁺] of bulk cerebrospinal fluid on ventilation. They found that less than half the ventilatory response to inhaled CO_2 could be explained by changes in [H⁺] of the perfusion fluid. Therefore, a large part of the effect of CO_2 on ventilation must result from stimulation of areas remote from the ventriculo-cisternal perfusion fluid, either from other areas in the brain or from peripheral chemoreceptors.

From these studies we can conclude that functionally, as well as anatomically, there are several chemosensitive sites in the brain where [H⁺] changes in the surrounding fluid alter ventilation. Some of these areas are located on the surface of the brain stem close to bulk cerebrospinal fluid, but others are remote from the ventriculo-cisternal system. Whether this chemosensitivity is a general characteristic of respiratory or other brain stem neurons or whether specific sensory endings exist is not known.

Dynamic Characteristics of Chemosensitive Areas. In addition to their contribution to steady state ventilation, the peripheral and central chemosensitive areas possess dynamic characteristics that determine their effectiveness in maintaining moment-to-moment stability of Po_2, Pco_2, and [H⁺] in arterial blood. The high blood flow to peripheral chemoreceptors[11] results in a rapid response to changes in composition of arterial blood. The central chemosensitive areas have a much lower blood flow and, in addition, an adjacent avascular volume of cerebrospinal fluid with which CO_2 must equilibrate; therefore, they respond more slowly to changes in arterial Pco_2.[38] The presence of two sensory systems with different response times is well suited to provide optimal stability. The peripheral chemoreceptors provide a moment-to-moment defense against changes in arterial Po_2, Pco_2 and [H⁺], whereas the central chemosensitive areas serve to dampen changes in the ventilatory drive arising from the peripheral chemoreceptors.

VENTILATORY RESPONSES

In the following section we shall discuss the way in which drives from the peripheral chemoreceptors and central chemosensitive areas interact to explain the acute responses and long-term adjustments of ventilation in response to changes in oxygenation or acid-base state of the organism.

Acute Ventilatory Responses. CARBON DIOXIDE. When the CO_2 concentration in inspired air is increased, ventilation increases owing to stimulation of both peripheral and central chemosensitive areas. The increase in ventilation lessens the change in $PaCO_2$ that would otherwise result. CO_2 inhalation is not a common experience in the course of human events, but it provides a useful tool for the physiologist to measure how effective the ventilatory control system is in maintaining acid-base homeostasis in blood and brain. The results are expressed as CO_2 response curves where ventilation is plotted as a function of arterial or alveolar PCO_2 (Fig. 23–6). The slope of the CO_2 response curve, which varies between 1 and 3 liters per min per mm Hg in a normal adult man, describes the ability of the system to minimize changes in PCO_2. The steeper the slope, the less is the rise in arterial PCO_2 for a given increase in inspired PCO_2 and the greater the ability to counteract ventilatory changes imposed by other stimuli. CO_2 response curves also provide a useful means of describing the effect of a variety of conditions on ventilation.[45] Sleep, depressant drugs and anesthetics all shift the curve to the right, often with a decrease in slope.[52, 56] Hormones and drugs such as norepinephrine, progesterone and salicylates displace the curve to the left.[37]

HYPOXIA. Acute oxygen lack stimulates ventilation solely by its effect on the carotid and aortic chemoreceptors. After removal or denervation of these structures, hypoxia not only fails to produce hyperpnea acutely but may actually depress ventilation.[12] Resting man at sea level is relatively unresponsive to appreciable decreases in inspired PO_2 (Figs. 23–2 and 23–7); not until alveolar PO_2 has dropped below 50 to 60 mm Hg[15] does ventilation begin to increase significantly. This PO_2 is much less than that necessary to stimulate peripheral chemoreceptors, for they are tonically active even at normal arterial oxygen tensions (Fig. 23–5).

The ventilatory stimulus from acute hypoxia mediated via the peripheral chemoreceptors is partially offset by the hypocapnia resulting from hyperventilation. The decreased PCO_2 reduces the contribution from both peripheral and central chemosensitive areas. The ventilatory response to acute hypoxia is, therefore, much less than if PCO_2 were kept constant (Fig. 23–7). The ability of the central H^+ receptors to counterbalance other ventilatory stimuli such as hypoxia is determined by their sensitivity to H^+ as indicated by the slope of the CO_2 response curve. The greater the slope of the CO_2 response curve the less ventilation will increase for a given stimulation of the peripheral chemoreceptors. Hence, during acute hypoxia, interaction between peripheral and central chemosensitive areas tends to maintain PCO_2 and, thereofre, the brain's $[H^+]$ close to normal at the expense of arterial PO_2.

METABOLIC $[H^+]$ CHANGES. Interaction between peripheral and central chemosensitive areas similarly determines the

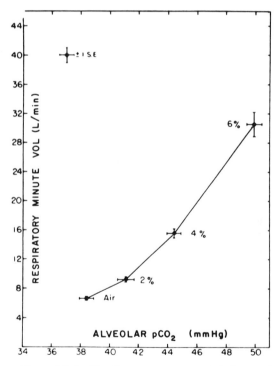

Figure 23–6 Ventilatory response to inhalation of varying concentrations of CO_2 in 21 per cent O_2 in N_2. Average response of 33 normal subjects. (From Lambertsen, *Anesthesiology*, 1960, *21*, 642–651.)

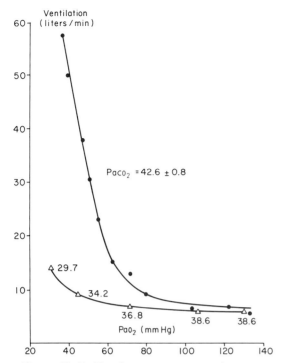

Figure 23–7 Ventilatory response to hypoxia in one human subject. The upper curve shows the ventilatory response when alveolar P_{CO_2} was kept constant at 42.6 mm Hg by adding CO_2 to inspired gas mixture. The lower curve shows the ventilatory response when alveolar P_{CO_2} was allowed to fall with hyperventilation. The numbers beside the lower curve refer to the measured alveolar P_{CO_2} at each point on the curve. (After Loeschcke and Gertz, *Pflügers Arch. ges. Physiol.*, 1958, 267, 460–477.)

magnitude of the ventilatory response to acute metabolic acid-base changes in the blood. In metabolic acidosis the increase in ventilation from peripheral chemoreceptor stimulation will be partly offset by a decrease in central drive caused by ensuing hypocapnia.[21, 53] The resulting increase in ventilation will be determined by the relative sensitivity of peripheral and central chemosensitive areas to $[H^+]$ changes. A high sensitivity of central chemosensitive areas to the CO_2-induced decrease in $[H^+]$ will minimize change in the brain $[H^+]$ while enhancing arterial acidosis. Opposite changes occur in acute metabolic alkalosis.

Chronic Ventilatory Adjustments. During acute acid-base derangements or hypoxia, blood $[H^+]$ and cerebrospinal fluid $[H^+]$ vary from their normal values of approximately 40 and 48 n moles per liter (*p*H's of 7.40 and 7.32). During primary

carbonic acid changes (hypo- and hypercapnia) both $[H^+]_{plasma}$ and $[H^+]_{CSF}$ change in the same direction; with noncarbonic acidosis or alkalosis, they change in opposite directions. When acid-base changes or hypoxia are sustained, ventilation continues to change in association with further alterations of $[H^+]$ in cerebrospinal fluid and blood.

Before discussing the mechanisms proposed to explain the chronic alterations in $[H^+]_{CSF}$, we shall describe the events that take place when ventilatory stimuli are sustained. In general, under chronic conditions $[H^+]_{CSF}$ has been found to deviate much less from normal than $[H^+]$ in blood.[43] Some authors have proposed that $[H^+]_{CSF}$ will be returned to its normal value during chronic acid-base derangements, but there is now abundant evidence that this is not so and that the chronic deviations in $[H^+]_{CSF}$, although small, are of importance in explaining chronic ventilatory adjustments.

That H^+, not CO_2, is the chemical stimulus to ventilation has been assumed in our discussion thus far. The quantitative justification for this assumption comes mainly from the work of Fencl *et al.*[18] in unanesthetized goats. Chronic states of metabolic acidosis or alkalosis were produced by feeding the animals ammonium chloride or sodium bicarbonate, and the ventilatory response to carbon dioxide was determined. There is no unique relationship between ventilation and $PaCO_2$ comparing the normal and chronically acid or alkaline states (Fig. 23–8A). When the associated values for $[H^+]_{CSF}$ are calculated, ventilation becomes an essentially unique function of $[H^+]_{CSF}$ (Fig. 23–8B), indicating that the primary stimulus is change in $[H^+]$ rather than change in PCO_2. The ability to account for most of the ventilatory response as a function of $[H^+]_{CSF}$ suggests that under the conditions of this study contribution from peripheral chemoreceptors was small.

CHRONIC HYPOXIA. One of the earliest adjustments during acclimatization to high altitude is an increase in ventilation initiated by hypoxic stimulation of the peripheral chemoreceptors. The resulting hypocapnic alkalosis decreases ventilatory drive so that the initial increase in ventilation and decrease in $PaCO_2$ are small. Within these first minutes the ventilatory response can be completely reversed by restoring

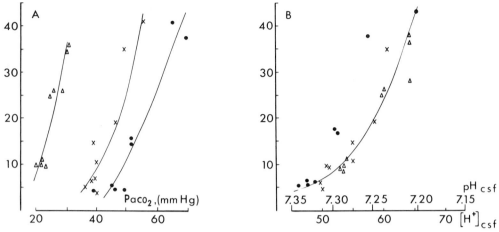

Figure 23–8 Ventilatory response of an unanesthetized goat to inhalation of varying concentrations of CO_2 during normal acid-base state ($\times$, $[HCO_3^-]_a = 28$ to 33 m moles/liter) during metabolic acidosis (Δ, $[HCO_3^-]_a = 9$ to 11 m moles/liter), and metabolic alkalosis ($\bullet$, $[HCO_3^-]_a = 35$ to 44 m moles/liter) of several days' duration. *A*, Alveolar ventilation is related to arterial P_{CO_2}. *B*, Alveolar ventilation is related to cerebrospinal fluid $[H^+]$ and *p*H. (After Fencl *et al.*, *Amer. J. Physiol.*, 1966, *210*, 459–472.)

the inspired P_{O_2} to its sea level value. If the sojourn at altitude is prolonged for hours or days, a progressive and considerably greater increase in ventilation and decrease in Pa_{CO_2} occur. This fall in Pa_{CO_2} is associated with an approximately equivalent increase in Pa_{O_2} that serves, along with continuing alkalosis in the blood,[55] to decrease the peripheral chemoreceptor drive to ventilation. If P_{O_2} is increased at this time, ventilation decreases slightly but remains well above the original sea level value.[33, 50, 55] This resetting of ventilation to a low P_{CO_2} indicates an increased central H^+ drive in the absence of hypoxia. Because central P_{CO_2} is also decreased,* an increased $[H^+]$ requires a fall in $[HCO_3^-]$ in brain interstitial fluid. This decrease in brain interstitial fluid $[HCO_3^-]$ during chronic hypoxia has been documented by

measurement of $[HCO_3^-]$ in cerebrospinal fluid ($[HCO_3^-]_{CSF}$)[48, 55, 63] (Fig. 23–9).

Although peripheral chemoreceptor stimulation by hypoxia has classically been regarded as the initiating event in the acclimatization process, more recent evidence indicates that a peripheral chemoreceptor drive is not essential. Permanent residents at high altitude usually possess minimal ventilatory response to acute hypoxia.[61] Nevertheless, they have lower Pa_{CO_2}'s at

---

*Mean brain tissue P_{CO_2} is several mm Hg higher than arterial P_{CO_2} and is slightly lower than jugular venous P_{CO_2}. The arterial-venous P_{CO_2} difference varies inversely with cerebral blood flow and is, therefore, affected by changes in arterial P_{O_2} and P_{CO_2} (see Chap. 17, Vol. II). We will assume that in the steady state the P_{CO_2} in bulk cerebrospinal fluid provides a reasonable measurement of the P_{CO_2} in the interstitial fluid surrounding areas of central chemosensitivity. Similarly, our use of $[HCO_3^-]_{CSF}$ assumes that in the steady state no concentration gradient for HCO_3^- exists between the sampling site and brain interstitial fluid.[18, 19]

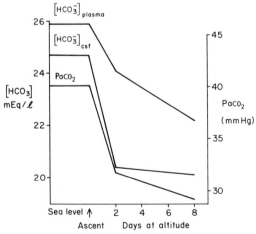

Figure 23–9 Average changes in $[HCO_3^-]_{plasma}$, $[HCO_3^-]_{CSF}$ and arterial P_{CO_2} in four normal men during sojourn at an altitude of 12,500 feet. (After Severinghaus *et al.*, *J. appl. Physiol.*, 1963, *18*, 1155–1166.)

altitude than at sea level,[62] as do awake goats that have been surgically deprived of their carotid chemoreceptors.[60] Following peripheral chemoreceptor denervation, rabbits made hypoxic for 24 hours showed decreases in $[HCO_3^-]_{CSF}$ similar to those seen in normal animals.[64] Ventilatory acclimatization in the absence of a drive from peripheral chemoreceptors may be explained by a slightly more acid cerebrospinal fluid in the high-altitude dweller as compared to the acclimatized lowlander.[63] How hypoxia might produce an increase in $[H^+]_{CSF}$ will be discussed under Regulation of $[H^+]$ and $[HCO_3^-]$ in Brain.

METABOLIC ACIDOSIS AND ALKALOSIS. When metabolic acid-base changes in arterial blood are sustained, the changes in ventilation that occur are primarily a result of changes in $[HCO_3^-]_{CSF}$. In chronic metabolic acidosis, $[HCO_3^-]_{CSF}$ decreases; in chronic metabolic alkalosis, it increases (Fig. 23–10).[18, 19] As described earlier, the acute change in $[H^+]_{CSF}$ (and, hence, of central ventilatory drive) is opposite in direction to that in arterial blood.[21, 53] With ventilatory acclimatization to a sustained metabolic acidosis, the fall in $[HCO_3^-]_{CSF}$ causes cerebrospinal fluid $[H^+]$ to change from the alkaline to the acid side of normal,[18, 19] resulting in a further increase in ventilation over that occurring with acute acidosis. This decrease in blood $[H^+]$ constitutes ventilatory compensation for the primary metabolic (noncarbonic) acidosis (see Chap. 27, Vol. II). In chronic metabolic alkalosis opposite changes take place.

CHRONIC HYPERCAPNIA. Hypercapnia can be produced by the addition of CO_2 to the inspired air, but in real life it is almost always the result of what is termed respiratory failure. A variety of disorders affecting either the ventilatory apparatus (the airways, lung tissue of chest wall) or the neuromuscular system controlling ventilation can cause CO_2-retention. An acute decrease in alveolar ventilation and increase in alveolar P_{CO_2} increases both the peripheral and the central drive to ventilation. The resulting ventilation represents a balance between the increased total ventilatory drive and the limitations placed upon the ventilatory apparatus by the primary disease process.

If hypercapnia is sustained, both $[HCO_3^-]_{plasma}$ and $[HCO_3^-]_{CSF}$ increase,[1, 4, 43, 66] decreasing arterial and cerebrospinal fluid $[H^+]$ and, therefore, ventilatory drive. As a result, P_{CO_2} increases further. In the steady state of chronic hypercapnia, $[H^+]_{CSF}$ is distinctly higher than normal,[1, 43, 49, 66] implying the presence of a greater than normal ventilatory drive. The rise in $[HCO_3^-]_{CSF}$ is a consequence of factors discussed in the following section, whereas $[HCO_3^-]_{plasma}$ is determined by the ability of the kidney to increasingly retain HCO_3 as Pa_{CO_2} increases.

What determines the new steady state level of P_{CO_2} is not clear, but sustained cerebrospinal fluid acidosis must be one of the contributing factors. Also, arterial acidosis may play a role, because renal HCO_3^- reabsorption fails to maintain a constant

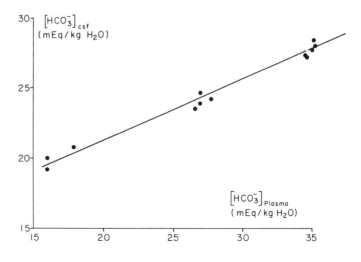

Figure 23–10 Relation between $[HCO_3^-]_{plasma}$ and $[HCO_3^-]_{CSF}$ in four normal men during experimental chronic metabolic acid-base changes in blood. The relation does not reflect the effect of pure metabolic changes, because Pa_{CO_2} was somewhat different in the three situations. (After Fencl *et al.*, *J. appl. Physiol.*, 1969, 27, 67–76.)

[H$^+$] when P$_{CO_2}$ becomes very high.[51] In addition, respiratory failure is usually complicated by arterial hypoxemia, which can also enhance ventilatory drive.

The role of hypoxemia as a factor determining the level of P$_{CO_2}$ in respiratory failure is demonstrated by the effect of oxygen administration on the ventilation of patients with sustained hypoxia and hypercapnia. When the patient is acclimatized to hypoxia the immediate effect of increasing Pa$_{O_2}$ is usually small, but over several hours P$_{CO_2}$ gradually increases as the patient deacclimatizes. If oxygen is administered during acute respiratory failure, when the peripheral chemoreceptor contribution to the ventilatory drive is relatively greater, the decrease in ventilation and rise in P$_{CO_2}$ may be more acute.

REGULATION OF [H$^+$] AND [HCO$_3^-$] IN BRAIN. How [H$^+$] in brain extracellular fluid is regulated has been the subject of much controversy and several symposia.[5, 58] Because the issue is as much in a state of flux as the ions on which it centers, we shall discuss only briefly some of the concepts and facts underlying current hypotheses concerning control of [H$^+$] and [HCO$_3^-$] in brain extracellular fluid. A more detailed discussion may be found in the recent review by Siesjö.[57]

An analysis of the factors that might determine [H$^+$] and [HCO$_3^-$] in brain extracellular fluid must consider the mechanisms affecting the distribution of the ions between blood and brain extracellular fluid across the blood-brain barrier. Although in most tissues interstitial fluid approximates an ultrafiltrate of plasma, this is not so for the brain. The tight cell layers separating blood and brain extracellular fluid provide a barrier that is poorly permeable to larger molecules and ions, although several large molecules and ions are actively transported across this barrier, e.g., PAH, Diodrast, Mg^{++}, Ca^{++} and K$^+$. In considering the role of the blood-brain barrier in determining [H$^+$] and [HCO$_3^-$] in brain extracellular fluid, we must examine the possible roles of active transport and passive ion distribution.

Measurement of the electrical potential difference between cerebrospinal fluid and blood in several species of mammals has shown the cerebrospinal fluid to be a few millivolts positive relative to blood under normal acid-base conditions.[24, 34] If H$^+$ and HCO$_3^-$ were in electrochemical equilibrium across the blood-brain barrier, this potential difference should result in a higher [HCO$_3^-$] and lower [H$^+$] in cerebrospinal fluid than in blood (see Nernst equation, Chap. 1, Vol. I). However, the reverse is true: the [H$^+$] is higher and the [HCO$_3^-$] is lower in cerebrospinal fluid than in blood. This electrochemical disequilibrium could be a consequence of active transport of H$^+$ or HCO$_3^-$ (or some other ions) across the blood-brain barrier,[47, 55] or it could result from metabolic production of H$^+$ by brain cells.

That H$^+$ production by the brain may play a role is illustrated by the response to oxygen lack. During acute[6] and chronic[63] hypoxia, increased anaerobic glycolysis by the brain adds lactic acid to the extracellular fluid. The added H$^+$ will react with HCO$_3^-$ in brain extracellular fluid to form CO$_2$, which then diffuses readily into blood, lowering [HCO$_3^-$]$_{CSF}$ independently of [HCO$_3^-$]$_{plasma}$ and producing electrochemical disequilibrium. The degree of disequilibrium for HCO$_3^-$ and H$^+$ would depend in part on the rate of lactic acid production by brain cells and the ease with which H$^+$ and lactate ions can cross the blood-brain barrier. Ventilatory acclimatization to chronic hypoxia may be partly explained by this mechanism.

Exploration of the changes occurring between cerebrospinal fluid and blood during chronic acid-base derangements might permit separation of the roles of active and passive processes in determining [H$^+$]$_{CSF}$. The relative stability of [H$^+$]$_{CSF}$ during sustained acid-base abnormalities might result, at least in part, from changes in the potential difference between cerebrospinal fluid and blood with changes in blood [H$^+$]. In dogs and rats the potential difference changes about 30 mv per 10-fold change in arterial [H$^+$], becoming more positive in cerebrospinal fluid with arterial acidosis and becoming negative relative to blood with arterial alkalosis.[24, 34] How alterations in blood [H$^+$] cause a change in the CSF-blood potential is not known; nor do we know whether potential changes in man are similar to those in dogs and rats. During arterial acidosis, as cerebrospinal fluid becomes more positive relative to blood, [HCO$_3^-$] in brain extracellular fluid would tend to increase. During metabolic (non-carbonic) acidosis, this potential change will counteract the effect of a low plasma $[HCO_3^-]$, helping to explain the stability of $[H^+]_{CSF}$ and $[HCO_3^-]_{CSF}$ relative to blood under these conditions (Fig. 23–10). During respiratory acidosis, the potential change is abetted by a rising plasma $[HCO_3^-]$, both serving to minimize the acidosis in brain extracellular fluid. Whether these potential changes are large enough to explain the stability of $[H^+]$ and $[HCO_3^-]$ in brain extracellular fluid during blood acid-base changes[57] or whether active transport also plays a role[47, 55] has not as yet been resolved.

In summary, three factors may play a role in determining $[H^+]$ of brain extracellular fluid during steady state derangements of acid-base balance: (i) a blood-brain barrier that is poorly permeable to ions, coupled with a potential difference between cerebrospinal fluid and blood that is altered by changes in blood $[H^+]$; (ii) active ion transport across the barrier; (iii) metabolic production of H^+ by brain cells. Considerable clarification is required before the relative importance of each of these mechanisms can be defined.

Exercise. Exercise produces the greatest sustained ventilatory response of which man is capable (Fig. 23–2). The minute volume during maximum work by a trained athlete may closely approach his maximum voluntary ventilation. During exercise, ventilation increases linearly with increasing oxygen consumption over a wide range of work loads (Fig. 23–2), and arterial Po_2 and Pco_2 change surprisingly little from normal resting values. Only at very high work levels do these relationships break down, resulting in relative hyperventilation for the amount of oxygen consumed.

At the onset of dynamic exercise, such as running or cycling, ventilation rises rapidly within a few seconds and then increases more slowly over several minutes to a relatively constant steady state level. With cessation of work, ventilation immediately falls precipitously, then declines slowly toward normal during the ensuing minutes. These fast and slow components to both onset and cessation of exercise provide the basis for the neurohumoral theory of ventilatory control during exercise, as reviewed by Dejours[14] (Fig. 23–11).

NEURAL FACTORS. The sudden increase

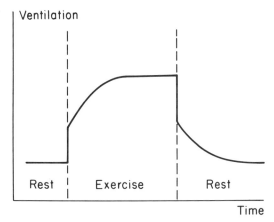

Figure 23–11 The changes in ventilation during and after exercise, according to Dejours *(Handb. Physiol.,* 1964, sec. 3, vol. I, 631–648.) Both at the onset and cessation of exercise a fast change in ventilation (neural component) is followed by a slow response (humoral component).

in ventilation with onset of work is considered too rapid to result from chemical changes in blood. Furthermore, occluding the circulation with tourniquets prior to exercise does not alter the response. Therefore, the early response is regarded as neural in origin, via either cortical input to the respiratory center or reflex pathways from muscles or moving joints.

In 1913 Krogh and Lindhard[35] pointed out that the initial ventilatory response was highly variable in magnitude and time of onset, sometimes even preceding the onset of exercise. They concluded that the initial increase in ventilation could not be caused solely by reflex input from exercising limbs. Still a considerable effort has been directed toward identifying receptors in the limbs that might mediate an increased ventilation. Joint receptors,[9] muscle spindles,[14] and "ergoreceptors"[31] have all been suggested as playing a role in exercise hyperpnea. Electrical stimulation of nerves or receptors in anesthetized animals and passive movement of the limbs result in increased ventilation. However, most studies of ventilatory drive during exercise report a lack of stimulation from limb receptors. Ventilation is not affected by pedaling frequency,[59] and differential blockade of fusimotor nerve fibers has indicated the absence of a contribution from muscle spindles.[30] Furthermore, recent careful studies reveal great variability in breathing patterns at the beginning of exercise,[2] supporting Krogh and Lindhard's original observations.[35] The role of conditioning in this response remains unresolved.

HUMORAL FACTORS. Possible blood-borne stimuli affecting ventilation during exercise include P_{CO_2}, P_{O_2}, $[H^+]$, catecholamines and temperature. A 1° C increase in body temperature may increase ventilation approximately 3.9 liters per min per m^2, but temperature rise in exercise is too slow to explain the major portion of the steady state response. The constancy of arterial P_{O_2} and P_{CO_2} would appear to rule out these factors as possible stimuli to work hyperpnea unless one postulates a sensitive negative feedback mechanism which counterbalances any change in stimulus. It is unlikely that arterial gas tensions can account for such delicate control. The frustrating constancy of arterial blood gas tensions has caused some investigators to postulate a contribution from unknown metabolites of exercising muscle, but such suggestions are based more on uneasy ignorance than physiologic fact. Similarly, because mixed venous P_{CO_2} increases almost in direct proportion to the increase in ventilation during exercise, a pulmonary arterial chemoreceptor has been sought but not convincingly demonstrated. The rise in mixed venous P_{CO_2} is more likely a consequence of the increased metabolism during work than a cause of the increased ventilation.

The possibility exists that the peripheral chemoreceptors might provide an increased ventilatory stimulus during exercise, even though mean arterial gas tensions and $[H^+]$ are equal to the resting values.[3, 65] Sympathetic stimulation might increase chemoreceptor discharge, perhaps by decreasing blood supply. Or the greatly enhanced respiratory oscillations in Pa_{O_2} and $[H^+]_a$ occurring during exercise might result in an enhanced peripheral drive in spite of normality of the mean blood stimulus. The few attempts to evaluate these hypotheses during active exercise have failed to reveal a definite contribution of the peripheral chemoreceptors to work hyperpnea.[3]

Having explored some of the factors that might contribute to work hyperpnea, we must conclude by stating that less is known about these stimuli and how they interact to explain the ventilatory response to exercise than about any other aspect of ventilatory control.

REFERENCES

1. Alroy, G. G., and Flenley, D. C. The acidity of the cerebrospinal fluid in man with particular reference to chronic ventilatory failure. *Clin. Sci.*, 1967, *33*, 335–343.
2. Beaver, W. L., and Wasserman, K. Transients in ventilation at start and end of exercise. *J. appl. Physiol.*, 1968, *25*, 390–399.
3. Biscoe, T. J. Carotid body: Structure and function. *Physiol. Rev.*, 1971, *51*, 437–495.
4. Bleich, H. L., Berkman, P. M., and Schwartz, W. B. The response of cerebrospinal fluid composition to sustained hypercapnia. *J. clin. Invest.*, 1964, *43*, 11–16.
5. Brooks, C. Mc., Kao, F. F., and Lloyd, B. B., eds. *Cerebrospinal fluid and the regulation of ventilation.* Oxford, Blackwell Scientific Publications, 1965.
6. Cohen, P. J., Alexander, S. C., Smith, T. C., Reivich, M., and Wollman, H. Effects of hypoxia and normocarbia on cerebral blood flow and metabolism in conscious man. *J. appl. Physiol.*, 1967, *23*, 183–189.
7. Comroe, J. H., Jr., and Schmidt, C. F. The part played by reflexes from the carotid body in the chemical regulation of respiration in the dog. *Amer. J. Physiol.*, 1938, *121*, 75–97.
8. Comroe, J. H., Jr. The location and function of the chemoreceptors of the aorta. *Amer. J. Physiol.*, 1939, *127*, 176–191.
9. Comroe, J. H., Jr., and Schmidt, C. F. Reflexes from the limbs as a factor in the hyperpnea of muscular exercise. *Amer. J. Physiol.*, 1943, *138*, 536–547.
10. Cozine, R. A., and Ngai, S. H. Medullary surface chemoreceptors and regulation of respiration in the cat. *J. appl. Physiol.*, 1967, *22*, 117–121.
11. Daly, M. de B., Lambertsen, C. J., and Schweitzer, A. Observations on the volume of blood flow and oxygen utilization of the carotid body in the cat. *J. Physiol. (Lond.)*, 1954, *125*, 67–89.
12. Davenport, H. W., Brewer, G., Chambers, A. H., and Goldschmidt, S. The respiratory responses to anoxemia of unanesthetized dogs with chronically denervated aortic and carotid chemoreceptors and their causes. *Amer. J. Physiol.*, 1947, *148*, 406–416.
13. De Castro, F. Sur la structure et l'innervation de la glande intercarotidienne (glomus caroticum) de l'homme et des mammiféres, et sur un nouveau système d'innervation autonome du nerf glossopharyngien. Études anatomiques et expérimentales. *Trab. Lab. Invest. biol. Univ. Madr.*, 1926, *24*, 365–432.
14. Dejours, P. Control of respiration in muscular exercise. *Handb. Physiol.*, 1964, sec. 3, vol. I, 631–648.
15. Dripps, R. D., and Comroe, J. H., Jr. The effect of the inhalation of high and low oxygen concentrations on respiration, pulse rate, ballisto-

cardiogram and arterial oxygen saturation (oximeter) of normal individuals. *Amer. J. Physiol.*, 1947, *149*, 277–291.

16. Duke, H. N., Green, J. H., and Neil, E.   Carotid chemoreceptor impulse activity during inhalation of carbon monoxide mixtures. *J. Physiol. (Lond.)*, 1952, *118*, 520–527.

17. Fay, F. S.   Oxygen consumption of the carotid body. *Amer. J. Physiol.*, 1970, *218*, 518–523.

18. Fencl, V., Miller, T. B., and Pappenheimer, J. R.   Studies on the respiratory response to disturbances of acid-base balance, with deductions concerning the ionic composition of cerebral interstitial fluid. *Amer. J. Physiol.*, 1966, *210*, 459–472.

19. Fencl, V., Vale, J. R., and Broch, J. A.   Respiration and cerebral blood flow in metabolic acidosis and alkalosis in humans. *J. appl. Physiol.*, 1969, *27*, 67–76.

20. Gemmill, C. L., and Reeves, D. L.   The effect of anoxemia in normal dogs before and after denervation of carotid sinuses. *Amer. J. Physiol.*, 1933, *105*, 487–495.

21. Gesell, R., and Hertzman, A. B.   The regulation of respiration. IV. Tissue acidity, blood acidity and pulmonary ventilation. A study of the effects of semipermeability of membranes and the buffering action of tissues with the continuous method of recording changes in acidity. *Amer. J. Physiol.*, 1926, *78*, 610–629.

22. Gray, B. A.   Response of the perfused carotid body to changes in pH and P_{CO_2}. *Resp. Physiol.*, 1968, *4*, 229–245.

23. Haldane, J. S., and Smith, J. L.   The physiological effects of air vitiated by respiration. *J. Path. Bact.*, 1892, *1*, 168–186.

24. Held, D., Fencl, V., and Pappenheimer, J. R.   Electrical potential of cerebrospinal fluid. *J. Neurophysiol.*, 1964, *27*, 942–959.

25. Heymans, J.-F., and Heymans, C.   Sur les modifications directes et sur la régulation réflexe de l'activité du centre respiratoire de la tête isolée du chien. *Arch. int. Pharmacodyn.*, 1927, *33*, 272–370.

26. Heymans, C., Bouckaert, J. J., and Dautrebande, L.   Sinus carotidien et réflexes respiratoires. II. Influences respiratoires réflexes de l'acidose, de l'alcalose, de l'anhydride carbonique, de l'ion hydrogène et de l'anoxémie. Sinus carotidien et échanges respiratoires dans les poumons et au delà des poumons. *Arch. int. Pharmacodyn.*, 1930, *39*, 400–450.

27. Heymans, C., and Neil, E.   *Reflexogenic areas of the cardiovascular system.* Boston, Little, Brown and Co., 1958.

28. Hornbein, T. F., Griffo, Z. J., and Roos, A.   Quantitation of chemoreceptor activity: Interrelation of hypoxia and hypercapnia. *J. Neurophysiol.*, 1961, *24*, 561–568.

29. Hornbein, T. F., and Roos, A.   Specificity of H ion concentration as a carotid chemoreceptor stimulus. *J. appl. Physiol.*, 1963, *18*, 580–584.

30. Hornbein, T. F., Sørensen, S. C., and Parks, C. R.   Role of muscle spindles in the lower extremities in breathing during bicycle exercise. *J. appl. Physiol.*, 1969, *27*, 476–479.

31. Kao, F. F.   An experimental study of the pathways involved in exercise hyperpnoea employing cross-circulation techniques. In: *The regulation of human respiration*, D. J. C. Cunningham and B. B. Lloyd, eds. Oxford, Blackwell Scientific Publications, 1963.

32. Katsaros, B., Loeschcke, H. H., Lerche, D., Schönthal, H., and Hahn, N.   Wirkung der Bikarbonat-alkalose auf die Lungenbelüftung beim Menschen. Bestimmung der Teilwirkungen von pH und CO_2 Druck auf die Ventilation und Vergleich mit den Ergebnissen bei Acidose. *Pflügers Arch. ges. Physiol.*, 1960, *271*, 732–747.

33. Kellogg, R. H., Pace, N., Archibald, E. R., and Vaughan, B. E.   Respiratory response to inspired CO_2 during acclimatization to an altitude of 12,470 feet. *J. appl. Physiol.*, 1957, *11*, 65–71.

34. Kjällquist, A., and Siesjö, B. K.   The CSF/blood potential in sustained acidosis and alkalosis in the rat. *Acta physiol. scand.*, 1967, *71*, 255–256.

35. Krogh, A., and Lindhard, J.   The regulation of respiration and circulation during the initial stages of muscular work. *J. Physiol. (Lond.)*, 1913, *47*, 112–136.

36. Lambertsen, C. J., Semple, S. J. G., Smyth, M. G., and Gelfand, R. H^+ and P_{CO_2} as chemical factors in respiratory and cerebral circulatory control. *J. appl. Physiol.*, 1961, *16*, 473–484.

37. Lambertsen, C. J.   Effects of drugs and hormones on the respiratory response to carbon dioxide. *Handb. Physiol.*, 1964, sec. 3, vol. I, 545–555.

38. Lambertsen, C. J., Gelfand, R., and Kemp, R. A.   Dynamic response characteristics of several CO_2-reactive components of the respiratory control system. In: *Cerebrospinal fluid and the regulation of ventilation*, C. McC. Brooks, F. F. Kao and B. B. Lloyd, eds. Oxford, Blackwell Scientific Publications, 1965.

39. Landgren, S., and Neil, E.   Chemoreceptor impulse activity following hemorrhage. *Acta physiol. scand.*, 1951, *23*, 158–167.

40. Laqueur, E., and Verzár, F.   Über die spezifische Wirkung der Kohlensäure auf das Atemzentrum. *Pflügers Arch. ges. Physiol.*, 1911, *143*, 395–427.

41. Leusen, I. R.   Chemosensitivity of the respiratory center. Influence of changes in the H^+ and total buffer concentrations in the cerebral ventricles on respiration. *Amer. J. Physiol.*, 1954, *176*, 45–51.

42. Mitchell, R. A., Loeschcke, H. H., Massion, W. H., and Severinghaus, J. W.   Respiratory responses mediated through superficial chemosensitive areas on the medulla. *J. appl. Physiol.*, 1963, *18*, 523–533.

43. Mitchell, R. A., Carman, C. T., Severinghaus, J. W., Richardson, B. W., Singer, M. M., and Shnider, S.   Stability of cerebrospinal fluid pH in chronic acid-base disturbances in blood. *J. appl. Physiol.*, 1965, *20*, 443–452.

44. Mitchell, R. A.   Respiration. *Ann. Rev. Physiol.*, 1970, *32*, 415–438.

45. Nielsen, M.   Untersuchungen über die Atemregulation beim Menschen. *Skand. Arch. Physiol.*, 1936, *74*, Suppl. 10, 83–208.

46. Nielsen, M., and Smith, H.   Studies on the regulation of respiration in acute hypoxia. *Acta physiol. scand.*, 1952, *24*, 293–313.

47. Pappenheimer, J. R., Fencl, V., Heisey, S. R., and

Held, D. Role of cerebral fluids in control of respiration as studied in unanesthetized goats. *Amer. J. Physiol.*, 1965, *208*, 436–450.

48. Pauli, H. G., Vorburger, C., and Reubi, F. Chronic derangements of cerebrospinal fluid acid-base components in man. *J. appl. Physiol.*, 1962, *17*, 993–998.

49. Posner, J. B., and Plum, F. Spinal fluid *p*H and neurologic symptoms in systemic acidosis. *New Engl. J. Med.*, 1967, *277*, 605–613.

50. Rahn, H., Stroud, R. C., Tenney, S. M., and Mithoefer, J. C. Adaptation to high altitude: Respiratory response to CO_2 and O_2. *J. appl. Physiol.*, 1953, *6*, 158–162.

51. Rector, F. C., Seldin, D. W., Roberts, A. D., Jr., and Smith, J. S. The role of plasma CO_2 tension and carbonic anhydrase activity in the renal reabsorption of bicarbonate. *J. clin. Invest.*, 1960, *39*, 1706–1721.

52. Reed, D. J., and Kellogg, R. H. Changes in respiratory response to CO_2 during natural sleep at sea level and at altitude. *J. appl. Physiol.*, 1958, *13*, 325–330.

53. Robin, E. D., Whaley, R. D., Crump, C. H., Bickelman, A. G., and Travis, D. M. Acid-base relations between spinal fluid and arterial blood with special reference to control of ventilation. *J. appl. Physiol.*, 1958, *13*, 385–392.

54. Schlaefke, M. E., See, W. R., and Loeschcke, H. H. Ventilatory response to alterations of H^+ ion concentration in small areas of the ventral medullary surface. *Resp. Physiol.*, 1970, *10*, 198–212.

55. Severinghaus, J. W., Mitchell, R. A., Richardson, B. W., and Singer, M. M. Respiratory control at high altitude suggesting active transport regulation of CSF pH. *J. appl. Physiol.*, 1963, *18*, 1155–1166.

56. Severinghaus, J. W., and Larson, C. P., Jr. Respiration in anesthesia. *Handb. Physiol.*, 1965, sec. 3, vol. II, 1219–1264.

57. Siesjö, B. K. The regulation of cerebrospinal fluid pH. *Kidney International*, 1972, *1*, 360–374.

58. Siesjö, B. K., and Sørensen, S. C., eds. *Ion homeostasis of the brain.* Copenhagen, Munksgård, 1971.

59. Sipple, J. H., and Gilbert, R. Influence of proprioceptor activity in the ventilatory response to exercise. *J. appl. Physiol.*, 1966, *21*, 143–146.

60. Sørensen, S. C., and Mines, A. H. Ventilatory responses to acute and chronic hypoxia in goats after sinus nerve section. *J. appl. Physiol.*, 1970, *28*, 832–835.

61. Sørensen, S. C., and Severinghaus, J. W. Respiratory sensitivity to acute hypoxia in man born at sea level living at high altitude. *J. appl. Physiol.*, 1968, *25*, 211–216.

62. Sørensen, S. C., and Severinghaus, J. W. Irreversible respiratory insensitivity to acute hypoxia in man born at high altitude. *J. appl. Physiol.*, 1968, *25*, 217–220.

63. Sørensen, S. C., Milledge, J. S., and Severinghaus, J. W. Cerebral anaerobic metabolism and ventilatory acclimatization to chronic hypoxia. *Fed. Proc.*, 1969, *28*, 337.

64. Sørensen, S. C. Ventilatory acclimatization to hypoxia in rabbits after denervation of peripheral chemoreceptors. *J. appl. Physiol.*, 1970, *28*, 836–839.

65. Torrance, R. W., ed. *Arterial chemoreceptors.* Oxford, Blackwell Scientific Publications, 1968.

66. Van Heijst, A. N. P., Maas, A. H. J., and Visser, B. F. L'équilibre acido-basique dans le sang et le liquide céphalo-rachidien dans l'hypercapnie chronique. *Entretiens Physiopathologie Respiratoire*, Nancy, 1964.

67. Winterstein, H. Die Regulierung der Atmung durch das Blut. *Pflügers Arch. ges. Physiol.*, 1911, *138*, 167–184.

68. Winterstein, H. Chemical control of pulmonary ventilation. III. The "reaction theory" of respiratory control. *New Engl. J. Med.*, 1956, *255*, 331–337.

CHAPTER 24 PASSIVE AND ACTIVE
TRANSPORT

by ARTHUR C. BROWN

INTRODUCTION[7, 9, 15]

A characteristic property of the living cell is its ability to maintain many substances at internal concentrations different from that in the surrounding medium. For particles able to pass through the cell wall, this finding is somewhat unexpected, since it appears to oppose the "natural" tendency of such particles to move down concentration gradients and eliminate concentration differences. When a cell dies, this ability is lost, and those substances which can pass through the cell wall equilibrate with the immediate surroundings.

The intracellular-extracellular concentration differences of many substances reflect the characteristics of the cell wall (rather than those of the cytoplasm). That region of the cell wall which sustains concentration and electrical gradients and manifests the transport mechanisms peculiar to living systems is called the cell membrane.

The forces responsible for transport through the cell membrane and an evaluation of their effect forms the subject matter of this chapter. The emphasis will be on the general physical basis of common characteristics and mechanisms, leaving details of membrane transport in specific organs to be covered in the appropriate chapters.

Role of Transport. One role of membrane transport in the economy of an individual cell is the maintenance of the proper intracellular composition. An appropriate intracellular environment is necessary for optimal function of the cell's biochemical apparatus. Also, transported particles may take part in other cellular functions; for example, the electrical and chemical energy created by the active separation of sodium from potassium is exploited by nerve and muscle cells for the rapid generation and propagation of action potentials.

To maintain a constant concentration of substances not produced or consumed within the cell, the average transport over an extended time must be zero; otherwise, the amount of substance in the cell would change progressively. If the substance is consumed or produced at a given rate, the net transport must proceed at an equal rate to assure stability of the cell's internal composition.

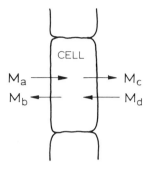

$$M_{cell} = M_a + M_d - (M_c + M_b)$$

$$M_{trans} = M_a - M_b = M_c - M_d$$

Figure 24–1 The fluxes which must be considered in cellular exchange. M_{cell} is the net flux entering the cell, and M_{trans} is the net transport through the tissue layer (when the internal cell concentration is in a steady state). Note that the intracellular concentration can remain constant ($M_{cell} = 0$) even if a transcellular net flux is present ($M_{trans} \neq 0$).

In some types of tissue, such as the capillary endothelium and the intestinal mucosa, membrane transport has an additional function related to the total economy of the body: *i.e.,* the transfer of substances across a tissue layer, resulting in movement of material from one part of the body to another. While not incompatible with intracellular stability (net membrane transport into the cell may still be zero), this function does imply an asymmetry of particle movement through the cell wall (see Fig. 24–1).

Definitions. FLUX. Normally, transport is measured in terms of *flux* or transfer of material across a unit area in a given time. Typical units for flux are micrograms per square centimeter per hour, or picomoles per square centimeter per second.

UNIDIRECTIONAL FLUX. Frequently it is advantageous to consider the net flux across a membrane as the difference between two unidirectional fluxes. In Figure 24–2, unidirectional flux from side 1 to side 2 across the membrane, noted as $M_{1,2}$ is defined as the rate at which particles from side 1 cross a unit area of the membrane and appear on side 2. Similarly, the unidirectional flux in the opposite direction, $M_{2,1}$, is the rate per unit area at which particles originating on side 2 reach side 1. Thus, the net transport from side 1 to side 2

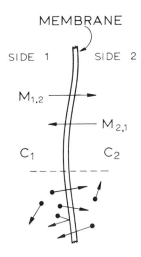

MEMBRANE

$M_{net} = M_{1,2} - M_{2,1}$

$P \doteq M_{1,2} / C_1 = M_{2,1} / C_2$

Figure 24–2 *Top*, unidirectional flux. *Bottom*, the statistical nature of diffusion. A net diffusion flux exists from side 1 to side 2 not because the individual particles on side 1 have a higher probability of penetrating the membrane than those on side 2, but simply because there are more of them on side 1.

is given by the difference between the two unidirectional fluxes, or

$$M_{net} = M_{1,2} - M_{2,1}$$

STEADY STATE. The term "steady state," frequently used in discussing transport, means simply "not changing with time." Thus, a flux or concentration has reached a steady state when it is constant and no longer varies with time.

EQUILIBRIUM. A substance that is transported through a membrane is in equilibrium if it fulfills the following conditions: (i) its net flux across the membrane is zero (although the unidirectional fluxes may be appreciable); (ii) all movement through the membrane is due to "passive" forces, such as concentration gradient or electrical potential.

PASSIVE. Movements or forces are defined as passive if they develop spontaneously and do not depend upon an energy supply linked to metabolism. Particle movements commonly treated in physical chemistry are passive, *e.g.*, diffusion, migration of ions in an electric field, osmosis.

ACTIVE. Movements or forces are active in general if they are directly linked to and utilize the energy created by biologic metabolism at the site of transfer. A more specific definition and description of active transport through biologic membranes is given in later sections of this chapter. These forces are common in biologic systems, but not in inanimate systems.

Modes of Membrane Transport.[13, 28] The rate at which a substance passes through a membrane depends upon the balance between two factors: The first is the magnitude of forces responsible for the movement, such as concentration gradient or electrical potential gradient. The second factor is the ease with which the particle passes through the membrane; this is expressed as permeability or conductance or by some other appropriate term. The ease of particle movement within a membrane depends in turn upon the mode of transport. For many substances, the detailed mechanism of movement within membranes is unproved or unknown, but there are several plausible hypotheses. These include (i) direct passage through the lipoprotein membrane, (ii) movement through membrane pores, and (iii) pinocytosis.

DIRECT PASSAGE.[32] For substances which can penetrate the membrane structure, transmembrane movement is not especially difficult. How easily the membrane structure can be entered is determined by the "solubility" of the transported substance in the lipoprotein matrix of the membrane. This property is measured by the concentration ratio between the membrane substance and the extracellular (or intracellular) aqueous solution; this ratio is called the partition coefficient. Substances with a high partition coefficient can establish relatively high concentrations within the membrane; therefore, high intramembrane concentration gradients and high fluxes can be attained. Substances with low partition coefficients are greatly impeded by the membrane, since there are not sufficient particles within it to lead to an appreciable flux (see Fig. 24–3).

In general, the lipid molecules limit the direct transport through the membrane substance; thus, substances moving by this method must have a high lipid-to-water partition coefficient. The major substances transported by this mechanism are lipid-soluble organic materials and dissolved

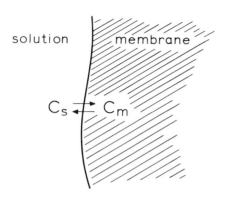

solution   membrane

$$C_s \rightleftharpoons C_m$$

$$\text{Partition Coefficient} = C_m / C_s$$

Figure 24-3 The partition coefficient.

gases. However, a particle which normally cannot penetrate the membrane surface may do so if it combines with an appropriate intramembrane molecule, much as soap permits the mixing of normally immiscible grease and water. If this intramembrane molecule is free to move within the membrane, it can shepherd the transported particle through the membrane and release it on the other side, functioning as a *carrier* within the membrane (Fig. 24–4). Such a

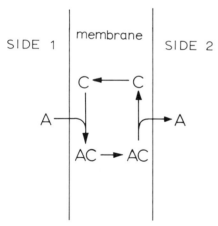

SIDE 1   membrane   SIDE 2

$$C \leftarrow C$$
$$A \quad \quad A$$
$$AC \rightarrow AC$$

Figure 24-4 Carrier-mediated transport of a particle, A, by an intramembrane molecule, C. If the reaction between A and C is passive, this mechanism is named *facilitated diffusion.* In contrast, if the carrier cannot return to side 1 without combining with another A (not shown), the mechanism is termed *exchange diffusion.* If the reaction between A and C involves metabolic free energy on one side of the membrane, then the carrier mechanism can form the basis for *active transport.*

mechanism could explain the ability of particles with low partition coefficients in lipids to pass through the membrane at an appreciable rate. Diffusion through a membrane which is aided by combination with carriers is termed *facilitated diffusion.*

TRANSPORT WITHIN PORES.[12, 16] The transport of non-lipid-soluble substances, including electrolytes, water and carbohydrates, might also be explained by the existence of small holes or pores through which they pass, rather than penetrating into the membrane structure itself. The rate of passage through a fenestrated membrane obviously depends partly on the particle size relative to the diameter of the pores. Thus, the relation between membrane permeability and particle size is a major experimental method for the study of movement through pores. In general, small particles are found to pass through membranes more rapidly than larger ones. For example, most membranes are more permeable to small univalent ions, such as potassium or chloride, than to the larger divalent ions, such as calcium, magnesium or sulfate. The size important here is probably the hydrated diameter, since water molecules bound to the dissolved particle effectively increase its size.

Often, instead of allowing particles of all sizes to pass, the membrane exhibits a cut-off, so that few particles larger than the critical size can move through the membrane. If particles pass through pores with a fixed range of internal diameters, the mean pore size can be calculated from the mean "cut-off" diameter, while the distribution of pore sizes about the mean can be calculated from the sharpness of the cut-off.

Additional information on pore size may be deduced from the rate of water movement through the membrane. Water would be expected to be confined to pores, since it is not lipid-soluble. Within the pores, water will move either because of hydrostatic pressure gradients or because of diffusion. If hydraulic flow in the pore is laminar, the water flux under a hydrostatic pressure gradient, Q_h, is directly proportional to the number of pores per unit membrane area (n) and the fourth power of the pore radius, r^4, and will be inversely proportional to the pore's length, L. Thus $Q_h \sim (n/L)r^4$. Diffusion flux, Q_d, will be directly proportional to n and to the cross sectional area presented by the pore, πr^2, and will be inversely proportional to L: thus $Q_d \sim (n/L)\pi r^2$. By measuring Q_h and Q_d, and taking their ratio, (n/L) is eliminated, permitting the calculation of r, the mean pore radius.

Finally, with electron microscopy, membrane pores have been demonstrated directly in some tissues and their sizes estimated.

Thus, three lines of evidence—influence of size of dissolved particles on transmembrane movement, rate of water fluxes, and direct electron microscopic observation—indicate that for many biological membranes, the major mode of transport for small, lipid-insoluble particles is through membrane pores (see Fig. 24–5).

PINOCYTOSIS.[19, 27] With electron microscopy, transported particles large enough to be seen sometimes are not uniformly distributed within the membrane but instead appear to be contained inside small intramembrane droplets. These droplets or *vesicles* appear to be formed by a process similar to phagocytosis: membrane invagination enclosing particles at the membrane surface. Then the vesicles are thought to move through the membrane, emptying their contents on the opposite side (see Fig. 24–6). This process has been named *pinocytosis.*

Highly specific binding sites at the membrane surface where invagination occurs have been postulated. By this mechanism, attracted particles could be concentrated within the vesicle and thus would be selectively transported. Such sites could account for the ability of pinocytosis to discriminate between the various particles in the external medium.

Pinocytosis is particularly adapted to the transport of large particles which would otherwise have difficulty in penetrating the membrane structure. The extent to which it

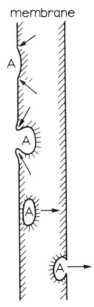

Figure 24–6 Successive stages in pinocytotic transport through a membrane.

contributes to the movement of smaller particles, such as electrolytes and simple organic molecules, is unknown.

PASSIVE MECHANISMS

Diffusion. The particles of a fluid (whether liquid or gas) are in constant motion. This motion is erratic and rather irresponsible; the particles lurch to and fro at random, continually changing both the magnitude and direction of their velocities. If the velocity of a single particle is tabulated at intervals (say, once a second), all directions of movement appear with equal frequency and the average magnitude of the velocity is independent of direction. Although such meanderings, so erratic and unpredictable at any single moment and so symmetrical when averaged, may seem of little consequence, this is not the case.

Consider a membrane separating two solutions with a solute concentration C_1 on side 1 and concentration C_2 on side 2 (Fig. 24–2). Since the solute particles are moving, some occasionally enter the membrane and pass completely through it. The number of particles which move through a unit area of membrane in a given time is directly proportional to the rate at which particles strike the membrane, and this in turn is proportional to the number of solute

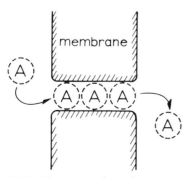

Figure 24–5 Movement through pores within a membrane.

particles near the membrane. Consequently, the unidirectional flux from a given side is directly proportional to the concentration on that side; the constant of proportionality is known as the permeability (P) of the membrane to that species of solute particles. Thus, $M_{1,2} = PC_1$. Similarly, the unidirectional flux in the opposite direction equals the concentration on the opposite side times the membrane permeability, or $M_{2,1} = PC_2$. The net transport rate is given by the difference between the two unidirectional fluxes, so

$$M_{net} = P(C_1 - C_2) \qquad (1)$$

This is the fundamental law describing passive diffusion through membranes.

The permeability of a membrane depends partly on its thickness: the farther the particle must traverse in its passage through the membrane, the lower its rate of transfer. In addition, permeability is determined by the energy necessary to pass through the membrane relative to the thermal energy of the diffusing particles: if considerable energy is necessary to pass through the membrane, relatively few particles will have the required energy; if the membrane presents only a low energy barrier to transport, the probability that an individual particle will have the necessary energy is considerably greater. This factor, which depends upon the structure of the membrane relative to the energy distribution of the particles in the solution, is expressed by the *diffusion coefficient*: a high diffusion coefficient implies low energy requirements for membrane passage; a low value implies that the membrane severely impedes particle movement. Thus,

$$P = \overline{D}/x_0 \qquad (2)$$

where $\overline{D}$ is the average diffusion coefficient within the membrane and x_0 is the membrane thickness. Combining equations (1) and (2) and noting that the total rate of particle transport is the product of flux and area (a) gives:

Total transport $= Ma = \overline{D}a\,(C_1 - C_2)/x_0$
$$\qquad (3)$$

Equation (3) implies that, to have rapid exchange without creating large concentration differences, a short diffusion distance and a large exposed area (as well as a high diffusion coefficient) are necessary. That

nature is not ignorant of equation (3) is evident from the architecture of those organ systems whose functions include the rapid distribution of material by diffusion, *e.g.*, the pulmonary alveoli, the intestinal mucosa, and the capillary beds of the circulatory system.

The differential form of the law of diffusion can be derived from the equations developed above. Consider the flux, not through the whole membrane, but rather through a thin lamina within the membrane. With the same reasoning as previously, the flux will be given by

$$M = -\overline{D}\,(\Delta C/\Delta x)$$

where ΔC is the concentration difference across the lamina and Δx is its thickness. The minus sign is necessary since the flux is from high concentration to low; in other words, if C is increasing with distance ($\Delta C/\Delta x$ positive), then the net flux is in the opposite (negative) direction.

If this lamina is taken progressively thinner, the ratio $\Delta C/\Delta x$ approaches the derivative $\partial C/\partial x$, since the derivative is defined as the limit of such a ratio. Also, $\overline{D}$ must approach D, the actual value of the diffusion coefficient at the place where the limit is taken. Thus,

$$M = -D\,(\partial C/\partial x) \qquad (4)$$

If the concentration varies in the y and z directions as well as the x direction, there will be components of the flux in all directions. Thus the directional derivative or gradient must be used in place of equation (4), so

$$M = -D\,\text{grad } C \qquad (5)$$

It must be emphasized that diffusion is not a force within the ordinary meaning of the term. An individual molecule has no more tendency to move in the direction of lower concentration than in the direction of higher concentration; net transport is due only to a volume element of high concentration having more molecules and thus losing more than a comparable volume having a low concentration. That a large number of actions, each individually random, can have a systematic and predictable effect is familiar to anyone with even a rudimentary knowledge of statistics.

Concentration, Activity and Partial Pressure. It has been assumed that the net effect of a number of particles acting simultaneously is simply the sum or statistical average of the effects of particles acting independently. Thus, it was possible to show that diffusion flux should be directly proportional to concentration gradient or difference, since concentration is a measure of the number of particles in a given volume. However, when the particles become crowded together, as in a highly concentrated solution, the assumption of independent action may no longer be valid, and the diffusion equations must be modified.

At the concentrations encountered in the body, surrounding a dissolved particle with other particles of the same species or of different species (other than molecules of the solvent, water) in general reduces flux rate below that predicted by the equations of the previous section. To account for this effect what is usually done is to preserve the form of the preceding equations by substituting for concentration a new variable, derived from it, called the *activity*. Thus, the flux within the solution is proportional to the activity gradient; the flow across a cellular membrane depends upon the difference in activity on the two sides.* The activity may be thought of as the effective concentration, as opposed to the actual concentration or content.

Sometimes the laws of diffusion are written using the activity coefficient, defined by the equation

$$A = \gamma C \qquad (6)$$

where A is the activity, C the concentration and γ the activity coefficient. Thus γ may be viewed as the fraction of the concentration which is effective in determining diffusion, chemical reactions, etc. Either γC or A may be substituted for concentration in the equations of diffusion.

At low concentrations, activity and concentration values are almost identical (or the activity coefficient is approximately 1), since when the solution is dilute, dissolved particles per unit volume are relatively few, so the probability of two or more dissolved particles interacting is correspondingly low. For dilute solutions, the diffusion equations may be used as they are written in the preceding section. The question then arises: At what concentration does a solution cease to be dilute and become concentrated? Obviously, the transition is continuous so no single value of concentration can be given; however, some general statements may be made.

Concentrations in biological fluids usually are less than 0.01 molar. At these relatively low levels, little error is made by assuming that concentration and activity are approximately equal, except in one of three conditions.

First, when the particles in solution have a net charge (ions), the interaction between them is considerably greater than if they are electrically neutral. The charged particles tend to be surrounded by others of opposite charge because of electrostatic attraction, although thermal motion of the particles prevents the establishing of rigid bonds. When an ion moves, it must break away from the surrounding particles or else drag them along; in either case its movement and therefore its rate of diffusion is decreased. Thus, as the solution becomes more concentrated and ionic interaction becomes increasingly prominent, the activity coefficient falls below 1. This effect in a sodium chloride solution is shown in Figure 24–7. For a NaCl solution isotonic with plasma, the activity is only about 75 per cent of the concentration. Adding additional ionic species, *e.g.*, $KHCO_3$, to this solution further reduces NaCl activity since K^+ and HCO_3^- ions contribute to the charged atmosphere surrounding Na^+ and Cl^-; for the same reason, $KHCO_3$ activity will be less than if NaCl were absent. In other words, the effects of various ions are cumulative, the reduction in activity coefficient of each individual species depending upon the total ionic strength of the solution. The ionic strength of blood plasma is about 244 mEq per liter, resulting in an activity coefficient of approximately 0.7 for all univalent ions; activity of polyvalent ions is reduced even more since their higher charge produces even greater interaction.

Activity is also reduced by sites of attraction which immobilize certain solute molecules. These sites may be either fixed (*e.g.*, along the capillary wall) or on another dissolved or suspended particle (*e.g.*, plasma protein molecule). If the attraction is weak or the sites few, the movement of particles is little affected, but if attraction is strong and sites are plentiful, the effect on particle movement is great. (The stronger of these forces are sometimes called reversible bonds.) Because of the continuum of bond strengths, it cannot be said when a simple attraction becomes a chemical reaction. In any case, the number of particles which can participate in other activities is reduced. This sometimes is called the "free concentration" to distinguish it from the actual concentration or "total content."

Gases are usually treated in a special manner. Their activity in an aqueous solution depends upon the concentration of dis-

---

*Also, it can be shown that the rate of a chemical reaction depends upon the activities and not the concentrations of the participating species; that is, to be accurate the law of mass action should be written in terms of activity rather than concentration.

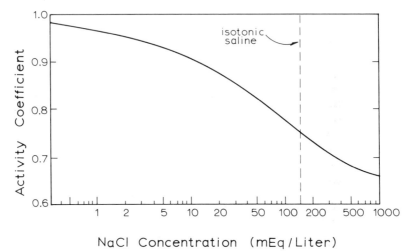

Figure 24–7 The activity coefficient at increasing sodium chloride concentrations (logarithmic scale on the concentration axis).

solved gas which is free to move. The "free" concentration frequently is difficult to measure directly, since much of the dissolved gas may be immobilized by binding (e.g., O_2 to hemoglobin) or transformed into another chemical form (e.g., CO_2 to bicarbonate). However, the activity can be evaluated indirectly, since it is directly proportional to the partial pressure of that gas with which the solution is in equilibrium.* Thus, partial pressure gradient may be thought of as the driving force for the diffusion of gas.

In summary, the diffusion equations should be modified: (i) if the particles are charged, by multiplying the concentration by the activity coefficient to account for the electrical interaction between particles; (ii) if some of the solute molecules are bound, by substituting the "free concentration" for the total concentration or content; (iii) if the solute molecules are dissolved gas, by substituting partial pressure for concentration.

Electrical Forces. When an electrical potential is established across a biologic membrane, an additional influence is superimposed upon the random thermal motion of charged particles. On the positive side of the membrane, positively charged par-

ticles are repelled, thus increasing their rate of passage through the membrane; on the other hand, negative particles are attracted and, thus, tend to remain on this side. On the negative side of the membrane, the opposite occurs: positive particles tend to remain and negative particles tend to cross the membrane. Thus, an electrical potential difference destroys the symmetry of movement characteristic of thermal agitation of charged particles by establishing a preferred direction of transport.

Consider a membrane with the potential V_1 on side 1 and V_2 on side 2 and with equal concentration, C, of some particular solute on both sides of the membrane (only uniform concentrations will be considered here to avoid complications of simultaneous transport via diffusion) (Fig. 24–8). When the potential is applied, particles flow through the membrane from side 1 to side 2 with a magnitude of

$$M_{net} = zmC(V_1 - V_2) \qquad (7)$$

Here, m is the electrical permeability of the membrane (analogous to P, the diffusion permeability). The quantity z is the net number of electronic charges on a solute particle; the sign of z indicates whether the particle is positive or negative; for example, for sodium ion, $z = +1$; for chloride ion, $z = -1$; for calcium ion, $z = +2$. Thus, the electrical flux through the membrane depends upon the driving force $(V_1 - V_2)$, the number of particles per unit volume, C,

---

*The use of partial pressure instead of fugacity, the analog of activity for gases, is valid only for ideal gases; however, in the normal physiologic range, little error is made by assuming that gases behave ideally.

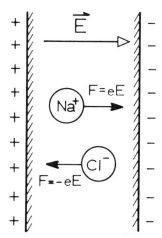

Figure 24-8 Movement of charged particles due to a transmembrane electrical potential. The potential is associated with an electric field, E, within the membrane, leading to a force of e (the net charge) times E. The force is in the same direction as E for positively charged particles and in the opposite direction for negative particles; however, the movement of both types of particles leads to a net current flow from the positive side of the membrane to the negative side.

the ease with which the particles are electrically propelled through the membrane, m, and the charge on an individual ion, z.

Much as in the development of the diffusion equation (2), the electrical permeability of the membrane may be written as the ratio of a property inherent in the molecular structure of the membrane, η, divided by the membrane thickness, Δx.

$$m = \eta/\Delta x \qquad (8)$$

The value of the η depends upon the relationship between the energy imparted to charged particles by the electrical potential gradient and the energy barrier imposed by the membrane structure. Thus, the η is for electrical migration the same membrane property expressed by D for diffusion. The membrane would be expected to present a similar resistance to moving particles whether their source of energy was thermal agitation or electrical potential. At a constant temperature, the ratio between η and D is constant and is given by equation 8(a).

$$\eta/D = F/RT \qquad (8a)$$

Here, F is Faraday's constant, the net charge per mol of univalent ions (96,496 coulombs); T is the absolute temperature (measured

in degrees Kelvin); and R is the gas constant, expressing the thermal energy per degree temperature (8.31 joules per mol per degree centigrade). Since the electrical energy gained by a charged particle in passing from one potential to another is directly proportional to its charge, the ratio F/RT indicates the ratio of the electrical energy (per volt) compared to the thermal energy as indicated by RT.

As in the derivation of equation (5), by substituting equation (8) into equation (7) and taking the limit as the region under consideration becomes progressively thinner, and by taking into account the direction of motion under an electrical potential difference, electrical flux can be written as

$$M = -\eta zC \, (\partial V/\partial x) \qquad (9)$$

Since the negative of the voltage gradient, $-(\partial V/\partial x)$, is equal to the electric field, E, equation (4) may be written as

$$M = \eta zCE \qquad (10)$$

Either equation (9) or (10) may be taken as the fundamental equation describing the movement of charged particles under the influence of an electric field.

A flux of charged particles represents an electric current. Specifically, the membrane current carried by any particular species of ions is given by

$$J = zFM \qquad (11)$$

where J is the current density (usually amperes per sq cm). Substituting into the above relation for M from equation (7) gives

$$J = z^2FmC(V_1 - V_2) \qquad (12)$$

The product "z^2FmC" may be abbreviated by the single term "g," called the specific membrane conductance for the particular ionic species. Thus

$$J = g(V_1 - V_2) \qquad (13)$$

This is the equivalent of Ohm's law for a biologic membrane since it states that current density is equal to the product of conductance and potential difference. When a number of ionic species traverse the membrane independently, the total conductivity is simply the sum of the conductivities of the individual ionic species.

Water Movement and Osmosis.[2,14,24,26,31,35] Water is not only the environment through which dissolved particles move while in solution, but is also itself capable of moving between various points within tissues and in particular of moving across membranes. Two types of forces are responsible for water movement.

The first type is ordinary mechanical force or pressure. If the pressures on each side of a membrane are not equal, water flows from the side of higher pressure to the side of lower pressure:

$$Q = P_h (Pr1 - Pr2) \qquad (14)$$

where Q is the net water flux from side 1 to side 2; P_h is the hydrostatic permeability of the membrane; and Pr1 and Pr2 are the respective pressures. If flow is through small holes or pores in the membrane, and if flow is laminar, P_h can be evaluated, by using Pousielle's law, from the length, radius and number of pores per unit area of membrane and the viscosity of water.

In addition, water has a second way of passing through membranes. Water molecules in the liquid state participate in the random thermal motion characteristic of all particles in the solution; thus, water can move by a mechanism similar to that responsible for solute diffusion. Suppose that a membrane, permeable to water, separates two aqueous solutions of different solute concentrations, C_1 on side 1 and C_2 on side 2, with $C_2 > C_1$. Because water has less tendency to move from a concentrated solution than from a dilute one, water will move from side 1 to side 2. The equation for water flux, like that for solute particle flux, may be written as

$$Q = P_{os}(a(H_2O)_1 - a(H_2O)_2) \qquad (15)$$

In this equation, $a(H_2O)_1$ and $a(H_2O)_2$ are the thermodynamic activities of water on the two sides of the membrane. Development of net water flux through a membrane because of a difference in water activity is usually called *osmosis*.

It must be pointed out that osmosis cannot be regarded as simple diffusion of water through a membrane. Measurements of water diffusion rate with isotopic tracers indicate that flux attributable to simple water diffusion alone is generally much less than that due to osmotic gradients. Why osmosis can produce such rapid water movement is not clear. Thus, although the source of energy responsible for diffusion and osmosis are similar—namely, the random thermal motion of particles in the solution—because of some mechanism not presently completely understood, a difference in water activity across a membrane usually leads to a greater water flow than can be explained by simple diffusion.

To evaluate osmotic flow from equation (15), the activity of water in the solution must be known. For dilute solutions, as discussed earlier, solute activity is adequately approximated by the concentration (or some modification, such as "free" concentration). However, in an aqueous solution, obviously water cannot be considered dilute. Thus, molar concentration is not an appropriate measure of water activity. There are several conventions for evaluating solvent activity; for dilute solutions, the most common is as follows:

The activity of any pure solvent is defined as unity (1). When solute particles are dissolved in the solvent, the activity of the solvent is reduced. For dilute solutions, the activity reduction is proportional to the number of dissolved particles, or

$$a(H_2O) = 1 - k\Sigma C \qquad (16)$$

C is the concentration of solute particles; k is a constant with an approximate value of 0.018 (its exact value is unimportant since it is usually incorporated into other constants in the flux equations); the summation sign, Σ, indicates that the concentrations of all species of dissolved particles are summed in evaluating activity. Thus, water activity is not determined by water concentration, but rather by the total concentration of all particles dissolved in the water. The greater the concentration of dissolved particles, the more water activity is reduced.

Since the coefficient k does not vary appreciably within physiologic limits, the osmotic activity of any solution is determined mainly by the summed concentrations of the dissolved solutes, ΣC. The summed molar concentration is called the *osmolarity* of the solution. Each distinct species of particle is counted separately. Thus, for a substance whose constituents separate upon going into solution, the concentrations of the individual parts must be included in calculating osmolarity. For example, for a 1 mM glucose solution, the osmolarity is 0.001 osmols or 1 milliosmol; for a 1 mM solution of NaCl, the osmotic activity is approximately 2 milliosmols; for a 1 mM solution of $CaCl_2$, the osmolarity is approximately 3 milliosmols. If the solutions become sufficiently concentrated so that the dissolved particles interact, the osmotic effect is less than that predicted from equation (16), because such particles do not act independently and cannot be simply summed as independent species. For concentrated solutions, ΣC must be multiplied by a factor called the osmotic coefficient. For example, a 154 mM NaCl solution (isotonic saline) has an osmotic activity of only 283 milliosmols, since its osmotic coefficient at this concentration is approximately 0.92; $(283 = 0.92 \times 2 \times 154)$.

Equation (16) is based upon the fact that, for an ideal solution, the activity of any given substance is directly proportional to its mol fraction. The mol fraction of any substance is the number of mols of that substance per unit volume divided by the total number of mols of all substances per unit volume. For a solute, A, with water as the solvent, the mol fraction f_A, is

$$f_A = \frac{m_A}{m_A + m_{H_2O}} \qquad (17)$$

where m_A and m_{H_2O} are the number of mols per unit volume of A and of water respectively. For a dilute solution in which $m_A \ll m_{H_2O}$, equation (17) may be approximated by

$$f_A \approx \frac{m_A}{m_{H_2O}} \approx \frac{C_A}{55.5} \approx 0.018\, C_A \qquad (17a)$$

The expressions on the right of (17a) follow from taking the unit volume equal to 1 liter, in which case m_A is simply the molar concentration, C_A, and m_{H_2O} is approximately 55.5 mols of H_2O per liter. Thus, in a dilute solution, solute concentration may be used as an index of solute activity, as was done in evaluating solute diffusion previously.

The mol fraction of the solvent, f_{H_2O}, is

$$f_{H_2O} = \frac{m_{H_2O}}{m_{H_2O} + m_A}$$

Again, using the approximation for a dilute solution in which $m_A \ll m_{H_2O}$

$$f_{H_2O} \approx 1 - \frac{m_A}{m_{H_2O}} \approx 1 - kC_A$$

Thus, for a pure solvent $(C_A = 0)$, $f_{H_2O} = 1$, while the decrease in activity of the solvent when a substance is dissolved in it depends upon the solute concentration. If several species of dissolved particles are present, their effects are additive in reducing solvent activity, leading to equation (16).

Substituting the value of water activity derived from equation (16) into equation (15) gives

$$Q = P'(\Sigma C_2 - \Sigma C_1) \qquad (18)$$

Here, P' equals $k \times P_{OS}$; this coefficient indicates the relation between osmolar difference and water flow. Note that equation (18) states that water moves from the side of low osmolarity to that of high osmolarity (since osmolarity and solvent activity of water are inversely related).

If a given solute can move freely across the membrane and come to equal concentrations on both sides, obviously it cannot contribute to sustained osmotic movement. Permanent osmotic effects are possible only from particles to which the membrane is impermeable or which cannot attain equal concentration for some other reason (such as electrical effects on charged particles or active transport). Thus, semipermeable membranes, which are permeable to water but not to some dissolved particles, can develop osmotic forces, but they are not the only types of membranes across which such forces can develop; any membrane able to prevent solute particles on either side from becoming equally concentrated can sustain permanent osmotic effects.

In evaluating osmotic water flow, only the concentrations of those substances which are out of concentration equality need be considered. Thus, only the concentrations of plasma proteins (and those charged particles which separate because of an electrical potential across the capillary wall) need be used to evaluate water movement across the capillary wall, since all other particles have the same concentration in plasma and interstitial fluid. However, in nerve or muscle cell membranes, the concentrations of sodium, potassium and chloride must be considered, even though the membrane is permeable to these ions, since they have greatly different concentrations intracellularly and extracellularly.

Osmotic differences do not always lead to water movement, since osmotic movement can be reduced by an opposing hydrostatic pressure. The hydrostatic pressure required to reduce net water flux to zero is a measure of the osmotic pressure across the membrane. Thus, *osmotic pressure* is the osmotic force across a membrane under conditions of no water flow; however, osmotic pressure is a fictional pressure in the sense that it cannot be measured directly with a manometer or any other pressure sensing device; its magnitude can be deduced only by measuring the hydrostatic pressure which neutralizes its effect.

From thermodynamic considerations, it can be shown that the relation between osmotic pressure and osmolarity is given by

$$\pi = RT\Sigma C \qquad (19)$$

where π is the osmotic pressure attributable to a solution with an osmolarity of ΣC, R is the gas constant, and T the absolute temperature. The net osmotic pressure developed across a membrane is the difference between the osmotic pressures on each side, or

$$\pi_{net} = \pi_2 - \pi_1 = RT (\Sigma C_2 - \Sigma C_1) \quad (19a)$$

As before, only solutes with differing concentrations on either side need be included in equation (19a).

Evaluating RT at 37° C. (310° K.) gives

$$RT = 25.4 \text{ atmospheres per osmol}$$
$$= 19{,}300 \text{ mm Hg per osmol}$$
$$= 19.3 \text{ mm Hg per milliosmol}$$

This value gives an idea of the large forces which can be developed by small differences in osmotic activity. That 180 grams of glucose (about 6 ounces) dissolved in a liter of water can support a column of distilled water over 300 meters high (about 1000 feet) seems preposterous but is true nonetheless.

The rate and direction of water movement through the membrane is determined by the combined effects of both hydrostatic and osmotic gradients. The net force causing water to leave side 1 is $(Pr1 - \pi_1)$. Thus water flows from the side with higher $(P - \pi)$ to the side with the lower. If the hydrostatic and osmotic forces just balance, so that

$$(Pr1 - \pi_1) = (Pr2 - \pi_2), \quad (20)$$

there is no net water transport.

When cells with a given internal osmotic activity are exposed to a solution with a different osmotic activity, water flows through the cell membrane. Obviously, such water movement cannot continue indefinitely. If the external solution is hypotonic, the resultant water movement into the cell has three effects: (i) the intracellular solution becomes less concentrated, decreasing its osmotic activity; (ii) the extracellular solution may become more concentrated, increasing its osmotic activity; (iii) because of the water influx, the cell enlarges, increasing its wall tension. The increased tension may increase internal hydrostatic pressure, although the relation between internal pressure and cell volume depends on the elastic characteristics of the cell wall. Thus, the first two, and perhaps the third, of these effects progressively

reduces water influx. If the cell can successfully withstand the mechanical stresses caused by its increased volume, equation (20) will eventually be satisfied, and net water movement will become zero. Similarly, if the cell is exposed to a hypertonic bathing solution, water leaves the cell, raising its internal osmotic activity, reducing the osmotic activity of the extracellular solution, and perhaps, reducing the internal hydrostatic pressure. Again, the process continues until equation (20) is satisfied or until the integrity of the cell wall is lost.

In summary, water movement through a membrane depends upon both hydrostatic pressure and osmotic force. The osmotic force may be expressed in terms of (i) water activity, (ii) osmolarity or (iii) osmotic pressure; the relations between these quantities are given by equations (16) and (19). Osmotic pressure is particularly useful in evaluating the effect of combined osmotic and hydrostatic forces.

Another aspect of the lowered activity of water due to dissolved particles is a reduction of the ease with which water molecules leave the solution surface and escape into the atmosphere. In other words, the partial pressure of water vapor is reduced as the osmolarity increases. Advantage is taken of this dependence of P_{H_2O} on solute concentration in the design of osmometers, instruments used to measure osmotic activity. One major type of osmometer equilibrates a drop of the unknown solution with a liquid of known vapor pressure, and converts the results to milliosmols. A second type is based on the fact that the temperature at which an aqueous solution and ice can exist in equilibrium (i.e., the freezing point) depends upon the solution vapor pressure; thus, freezing point depression can be used to measure osmolarity.

Effect of Water Flow on Particle Movement; Solvent Drag and Bulk Flow.[1, 33] When particles are transported through a membrane, osmotic forces may be developed which make water follow. The converse effect may also occur: the net flux of water through a membrane may move dissolved particles in the same direction.

When solute particles diffuse through a membrane, the fixed membrane structure slows particle motion, thus reducing the flux rate. However, when water is also moving through the membrane in the same direction, the water flux tends to carry the dissolved particles along. Similarly, water movement in the direction opposite to solute diffusion reduces the flux rate. Because this effect depends upon the difference between solute and solvent velocities, it is called *solvent drag*. The influence of solvent drag on particle motion depends on the relative magnitudes of dissolved particle interaction with the membrane and dissolved particle interaction with the water moving in the membrane.

If solute particles and water move through the membrane by separate pathways, their interaction is small and solvent drag effects do not occur. On the other hand, if the solute particles and water share a common transport pathway through the membrane, solvent drag effect upon solute transport may be large. In addition to solvent drag effects, solute movement is also influenced by the changes in concentration which occur along with a net movement of water through a membrane. When water leaves one side of a membrane, the concentration of solute particles on that side increases and that on the other side decreases. Thus, the movement of water establishes concentration gradients tending to move particles in the same direction as net water flux.

The sum of solvent drag and concentration effects may be so great that it dominates the movement of a given species of dissolved particles. In this case, the particles are swept through the membrane as if simply carried along in aqueous solution. Such particle movement is called *bulk flow* and is characterized by the relation

$$M = QC$$

where Q is net water flux, and C is the solute concentration on the side of the membrane from which the water movement originated.

Bulk flow is characteristic of extremely permeable membranes where the interaction forces between particles and membrane are small, *e.g.*, in the systemic capillaries and the renal glomerulus.

ACTIVE TRANSPORT

Definition of Active Transport.[5] The mechanisms of membrane transport discussed in the preceding section have one property in common: they can cause a net flux of material only by decreasing or dissipating the total available (free) energy of the system. If no other mechanisms are present, the system will spontaneously decrease in free energy under the impetus of these forces until the minimum value is reached. At this point, the concentrations and free energy cease changing and the net fluxes of all particles are zero. In other words, the energy decreases until a thermodynamic equilibrium is reached. To call these influences on transport "passive" is not to deprecate their vigor or independence, but rather to indicate that they require no external energy. They occur in both living and nonliving material; the laws that describe them arise from purely physical properties of moving particles in the liquid state when interacting with a fixed structural matrix or membrane. Diffusion, electrostatic movement, osmosis and solvent drag are all passive mechanisms, although they do not exhaust the catalog of this type of transport.

The asymptotic approach to thermodynamic equilibrium is not characteristic of living organisms, and, indeed, is inconsistent with life as we know it. All living material can avoid, at least briefly, this energetically degraded condition. Since all matter has the mechanisms of passive transport, it is clear that living cells can maintain a steady state out of equilibrium only by supplying free energy at the same rate at which it is spontaneously degraded; the source of this energy is the biochemical metabolism of energetic substrates.

Two mechanisms are used by living cells to hold materials out of equilibrium across membranes. One is exemplified by the transport of O_2 and CO_2. Since O_2 is consumed within the cell, its partial pressure is lowered below that of the surrounding blood and interstitial fluid, thereby maintaining a flux of O_2 into the cell. In a similar manner, CO_2 is produced in the cell, causing a net efflux. The combination of O_2 with metabolic substrates makes free energy available at the expense of substrate energy; otherwise the reaction would not proceed spontaneously. Part of this energy is tapped to provide the gradients for O_2 and CO_2 diffusion. This mechanism is similar to that permitting the cell's internal metabolic mechanisms to function; that is, the tapping of substrate energy allows the biochemical reactions within the cell to be displaced from equilibrium, making it possible for a substance to proceed along (or around) biochemical pathways at a finite rate.

The second mechanism, and the one which concerns us, is the coupling of metabolic energy to transport within the membrane, *i.e., active membrane transport.* This term has been defined in various ways. To avoid confusion we shall classify a force influencing membrane transport of particles as active only if it meets the following criteria: (i) the force is located within the membrane; (ii) the force *directly* influences particle motion; (iii) the force tends to increase the free energy of the particle as it passes through the membrane; (iv) the force is established by and maintained through the consumption of free energy made available by metabolism. Thus, an *active* force may be viewed as a means of transferring energy released by biochemical reactions to particle movement through the membrane.

An example of active transport encountered previously is the movement of sodium through the axon membrane (Chap. 2, Vol. I). Since sodium moves out of the axoplasm against both a concentration gradient and an electric field, an additional force clearly must be invoked to explain its movement. This active force is provided by the coupled sodium-potassium "pump." However, the movements of chloride ion and water following sodium movement is not active since they are only *indirectly* linked to an active force. Thus, chloride moves only by the passive electrical forces established by the exchange of sodium for potassium and the subsequent potassium diffusion, while water moves passively only because of the osmotic gradients consequent to particle interchange. For the same reasons, sodium movement during the initial part of an action potential is not ascribed to active forces, because it is due to the passive or natural forces resulting from its concentration and electrical potential gradients. Thus, across any membrane, movement directly linked to active transport may exist simultaneously with passive movement linked only indirectly to active transport and with passive movement totally unconnected with active transport; also, the motion of the same particle may be dominated at one time by active forces, and, at another time, by passive forces.

In summary, a variety of forces influence the movement of particles across living membranes. Some are classified as passive, meaning that they reduce the system's free energy, require no outside energy supply, and occur in inanimate as well as living membranes. Other forces are classified as active, as defined above, and are characteristic of many biologic membranes. The detailed properties of active transport and the membrane mechanisms which may be responsible for energy coupling are the subjects of the remainder of this chapter.

General Characteristics of Active Transport. The molecular mechanisms of active particle transport through membranes may be simple modifications of similar passive mechanisms. However, the availability of and the dependence upon external energy gives active transport several characteristics not present in passive movement.

DEPENDENCE UPON METABOLIC SUBSTRATES.[20, 38] The energy which moves particles through the membrane comes from the free chemical energy of organic substrates, for example, glucose. Most sustained active transport is aerobic; thus, oxygen is consumed so that the withdrawal of either the substrate or oxygen eventually stops active transport (whereas passive transport would be relatively unchanged as long as the membrane remains intact).

The dependence of active transport upon energy sources is seen in the mammalian intestine. Normally, the transporting cells in the intestinal epithelium are supplied with oxygen through the mucosal capillaries. Active transport can be maintained even when intestinal segments are removed from the body and placed in a test tube, but only if oxygen-saturated solution is vigorously perfused by the mucosal surface. For optimal transport, the perfusing solution must contain glucose. Similar behavior is seen in the sodium-potassium pump in nerve axon. If nitrogen is flushed through the solution surrounding an excised axon to wash out the dissolved oxygen, the axon soon ceases to pump ions, and eventually the internal solutions equilibrate with the external solutions and the membrane potential falls to zero.

SENSITIVITY TO METABOLIC POISONS. Because of their direct dependence upon cellular metabolism, active transport forces are particularly susceptible to metabolic poisons, *i.e.,* agents which inhibit or divert metabolic reactions as in the action of DNP (dinitrophenol) on the rate of sodium extrusion from nerve (Chap. 2, Vol. I). DNP uncouples substrate utilization from oxidative phosphorylation so that less energy is available for active transport. (This observation also indicates that the products of oxidative phosphorylation must be intermediates in the transfer of free energy to active transport in this system.)

ABILITY TO MAINTAIN CONCENTRATION DIFFERENCES IN THE FACE OF PASSIVE GRADIENTS. A passive gradient across membranes or cell layers not possessing active forces causes particle flux until a passive equilibrium is reached. An external free energy supply allows those membranes which can produce active forces to resist successfully the establishment of a passive equilibrium and maintain indefinitely concentration differences in the face of opposing passive forces.

Consider a membrane which separates two solutions of initially identical composition (Fig. 24–9). If the membrane begins to transport some species of particles actively, the concentration on one side decreases and that on the other increases. This concentration difference has two effects. First, it leads to a passive diffusion flux in the direction opposite to the active transport;

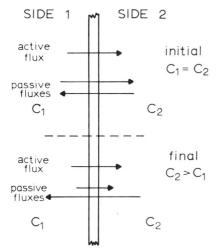

Figure 24–9 Transition from an equilibrium ($C_1 = C_2$) to a steady state in which $C_2 > C_1$ because of active transport. The steady state is reached when the concentration difference leads to a net passive flux equal in magnitude and opposite in direction to the active flux.

and, secondly, the minimum energy requirement for active transport goes up, since more free energy is required to transport particles from dilute to concentrated solutions than between solutions of equal concentration. As long as active transport flux is greater than opposing passive diffusion flux, concentration difference continues to rise. This greater concentration difference increases the passive flux and, because of increased energy requirements, may reduce active flux. Eventually, these fluxes become equal, after which the solution composition remains stable. The concentration ratio so established is called the maximum concentrating ability of the membrane or tissue; it is a measure of the relative magnitudes of active and passive forces. (For simplicity, diffusion is assumed above to be the only important mode of passive movement. The existence of other passive forces does not alter the general conclusion, although the analysis is more complex.)

ABILITY TO MAINTAIN FLUX IN THE ABSENCE OF PASSIVE GRADIENTS.[37] Just as an active transport force is able to maintain a steady state by balancing the effect of opposing passive forces, so can it, in the absence of passive opposition, establish a net flux across a membrane. For example, if plasma serum is introduced into the small intestine, it eventually becomes completely absorbed, although obviously no concentration gradient is present initially. (Note that this does not imply that every serum component is actively transported from the intestinal lumen, since transport of one or several particle species may create forces— e.g., electrical gradients, osmotic gradients —which cause the remaining particles to be absorbed passively.) Thus active transport can lead to a net flux in the absence of or even when opposed by passive forces.

SATURATION. For simple passive diffusion, the unidirectional flux increases linearly in direct proportion to the concentration. In contrast, an actively transporting system would be expected to be limited by the rate at which it can supply energy to the transported particles. At low particle concentrations, this maximum energy limitation may not be important. But at higher concentrations, its influence upon the membrane flux becomes increasingly significant. In other words, the flux rate would be concentration limited at low concentrations, but transport mechanism limited at high concentrations. Thus a plot of flux versus concentration levels off as the active transport mechanism becomes saturated. Such saturation is characteristic of many actively transporting systems.

Saturation is not unique to active transport systems. Although simple passive diffusion does not reach a saturation point, more complex passive processes may well exhibit flux maxima. For example, passive diffusion facilitated by intramembrane carriers, as discussed earlier, is limited by the number of carriers and the rate at which they can shuttle across the membrane.

Mechanism of Active Movement. To establish an actively transporting system, nature must solve two difficult problems: (i) coupling metabolically derived energy to moving particles and (ii) supplying this energy primarily to particles moving in one direction, since coupling energy to all particles traversing a membrane does not result in active transport. Thus, the particles traveling in the direction promoted by active transport must receive energy which is unavailable to or hinders particles moving in the opposite direction.

Because the molecular events within a membrane are difficult to investigate, direct experimental evidence establishing the detailed mechanism of active transport is sparse. However, several plausible hypotheses have been postulated, each supported to some extent by indirect evidence. These are (i) carrier-mediated active movement, (ii) oriented binding sites within the membrane, and (iii) pinocytosis.

CARRIER-MEDIATED ACTIVE TRANSPORT.[18] The utility of a carrier molecule in shepherding particles through a membrane was noted earlier. The particle is bound to the carrier on one side of the membrane, moved through the membrane, and released from the carrier into the solution on the other side. The carrier then returns to its original site (see Fig. 24–4). Writing A for the transported particle and C for the carrier molecule, the reaction between them is

$$AC \rightleftharpoons A + C \qquad (21)$$

But how can a reaction, which on one side of the membrane goes spontaneously to the left, binding the particle, on the other side go spontaneously to the right, releasing the

particle? A localized catalyst alone cannot be responsible, because catalysis can alter only the rate, not the predominant direction or equilibrium constant of a reaction. Instead, a reaction will proceed spontaneously in the direction in which free energy is released. Thus, the direction of the reaction can be reversed only by altering the free energy evolved. How this could be accomplished is exemplified by the following:

Assume that the particle and carrier combine spontaneously, forming AC on one side of the membrane, and that AC then diffuses through the membrane. On the other side, conditions are such that AC dissociates when stochiometrically linked with a second reaction in which a high energy compound is degraded to a lower energy compound, say ATP $\longrightarrow$ ADP; this energy released by the second reaction allows separation of the particle from the carrier. Thus, the reaction is spontaneous on one side of the membrane and utilizes a coupled, energy-evolving, auxiliary reaction to reverse on the other side of the membrane (see Fig. 24–10).

Thus, the particle is transferred from the solution on one side of the membrane to that on the other, possibly against passive forces, fueled by energy originally derived from metabolism—in other words, by active transport. Note that energy is used not in conjunction with movement within the membrane, but rather in conjunction with release from the carrier.

DIRECTIONAL SITES.[3] Mobile carriers are not necessary for active transport; binding sites fixed in the membrane structure also could lead to active movement, if they could impart free energy to particles moving in a preferred direction.

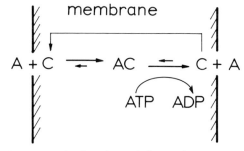

Figure 24–10 Coupling of chemical energy to particle-carrier separation to produce active transport.

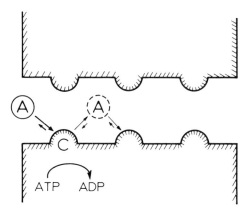

Figure 24-11 Coupling of chemical energy to a fixed attraction site to produce active transport.

For example, consider a sequence of binding sites lining a channel or pore through the membrane and having the following three properties (see Fig. 24-11): (i) The particle can become bound to the site from either the right or the left side. (ii) On one side, say the left, the binding is firm, but on the other, the right, it is weak. From equation (21), with C designating the binding site rather than the carrier, it can be seen that if the particle is close to the site and on its right, the reaction tends to go to the right, whereas if the particle is on the left, the reaction tends to go to the left. (iii) The bond angle can fluctuate between the two positions.

Consider next a particle as it enters the channel from the left. When the particle nears a binding site, it is likely to be bound since strong bonds are formed on the left. When the bond configuration changes spontaneously, moving the particle to the right, the particle probably will be released, because of the weak binding on the right. Next the particle will be attracted to the adjacent site; in this manner the particle is "handed" from site to site and moves through the membrane from left to right.

But how can the molecular structure of the site change in this way, first binding and then releasing the transported particle? Such an alteration cannot be passive, since it implies a change in the equilibrium constant or, in other words, a free energy change—hence, active transport.

PINOCYTOSIS. The formation of vesicles and their movement within the membrane involve modifications of part of the mem-

brane structure. Such modifications require energy. Thus, pinocytotic active transport is linked directly to energy supplying reactions in the production and movement of vesicles.

Control of Active Transport. The rate of active transport in a given tissue is not necessarily steady or constant, but usually varies to meet the changing needs of the body. Thus, following a burst of action potentials, active sodium extrusion from nerve axoplasm increases relative to the quiescent rate, to remove the accumulated sodium; the rate of absorption of ingested particles from the intestinal lumen into the capillaries varies with autonomic nerve activity; the renal excretion of several substances depends upon their plasma concentrations, through mechanisms depending in part upon active transport.

Knowledge of the processes involved in the control of active transport is still relatively sparse. However, several mechanisms may be postulated, each supported to some extent by experimental evidence. These mechanisms need not be mutually exclusive; frequently, the net transport rate may result from the balance of several influences.

DIRECT CHEMICAL AND HORMONAL CONTROL.[4] Evidence from some tissues indicates that the velocity of active transport reactions may be controlled by chemicals either circulating in the blood or locally produced. For example, the effect of aldosterone on electrolyte balance is probably due directly to its stimulating action on the sodium-potassium pump, particularly in renal tubule cells. The action of various hormones on the glands of the gastrointestinal tract, e.g., promotion of hydrochloric acid secretion by gastrin, likely are additional instances of such control. Exogenous chemicals may be effective also; for example, the drug Dilantin (diphenylhydantoin) appears able to mimic aldosterone, whereas the cardiac glycosides (such as digitalis) have been shown to have the opposite effect, inhibiting the sodium-potassium pump in cardiac muscle and other tissues.

CONTROL THROUGH PASSIVE PERMEABILITY.[6, 11, 25] When a particle moves through a tissue layer in which the active transport pump is confined to one surface, the flux rate can be effectively controlled by

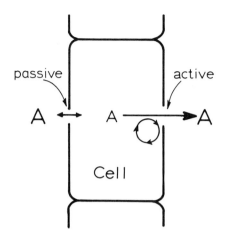

Figure 24–12 Control on one cell surface of active transport rate due to changes in the intracellular concentration established through the passive exchange on the opposite surface.

varying the passive permeability of the other surface. Figure 24–12 shows a possible configuration with the particle A being actively extruded from one side of a representative cell while entering the cell passively on the other side. If the passive permeability is high, the active pump operates near its maximum rate. However, if the passive permeability is low, active pumping tends to exhaust the intracellular supply of A; thus, active pumping is limited by the rate of passive intracellular replenishment of A. In this way, changes in passive permeability can influence the rate of active transport.

CIRCULATORY CONTROL. The cardiovascular system can influence the rate of active transport in two ways. First, for active movement to proceed, metabolic substrates must be furnished and metabolic end products must be carried away. Since most active transport energy depends eventually on aerobic reactions, a continuing supply of oxygen is also necessary. Substrate distribution and end product removal are the primary functions of the circulatory system. Thus, the rate of tissue perfusion can influence active transport by limiting the metabolic energy available to the transporting cells.

The second means of circulatory influence upon active transport is illustrated by the function of several types of glands, such as the salivary, sweat and mammary glands. The secretions of these glands consist primarily of substances simply extracted from surrounding interstitial fluid, frequently through active transport. The transport clearly cannot proceed more rapidly than permitted by the rate at which circulating blood restores the local interstitial supply of secreted particles. Similarly, the circulation can sometimes promote active movement by reducing the local accumulation of transported material (for example, see the discussion of the role of the circulation upon intestinal absorption).

Thus the circulation affects active movement both through metabolic mechanisms and by reducing local concentration gradients of transported particles which might otherwise inhibit active movement.

INTRACELLULAR CONTROL. The mechanisms discussed above are particularly suitable for simultaneous control of many cells. However, in certain tissue, such as brain or spinal cord, adjacent cells may have different rates of activity, and thus widely varying requirements for active transport. For such tissue, a gross control mechanism which affects all cells similarly is clearly inappropriate. Instead, there must exist a method of adjusting the active transport of an individual cell to fit its own activity level.

At present, little is known about the details of such a mechanism. One hypothesis is that the active pumping rate of a given substance can be controlled by its intracellular concentration. For example, the sodium influx associated with action potentials tends to raise the intracellular sodium concentration of active cells. If increased sodium concentration stimulates active sodium extrusion, either directly or through the reduction in membrane potential associated with the ionic redistribution, active transport could be raised to accommodate the firing frequency. Such dependence upon concentration is a homeostatic or negative feedback mechanism, since any intracellular deviation from some ideal concentration would act upon the pumping rate in a manner to restore the ideal level.

QUANTITATIVE ASPECTS OF MEMBRANE TRANSPORT[30]

Simple Passive Movement. The complete description of exchange across membranes must include, besides a list of the forces influencing particle movement, the quantitative relations between these forces and the resulting fluxes. Ideally, this description would result in an equation accurately predicting particle flux as a function of concentration difference, membrane electrical potential, solvent drag force, active transport force, etc. Such an equation would permit quantitative as well as qualitative insight into the mechanism of membrane transport.

However, the physical chemistry of membrane transport is not sufficiently developed to establish such an equation. Those influences of greatest interest to physiologists, such as active transport, are the least well understood. Thus, the major quantitative approach has been to derive flux relations from those forces whose theory is understood; these relations are then compared with those observed experimentally to determine what additional forces need be postulated.

The remainder of this section covers the theoretical development of two of these influences: (i) passive diffusion caused by a concentration gradient and (ii) movement of ions caused by an electrical potential gradient. The development will also be restricted to steady state fluxes, so that concentration distribution within the membrane does not change with time. Although phrased in terms of "membrane" transport, the subsequent equations and relations can be applied equally well to transport across complete tissue layers, such as capillary wall or intestinal epithelium.

The equations describing flux caused by diffusion alone and movement in an electric field alone have been derived previously in differential form as equations (4) and (9), respectively. It is natural to assume that when both influences are present simultaneously, the net flux is simply the sum of the fluxes given by the two equations. If this assumption is valid, so that only the forces from electrical and concentration gradients are causing movement and these can be combined in a linear manner, transport is called "simple passive." Thus, the net flux for simple passive movement within a membrane is given by (employing equation 8a)*

$$M = -D\left(\frac{dC}{dx} + z\beta C\frac{dV}{dx}\right) \qquad (22)$$

Here M is the net flux resulting from the combined effects of diffusion and electric migration, and β denotes the ratio F/RT. This may be condensed as equation (23).

$$M = -De^{-z\beta V}\frac{d}{dx}(Ce^{z\beta V}) \qquad (23)$$

The quantity in parentheses, $Ce^{z\beta V}$, is frequently called the electrochemical activity; it is, in a sense, the "effective" concentration under the influence of the potential, V. Its natural logarithm when multiplied by RT is the electrochemical potential, μ, which represents the work necessary to accumulate a concentration, C, at electrical potential, V, starting from some standard state. Thus,

$$\mu = RT \ln (Ce^{z\beta V}) + \mu_o$$
$$= RT \ln C + zFV + \mu_o \qquad (24)$$

where μ_o is the electrochemical potential in the standard state.

The further development from these equations depends upon the specific circumstances in which it is to be applied.

NO NET FLUX; NERNST EQUATION. Suppose that an ionic species is in equilibrium so that its net flux across a membrane is zero. Thus, from equation (23)

$$0 = -De^{-z\beta V}\frac{d}{dx}(Ce^{z\beta V})$$

Since the product of the three terms, D, $e^{-z\beta V}$ and $d(Ce^{z\beta V})/dx$ in the above equation is zero, at least one of the individual terms must be zero also. The terms D and $e^{-z\beta V}$ are not zero since the membrane is assumed to be permeable and the potential finite, so it is the derivative $d(Ce^{z\beta V})/dx$ which must be zero. But the only function whose derivative is always zero is a constant. Thus the electrochemical activity, $Ce^{z\beta V}$, must be constant everywhere; and, thus it must be the same each side of the membrane:

$$C_1e^{z\beta V_1} = C_2e^{z\beta V_2} \qquad (25)$$

In this equation C_1, V_1 and C_2, V_2 represent the concentrations and the potential on side 1 and side 2 of the membrane.

Equation (25) can be rewritten in more familiar form by taking the logarithm of both sides and rearranging terms. Using the natural logarithm (ln) gives

$$V_2 - V_1 = \frac{1}{z\beta}\ln(C_1/C_2) \qquad (26a)$$

---

*In equation (22) ordinary derivatives (d/dx) are substituted for the partial derivatives ($\partial/\partial x$) of equations (4) and (9), since we are confining ourselves to the one dimensional, steady state case. Thus C and V are functions only of x, the distance into the membrane.

while taking the logarithm to the base 10 (log) gives

$$V_2 - V_1 = \frac{1}{z\beta \log e}\log(C_1/C_2) \quad (26b)$$

Using the value of $1/\beta$ at 37° C. and noting that $1/\log e = 2.3$, these equations can be written as

$$V_2 - V_1 = \frac{27 \text{ mv}}{z}\ln(C_1/C_2)$$
$$T = 37° \text{ C.} \quad (26c)$$

$$V_2 - V_1 = \frac{61 \text{ mv}}{z}\log(C_1/C_2)$$
$$T = 37° \text{ C.} \quad (26d)$$

Equations (26a-d) are a particular form of the general relation usually named the *Nernst equation.*

The Nernst equation has three major applications in the study of membrane and tissue transport: (i) development of a criterion for passive equilibrium, (ii) prediction of the membrane potential, and (iii) derivation of the Gibbs-Donnan relation.

CRITERION FOR PASSIVE EQUILIBRIUM. If the electrical potential difference and the concentration of a dissolved species of particles (which have no net flux) are measured on each side of a membrane or tissue layer, and the experimental measurements are found to fit equation (25), it can be concluded that electrical and diffusion forces alone explain the observed concentration distribution. If the results do not fit the equation, an additional force acting on the particles as they move through the membrane (*e.g.*, active transport) must be postulated to explain the deviation from the predicted results (assuming, of course, that the membrane is permeable to this species of particles). Specifically, if the left side of equation (25) is greater than the right, a net diffusion-electrical force must cause particles to move from side 1 of the membrane to side 2; and, since net flux is zero, an opposing force must also exist. Similarly, if $C_2 e^{z\beta V_2} > C_1 e^{z\beta V_1}$, an additional force tending to move particles from side 1 to side 2 must be present (see Fig. 24–13).

In the terminology of Chapter 1, Volume 1, the Nernst equation and equation (25) will be satisfied only if the calculated ionic equilibrium potential is equal to the experimentally measured membrane potential. Thus, the Nernst equation can be used to

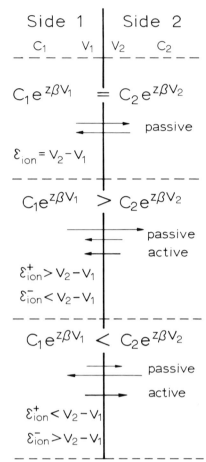

Figure 24–13 Fluxes and electrochemical activities. If the electrochemical activity of a given ion is the same on both sides of a membrane (or the ionic equilibrium potential, $\mathcal{E}_{ion}$, is equal to the observed membrane or transtissue potential, $V_2 - V_1$), the passive unidirectional fluxes would be expected to be equal. If the electrochemical activity is not the same on the two sides but the net flux is zero, an additional influence, here noted as active, must be present to counteract the imbalance in passive electrochemical diffusion. This figure also indicates the relation between the ionic equilibrium potential for positive ions, $\mathcal{E}_{ion}^-$, and negative ions, $\mathcal{E}_{ion}^-$, and the potential difference in these circumstances.

determine the necessity for and the direction of membrane forces other than simple passive diffusion and electrical migration.

PREDICTION OF THE MEMBRANE POTENTIAL. In general, the electrical potential across a membrane is determined by the complicated relations between the forces and fluxes for all the ionic species present. However, when the *passive flux of a single ionic species* dominates membrane exchange, the membrane potential approaches that predicted by the Nernst relation for that species (which is the same as the ionic equilibrium potential of Chapter 1, Volume I). For this reason the axon membrane potential rises toward $(1/\beta) \ln ([Na]_{out}/[Na]_{in})$ during the initial phase of the action potential when sodium movement dominates, and declines toward $(1/\beta)\ln([K]_{out}/[K]_{in})$ during the falling phase of the action potential when membrane current is carried mainly by potassium ions.

GIBBS-DONNAN EQUILIBRIUM. An electrical potential can be maintained across a membrane by several mechanisms. Ions can be pumped through a membrane, thereby causing separation of electrical charges; such a pump is called "electrogenic" since it is the direct cause of the membrane potential. A pump also can cause two simultaneous active fluxes without net current flow ("nonelectrogenic"), with the membrane potential resulting from the passive redistribution of the pumped ions. In both of these cases, continuous active transport is required to sustain the membrane potential.

However, a membrane potential can be maintained by purely passive means, *i.e.*, the simultaneous presence of both charged particles to which the membrane is impermeable and of other particles which can pass through the membrane. If the permeating particles are influenced only by simple passive forces, the resulting concentration distribution is called a *Gibbs-Donnan equilibrium*. In such an equilibrium, the direction of the potential change depends upon the charge of the nonpermeating ion. A positive potential will develop on the side with nonpermeating positive ions or a negative potential on that side with nonpermeating negative ions. The magnitude of the potential depends upon the ratio of concentrations of permeating and nonpermeating ions, as shown by the following example:

Suppose that two permeating ions, sodium and chloride, are passively distributed across a membrane, but that the solution on one side (here noted as side 2) also contains the nonpermeating anion, A^- (Fig. 24–14). Since only simple passive forces are present, the permeating ions must obey the Nernst relation:

$$[Na^+]_1 e^{\beta V_1} = [Na^+]_2 e^{\beta V_2} \qquad (27a)$$

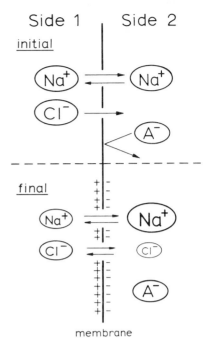

Side 1 | Side 2

Figure 24–14 Development of a membrane potential due to a nonpermeating ion. Initially, equal concentrations of NaCl and NaA are placed on each side of a membrane. If Na^+ and Cl^- can pass through the membrane but A^- cannot, the initial net flux of Cl^- will cause side 2 to become negative with respect to side 1. The membrane potential developed will in turn cause Na^+ to redistribute passively according to the Nernst relation and will progressively reduce net Cl^- flux. Final equilibration will take place when equation (31) is satisfied.

and

$$[Cl^-]_1 e^{-\beta V_1} = [Cl^-]_2 e^{-\beta V_2} \qquad (27b)$$

Also, the requirement for electrical neutrality of the solutions on the two sides leads to two additional equations:

$$[Na^+]_1 = [Cl^-]_1 \qquad (28a)$$

and

$$[Na^+]_2 = [Cl^-]_2 + [A^-]_2 \qquad (28b)$$

These four equations can be solved as follows to find the concentration distribution and electrical potential:

Multiply equations (27a) and (27b) to get the relation:

$$[Na^+]_1[Cl^-]_1 = [Na^+]_2[Cl^-]_2 \qquad (29)$$

This equation often is formally called the *Donnan rule* (or the Gibbs-Donnan rule), although it is merely a simple consequence of the Nernst equation. Using equations (28a) and (28b) to eliminate the sodium concentrations from equation (29) and rearranging terms gives

$$\left(\frac{[Cl^-]_1}{[Cl^-]_2}\right)^2 = 1 + \frac{[A^-]_2}{[Cl^-]_2} \qquad (30)$$

The chloride concentration ratio, as calculated from equation (30), is sufficient to determine the sodium concentration ratio and the membrane potential by using equations (27b) and (29), thereby giving the relation

$$e^{\beta(V_1 - V_2)} = \frac{[Na^+]_2}{[Na^+]_1} = \frac{[Cl^-]_1}{[Cl^-]_2}$$

$$(31)$$

$$= \left[1 + \frac{[A^-]_2}{[Cl^-]_2}\right]^{1/2}$$

Thus, nonpermeating anions, A^-, on side 2 cause sodium accumulation, a chloride depletion, and a reduced potential on side 2 relative to side 1; the magnitude of this effect is a function of the ratio of nonpermeating to permeating anions, $[A^-]_2/[Cl^-]_2$.

An important example of the Gibbs-Donnan relation is the equilibrium across the capillary walls as modified by the negatively charged plasma proteins. The charge carried by these proteins represents about 10 per cent of the total negative charge of the plasma anions. Thus, in those capillary beds impermeable to plasma proteins, plasma is slightly negative (about 1.3 mV) relative to the interstitial fluid; the interstitial concentrations of sodium and other univalent positive ions are less than their plasma values by a factor of about 0.95, while the concentrations of chloride and other univalent negative ions are more by a factor of about 1.05.

NET FLUX; USSING EQUATION.[36] Using the Nernst equation to distinguish between passive transport and active transport is limited to circumstance for which the equation was derived, i.e., zero net flux. However, the existence of a net flux is normal in many organ systems, such as the gastrointestinal mucosa or the renal tubule. Thus, several more general criteria have been derived.

Equation (22) usually is taken as the starting point. By dividing both sides by $(-De^{-z\beta V})$ and noting that M is constant

within the membrane in a steady state, this equation can be integrated to give

$$M\left[\int_{side\ 1}^{side\ 2}(-e^{z\beta V}/D)dx\right] = C_2 e^{z\beta V_2} - C_1 e^{z\beta V_1} \qquad (32)$$

Unfortunately, the integral in brackets cannot be evaluated unless the diffusion coefficient, D, and the electrical potential, V, are known at all points within the membrane, and such data have never been obtained for biologic membranes. Sometimes, D is assumed to be constant and V to be a linear function of distance within the membrane; the resulting relation is known as the Goldman constant field equation. However, such assumptions cannot be directly verified.

Alternatively, by considering the unidirectional fluxes through the membrane, a relation can be developed in which the integral term does not appear. Such a relation was derived for biologic membranes by H. H. Ussing and, thus is called the *Ussing criterion* for passive transport.

Consider the application of equation (32) to the unidirectional flux $M_{2,1}$ (particles passing from side 2 to side 1). On side 1, the concentration of particles which contribute to this flux must be zero; thus for $M_{2,1}$, $C_1 = 0$, so that equation (32) becomes

$$M_{2,1}\left[\int_{side\ 1}^{side\ 2}(-e^{z\beta V}/D)dx\right] = C_2 e^{z\beta V_2} \qquad (33a)$$

Similarly, for the particles which contribute to the flux in the opposite direction, $M_{1,2}$, the concentration, C_2, may be considered zero. Thus

$$M_{1,2}\left[\int_{side\ 1}^{side\ 2}(-e^{z\beta V}/D)dx\right] = -C_1 e^{z\beta V_1} \qquad (33b)$$

Taking the ratio of equations (33a) and (33b) eliminates the integral term, giving the Ussing equation for the magnitude of the flux ratio:

$$\left|\frac{M_{2,1}}{M_{1,2}}\right| = \frac{C_2}{C_1}e^{z\beta(V_2 - V_1)} \qquad (34)$$

Thus, even though the magnitude of neither of the unidirectional fluxes (or the net flux) can be predicted, the ratio of these fluxes is predictable and can be observed experimentally. The concentrations, C_1 and C_2, and the membrane potential, $V_2 - V_1$, can be measured directly, and the unidirectional fluxes can be evaluated by placing a radioactive isotope of the species whose flux is to be determined on one side of the membrane and then measuring its rate of appearance on the other.

If the results agree with equation (34), simple passive forces are sufficient to account for observed fluxes. Agreement with this equation does not rule out the possibility of other forces; however, it does imply that other forces are unnecessary to explain the data. If the equation is not satisfied, an additional force must be postulated; the direction of this force can be derived from considerations discussed in connection with the Nernst equation.

MOVEMENT AGAINST A PASSIVE GRADIENT. If there is a difference in electrochemical activity, $Ce^{z\beta V}$, across a membrane, passive forces tend to cause a net flux from the side of higher activity to that of lower activity, since such a flux is a consequence of the spontaneous tendency to reduce the total chemical potential energy, μ. Thus a net flux from a region of high to one of low electrochemical activity (or potential) requires no external energy. However, if a net flux occurs in the opposite direction (i.e., from low to high electrochemical activity), the total energy, inherent in the concentration and the electrical potential, is increasing, as can be seen from equation (24); therefore additional energy must be supplied to sustain this flux. Thus, movement against an electrochemical activity gradient requires an active transport mechanism (assuming that other passive forces, such as solvent drag, have been taken into account).

SUMMARY OF MATHEMATICAL CRITERIA FOR ACTIVE TRANSPORT. If the net flux through a membrane occurs against an electrochemical gradient, active transport is indicated. If the net flux is in the direction of the electrochemical gradient and if in addition the unidirectional flux ratio fits Ussing's equation, no forces other than simple passive diffusion in an electric field need be postulated. If the net flux is in the direction of the electrochemical gradient but Ussing's equation does not hold, then active transport is not necessary to account for the energy change (although it may be present anyway), but the simple passive mechanisms which led to equation (34) are obviously insufficient to account for the observed fluxes; in this case more detailed investigation into the forces responsible is necessary.

ANNOTATED BIBLIOGRAPHY

A comprehensive coverage of membrane exchange and active transport can be found in Harris, *Transport and Accumulation in Biological Systems*.[13] Also, H. Davson, in his *Textbook of General Physiology*,[8] discusses a number of active transport systems. Much of this material as well as the pharmacology of membrane exchange is covered in two papers by A. M. Shanes, "Electrochemical aspects of physiological and pharmacological action in excitable cells," I and II.[34]

A discussion of the biochemical mechanisms of metabolic energy coupling to sodium ion movement is presented in the review of J. D. Judah and K. Ahmed, "The biochemistry of sodium transport."[20] Particular aspects of membrane exchange have been discussed in two symposia: "Borderline problems around the field of active transport,"[10] and *The cellular functions of membrane transport*.[17]

The quantitative development of transport equations based on classical thermodynamics is exemplified in an important paper by Ussing, "The distinction by means of tracers between active transport and diffusion."[36] An attempt to employ the modern theory of irreversible thermodynamics in transport problems is developed by Kedem and Katchalsky in "A physical interpretation of the phenomenological coefficients of membrane permeability,"[21] and by Kimizuka and Koketsu in "Ion transport through cell membrane."[22] Alternative quantitative approaches which have been attempted include stochastic (probabilistic) models, e.g., Hodgkin and Keynes, "The potassium permeability of a giant nerve fiber"[16]; chemical kinetics models, e.g., Kirschner, "On the mechanism of active sodium transport across the frog skin"[23]; and the treatment of the molecular structure of the membrane as a sequence of energy barriers, e.g., Parlin and Eyring, "Membrane permeability and electrical potential."[29]

REFERENCES

1. Andersen, B., and Ussing, H. H. Solvent drag on non-electrolytes during osmotic flow through isolated toad skin and its response to antidiuretic hormone. *Acta physiol. scand.*, 1957, 39, 228–239.
2. Chinard, F. P. Derivation of an expression for the rate of formation of glomerular fluid (GFR). Applicability of certain physical and physicochemical concepts. *Amer. J. Physiol.*, 1952, *171*, 578–586.
3. Christensen, H. N. Transport by membrane-bound sites or by free shuttling carriers? In:

Membrane transport and metabolism, A. Kleinzeller and A. Kotyk, eds. Prague, Czech. Acad. Sci., 1961.

4. Crabbé, J. Stimulation of active sodium transport across the isolated toad bladder after injection of aldosterone to the animal. *Endocrinology*, 1961, *69*, 673–682.

5. Csaky, T. Z. Transport through biological membranes. *Ann. Rev. Physiol.*, 1965, *27*, 415–450.

6. Curran, P. F., Herrera, F. C., and Flanigan, W. J. The effect of Ca and antidiruetic hormone on Na transport across frog skin. II. Sites and mechanisms of action. *J. gen. Physiol.*, 1963, *46*, 1011–1027.

7. Danielli, J. F. Structure of the cell surface. *Circulation*, 1962, *26*, 1163–1166.

8. Davson, H. *Textbook of general physiology*, 3rd ed. Boston, Little, Brown and Co., 1964.

9. Davson, H., and Danielli, J. F. *The permeability of natural membranes.* Cambridge, Cambridge University Press, 1952.

10. Mommaerts, W. F. H. M., Csaky, T. Z., Hokin, L. E., Hokin, M. R., Tosteson, D. C., Zabin, I., and Woodbury, J. W. Borderline problems around the field of active transport. *Fed. Proc.*, 1963, *22*, 1–35.

11. Frazier, H. S., Dempsey, E. F., and Leaf, A. Movement of sodium across the mucosal surface of the isolated toad bladder and its modification by vasopressin. *J. gen. Physiol.*, 1962, *45*, 529–543.

12. Goldstein, D. A., and Solomon, A. K. Determination of equivalent pore radius for human red cells by osmotic pressure movement. *J. gen. Physiol.*, 1960, *44*, 1–17.

13. Harris, E. J. *Transport and accumulation in biological systems*, 3rd ed. Baltimore, University Park Press, 1972.

14. Hays, R. M., and Leaf, A. J. Studies on the movement of water through the isolated toad bladder and its modification by vasopressin. *J. gen. Physiol.*, 1962, *45*, 905–919.

15. Hillier, J., and Hoffman, J. F. On the ultrastructure of the plasma membrane as determined by the electron microscope. *J. cell. comp. Physiol.*, 1953, *42*, 203–220.

16. Hodgkin, A. L., and Keynes, R. A. The potassium permeability of a giant nerve fiber. *J. Physiol.*, (Lond.), 1955, *128*, 61–81.

17. Hoffman, J., ed. *The cellular functions of membrane transport.* Englewood Cliffs, N. J., Prentice-Hall, 1964.

18. Hokin, L. E., and Hokin, M. R. Studies on the carrier function of phosphatidic acid in sodium transport. I. The turnover of phosphatidic acid and phosphoinositide in the avian salt gland on stimulation of secretion. *J. gen. Physiol.*, 1960, *44*, 61–85.

19. Holter, H. Pinocytosis. In: *Enzymes and drug action.* Ciba Found. Symp. G. W. Wolstenholme, ed. Boston, Little, Brown and Co., 1962.

20. Judah, J. D., and Ahmed, K. The biochemistry of sodium transport. *Biol. Rev.*, 1964, *39*, 160–193.

21. Kedem, O., and Katchalsky, A. A physical interpretation of the phenomenological coefficients of membrane permeability. *J. gen. Physiol.*, 1961, *45*, 143–179.

22. Kimizuka, H., and Koketsu, K. Ion transport through cell membrane. *J. theor. Biol.*, 1964, *6*, 290–305.

23. Kirschner, L. B. On the mechanism of active sodium transport across the frog skin. *J. cell. comp. Physiol.*, 1955, *45*, 61–87.

24. Koefoed-Johnsen, V., and Ussing, H. H. The contributions of diffusion and flow to the passage of D_2O through living membranes. Effect of neurohypophyseal hormone on isolated anuran skin. *Acta physiol. scand.*, 1953, *28*, 60–76.

25. Leaf, A., and Hays, R. M. Permeability of the isolated toad bladder to solutes and its modification by vasopressin. *J. gen. Physiol.*, 1962, *45*, 921–932.

26. Mauro, A. Nature of solvent transfer in osmosis. *Science*, 1957, *126*, 252–253.

27. Palade, G. E. Transport in quanta across the endothelium of blood capillaries. *Anat. Rec.*, 1960, *136*, 254.

28. Pappenheimer, J. R. Passage of molecules through capillary walls. *Physiol. Rev.*, 1955, *33*, 387–423.

29. Parlin, R. B., and Eyring, H. Membrane permeability and electrical potential. In: *Ion transport across membranes*, H. T. Clarke and D. Nachmansohn, eds. New York, Academic Press, 1954.

30. Patlak, C. S. Energy expenditure by active transport mechanisms. II. Further generalizations. *Biophys. J.*, 1961, *1*, 419–427.

31. Ponder, E. The cell membrane and its properties. In: *The cell*, vol. II, J. Brachet and A. E. Mirsky, eds. New York, Academic Press, 1961.

32. Renkin, E. M. Capillary permeability to lipid-soluble molecules. *Amer. J. Physiol.*, 1952, *168*, 538–545.

33. Rosenberg, T., and Wilbrandt, W. Uphill transport induced by counterflow. *J. gen. Physiol.*, 1957, *41*, 289–296.

34. Shanes, A. M. Electrochemical aspects of physiological and pharmacological action in excitable cells. Part I and Part II. *Pharmacol. Rev.*, 1958, *10*, 59–273.

35. Tosteson, D. C., and Hoffman, J. F. Regulation of cell volume by active cation transport in high and low potassium sheep red cells. *J. gen. Physiol.*, 1960, *44*, 169–194.

36. Ussing, H. H. The distinction by means of tracers between active transport and diffusion. The transfer of ioside across the isolated frog skin. *Acta physiol. scand.*, 1949, *19*, 43–56.

37. Ussing, H. H., and Zerahn, K. Active transport of sodium as the source of electric current in the short-circuited isolated frog skin. *Acta physiol. scand.*, 1951, *23*, 110–127.

38. Zerahn, K. Oxygen consumption and active sodium transport in the isolated and short-circuited frog skin. *Acta physiol. scand.*, 1956, *36*, 300–310.

by ALAN KOCH

Regulation of the body fluids is accomplished primarily by the kidneys. These two organs, which together weigh only about 300 g, directly control the volume and composition of the extracellular fluid, and exert indirect control over the intracellular fluid. This regulatory task is fulfilled so successfully that, over a wide range of water and solute intake, the volume and the composition of the fluids in the body are held remarkably constant. In order to effect this control, a great many different operations are required. The multiplicity of operations and the versatility of each lead to great diversity of renal function. This apparent complexity, however, can be explained by the electrochemistry of fluids, the permeability characteristics of renal structures, and metabolic pumps such as the ones already discussed in Chapter 1, Volume I and Chapter 24, Volume II.

Properties of Solutions. A *solution* consists of a fluid medium, called the *solvent*, in which are distributed a number of particles, the *solute*. The concentration (C or [S]) of a substance in a solution is defined as the quantity present (Q) in moles, millimoles, or micromoles, divided by the volume (V) in liters or cubic centimeters through which it is distributed. The concept of concentration is frequently generalized to that of mole fraction. The *mole fraction* of a substance in a solution is defined as the number of molecules of that substance divided by the total number of molecules present in the solution.

As will be remembered from physical chemistry, the mole fraction of a solvent, and therefore of the total solute, can be measured by determination of the colligative properties of the solution. These properties include the lowering of the vapor pressure, elevation of the boiling point, lowering of the freezing point, and osmotic pressure. The changes are all measures of the same thing, the tendency of water molecules to escape from the solution. The physical chemistry of solutions is discussed lucidly by Moore.[26]

The relationship between the quantity present, the volume through which it is distributed, and the resulting concentration is given by:

$$C = \frac{Q}{V}$$

This relationship, which has been developed for a static system, can also be applied to a system, like the kidneys, in which fluids are moving. Only the steady state situation, *i.e.*, the situation in which the concentration in the flowing fluid is not changing with time, will be considered. The *mass flow* is the quantity of solute (Q), in millimoles or milligrams, which passes a cross section every minute and is symbolized by $\dot{Q}$.[*] The *volume flow* is the volume of the solvent which passes a cross section every minute and is symbolized by $\dot{V}$. In a given period, the quantity which passes a cross section is $\dot{Q}\Delta t$, and the volume through which it is distributed is the volume which passes, $\dot{V}\Delta t$. The concentration of material is still Q/V; hence, $C = \dot{Q}\Delta t / \dot{V}\Delta t$, and the equation is

$$C = \frac{\dot{Q}}{\dot{V}}$$

Terminology. Unfortunately, renal terminology arose independently, without reference to physical terminology. Concentrations were considered fundamental quantities, and mass or mass flows were derived from them. In addition, no explicit distinction is customarily made between mass and mass flow or between volume and volume flow. In standard renal notation, the plasma concentration is denoted by P, and a subscript is used to indicate the substance. The urinary concentrations are denoted by U, again with a subscript to indicate the substance. The rate of urine flow is denoted by V.

In this chapter, a different system of notation is used. The mass of a substance will be denoted by Q, the mass flow will be denoted by $\dot{Q}$, and $\dot{V}$ will indicate volume flow. Square brackets will be used to indicate concentration, and a subscript will tell where the concentration is measured. Table 25–1 gives the conversions between the two systems.

Functional Anatomy. GROSS STRUCTURE (Fig. 25–1). The medial side of the kidney contains a deep sinus through which the ureter and all the blood vessels enter or leave the renal parenchyma. Just outside the renal sinus, the ureter expands into the extrarenal pelvis, which continues inside the border of the sinus as the intrarenal pelvis. The intrarenal pelvis divides into two major calyces, each of which subdivides into two or three minor calyces. Each minor calyx terminates around the base of one or two papillae. Formed urine is delivered into the collecting duct system in the renal papillae, passes through the papillary duct system into a minor calyx, and thence goes eventually to the bladder.

---

[*]Pronounced "Q dot"; the dot to express rate is a symbol going back to Sir Isaac Newton.

Table 25-1 *Correlation of Terms Describing Renal Function*

STANDARD RENAL NOMENCLATURE	MEANING	NOMENCLATURE USED HERE	UNITS
P_{Na}	Plasma concentration of Na^+	$[Na]_p$	$\mu M/cm^3$ or $mM/liter$
U_G	Urinary concentration of glucose	$[G]_u$	$\mu M/cm^3$, $mM/liter$, or mg/cm^3
V	Rate of urine flow	$\dot{V}_u$	$cm^3/minute$
$U_{Cl}V$	Rate of excretion Cl^-	$\dot{Q}_{Cl_u}$	$\mu M/minute$
C_{PAH} or RPF	The rate of flow of plasma into the kidney, measured by the clearance of PAH	$\dot{V}_p$	$cm^3/minute$
C_{in} or GFR	The rate of flow of glomerular filtrate into the nephrons, measured by the clearance of inulin	$\dot{V}_g$	$cm^3/minute$
L_K or $C_{in} \times P_k$	The filtered load of K^+; *i.e.*, the rate at which K^+ is filtered at the glomerulus	$\dot{Q}_{K_g}$	$\mu M/minute$
$\dfrac{C_{urea}}{C_{in}}$	The clearance ratio of urea; *i.e.*, the fraction of the filtered urea which is excreted	$\dfrac{\dot{Q}_{urea_u}}{\dot{Q}_{urea_g}}$	no units

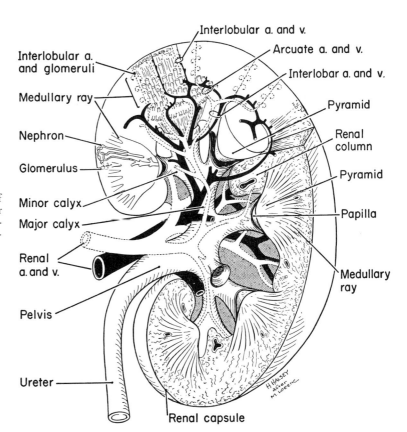

Figure 25-1 Gross structure of the kidney, sagittal section. (After Smith, *Principles of renal physiology.* New York, Oxford University Press, 1956.)

Interlobular a. and v.

Arcuate a. and v.

Interlobar a. and v.

Interlobular a. and glomeruli

Medullary ray

Nephron

Glomerulus

Minor calyx

Major calyx

Renal a. and v.

Pelvis

Ureter

Pyramid

Renal column

Pyramid

Papilla

Medullary ray

Renal capsule

A medullary and a cortical type of tissue can be distinguished when the kidney is sectioned longitudinally. The triangularly shaped papillae give rise to medullary pyramids which, in turn, give rise to medullary rays; the over-all impression is one of a decrease in the density of medullary substance as the cortex is approached. Cortical substance lines the surface of the organ and, between medullary pyramids, dips in toward the medulla in the renal columns.

MICROSCOPIC STRUCTURE (Fig. 25–2).[24] The unit of structure and function in the kidney is the *nephron*. Renal parenchyma is composed of a great many nephrons, each with its associated blood supply. Urine is formed in the nephrons, and total renal function can be viewed as a summation of the function of about two million extremely similar but distinct units. A nephron is composed of two major sections, a *glomerulus* and a *tubule*. The glomerulus consists of Bowman's capsule, the spherical blind end of the tubule, and of coiled capillaries. These capillaries lie within an invagination of the capsule, which is formed of squamous epithelium.

The tubular portion of the nephron begins at the glomerulus and undergoes several convolutions in this region, traveling generally outward toward the cortex. The tubule then straightens and descends in a straight line toward the medulla. The convoluted portion and the first part of the descending portion constitute the *proximal tubule*. Near the end of the descent the walls of the tubule become exceedingly thin. The thin-walled portion is termed the *thin segment* of the loop of Henle. After a sharp hairpin turn the tubule travels back toward its associated glomerulus in the cortex. Along this ascending limb the walls become thick again. The point of thickening marks the beginning of the *distal tubule*. In the region of the glomerulus, the tubule undergoes several more convolutions before emp-

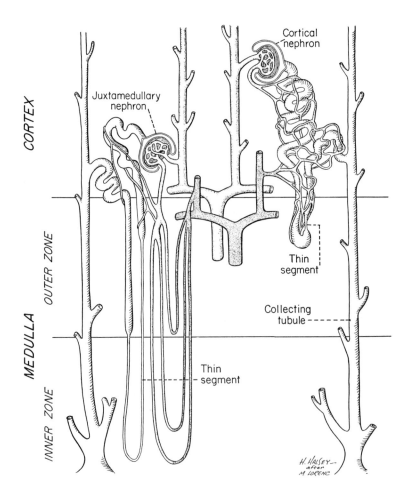

Figure 25–2 Structural differences between cortical and juxtamedullary nephrons. (After Smith, *The kidney, its structure and function in health and disease.* New York, Oxford University Press, 1951.)

tying into the system of *collecting ducts.* These ducts travel in straight lines through the medulla; they accept fluid from several nephrons, coalesce, and then enter the renal papillae. Cortical substance is thus composed of glomeruli and the proximal and distal convoluted portions of the tubules; medullary substance is composed of the descending and ascending limbs of the tubules and the collecting duct system.

Short *afferent arterioles*, each feeding a glomerulus, arise from the intralobular arteries. Upon entering the corpuscle, the afferent arteriole arborizes into six to ten glomerular capillaries, which lie close to the invaginated surface of Bowman's capsule. The capillaries then recombine to form the *efferent arteriole*, which leaves the glomerulus and goes to the tubular portion of the nephron. Upon reaching the proximal portion of the tubule, the efferent arteriole arborizes into a second group of capillaries, the *peritubular capillaries*, which wind around the tubule. These capillaries traverse the entire length of the renal tubule, following both its descending course into the medulla and its ascending return. At this point, the capillaries coalesce to form the renal venules.

The kidney is abundantly supplied with nerves which travel beside the major arteries and apparently innervate the arteriolar vasculature. There is no evidence that these nerves supply the nephron.

All tubules are not quite alike and all vascular supplies are not identical. Nephrons with glomeruli lying in the outer two-thirds of the cortex tend to have very short descending and ascending limbs and only vestigial loops of Henle. The efferent arterioles of these nephrons are very short, and the peritubular ramification occurs immediately and extensively. Nephrons whose glomeruli lie in the deeper third of the cortex have a rather different structure. These nephrons have long descending and ascending limbs, which penetrate deeply into the renal pyramids and possess well-developed loops. The efferent arterioles of these nephrons tend to be long and, instead of ramifying extensively into peritubular capillaries, give rise to one or two long straight vessels, the *vasa recta*. These vessels follow the course of the nephron into and out of the medullary pyramids and do not appear to break up into true capillaries.

Précis of Renal Function. The quantity of blood entering the kidneys every minute represents one-fourth to one-fifth of the resting cardiac output. As the blood flows through the glomerular capillaries, about one-fifth of the plasma water passes through the membranes of the capillaries and glomerulus to enter the proximal portion of the renal tubule. The blood remaining in the vascular system enters the efferent arterioles and perfuses the tubules via the peritubular capillaries. The plasma water removed from the blood is termed the *glomerular filtrate*; the process of removal is *glomerular filtration.* Glomerular filtrate is an ultrafiltrate and normally contains no erythrocytes and little or no plasma protein. Other molecules which are sufficiently small to be in true solution pass freely through the glomerular membranes. All major ions, glucose, amino acids and urea appear in the glomerular filtrate at approximately the concentration at which they exist in the plasma.

In the tubule, both solute and water transport take place. Materials are transported across the tubular epithelium from the lumen of the tubule to the interstitial fluid surrounding the nephron and thence to the blood in the peritubular capillaries. This process is called *reabsorption* and results in the return of filtered material to the blood stream. Materials are also transported from the peritubular blood to the interstitial fluid, across the tubular epithelium, and into the lumen. This process is called *secretion* and results in an excretion which is more rapid than would be possible solely through glomerular filtration. The terms reabsorption and secretion denote direction rather than a difference in mechanism. All the glucose that is filtered may return to the circulation and remain in the body as a result of complete reabsorption. Conversely, all the para-aminohippuric acid (PAH) that enters the kidneys may leave in the urine, even though only one-fifth of the material is filtered. This complete excretion results from the efficient secretory system for PAH.

The proximal tubule reabsorbs physiologically important solute material and secretes organic substances that are destined for excretion. The filtered glucose and amino acids are reabsorbed, as are most of the filtered Na^+, Cl^-, and HCO_3^-. As solute particles are removed by active transport processes, the osmotic gradient produced causes reabsorption of water as well. The distal tubule and collecting duct are engaged primarily in the precise regulation of acid-base and ionic balance. Since much of the filtered solute and water is reabsorbed proximally, both $\dot{V}$ and $\dot{Q}$ are relatively small in the distal tubule. Final regulation of solute excretion and final modification of urinary solute concentration to form hypertonic urine occurs in the collecting ducts.

FLUID DYNAMICS

Two roles are played by the blood that enters the kidney. The first is to supply oxygen and metabolites to enable the kidney to function; this is the role filled by the blood in any organ. In the kidney, however, the blood must also supply the water and solute material which the kidney will process. Because of this latter role, fluid dynamics in the kidney differs from that in any other organ.

The volume of blood flowing into the kidney exceeds by far the amount needed to meet its requirements for oxygen and metabolites. Two corollaries of this high volume flow may be pointed out. First, the extraction of oxygen and metabolites is normally extremely low. Second, in an emergency such as hemorrhage, the reduction of renal blood flow which occurs causes an increase, or lessens a decrease, in the blood flow in other regions. Normally 1 to 1.5 liters of blood enters the two kidneys each minute. After hemorrhage or severe injury, the flow may be reduced to as little as 250 ml per min.

Physiologic Analysis of Blood Flow. The study of renal fluid dynamics involves an application of the principle that flow is equal to the pressure drop divided by vascular resistance (Ohm's law). This principle, as applied to renal fluid circulation, is illustrated in Figure 25–3, in which each screw clamp corresponds to a resistance to flow. The amount of resistance is shown by the degree of constriction of each clamp. Screw clamp R_1 indicates the resistance to flow from the renal artery to the afferent arterioles; R_2, through the afferent arterioles; R_3, through the efferent arterioles; R_4, through the peritubular capillaries; R_5, through the glomerular membrane; and R_6, through the entire venous bed. As fluid travels through the tubular portion of the nephron, it is removed and returned to the peritubular capillaries. This transfer is driven by the transport of solute and, in Figure 25–3, is indicated by the centrifugal pumps in the reabsorptive channels.

Two conclusions can be made from Figure 25–3. If the pressures at each point and all the resistances were known, the total flow and the flow in each limb could be calculated. Unfortunately, the resistances are not known but must be computed from the flows which have been estimated by other means. The main point is that *the total plasma flow is split at the glomerulus, some fluid being filtered to enter the tubular lumen and some continuing in the vascular system.* The relative size of the two flows is determined by the relative resistances in the two parallel branches. The ratio of the rate of glomerular filtration to the total rate of plasma flow is determined by the venous resistance and the resistances in the two limbs. This ratio is referred to as the *filtration fraction.*

Filtration rate and renal plasma flow may be varied widely by both extrarenal and intrarenal factors. The glomerular filtration rate is determined by the resistance to flow

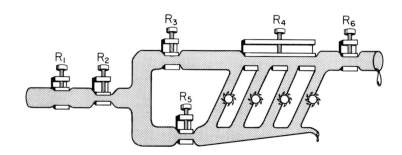

Figure 25–3 Analogue of fluid flow in the kidney. Resistances are designated as R, such as R_1 in the distributing arteries, R_2 in the afferent arteriole, R_3 in the efferent arteriole, R_4 in the peritubular capillary, R_5 across the glomerular membrane, and R_6 in the venous bed. The turbine wheels signify the active transport of solute.

across the glomerular capillary wall and the effective filtration pressure. Since protein does not pass the glomerular membrane freely, work must be done to separate the protein from the solution being filtered. The effective filtration pressure, therefore, is the pressure difference across the glomerular capillary wall less the osmotic pressure corresponding to concentration of nonpermeating solute in plasma:

filtration pressure = pressure in glomerular capillaries, minus pressure in capsular space, minus osmotic pressure of plasma protein.

The osmotic pressure term (π) is approximately equal to RT (R = universal gas constant; T = temperature) times the molar concentration of protein in the plasma. The protein concentration is about 1 mM per liter, which corresponds to a value for π of a little less than 20 mm Hg. At the normal mean arterial pressure of 100 mm Hg, capillary pressure is probably about 70 mm Hg, and intracapsular pressure is about 20 mm Hg. Thus, the filtration pressure is about 30 mm Hg (70 − 20 − 20).

Both capillary and intracapsular pressures vary with arterial pressure. Although the precise relations are unknown, it is unlikely that either are linear functions of the arterial pressure. Hence, filtration pressure would not be expected to bear a constant relation to arterial pressure. Indeed, it does not. Glomerular filtration starts at arterial pressures around 30 mm Hg, rises rapidly in the region between 30 and 90 mm Hg, and then rises more slowly or may even stay constant over the physiologic range of 90 to 180 mm Hg (see Fig. 25–4). Presumably, no filtration occurs at very low pressures because the capsular back pressure and protein osmotic pressure have not been exceeded by the glomerular capillary pressure. The nonlinear pressure-flow relationship, seen after filtration has been instituted, probably reflects the manner in which glomerular capillary and intracapsular pressures change with arterial pressure.

The volume flow of plasma through the kidney depends directly on the driving pressure and inversely on the vascular resistance. Since renal venous pressure is normally zero, the driving pressure must

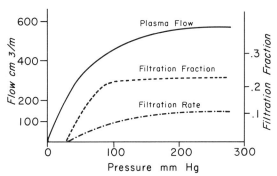

Figure 25–4 Pressure-flow relations in a normal kidney. Filtration fraction is the ratio between filtration rate and plasma flow.

be the arterial pressure. So long as resistances are constant, flow increases linearly with pressure. As arterial pressure is increased above 50 mm Hg, vascular resistance increases so that the flow fails to increase as much as it would have (Fig. 25–4).

Control of Flow Rate; Autoregulation. The kidney is richly supplied with sympathetic nerves, all of which probably function to innervate vascular smooth muscle. The vasculature of the kidney is very labile, and in emergencies, such as hemorrhage or shock, much of the blood that normally perfuses the kidney is diverted to other regions. This would seem a reasonable mechanism for the body to possess since, in an emergency, the body can better afford to lose renal function for two or three hours than brain or cardiac function. Unfortunately, the response may outdo itself and after severe blood loss or shock, the kidneys may become ischemic. A protracted period of renal ischemia frequently initiates a degenerative process which culminates in renal failure with very low blood flow, little or no glomerular filtration, and tubular cell necrosis. Hence, after injuries in which the patient has undergone a period of shock, it is necessary to be on guard against renal failure.

In addition to this active, central control of the renal vasculature, the kidney also exhibits a form of self-control or *autoregulation*. This can be seen in the isolated, perfused kidney in which the pressure-flow relations described above, and illustrated in Figure 25–4, are still evident, although less pronounced. Thus, the continued existence of these nonlinear relations after the removal of nervous control would seem to indicate a regulating component characteristic of the kidney itself. The tendency for blood flow to level off with increasing pressure means that the renal vascular resistance increases somewhere as a function of pressure. The tendency of filtration rate to level off with increasing arterial pressure presumably is related, not to a change in filtration resistance, but to a change in the relationships between arterial pressure and the pressures involved in glomerular filtration. Thus, the two facets of autoregulation come about in different manners. Figure 25–3 shows that the change of a single resistance with pressure might account for the whole phenomenon. If either the preglomerular resistance, R_2, which corresponds to the resistance in the afferent arterioles, or the venous resistance, R_6, were to increase, total vascular resistance would increase and the filtration fraction would remain constant.

Two major theories have been advanced to explain autoregulation. Proponents of one theory have selected the afferent arterioles as the place where resistance changes as a function of pressure, and proponents of the other have chosen the venous bed. Those who favor the first theory advocate the presence of a myogenic reflex. A pressure sensor located within the kidney is postulated, and pressure variations in the region of this sensor result in contractions of vascular smooth muscle. Thus, a vessel's diameter is smaller at a higher pressure than at a lower pressure, and the resistance is increased.[41] The resistance is presumed to change in the afferent arterioles.

The second theory states that the resistance changes in the venous portion of the vascular bed. An increase in resistance results from an increase in extravascular pressure and a consequent decrease in the transmural pressure. Extravascular pressure increases because of an increase in vascular pressure at the level of the capillaries and a resulting increase in transcapillary filtration. The events required for and predicted by the tissue pressure hypothesis are known to take place. Thus, autoregulation by tissue pressure does occur.[16, 33] The recent findings of Başar et al.[1] suggest the presence of both a mechanical and a second, presumably myogenic, component to autoregulation.

Renal Clearances. This method, first used extensively by Smith, has quantified the study of normal and pathologic renal mechanisms. A clearance value expresses the degree to which a substance is removed from the blood by excretion into urine. This value is expressed not as a percentage but rather as the number of cubic centimeters which would be "cleared" were all of the substances removed. It is therefore the virtual rather than the actual amount of blood cleared. A good way of looking at clearance is that it is the number of cubic centimeters of blood which would have to be presented to the nephron to provide the amount of substance actually found in the urine, in a unit time.

$$\text{Clearance} = \frac{\dot{Q}_u}{C_p} \quad \text{or} \quad \frac{UV}{P} \quad \text{or}$$

$$\frac{\text{amount of substance in urine per minute}}{\text{concentration of substance in plasma}}$$

The units of clearance are those of volume flow, usually cubic centimeters per minute.

The clearances of two particular substances are especially important. The clearance of para-aminohippuric acid (PAH) is a measure of renal plasma flow; the clearance of inulin is a measure of the filtration rate. Knowledge of these volume flows allows calculation of some of the important relationships in renal fluid dynamics; it also enables us to know the mass flows of material entering the kidney or being filtered or being presented to the peritubular borders of tubular cells. The mass flow of a substance is the product of the volume flow of solute and the concentration of the material of interest.

MEASUREMENT OF RENAL PLASMA FLOW. Blood flow could be measured by interrupting the renal artery or vein and inserting a flowmeter. Although this procedure has been followed experimentally, it is obviously not applicable to man, and even in work on animals it is too difficult for routine determinations. Indirect methods are therefore used. The kidney is regarded as a Y tube of the type illustrated in Figure 25–5b. If renal tissue neither creates nor destroys a material, the rate at which the material enters the kidney through the renal artery must equal the rate at which it leaves by the two available routes, the renal vein and the urine. If the arterial and venous concentrations of such a substance and the rate at which it is excreted by the kidney are

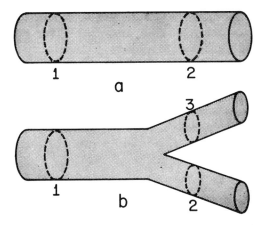

Figure 25–5 *a*, Straight tube, $\dot{Q}_1$ must equal $\dot{Q}_2$.
b. Branched tube, $\dot{Q}_1$ must equal $\dot{Q}_2 + \dot{Q}_3$.

known, the rate at which plasma must enter the kidney can be calculated. This procedure is already familiar as the Fick principle (Chap. 3, Vol. II). Thus, if the renal arterial plasma concentration of a substance is 3 mg per liter and the renal venous plasma concentration is 2 mg per liter, 1 liter of plasma must pass through the kidney each minute in order to furnish sufficient material to account for an excretion rate of 1 mg per minute. For substance S, the plasma flow ($\dot{V}_p$) is given by the formula:

$$\dot{V}_p = \frac{\dot{Q}_{su}}{[S]_a - [S]_v} \quad \text{or} \quad RPF = \frac{UV}{P_a - P_v}$$

$$\text{or} \quad \frac{\text{amount in urine per unit time}}{\text{loss per liter of plasma}}$$

All that is necessary is to collect urine in a given period of time and, sometime during this period, to obtain a sample of arterial plasma (systemic venous plasma will usually have about the same concentration) and a sample of renal venous plasma. The latter sample can be obtained by catheterization of the renal vein.

The method described is simple and applicable to any substance that is neither destroyed nor created in the kidney. Renal venous catheterization is obviously the most difficult procedure, but this can be avoided by selecting a substance which, once it has entered the kidney, is entirely excreted in the urine. The two agents most commonly used are iodopyracet (Diodrast) and PAH. The renal venous plasma concentration of either of these agents is zero, and no renal venous plasma sample is

needed. The rate at which the material enters the kidney is equal to the rate at which it is excreted, and the flow can be computed from the clearance of PAH:

$$\dot{V}_p = \frac{\dot{Q}_{PAHu}}{[PAH]_p} \quad \text{or} \quad RPF = \frac{U_{PAH}V}{P_{PAH}}$$

Any substance which is abstracted completely from renal plasma enters the nephron by two routes: (i) by filtration at the glomerulus; (ii) by transport across the tubular cells of the portion of the substance remaining in the plasma that passes along the peritubular capillaries. Further, the transcellular transport mechanism must be exceedingly active if the last trace of the material is to be removed from the plasma. A class of cyclic organic acids, including the two mentioned above, appears to be handled in the kidney in approximately this manner.

Even at low plasma concentrations of PAH (or other compounds of this group) about 10 per cent of the material is still present in renal venous plasma. Hence, the plasma flow computed from the PAH clearance is about 10 per cent lower than the actual flow.

A number of reasons exist for this incomplete extraction. The tubular cells may not extract all of the PAH in the perfusing blood, although they certainly extract almost all of it. Apparently, some back-diffusion of PAH from more distal regions of the nephron back into peritubular blood occurs. There is some flow through connective tissue. Finally, blood that perfuses the juxtaglomerular nephrons tends to go directly to the medulla rather than perfuse the proximal tubular cells of those nephrons. Currently, the significance of each of these factors is unknown, but the assumption is that the last factor is dominant. Using this assumption, a partition of renal blood flow can be made into that flow which perfuses the cortex, obtained from the clearance of PAH, and the flow which perfuses the medulla, obtained as the difference between the clearance of PAH and the total renal plasma flow. The estimates of medullary flow so obtained are in reasonable agreement with more direct measurements of renal medullary flow.[19, 30]

MEASUREMENT OF GLOMERULAR FILTRATION RATE. Glomerular fluid cannot be sampled except in animal experiments, and even there the procedure is extremely difficult; hence clearance techniques are used. Consider a substance that is neither secreted nor reabsorbed by renal tubular cells (Fig. 25–5a). In the steady state, as much material leaves the nephron each minute as enters it; $\dot{Q}_g = \dot{Q}_u$. Since the concentration of this material in glomerular fluid is approximately the same as the concentration in plasma, knowledge of its rate of excretion and plasma concentration is sufficient to allow the computation of the rate of glomerular filtration. If 210 mg of the measuring material is excreted each minute, the same amount has entered the nephrons. If the plasma concentration is 2 mg per ml, 105 ml of fluid must enter the glomeruli each minute in order to account for the mass flow of the measuring material. We have already defined a clearance in such terms. The glomerular filtration rate can be calculated from the clearance of inulin.

$$V_g = \frac{\dot{Q}_{In_u}}{[In]_p} = GFR = C_{In} = \frac{U_{In}V}{P_{In}}$$

Several materials are suitable for the estimation of filtration rate. As stated, the substance cannot be reabsorbed or secreted, nor can it be created or destroyed by the nephrons. Further, its concentration in the glomerular filtrate should be proportional to the concentration in plasma. Other criteria which must be met are given by Smith.[35] The substance which fulfills these criteria best is *inulin*, a fructose polysaccharide derived from Jerusalem artichokes. In all species in which the inulin clearance has been examined, it appears to be an accurate measure of filtration. Other substances which may be used include ferrocyanide and sucrose in man and dog and creatinine in dog.

NORMAL VALUES OF FILTRATION RATE AND PLASMA FLOW. Renal mass and, consequently, the values for the filtration rate and plasma flow correlate fairly well with the body surface area. For this reason, values for the plasma flow and filtration rate in man are generally expressed in terms of surface area and then corrected to the value of a hypothetical man with a surface area of 1.73 m². In animals, this correction is frequently neglected. In men, plasma flow averages 655 ml per min, and the filtration rate 127 ml per min. In women, plasma flow averages 600 ml per min, and filtration rate 118 ml per min. Both volume flows decrease with increasing age and fall in an approximately linear fashion from the stated values at the age of about 30 years to approximately half those values by the age of 75. The measured values vary considerably in the absence of clinically detectable illness.

TUBULAR TRANSPORT

The main activity of the kidney is to transport solute materials and water across tubular cells. Such transport is termed *reabsorption* when its direction is from the tubular lumen to the interstitial fluid; it is termed *secretion* when its direction is from the interstitial fluid to the tubular lumen. The two terms denote the direction of the transport and have no implications as to the cellular basis of the transport. Transtubular transport is said to be *passive* when it can be accounted for on the basis of a *passive flux equation* (see below) and *active* when this equation does not account for the observed transport.

Techniques of Study. The experiments that are performed to analyze renal function fall into two categories: steady state studies and transient analysis. Thus far the steady state studies have been much more extensively used and the bulk of our current understanding comes from clearance experiments, analysis of tissue slices and tubular puncture experiments.

STEADY STATE EXPERIMENTS. *Clearance Experiments.*[35] The analysis of tubular transport of both solutes and water by means of clearance experiments was the beginning of the quantitative analysis of renal function. A clearance experiment studying the tubular transport of glucose, for example, might be conducted in the following way: A healthy dog would be anesthetized and catheterized so that urine would be collected as soon as it reached the bladder. A convenient systemic artery would be isolated for subsequent blood collection. An intravenous infusion containing inulin, PAH and glucose would be instituted and the experimenter would then wait until renal function and the plasma concentrations of these solutes had become stable. At that time, a urine collection period of perhaps 20 min would be started and, at the midpoint of it, an arterial blood sample would be collected. The volume of urine formed during the collection period would be recorded. If the experiment is to find the effect of changing the concentration of glucose in plasma, the infusion would be changed and the experimenter would wait for the attainment of the new steady state. Then another urine collection period would be taken with another blood sample. At the end of the experiment, the concentrations of inulin, PAH and glucose would be determined in the collected plasma and urine. From these primary data, the rate of urinary excretion of glucose, the rate at which it was filtered, and the rate at which it was presented to the kidney can be determined. The net rate of transtubular transport is the difference between the rate at which the material was filtered and the rate at which it was excreted. The clearance experiment gives the overall results of renal function but does not lend itself readily to detailed analysis of mechanism.

One can determine the clearance of glucose or of sodium just as well as the clearance of inulin or of PAH. However, only in the latter two cases is any physical meaning attached to the term. The clearance of glucose is related in some manner to the excretion of glucose, but this information alone does not tell how the kidney handles glucose. A derived computation, the *clearance ratio*, does give important information about the way a substance is handled by the kidney. The clearance ratio is the ratio of the clearance of a substance to the clearance of inulin. The clearance of inulin gives the amount of glomerular filtration. If that is multiplied by the plasma concentration of the substance under investigation, the amount of the substance entering the tubule, the *filtered load*, is known. If the substance, in addition to being filtered, is also secreted by the tubule, the clearance of the substance will be higher than that of inulin. If the substance is reabsorbed, its clearance will obviously be lower.* When the clearance ratio is less than one, *net re-*

absorption occurs, but it cannot be stated whether secretion also takes place. If filtration and reabsorption are the only processes operating, the clearance ratio gives a quantitative measure of the reabsorptive process; if secretion also occurs, a theoretical possibility, this ratio gives a minimum value for the reabsorptive process. When the clearance ratio is greater than one, *net* secretion occurs. When filtration and secretion are the only processes operating, the clearance ratio gives a quantitative measure of the secretory process. The clearance ratio is one for inulin or for any substance which is handled like inulin.

Tissue Slice Experiments. The experiment of removing a functioning kidney and analyzing different regions of it for various materials has been used since the turn of the century, both for histological and chemical determinations. As an example, Ljungberg in 1947[21] excised kidneys from cats and separated them into cortical slices and slices from progressively deeper in the medulla. He then stained these slices with silver in order to obtain an estimate of chloride content. He found extraordinarily large amounts of chloride in the renal medulla. This observation remained largely unnoticed until the work of Wirz, Hargitay and Kuhn[14, 48] called attention to the osmotic gradient through the renal medulla (see below). Analysis of tissue slices from different regions and calculation of the stable spatial gradients of solute concentration has led to important contributions in the understanding of renal concentrating and diluting mechanisms.

Micropuncture Experiments. In 1924, Wearn and Richards[42] published the results of experiments in which they had inserted fine glass capillaries into the capsular space and collected glomerular filtrate from single nephrons. The technique was extremely difficult because it required a fine experimental hand and the subsequent analysis of minute quantities of fluid. However, by this *tour de force*, they established the composition of glomerular fluid and the nature of glomerular function. Later, Walker *et al.*[40] showed that the total solute concentration remains constant in the proximal tubule, but that glucose and water are reabsorbed. Recently, these methods and others have been extended and refined to include precise localization of the point of collection,

---

*These relationships may be developed as follows: The clearance of a substance is $\dot{Q}_u/C_p$, and the clearance of inulin is numerically equal to the filtration rate. Hence the clearance ratio is $\dot{Q}_u/C_p\dot{V}_g$. But $C_p\dot{V}_g$ is the filtered load of the material ($\dot{Q}_g$); therefore,

$$\text{Clearance Ratio} = \frac{\dot{Q}_u}{\dot{Q}_g} = \frac{UV}{PXGFR}$$

$$= \frac{\text{rate of excretion}}{\text{rate of entry at glomerulus}}$$

stop-flow microperfusion,* free-flow microperfusion,† and measurement of transcellular and transtubular potentials.[44] As in clearance experiments, estimates of the movement of water and of solute are obtainable from the behavior of inulin. Indeed, both the inulin clearance and the clearance ratio have their analogs in single tubular studies. The fraction of fluid that has been reabsorbed between the glomerulus and the site of puncture can be determined from the ratio of inulin concentrations in tubular fluid and plasma. Since inulin is neither reabsorbed nor secreted, steady state conditions require that the mass flow of inulin past any site in the tubule be identical. Changes in the concentration of inulin from its value in plasma must therefore reflect tubular reabsorption of water. Thus, at the site where the ratio of inulin concentrations in tubular fluid and water is two, half the filtered water must have been reabsorbed. In considering the transport of solutes, the tubular fluid to plasma concentration ratio divided by the tubular fluid to plasma ratio of inulin has the same significance as the clearance ratio. In fact, it is the clearance ratio for the site of puncture. When discussing micropuncture studies, we shall call this ratio of ratios the *tubular clearance ratio*.

At this point, micropuncture experiments, either for chemical analysis or for the measurement of electric potentials, give the most detailed and precise data available. In the treatment of the individual solutes below, considerable emphasis will be given to such experiments.

TRANSIENT ANALYSIS. Chinard[4] has injected materials into the renal artery "instantaneously," and analyzed the patterns of their appearance in both renal venous blood and urine. He has shown that glucose is transported across renal cells without entering into cellular metabolism. In addition,

---

*Process by which a single nephron is isolated by means of droplets of oil placed at two different points and a defined solution is placed between the oil droplets, allowed to remain there for a period of time and then removed for analysis.

†Process by which a portion of a single tubule is isolated by oil droplets and a defined fluid is pumped into the proximal portion of the nephron segment at controlled rate and pressure and collected from the distal portion of the segment for analysis.

he has reported many provocative findings pertaining to the handling of electrolytes by the kidney. More recently, another method has been developed by Malvin et al.[23] and used by many workers. In so-called "stop-flow" experiments, the ureter is blocked for 4 to 8 min. It is then opened and the accumulated tubular fluid is collected serially. The first samples are presumed to have come from the most distal portions of the nephron; later samples are presumed to have been "processed" by more proximal portions. Although the techniques of transient analysis are inherently capable of yielding more information than those of steady state analysis, they are much more difficult to analyze, and thus far they have provided mostly corroborative evidence.

Principles of Transport. PASSIVE TRANSPORT.[18] The principles governing transport of solute across membranes, discussed fully in Chapter 1, Volume I, and Chapter 24, Volume II, will be briefly recapitulated here. The passive transport of a component of a solution across a membrane may result from one of two distinct types of motion. Movement of the whole solution through the membrane carries with it all specific components; this is termed *bulk flow*. Movement of the component through the solution lodged in the membrane and thus through the membrane itself is termed *diffusion*. Diffusion will occur regardless of whether the fluid in the membrane is moving or stationary. In general, both kinds of movement may exist, and the total movement of a component through the membrane is the sum of its bulk flow and its diffusion. Two factors are important in the diffusion term of the transport equation, the concentration gradient and the voltage gradient. To decide whether a specific substance is transported passively it is necessary to compare its transport quantitatively with that predicted by the equation derived below.

The bulk flow of a component in a solution is the product of the velocity of the solution with respect to the membrane and the concentration of the solute as it passes through the solution. If the solute is relatively impermeable, its concentration may be lower as it passes through the membrane than in the bulk solution whence it comes. A measure of this reduction of concentration is given by the reflection coefficient (σ). When σ is one, all the material that approaches a membrane bounces back and there is no bulk flow of the solute. When σ is zero, there is no separation of solute from solvent and the concentration of the solute as it passes through the membrane is the same as in the bulk solution. The flux, in moles per cm² per sec is given by the expression:

$$M = Cv(1 - \sigma)$$

where v is the velocity. When the reflection coefficient is zero, this expression is the same as $\dot{Q} = C\dot{V}$ given above, because the integral of the flux over the total cross-sectional area is the mass flow and the integral of the velocity over the total cross-sectional area is the volume flow.

The diffusion term generally includes the gradients of both concentration and voltage; these gradients will produce movement of the component through the solution as a whole. In one dimension, the gradient (grad) of a function may be defined as the rate of change of the function with distance. Movement through the solution is in the direction of the negative of the gradient (see Chap. 1, Vol. I).

Consider two different solutions of glucose separated by a membrane. Grad C is equal to $(C_1 - C_2)/\delta$, where δ is the thickness of the membrane. If the membrane is permeable to glucose, glucose will move from the side of higher concentration to the side of lower concentration, *i.e.*, it will move down its gradient. The flux is given by the expression:

$$M = D\frac{C_2 - C_1}{\delta} = \frac{D}{\delta}(C_2 - C_1) = P(C_2 - C_1)$$

where D is the diffusion coefficient, and P the permeability. Permeability includes both the diffusivity within the membrane and the membrane thickness. In the same manner, if two identical solutions of NaCl are separated by a membrane which is permeable only to Na^+, and if a voltage gradient (grad $\mathscr{E}$) is impressed across the membrane by an external source, Na^+ will move down the voltage gradient. The flux of Na^+ will be from the side of higher electric potential to the side of lower electric potential. The magnitude of the flux will be proportional to the charge on the particle, to grad $\mathscr{E}$, and to the number of particles present which can move. The expression for net flux under these conditions is:

$$M = -D\frac{ZF}{RT} C \text{ grad } \mathscr{E}$$

where Z is the charge of the particle, F is the Faraday, R is the universal gas constant and T is the absolute temperature. The diffusion term is the sum of the concentration and electrical terms:

$$M_{\text{diffusion}} = -D\left(\text{grad } C + \frac{ZF}{RT} C \text{ grad } \mathscr{E}\right)$$

The total passive transport of a substance with respect to the membrane is the sum of the bulk flow and the diffusion terms.

$$M = -D\left(\text{grad } C + \frac{ZF}{RT}C \text{ grad } \mathscr{E}\right) + Cv(1 - \sigma)$$

In Chapter 24, Volume II, we saw how different conditions gave rise to different solutions of this equation. Thus, in the steady state, when the volume flow is zero, this equation gives rise to the Ussing equation for passive transport. For a cation, this is:

$$\frac{M_{12}}{M_{21}} = \frac{C_1}{C_2} \exp \frac{ZF}{RT} (\mathscr{E}_1 - \mathscr{E}_2)$$

where M_{12} is the one way flux from compartment 1 to compartment 2 and M_{21} is the reverse flux. At equilibrium, when the volume flow and the net flux are both zero, the flux equation can be integrated to give the familiar Nernst equation, which, for a cation is:

$$\mathscr{E}_1 - \mathscr{E}_2 = \frac{RT}{ZF} \ln \frac{C_1}{C_2}$$

In the kidney, we are often interested in solutions that include the bulk flow term. In experiments utilizing stopped flow microperfusion, a solution is put into a portion of the tubule and left there until concentrations within the tubular fluid come to equilibrium. At that time, no net movement of the solute occurs and reabsorption by bulk flow is balanced by secretory diffusion. The solutions that are obtained are analogous to the Nernst equation (see above). When there is no voltage gradient, or when the material is uncharged, the equilibrium condition of no net transport leads to a relation between the concentrations of the solute in the two compartments and the bulk flow for compartment 1 to compartment 2:

$$\frac{C_2}{C_1} = \exp \frac{v (1 - \sigma)}{P}$$

The steady state solution for when there is net transport is more appropriate in free-flow micropuncture experiments. This solution, for an uncharged molecule, is:

$$C_1 - C_2 e^{\frac{v}{P}(1-\sigma)} = \frac{J}{v(1-\sigma)} (1 - e^{\frac{v}{P}(1-\sigma)})$$

When voltage terms are required as well as concentration and bulk flow terms, even more complicated solutions are needed, but they can readily be derived from the passive transport equation.

ACTIVE TRANSPORT. Transport is said to be active when, in addition to the passive terms, another term is required to describe it. Detectable active transport thus requires the combination of a pump that will move material across the membrane and relative impermeability that will prevent rapid diffusion of material back to the site from which it came.

In the kidney, many of the materials that are transported show significant passive permeability which makes the detection of active transport more difficult. In addition, concentration differences, voltage differences and bulk flow terms may all be involved, and hence the number of separate measurements required to test whether the movement of a given solute is passive can be prohibitively large.

The design of the kidney allows the severity of the problem to be reduced. The proximal tubule is probably the only region where bulk flow has a significant effect on solute movement. The rate of water reabsorption is highest and the reflection coefficients are lowest in that portion of the nephron. Since, in the mammalian nephron, the potential difference between the tubular fluid and the blood is nearly zero across proximal tubular cells (see below), we have only to balance bulk flow effects against concentration gradient effects. In the more distal portions of the nephron, reflection coefficients appear to be much higher and the criteria for active transport can be obtained by balancing the electrical and concentration gradient effects. Even this simplification is not always sufficient, and the decision as to whether a substance is actively or passively transported must often be based on more indirect evidence.

Fulfillment of four ancillary criteria provides strong presumptive evidence for active transport. These criteria are the demonstration of (i) a maximum rate of tubular transport, (ii) competition between similar molecular species, (iii) inhibition by metabolic inhibitors, and (iv) in certain cases, failure to vary in the expected manner with such variables as urine flow or pH.

The first two criteria imply a specific combination between the transported substance and the transport system at a limited number of sites. When all sites are occupied, a maximum rate of transport has been attained and further elevation of the plasma concentration of the material will not increase the rate of transport. If some sites are occupied with one molecular species, correspondingly fewer sites are available for transport of a similar compound handled

by the same system. The third criterion implies that the transport requires energy derived from cellular metabolism. The fourth criterion is a portion of the definition given above of active transport. Glucose transport, for example, is not changed by wide variations in the rate of urine flow or the pH of urine. Reabsorption of urea, on the other hand, is closely related to the reabsorption of water, a finding that suggests that bulk flow may play a significant role.

Characteristics of Tubular Transport. We are interested in two different types of transport characteristics. If a fraction of a single tubule is isolated and perfused with a known solution, the pertinent fluxes and electrical potentials are suitable data for insertion into the passive flux equations and for a detailed analysis of tubular transport mechanisms. A cellular physiologist would follow this approach. On the other hand, data from a clearance experiment tell how the kidney as a whole acts in the regulation of body fluid composition. A practicing physician might be more interested in this type of information. The overall clearance characteristics are formed from a combination of the local transport properties, the distribution of transport sites within the nephron and the way conditions within tubular fluid change in different parts of the nephron.

THE LOCAL PROCESS. *Concentration Dependent Transport.* Current concepts of carrier-mediated transport lead one to expect that any active transport system must ultimately have a maximum velocity, and, indeed, this expectation is the basis for the ancillary criteria used to detect active transport. At a local level, this view leads to the prediction of a relation between concentration of material and transport rate that depends on concentration at low concentrations, but which saturates at a constant rate at high concentrations. When small amounts are delivered to the site of transport, all or almost all of the material will be transported. Transport sites in the early portion of the nephron may be saturated, and those farther downstream may be working at very low substrate levels. As the amount presented to the transport site is increased, saturation occurs farther and farther down the nephron until, finally, all of the transport sites throughout the nephron are saturated. This maximal attainable rate of transport is called the *transport maximum* (Tm). The local transport characteristics of many organic solutes exhibit this behavior.

Flow Dependent Transport. A second type of local transport characteristic is seen in the kidney. For many ions, the rate of local transport is a linear function of the mass flow of material past the site of transport. Thus the transport rate will change, not only with a change in concentration of the material, but also with a change in the rate of fluid flow. The physical mechanism by which this occurs is unclear. Transport systems that are known to behave in this manner are all reabsorptive, and the consequence of such reabsorption is that a fixed fraction of the material entering any segment of the nephron is reabsorbed in that segment. The proximal reabsorption of Na^+, Cl^-, and HCO_3^- behaves in this manner.

Both types of systems may be subject to regulation. H^+ secretion acts like a concentration dependent system where the dependence is on cellular H^+ concentration. Changes in cellular pH modify the maximum transport rate in this system. Proximal Na^+ reabsorption is a classic flow dependent system, and the fraction of the solute that is reabsorbed in any segment depends on the interstitial fluid volume.

INTEGRATION OF THE LOCAL PROCESS. Tubular puncture data of solute flow as a function of length along the nephron give rise to one of three types of relations. If the transport activity is restricted to a very small portion of the nephron, the tubular clearance ratio changes in an almost step-like manner there and remains constant elsewhere. The secretion of PAH occurs only in the middle third of the proximal tubule and generates such a relation. Transport systems which are distributed throughout an appreciable portion of the nephron and which show flow dependent properties generate exponential reabsorption. A plot of the logarithm of the tubular clearance ratio as a function of distance along the tubule gives a straight line with a slope that depends on the fractional reabsorption. Na^+ transport gives this sort of relation. Finally, concentration-dependent systems that are distributed show reabsorption in their more distal portions that is more rapid than the exponential relation of flow-dependent systems. The plot of log tubular

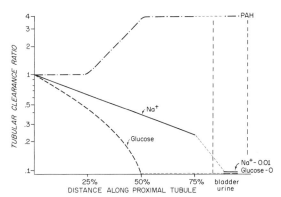

Figure 25–6 Tubular clearance ratio as a function of distance along the proximal tubule for three types of materials. *Dotted and dashed line,* PAH-tubular secretion is indicated by the elevation of the tubular clearance ratio in the proximal tubule. No further transport occurs, and the tubular clearance ratio remains constant in the more distal portions of the nephron. *Solid line,* Na^+-exponential reabsorption distributed throughout the proximal tubule. Further reabsorption occurs more distally so that the bladder clearance ratio is lower than the most distal proximal tubular clearance ratio. *Dashed line,* Glucose–uniform concentration-dependent reabsorption distributed throughout the proximal tubule.

clearance ratio against distance for such systems curves downward. Glucose reabsorption shows this character. Three representative curves are shown in Figure 25–6.

The properties observed in clearance experiments depend on the local properties, the way they are distributed and the number of different types of transport to which any substance is subjected. Na^+ is reabsorbed in the proximal tubule, the thin limb, the distal tubule, and the collecting duct in a variety of different ways. The dominant process, numerically, is proximal reabsorption and this is a flow dependent process. Thus the manner of expressing total Na^+ reabsorption that seems most useful is in terms of percentage of the filtered load that is reabsorbed, *i.e.,* clearance ratio. For fixed reabsorptive conditions the clearance ratio of Na^+ is independent of the filtered load, and changes in the clearance ratio reflect changes in the activity of the transport system. Urea is reabsorbed in some portions of the tubule and secreted in others, the reactions being strongly influenced by passive diffusion properties and, perhaps, by a weak active transport system. The clearance ratio of urea depends on the filtered load of urea, on the pre-existing dietary condition of the animal and on urine flow; all of these dependencies reflect the variety of processes involved in urea transport. Glucose and PAH are transported by concentration-dependent systems in the proximal tubule, and any appreciable transtubular flux of these solutes occurs only at these transport sites. Thus the properties of these latter transport systems are clearly seen in the clearance relations.

In the case of glucose, reabsorption is essentially complete at low filtered loads and there is no urinary excretion. The clearance ratio is therefore zero. As the load is increased above Tm, the increment in excretion is just equal to the increment in the filtered load. At all plasma concentrations above that which produces a filtered load equal to Tm, the rate of excretion is given by the expression $Qu = Qg - Tm = C_p \cdot \dot{V}g - Tm$. The clearance ratio rises from zero and approaches one asymptotically. The plasma concentration at which Tm is reached is called the *threshold concentration* and is characteristic of the substance, and, of course, the filtration rate. In the case of PAH, which is secreted by the tubules, the material that appears in the urine is composed of two moieties, one which entered the lumen at the glomerulus and one which entered as the result of transtubular transport. The former, which is filtered, is always proportional to the plasma concentration (barring peculiarities in protein binding), whereas the latter depends on transport. When the plasma concentration is low, all of the material

that escapes glomerular filtration is transported. So long as the plasma concentration remains well below that resulting in Tm, the amount of solute contributed to the urine by both filtration and transport increases in proportion to the plasma concentration. In these circumstances, the clearance ratio remains constant and above one. If plasma concentration is such that the amount presented to the tubules is above Tm,* the amount secreted remains constant. Although the rate of excretion in urine still increases as the plasma concentration increases, the only cause for this greater excretion is the increase in the filtered load. Since the total excretion no longer increases at the same rate as the plasma concentration does, the clearance ratio falls. It declines from its initial high value (which is characteristic for the substance) toward one. The relations between the filtered load, the amount of tubular transport, the renal excretion and the clearance ratio for avid concentration dependent systems with no passive permeability are illustrated in Figure 25–7.

The Major Processes. PASSIVE PROCESSES. The reabsorption of water secondary to the reabsorption of the major tubular solutes is an obvious and important passive transcellular movement. In addition, four

---

*Remember that the amount presented to the tubules is equal to the amount which enters the kidney each minute minus the amount which is filtered at the glomerulus. The tubular load of PAH = $\dot{V}p(PAH)p - \dot{V}g(PAH)p = (PAH)p(\dot{V}p - \dot{V}g)$.

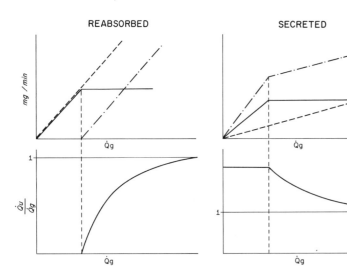

REABSORBED   SECRETED

Figure 25–7 Relationships between measures of tubular transport activity for concentration-dependent transport systems. *Upper graphs,* dashed line, filtered load; solid line, tubular transport; dotted and dashed line, excretion. *Lower graphs,* clearance ratio as a function of filtered load. *Left hand graphs,* reabsorbed material; *right hand graphs,* secreted material.

other conditions exist in the kidney which induce significant transcellular movement: (i) solute movement secondary to bulk flow, (ii) movement of both water and solute across tubular cells when they pass through a region of interstitial fluid in which the concentration of solutes is markedly different from that in systemic plasma, (iii) movement in response to strong voltage gradients and (iv) nonionic diffusion.

Movement Secondary to Bulk Flow. This mechanism is probably important only in the proximal tubule. Most of the filtered Na^+, Cl^-, HCO_3^- and water are reabsorbed in this region. Both passive and active mechanisms are probably responsible for the reabsorption of the filtered ions, but it is the reabsorption of these ions that drives water reabsorption. Proximal tubular epithelium is freely permeable to water and, as solute is reabsorbed, water is also reabsorbed in osmotically equivalent quantities. Since the reflection coefficients of the major ions are relatively low in the proximal tubule, this reabsorption of water carries its dissolved solutes along with it. The overall result is the massive reabsorption of nearly unmodified glomerular filtrate. Superimposed on this massive reabsorption are various active transport processes so that proximal tubular function is more complex than simply the reabsorption of glomerular filtrate. However, some fraction of the filtered Na^+ and Cl^- and much of the filtered K^+ and Ca^{++} are probably reabsorbed secondary to bulk flow.

An additional driving force is needed to get the bulk reabsorption started. An active exchange mechanism takes place wherein luminal Na^+ is reabsorbed in exchange for cellular H^+, and in the proximal tubule this process leads to net reabsorption of Na^+ and HCO_3^- (see below). Apparently, another active process leads to the reabsorption of Na^+ and Cl^-. In addition, there is active reabsorption of many organic molecules. These processes would initiate a reabsorptive water flow which, in turn, would carry more of these same ions with them across the tubular cells. In addition, recent measurements have shown that there is a detectable hydrostatic pressure difference between the tubular fluid and its surrounding blood.[2a] This pressure difference constitutes a net force driving water from the tubule, across the tubular cells into the blood. Although the oxygen consumption of the kidney is extraordinarily high, it is not high enough. Active Na^+ transport is coupled stoichiometrically to oxygen consumption, and in other tissues 15 to 18 moles of Na^+ is transported per mole of oxygen. In the kidney, almost twice that amount of Na^+ is transported.[37] Although other possible explanations of this anomaly exist, including a significant use of anaerobic oxidation by renal tissue, one explanation is that a significant portion of the filtered Na^+ is reabsorbed by a passive process.

Movement Across Tubular Cells Passing Through Unusual Interstitial Fluid Environments. The thin limb of the loop of Henle dips down into the medulla, makes a sharp turn and then comes back toward the cortex. The collecting duct also passes through the medulla. As we shall discuss below, the interstitial fluid in the renal medulla is markedly hypertonic to systemic plasma, so that as a tube, either descending limb or collecting duct, passes down into the renal medulla, the tubular fluid passes interstitial fluid of successively higher and higher solute concentration. Passive reabsorption of water and passive secretion of solutes take place in descending limbs. In the ascending limb, tubular fluid passes interstitial fluid of successively lower solute concentration and the reverse processes take place. Because the concentration of urea, relative to plasma, is much higher than that of any other solute in the medulla, passive secretion of urea is probably the most important result of the first pass of the nephron through the medulla. Passive reabsorption of water to result in a final urine that is hypertonic to systemic plasma — but not to the medulla — and passive reabsorption of urea to a concentration that is much higher than that found in systemic plasma are the major results of the action of this concentrated medullary interstitial fluid on the tubular fluid in the collecting duct.

Movement in Response to Strong Voltage Gradients. Two measurements of voltage are important in the determination of the processes involved in the tubular transport of ions. One of these is the electric potential between the interior of the cell and the interstitial fluid — *the transcellular potential.* The second is that potential measured between the tubular lumen and the interstitial fluid — *the transtubular potential.* Although both potentials change with varying conditions, some representative values can be given for each. The transcellular potentials from proximal tubular cells are about 75 mv, cell fluid negative to interstitial fluid.[8] Fewer measurements have been made from distal tubular cells, but they appear to be in about the same range.[36] In mammals, transtubular potentials from the proximal tubule are quite low. No proximal transtubular potential seems to exist in the rodent,[7] although a small potential may be found in other mammals.[3] A negative transtubular potential occurs in the first portion of the distal tubule,[45] the straight portion that ascends through the outer medulla. A significant negative transtubular potential is present in the distal convoluted tubule. The value of this potential depends quite strongly on the composition of distal tubular fluid and may vary from as small as 10 mv to over 100 mv.[32] A value for "normal conditions" might be taken as 60 mv. Finally, the collecting duct shows a negative transtubular potential which is about 25 mv high in the collecting duct.[43] As these values suggest, the effects of the transtubular potential on ion movement are greatest in the distal convoluted tubule. In response to that driving force, there is a net passive secretion of cations in the cortical portion of the distal tubule. Distal tubular secretion of K^+ and of H^+ both take place, and both have great significance to ionic and acid-base balance of the body. Both are driven by the transtubular potential difference in this portion of the nephron. The negative transtubular potential in the collecting duct has less significance. Under some conditions, passive secretion of cations and passive reabsorption of anions may occur in the collecting duct. The negative transtubular potential found in the distal portions of the nephron shows that distal Na^+ reabsorption must be an active process. This negative potential aids in the distal reabsorption of Cl^-, but is not sufficient to account for the lowering of Cl^- concentration that is observed in distal tubular fluid.

Nonionic Diffusion.[25] Tubular epithelium is much more permeable to the un-ionized than to the ionized species of a number of weak acids and bases, *e.g.*, ammonia,[29] quinacrine (Atabrine), salicylic acid and barbiturates. For substances such as these, the un-ionized molecule diffuses across tubular cells until it is in equilibrium between cell fluid, tubular fluid and peritubular blood. In each fluid compartment, ionization occurs according to the *p*H of that compartment. Thus ammonia is trapped in acid tubular fluid, mostly in the form of NH_4^+, but not much ammonia is trapped in alkaline tubular fluid. This process occurs throughout the length of the nephron, but the portion of the nephron that is important in determining the excretion of these materials is the terminal portion, the collecting duct. In the urine that has left the collecting duct, the concentration of the permeant species will be the same as that in renal tissue or in renal venous blood. The concentration of the ionized species will be determined by the concentration of the un-ionized species and the *p*H of the urine. Acidifying the urine will enhance the total concentration of weak bases such as ammonia or quinacrine, whereas alkalinizing the urine will enhance the concentration of weak acids, such as salicylic acid.

The ability to modify the total urinary concentration of solutes which are handled by nonionic diffusion is used in the therapy of acute drug poisoning. Salicylate intoxication is routinely treated by the administration of HCO_3^- or lactate in order to alkalinize the urine. Under these conditions, a much higher concentration of salicylate appears in the urine and, since the urine flow is also elevated, the rate of excretion of salicylate is much enhanced. The same technique is used after intoxication with barbiturates.

One final passive phenomenon should be mentioned. Even though there are a large number of independent transport processes, both active and passive, taking place along the nephron, there is an overall constraint of electroneutrality. If passive movements were to lag behind active movements and a net charge transfer were to take place, the transtubular potential would change until the potential drove the passive movements at a rate that reduced net charge transfer to zero across the nephron.* Active transport processes that are really directed toward the movement of a single ion, such as Na^+ reabsorption, or H^+ secretion, have the effect of producing either coupled cation and anion transport or cation exchange.

---

*The reabsorption of as little as one part in a million of the filtered Na without associated anion reabsorption or cation secretion would lead to an increase in the potential across the tubular cell of at least 1 mv per min. By the age of 20, one could then generate about 10,000 volts and micturition would be both impractical and dangerous.

ACTIVE PROCESSES. Although many materials are actively transported, five different processes will describe the major ionic movements and give prototypes for the active transport of organic materials.

Glucose Reabsorption. Glucose transport was one of the first renal transport systems to be well delineated; the work was done by Shannon and Fisher.[34] Both stop-flow experiments and micropuncture studies[40, 47] indicate that glucose reabsorption occurs in the proximal tubule and normally proceeds there essentially to completion. Any glucose escaping reabsorption in this region is destined for excretion.

The maximum rate of transport is 375 mg per min ± 80 mg in men and slightly lower in women. Endocrine imbalances may alter Tm_G slightly, but in the main it is quite constant and depends only on the number of nephrons functioning. For this reason, measurement of Tm_G is sometimes used clinically to estimate the number of functioning nephrons. This measurement is called "tubular reabsorptive capacity."

The best tubular puncture data for glucose are still from the work of Walker *et al.*[40] who did much of the pioneering work in 1941. The curve for glucose illustrated as the dashed lines in Figure 25–6 is taken from their work and is roughly consistent with uniform reabsorptive activity of a concentration dependent system through the first half of the proximal tubule.

If Tm_G is 375 mg per min and the filtration rate is 120 ml per min, then the renal threshold for glucose would be about 300 mg of glucose per 100 ml of plasma. Since plasma glucose is normally under 100 mg per 100 ml, one can predict that normal urine should be glucose free and that the concentration of glucose in plasma is not normally controlled by the kidney.

Even though Tm_G is attained at plasma concentration of 300 mg per 100 ml, some glucose appears in the urine when the plasma concentration is 200 mg per 100 ml.

This could be interpreted to mean that the combination between glucose and the transport site is relatively loose. However, this is apparently not the case. Glucose reabsorption is apparently quite avid in each individual nephron. But there is statistical dispersion of both glomerular filtering capacity and tubular reabsorptive capacity. Thus a nephron in which a "large" glomer-

ulus is attached to a "small" proximal tubule reaches its individual Tm at a relatively low plasma concentration. Conversely, if a "small" glomerulus is attached to a "large" proximal tubule, the nephron will not reach its Tm until relatively high plasma concentrations of glucose are reached. Recently, Oliver and associates have shown that the distribution of the ratio of glomerular activity to reabsorptive activity, which is calculated on the assumption that each individual nephron reabsorbs completely at loads below Tm, agrees closely with distribution of sizes determined directly.[28]

Para-aminohippuric Acid Secretion. PAH is bound to plasma proteins; hence its concentration is not the same in glomerular filtrate and in plasma. Instead, the concentration in the glomerular filtrate is the same as the concentration of unbound PAH in plasma water. In man, this value is taken to be 78 per cent of the total plasma concentration. So avid is the tubular transport system, or so rapid is the binding reaction, that tubular cells remove essentially all of the PAH — both bound and unbound — from the blood perfusing them, so long as the amount presented per minute is less than Tm. This secretory system is located in the middle third of the proximal tubule. The data in Figure 25–6 are taken from the work of Courtney *et al.*[5] and show that secretion of PAH occurs only in this region.

The Tm of PAH is 80 mg per min ± 17 mg in men and somewhat lower in women. The value is quite constant and is the basis for another clinical test of tubular functions that is similar to the use of Tm_G. Tm_{PAH} is considered an estimate of "tubular secretory capacity."

A large number of other compounds are transported by the same system. All are aromatic organic acids, the most important being penicillin. Before supplies of penicillin became plentiful, considerable effort was expended in attempting to inhibit this system. It was found that the transport involves two steps. The first is a specific combination with a "carrier" molecule; the second is transport of the resulting complex. Chemical reactions leading to the decomposition and release of the acid occur at the luminal border. Tubular secretion is competitively inhibited either by a compound which combines well with the carrier but which is transported only slowly, or, better, by a compound which combines well with the carrier but which cannot undergo the biochemical transformation necessary for decomposition. Probenecid (Benemid), which falls in the latter class, was developed as a result of these investigations. Penicillin is now cheaper than probenecid, and the original purpose of the drug is largely forgotten. However, it also inhibits the renal reabsorption of uric acid and is now widely used in the treatment of chronic gout.

Na and Cl Reabsorption. Na^+ is reabsorbed both in association with the reabsorption of anions and with the secretion of cations. There is undoubtedly some active transport of Na^+ and Cl^- in the proximal tubule, although little is known about the characteristics of this system.[13]

Active reabsorption of salt occurs in the first portion of the distal tubule. This is the portion of the nephron that passes through the outer zone of the medulla. The transtubular potential in this region is about +9 mV, and since the concentration of Cl^- falls in tubular fluid, an active transport system directed toward Cl^- must be present.[2b] Tests of the flux equation indicate that Na^+ reabsorption in this region can be accounted for in terms of the passive forces present. This active solute reabsorptive system is one of the driving forces for the concentration process, and inhibition of this system reduces the ability of the kidney to produce either a hypertonic or a hypotonic urine. The clinical diuretic furosemide acts on this system.

Active systems for both Na^+ and Cl^- are present in the distal tubule and the collecting duct. In the distal tubule Na^+ reabsorption takes place from a solution in which the Na^+ concentration is about one-half that of plasma. Since both the chemical and the electrical forces would tend to produce cation secretion, the continuing reabsorption of Na^+ must be the result of an active transport system. The concentration of Na^+ in tubular fluid can be reduced to very low values in the collecting duct and this also requires active transport. Cl^- reabsorption is favored by the transtubular potential in the distal tubule, but an active process must also exist, because the concentration of Cl^- may fall below that predicted by the Nernst equation at a time when Cl^- reabsorp-

tion is still taking place.[32] The active transport systems for Na^+ and Cl^- in the distal portions of the nephron are coupled in some way because agents that inhibit one system inhibit the other. The benzothiazide diuretic drugs act in this region. The main difference between the events in the distal tubule and the collecting duct is that, in the collecting duct, Na^+ and Cl^- concentrations may fall to very low values, whereas in the distal tubule they rarely fall below one-half their concentration in plasma. This difference probably stems from a reduced passive permeability and therefore a reduced back-diffusion in the collecting duct. A single system is probably distributed through the distal tubule and the collecting duct that leads to active reabsorption of both Na^+ and Cl^-. This system works on a small fraction of the filtered load, because about 90 per cent of the filtered salt has been reabsorbed more proximally. However, this is the system that is responsible for the final conservation of salt and is involved in regulation of body Na^+ and Cl^- content.

Hydrogen Ion Secretion. Hydrogen ion secretion takes place in the proximal tubule, the distal tubule and the collecting duct. The driving force for H^+ secretion in the distal portions is the transtubular voltage. In the proximal tubule, however, H^+ is actively secreted in exchange for the reabsorption of Na^+.[31] The source of the cellular H^+ is the hydration reaction of carbon dioxide within proximal tubular cells. Carbon dioxide combines with water within the cell to produce carbonic acid, which then ionizes into cellular H^+ and cellular HCO_3^-. The hydrogen ion is secreted into the tubular lumen, and the reabsorbed Na^+ is put into the peritubular blood in association with the HCO_3^- that was formed within the cell. Thus the secretion of each mole of cellular H^+ leads to the appearance of one mole of Na^+ and one mole of HCO_3^- in peritubular blood. In the proximal tubule, the secretion of H^+ leads to the destruction of HCO_3^-. Luminal HCO_3^- combines with the secreted H^+ to form H_2CO_3. The H_2CO_3, formed in the tubular lumen, is dehydrated to form CO_2 and water, and this reaction is catalyzed by the enzyme carbonic anhydrase, which is located on the luminal border of proximal tubular cells. The CO_2 diffuses

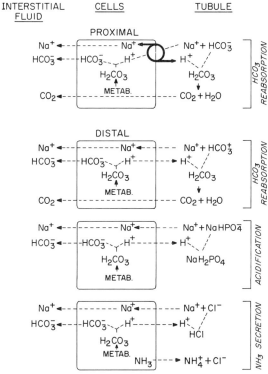

INTERSTITIAL FLUID | CELLS | TUBULE

Figure 25–8 Summary of effects of hydrogen ion exchange. Broken lines indicate passive transport; solid lines, active transport.

back into the cell, where it is again rehydrated and used as the source for another H^+ to be secreted. Ninety per cent or more of the filtered HCO_3^- is reabsorbed in the proximal tubule as the result of Na^+ for H^+ exchange. Figure 25–8 illustrates the action of this system.

K Reabsorption. Potassium ion is probably actively reabsorbed in both the proximal and the distal tubule.[22] This is a relatively weak transport system. The passive movements of K^+ in the distal tubule (where it is secreted because of the transtubular voltage) and the collecting duct (where it is reabsorbed *pari passu* with Na^+) are normally the important processes for regulation of K^+ balance.

REGIONAL DESCRIPTION OF RENAL FUNCTION

Large volumes of fluid enter the nephron at the glomerulus. The composition of this fluid is very similar to the composition of plasma except that glomerular filtrate has a very low protein concentration and slight

modification of the concentration of ions present occurs because of the Donnan equilibrium which arises from this protein separation. Organic solutes such as glucose are reabsorbed in the proximal tubule, and some substances such as PAH are secreted. Na^+ for H^+ exchange starts early in the proximal tubule and leads to the reabsorption of about 90 per cent of the filtered HCO_3^-. Because of this extensive reabsorption of HCO_3^-, the *p*H of proximal tubular fluid falls. Much of the NH_3 which has diffused into proximal tubular fluid is acidified to form the less permeable NH_4^+ so that a significant amount of the NH_4^+ that is excreted enters the tubular fluid in the proximal tubule. K^+ reabsorption takes place so that the concentration of K^+ in proximal tubular fluid may be lower than that in plasma. Na^+ and Cl^- are reabsorbed throughout the length of the proximal tubule, probably by both passive and active processes. Proximal tubular epithelium is permeable to water and, as the solutes are reabsorbed, water is reabsorbed with them. This water carries with it some of the dissolved solute. As water is reabsorbed, the concentration of urea rises in luminal fluid. The concentration difference of urea across the tubular cells drives urea reabsorption so that much of the filtered urea is also reabsorbed in the proximal tubule. By the end of the proximal tubule, about 80 per cent of the filtered Na^+ and Cl^- and K^+ have been reabsorbed, about 90 per cent of the filtered HCO_3^- has been reabsorbed, some NH_4^+ has been added and about 80 per cent of the filtered water has been reabsorbed. For many of the organic molecules, whatever tubular transport is to take place has already taken place.

Tubular fluid then passes into the thin limb and descends into the renal medulla. As it dives deeper into the medulla, it passes interstitial fluid of successively higher Na^+ and Cl^- concentration and sucessively higher urea concentration. Urea diffuses into the thin limb, as might Na^+ and Cl^-. Water diffuses out to some extent, although much of the equilibration of tubular fluid with hypertonic interstitial fluid comes from passive secretion of solute particles. At the tip of the loop of Henle, tubular fluid, interstitial fluid, and blood are all at the same total solute concentration and markedly hypertonic to systemic plasma. As the fluid passes up the thin limb, in the ascending portion, there is probably some reversal of the exchanges that occurred in the descending portion. Current evidence suggests that the exchanges that occur in the thin limb are not very important.

The first portion of the distal tubule starts when the nephron enters the outer zone of the renal medulla. Here, Na^+ and Cl^- are actively transported out of the nephron but, unlike the proximal tubule, water does not follow. This reabsorption of solute without water creates the concentrated medullary interstitial fluid. About 10 per cent of the filtered Na^+ and Cl^- are reabsorbed in this region.

The fluid that enters the distal convoluted tubule in the cortex is dilute to systemic plasma, and this osmotic difference is a driving force for the reabsorption of water from the distal convoluted tubule. The permeability of the distal tubule and collecting duct to water varies with the concentration, in plasma, of a hormone which is made in the hypothalamus and stored in the posterior pituitary. This hormone is officially designated *vasopressin*, but is often referred to as *antidiuretic hormone* (ADH). In the presence of ADH, the distal convoluted tubule is permeable to water and some water reabsorption takes place. In the absence of this hormone, little or no water reabsorption takes place from the distal convoluted tubule. The major solutes in tubular fluid at the beginning of the distal convoluted tubule are urea and Na^+ and Cl^-. More urea is present than was in the tubule at the end of the proximal tubule; this is a reflection of medullary urea secretion. Very little HCO_3^- or K^+ remains in tubular fluid. Since the HCO_3^- concentration is low, distal tubular fluid is acid to systemic plasma, and therefore an appreciable fraction of the NH_4^+ that will be excreted is present in luminal fluid. Na^+ and Cl^- are reabsorbed in the distal tubule, and their concentrations may fall to low levels in tubular fluid by the end of the distal tubule. The strong transtubular potential brings about a passive secretion of K^+, and most of the K^+ that appears in the urine enters the lumen across distal tubular cells. The transtubular potential also brings about secretion of H^+. This distal H^+ secretion has three consequences. It normally finishes the reabsorption of the filtered HCO_3^-. Even though this is probably a passive process, the re-

sults are similar to the active secretion of H^+ in the proximal tubule in that the H^+ reacts with luminal HCO_3^- to form H_2CO_3. The secretion of one H^+ which was derived from cellular CO_2 and water leads to the appearance in peritubular blood of one HCO_3^-. Distal reabsorption of Na^+ takes place too, so the result is the loss of one molecule of Na^+ and HCO_3^- from tubular fluid concomitant with the appearance of one molecule of Na^+ and HCO_3^- in peritubular blood. Distal H^+ secretion further acidifies the urine and, as luminal pH falls, more NH_3 is trapped in tubular fluid as NH_4^+.

The fluid that enters the collecting duct contains Na^+, K^+, and NH_4^+ as its major cations and Cl^- as its major anion. About one-half the solute in the tubular fluid is urea, and the solution is acid to plasma. In the presence of ADH, sufficient water leaves the distal tubule, making the fluid near isotonic.* In its absence, the solute concentration of tubular fluid is less than one-half that of systemic plasma. Final, avid Na^+ and Cl^- reabsorption takes place, and the concentrations of these ions may fall to extremely low values. Normally K^+ is reabsorbed in the collecting duct. The reabsorptions of Na^+ and K^+ in the collecting duct are in proportion to their concentrations in tubular fluid, so that the greater the amount of K^+ secretion that occurred in the distal tubule or the less the amount of collecting duct Na^+ reabsorption, the greater the K^+ excretion. Further H^+ secretion takes place which further acidifies the tubular fluid. This further acidification converts luminal NH_3 to NH_4^+ so that more NH_3 diffuses into tubular fluid and more NH_4^+ is trapped. Finally, urea is reabsorbed from the collecting duct. The concentration of urea in bladder urine is the same as the concentration in the interstitial fluid at the tip of the renal papillae. Thus the collecting duct reabsorption of urea is

---

*The degree to which tubular fluid equilibrates osmotically with plasma in the distal convoluted tubule differs in different species. In the rat, when adequate amounts of ADH are present, equilibration takes place early in the distal convoluted tubule and tubular fluid in the latter half of the distal convoluted tubule is isotonic to systemic plasma. In the dog and the monkey, water efflux is slower and, even in the presence of large amounts of ADH, tubular fluid is still slightly hypotonic when it leaves the distal tubule.

possibly just a consequence of passive diffusion, and the concentration of urea is elevated because of the abstraction of water. However, some active urea reabsorption from the collecting duct is also possible.* Although, in the presence of ADH, there may be passive diffusion of urea from collecting duct to interstitial fluid, the question remains, "How did the urea get so concentrated in the medulla in the first place?"

REGULATION OF PLASMA COMPOSITION

The kidney has so far been viewed as a machine with a set of transport properties built into it. The end result of these properties is that renal function maintains body composition rather constant. The manner in which this control of body composition takes place is of some interest. A dominant principle in the renal control of plasma composition is the *high turnover of plasma through the kidney.* In the normal man, with an extracellular space of about 15 liters and a filtration rate of 100 ml per min, half of the extracellular fluid in the body is filtered every 75 min. To see the effect of this high turnover rate, suppose the tubular reabsorption of sulfate was decreased by 10 per cent one morning. Sulfate would be excreted in the urine until the plasma sulfate concentration reached a new equilibrium level. This would be at 90 per cent of the original concentration. The concentration of sulfate would drop halfway toward its new level within two hours, and by eight hours, the end of a normal working day, the concentration would be essentially at its new equilibrium value. On the other hand,

if the tubular reabsorption were to increase, urinary excretion would fall and plasma sulfate concentration would increase as sulfate and protein sulfhydryl were taken into the body. Little or no sulfate would be excreted until the plasma concentration had reached its new equilibrium value. Here, the time course would be determined, not by the characteristics of the kidney, but by the rate at which sulfate is taken into the body and produced from protein mercaptan.

Two other principles — *over-all body balance* and *glomerulotubular balance*† — are involved in the control of body balance by the kidney. According to the first principle — that of over-all body balance — a person excretes material at the same rate at which it is taken in, over a long period. Although this is simply a statement of steady state conditions, it suggests the existence of both a method for sensing changes in net balance and a mechanism for modifying renal characteristics. We shall look at the ways that the renal characteristics are changed below.

Just above, we discussed what would happen if the intake of sulfate remained constant and the renal characteristics changed. As an illustration of glomerulotubular balance, let us now observe the same system in the more realistic situation in which the intake of sulfate varies, but the renal characteristics remain constant. As the intake rises, more sulfate is presented to the tubules. The Tm is eventually exceeded and the excess sulfate enters the urine, thereby reducing the plasma concentration toward its previous level. A reduction of sulfate intake reduces the plasma concentration and reduces excretion. Only after more sulfate has been taken into the body or produced and the plasma concentration is again brought to normal levels will urinary excretion take place. For a reabsorptive system with a sharp renal threshold, this relationship

---

*Under some conditions, collecting duct urea concentration can be less than medullary concentrations at a time when urea is being reabsorbed from the collecting duct.[2, 20] In addition, urea seems to be too intimately involved in the concentrating mechanism simply to be a passive attendant. The maximum solute concentration that can be attained in urine increases with increasing urea concentration. Although most solutes are concentrated in the interstitial fluid of the medulla, the ratio of the concentration in a medullary slice to the concentration in plasma is normally far greater for urea than for any other solute. This ratio falls after protein deprivation and, concomitantly, the ability to produce a highly concentrated urine also decreases.

---

†*Glomerulotubular balance*, used here in the old sense, was coined by Homer Smith. Recently, the same phrase has been widely used to express an exact proportionality between the amount of Na^+ passing a region of the proximal tubule and the tubular reabsorption from that region. As discussed above, this problem is centrally involved in investigation of the mechanisms of proximal ion transport. However, no good alternative phrase exists for the old meaning of glomerulotubular balance, so we shall use it here.

may be stated fairly precisely. Averaged over a period of time, the intake will equal the outgo:

Plasma concentration × filtration rate =
 Tm + excretion = Tm + intake

For secretory systems, analogous relations exist. This sort of relation provides the basis for the clinical evaluation of kidney function from the blood urea nitrogen concentration. Even though there is not a sharp Tm in this case, blood urea becomes very high only when the filtration rate is markedly reduced.

The effect of glomerulotubular balance enters into the control of most solutes. For many of them, notably the important electrolytes and water, there are other control systems superimposed that modify the renal characteristics. Together, these two types of systems normally maintain plasma composition, and thus total body fluid composition, quite constant.

Na and Cl. A normal American diet contains about 100 mEq of Na^+ and Cl^- per day. Thus we normally excrete about 100 mEq per day. This constitutes only about 1 per cent of the filtered load. Most of the reabsorption takes place in the proximal tubule, where the characteristics are flow dependent, meaning that variations in $\dot{V}_g$ or in plasma concentration do not have very much effect on excretory rate. Ninety-nine per cent of the increment in filtered load that occurs after elevation of filtration rate is reabsorbed. The fractional reabsorption, however, depends very strikingly on the extracellular fluid volume. Elevation of extracellular volume reduces proximal fractional reabsorption, and lowering of extracellular volume increases it. Not all of these changes are reflected in urinary excretion, because distal processes may increase their activity in response to increased distal loads after the reduction of proximal reabsorption. The distal convoluted tubule and collecting duct system also modifies its activity in response to changes in extracellular volume, and this latter region is probably where the important regulatory effects occur. After expansion of the extracellular fluid, Na^+ and Cl^- excretions increase and the gain of the regulatory system can be estimated from the time it takes to excrete the infused saline. In the dog, half the infused saline is ex-

creted in about 2½ hours, half of the remaining saline, or a quarter of the initial load, in the next 2½ hours and so on. The response in man is slower, and it takes about one day for half of the infused load to be excreted. This mild response of man is part of the reason that man develops edema relatively easily.

Two other phenomena enter into the regulation of Na^+ excretion. Na^+ is the primary cation in extracellular fluid, and when large amounts of anion have to be excreted most of the cation that accompanies the anion is Na^+. In uncorrected diabetes mellitus, the faulty glucose metabolism leads to the production of large amounts of organic acids. The major acid metabolites, β-hydroxybutyric acid and acetoacetic acid, have low renal thresholds, and hence large amounts of these acids appear in the urine. Na^+ is the major cation that accompanies them and large amounts of Na^+ can be lost in the urine regardless of what the extracellular volume control system is trying to do. Na^+ excretion is also increased whenever large amounts of any nonreabsorbable solute are excreted. If sufficiently large amounts of inulin, for example, are given to an animal, that inulin makes up 20 per cent of the total solute in plasma; proximal Na^+ reabsorption leads to a reduction in the concentration of Na^+ in proximal tubular fluid. This process can best be illustrated by a numerical example. Normally, the Na^+ concentration in plasma, tubular fluid and proximal reabsorbate are all about 150 mEq per liter. However, after inulin, the Na^+ concentration in plasma and at the beginning of the proximal tubule may be reduced to 120 mEq per liter, with inulin making up the rest of the solute concentration. As Na^+ and Cl^- are reabsorbed in the proximal tubule, water will still follow in osmotically equivalent quantities and the concentration of the proximal reabsorbate will still be 150 mEq per liter. This reabsorption lowers the concentration of Na^+ in the remaining luminal fluid so that, halfway down the proximal tubule, the Na^+ concentration in luminal fluid may be as low as 60 mEq per liter. In these conditions, significant passive diffusion back into the lumen takes place. The overall result is that *net* proximal Na^+ and Cl^- reabsorption are reduced.

K. Most of the K^+ that appears in the

urine is added to tubular fluid in the distal convoluted tubule. The driving force for this secretion is the transtubular potential. As might be expected, variation in the transtubular potential, in the cellular K^+ concentration and in the volume flow through the distal tubule all modify the secretion. Conditions in which the transtubular potential is strongly negative, *e.g.*, when large amounts of impermeable anions are present, and conditions which elevate cellular K^+ concentration lead to increased tubular secretion of K^+. A higher volume flow through the distal tubule, by increasing the concentration difference between cell and luminal fluids, also increases K^+ secretion. Finally, the presence of HCO_3^- in distal convoluted tubular fluid has a marked enhancing effect on K^+ secretion. It is not presently known whether this effect stems from an elevation of cellular K^+ concentration or from a change in the transtubular potential or whether it is a specific effect. Normally, K^+ is reabsorbed in the collecting duct along with the final Na^+ reabsorption.

In a real sense then, the body regulates K^+ excretion to maintain a constant concentration in distal tubular cells. If the plasma concentration of K^+ increases, active transport mechanisms on the *peritubular* side of the tubular cells take up more K^+ and the cellular concentration increases. This leads to enhanced K^+ excretion and ultimately brings the K^+ concentration back to normal in plasma. Other effects, such as the presence of an impermeable anion, may temporarily alter K^+ excretion, but when an elevated K^+ secretion leads to cellular K^+ depletion, K^+ secretion falls. Agents that inhibit the active K^+ uptake by cells, such as the cardiac glycoside, will reduce K^+ secretion if the distal tubular volume flow and transtubular potential stay constant. However, these same agents may increase K^+ secretion if their effect on volume flow is greater than their effect on cellular uptake. The final result is that, when K^+ secretion is initially low and, therefore, not much K^+ has to be taken up from the luminal border of the cell to keep up with the passive secretion, the cardiac glycosides and many diuretic agents have their major effect on distal volume flow and produce an increase in K^+ excretion. When, however, K^+ secretion is initially high, the rate at which K^+ flows through the cell rapidly depletes cellular K^+ when the luminal pump is inhibited and these same agents will reduce K^+ excretion. Inhibition of the luminal K^+ uptake process, however, always reduces the ratio of K^+ to Na^+ in the tubular fluid that enters the collecting duct.

Acid-Base Balance. The sort of diet one lives on is a major determinant of the daily acid-base problems the kidney must handle. If one lives on a high protein diet, one ingests a large amount of sulfur in the form of sulfhydryl groups. During metabolism, the sulfur is oxidized to $SO_4^=$ and this process tends to lead to metabolic acidosis. Conversely, a vegetarian takes in large amounts of lactate or acetate and the metabolism of these anions tends to lead to a metabolic alkalosis. The kidney is the organ that corrects these dietary tendencies toward acid-base imbalance. In addition, the kidney is capable of compensating some respiratory imbalances. The basic mechanism used is the secretion of H^+. It will be recalled that H^+ is secreted actively in the proximal tubule and passively in the distal convoluted tubule. Most of the filtered HCO_3^- is reabsorbed in the proximal tubule as a result of proximal H^+ secretion. This process has been discussed above and is illustrated in Figure 25–8. Even though distal H^+ secretion is passive, the end result is the same as that of the more proximal process. Every H^+ that is secreted into the distal lumen leaves an HCO_3^- within the tubular cell. The electroneutrality principle assures us that the H^+ that is secreted is, one way or another, balanced by the reabsorption of a cation, usually Na^+. Thus the passive secretion of one mole of H^+, like the active process, ultimately leads to the deposition, in peritubular blood, of one mole of HCO_3^- and one mole of Na^+. So long as tubular HCO_3^- is present in the lumen and being destroyed by H^+ secretion, the result of H^+ secretion is just the conservation of the filtered HCO_3^-. However, the continued secretion of H^+ afterward, which leads to the production of titratable acid and the trapping of NH_4^+ in tubular fluid, also puts Na^+ and HCO_3^- in peritubular blood. *Thus the excretion of titratable acid and ammonia is a measure of the rate of production of new HCO_3^- and, whenever the kidney excretes an acid urine, more HCO_3^- leaves the kidney in the renal vein than enters it in the renal artery.* Normally, the new HCO_3^- that is produced just balances the acids that are produced metabolically and one maintains acid-base balance in the face of an acid-producing diet. Let us look at the way this system works in three unbalanced states.

A *primary metabolic acidosis* is characterized by a reduction in plasma HCO_3^- concentration. This reduction might come about from excess metabolic production of acid, as in diabetes mellitus. Although there will tend to be a compensatory response of hyperventilation which will reduce the magnitude of the plasma pH change, correction of the problem depends on the kidney. Less HCO_3^- is filtered and hence HCO_3^- reabsorption is completed higher in the tubule. More of the distal tubule is involved in titratable acid and NH_4^+ excretion, which means that more new HCO_3^- is made by the kidney. Over a period of time, the new HCO_3^- made in the kidney will build plasma HCO_3^- concentration back.

A *primary metabolic alkalosis* comes about from an increase in plasma HCO_3^- concentration. In this condition, more HCO_3^- is filtered, HCO_3^- reabsorption is not completed until further down in the nephron and less of the tubule is involved in titratable acid and NH_4^+ excretion. Over a period of time, HCO_3^- will fall in plasma, even if there is no frank HCO_3^- excretion whenever the rates of titratable acid and NH_4^+ excretion are less than the rates of metabolic production of new acids. In extreme metabolic alkalosis, the whole nephron may be involved in HCO_3^- reabsorption, or, if the plasma HCO_3^- is sufficiently high, HCO_3^- diuresis may occur. In these conditions, if one calculates total H^+ secretion (which, in the case of HCO_3^- diuresis, is simply the rate of HCO_3^- reabsorption), one usually finds that H^+ secretion is elevated above normal. However, the measure that counts for acid-base balance is not total H^+ secretion, but rather the H^+ secretion that is not associated with HCO_3^- reabsorption. This latter measure is, of course, zero.

A *primary respiratory acidosis* occurs when the arterial Pco_2 tension is elevated. As will be discussed below, elevation of Pco_2 increases the total renal H^+ secretory capacity. As a result of this increased secretory capacity, the filtered HCO_3^- is reabsorbed higher up in the nephron and more of the nephron is left for the production of new HCO_3^-. Plasma HCO_3^- rises and finally comes to equilibrium when the elevated filtered load of HCO_3^- uses up enough of the nephron so that just sufficient distal H^+ secretion is left to balance dietary acid production. This elevation of plasma HCO_3^- in response to elevated Pco_2 is the renal compensation to a respiratory acidosis.

The controls of renal H^+ secretion are just those that would be expected. Since the H^+ is derived from the hydration of cellular CO_2, elevation of Pco_2 enhances the rate of H^+ secretion and lowering of Pco_2 lowers the rate. Further, the distal secretory process is modified by alterations in the transtubular potential in the distal convoluted tubule. The presence of large amounts of non-reabsorbable anion or the stimulation of distal Na^+ reabsorption thus enhances distal H^+ secretion. The amount of titratable acid that is ultimately excreted, however, is limited by the buffer capacity of the urine. Tubular cells are permeable to H^+, and back-diffusion of H^+ occurs. This leads to a limiting pH of the urine, and in man this pH is about 4.5. When the urine is slightly buffered, only a few μmoles per minute of H^+ brings the pH of the tubular fluid down to 5 and further net H^+ secretion stops. However, in highly buffered urines, several hundred μmoles per minute of H^+ secretion may lower tubular fluid pH only to 6, for example, and the full secretory capacity of the nephron can be used. Finally, inhibition of the enzyme carbonic anhydrase which is situated on the luminal border of proximal tubular cells, inhibits the active proximal system. Carbonic anhydrase inhibitors have been used as diuretic agents because of this property, although they are now largely superseded by more modern drugs.

Water.[11] Most of the water filtered from the plasma in the glomeruli is reabsorbed in the tubules; only 1 to 2 per cent of the amount filtered is normally excreted in the urine. Both urine flow and the concentration of solute particles in the urine may vary widely; the rate of urine flow may be less than 1 per cent or more than 50 per cent of the rate of glomerular filtration, and the concentration of solute may vary from almost zero to three or four times the concentration in the glomerular filtrate.

The concentration of solute in the urine depends both on the concentration of antidiuretic hormone (ADH) in the plasma perfusing the kidney and on the rate of urine flow. ADH synthesis in the hypothalamus, its passage from the hypothalamus to the hypophysis, and its release into the blood stream are discussed in Chapter 8, Volume III.

When plasma entering the kidney contains a high concentration of ADH, the solute concentration is higher in the urine than in plasma; urine is *hypertonic* to plasma. When little or no ADH enters the kidney, the concentration of solute is much lower in the urine than in plasma; urine is *hypotonic* to plasma. The rate at which ADH is released depends on the solute concentration in circulating plasma. If the concentration is high, the hypophysis releases ADH into the blood. When the hormone reaches the kidney, water reabsorption is promoted and the urine becomes hypertonic. The water returned to the body reduces the concentration of the solute in body fluids. Conversely, if the concentration of solute in plasma is low, ADH is not released. In these circumstances, hypotonic urine is excreted, and large amounts of water are lost from the body. This loss tends to concentrate body fluids.

Within the limits set by the concentration of ADH in the plasma entering the kidney, the concentration of solute in the formed urine depends on the rate of urine flow. At low rates of flow, solute concentration is relatively high; it decreases progressively as urine flow increases. Urinary concentration of solute as a function of both rate of urinary flow and release of ADH is depicted in Figure 25–9.

The mechanisms responsible for the production of dilute or concentrated urine have been alluded to throughout this chapter. A unified description of these processes will now be given. The collective description of these processes is termed the *renal medullary countercurrent theory*.

The fundamental process is the active reabsorption of Na^+ and Cl^- from the ascending portion of the distal tubule in the outer zone of the medulla. As a consequence, the recent findings on solute concentrations in different regions of the kidney (to be detailed below) and the older data on the concentration and dilution of urine all fit into a single unified theory.

The most important findings which gave

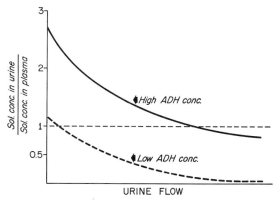

Figure 25–9 Relationship between urine flow and the ratio of urinary to plasma solute concentration. *Solid line*, dehydration; *dashed line*, hydration.

rise to the countercurrent theory are as follows: (i) The mole fraction of solute is higher in slices of kidney taken from the medulla than in slices taken from the cortex. The deeper the site in the medulla from which the sample is derived, the higher the mole fraction of the solute.[39] When ADH is present in adequate amounts, the solute concentration of the formed urine approximates that at the tip of the renal papillae. During water diuresis, when ADH is absent, the solute concentrations of the papillary tips and formed urine are disparate.[46] These findings are shown in Figure 25–10.

(ii) Fluid can be collected from certain portions of the nephron by inserting small glass pipettes into individual tubules. Fluid collected from proximal tubules is always isotonic to systemic plasma. Fluid collected from the turn in the thin segment, at the tip of the papilla, is always hypertonic to systemic plasma, but of the same solute concentration as blood collected from blood vessels in that region and as papillary slices. Fluid collected from the first portion of the distal convoluted tubule is always hypotonic to systemic plasma. When ADH is present, the fluid in the distal portions of the distal tubule is isotonic to systemic plasma. In the collecting duct, the solute concentration of tubular fluid continues to rise and the urine formed is hypertonic to systemic plasma. When ADH is not present, the solute concentration of the already dilute tubular fluid decreases as it passes through the distal tubule and the collecting duct. Thus, the urine formed is dilute. The solute concentration of luminal fluid as a

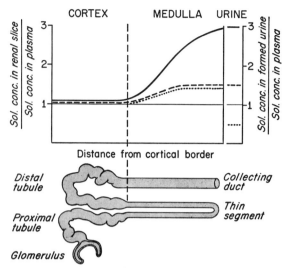

Distance from cortical border

Distal tubule

Collecting duct

Thin segment

Proximal tubule

Glomerulus

Figure 25–10 Solute concentration in slices of kidney as a function of depth of juxtamedullary nephron is also shown to indicate structures present at each level. Concentrations are shown for three conditions: low urine flow with adequate ADH (*solid line*), high urine flow with adequate ADH (*dashed line*), and high urine flow with inadequate ADH (*dotted line*). It should be remembered that a renal slice contains both tubular contents and interstitial fluid. (Data from Ullrich and Jarausch, *Pflügers Arch. ges. Physiol.*, 1956, *262*, 537–550, and Ullrich, Drenckhahn and Jarausch, *Pflügers Arch. ges. Physiol.*, 1955, *261*, 62–77.)

function of distance along the distal tubule is depicted in Figure 25–11.

ADH increases the permeability of the distal portions of the nephron to water and to urea.[27] This has been assumed for many years on the basis of distal tubular punctures[17] and of a like effect on frog skin. Recently, direct measurements of diffusion permeability and of water flux in response to an osmotic driving force have shown increases in flux in the collecting ducts. The overall picture is that ADH increases the permeability of distal tubular and collecting duct cells primarily to water and, to a limited extent, to solute materials.

The theory which fits these data together can be expressed as follows: In the proximal tubule, considerable solute reabsorption takes place. The membranes of the proximal tubular cells are freely permeable to water and, as this solute reabsorption occurs, an osmotically equivalent amount of water is reabsorbed. Since the water permeability is high, no appreciable gradient of water

concentration develops across proximal tubular cells. The fluid then enters the descending limb of the thin segment and progresses down through the renal medulla. As the fluid descends deeper into the medulla, the interstitial fluid opposite the tubular fluid has an increasingly high solute concentration. This portion of the nephron is permeable both to water and to solute materials. Water is reabsorbed from tubular fluid and Na^+ and urea enter so that, at the tip of the papillae, the solute concentration of tubular fluid is the same as that of the surrounding interstitial fluid, but considerably higher than that of systemic plasma. As the fluid traverses the first portion of the distal tubule, conditions change. Here, the membranes of the tubular cells are impermeable to water. In this region, an active transport process is present. Solute is transported out of the tubular fluid into the medullary interstitium and the solute concentration of the tubular fluid is reduced while that of the medullary interstitial fluid is increased. Tubular fluid enters the early portion of the distal convoluted tubule hypotonic to systemic plasma. In the absence of ADH, neither the distal tubule nor the collecting duct is very permeable to water and most of the fluid that enters the distal tubule is destined for excretion. Indeed, under these conditions, which are those of a water diuresis, the solute concentration of the formed urine may be even lower than the solute concentration of the fluid entering the distal tubule. This is a reflection of the continuing reabsorption of solute in the latter portions of the nephron.[14]

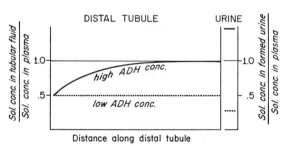

Distance along distal tubule

Figure 25–11 Solute concentration of tubular fluid as a function of length along the distal tubule, when adequate amounts of ADH are present (*solid line*) and during water diuresis (*dotted line*). (Data from Wirz. *Helv. physiol. acta*, 1956, *14*, 353–362, and Gottschalk and Mylle, *Amer. J. Physiol.*, 1959, *196*, 927–936.)

When ADH is present, the permeability of the latter portions of the nephron to water is great. On entering the distal convoluted tubule, water is immediately reabsorbed into the cortical interstitial fluid. Normally, in the first half of the distal convoluted tubule, sufficient water has been reabsorbed to bring the tubular fluid into osmotic equilibrium with cortical interstitial fluid, which has, of course, the same solute concentration as systemic plasma. This tubular fluid then enters the collecting duct and passes through the renal medulla once more. As when going through the descending limb of the thin segment, water passes out of the collecting duct into the concentrated interstitial fluid, producing a concentrated urine.

Collecting duct fluid equilibrates with the interstitial fluid at each level of the medulla and the final solute concentration of the formed urine is the same as that of the interstitial fluid at the tips of the papillae. Thus, water without an equivalent amount of solute is lost from both the descending limb of the thin segment and the collecting duct. This water enters the medullary interstitium. Solute without water is lost from the ascending portion of the distal tubule. Both this solute and water must leave the medullary interstitial compartment via the blood vessels which perfuse this region. Considerable water without equivalent solute is also lost from the distal tubule in the cortex. Here, however, the fluid enters the cortical interstitial compartment, is rapidly carried away by the rich cortical blood supply, and does not affect the balance of solute and water

transport taking place in the medulla. Figure 25–12, showing the solute concentration of the tubular fluid and the interstitial fluid outside the tubule as a function of distance along the tubule for a straightened-out nephron, depicts the transcellular water gradient. One can see that the solute-free water reabsorption that occurs in the early distal tubule, in the descending limb of the thin segment, and in the collecting duct, would be expected on the basis of permeability of these membranes to water, but the production of a hypotonic tubular fluid in the ascending limb of the thin segment requires an active transport process.

Two questions remain to be answered: (i) What is the nature of the active process involved? Active transport of Na^+ out of the ascending limb is involved, but there may be other active processes also. The findings that normal urea concentrations are required for normal concentrating ability and that urea normally is concentrated in the medulla, to a greater extent than any other solute, point to a unique role for this substance. High concentrations of a solute other than Na^+ and Cl^- must be present in the interstitial fluid to permit balancing of solute and water transport into and out of the medulla. (ii) Why does the blood flowing through the medulla not wash all the extra solute away (or supply sufficient water to bring the solute concentration down to isotonic levels)? The answer to this question comes out of the peculiar geometric arrangement of the blood vessels in this region. These vessels, termed the *vasa recta*, are long and straight and dive deeply into the medulla, take a sharp hairpin turn,

Figure 25–12 Solute concentration in tubular fluid and in interstitial fluid opposite the tubule as a function of length along the nephron. Interstitial fluid outside the nephron (*solid line*); tubular fluid in the presence of adequate ADH (*dashed line*) and during water diuresis (*dotted line*).

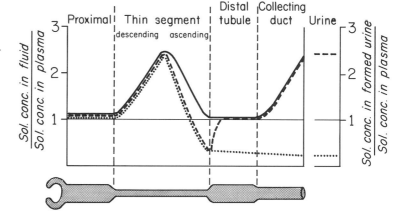

and return. Indeed, all of the tubular structures tend to be packed tightly together so that ascending and descending tubes are in intimate contact. As the blood passes through the descending portions of the vasa recta, it loses water to the medullary interstitial fluid or gains solute material from it or both. As the blood later traverses the ascending portions of the vasa recta, passing through regions where the solute concentration progressively decreases, it regains the fluid it lost on the way down. In addition, the solute and water that have entered the medulla from the nephron are also picked up. This is a true countercurrent exchange system; as such, it minimizes the amount of water and solute transfer that can take place for a given blood flow.

Finally, one can see how the theory described here accounts for the observed relationship between urine flow and solute concentration (Fig. 25–9). In the steady state, the amount of water that enters the medullary interstitial compartment must equal the amount removed per unit time. The rate of fluid leaving the medulla is determined by the active solute transport process but, for present purposes, can be considered as constant. The rate that water enters the medulla from the collecting duct is that which produces osmotic equilibrium at each level in the duct. If the volume flow through a section of the collecting duct is doubled while the interstitial concentration remains the same, twice as much fluid leaves that section to enter the interstitial compartment. Hence, if the volume flow entering the collecting duct is increased, more water enters the medullary interstitium for a time than leaves it. This lowers the solute concentration of the interstitial fluid and the solute concentration of urine produced is correspondingly lowered. The concentration–urine flow relation illustrated in Figure 25–9 is a reflection of the progressive dilution of the medullary interstitium with increasing volume flow entering the collecting duct.

DIURESIS

The condition in which there is a high rate of urine flow is called "diuresis." Two general types exist: *water diuresis* and *osmotic diuresis*. A water diuresis occurs when the blood contains slight amounts of ADH. Solute transport is affected only minimally, but an abnormally high rate of water excretion is observed. The total solute concentration is very low and may be as low as one-tenth that of plasma. Ingestion of water causes a water diuresis by inhibiting the release of ADH, as does ingestion of alcohol. *Diabetes insipidus* is a state of permanent water diuresis caused by destruction of the posterior pituitary or of the supraventricular nucleus of the hypothalamus. Administration of exogenous ADH reduces a water diuresis.

Osmotic diuresis results from an increase in the rate of solute excretion. Since water excretion varies directly with solute excretion, an increase in the latter will cause an increase in the former. Generally, osmotic diuresis is induced either without change in water balance or in association with mild dehydration. Adequate ADH is present to insure the formation of urine with a solute concentration the same as that at the tips of renal papillae. The sole cause for the increase in urine flow is an increase in solute excretion. Excessive excretion of any solute will produce an osmotic diuresis. Thus, in the diabetic patient, the concentration of glucose in the plasma is high, and the load of glucose filtered at the glomerulus is greater than Tm_G. Large amounts of glucose may be excreted in the urine. In addition, organic acids are incompletely metabolized, and these compounds must also be excreted. The consequent elevation of solute excretion brings about an elevation of urine flow. After large doses of urea have been administered, significant quantities may escape into the urine, producing osmotic diuresis. Inhibition of Na^+ reabsorption by diuretic agents produces an osmotic diuresis in which the high rate of excretion of Na^+ and its attendant anion are responsible for the increase in urine flow.

Clinically useful diuretic agents decrease Na^+ reabsorption and thus lead to an increase in the rate of Na^+ excretion. This increase in the rate of Na^+ excretion is the primary and desired effect, and the increase in urine flow is an automatic consequence.

ENDOCRINE CONTROL OF RENAL FUNCTION

The kidneys are autonomous to a remarkable degree. Solute and water excretion vary widely with the composition of the plasma. This intrinsic regulation of renal function, however, can be modified by hormones. Regulation of renal function is a defense in depth; a mean value for the transport is established by the endocrine milieu, and moment-to-moment variations in transport around this mean result from local variations in the rate at which material is presented to tubular cells. We normally think only of these rapid changes; and, indeed, as long as the external environment is not varied too greatly, the slower hormonally induced changes in renal function may not appear. When severe stress is laid on the animal's regulatory systems, however, hormonal regulation is brought into play. Four different effects may be mentioned: those induced by ADH, by parathyroid hormone, by the renotrophic hormones, and by aldosterone. ADH has been discussed above; its effect, unlike that of other hormones, is rapid. The effect of ADH on tubular permeability of water is immediate, and changes in the ADH concentration in the blood lag only 10 to 15 min behind changes in the solute concentration.

Increases in the concentration of circulating parathyroid hormone increase the rate of phosphate excretion by reducing net tubular reabsorption. The release of parathyroid hormone is regulated by the plasma concentration of calcium; as the calcium concentration falls, the amount of circulating hormone rises. Plasma is nearly a saturated solution of calcium phosphate, and the product of the calcium and the phosphate concentrations is about constant. As more parathyroid hormone is released, more phosphate is excreted. The plasma phosphate concentration then falls, and the plasma calcium concentration rises. This last change reduces the release of the hormone. The kidney and the parathyroid gland work together to control calcium and phosphate balance.

Hormones of the third group modify cellular metabolism throughout the body, and renal cells are only one of their points of action. Somatotropin from the anterior pituitary, corticosterone and cortisol from the adrenal cortex, and thyroid hormone all exhibit a renotrophic action. That is, they increase the total amount of renal tissue, the plasma flow rate, the filtration rate and all renal function. This effect is not regulatory, but is exerted because renal cells, like all other cells, respond to these hormones.

Aldosterone, which is released from the adrenal cortex, has a striking effect on the renal transport of Na^+ and K^+. This effect develops over days or weeks and is of long duration. If excessive aldosterone is present, an excessive amount of Na^+ is retained in the body. K^+ secretion is also hyperactive, and an excessive amount of this ion is lost. Absence of the hormone leads to opposite changes. In adrenal insufficiency, the loss of Na^+ leads to a large loss of extracellular fluid. Circulatory collapse eventually ensues. Striking as the effects of aldosterone on renal ion transport are, this hormone probably is not primarily concerned with the regulation of renal function, for it seems to enhance the rate of active Na^+ extrusion from all cells in the body.[49] Probably the renal effects of aldosterone are indirect and the result of changes in renal cell composition consequent to changes in the activity of a general maintenance pump on the peritubular border.

Attention has been directed to the renal action of aldosterone for two reasons. First, the renal effect is the most obvious one in the body. As discussed in Chapter 1, Volume I, a 5 per cent change in Na^+ pumping is barely detectable in muscle or brain tissue. In the kidney, however, a 5 per cent increase in Na^+ reabsorption results in striking retention of this ion. The situation is simply a reflection of the extremely high flux rates in the kidney and of the fact that Na^+ excretion is always very small in comparison with Na^+ reabsorption. Second, renal ion transport modifies the release of aldosterone itself.

Increase in blood volume decreases and decrease in blood volume increases the rate at which aldosterone is released. At least some of these changes are mediated through a system involving the kidney. Apparently, low arterial pressure is sensed by the juxtaglomerular cells in the kidney and, as a result, renin is released. Renin catalyzes the release of angiotensin II, which stimulates the release of aldosterone.[6] Thus, when blood volume is decreased, an increased amount of circulating aldosterone enhances renal Na^+ reabsorption. With normal dietary intake of Na^+ and water, Na^+ is retained, extracellular fluid is accumulated and blood volume is returned toward normal. The kidney and adrenal cortex thereby constitute a long-term controlling system which maintains the cation composition and volume of the plasma.

REFERENCES

1. Başar, E., Tischner, H., and Weiss, C. Untersuchungen zur Dynamik druckinduzierter Änderungen des Strömungswiderstandes der autoregulierenden, isolierten, Rattenniere. *Pflügers Arch. ges. Physiol.*, 1968, 299, 191–213.

2. Bray, G. A., and Preston, A. S. Effect of urea on urine concentration in the rat. *J. clin. Invest.*, 1961, 40, 1952–1959.

2a. Brenner, B. M., Troy, J. L., and Daugharty, T. M. Pressures in cortical structures of the rat kidney. *Amer. J. Physiol.*, 1972, 222, 246–251.

2b. Burg, M. B., and Green, N. Function of the thick ascending limb of Henle's loop. *Amer. J. Physiol.*, 1973, 224, 659–668.

3. Burg, M. B., Issacson, L., Grantham, J., and Orloff, J. Electrical properties of isolated perfused rabbit renal tubules. *Amer. J. Physiol.*, 1968, 215, 788–794.

4. Chinard, F. P. Comparative renal excretions of glomerular substances following 'instantaneous' injection into a renal artery. *Amer. J. Physiol.*, 1955, 180, 617–619.

5. Courtney, M. A., Mylle, M., Lassiter, W. E., and Gottschalk, C. W. Renal tubular transport of water, solute, and PAH in rats loaded with isotonic saline. *Amer. J. Physiol.*, 1965, 209, 1199–1205.

6. Davis, J. O. The control of aldosterone secretion. *Physiologist*, 1962, 5, 65–86.

7. Frömter, E., and Hegel, U. Transtubuläre Potentialdifferenzen an proximalen und distalen Tubuli der Rattenniere. *Pflügers Arch. ges. Physiol.*, 1966, 291, 107–120.

8. Giebisch, G. Electrical potential measurements on single nephrons of necturus. *J. cell. comp. Physiol.*, 1958, 51, 221–239.

9. Giebisch, G. Measurements of electrical potential differences on single nephrons of the perfused *necturus* kidney. *J. gen. Physiol.*, 1961, 44, 659–678.

10. Giebisch, G., Klose, R. M., and Windhager, E. E. Micropuncture study of hypertonic sodium chloride loading in the rat. *Amer. J. Physiol.*, 1964, 206, 687–693.

11. Gottschalk, C. W. Micropuncture studies of tubular function in the mammalian kidney. *Physiologist*, 1961 4(1), 35–55.

12. Gottschalk, C. W., and Mylle, M. Micropuncture study of the mammalian urinary concentrating mechanism: evidence for the countercurrent hypothesis. *Amer. J. Physiol.*, 1959, 196, 927–936.

13. Grantham, J. J., Burg, M. B., and Orloff, J. The nature of transtubular Na and K transport in isolated rabbit renal collecting tubules. *J. clin. Invest.*, 1970, 49, 1815–1826.

14. Hargitay, B., and Kuhn, W. Das Multiplikationsprinzipals als Grundlage der Harnkonzentrierung in der Niere. *Z. Elektrochem.*, 1951, 55, 539–558.

15. Hilger, H. H., Klümper, J. D., and Ullrich, K. J. Wasserrückresorption und Ionentransport durch die Sammelrohrzellen der Säugetierniere (Mikroanalytische Untersuchungen). *Pflügers Arch. ges. Physiol.*, 1958, 267, 218–237.

16. Hinshaw, L. B., Day, S. B., and Carlson, C. H. Tissue pressure as a causal factor in the autoregulation of blood flow in the isolated perfused kidney. *Amer. J. Physiol.*, 1959, 197, 309–312.

17. Jaenike, J. R. The influence of vasopressin on the permeability of the mammalian collecting duct to urea. *J. clin. Invest.*, 1961, 40, 144–151.

18. Koch, A. Transport equations and criteria for active transport. *Amer. Zoologist*, 1970, 10, 331–346.

19. Kramer, K., Thurau, K., and Deetjen, P. Hämodynamik des Nierenmarks. I. Mitteilung: Capilläre Passagzeit, Blutvolumen, Durchblutung, Gewebshämatokrit und O_2-Verbrauch des Nierenmarks in situ. *Pflügers Arch. ges. Physiol.*, 1960, 270, 251–269.

20. Lassiter, W. E., Gottschalk, C. W., and Mylle, M. Micropuncture study of net transtubular movement of water and urea in nondiuretic mammalian kidney. *Amer. J. Physiol.*, 1961, 200, 1139–1147.

21. Ljungberg, E. On the reabsorption of chlorides in the kidney of the rabbit. *Acta med. scand.*, 1947, 127 (Suppl. 186), 1–189.

22. Malnic, G., Klose, R. M., and Giebisch, G. Micropuncture study of renal potassium excretion in the rat. *Amer. J. Physiol.*, 1964, 206, 674–686.

23. Malvin, R. L., Wilde, W. S., and Sullivan, L. P. Localization of nephron transport by stop flow analysis. *Amer. J. Physiol.*, 1958, 194, 135–142.

24. Maximow, A., and Bloom, W. *Textbook of histology*, 7th ed. Philadelphia, W. B. Saunders Co., 1957.

25. Milne, M. D., Scribner, B. H., and Crawford, M. A. Non-ionic diffusion and the excretion of weak acids and bases. *Amer. J. Med.*, 1958, 24, 709–729.

26. Moore, W. J. *Physical chemistry*, 2nd ed. New York, Prentice-Hall, Inc., 1955.

27. Morgan, T., and Berliner, R. W. Permeability of the loop of Henle, vasa recta, and collecting duct to water, urea, and sodium. *Amer. J. Physiol.*, 1968, 215, 108–115.

28. Oliver, J., and MacDowell, M. The structural and functional aspects of the handling of glucose by the nephrons and the kidney and their correlation by means of structural-functional equivalents. *J. clin. Invest.*, 1961, 40, 1093–1112.

29. Orloff, J., and Berliner, R. W. The mechanism of the excretion of ammonia in the dog. *J. clin. Invest.*, 1956, 35, 223–235.

30. Pilkington, L. A., Binder, R., de Haas, J. C. M., and Pitts, R. F. Intrarenal distribution of blood flow. *Amer. J. Physiol.*, 1965, 208, 1107–1113.

31. Rector, F. C., Carter, N. W., and Seldin, D. W. The mechanism of bicarbonate reabsorption in the proximal and distal tubules of the kidney. *J. clin. Invest.*, 1965, 44, 278–290.

32. Rector, F. C., and Clapp, J. R. Evidence for active chloride reabsorption in the distal renal tubule of the rat. *J. clin. Invest.*, 1962, 41, 101–107.

33. Scher, A. M. Mechanism of autoregulation of renal blood flow. *Nature (Lond.)*, 1959, 184, 1322–1323.

34. Shannon, J. A., and Fisher, S. The renal tubular reabsorption of glucose in the normal dog. *Amer. J. Physiol.*, 1938, 122, 765–774.

35. Smith, H. W. *The kidney, structure and function in health and disease*. New York, Oxford University Press, 1951.

36. Solomon, S. Transtubular potential differences of rat kidney. *J. cell. comp. Physiol.*, 1957, 49, 351–365.

37. Torelli, G., Milla, E., Faelli, A., and Constantini, S. Energy requirement for sodium reabsorption in the *in vivo* rabbit kidney. *Amer. J. Physiol.*, 1966, *211*, 576–580.

38. Ullrich, K. J., Drenckhahn, F. O., and Jarausch, K. H. Untersuchungen zum Problem der Harnkonzentrierung und -verdünnung. *Pflügers Arch. ges. Physiol.*, 1955, *261*, 62–77.

39. Ullrich, K. J., and Jarausch, K. H. Untersuchungen zum Problem der Harnkonzentrierung und Harnverdünnung. *Pflügers Arch. ges. Physiol.*, 1956, *262*, 537–550.

40. Walker, A. M., Bott, P. A., Oliver, J., and Mac-Dowell, M. C. The collection and analysis of fluid from single nephrons of the mammalian kidney. *Amer. J. Physiol.*, 1941, *134*, 580–595.

41. Waugh, W. H., and Shanks, R. G. Cause of genuine autoregulation of the renal circulation. *Circulat. Res.*, 1960, 8, 871–888.

42. Wearn, J. T., and Richards, A. N. Observations on the composition of glomerular urine with particular reference to the problem of reabsorption in the renal tubules. *Amer. J. Physiol.*, 1924, *71*, 209–227.

43. Windhager, E. E. Electrophysiological study of renal papilla of golden hamsters. *Amer. J. Physiol.*, 1964, *206*, 694–700.

44. Windhager, E. E. *Micropuncture techniques and nephron function.* New York, Appleton-Century-Crofts, 1968.

45. Windhager, E. E., and Giebisch, G. Electrophysiology of the nephron. *Physiol. Rev.*, 1965, *45*, 214–244.

46. Wirz, H. Der osmotische Druck in den corticalen Tubuli der Rattenniere. *Helv. physiol. pharm. Acta*, 1956, *14*, 353–362.

47. Wirz, H., and Bott, P. A. Potassium and reducing substances in proximal tubule fluid of the rat kidney. *Proc. Soc. exp. Biol. (N. Y.)*, 1954, 87, 405–407.

48. Wirz, H., Hargitay, B., and Kuhn, W. Lokalisation des Konzentrierungsprozesses in der Niere durch direkte Kryoscopie. *Helv. physiol. pharm. Acta.*, 1951, 9, 196–207.

49. Woodbury, D. M., and Koch, A. Effects of aldosterone and desoxycorticosterone on tissue electrolytes. *Proc. Soc. exp. Biol. (N. Y.)*, 1957, 94, 720–723.

CHAPTER 26 PHYSIOLOGY OF BODY FLUIDS

by DIXON M. WOODBURY

The concepts of water and ion distribution at the cellular level, developed in Chapter 1, Volume I, apply as well to the distribution of water and electrolytes between the various fluid compartments (see below) of tissues, organs and the whole body. The present chapter deals with the volume and the composition of the body fluids and with the factors which regulate and maintain them within narrow limits.

Because of their importance to many aspects of medicine, a sound knowledge and thorough understanding of the principles determining the dynamic state of body fluids and electrolytes are essential. Whether a physician is concerned with excessive electrolyte or water losses or both caused by diarrhea, cholera, suction drainage of gastrointestinal fluids during bowel surgery, or diabetes insipidus, or with retention of fluid and electrolytes which results from heart failure or liver disease, his therapeutic approach must always be based on fundamental principles of water and electrolyte distribution. Ignorance of such physiologic principles may cost lives and may jeopardize the results of the finest surgical skills.

Water is the solvent of the body fluids in which are dissolved numerous inorganic and organic solutes. The organic solutes are mainly foodstuffs and products of metabolism which are constantly moving in and out of cells, tissues or the body. Water and the inorganic substances constitute the stable *milieu interieur* of Claude Bernard and are the chief concern of this chapter. Water is distributed passively along with the electrolytes. The *milieu interieur* is maintained in a dynamic steady state, not at equilibrium, by the constant expenditure of energy derived from cellular metabolism. There are many extensive treatments of fluid and electrolyte physiology.[5, 10, 19, 41, 49, 56, 64, 73]

VOLUME AND COMPOSITION OF BODY FLUIDS

Water is the main volume-occupying substance (about 70 per cent of the body weight) in which the major cations (sodium, potassium, hydrogen, calcium and magnesium) and anions (chloride, bicarbonate and protein) of the body are dissolved. The distribution of water delineates the compartments in which the various biochemical reactions and ionic movements and exchanges between various water and solid compartments (*e.g.*, bone) of the body take place.

Water Content and Distribution. In the organism as a whole, and in specific organs and tissues, the body fluids may be divided into two main compartments, the *extracellular* and the *intracellular*. The boundary between them is the cell membrane (Chap. 1, Vol. I). The *intracellular phase* is that portion of the total body water with its dissolved solutes which lies within cell membranes and is the site of all the metabolic processes of the body. The *extracellular phase* lies outside the cell membranes and is the compartment which

provides a constant external environment for the cells.*

The extracellular fluid compartment is divided into a number of subcompartments. The *blood plasma* constitutes a major subcompartment of the extracellular fluid. Supporting tissues, collagen, connective tissue and bone constitute another subcompartment. Although produced by cells, they are deposited as solid materials extracellularly, and thus are part of the extracellular space. Secretory cells produce solutions similar in composition to those of the extracellular fluid, *e.g.*, cerebrospinal fluid. These solutions are secreted into portions of the body separated from the main extracellular space by a continuous layer of epithelial cells held together by tight junctions. These fluids are formed by active transport of Na^+ and occasionally Cl^- across the epithelial cell layer; water moves passively with the transported Na^+. Thus the fluid formed has a composition that is determined by the characteristics of the transport system. These modified extracellular fluids are called *transcellular fluids;* the volumes they occupy are the *transcellular subcompartment* of the extracellular space. They include the digestive secretions and the cerebrospinal, intraocular, pleural, pericardial, peritoneal and synovial fluids, the luminal fluid of the thyroid, the cochlear endolymph and the secretions of sweat glands and other glands. The final subcompartment of the extracellular space is the *interstitial fluid*. It is the fluid interposed between the rapidly circulating plasma and the cells. Interstitial fluid flows slowly through tissue interstices and bathes the cells, but is in rapid equilibrium with the blood plasma. Lymph is a small part of the interstitial space. Transcapillary fluid movements and formation of interstitial fluid and lymph are discussed in Chapter 9, Volume II.

The intracellular fluid, like the extracellular, is nonhomogeneous even within

---

*This function was once served by the vast extracellular medium of the sea in ancient times when organisms were unicellular. The ionic composition of the extracellular fluid in present day animals reflects qualitatively the composition of the Cambrian ocean. In the Cambrian Period, when multicellular organisms are thought to have developed, the sea water was incorporated into the organism as extracellular fluid.

single cell types, since many anatomic subdivisions exist in the cell. Striking differences in water content and ionic composition have been demonstrated between the cytoplasm, nucleus, mitochondria and endoplasmic reticulum (microsomes) of various cell types. Despite the nonhomogeneity of the intracellular fluids, the concept of a single intracellular compartment is useful in describing the fluid and electrolyte balance of the whole organism.

MEASUREMENT OF BODY FLUID COMPARTMENTS. Before the composition, volume and fluid dynamics of the body compartments, organs and tissues can be discussed, it is necessary to understand how the various compartments are measured. The volume of water in each compartment cannot be measured directly. Indirect methods of measurement, called dilution techniques, are based on the relationship between the quantity of a substance present, the volume through which the substance is distributed, and the final concentration attained. The equation for this relationship is expressed as follows:

$$[C] = \frac{Q}{V}; \text{ hence, } V = \frac{Q}{[C]}$$

where V is the volume (in milliliters or liters) through which the quantity Q (g, kg or mEq) is distributed, yielding concentration [C] (g per ml or per liter; or mEq per ml or per liter). The dilution principle may be illustrated by an example. The volume of a beaker can be determined, without actually measuring its content in a graduated cylinder, by adding a known amount of material (Q)—for example, a dye—to the beaker, and then determining in a colorimeter its final concentration [C] after thorough mixing to insure uniform distribution. If 25 mg of dye is added and the final concentration is 0.05 mg per ml, then the volume of the beaker (V) from the above equation is:

$$V = \frac{25 \text{ mg}}{0.05 \text{ mg per ml}} = 500 \text{ ml}$$

Measurement of body fluid compartments by the dilution principle is only slightly more difficult. The method requires that the injected solute be evenly distributed but confined within the body fluid compart-

ment to be measured. If the solute leaves the compartment by excretion in the urine, by transfer into another compartment where it exists in a different concentration, or if it is metabolized, then a correction for the loss must be made. The amount lost from the space to be measured is subtracted from the quantity administered:

Volume of distribution =
$$\frac{\text{Quantity administered} - \text{Quantity removed}}{\text{Concentration}}.$$

Volume of distribution is defined as the volume of a solution having the same concentration as that in the plasma water, which would contain the quantity of test substance left in the body. Other requirements for the substances used are that they be nontoxic, be free of pharmacologic activity, and be readily determined analytically. The volume of distribution obtained from application of the method is usually calculated in liters and is generally expressed as a percentage of body weight in kilograms.

Two principal methods, with many variations,[74] are in use for measurements of body space by the dilution technique: (i) the *infusion—equilibration method,* and (ii) the *kinetic method, with or without extrapolation.*

The Infusion—Equilibration Method. This is used for substances such as inulin which are rapidly excreted (Fig. 26–1A). A priming dose of the substance is given to saturate the system; then the material is infused slowly to maintain a constant plasma concentration. After a time sufficient to insure equilibration, a series of plasma samples is taken and the concentration of indicator is measured in each sample. When the plasma level becomes constant, the infusion is stopped and urine is collected until all the substance which was present in the body at the time the infusion ended has been excreted. This amount represents the quantity (Q) of material in the body at the end of the infusion. If this quantity is divided by the plasma concentration at the end of the perfusion, the volume of distribution of the indicator is obtained.

The Kinetic Method. A single dose of an indicator substance is injected and plasma concentrations of the substance are determined at various intervals after injection (Fig. 26–1B). Semilogarithmic plots (log concentration versus

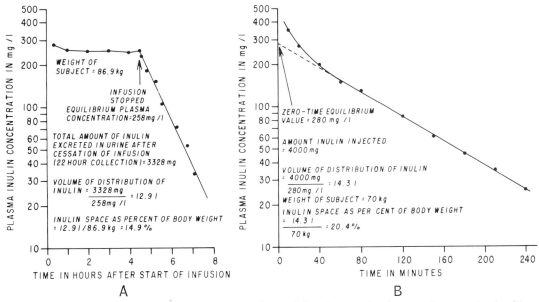

Figure 26–1 Two methods for determining the volume of distribution of indicator substances by the dilution method.

time) show an initial nonlinear decline of concentration but thereafter become a straight line. Extrapolation of the linear segment of the regression line to zero time gives the theoretical concentration if the injected substance were instantaneously distributed. The zero-time equilibrium concentration divided into the total quantity of substance given yields the volume of distribution of the indicator substance. Also, the volume of distribution at any point along the falling curve can be obtained by dividing the amount retained in the body at the time desired by the concentration in the plasma at the same time. The latter method can be used only if the material is lost in the urine and can be measured. If the substance is metabolized in the body, then the calculated volume of distribution is erroneous. The correct volume can be obtained by extrapolation to zero time as described above, if the rate of loss from the body is constant. Examples of the infusion – equilibration and kinetic methods with sample calculations are shown in Figure 26–1A and B.

TOTAL BODY WATER. Total body water constitutes a very high and constant proportion of the body weight in the "average" lean man. Among individuals, the ratio of water to body weight varies inversely with the amount of fatty tissue, which contains practically no water. However, the relation between total body water and the

weight of fat-free tissue only (*lean body mass*) is remarkably constant (Fig. 26–2) at 73.2 per cent.[45] In the newborn, the ratio is 82 per cent.

The value of 73.2 per cent water is so constant that it can be used to calculate the fat content of man as follows:[45]

$$\text{Percentage of fat} = 100 - \frac{\text{Percentage of water in whole body}}{0.732}$$

Body fat, calculated from the specific gravity (G) of the body, agrees with the above calculation. One formula for fat content empirically derived from body specific gravity data is as follows:[4]

$$\text{Percentage of fat} = 100\left(\frac{5.548}{G} - 5.044\right)$$

Once fat content is obtained by the specific gravity method, then the percentage of body water can be obtained from the formula above. The results so obtained agree well with those obtained by desiccation and dilution techniques.[44] Calculation of fat content with these formulae, together with the data on total body water, supplemented by measurements of the metabolic balance, allow for an accurate study of changes in body composition (fat, water, and

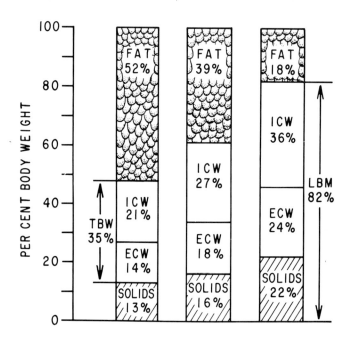

Figure 26–2 The influence of fat content on the percentage body weight of the various fluid compartments. When fat content is high (left), the percentage of total body water, intracellular water, extracellular water and body solids is low; conversely, when fat content is low (right), the percentage values for water and solids are high. If the values for water and solids are compared on a fat-free basis (lean body mass), the percentages of water and solids are constant (73 per cent total body water, 27 per cent solids, in adults). *TBW* = total body water, *ICW* = intracellular water, *ECW* = extracellular water, *LBM* = lean body mass.

fat-free solids) under various physiologic and pathologic conditions.

The measurement of total body water by a dilution technique requires an indicator substance which diffuses rapidly through all body water, including the transcellular component. Deuterium oxide (D_2O), tritium oxide (HTO), antipyrine, and N-acetyl-4-amino antipyrine (NAAP) all yield values which correlate well with those obtained from desiccation or specific gravity measurements in man.[6, 18, 44, 50, 54, 59, 75]

The hydrogen isotopes contained in these molecules exchange with hydrogen atoms in body water and with the exchangeable hydrogen atoms in organic molecules. Over the interval required for the equilibration of D_2O or HTO with body water these errors amount to a water equivalent of only 0.5 to 2.0 per cent of body weight in man[48] and, consequently, are insignificant. Antipyrine and NAAP are distributed throughout total body water and diffuse rapidly across cell membranes. However, antipyrine is metabolized by the body and also is excreted in the urine; hence, correction must be made for these losses by multiple sampling and use of the kinetic falling plasma method. NAAP is metabolized only negligibly by tissues and is excreted slowly in the urine. Plasma binding of antipyrine and NAAP is small and generally does not require correction.

A summary of current estimates of total body water obtained by the various methods is presented in Table 26–1.[17]

During the first year of life water content decreases principally because of contraction of the extracellular fluid volume. Thereafter, it decreases more slowly with age. After puberty, water content is higher for males than females because of differences in body fat content.

The body water is distributed unequally in the various tissues. Table 26–2 summarizes some of these differences. The exchange rate between tissue and plasma water (time for D_2O equilibration) is also indicated. Most of the water in the body is in muscle and skin, whereas the skeleton and adipose tissue contain the lowest percentages. The exchange rate of water is characteristic for each tissue; apparently each tissue regulates its own water exchange. The rates of exchange, for example, of visceral tissues are faster than those for bone and muscle. Factors which regulate movement of water in and out of cells and tissues are considered below.

EXTRACELLULAR FLUID. The varied results derived from anatomic, dilution, metabolic balance, and tissue studies of the extracellular space necessitate division of this space into various subcompartments. From anatomic and physiologic considera-

Table 26–1 *Total Body Water in Normal Man as Measured by Different Methods*

GROUP	METHOD	TOTAL BODY WATER (PER CENT OF BODY WEIGHT)			
		0 to 1 Month	1 to 12 Months	1 to 10 Years	
Children	D₂O AP	75.7 (n = 20)	64.5 (n = 15)	61.7 (n = 24)	
		10 to 16 Years	17 to 39 Years	40 to 59 Years	Over 60 Years
Adolescent and adult males	D₂O, HTO AP, G	58.9 (n = 11)	60.6 (n = 15)	54.7 (n = 127)	51.5 (n = 20)
Adolescent and adult females	D₂O, AP G	57.3 (n = 7)	50.2 (n = 61)	46.7 (n = 38)	45.5 (n = 14)

Values are means of all methods and the n values in parentheses are the number of subjects in each group. D_2O = deuterium oxide, AP = antipyrine and its derivatives, HTO = tritium oxide, and G = specific gravity. (Values taken from Edelman and Leibman, *Amer. J. Med.*, 1959, *27*, 256–277.)

tions the extracellular fluid is made up of the following subdivisions which were described above: plasma, interstitial and lymph fluid, connective tissue and cartilage, bone, and transcellular fluids. No single substance can be used to measure the whole fluid extracellular volume; therefore, measures of several substances are necessary.

Anatomic Considerations. The histochemical characterization of the extracellular fluid and the electronmicroscopic appearance of the capillaries are described in Chapter 9, Volume II.

The volume of the extracellular fluid has been measured by various histologic techniques.[41] In sections of adult muscle the area not occupied by cells is 14.5 to 23 per cent of total muscle area; in young chick muscle it is about 50 per cent. In liver the histologically determined extracellular space is 24 per cent,[63] and in thyroid the stromal (nonfollicular) volume varies between 10 and 20 per cent.[9, 28] These measurements agree well with the extracellular volumes determined by various indicator substances.

Blood and Plasma Volume.[38, 58] Plasma volume may be measured by either of two dilution techniques. In the first, substances used neither leave the vascular bed nor penetrate the erythrocytes; in the second, substances confined only to the erythro-

Table 26–2 *Distribution of Water and Kinetics of Water Movement in Various Tissues*

TISSUE	PER CENT WATER	PER CENT BODY WEIGHT	LITERS WATER IN 70 KG MAN	TIME FOR D₂O EQUILIBRATION (MINUTES)
Skin	72.0	18	9.07	120–180
Muscle	75.7	41.7	22.10	38
Skeleton	31.0	15.9	3.45	120–180
Brain	74.8	2.0	1.05	2
Liver	68.3	2.3	1.10	10–20
Heart	79.2	0.5	0.28	
Lungs	79.0	0.7	0.39	
Kidneys	82.7	0.4	0.23	
Spleen	75.8	0.2	0.11	
Blood	83	7.7	4.47	Erythrocytes 1/60
Intestine	74.5	1.8	0.94	Gastric juice, 20 to 30
Adipose	10	9.0	0.63	
Total body	62	100	43.4	180

(Water values from Skelton, *Arch. intern. Med.*, 1927, *40*, 140–152; D₂O equilibration values from Edelman, *Amer. J. Physiol.*, 1952, *171*, 279–296.)

cytes are used. The two most useful indicators that are confined to the plasma are Evans blue (T 1824) and radioiodinated human serum albumin (RISA). Both are bound to plasma albumin;* Evans blue is bound *in vivo* after injection of the material; radioiodine is prebound *in vitro* to albumin before injection. The volume of distribution measured is thus that of albumin. The kinetic-extrapolation method for analyzing the data is generally used. Total blood volume can then be obtained from the plasma volume and the hematocrit by the following formula:

$$\text{Blood volume} = \text{Plasma volume} \times \frac{100}{100 - \text{Hematocrit}}$$

An iron-dextran complex has also been used by a micromethod to estimate plasma volume.

The second method for measuring plasma volume is based on the fact that radioisotopes (P^{32}, $Fe^{55, 59}$ or Cr^{51}) penetrate red cells and become incorporated or firmly bound. The tagged cells are injected intravenously and their volume of distribution is measured. Plasma volume is then calculated from the measured red cell volume and hematocrit. The most accurate method for obtaining total blood volume is separate determination of plasma and erythrocyte volumes and addition of the two volumes together. This eliminates need for the large vein hematocrit value which does not accurately portray the whole body hematocrit.

In man the average value for total blood volume is about 5.7 liters (range 4.09 to 7.76), which corresponds to an average of 7.7 per cent of body weight. Adult women probably have a lower average blood volume per unit body weight than do adult men, but the difference is not conclusive since sufficient data based on the best methodology are not available.

---

*Since plasma albumin with its attached indicators leaks out of the circulation into the interstitial fluid, the plasma volume is slightly overestimated. Plasma volumes measured with radioiodinated gamma globulin and fibrinogen, proteins which do not generally leak out of the vascular system, are lower by 2 to 12 per cent than those determined by Evans blue and RISA.

In adult males, the average values for plasma volume determined by four different methods tend to be slightly larger than those in females, but the overlap is considerable. The range of values for plasma volume is from 3.1 to 5.8 per cent of body weight for adult males, mean 4.2, and 2.7 to 5.2 per cent for adult females, mean 3.7.

Extracellular Volumes. The remainder of the extracellular fluid consists of the following components: interstitial fluid and lymph, connective tissue, bone and transcellular fluids. The extracellular fluid in connective tissue appears to be similar to that found elsewhere. The volume of interstitial fluid and lymph cannot be obtained by dilution techniques, since no substance is known that distributes exclusively in this compartment. However, the volumes of distribution of many substances closely approximate the combined volume of the plasma and interstitial-lymph space. Therefore, the interstitial-lymph space may be computed by subtracting the plasma volume from these values. Such computed values may not be exact because some of the indicators penetrate into portions or all of the water of the connective tissue subdivision of extracellular fluid.[12, 37, 43, 72] Hence, a correction for the volume occupied by the substance in connective tissue water must be subtracted from the total volume of distribution of the indicator in order to obtain the true interstitial volume.

The substances which have been used for measuring the combined plasma-interstitial-lymph space plus a portion of the connective tissue space are of two types: (i) saccharides—inulin, raffinose, sucrose and mannitol, and (ii) ions—thiosulfate, sulfate, ethanesulfonate, thiocyanate, chloride, bromide and sodium. Most of these substances are available labeled with radioactive isotopes; either the stable or radioactive forms may be used. The diffusion rates are in the order of inulin < raffinose and sucrose < mannitol and the inorganic ions; the molecular weights are in the reverse order.[62] Often the volume of distribution of inulin (5000 MW) is less than that of the smaller saccharides and ions. This has been attributed in part to the fact that lymph flow influences the steady-state distribution of extracellular markers.[36] The amount of this influence depends on the relative magnitude of the lymph flow when compared to the permeability coefficient–capillary area product. Inulin has a lower capillary permeability coefficient than do the smaller markers; hence for the same lymph flow its volume of distribution is less. In the heart the inulin space may amount to a difference from sucrose space of 14 per cent.

Two phases of inulin, sucrose or thiosulfate penetration into the extracellular space have been distinguished.[12, 43] A rapidly equilibrating phase with a half-time of 20 min or less has been identified with penetration into the plasma-interstitial fluid-lymph space. A slowly equilibrating phase, with a half-time of between 5 and 9 hours, is more difficult to identify. It has been suggested that this phase corresponds to the uptake of inulin in dense connective tissue, but more recent evidence suggests that it corresponds to the accumulation of localized clumps of inulin, probably within macrophages (Fig. 26–3). Thus, after the passage of time, inulin loses one of the characteristics required of an indicator substance, namely, that of homogeneous distribution. After 12 hours, inulin, and perhaps other saccharides such as sucrose, give erroneously high measures

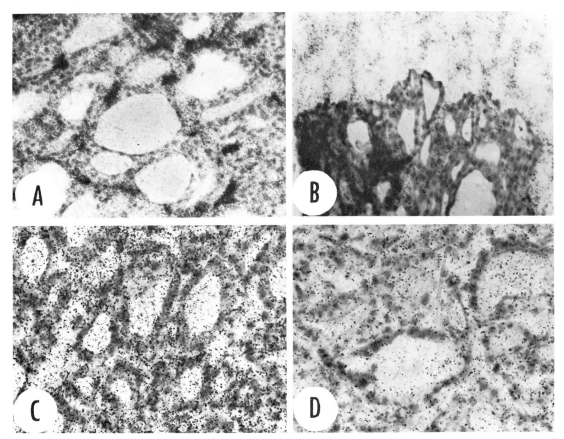

Figure 26–3 H³-inulin, S³⁵O₄ and Cl³⁶ radioautographs of frozen section of rat thyroid. A, Radioautograph 24 hours after administration of H³-inulin. Note localized clumping of the inulin in the stroma (interstitium) outside the follicles. The follicular cells and lumen contain no radioactivity except that contributed by background fogging of the emulsion. B, Radioautograph 2 hours after administration of H³-inulin. Note the uniform distribution of the inulin in the interstitial space and its absence in the follicular cells and lumen. The upper portion of the picture shows the radioactive tracks in the emulsion without the superimposed thyroid section seen in the lower portion. C, Radioautograph 2 hours after administration of S³⁵O₄. The sulfate is distributed uniformly in the stromal and follicular luminal spaces; minute amounts are also present in the follicular cells. D, Radioautograph 2 hours after administration of Cl³⁶. The chloride is uniformly distributed in the stromal and follicular luminal spaces; smaller but still appreciable amounts are also present in the follicular cells. (From Chow *et al.*, *Endocrinology*, 1965, 77, 818–824.)

of extracellular space. Sulfate and mannitol spaces do not increase with increasing time in nephrectomized rats and appear to be suitable for long-term measures of extracellular space.

The "rapidly equilibrating space" measured thus consists of the plasma and interstitial-lymph spaces, plus that portion (60 to 80 per cent) of connective tissue water which is penetrated by inulin, sucrose and particularly the smaller molecules—thiosulfate, sulfate and mannitol—within less than 12 hours.[12, 15, 43, 72] The remaining portion of connective tissue water is probably intracellular, inasmuch as intracellular potassium concentration calculated on the basis of the difference between total connective tissue water and inulin space gives a value comparable to that found for other tissue cells. Consequently, interstitial water of connective tissue cannot be distinguished from the interstitial water of other tissues. For example, separation of the extracellular space of muscle into two subcompartments for water is not necessary. However, compartmentalization is necessary in the case of the ionic composition of connective tissue as discussed later.

Chloride was first used to estimate extracellular space by Fenn[20] and by Hastings and co-workers.[30, 31, 41] More recently Cotlove and Hogben[13] concluded that Cl^- space is a valid measure of total extracellular space including transcellular volumes. These authors believe that Cl^- is actively transported out of most cells and that, except for erythrocytes, most cells contain negligible Cl^-. However, their conclusion that Cl^- is actively extruded from all cells is based on the assumption that the slow component of inulin penetration represents entry into dense connective tissue, an assumption that is probably incorrect. Since most evidence demonstrates that Cl^- is passively distributed across cell membranes according to the $\mathcal{E}_s$, Cl^- space is only a rough approximation of total extracellular volume.

The *transcellular fluids* are formed by active transport mechanisms across epithelial cells. The cells modify the extracellular fluid to form the specialized secretion fluids. Total volume occupied by transcellular fluids is small: about 15.3 ml per kg body weight in man.[17] About half of this (7.4 ml

per kg) is in the gastrointestinal lumen; cerebrospinal fluid constitutes 2.8 ml per kg and biliary fluid, 2.1 ml per kg. As previously mentioned, deuterium oxide, tritium oxide and antipyrine distribute rapidly and completely in the transcellular fluids; hence, transcellular fluid volume is included in the total body water measured by these substances. Small-molecule indicators of extracellular fluid volume enter the transcellular fluids (except cerebrospinal fluid) but large-molecule indicators do not measure these spaces well.

Physiologic Considerations of Extracellular Fluids. The volume of the whole body extracellular space (Fig. 26–4) changes with age. For example, the corrected bromide space in neonatal infants is about 360 ml per kg body weight for the first 48 hours and decreases progressively to a value of 267 ml per kg at one year to 250 ml per kg in late childhood, and finally to the adult value of 220 to 240 ml per kg. Similar marked decreases in extracellular space with age occur in special tissues, particularly in muscle, heart and brain. In fetuses and premature infants extracellular volume constitutes the largest phase of total body water. With growth the extracellular fluid is replaced by intracellular fluid and cell solids.

The water content of bone, cartilage, tendon and connective tissue is summarized in Figure 26–4. In the whole body, total bone water constitutes only 49 ml per kg of body weight of which 28 ml per kg is extracellular and 21 ml per kg is bone matrix water.[15] In connective tissue, 70 per cent of the water is extracellular (41 ml per kg) and only 30 per cent (19 ml per kg) is inside the cell membrane. Much is bound to the bone crystals and hence slow (180 min) to equilibrate with tracers such as D_2O (Table 26–2). The amount of bone water is little altered by physiologic conditions, but decreases with age as deposition of bone mineral occurs.

INTRACELLULAR FLUID. Intracellular fluid volume cannot be measured directly by dilution, because there is no substance that distributes only in this compartment. It is obtained by subtracting the volume of the extracellular fluid from the total body or tissue water; any errors of extracellular volume measurement are magnified in determining intracellular volume "by difference." The volume of intracellular fluids

Figure 26–4 Schematic drawing depicting the relation between the water and solid compartments of the body. The information below the anatomic diagram gives the names and volumes of the various anatomic compartments. The figures in parentheses are the volumes in grams per kg body weight. The various substances which measure the physiologic spaces of the body water compartments are listed above the diagram. The figures after each substance are the volumes of distribution in ml per kg body weight. Both Na^+ and Cl^- penetrate cells as shown by their volumes of distribution. The amounts in cells as depicted on the chart represent penetration into all cells, including erythrocytes, and not just bone and connective tissue cells as illustrated. (After Elkinton and Danowski, *The body fluids.* Baltimore, Williams & Wilkins Co., 1955; and after Cotlove and Hogben, Chap. 27 in *Mineral metabolism,* Comar and Bronner, eds. New York, Academic Press, 1962.)

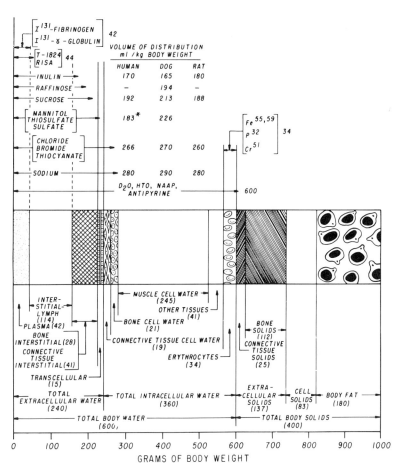

is variable, but usually amounts to 30 to 40 per cent of body weight (see Fig. 26–4).

Ionic Composition of Body Fluids. The predominant cations of the body fluids are the monovalent elements, Na^+ and K^+. The divalent cations Mg^{++} and Ca^{++} exist in body fluids only in relatively low concentrations. K^+ and Mg^{++} are the main intracellular cations, Na^+ and Ca^{++} the main extracellular cations. Na^+ and K^+ are widely spread throughout all subdivisions of the earth, the lithosphere, the hydrosphere, the atmosphere and the biosphere. The quantitative distinctiveness of the biosphere lies in the fact that its chemical composition is determined by the ability of organisms to concentrate elements and to synthesize new compounds.[61]

The chief anions of the body fluids are Cl^-, HCO_3^-, phosphate, organic ions and polyvalent proteins. The main intracellular anions are phosphate, protein and organic

ions; and Cl^- and HCO_3^- are the predominant extracellular anions.* The sum of the concentrations of the cations equals the sum of the concentrations of the anions in each compartment, making the solutions electrically neutral. Inasmuch as all body compartments are in osmotic balance the total solute concentrations in the various compartments are the same. Total body water is, therefore, *determined by the total quantity of solutes in the body.* The distribution of water between the compartments is determined by the quantity of solute in each compartment. It is important, therefore, to study the total body stores of the major ions, distribution of the ions in

---

*In some tissues the organic amino acid anions, glutamate and aspartate, are present in sufficient quantities to contribute significantly to the anionic content of cells. Similarly, lysine and arginine may contribute to the cation content of cells.

various compartments of the body, and the factors which affect their distribution.

METHODS OF MEASUREMENT. Although the total body content of ions can be determined only by direct analysis of dead animals, the pools of ions readily available for maintenance of homeostasis, i.e., the total exchangeable ion content, can be determined by the *isotope dilution* method. This is simply another application of the relations between V, Q and [C]. Isotopes distribute to the same extent in the body fluids as do their normal counterparts.

SODIUM.[21] The total amount of Na^+ in the body depends upon the balance between intake and output. Since the average intake far exceeds minimum need, Na^+ balance must be rigidly controlled; this control is discussed below. The urine is the main route for excretion of Na^+ (90 to 95 per cent), but small amounts are lost in stool and in sweat.

The total body Na^+ content of the adult human male is 58 mEq per kg or about 4100 mEq in a 70 kg man. Its distribution is indicated in Table 26–3. Determination of total body Na^+ content in man by whole body neutron activation, however, yields a lower value than that obtained by chemical analyses of cadavers. A value of 49.6 mEq per kg has been found by this technique.[7] Of the total body Na^+ measured, about two-thirds is present in extracellular fluid, where it is the predominant cation. Between 5 and 10 per cent of the body Na^+ is within cells. The remaining 25 per cent is found in the crystal lattice of bone.

The amount of exchangeable Na^+ (in man) as determined by isotope dilution is much smaller than that determined by chemical analysis and averages about 41.0 mEq per kg of body weight. The nonexchangeable Na^+ fraction is located in bone, which makes up from 40 to 45 per cent of body weight. Between 20 and 35 per cent of bone Na^+ exchanges with radiosodium in 24 hours, after which the exchangeable Na^+ pool size increases about 1 per cent per day because of slow penetration of isotope into bone.[17]

The rapidly exchangeable (24-hour) Na^+ content of bone (6.4 mEq per kg of body weight) consists of all that in the bone extracellular space (3.8 mEq per kg of body weight) and some

Table 26–3 *Distribution of Sodium, Potassium, Magnesium, Chloride and Bicarbonate in the Various Body Compartments of Man*

COMPARTMENT	SODIUM	POTASSIUM	MAGNESIUM	CHLORIDE	BICARBONATE
	(mEq per kg Body Weight)				
Plasma	6.5	0.2	0.08	4.5	1.1
Interstitial–lymph	16.8	0.5	0.12	12.3	2.9
Dense connective tissue and cartilage	6.8	0.2	1.0	5.2°	1.7
Exchangeable bone	6.4	0.6	1.0	4.3	0.8
Nonexchangeable bone	14.8	0.5	19.5	—	—
Transcellular	1.5	0.5	0.07	1.5	0.3
Total extracellular	52.8	2.5	21.8	27.8	6.8
Total intracellular	5.2	51.3	8.2	5.2	5.9
Total body	58.0	53.8	30.0	33.0	12.7
Total exchangeable	41.0	52.8	3.4, 4.9, 10†	33.0	12.7
Total body intracellular concentration (mEq per liter icw)	14.4	143	22.8	14.4	16.4

Values for connective tissue and bone include only those portions that are considered extracellular and solids. (After Edelman and Leibman, *Amer. J. Med.*, 1959, 27, 256–277; Wacker and Vallee, *New Engl. J. Med.*, 1958, 259, 431–438; idem, 475–482; MacIntyre et al., *Clin. Sci.*, 1961, 20, 297–305; and Freeman and Fenn, *Amer. J. Physiol.*, 1953, 174, 422–430.)

°A fraction of skin chloride appears to be nonexchangeable as determined by radiochloride studies.

†Equilibrated for 24, 48 and 89 hours respectively. Total exchangeable magnesium is a function of time of equilibration.

(2.6 mEq per kg of body weight) of that adsorbed to the outer surface of the hydroxyapatite crystal structure of bone. Na^+ and other ions are adsorbed in layers on the surface of the hydroxyapatite bone crystal in large amounts. The outer mineral layers of the bone crystal surface contain Na^+ which has been recently deposited and is completely exchangeable, whereas the inner mineral layers contain Na^+ which has been deposited for a long time and which is very slowly exchanged.

The exchangeable Na^+ of bone is of interest because it has been shown that in many situations involving loss of body Na^+, this cation is mobilized from the surface layers of bone. Also, a rise in extracellular Na^+ concentration is mitigated by entrance of some of the excess Na^+ into the bone stores of the body. Acidosis can also mobilize Na^+ from bone.

The total exchangeable Na^+ content of infants (76 mEq per kg of body weight) is considerably higher than that of adults and is equal to the total sodium content determined by chemical analysis.[64] This means that all the bone Na^+ in infants is readily exchangeable with radiosodium. The higher value of total Na^+ in the young is due to the very large extracellular space of the infant; the amount decreases *pari passu* with decrease of extracellular volume.

Low values for total exchangeable Na^+ have been observed in adrenal insufficiency and high values in edematous states and hypertension.

POTASSIUM.[76] The amount of K^+ in the body depends on the balance between intake and output. The K^+ content remains nearly constant in healthy adult life with only minor fluctuations from day to day. K^+ is an integral part of protoplasm and in the developing organism the over-all K^+ balance is positive. The turnover rate of K^+ is higher in infants than in adults. Young rats do not grow normally on a K^+-deficient diet. The normal intake of K^+ is dietary and is derived from both animal and plant sources, both of which are rich in K^+. The normal dietary intake of K^+ ranges from 50 to 150 mEq per day; this amount is also excreted daily. Most of the output is in the urine, but some is excreted in feces (9 mEq per day); a very small amount is excreted in sweat.

Content and Distribution. The total K^+ content of the human adult male is 50 to 54 mEq per kg of body weight.[22] It is slightly lower in the female. The distribution of K^+ in the various compartments of the body is summarized in Table 26–3. Only 5 per cent of the total K^+ is in extracellular fluids; the bulk is located intracellularly. Almost 90 per cent of the intracellular K^+ is found in muscle cells. The intracellular K^+ concentration averaged over the whole body is 143 mEq per liter of cell water. This value is close to the 150 mEq per liter shown for muscle in Table 26–5.

The rate of K^+ exchange between extracellular fluid and cells is rapid as compared to that of Na^+, but varies with the type of tissue. For example, the time for equilibration between plasma K^+ and tissue is 2 min for kidney; 8 to 10 min for intestine and lung; 90 to 100 min for liver, skin and spleen; 600 min for muscle mass; and greater than 62 hours for erythrocytes and brain. In all these tissues except bone and erythrocytes, the K^+ is completely exchangeable.[26, 76] Bone contains moderate amounts of K^+ (Table 26–3), of which only 58 per cent is exchangeable in 40 hours. Thus nonexchangeable K^+ in bone represents about 1 per cent of total body K^+.[29] Also, about 1 per cent of erythrocyte K^+ is nonexchangeable in 40 hours, the usual time for measuring exchangeable K^+ with $^{42}K^+$. Since 80 to 90 per cent of the total body K^+ is in muscle, the quantity of exchangeable K^+ is a reasonably good measure of lean muscle mass. Although the cellular concentration of K^+ remains fairly constant with increasing age, total exchangeable K^+ is lower in infants and children than in adults. This is a reflection of the large extracellular space characteristic of children.[65, 66]

K^+ Deficiency. Cellular stores of K^+ are depleted in a number of clinical conditions which result in excessive loss of the ion from the body. These losses occur either via the urine (as in adrenocortical hyperfunction, renal disease of various kinds, diuretic therapy and diabetic acidosis) or via the gastrointestinal tract (as in vomiting, diarrhea and continuous aspiration of intestinal fluids during surgery). The latter conditions result in rapid K^+ deficiency because the transcellular fluids which enter the gastrointestinal tract contain K^+ in concentrations 2 to 5 times higher than that of extracellular fluid. Reduction of K^+ stores causes characteristic clinical effects, mainly manifested in muscle tissue. In skeletal muscle, weakness develops and

may progress to flaccid paralysis. Weakness in gastrointestinal smooth muscle leads to diarrhea, distension, and paralytic ileus; in vascular smooth muscle, weakness leads to hypotension. Loss of K^+ from cardiac muscle leads to tachycardia, arrhythmias and electrocardiographic changes characterized by a prolonged Q-T interval, T wave inversion and the appearance of U waves.

The reverse of these changes occurs in plasma and in total body K^+ during K^+ repletion. Increases in plasma K^+ levels occur in anuria, adrenocortical insufficiency, and excessive therapy with K^+ solutions. The rise in plasma K^+ is accompanied by an increase in cellular concentration, but the ratio of $[K^+]_i/[K^+]_o$ is decreased and transmembrane potential is reduced. If plasma K^+ continues to rise, progressive signs of K^+ intoxication appear. Hyperkalemia is even more dangerous than hypokalemia. Muscle weakness and central nervous system changes due to membrane depolarization are a striking feature of this condition. Cardiac effects include arrhythmias and progressive changes in the electrocardiogram.

Regulation of Potassium Distribution. Observations in experimental animals and man have shown that the extracellular concentration of K^+ is increased by acidosis and decreased by alkalosis, regardless of how the pH change is produced. In the case of acidosis, K^+ is lost from the cells, particularly those of muscle, and this causes plasma K^+ concentration to rise, increasing K^+ excretion in the urine. Therefore, chronic acidosis results in a reduction of total body K^+. In alkalosis, K^+ moves into cells from the extracellular fluid, plasma K^+ decreases, but urinary K^+ excretion increases. This increase in urinary K^+ excretion is the result of local conditions in the kidney which strongly favor K^+ secretion. Thus, in both acidosis and alkalosis, there is a tendency for K^+ to flow from muscle cells, through the plasma, into the kidney, and out in the urine. Just as changes in acid-base status modify K^+ balance, changes in K^+ balance modify acid-base status. When excessive K^+ is administered, plasma K^+ concentration is elevated and active uptake of K^+ by cells is stimulated. Since active K^+ influx and Na^+ efflux are coupled (Chap. 1, Vol. I), cellular Na^+ concentration falls. H^+ efflux from cells is probably also coupled into the Na^+–K^+ pump, and its transport from cells to extracellular fluid is also increased. Thus the elevation of plasma K^+ concentration produces a decrease in cellular Na^+ concentration, an increase in cellular K^+ concentration, and an elevation in cellular pH (Table 26–4). A reflection of the

Table 26–4 *Effect of Elevated Plasma Potassium Concentration on Inulin, Sodium and Chloride Spaces, and on Intracellular Electrolyte Concentrations of Rat Skeletal Muscle**[*]

		CONTROL	NEPHRECTOMY
Interstitial	Volume (inulin space) – per cent	11.6	11.8
	$[Na^+]_e$	154.0	149.0
	$[K^+]_e$ mEq/liter water	4.2	9.3
	$[Cl^-]_e$	116.0	113.0
	pH_e	7.45	7.35
Intracellular	Volume – per cent	65.1	65.6
	$[Na^+]_i$	17.4	12.3
	$[K^+]_i$ mEq/liter water	168.0	179.0
	$[Cl^-]_i$	4.6	6.4
Space	Na^+ per cent	19.0	17.2
	Cl^- per cent	14.1	15.6
Potentials	$\mathscr{E}_s$	−90.9	−77.3
	$\mathscr{E}_{K^+}$ millivolts	−98.4	−79.0
	$\mathscr{E}_{Cl^-}$	−86.0	−77.0
	$\mathscr{E}_{Goldman}$†	−90.0	−75.2

[*]Data obtained from observations of Williams *et al.*[77]

†$\mathscr{E}_{Goldman} = \dfrac{RT}{nF} \ln \dfrac{[K^+]_o + P[Na]_e}{[K^+]_i + P[Na^+]_i}$, where $P = \dfrac{P_{Na}}{P_K} = 0.01$ for skeletal muscle.

transport of H^+ out of the cell is seen in the decrease in extracellular pH. However, since extracellular K^+ concentration increases to a greater extent than does intracellular K^+ concentration, $\dfrac{[K^+]_i}{[K^+]_o}$ decreases and so does $\mathscr{E}_s$ (see Table 26–4). The values for $\mathscr{E}_{Cl^-}$ and $\mathscr{E}_{Goldman}$ calculated from the electrolyte data agree well with the observed $\mathscr{E}_s$. Thus the increased $[K^+]_i$ and decreased $[Na^+]_i$ which result from stimulation of the Na^+–K^+ pump by the high plasma K^+ appear to be regulatory responses to maintain cellular $[Na^+]$ and $[K^+]$ in face of a large depolarization. If the pump had not been stimulated by the increase in $[K^+]_o$, the depolarization of the muscle would have been greater by about 2 mV. The clinically more important condition of K^+ deficiency produces effects on extracellular and intracellular acid–base balance opposite from those produced by excessive K^+ administration. If body K^+ stores are depleted by reduced intake or excessive losses, an extracellular hypokalemic alkalosis develops. In the intracellular compartment, the concentration of Na^+ increases and of K^+ decreases and metabolic acidosis results (Fig. 26–5). Summaries of the evidence that cells are acidotic in K^+ deficiency and methods of measuring cell pH have been presented by a number of workers.[11, 16, 24, 34, 35, 68, 69, 70]

The K^+ content of cells is maintained constant in practically all tissues at about 150 mEq per liter by the Na^+–K^+ active transport system. As a result, the level of K^+ and Na^+ in cells is influenced by any condition or agent which alters the active transport system. The main regulator of cellular K^+ and Na^+ concentrations appears to be aldosterone, which acts by increasing active cation transport across cells.[56, 78] The initial effect of the steroid is to increase K^+ and decrease Na^+ levels in all cells. The result of the aldosterone-induced increase in K^+ concentration of renal tubular cells is an increase in the urinary excretion of this cation. If the increased aldosterone output is maintained, then cellular K^+ deficiency results from chronic loss in the urine and the sequelae already discussed appear. Adrenocortical hypofunction causes opposite changes from hyperfunction. Initially, cellular K^+ is decreased and Na^+ increased, but as tubular cell K^+ decreases,

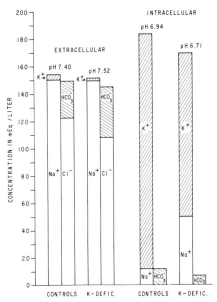

Figure 26–5 Effect of chronic potassium deficiency on extracellular and intracellular pH and electrolyte values in rat skeletal muscle. Potassium lack produces an extracellular hypokalemic, hypochloremic, metabolic alkalosis, and an intracellular metabolic acidosis characterized by marked loss of K^+ and gain of Na^+. Note the decrease in the sum of Na^+ plus K^+ in the acidotic cells. As a result of the acidosis, total anionic charge on the protein in the cells is reduced; hence total cation concentration is decreased. (Adapted from Irvine *et al.*, *Clin. Sci.*, 1961, *20*, 1–18.)

K^+ excretion in the urine is reduced and plasma K^+ increases. The sequence of events is then the same as that discussed above for excessive K^+ administration and an extracellular hyperkalemic metabolic acidosis develops.

Loss of K^+ from cells may also result from massive breakdown or destruction of tissue. Each gram of nitrogen in protoplasm is associated with 2.4 mEq of K^+ and this much is lost for every gram of nitrogen broken down. Repair or formation of new tissue is associated with K^+ uptake into cells.

The subcellular distribution of K^+ is characterized by heterogeneity. For example, mitochondria concentrate K^+ by a process which requires energy. *In vitro* incubation studies indicate, however, that the ratio of mitochondrial to suspending medium K^+ is low (about 2:1). In addition, some nuclei possess a low membrane potential (inside negative) and also concentrate K^+ slightly above the outside medium.[32] Depending on the pH of the medium, microsomes appear to "bind" K^+ in much the same manner as cationic exchange resins. Although many of the subcellular particles contain K^+ at concentrations higher than that of the surrounding cytoplasm, 65 to 85 per cent of cellular K^+ is in the cytoplasm proper.[33] Brain mitochondria contain 30 per cent or more of the total tissue K^+.[32] Possibly the slowly exchanging fraction of K^+ which has been described for this tissue is due to slow penetration of K^+ into mitochondria.

MAGNESIUM.[27, 67, 71] Mg^{++} is predominantly an intracellular cation which is present in quantities second only to K^+. The element is important for living organisms and, in its absence, deficiency symptoms develop. In many respects, movements of Mg^{++} into and out of cells resemble those of K^+. Mg^{++} activates many enzymes, plays an essential role in neuromuscular function and is an important constituent of bone.

Mg^{++} is relatively abundant in foods; green plants contain large amounts inasmuch as the element is an integral part of the chlorophyll molecule. The daily intake of Mg^{++} in an average diet is about 25 mEq (20 to 40), of which about one-third is ordinarily absorbed, mostly from the upper small bowel. Vitamin D increases absorption of Mg^{++} from the gut. The daily requirement has not been established, but about 18 to 20 mEq per day for adults, about 12.5 mEq per day for infants and about 33 mEq per day during pregnancy and lactation appear to be adequate to maintain Mg^{++} balance. In normal adults, the 5 to 12 mEq which is absorbed daily is balanced by urinary excretion by the processes of glomerular filtration of the unbound form and of tubular reabsorption. There is as yet no evidence for net tubular secretion of Mg^{++}.

Content and Distribution. The total content of Mg^{++} in the body averages 30 mEq per kg of body weight; its distribution is indicated in Table 26–3. Mg^{++} distribution is similar to that of K^+ except that a large amount is present in bone. It is present in the hydration shell and in the surface-bound ion layer but probably not in the crystal lattice interior. About two-thirds of the total body Mg^{++} is located in bone, 4 per cent in the extracellular fluid and connective tissue, and the remainder in cellular fluid. The highest amounts are found in skeletal muscle, brain and liver. On the assumption that Mg^{++} is not bound in cells,

the calculated intracellular concentration in muscle is about 34 mEq per liter of cell water.

The interstitial concentration of Mg^{++} is lower than its concentration in plasma because about 35 per cent of the plasma Mg^{++} is bound to protein. Thus, if the plasma concentration averages 2.0 mEq per liter the interstitial concentration (after correction for the Donnan effect) averages 1.30 mEq per liter. The ratio of Mg^{++} across the cell membrane is

$$\frac{34 \text{ mEq per liter}}{1.30 \text{ mEq per liter}} = 26.2$$

Since Mg^{++} is doubly charged, a membrane potential of -90 mV predicts a passive ratio across the membrane of 1000, rather than the 32 that would be predicted for a monovalent ion. Since the ratio is about one-fortieth that predicted by passive distribution, Mg^{++} must be pumped out of muscle cells. Any cellular binding that may occur would make this argument even stronger.

When Mg^{28} is injected and the turnover rates in the body pool are analyzed, four compartments can be distinguished.[1, 40, 57] A rapid compartment with a half-time of 1 hour contains less than 5 per cent of the body Mg^{++}; this probably corresponds to the distribution throughout the extracellular fluid. An intermediate compartment with a half-time of 3 hours contains slightly more Mg^{++} and probably corresponds to its distribution through skin, connective tissue, liver, intestine and heart. A slower compartment with a half-time of 1 day contains about 20 per cent of the body Mg^{++}; this represents exchange with muscle cells and erythrocytes. Finally, about two-thirds of the body Mg^{++} exchanges very slowly with a half-time of about 25 days. This is the so-called unexchangeable Mg^{++} situated on the inner surface layers of bone.

Magnesium Deficiency and Excess. The main symptoms of Mg^{++} deficiency are tetany, which is indistinguishable from hypocalcemic tetany, and occasionally convulsions. Chronic deficiency in animals causes alopecia, skin lesions, hyperemia and fibrotic changes in the blood vessels. Both plasma and skeletal muscle Mg^{++} levels decrease markedly in humans,[40] but soft tissue Mg^{++} does not decrease in experimental animals.[3, 14] Total 24-hour exchangeable Mg^{++} is also decreased[28] because of the loss from the slowly exchangeable fraction in muscle[40] and bone.[3] In addition, Na^+ is gained and K^+ is lost from muscle in Mg^{++} deficiency. This cannot be prevented by a fairly large intake of K^+. The plasma K^+ level does not change and a metabolic alkalosis does not develop. The muscle K^+ deficit and Na^+ excess may be due to inhibition of Na^+-K^+-activated ATPase, which is also Mg^{++} dependent, as a result of the Mg^{++} deficiency. Hypercalcemia and hypophosphatemia also occur in Mg^{++} deficiency and are most likely due to stimulation of parathyroid hormone release as a result of the Mg^{++} deficiency. Administration of Mg^{++} corrects these deficiencies, but if excessive amounts are given, the following effects are noted: central nervous system depression and anesthesia, neuromuscular and ganglionic blockade, vasodilation and depression of the myocardium. Electrocardiographic changes similar to those produced by K^+ excess are also noted.

Regulation of Magnesium Distribution. The factors regulating Mg^{++} levels in cells and plasma are not known. However, they appear to be correlated with aldosterone secretion, as is K^+. In Mg^{++} deficiency aldosterone secretion is increased. As might be anticipated from the similarity of Mg^{++} to Ca^{++}, parathormone mobilizes Mg^{++} from bone and increases its excretion in the urine. However, unlike Ca^{++}, plasma Mg^{++} levels are decreased. The relation between acid–base changes and Mg^{++} metabolism has not been clearly elucidated, but the available data indicate that the movements of Mg^{++} in acid–base derangements are like those of K^+.

CALCIUM. The physiology of Ca^{++} metabolism is discussed in Chapter 7, Volume I.

CHLORIDE.[13] Cl^-, like Na^+, is confined mainly to the extracellular fluid and is the predominant anion of this fluid. It is present in cells, but unlike Na^+, which is actively transported out of cells, the concentration of Cl^- is usually determined by the membrane potential. In gastric and intestinal mucosa, thyroid gland, glial cells, and also probably in other secretory cells, however, Cl^- is actively transported (see Chaps. 24 and 28, Vol. II, and Chap. 2, Vol. III).

Content and Distribution. The total body Cl^- content in man and experimental animals averages about 33 mEq per kg of body weight. The distribution of Cl^- in the various fluid compartments is presented in Table 26–3. About 50 per cent of the Cl^- is found in plasma and interstitial fluids. The content of Cl^- in connective tissue is high and represents about 16 per cent of the total, as compared to 12 per cent of the total body Na^+ found in connective tissue. Thus, some of the Cl^- in connective tissue is either intracellular, bound to extracellular proteins or, more likely, both. Since not all of the Cl^- present in connective tissue exchanges with radiochloride, some binding may occur.

Transcellular Cl^- is a small, but important, fraction of total body Cl^-. The Cl^- concentration in such fluids varies from 20 mEq per liter or less in sweat or colonic fluid to more than 150 mEq per liter in actively secreted gastric acid. Approximately 40 per cent of transcellular fluid Cl^- is in the gastrointestinal tract and about 25 per cent is in cerebrospinal fluid.

Since Cl^- is passively distributed across the majority of cell membranes, most of the intracellular Cl^- is present in cells with low membrane potentials. These include smooth muscle, exocrine glands, liver, connective tissue and erythrocytes. Erythrocytes contain approximately 40 per cent of this amount. Muscle cells contain an additional 20 per cent of the intracellular Cl^-.

Values for exchangeable Cl^- reach about the same value as that obtained by chemical analysis if sufficient time is allowed for equilibration (24 to 48 hours). Most of the Cl^- is rapidly exchangeable and, in rats, only the Cl^- in skin, testes and brain is not completely exchanged with radiochloride within 30 min. After 24 hours, only the skin Cl^- is not equilibrated.

Regulation of Chloride Distribution. Except for acid–base derangements, changes in body Cl^- content are influenced by the same factors and in the same direction as are changes in body Na^+ content. Changes in Cl^- levels secondary to acid–base distortions are covered in Chapter 27, Volume II. Where Cl^- distribution is passive, its movement is governed by changes in the membrane voltage and thus, ultimately, by the activity of the sodium pump. In the case of tissues that actively transport Cl^-, alterations of Cl^- distribution can be produced by inhibitors of the transport process such as perchlorate, which appears to compete with Cl^- for the carrier.

BICARBONATE. The presence of HCO_3^- in the body is dependent on two factors: (i) the metabolic production of CO_2 by cells and (ii) the excess of cations, such as Na^+,

K^+, Ca^{++} and Mg^{++}, over nonlabile anions, such as Cl^-, $SO_4^=$, $H_2PO_4^-$, $HPO_4^=$, and protein. HCO_3^- is a labile anion derived from the hydration of metabolically produced CO_2. The amount of HCO_3^- in body fluids is equal to the difference between the sum of the fixed cations and the sum of the fixed anions in any fluid. An excess of cations over anions causes the formation of HCO_3^- from the hydration of CO_2 and an excess of anions causes conversion of HCO_3^- to carbonic acid with liberation of CO_2. The role of HCO_3^- in regulation of body acid–base balance is discussed in Chapter 27, Volume II.

The total body HCO_3^- averages 13 mEq per kg of body weight. The total acid-releaseable CO_2 of the body, however, is much greater than this and amounts to approximately 76 mEq per kg of body weight.[23] The difference between these values is contributed by carbonate from the crystal lattice of bone.

The distribution of HCO_3^- in the various body fluids is indicated in Table 26–3. About half the HCO_3^- is in the extracellular compartment. The over-all average concentration in transcellular fluids is about the same as that in plasma, but the values in individual fluids vary considerably. The concentration is high in pancreatic fluid and aqueous humor but absent or extremely low in gastric acid. In transcellular fluids with high HCO_3^- concentrations, e.g., aqueous humor and pancreatic juice, this anion is actively secreted; it appears that carbonic anhydrase and $(HCO_3^- \text{-} Mg^{++})$-activated ATPase are involved in the secretory process.

The other half of the total body HCO_3^- is present in cells. Muscle cells contain HCO_3^- at a concentration of about 12 mEq per liter (Table 26–5) and most other cells contain it at higher concentrations. Thus, the concentration ratio of HCO_3^- across the muscle cell membrane $\left(\dfrac{[HCO_3^-]_o}{[HCO_3^-]_i} = \dfrac{26}{12} \right)$ is 2.2. This value is far from the value of 30 predicted by the membrane potential. Either HCO_3^- must be transported actively into cells or H^+ must be transported actively out of them. The nature of and evidence for

Table 26–5 *Intracellular Concentrations of Cations and Anions in Some Representative Tissues*

	Ion	Skeletal Muscle mEq/l Intracellular Water	Skeletal Muscle mM/l Intracellular Water	Cardiac Muscle mEq/l Intracellular Water	Liver mEq/l Intracellular Water	Thyroid Gland mEq/l Intracellular Water	Erythrocytes mEq/l Intracellular Water	Erythrocytes mM/l Intracellular Water
Cations	Na^+	12	12	7	29	42	19	19
	K^+	150	150	134	166	147	136	136
	Mg^{++}	34	17	28	29	?	6	3
	Ca^{++}	4	2	4	2	?	0	0
	pH	6.90		7.10	7.23	7.17	7.28	
	Total	200	181	173	226		161	158
Anions	Cl^-	4	4	4	19	19	78	78
	HCO_3^-	12	12	12	16	14	18	18
	$HPO_4^= \text{–} H_2PO_4^-$	40	17				4	2
	Organic	90	84				25	23
	Protein	54	6				36	5
	Total	200	123				161	126
Total mOsm/l			304					284
E. C. F. volume		8.2%		19.5%	10.5%	20%		
$\mathscr{E}_s$ (millivolts)		−90		−89	−43	−50	−10	

active transport of H^+ or of HCO_3^- are discussed more fully in Chapter 1, Volume I, and Chapter 27, Volume II.

The average cell body concentration of HCO_3^-, 16 mEq per liter, is not far from the values found in muscle, brain and heart. The over-all cell body pH is 7.23 as calculated from this value and from the assumptions that (i) Pco_2 is in equilibrium across all cell membranes and (ii) that the hydration reaction of $CO_2 \rightleftarrows H_2CO_3$ is in equilibrium everywhere. Values of over-all cellular pH of between 6.9 and 7.0 have been measured from the distribution of the weak acid 5,5-dimethyl-2,4-oxazolidinedione (DMO);[51, 69] such observations have been made in muscle, brain and thyroid tissue. The interpretation appears to be that the reaction between CO_2 and H_2CO_3 is not always at equilibrium in cells. The cellular concentration of H_2CO_3 is higher and the cellular pH is lower than would be predicted on the assumption of equilibrium. In skeletal muscle, the cells do not contain carbonic anhydrase and this disequilibrium is greater than normally found in brain or thyroid tissue, which contain this enzyme. After inhibition of carbonic anhydrase, the disequilibrium is enhanced in these latter tissues.

Present data indicate that all of the HCO_3^- in the body is exchangeable. Factors affecting the distribution, excretion and transcellular movement of HCO_3^- are discussed in Chapters 25 and 27, Volume II.

Fluid Measurement and Composition in Isolated Tissues. Methods for measurement of whole body distribution of water and ions in the various fluid compartments of the body have already been described. These methods involve a knowledge of total body water and ionic content, the volume of the extracellular fluid, and the ionic composition of plasma. The same principles apply in isolated tissues. The extracellular volume, measured with one of the polysaccharides or with sulfate, the water and ionic composition of the plasma, and the total tissue water and electrolytes must be determined. From such raw data, the extracellular and intracellular values can be derived.

PLASMA. The values for water and electrolyte content are determined by standard techniques: the water content by desiccation, Na^+ and K^+ by flame photometry, and Cl^- by any of several methods, the most accurate of which is the automatic electrometric titration of silver. Since ions are dissolved only in the aqueous phase of plasma, concentrations of electrolytes are expressed in terms of plasma water. The concentrations of the various ions in plasma water are shown in Table 26–6. The solid content of plasma is made up mainly of negatively charged protein, which constitutes 6 to 8 per cent of plasma weight. Although the protein occupies a large volume, it makes up only a small portion of the anionic charge in plasma and contributes only a small fraction to the total plasma osmotic concentration. However, the protein is important in maintaining the differential concentration

Table 26–6 *Concentration of Cations and Anions in Plasma Water and Interstitial Fluid*

	ION	PLASMA WATER* mEq/l	mM/l	INTERSTITIAL FLUID† mEq/l	mM/l
Cations	Na^+	153.2	153.2	145.1	145.1
	K^+	4.3	4.3	4.1	4.1
	Ca^{++}	3.8	1.9‡	3.4	1.7
	Mg^{++}	1.4	0.7§	1.3	0.65
	Total	162.7	160.1	153.9	151.6
Anions	Cl^-	111.5	111.5	118	118
	HCO_3^-	25.7	25.7	27	27
	$H_2PO_4^- - HPO_4^=$	2.2	0.66	2.3	0.7
	Other	6.3	5.9	6.6	6.2
	Protein	17.0	1.5	0	0
	Total	162.7	145.3	153.9	151.9
Total mOsm per liter			305.4		303.5

*Plasma water content assumed to be 93 per cent.
†Gibbs-Donnan factors used as multipliers are 0.95 for monovalent cations, 0.9 for divalent cations, 1.05 for monovalent anions and 1.10 for divalent anions.
‡Total Ca is 2.7 mM; ionized Ca is about 70 per cent of total Ca.
§Total Mg is 1 mM; ionized Mg is about 65 per cent of total Mg.

of ions between plasma and interstitial fluid via the Gibbs–Donnan distribution.

INTERSTITIAL FLUID. The concentration of any ion in the interstitial fluid is determined by its concentration in plasma and its Gibbs–Donnan distribution ratio. Capillary membranes are not freely permeable to protein, and hence interstitial fluid, unlike plasma, is nearly devoid of protein.[42] In any system in which two fluid compartments are separated by a membrane permeable to water and to all but one of the charged solutes (e.g., plasma protein), a potential difference develops across the membrane which affects the distribution of the ions present. The sequence of events leading to equilibration is similar to that described in Figure 1–3, Chapter 1, Volume I. The concentration of Cl^- in the plasma water must be less than in interstitial fluid because some of the negative charges are carried by proteins in plasma. The potential arises from the flow of Cl^- ions down their concentration gradient from interstitial fluid to plasma, thereby charging the capillary membrane negatively on the plasma side. In turn, Na^+ ions flow down the voltage gradient from interstitial fluid to plasma, i.e., they accompany Cl^-. Hence there will be a net flow of Na^+ and Cl^- into plasma until the concentration of Na^+ has increased enough to provide an equal counterflow. Thus, in the case of the capillary membrane, cations are less concentrated and anions are more concentrated in the interstitial fluid than in the vascular fluid. The total cation concentration balances the total anion concentration in each compartment. If equilibrium has been attained, one can use the Nernst equation to calculate the ratio of concentrations from the voltage or vice versa (see Chap. 1, Vol. I, and Chap. 24, Vol. II). In addition, if water is to be in equilibrium, a hydrostatic pressure difference must be present across the membrane. In capillaries, the vascular side must be at a higher pressure than the interstitial side. The concentrations of permeant monovalent cations is about 5 per cent less and of permeant anions about 5 per cent more in interstitial fluid than in the capillaries from which this fluid derives, and the transcapillary voltage is about 1.3 mV. For divalent ions, there is about a 10 per cent difference in concentrations. The average concentrations of some of the ions in interstitial fluid as compared to their concentrations in plasma water are presented in Table 26–6. It is of interest to note that the osmolar concentration of plasma protein must be about 1.5 mMol per liter to produce the observed osmotic pressure of 25 mm Hg, while the equivalent concentration must be 17 mEq per liter (Table 26–6) to produce the observed Gibbs–Donnan factor of 1.05 ($[Cl^-]_e/[Cl^-]_p$) for monovalent anions and 0.95 ($[Na^+]_e/[Na^+]_p$) for monovalent cations; i.e., the average valence of plasma protein is about −10.

INTRACELLULAR FLUID. In order to calculate the concentrations of ions in cell water, not only must tissue water and electrolyte content be known, but the volume of interstitial fluid present in the tissue must be determined. The volume of distribution of inulin is a measure of the latter. The calculation for muscle intracellular Na^+ is as follows: Total muscle water = 0.774 kg per kg of fat-free wet muscle; total muscle inulin = 18.04 mg per kg of fat-free wet muscle; total muscle Na^+ = 20.7 mEq per kg of fat-free muscle. The concentration of inulin in interstitial water is 220 mg per kg of water and the concentration of Na^+ is 147 mEq per kg of water. The volume of interstitial fluid must be that volume required to account for the measured inulin or 18.04/220 = 0.082 kg per kg of fat-free wet muscle. This interstitial space of 8.2 per cent contains 0.082 × 147 = 12.1 mEq per kg of fat-free wet muscle of Na^+. Hence 20.7 − 12.1 = 8.6 mEq of Na^+ that must be intracellular in every kg of muscle. Since 0.082 kg of water is interstitial, 0.774 − 0.082 = 0.692 kg water which must be intracellular in every kg of muscle. The concentration of Na^+ within the cells is the quantity of intracellular Na^+ divided by the volume of intracellular water: $[Na^+]_i$ = 8.6 mEq per kg/0.692 kg per kg = 12.4 mEq per kg intracellular water. Values for the intracellular concentration of other solutes are calculated in the same manner.

Intracellular cations and anions of some representative tissues are summarized in Table 26–5. The intracellular values for each tissue appear to be characteristic of that tissue. It is evident, therefore, that each tissue has its own intrinsic mechanism for regulation of its internal composition. Note that HCO_3^- is higher than would be predicted by the transmembrane voltage. This discrepancy is due to the continuous metabolic production of CO_2 and its subsequent hydration within the cell which keeps HCO_3^- from ever attaining equilibrium. Rather, a steady state is reached in which intracellular HCO_3^- is higher than the equilibrium value. The dependence of the intracellular concentration of Cl^- on the membrane potential is dramatically illustrated by comparing the difference in concentrations between muscle and red blood cells. In muscle, where $\mathscr{E}_s$ is about −90 mV, intracellular Cl^- is only 4 mEq per liter. In red blood cells, where the $\mathscr{E}_s$ is less than −10 mV, intracellular Cl^- is 78 mEq per liter.

It is difficult to determine the total osmolar concentration in cellular fluid, because the total charge on the major anion, protein, is not known. However, if osmolar equilibrium with extracellular fluid is assumed (see below), one can estimate protein valence. The total molar concentration of protein is the difference between the total solute concentration in cells (equal to that in plasma) and the total solute concentration of the determined ions in those cells. The charge on that protein is the difference between the total concentration of determined cations and the total concentration of determined anions. The valence is calculated as the charge per mol of protein. Calculation from the muscle data in Table 26–5 yields a value of almost 3 for the valence of the protein.

EXCHANGES OF WATER AND ELECTROLYTES BETWEEN BODY COMPARTMENTS

The dynamics of fluid and electrolyte movements across the capillary endothelium and the cell membrane are an important aspect of body fluid physiology. Such movements are the basis for nourishment of cells and the maintenance of osmotic equilibrium between vascular, interstitial and intracellular spaces.

Plasma and Interstitial Fluids.[8, 46, 53] Water is continuously exchanged between the intravascular and interstitial fluids through the walls of capillaries. The fluid exchange depends upon the physical factors of diffusion and filtration and upon the colloid osmotic (oncotic) pressure of the plasma. Although there has been considerable discussion on the relative contributions of diffusion and bulk filtration in capillary exchange, the over-all result is that capillary exchange follows the general pattern first suggested by Starling. The dominant force tending to move material from the vascular compartment into the interstitial compartment is the hydrostatic pressure of the blood. Since protein crosses the capillary membrane very slowly, its concentration in interstitial fluid is much lower than its concentration in plasma. The osmotic pressure corresponding to the difference in concentrations of protein constitutes the dominant force tending to bring material back from the interstitial compartment to the blood. At the arterial end of a capillary, the hydrostatic pressure is greater than the oncotic pressure and there is a net bulk movement of water, electrolytes and the small organic molecules from plasma into interstitial fluid. At the venous end of the capillary, the hydrostatic pressure has fallen and is below the oncotic pressure, and there is net bulk movement of water, electrolytes and small organic molecules back into blood. These transcapillary movements must balance in the steady state. Thus, an increase in vascular pressure or a decrease in plasma protein concentration (such as occurs in nephrosis or after severe burns) leads to a net transfer of fluid from vascular to interstitial fluid. The result is edema.

Although water and the dissolved electrolytes appear to cross capillaries only 'hrough a small portion of the capillary membrane, the respiratory gases appear to diffuse through the whole surface. This accounts for the rapidity of gas transport compared to that of water and electrolytes.

Interstitial and Intracellular Fluids. With a few exceptions, such as cells in the ascending limb of renal tubules and cells lining the ducts of sweat and salivary glands, all cells appear to be freely permeable to water. Hence, the total solute concentration of cell fluid is the same as that of interstitial fluid, and water is exchanged quickly across cell membranes. The mechanisms of exchange of solutes between interstitial and intracellular fluids have already been discussed.

An interesting problem arises from the fact that cells contain considerable amounts of nonpermeating anion. Since interstitial fluid contains little or no protein, a Donnan equilibrium is set up across the cell membrane. The situation is similar to that across the capillary membrane with one important difference: cell membranes have remarkably low tensile strengths and are unable to support the pressure difference necessary to balance these forces. Active cation transport resolves the apparent difficulty. The low permeability and active transport of Na^+ out of cells produces what is in effect a nonpermeating cation. This effect balances that of the nonpermeating anion within the cell. Consequently, conditions which increase the passive permeability of cells to Na^+ or decrease the rate of active extrusion should unbalance this double Donnan system and lead to entry of water into cells. This has been found to be the case for ouabain, a selective inhibitor of the $Na^+ - K^+$ pump. The converse, that increased pumping should lead to a shift of water out of the cells, by increasing the effective impermeability to Na^+ is also true.

If the active transport system of a cell is very efficient and essentially all of the interstitial Na^+ can be considered impermeant, then the concentration of impermeant ion is higher outside than inside the cell. Since pressure differences cannot balance the osmotic forces, it would be expected that, under these conditions, cells would be slightly hypertonic. Robinson[52] found that there are circumstances in which cells do appear to be hypertonic. That the process which renders the cells hypertonic should be the active transport of solute material out of them is a fine physiologic paradox.[39]

Alterations of Fluid and Electrolyte Balance. Examples of various clinical states characterized by changes in fluid volume or osmolal concentration induced by alterations in Na^+ or water content are presented in Figure 26–6. Volume is indicated on the abscissa and osmolal concentration on the ordinate. The total quantity of solute material in a compartment is represented by the area.

Calculation of the shift in fluid compartments is based on the principle used throughout this chapter that $Q/V = [C]$. When only water has been added to or subtracted from a compartment or the body as a whole, the product of the concentration of a solute in a compartment and the compartmental volume is constant, and

$$C_{initial} \ V_{initial} = C_{final} \ V_{final}.$$

When solute material is added, the change in total quantity must first be calculated and then the relation between concentrations and volumes deduced.*

Changes in extracellular Na^+ concentration accurately reflect changes in total osmolality; therefore, plasma Na^+ level is used to predict volume and concentration changes in conditions of altered fluid and electrolyte balance.[79]

EXCESSIVE INTAKE OF WATER (HYDRATION). Hydration may result from excessive ingestion or decreased loss of water or from administration of vasopressin (antidiuretic hormone, ADH). The effect of oral ingestion of water on the volume and

---

*Normally, intracellular solutes are regarded as remaining intracellular and extracellular solutes are regarded as remaining extracellular. Calculation of fluid balances in this manner is generally close enough to be adequate for clinical purposes. However, there may be secondary solute transfer across cell membranes, e.g., in severe dehydration.

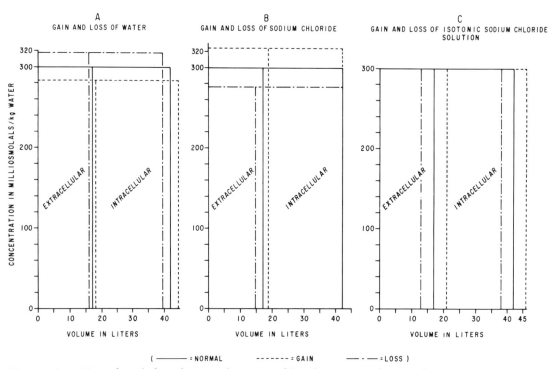

Figure 26–6 Examples of clinical states characterized by changes in volume and osmolar concentration of extracellular and intracellular compartments. *Ordinate,* Concentration in mOsm per kg water. *Abscissa,* Volume of the compartment in liters. Solid lines represent the initial normal state, dashed lines the final state after *addition* of fluid or solute, and dot-dashed lines the final state after *loss* of water or solute. (After Darrow and Yannet, *J. clin. Invest.,* 1935, *14,* 266–275.)

composition of the body compartments is shown in Figure 26–6A. After absorption from the intestinal tract, water distributes rapidly and uniformly throughout the body in proportion to the initial volumes of the extracellular and intracellular compartments.

Excessive water intake may produce the syndrome of water intoxication. The symptoms probably result from disturbed cellular metabolism secondary to decreased cellular electrolyte concentration. The disturbances are mainly related to the central nervous system: disorientation, convulsions and coma. Other disturbances are gastrointestinal dysfunction, muscular weakness and cardiac arrhythmias.

EXCESSIVE LOSS OF WATER (DEHYDRATION). Water loss by excessive evaporation through the lungs and skin or by lowered intake causes volume and composition changes opposite to those produced by hydration (Fig. 26–6A). The total osmolality of the body fluids is increased and the volumes of the intracellular and extracellular fluid compartments are decreased in proportion to their original volume.

EXCESSIVE INTAKE OF SOLUTE. Intravenous infusion of a strongly hypertonic solution of NaCl leads to the following sequence of events. The quantity of solute in the extracellular compartment increases and the volume also increases initially by the amount infused. Next, water shifts from the intracellular to the hypertonic extracellular compartment. The final result is that solute concentrations are equalized and that the extracellular compartment has gained in volume in addition to the amount infused. The intracellular compartment has shrunk. The total solute in the extracellular compartment has increased by the amount of solute infused, but total cellular solute has remained unchanged. The solute concentration in both compartments has risen. Actually all these processes occur simultaneously (Fig. 26–6B).

Example: A 70 kg man was given an infusion of 1 liter of a NaCl solution containing 1000 mOsm (500 mEq per liter).
Initial intracellular volume = 25 liters.
Initial extracellular volume = 17 liters.
Initial solute concentration in each compartment = 300 mOsm per liter.
Quantity of solute in whole body initially = (25 liters + 17 liters) 300 mOsm per liter = 12,600 mOsm.

Quantity of solute in whole body after infusion = 12,600 mOsm + 1000 mOsm = 13,600 mOsm.
Volume of water in whole body after infusion = (25 liters + 17 liters + 1 liter) = 43 liters.
Concentration in each compartment after infusion = $\frac{13{,}600 \text{ mOsm}}{43 \text{ liters}}$ = 316 mOsm per liter.
Quantity of intracellular solute initially and after infusion = 25 liters × 300 mOsm per liter = 7500 mOsm.
Volume of intracellular compartment after infusion = $\frac{7500 \text{ mOsm}}{316 \text{ mOsm per liter}}$ = 23.7 liters.
Volume of extracellular compartment after infusion = 43 liters − 23.7 liters = 19.3 liters.
Thus, 19.3 liters − (17 liters + 1 liter) = 1.3 liters of water which have shifted from the intracellular to the extracellular compartment.

EXCESSIVE LOSSES OF SOLUTE. The changes which result from loss of NaCl without water from the extracellular fluid are opposite from those induced by excessive intake. Body NaCl may be depleted by reduced intake, adrenocortical insufficiency, peritoneal dialysis with 5 per cent dextrose solutions,[79] or by excessive parenteral therapy with NaCl-free solutions. Extracellular fluid becomes hypotonic and water transfers into cells. Cellular hydration and extracellular dehydration result (see Fig. 26–6B).

ISOTONIC EXPANSION OF VOLUME. If isotonic (0.9 per cent) NaCl is infused intravenously, the volume of the extracellular fluid is expanded without change in osmolality. Hence, there is no shift of water between extracellular and intracellular compartments and the intracellular compartment remains unchanged. Clinically, this condition is termed *edema* (Fig. 26–6C).

ISOTONIC CONTRACTION OF VOLUME. If both Na$^+$ and water are lost from the body such that plasma Na$^+$ concentration is not altered, the volume of the extracellular fluid is reduced with no change in the intracellular compartment. This state occurs in conditions such as hemorrhage, extensive burns, and mild loss of electrolytes from the gastrointestinal tract (Fig. 26–6C).

In the case of hemorrhage or severe burns, the water and electrolytes are frequently replaced without replacement of the plasma protein that has also been lost. This lowers the plasma protein concentration and upsets the fluid balance between the vascular and interstitial compartments. Severe edema may result.

EXCHANGES OF WATER AND ELECTROLYTES BETWEEN BODY AND EXTERNAL ENVIRONMENT

The body surface is a site of exchange of matter and energy between the organism and its environment. Heat is mainly exchanged across the body surface, and water and electrolytes are mainly exchanged across the lungs, alimentary tract and the kidney. The exchanges take place across specialized cell membranes between extracellular fluid and the environment.

Fluid Intake. The daily net turnover of water averages between 3 and 6 per cent of total body water, that is, between 1.5 and 3.0 liters.

The total volume of body fluid results from the balance between intake and output of water. Although solid foods provide a considerable quantity of water (both preformed and as water of oxidation), variation in water intake depends largely upon the volume of liquid ingested. Physiologic regulation of the volume imbibed depends on thirst, which is discussed in Chapters 12 and 18, Volume I.

Fluid Absorption. The stomach absorbs practically no water; the intestinal mucosa absorbs ingested water and electrolytes as well as the digestive secretions. Since these secretions amount to more than 6 liters per day, interference with this reabsorptive process can rapidly lead to dehydration and Na^+ depletion. Because gastrointestinal secretions are higher in K^+ concentration than is plasma, continued loss of gastrointestinal secretions, as in diarrhea or vomiting, leads to K^+ deficiency. The main force for the movement of water is active Na^+ absorption throughout the whole length of the intestinal epithelium (see Chap. 3, Vol. III). The intestine also appears capable of actively absorbing Cl^-. Water movement is passive, secondary to solute movement. In the small intestine, the passive permeabilities to both water and ions are high. Appreciable concentration gradients are not maintained, but bulk absorption of large amounts of solute and water take place. In the colon, the passive permeabilities are lower and concentration gradients are produced and maintained. Thus colonic contents may have very low Na^+ concentrations. Because of the high permeability of the small intestine to water, chyme rapidly equilibrates with plasma. Thus, if hypertonic solutions are placed in the small intestine, water may first enter the intestinal lumen from the perfusing blood. Only later, as active solute transport continues, is the volume reduced.

Absorption of water and electrolytes is also influenced by adrenocortical hormones. Adrenocortical insufficiency results in a decreased rate of solute and water absorption. Deoxycorticosterone, ACTH and cortisone decrease the amount of Na^+ and Cl^- and increase the amount of K^+ left in the gastrointestinal tract.

Fluid Output. Most of the fluid loss from the body is "obligated" by renal osmotic forces and by cutaneous, pulmonary and oral evaporation. However, the variable or "facultative" loss of fluid depends upon excretion of water through the renal and glandular epithelia. The major regulation of body fluids is directed toward the renal output. The kidney helps to stabilize the osmolarity and volume of body fluids by eliminating or conserving water or Na^+. These functions are regulated by two independent neurohumoral systems, the "antidiuretic system" and the "antinatriuretic system."

INSENSIBLE WATER LOSS. Water without electrolyte is lost continuously through the skin and lungs by evaporation. Insensible water loss from both skin and lungs depends on the ambient temperature and the relative humidity; loss from the lungs is also modified by changes in respiratory rate.

SWEATING.[64] Sweating is an active secretory process concerned mainly with temperature regulation. Thus, although sweating affects both salt and water, it is not directed toward fluid regulation. Most of the sweat glands are activated when demands for heat loss are increased, as in exercise or in elevated environmental temperature. Sweating rates as high as 70 ml per min over short periods or 1500 ml per hour over several hours have been reported. Such high rates of sweating are not maintained; a decline occurs even though the water which is lost is replaced. Sweating is reduced when salt in excess of that

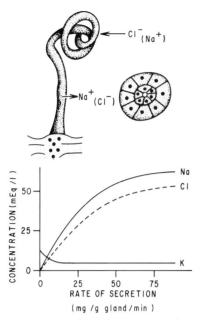

Figure 26–7 Histologic structure and functional characteristics of sweat glands. Graph at the bottom shows that the concentration of Na^+ and Cl^- in sweat is determined by the rate of secretion. (After Ussing et al., *The alkali metal ions in biology.* Berlin, Springer-Verlag, 1960.)

lost is ingested or injected. Similarly, sweating may be reduced in severe dehydration, although moderate water depletion does not influence the rate of sweating.

The secretion of the sweat glands is an active process which takes place in two stages as illustrated in Figure 26–7. Sweat secretion represents one process by which transcellular fluids are formed, and the general principles discussed here for sweat also apply with only slight modifications to the secretory process in the other duct-possessing glands such as salivary, pancreatic and lacrimal glands and in bile secretion. The first stage is the formation of a primary secretion in the acini of the gland. The second stage is the modification of the primary fluid by reabsorption of salts and water in the ducts.

The primary secretion of sweat is thought to be formed by active transport of either Na^+ or Cl^- or both. As a result of the osmotic and voltage gradients created by the active solute transport, water and the unpumped ions move passively into the lumen. The primary secretion is approximately isotonic with plasma and the sum of the Na^+ and K^+ concentrations is equal

to the sum of their concentrations in plasma. As the primary secretion flows down the sweat gland ducts, Na^+ but not K^+ is reabsorbed by an active process with a limited maximal capacity; Cl^- is reabsorbed passively secondary to the active Na^+ movement. Permeability of the duct wall epithelium to water is low; hence, reabsorption of water is limited but does occur secondary to active solute transport. As a result of these two processes—primary active solute and passive water secretion followed by active solute and passive, limited water reabsorption—the concentrations of Na^+ and Cl^- in the sweat vary with the secretory rate but are always lower than their plasma concentrations. Therefore, the final secretory fluid is always hypotonic. The concentration of K^+ appears to be independent of the secretory rate, except that with low rates the concentration rises (Fig. 26–7). At high flow rates, the rate of passage of the primary secretion down the duct lumen exceeds the maximal sodium reabsorptive capacity; consequently, less Na^+ is reabsorbed. The faster the secretory rate of the primary secretion the higher the Na^+ (and Cl^-) concentration in the sweat (Fig. 26–7). The body adapts to excessive sweating, generally caused by heavy work in hot environments, by a reduction in secretory rate, a decrease in sweat sodium, and an increase in sweat potassium concentrations. This adaptation, however, occurs only when sodium is lost because of inadequate sodium replacement. The adaptation results in conservation of sodium and is undoubtedly due to aldosterone release secondary to sodium depletion. The effects of aldosterone and adaptation on sweat electrolyte concentrations are the same.

The magnitude of electrolyte and water losses from excessive sweating can be illustrated by the following example.[64] During maximal sweating, 11 to 15 liters of sweat may be produced in 24 hours. This represents a water turnover of 25 per cent of total body water per day. If the average concentration of Na^+ in the sweat is 60 mEq per liter, the 24-hour Na^+ turnover represents 20 to 30 per cent of total exchangeable Na^+. These turnover rates are so large that maintenance of normal homeostasis, even for short periods, rests almost entirely on a balanced replacement of water and NaCl losses.

OTHER MEANS OF FLUID EXCRETION.[64] Small amounts of water (approximately 100 to 200 ml per day) and electrolytes are excreted with the feces and with glandular secretory products, such as salivary, lacrimal and genital secretions. During lactation, water and small amounts of electrolytes are lost through the mammary glands. The secretion of the transcellular fluids of these glands follows the same principles as described above for the sweat glands, that is, primary secretion followed by reabsorption. However, unlike sweat, the secretory fluid of many of these glands contains HCO_3^- in addition to Na^+, Cl^- and K^+.

Normal Fluid Balance. Despite wide variations in daily intake and output of water, total body fluid volume remains remarkably constant. In the adult, under normal conditions and over extended periods, the total intake must necessarily equal the total output. The several sources of water intake are (i) water ingested orally, (ii) water in food and (iii) water derived from oxidation of food. The water content of food contributes a considerable proportion of total intake, more than is often realized. For example, meat is 70 per cent water and certain vegetables and fruits are nearly 100 per cent water. Daily average values for the quantities of water taken into the body and excreted by various avenues are given in Table 26–7.

The requirement for water generally parallels energy metabolism, and it has been estimated to be approximately 1 ml of water per calorie. The fluctuations in the total volume of body fluid which occur during water loss and water replacement take place predominantly in the extracellular fluid. In infants, compared with adults, the total turnover of water represents a greater percentage of the volume of extracellular fluid; therefore, in infancy any excessive loss of fluid (diarrhea or vomiting) leads more rapidly to serious disturbances than in the adult. Daily changes in the total water content of the body are reflected by changes in body weight. This fact is utilized in the clinical estimation of water balance.

REGULATION OF VOLUME AND OSMOLARITY OF EXTRACELLULAR FLUID[25, 55, 60]

That the total volume and osmolarity of the intracellular and extracellular compartments are rigidly controlled is evident from the small fluctuations which occur despite wide variation in dietary intake. Body weight may vary only 0.5 kg per day if caloric balance exists and osmolarity is held constant to within 2 to 3 per cent. Intracellular volume and concentration are maintained constant by the factors already discussed. Regulation of these variables in the extracellular compartment is accomplished by control of both intake and output. The intake of water and salt is important for control of volume and osmolarity because deficits of Na^+ and water cannot be corrected solely by renal conservation mechanisms. *Excessive quantities in the body can be excreted, but deficits must be replaced by dietary intake.* The thirst mechanism plays an integral part in maintaining water homeostasis. Loss of water or gain of Na^+ induces thirst and the resulting increased intake of water, along with the renal compensating mechanisms, restores the altered volume and concentration of the extracellular fluid to normal. Since, under ordinary conditions, the intake of both Na^+ and K^+ far exceeds the minimum need, regulation of salt intake normally plays very little part in maintaining the stability of extracellular fluid volume and composition.

The major control of these variables is accomplished by the kidney. Two aspects must be considered: the factors regulating volume and the factors regulating osmolarity (mainly Na^+ concentration). The *osmolarity* of the extracellular fluid is regulated by the control of free water excretion, whereas *volume* is regulated by the control of both Na^+ and free water excretion.

Table 26–7 *Average Daily Water Balance*

WATER INTAKE (APPROXIMATE)		WATER EXCRETION (APPROXIMATE)	
Drinking water	1200 ml	Urine	1400 ml
Water content of food	1000 ml	Insensible water loss	900 ml
Water of oxidation	300 ml	Stool	200 ml
	2500 ml		2500 ml

Osmolar Regulation. (See also Chap. 12, Vol. I, and Chap. 8, Vol. III). Osmolar changes in the extracellular fluid are sensed by osmoreceptors which regulate the release of ADH. Under normal conditions, the release of ADH is controlled almost entirely by changes in osmolarity. The osmoreceptor cells are located in or near the supraoptic and paraventricular nuclei of the hypothalamus. They are thought to act like osmometers, shrinking when the extracellular fluid is hypertonic and swelling when it is hypotonic. Shrinkage of the cells stimulates neural tracts leading from the osmoreceptors to the median eminence and the neurohypophysis via the supraoptic–neurohypophyseal tract; stimulation through this tract increases the release of ADH. ADH increases the permeability of the distal portions of the nephron to water and the urine is rendered hypertonic. The excretion of hypertonic urine tends to bring the plasma back to its normal osmolarity. This recovery is facilitated by increased intake of water. Conversely, swelling of the osmoreceptors inhibits the release of ADH. As the circulating ADH is destroyed, the distal portions of the nephrons become impermeable to water and osmotically free water is excreted into the urine. This tends to elevate plasma osmolarity and bring it back to normal.

Volume Regulation. Survival of the animal depends on the relative constancy of blood volume. Volume regulation involves a receptor–effector system which has the following components, each of which will be discussed in turn: (i) receptors which detect changes in volume of some portion of the extracellular fluid; (ii) an afferent limb; (iii) effectors which carry out the required adjustments to the volume change. Volume is regulated by two systems, one acting through aldosterone, the other through ADH. Normally the aldosterone system is the more important.

Aldosterone secretion is enhanced by adrenocorticotropic hormone, and removal of the adenohypophysis results in reduced secretion. However, complete regulation of aldosterone release involves factors in addition to adrenocorticotropin. Thus, aldosterone secretion is also augmented by acute hemorrhage, acute suprarenal aortic constriction, low and high output cardiac failure, and vena caval constric-

tion. In addition to these acutely induced volume changes, changes in volume resulting from alterations in Na^+ and K^+ balance also influence aldosterone release.

Changes in aldosterone modify active Na^+ reabsorption in the kidney in the direction to bring extracellular volume back toward normal. Renin, a hormone secreted by the juxtaglomerular apparatus of the kidney, is the main regulator of aldosterone release. The juxtaglomerular apparatus (JGA) consists of the juxtaglomerular cells and the macula densa. Evidence for renin secretion by the JGA is based on the following observations: (i) There is a consistent correlation between the granule content of the JGA cells and the renin content of kidney. For example, renal artery constriction and sodium chloride depletion increase the juxtaglomerular cell granularity and renal renin content, whereas desoxycorticosterone acetate in combination with sodium chloride or a high-sodium chloride intake alone diminishes both granularity and renin content. (ii) Microdissection studies have localized the production of renin to the JGA. (iii) The fluorescent antibody technique has demonstrated specific staining of juxtaglomerular cells of hog, dog and rabbit after treatment with fluorescent antirenin antibodies. (iv) Cell cultures from human renal cortex produce renin only when JGA granules are present.

The factors regulating the release of renin from the JGA, although still somewhat controversial, appear to be as follows: renal perfusion pressure, sodium delivery to the macula densa, adrenergic activity, humoral release of catecholamines, plasma potassium concentration, humoral release of angiotensin, antidiuretic hormone, extracellular fluid volume, adrenocorticotropic hormone and the plasma sodium concentration. However, most of the experimental evidence suggests that alterations in the sensing of renal vascular perfusion pressure by the afferent renal artery (baroreceptor hypothesis), the delivery of sodium to the distal tubule (macula densa hypothesis) and alterations in adrenergic activity are the major factors regulating renin release from the JGA. Extensive observation suggests that an increase in renal perfusion pressure suppresses renin secretory production from juxtaglomerular cells and vice versa. In addition, because of the proximity of the early distal tubule and the afferent arteriole of a given nephron, it was postulated that the amount of Na^+ delivered to the distal tubule regulated renin release via an effect on the macula densa cells of the JGA and that this provided the chief intrarenal regulatory mechanisms for renin release and regulation of Na^+ excretion and renal blood flow. The majority of available data suggests an inverse relation between distal tubular sodium delivery and renin release. How this occurs is not clear, but it is probably related primarily to the sodium flux rate across the macula densa into the interstitium surrounding the juxtaglomerular granular cells. The amount of sodium transferred to the granular cells is most likely regulated by the degree of contact between the distal tubule and the granular cells. Thus a smaller sodium load in the distal tubule appears to be accompanied by a decrease in its volume and therefore decreased contact with granular cells and less sodium transfer to these cells; renin release would thereby increase. A large sodium load to the distal tubule would have the opposite effect.[2]

Renin release can also be enhanced by adrenergic agents or sympathetic nervous system stimulation by a mechanism that appears to involve the intracellular release of cyclic adenosine monophosphate.

Renin is a proteolytic enzyme which, after being released into the blood, acts on renin substrate, an α-2-globulin, in the plasma to produce angiotensin I, an inactive decapeptide. Converting enzyme then converts angiotensin I to angiotensin II in the capillary bed of the lungs. Angiotensin II is a potent octapeptide which then acts on the zona glomerulosa of the adrenal cortex to stimulate the release of aldosterone. Aldosterone, in turn, acts on the renal tubules to promote Na^+ reabsorption.

In addition to its role in regulating aldosterone release, there is evidence that the renin-angiotensin system controls ADH release in part. The level of angiotensin II in the plasma regulates ADH release from the osmoreceptors in the brain in a direct proportion, probably by acting on receptors in the paraventricular nucleus lying close to the ventricular surface.

Volume regulation is also mediated through the ADH system. The observations of Gauer and Henry[25] (see also review by Share and Claybaugh[55]) suggest that moderate changes of blood volume have their prime effects on the low pressure side of the circulation. The large veins, right heart, pulmonary circulation and the left atrium contain 80 to 85 per cent of the total blood volume and are 100 to 200 times as distensible as the arterial side of the circulation. It appears, therefore, that receptors located in the walls of this low pressure system could best measure the "fullness of the blood stream."[47] Stretch or baroreceptors which sense changes in volume and relay information to the hypothalamo–neurohypophyseal–ADH system are mainly located in the left atrium and are sensitive to distension of the atrial wall. However, there is some evidence that receptors in the carotid sinus and aortic arch on the arterial side might be involved to a limited extent. When blood volume increases, the atrial wall is stretched and receptors in the wall are activated. This results in inhibition of ADH release and water diuresis ensues. The elevated blood volume is thus restored to normal. The afferent path from the volume receptors is via the vagus nerve to the medulla and thence by unknown pathways to the supraoptic region.

Some evidence exists that a "natriuretic hormone" is released when either blood volume or extracellular fluid volume is expanded; it then causes increased excretion of Na^+ in the urine. The site of production of this presumed hormone is not known, but it is thought to act by inhibiting tubular reabsorption of Na^+. The sites of the volume receptors are not known but have been postulated to be in the heart, thorax and/or liver. Receptors in the brain have also been postulated.

Two reservations must be stated about volume regulation through the ADH system. First, it is an emergency system which is called into play only in severe derangements. When it is active, osmolar regulation is sacrificed and, after severe fluid loss, extracellular fluid may become increasingly hypotonic. Second, since the regulation uses changes in free water, most of the volume change takes place in the cells, next in the interstitial fluid, and the least in the blood volume. Indeed, if one considers the factors involved in the exchange of fluid across capillaries, it becomes evident that if water is to be lost and plasma protein to be concentrated, the requirement for *any* sustained loss of volume from the vascular system is that tissue interstitial pressure fall sufficiently to balance the increase in plasma oncotic pressure.

It appears that the renin–angiotensin system provides a reasonable basis for the physiologic regulation of aldosterone secretion. The alterations of aldosterone secretion which result from changes in daily Na^+ intake are probably mediated through this negative feedback system and provide one of the important means for regulation of the extracellular fluid volume and composition. The release of ADH is regulated by changes in osmolarity of plasma and, under exceptional circumstances, by changes in volume. This system provides a sensitive mechanism for the regulation of body solute concentration. The ADH system can also be strongly influenced by higher nervous activity.

REFERENCES

1. Aikawa, J. K., Gordon, G. S., and Rhoades, E. L. Magnesium metabolism in human beings: Studies with Mg^{28}. *J. appl. Physiol.*, 1960, *15*, 503–507.
2. Barajas, L. Renin secretion: An anatomical basis for tubular control. *Science*, 1971, *172*, 485–487.
3. Barnes, B. A., and Mendelson, J. The measurement of exchangeable magnesium in dogs. *Metabolism*, 1963, *12*, 184–193.
4. Behnke, A. R., Jr., Feen, B. G., and Welham, W. C. The specific gravity of healthy men. *J. Amer. med. Ass.*, 1942, *118*, 495–498.
5. Bittar, E. E., ed. *Membranes and ion transport.* New York, Wiley-Interscience, 1970.
6. Brodie, B. B., Berger, E. Y., Axelrod, J., Dunning, M. F., Porosowska, Y., and Steele, J. M. Use of N-acetyl 4-aminoantipyrine (NAAP) in measurement of total body water. *Proc. Soc. exp. Biol. (N.Y.)*, 1951, *77*, 794–798.
7. Chamberlain, M. J., Fremlin, J. H., Peters, D. K., and Philip, H. Total body sodium by whole body neutron activation in the living subject: Further evidence for non-exchangeable sodium pool. *Brit. med. J.*, 1968, *2*, 583–585.
8. Chinard, F. P. Starling's hypothesis in the formation of edema. *Bull. N.Y. Acad. Med.*, 1962, *38*, 375–389.
9. Chow, S. Y., Jee, W., Taylor, G. N., and Woodbury, D. M. Radioautographic studies of inulin, sulfate and chloride in rat and guinea pig thyroid glands. *Endocrinology*, 1965, *77*, 818–824.
10. Comar, C. L., and Bronner, F., eds. *Mineral metabolism.* New York, Academic Press, 1962.
11. Cooke, R. E., Segar, W. E., Cheek, D. B., Coville, F. E., and Darrow, D. C. The extrarenal correction of alkalosis associated with potassium deficiency. *J. clin. Invest.*, 1952, *31*, 798–805.
12. Cotlove, E. Mechanism and extent of distribution of inulin and sucrose in chloride space of tissues. *Amer. J. Physiol.*, 1954, *176*, 396–410.
13. Cotlove, E., and Hogben, C. A. M. Chloride. In: *Mineral metabolism*, vol. II, part B, C. L. Comar and F. Bronner, eds. New York, Academic Press, 1962.
14. Cotlove, E., Holliday, M. A., Schwartz, R., and Wallace, W. M. Effects of electrolyte depletion and acid-base disturbance on muscle cations. *Amer. J. Physiol.*, 1951, *167*, 665–675.
15. Dosekun, F. O. The effect of alterations of plasma sodium on the sodium and potassium content of bone in the rat. *J. Physiol. (Lond.)*, 1959, *147*, 115–123.
16. Eckel, R. E., and Sperlakis, N. Membrane potentials in K-deficient muscle. *Amer. J. Physiol.*, 1963, *205*, 307–312.
17. Edelman, I. S., and Leibman, J. Anatomy of body water and electrolytes. *Amer. J. Med.*, 1959, *27*, 256–277.
18. Edelman, I. S., Olney, J. M., James, A. H., Brooks, L., and Moore, F. D. Body composition: Studies in the human being by the dilution principle. *Science*, 1952, *115*, 447–454.
19. Elkinton, J. R., and Danowski, T. S. *The body fluids.* Baltimore, Williams and Wilkins Co., 1955.
20. Fenn, W. O. Electrolytes in muscle. *Physiol. Rev.*, 1936, *16*, 450–487.
21. Forbes, G. B. Sodium. In: *Mineral metabolism*, vol. II, part B, C. L. Comar and F. Bronner, eds. New York, Academic Press, 1962.
22. Forbes, G. B., and Lewis, A. M. Total sodium, potassium and chloride in adult man. *J. clin. Invest.*, 1956, *35*, 596–600.
23. Freeman, F. H., and Fenn, W. O. Changes in carbon dioxide stores of rats due to atmospheres low in oxygen or high in carbon dioxide. *Amer. J. Physiol.*, 1953, *174*, 422–430.
24. Gardner, L. I., MacLachlan, E. A., and Berman, H. Effect of potassium deficiency on carbon dioxide, cation, and phosphate content of muscle. *J. gen. Physiol.*, 1952, *36*, 153–159.
25. Gauer, O. H., and Henry, J. P. Circulatory basis of fluid volume control. *Physiol. Rev.*, 1963, *43*, 423–481.
26. Ginsburg, J. M., and Wilde, W. S. Distribution kinetics of intravenous radiopotassium. *Amer. J. Physiol.*, 1954, *179*, 63–75.
27. Gitelman, H. J., and Welt, L. G. Magnesium deficiency. *Ann. Rev. Med.*, 1969, *20*, 233–242.
28. Halmi, N. S., Stuelke, R. G., and Schnell, M. D. Radioiodide in the thyroid and in other organs of rats treated with large doses of perchlorate. *Endocrinology*, 1956, *58*, 634–650.
29. Hartsuck, J. M., Johnson, J. E., and Moore, F. D. Potassium in bone: Evidence for a nonexchangeable fraction. *Metabolism*, 1969, *18*, 33–37.
30. Hastings, A. B. The electrolytes of tissues and body fluids. *Harvey Lect.*, 1941, *36*, 91–125.
31. Hastings, A. B., and Eichelberger, L. The exchange of salt and water between muscle and blood. I. The effect of an increase in total body water produced by the intravenous injection of isotonic salt solutions. *J. biol. Chem.*, 1937, *117*, 73–93.

32. Holland, W. C., and Auditore, G. V. Distribution of potassium in liver, kidney and brain of the rat and guinea pig. *Amer. J. Physiol.*, 1955, *183*, 309–313.

33. Huddart, H. The subcellular distribution of potassium and sodium in some skeletal muscles. *Comp. Biochem. Physiol.*, 1971, *38A*, 715–721.

34. Irvine, R. O. H., and Dow, J. W. Potassium depletion: Effects on intracellular *p*H and electrolyte distribution in skeletal and cardiac muscle. *Aust. Ann. Med.*, 1968, *17*, 206–213.

35. Irvine, R. O. H., Saunders, S. J., Milne, M. D., and Crawford, M. A. Gradients of potassium and hydrogen ion in potassium-deficient voluntary muscle. *Clin. Sci.*, 1961, *20*, 1–18.

36. Johnson, J. A. Capillary permeability, extracellular space estimation and lymph flow. *Amer. J. Physiol.*, 1966, *211*, 1261–1263.

37. Kruhøffer, P. Inulin as an indicator for the extracellular space. *Acta physiol. scand.*, 1946, *11*, 16–36.

38. Lawson, H. C. The volume of blood—a critical examination of methods for its measurement. *Handb. Physiol.*, 1962, sec. 2, vol. I, 23–49.

39. Leaf, A. Regulation of intracellular fluid volume and disease. *Amer. J. Med.*, 1970, *49*, 291–295.

40. MacIntyre, I., Hanna, S., Booth, C. C., and Read, A. E. Intracellular magnesium deficiency in man. *Clin. Sci.*, 1961, *20*, 297–305.

41. Manery, J. F. Water and electrolyte metabolism. *Physiol. Rev.*, 1954, *34*, 334–417.

42. Maurer, F. W. Isolation and analysis of extracellular muscle fluid from the frog. *Amer. J. Physiol.*, 1938, *124*, 546–557.

43. Nichols, G., Jr., Nichols, N., Weil, W. B., and Wallace, W. M. The direct measurement of the extracellular phase of tissues. *J. clin. Invest.*, 1953, *32*, 1299–1308.

44. Pace, N., Kline, L., Schachman, H. K., and Harfenist, M. Studies on body composition. IV. Use of radioactive hydrogen for measurement *in vivo* of total body water. *J. biol. Chem.*, 1947, *168*, 459–469.

45. Pace, N., and Rathbun, E. N. Studies on body composition. III. The body water and chemically combined nitrogen content in relation to fat content. *J. biol. Chem.*, 1945, *158*, 685–691.

46. Pappenheimer, J. R. Passage of molecules through capillary walls. *Physiol. Rev.*, 1953, *33*, 387–423.

47. Peters, J. P. *Body water.* Springfield, Ill., Charles C Thomas, 1935.

48. Pinson, E. A. Water exchanges and barriers as studied by the use of hydrogen isotopes. *Physiol. Rev.*, 1952, *32*, 123–134.

49. Pitts, R. F. *Physiology of the kidney and body fluids.* Chicago, Year Book Publishers, Inc., 1968.

50. Prentice, T. C., Siri, W., Berlin, N. I., Hyde, G. M., Parsons, R. J., Joiner, E. E., and Lawrence, J. H. Studies of total body water with tritium. *J. clin. Invest.*, 1952, *31*, 412–418.

51. Robin, E. D., Wilson, R. J., and Bromberg, P. A. Intracellular acid-base relations and intracellular buffers. *Ann. N.Y. Acad. Sci.*, 1961, *92*, 539–546.

52. Robinson, J. R. The active transport of water in living systems. *Biol. Rev.*, 1953, *28*, 158–194.

53. Robinson, J. R. Body fluid dynamics. In: *Mineral metabolism,* vol. I, part A, C. L. Comar and F. Bronner, eds. New York, Academic Press, 1960.

54. Schloerb, P. R., Friis-Hansen, B. J., Edelman, I. S., Solomon, A. K., and Moore, F. D. The measurement of total body water in the human subject by deuterium oxide dilution; with a consideration of the dynamics of deuterium distribution. *J. clin. Invest.*, 1950, *29*, 1296–1310.

55. Share, L., and Claybaugh, J. R. Regulation of body fluids. *Ann. Rev. Physiol.*, 1972, *34*, 235–260.

56. Sharp, G. W. G., and Leaf, A. Mechanism of action of aldosterone. *Physiol. Rev.*, 1966, *46*, 593–632.

57. Silver L., Robertson, J. S., and Dahl, L. K. Magnesium turnover in the human studied with Mg^{28}. *J. clin. Invest.*, 1960, *39*, 420–425.

58. Sjostrand, T. Blood volume. *Handb. Physiol.*, 1962, sec. 2, vol. I, 51–62.

59. Soberman, R., Brodie, B. B., Levy, B. B., Axelrod, J., Hollander, V., and Steele, J. M. The use of antipyrine in the measurement of total body water in man. *J. biol. Chem.*, 1949, *179*, 31–42.

60. Stein, J. H., and Ferris, T. F. The physiology of renin. *Arch. intern. Med.*, 1973, *131*, 860–872.

61. Steinbach, H. B. The prevalence of K. *Perspect. Biol. Med.*, 1962, *5*, 338–355.

62. Swan, R. C., Madisso, H., and Pitts, R. F. Measurement of extracellular fluid volume in nephrectomized dogs. *J. clin. Invest.*, 1954, *33*, 1447–1456.

63. Truax, F. L. The equality of the chloride space and the extracellular space of rat liver. *Amer. J. Physiol.*, 1939, *126*, 402–408.

64. Ussing, H. H., Kruhøffer, P., Thaysen, J. H., and Thorn, N. A. *The alkali metal ions in biology.* Berlin, Springer-Verlag, 1960.

65. Vernadakis, A., and Woodbury, D. M. Electrolyte and amino acid changes in rat brain during maturation. *Amer. J. Physiol.*, 1962, *203*, 748–752.

66. Vernadakis, A., and Woodbury, D. M. Electrolyte and nitrogen changes in skeletal muscle of developing rats. *Amer. J. Physiol.*, 1964, *206*, 1365–1368.

67. Wacker, W. E. C., and Parisi, A. F. Magnesium metabolism. *New Engl. J. Med.*, 1968, *278*, 658–663, 712–717, 772–776.

68. Waddell, W. J. Intracellular *p*H. *Physiol. Rev.*, 1969, *49*, 285–329.

69. Waddell, W. J., and Butler, T. C. Calculation of intracellular *p*H from the distribution of 5,5-dimethyl-2,4-oxazolidinedione (DMO). Application to skeletal muscle of the dog. *J. clin. Invest.*, 1959, *38*, 720–729.

70. Wallace, W. M., and Hastings, A. B. The distribution of the bicarbonate ion in mammalian muscle. *J. biol. Chem.*, 1942, *144*, 637–649.

71. Walser, M. Magnesium metabolism. *Ergebn. Physiol.*, 1967, *59*, 185–296.

72. Weil, W. B., Jr., and Wallace, W. M. The effect of alterations in extracellular fluid on the

composition of connective tissue. *Pediatrics,* 1960, *26,* 915–924.

73. Welt, L. G. *Clinical disorders of hydration and acid-base equilibrium.* 2nd ed. Boston, Little, Brown & Co., 1959.

74. White, H. L., and Rolf, D. Comparison of various procedures for determining sucrose and inulin space in the dog. *J. clin. Invest.,* 1958, *37,* 8–19.

75. Widdowson, E. M., McCance, R. A., and Spray, C. M. The chemical composition of the human body. *Clin. Sci.,* 1951, *10,* 113–125.

76. Wilde, W. S. Potassium. In: *Mineral metabolism,* vol. II, part B, C. L. Comar and F. Bronner, eds. New York, Academic Press, 1962.

77. Williams, J. A., Withrow, J. W., and Woodbury, D. M. Effects of nephrectomy and KCl on transmembrane potentials, intracellular electrolytes, and cell pH of rat muscle and liver *in vivo. J. Physiol. (Lond.),* 1971, *212,* 117–128.

78. Woodbury, D. M., and Koch, A. Effects of aldosterone and desoxycorticosterone on tissue electrolytes. *Proc. Soc. exp. Biol. (N.Y.),* 1957, *94,* 720–723.

79. Wynn, V. The osmotic behaviour of the body cells in man. *Lancet,* 1957, *273,* 1212–1217.

CHAPTER 27 BODY ACID-BASE STATE AND ITS REGULATION

by J. WALTER WOODBURY

For continued survival, an animal requires a steady intake of food, oxygen and other nutrients and a steady output of waste products, carbon dioxide and water. The intake and output of these substances are closely regulated to the needs of the body, and their total body contents remain constant over long periods. The lungs, kidneys, digestive tract and skin are the primary organs of intake and excretion; the activities of these organ systems are closely integrated, giving over-all regulation of body content of the many substances dissolved in the body fluids, e.g., sodium, potassium, chloride, hydrogen, bicarbonate, calcium, and phosphate ions. Bodily function is extremely dependent on the hydrogen ion concentration ($[H^+]$) of the cells and extracellular fluids; this concentration is closely regulated by the combined activities of the lungs and kidneys. Although the hydrogen ion concentration in the body fluids is a million times smaller than the concentration of sodium ions, $[H^+]$ must be closely regulated because of the extreme reactivity of H^+, particularly with proteins. Even at the low $[H^+]$ found in the body, slight changes dramatically affect enzyme activity because the association or dissociation of a hydrogen ion with an enzyme changes the charge distribution throughout the molecule and thus affects its rate of combination with substrate molecules. Hydrogen ions are much more reactive than sodium or potassium ions because the hydrated hydrogen ion (H_3O^+) is much smaller than the hydrated Na^+ and K^+ and hence is more strongly attracted to negatively charged regions of molecules; i.e., H^+ are generally much more tightly bound than Na^+ or K^+.

This chapter describes the mechanisms limiting the changes in the body's hydrogen ion concentration in response to changes in total H^+ content and the regulatory mechanisms which reduce the changes in content and eventually restore the balanced condition. These buffering and compensatory changes are complex, but a careful, step-by-step study of the components of the bodily response to acid-base changes can give a clear comprehension of the concepts involved and of the gaps in present knowledge of the subject. The fundamentals of acid-base regulation were elucidated by Van Slyke and his co-workers in the 1920's.

Davenport's *The ABC of Acid-base Chemistry*[19] is a detailed account of the chemical fundamentals (partial pressure, pH, equilibria) of transport of CO_2 by blood and a simplified account of bodily compensatory processes. There are many recent books, symposia and programmed books[5, 17, 25, 33, 37, 39, 56, 68] on acid-base regulation, giving considerable choice of viewpoint and specific subject matter. The presentation of acid-base data on the $[HCO_3^-]$-pH diagram, as espoused by Davenport and used by many earlier workers,[61, 63] simplifies interpretation of the data, compactly summarizes many regulatory processes and is used throughout this chapter. Detailed understanding of the concept of buffering requires the use of the equations for chemical equilibrium and also of buffering. However, their use is kept to a minimum and is always carefully explained. The primary emphasis is on interpretation of acid-base phenomena from plots of points representing the pH, $[HCO_3^-]$, and P_{CO_2} of blood plasma on the $[HCO_3^-]$-pH diagram. The reader seeking a primarily verbal description may find the books by Frisell[25] and Hills[33] helpful.

The sequence of development of the material is (i) buffering, (ii) the CO_2 system, (iii) titration of body fluid compartments, (iv) physiological compensatory mechanisms, (v) regulation of acid-base state by the body and (vi) problems on the aforementioned topics and their answers.

Hydrogen Ion Balance. The body's total H^+ in normal conditions is such that the $[H^+]$ of the plasma is close to 40 nanomols per liter (40×10^{-9} mols per liter, pH = 7.40) and that of cells is about 100 nmols per liter (pH = 7.0). An increased intake or production of acid in the body increases these concentrations, but much less than might be expected from the amount of acid added; i.e., most of the added H^+ combine with bicarbonate, protein and other substances in the body, disappear as separate ions, and thus do not contribute to the concentration of H^+ ions in the body. Nevertheless, the restoration of normal conditions requires that exactly the amount of acid that was added be removed from the body. In those clinical situations in which the doctor must act to correct acid-base disturbances, he must be able to estimate how much excess acid or base the body has. In theory, the most accurate way of obtaining such information is from a balance study, *i.e.*, a careful accounting of the intake, production and excretion of acid:

Increase = Income + Production – Outgo

However, such measurements are technically difficult, and some of them must be made *before* the patient becomes ill. However, the clinician usually must rely on measurements made after the patient becomes ill. The only measurements made routinely in the clinic on a patient's acid-base status are on the blood. This information together with other clinical information on the nature of the disease is used to diagnose the primary cause of the acid-base disturbance and the nature and amount of corrective therapy required, *e.g.*, intravenous infusion or oral administration of base or acid. Estimates of body H^+ unbalance derived from measurements on blood can be reasonably accurate if the principles involved are understood, *e.g.*, the buffering properties of the body as a whole and the time course of the transient changes in H^+ content of the various body compartments.

Although the fundamentals of acid-base chemistry of the blood were worked out 30 to 40 years ago, the conceptual framework presented in this chapter[69] represents a considerable enlargement of the usefulness of the $[HCO_3^-]$-pH diagram in the interpretation of acid-base disturbances; well established facts are integrated with the newer knowledge of membrane ion transport, interchange of carbonate between bone and blood and renal and respiratory functions to provide a cohesive view of acid-base balance in the body. Further, the distinction between transient and steady-state H^+ ion unbalances is considered and new ways to distinguish between them are developed and explained.

Carbonic and Noncarbonic Acids. There are two classes of acids of physiologic importance, carbonic acid and noncarbonic acids. These are distinguished on the basis of their mode of excretion: Carbonic acid is excreted via the lungs in the dehydrated form, CO_2. Noncarbonic acids are any other acids which do not dehydrate to form a volatile product and must, therefore, be excreted via the kidneys. Carbon dioxide and water are the most abundant end-products of me-

tabolism and CO_2 is excreted, on the average, as rapidly as formed. The metabolism of exogenous protein produces sulfuric and phosphoric acids which must be excreted by the kidneys as rapidly as formed. Other noncarbonic acids are the products of intermediary metabolism, *e.g.*, lactic and acetoacetic acids. These are normally metabolized as rapidly as produced, but unbalances occur in exercise and other forms of tissue hypoxia and in diabetes.

Résumé of Compensatory Responses. The contents of this chapter can be outlined by briefly describing the sequence of changes which occur in the body in response to a sudden change in the body's H^+ balance. The response depends on whether the added acid is carbonic or noncarbonic, but the principles involved are better illustrated by considering what happens when the body has an excess of noncarbonic acid. In the clinic, such changes are commonly found to be associated with various types of pathologic conditions and illness.

The change in $[H^+]$ of the blood plasma after an intravenous infusion of a strong acid such as HCl follows an ever slowing time course: (i) The added acid largely disappears, first reacting with the hemoglobin, plasma proteins and bicarbonate of blood, but a small fraction of the added H^+ remains in the ionized form and raises $[H^+]$. This phase takes a few circulation times (a few minutes). As the blood circulates through the lungs, the excess CO_2 produced by the reaction of H^+ with HCO_3^- is excreted, reducing the $[H^+]$ slightly. (ii) The $[H^+]$ of the blood is further reduced in the tissues by diffusion of HCO_3^- (bicarbonate) ions into the blood from the interstitial spaces. The CO_2 thus produced by combination with H^+ is eliminated in the lungs. This interstitial buffering is nearly complete in 15 min but there is a slower component that may take up to 2 hours. (iii) Some of the excess H^+ enters muscle cells, reducing the excess in the extracellular space. The H^+ which enters cells largely disappears by combining with cellular proteins, phosphates and bicarbonate. The time course and extent of this process is not accurately known. (iv) The raised $[H^+]$ of the blood increases respiratory minute volume, which further tends to eliminate CO_2 via the lungs. This removal of CO_2 reduces the $[H^+]$ of the blood and thus compensates, partially, for the excess noncarbonic acid in the blood. The time scale is minutes to hours. (v) Over a period of hours and days, carbonate ($CO_3^=$) is released from bone and combines with some of the extra H^+ ions to form HCO_3^-. (vi) If kidney function is normal, *i.e.*, if the original disturbance is not due to a pathologic change in the kidney, the rate of acid secretion by the kidneys increases and the excess acid is excreted in about a week.

$[H^+]$ and pH. A current trend in the clinical literature is to express acidity in terms of $[H^+]$ rather than pH. By definition, $pH = -\log a_H$, where a_H is the hydrogen activity. $[H^+]$ is related to a_H by an unknown multiplicative factor and since pH is the quantity actually measured, a_H but not $[H^+]$ can be calculated. When $[H^+]$ is given, it is assumed equal to the hydrogen activity (see discussion of pH measurement). The use of pH terminology was originally introduced and permits convenient specification of the wide ranges of $[H^+]$ encountered in chemical experiments, *e.g.*, from 1 to 10^{-14} mols per liter, pH's of 0 to 14. The range of blood $[H^+]$ compatible with life is about 20 to 160 nmols per liter (pH 6.8 to 7.7), sufficiently small that $[H^+]$ can be used conveniently. Since concentration is a familiar concept, using $[H^+]$ routinely rather than pH would seem natural. However, as pointed out by Van Slyke[58] in 1922, pH must be used to give a clear, precise meaning to the general concept of buffering as shown in the discussion of titration curves. From this viewpoint, it is most convenient to interpret acid-base data using the $[HCO_3^-]$-pH diagram.

BUFFERS AND BUFFERING[13]

Electrolytes. Substances which ionize in water solution and hence can conduct an electrical current are called electrolytes. The ionization process consists of the donation of all valence shell electrons by the cation to complete the valence shell of the anion. Electrolytes dissolve in water solutions because their charged ions have higher binding energies to the dipolar water molecules than they do to each other. *Strong electrolytes* are completely ionized in solution, *e.g.*, the reaction $NaCl \rightleftharpoons Na^+ + Cl^-$ goes completely to the right. Although strong electrolytes are completely dissociated, they act as if they were only about 80 per cent ionized because of the mutual electrostatic attractions of the anions and cations; *i.e.*, activities are about 80 per cent of concentrations. Usually, in physiologic systems, it is sufficiently accurate to use the terms "concentration" and "activity" interchangeably. Most soluble salts and inorganic acids and bases are strong electrolytes. The ions found in plasma are formed from strong electrolytes, with a few exceptions, *e.g.*, calcium, phosphate, protein.

Weak electrolytes ionize only partially; the anions and cations have about the same affinity for each other as they have for water. Thus, the concentrations of neutral molecules and ions in solution are comparable. The most important weak electrolytes are weak acids and bases, such as metabolic acids and ammonia.

Acids and Bases. Brønsted defined an acid as a substance which can supply H^+ (protons) and a base as a substance which can accept H^+. The reaction $HCl \rightleftharpoons H^+ + Cl^-$ is characterized in the Brønsted formulation by calling HCl an acid since it can supply H^+; Cl^- is called the conjugate base since it can, at high $[H^+]$, accept H^+. A strong acid has a weak conjugate base. Similarly, a strong base has a weak conjugate acid. Table 27–1 gives several acids and their conjugate bases. Substances of particular interest are those which are more or less equally divided between the acidic and basic forms at body $[H^+]$. Thus, the imidazole side groups of hemoglobin undergo the following reaction: $HHb \rightleftharpoons H^+ + Hb^-$; the acid, HHb, dissociates to form H^+ and the conjugate base, Hb^-. HHb and Hb^- occur in about equal concentrations in the blood cells. This system acts like a base in respect to added acid and like an acid to added alkali.

Buffering. The property called buffering

Table 27–1 *Acids and Conjugate Bases*

ACID = H^+ + CONJUGATE BASE
$HCl = H^+ + Cl^-$
$HCH_3COO = H^+ + CH_3COO^-$
$H_2CO_3 = H^+ + HCO_3^-$
$HCO_3^- = H^+ + CO_3^=$
$NH_4^+ = H^+ + NH_3$
$HOH = H^+ + OH^-$

arises from the properties of weak acids and bases, *i.e.*, those that are incompletely dissociated at the [H+] under consideration. A solution consisting of roughly equal parts of a weak acid and a strong salt of the conjugate base anion (*e.g.*, acetic acid and Na+ acetate) contains molecules that can both supply and absorb H+ ions. Hence, adding strong acid or alkali produces a smaller change in [H+] than would occur if the weak acid and conjugate base were not present. The potential number of H+ is not changed but the number in the ionized form is reduced. This "masking" of H+ is called *buffering*. Nearly all of the H+ added in the form of strong acid combine with the ionized form of the weak acid. The amount of the buffer anion (or basic form of the weak acid) which is bound is nearly equal to the amount of acid added, as long as the added amount is less than that of the buffer anion present. *The near equality of added strong acid and amount of buffer anion disappearing is assumed in all the calculations given below. This assumption is, in fact, the central one in making buffer calculations.*

DISSOCIATION CONSTANT. The law of mass action describes the equilibrium condition of a chemical reaction. The condition of equilibrium for the reaction, $HA \rightleftharpoons H^+ + A^-$, where A^- is generally an organic anion in biologic systems, is given by the equation

$$K_a' = \frac{[H^+]\,[A^-]}{[HA]} \qquad (1)$$

where K_a' is the equilibrium or ionization constant characteristic of the particular reaction. The subscript "a" refers to "acid," and the prime sign indicates that concentrations rather than activities are used; hence K_a' varies somewhat with ionic strength. Since the ionic strength of body fluids is closely regulated, K_a' normally has one value for each reaction. (A similar equation holds for weak bases, poorly ionizing substances in which the basic form has no charge, *e.g.*, NH_4^+. Since the principles involved are identical, the behavior of weak bases will not be specifically discussed.) By convention, concentrations are in mols per liter (molar, M) or millimols per liter (millimolar, mM). In the body, the principal weak acid buffering substances are proteins, particularly hemoglobin and

organic and inorganic phosphates. Bicarbonate also acts as a buffer at body *p*H if the partial pressure of CO_2 (P_{CO_2}) is held approximately constant. Since the total amounts and concentrations of body buffers remain relatively constant in the steady state and since the body's total buffer concentration, [B], must consist of the sums of the concentrations of the acid and conjugate base forms, [B] = [HA] + [A−]. If [HA] = [A−], [HA]/[A−] = 1 and from equation (1), [H+] = K_a'. Maximum buffering action occurs when [H+] is near the value of K_a' so that weak acids and bases with dissociation constants of about 10^{-7} mols per liter are most effective at moderating [H+] changes due to the addition of strong acid or alkali to the body fluids.

CALCULATION OF [H+] IN BUFFERED SOLUTIONS. Equation (1) can be used to calculate the changes in [H+] produced by adding a given amount of strong acid. Rearrangement gives

$$[H^+] = K_a' \frac{[HA]}{[A^-]} \qquad (2)$$

However, [HA] + [A−] = [B]. Eliminating [HA] in equation (2) gives the relationship between [H+] and [A−]:

$$[H^+] = K_a' \frac{([B] - [A^-])}{[A^-]} \qquad (3)$$

The assumption stated above that all but a small fraction of added acid disappears by combining with A^- gives the relationship between [H+] and the amount of acid added.

Example: Since skeletal muscle contains considerable concentrations of organic and inorganic phosphates, it is of interest to calculate the buffer properties of a Na_2HPO_4, NaH_2PO_4 solution containing 20 mmols per liter of phosphate. The reaction is $H_2PO_4^- \rightleftharpoons H^+ + HPO_4^=$ and so

$$K_a' = \frac{[H^+] \times [HPO_4^=]}{[H_2PO_4^-]}$$

At the ionic strength of body fluids, $K_a' = 1.5 \times 10^{-7}$ mols per liter (150 nmols per liter). Suppose that initially $[HPO_4^=] = [H_2PO_4^-] = 10$ mmols per liter so that $[H^+] = K_a' = 150$ nmols per liter [equation (2)]. What is the final [H+] of the solution if 5 mEq of concentrated HCl of negligible volume is added to a liter of the solution? What fraction of the added acid remains ionized?

Solution: The reaction is $H^+, Cl^- + 2\,Na^+,$ $HPO_4^= \rightleftharpoons Na^+, Cl^- + Na^+, H_2PO_4^-$. If all of the added HCl combines with $HPO_4^=$, then $[HPO_4^=]$ decreases from 10 to 5 mmols per liter and $[H_2PO_4^-]$ increases from 10 to 15 mmols per liter. Substituting these values in equation (2) gives

$$[H^+] = K'_a\,\frac{[H_2PO_4^-]}{[HPO_4^=]} = 150\frac{15}{5}$$
$$= 450 \text{ nmols per liter.}$$

Thus the $[H^+]$ has increased approximately $450 - 150 = 300$ nmols per liter. Five mmols per liter = 5,000,000 nmols of H^+ per liter were added and only 300 nmols (one in 17,000) remained in the ionic form after buffering. This calculation illustrates vividly the validity of the assumption that the change in buffer anion is equal to the added acid; the error is negligible, as can be seen by noting that the actual $[H_2PO_4^-]$ after adding acid is close to 15,000,000 − 300 = 14,999,700 nmols per liter and $[HPO_4^=]$ is close to 5,000,300 nmols per liter. The ratio of these two quantities is 2.99976, a number negligibly different from 15/5 = 3. Similarly the change in H^+ ion concentration is 299.964 instead of 300 nmols per liter.

Dilution. Diluting a solution containing buffer has little effect on $[H^+]$. For example, if the solution in the example were diluted with an equal volume of water, all concentrations would be halved but only momentarily since this is not an equilibrium condition. Consequently, nearly 75 nmols of $H_2PO_4^-$ dissociate to raise the $[H^+]$ from 75 to 150 nmols per liter. This small change in $[H_2PO_4^-]$ has a negligible effect on $[H_2PO_4^-]/[HPO_4^=]$ and hence on $[H^+]$.

TITRATION CURVES. The buffering properties of weak acids and bases are illustrated by graphs, called titration curves, showing the relationship between the amount of acid added and the resulting change in $[H^+]$ of the solution. Contrary to the usual convention in making graphs, it is customary to plot the independent variable, the amount of added acid, on the ordinate and the pH on the abscissa because the slope of the resulting curve is directly related to buffer value instead of inversely related as would be true of the conventional plot. Figure 27–1 shows two ways of plotting a titration curve, using $[H^+]$ and pH as the ordinate for the $HPO_4^=, H_2PO_4^-$ buffer sys-

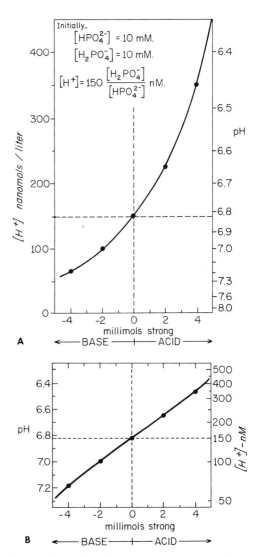

Figure 27–1 Titration curves of 1 liter of a 20 mmol per liter solution of NaH_2PO_4, Na_2HPO_4. Initially, $[H_2PO_4^-] = [HPO_4^=] = 10$ mmols per liter and $[H^+] = 150$ nEq per liter. Points are from Table 27–2. Change in volume due to added acid is small enough to be neglected. *A*, $[H^+]$ plotted on linear ordinate scale (left) and corresponding pH (right). *B*, pH plotted linearly (left) with corresponding $[H^+]$ (right). Note that titration curve plotted this way is nearly linear.

tem considered in the example. Additional points for the plot are obtained in the same manner; for different amounts of added acid or base, the resulting $[H^+]$ is calculated from $[H_2PO_4^-]/[HPO_4^=]$ (equation 3). The pertinent values, including pH, are given in Table 27–2 and are plotted in Figures 27–1 and 27–2. In Figure 27–1A, the amount of acid or base added has been plotted on the abscissa with the resulting $[H^+]$ in the

Table 27–2 *Titration of a 1 Liter Solution of NaH_2PO_4–Na_2HPO_4:*
H^+, $Cl^- + 2\ Na^+$, $HPO_4^= \rightleftharpoons Na^+\ Cl^- + Na^+$, $H_2PO_4^-\ Na^+$, $OH^- + Na^+$, $H_2PO_4^- \rightleftharpoons HOH + 2\ Na^+$, $HPO_4^=$ Initially $[H_2PO_4^-] = [HPO_4^=] = 10$ mmols per liter

	ADDED° MMOLS PER LITER	$[H_2PO_4^-]$ MMOLS PER LITER	$[HPO_4^=]$ MMOLS PER LITER	$\dfrac{[H_2PO_4^-]}{[HPO_4^=]}$	$[H^+]$† NMOLS PER LITER	LOG $\dfrac{[H_2PO_4^-]}{[HPO_4^=]}$	pH §
	−10	0.0366	19.9634	0.00183	0.274‡	−2.74	9.56
B	− 9	1.0	19.0	0.053	7.9	−1.28	8.10
A	− 8	2.0	18.0	0.11	16.6	−0.96	7.78
S	− 6	4.0	16.0	0.25	37.5	−0.60	7.42
E	− 4	6.0	14.0	0.43	65	−0.37	7.19
	− 2	8.0	12.0	0.66	100	−0.18	7.00
	0	10.0	10.0	1.0	150	0.00	6.82
	2	12.0	8.0	1.5	225	0.18	6.65
A	4	14.0	6.0	2.33	350	0.37	6.46
C	6	16.0	4.0	4.0	600	0.60	6.22
I	8	18.0	2.0	9.0	1350	0.95	5.87
D	9	19.0	1.0	19.0	2850	1.28	5.54
	10	19.9454	0.0546	365	54600‡	2.56	4.26

°Concentrated strong acid or base of negligible volume is added, *i.e.*, 5 ml of 2 M acid contains 10 mmols. Except for additions of ±10 mmols, change in $[H_2PO_4^-]$ = amount acid added.

†$[H^+]$ is given in units of nmols per liter (10^{-9} mols per liter) $[H^+] = 150\ \dfrac{[H_2PO_4^-]}{[HPO_4^=]}$.

‡Equation (3) cannot be used to calculate these values.

§Since $[H^+]$ is in nmols per liter, $pH = 9 - \log [H^+] = 6.82 + \log \dfrac{[HPO_4^=]}{[H_2PO_4^-]}$.

solution on the ordinate. Note that the units on the ordinate are a million times smaller (nmols per liter) than those on the abscissa (mmols per liter). This scale difference serves to emphasize the small fraction of added H^+ that remains in the ionized form after buffering. The relationship is not linear: Adding 2 mmols of strong acid per liter increases the $[H^+]$ from 150 to 225 nmols per liter, $3/2$ times the original value of 150 nmols per liter; adding 2 mmols of base per liter decreases $[H^+]$ to 100 nmols per liter, a decrease to $2/3$ the original. Thus, the fractional or percentage changes in $[H^+]$ are related since successive 2 mmol per liter increments increase $[H^+]$ by 50 per cent, *i.e.*, from 100 to 150 and 150 to 225 nmols per liter. This relationship holds only over the center range of acid or base additions (−4 to +4 mmols) but shows that using log $[H^+]$ or pH instead of $[H^+]$ as one of the coordinates will give a linear titration curve over part of the range.

Figures 27–1B and 27–2 show titration curves of the same solution with pH as one coordinate. In Figure 27–1B, the scale on the abscissa is the same as in Figure 27–1A, but the ordinate is pH. The same range

of $[H^+]$ values is covered in both parts of Figure 27–1. Over the same range of added acid, the titration curve is almost straight when pH is the ordinate. In Figure 27–2,

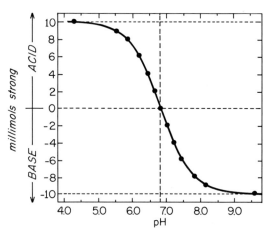

Figure 27–2 Titration curve of 1 liter of a 20 mmol per liter NaH_2PO_4, Na_2HPO_4 solution. Same as Figure 27–1B except that ordinate and abscissa have been interchanged and a broader range of pH's is covered. Slope of the curve at any point is the negative of the buffer value. Buffer value is maximum, having a value of 11.5 sl, at vertical dashed line. Points are from Table 27–2.

pH is plotted on the abscissa and amount of acid on the ordinate, the same as in Figure 27–1B except for an interchange of coordinate axes and a wider range of values. Over this wider range, it can be seen that the titration curve is S-shaped, the pH changing rapidly for small additions of acid or base at the high and low ends as expected when the buffer is mostly converted to the acidic or basic forms.

Buffer Value. When amount of acid (or base) added is plotted against pH, the slope of the titration curve at any point is a direct measure of the buffering capability of a solution at that pH. The slope is the amount of acid or base that must be added to produce a pH change of 1 unit (a.10-fold change in $[H^+]$) if the titration curve were a straight line. *The buffer value of a solution at any pH is defined as the negative of the slope of the titration curve at that pH* (cf. Fig. 27–2). The steeper the curve, the greater the amount of acid which must be added to produce a given change in pH. This is the reason that pH is plotted on the abscissa; buffer value equals the magnitude of the slope instead of its reciprocal. Since buffering is the body's first line of defense against the pH changes caused by added acid or alkali, it is important—and natural—to represent the body's acid-base state in a way that can relate it to titration curves. Hence, *the linear midregion of the titration curve is a preemptory reason for using pH instead of $[H^+]$ in describing acid-base behavior in buffered systems such as the body.*

The maximum buffer value of the 20 mmols per liter phosphate solution whose titration curve is shown in Figures 27–1B and 27–2 can be accurately calculated from Table 27–2. The addition of −2 mmols acid per liter (addition of base) increases the pH of the solution from 6.82 to 7.00; the addition of 2 mmols of acid per liter decreases the pH to 6.65. Hence, the buffer value defined as the negative of the slope of the titration curve is

$$\text{buffer value} = -\frac{(-2-2) \text{ mmols per liter}}{(7.00 - 6.65) \text{ pH units}}$$
$$= \frac{4.0 \text{ mmols}}{0.35 \text{ liter} \times \text{pH units}}$$
$$= 11.4 \text{ mmols per (liter} \times \text{pH)}$$

In words, 11.4 mmols of acid per liter must be added to the 20 mmols per liter phosphate solution to decrease the pH by 1 unit if the titration curve remained straight over a range of one pH unit or more.

The units of buffer value (mmols base/liter), per pH unit are clumsy and lengthy and, since they are used frequently throughout this chapter, it is useful to define a new unit of buffer value. In view of D. D. Van Slyke's major pioneering contributions in this field, especially his analysis of buffering,[58] it is appropriate and fitting to call this new unit of buffer value a Van Slyke or more simply a *slyke*, abbreviated *sl*, e.g., the maximum buffer value of a 20 mmols per liter phosphate solution is 11.5 slykes or 11.5 sl.

Another useful concept in describing the properties of buffered solutions is Van Slyke's *molar buffer value*,[58] defined as the buffer value of a 1 molar (1 mol per liter) solution, i.e., the number of mols (mmols) of acid required to reduce the pH of a 1 molar (1 mmolar) solution by 1 pH unit. In the example given above, the buffer value of a 20 mmols per liter phosphate solution is 11.5 sl, so that the molar buffer value is 11.5 (mmols per liter) per pH divided by 20 mmols per liter = 11.5/20 per pH = 0.57 per pH unit. More accurately, the maximum buffer value of any weak acid or base is 0.575 per pH unit for each independently ionizing group on the molecule regardless of its structure or K'_a.

Body Buffers. Considerable amounts of acid or base can be added to the body without producing lethal changes in pH because of the body buffer stores. Most of the body buffers are located intracellularly, the major exceptions being plasma protein and the small amount of phosphate in extracellular fluid. However, the hemoglobin of red cells, though intracellular, is more conveniently regarded as extracellular because it is confined to the blood and because it is readily available for buffering of extracellular acids. The buffer substances in cell fluids are the various phosphates, previously mentioned, and organic cations and anions, including proteins, which have ionizable groups with dissociation constants near 10^{-7}, i.e., between 10^{-6} and 10^{-8}. (See Chap. 26, Vol. II.) Since skeletal muscle is about half of the cellular mass, most intracellular buffering presumably occurs in muscle. Carbon dioxide readily penetrates cell membranes and thus carbonic acid is buffered by the whole body. On the other hand, noncarbonic acids penetrate cell membranes slowly so that an excess of acid in the extracellular fluid is initially buffered there. Hence, the buffering properties of hemoglobin must be known to calculate the amount of acid that has been added to the extracellular fluid by a disease process from the deviation in blood pH from its normal value of about 7.4.

HEMOGLOBIN.[47] Hemoglobin is specialized for carrying O_2 and CO_2 and also has a high buffer value over the normal range of body pH values. As was seen in Chapter 21, Volume II, these properties interact (Bohr effect) to reduce the changes that occur in the pH of blood as it releases O_2 and takes up CO_2 in the tissues and in the reverse process in the lungs. Hemoglobin is a tetrahedral molecule consisting of four rather tortuous globin chains each enfolding a heme group. There are two types of chains, alpha and beta, which are only slightly different from each other in structure. A hemoglobin molecule consists of a pair of alpha and a pair of beta chains, the members of each pair being placed oppositely across an axis of symmetry. Each chain is capable of taking up one molecule of O_2, which attaches to the iron in the center of the heme group. Complete oxygenation is associated with a displacement of the two beta chains toward each other.[47] The molecular weight of each chain is about 16,700,[20] one-fourth that of the whole molecule. It is customary in acid-base chemistry to take the oxygen combining weight, the weight of one chain, as the molecular or combining weight of hemoglobin. This custom will be followed here.

There are several kinds of dissociable groups in proteins which act as acids or bases of different strengths, e.g., free carboxyl, free amino, guanidino and imidazole groups. Of these, only the latter has a dissociation constant in the range of normal $[H^+]$ of blood; e.g., the dissociation constant of the carboxyl group of histidine is more than 10^{-2} mols per liter; that of the imidazole ring is about 10^{-6} mols per liter and that of the amino group is less than 10^{-9} mols per liter.[20] It must be kept in mind, however, that incorporating an amino acid into a protein can greatly modify the ionization constants of any remaining free groups in a manner which depends on the adjacent amino acid subunits.

Figure 27–3 shows idealized titration

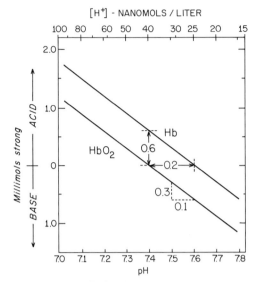

Figure 27–3 Idealized titration curves of 1 mmol per liter solutions of oxyhemoglobin (HbO_2) and hemoglobin (Hb). The molar buffer value of hemoglobin in this range is $0.3/0.1 = 3$ per pH. Oxygenation of Hb decreases pH by 0.2 unit or requires addition of 0.6 mmols of alkali to prevent change in pH. (After Wyman and German, *J. biol. Chem.*, 1937, *117*, 533–550; and Wyman, *Advanc. Protein Chem.*, 1948, *4*, 407–513.)

curves of 1 mmol per liter solutions of oxyhemoglobin and hemoglobin over the pH range 7 to 7.8.[26] The curves are nearly straight and parallel over this range and will be considered as such. These curves are more nearly linear than the titration curve of phosphate (Fig. 27–2) because of the existence of several ionizable groups with slightly different K_a' values; as the buffer value of one group falls off when pH is changed, the buffer value of another group with a different K_a' may be increasing.

The molar buffer value of hemoglobin is given by the magnitude of the slope of the curves in Figure 27–3 and is 3 per pH. This buffer value is almost entirely due to imidazole rings of histidine subunits in the hemoglobin. Each of the four chains of hemoglobin has nine histidine subunits. Since the maximum molar buffer value of a single ionizable group is 0.575 per pH, the molar buffer value of 3 per pH must arise from a minimum of $3/0.575 = 5.2$ dissociable groups. Hence, at least six and probably all nine of the histidines in a chain act as buffers in the 7 to 7.8 pH range since their dissociation constants are unlikely to be equal and the buffer value contributed by each group will be less than the maximal value of 0.575. The reaction of the imidazole ring of histidine with hydrogen ions is

$$H^+ + N\overset{\displaystyle C\!\!-\!\!H}{\diagup\diagdown} NH \rightleftharpoons$$

(basic form)

(acidic form)

the neutral form being the hydrogen ion acceptor (conjugate base).

H+ Liberating Effect of Hemoglobin Oxygenation. Oxygenation of hemoglobin acts as though it changes the dissociation constant of one of the nine histidines in a hemoglobin chain, making it a stronger acid; *i.e.*, at a given *p*H, the fraction of histidines in the uncharged, basic form increases by dissociating H+. The result is that 0.6 mmol of strong base must be added for each mmol of hemoglobin to prevent a *p*H shift on oxygenation (Fig. 27–3). Conversely, if no base is added the *p*H shifts 0.2 unit downward when hemoglobin is oxygenated. The size of these changes is reduced, however, by carbaminohemoglobin formation if CO_2 is present in the system[53] (see below).

THE CARBON DIOXIDE SYSTEM[24]

There are two distinct aspects to the role of carbon dioxide in the body: (i) Carbon dioxide produced by metabolism acts as a strong acid at body [H+] because of the reactions $H_2O + CO_2 \rightleftharpoons H_2CO_3 \rightleftharpoons H^+ + HCO_3^-$. Carbon dioxide as such is not an acid since it contains no protons; H_2CO_3 is. On the average, the rate of CO_2 excretion must equal its rate of production. (ii) The partial pressure of CO_2 (P_{CO_2}) in the alveoli, blood and tissues is determined by the ventilatory rate of the lungs. The normal maintenance of a fixed alveolar P_{CO_2} ($P_{A_{CO_2}}$) means that

the CO_2 system acts like an excellent buffer with respect to noncarbonic acids or bases which may be added to the blood by diet, metabolism or disease.

Carbon Dioxide System: Equilibria and Kinetics. DISSOLVED CARBON DIOXIDE. All gases dissolve to some extent in water. At pressures less than 1 atmosphere the concentration of dissolved gas is proportional to the partial pressure of the gas in a vapor phase with which the water is equilibrated, *e.g.*, by vigorous bubbling of the gas through the solution. The partial pressure of CO_2 in blood is thus defined as the pressure of CO_2 in a vapor phase with which the blood is equilibrated. Since arterial blood is usually equilibrated with alveolar air, the P_{CO_2} of arterial blood (Pa_{CO_2}) is equal to $P_{A_{CO_2}}$, the partial pressure of CO_2 in alveolar air. The concentration of CO_2 dissolved in blood plasma, denoted by $[CO_2]_{dis}$, is proportional to the partial pressure of CO_2 in the plasma. If Pa_{CO_2} is given in Torr (mm Hg), then $[CO_2]$ in mmols per liter is[62]

$$[CO_2]_{dis} = 0.03 \; P_{CO_2}$$

The P_{CO_2} of alveolar air is about 40 Torr; hence $[CO_2]_p = 0.03 \times 40 = 1.2$ mmols per liter for arterial plasma. The value is somewhat higher for venous plasma.

HYDRATION OF CARBON DIOXIDE. Carbon dioxide reacts reversibly with water to form carbonic acid, H_2CO_3. The equilibrium of the reaction $H_2O + CO_2 \rightleftharpoons H_2CO_3$ is far to the left and is attained slowly. At body temperature there are about 500 CO_2 molecules in solution for every H_2CO_3 molecule. The equilibrium condition for this hydration reaction is $K_h = [H_2CO_3]/[CO_2] \times [H_2O]$. Since the concentration of water in plasma remains nearly constant, the product $K_h \times [H_2O]$ can be regarded as a constant. Defining $K_h' = K_h[H_2O]$, the equilibrium conduction for the hydration of CO_2 becomes $K_h' = [H_2CO_3]/[CO_2]$. Thus $[H_2CO_3] = K_h' [CO_2]$, *i.e.*, the concentration of carbonic acid in water is directly proportional to the concentration of dissolved CO_2, which in turn is proportional to P_{CO_2}. Since both $[H_2CO_3]$ and $[CO_2]_{dis}$ are proportional to P_{CO_2}, it is customary to combine them: $[CO_2] = [CO_2]_{dis} + [H_2CO_3]$. At equilibrium, $[H_2CO_3]$ is only about 0.002 of $[CO_2]_{dis}$[20] and so the inclusion of $[H_2CO_3]$ has no appreciable effect on the proportionality constant between [CO$_2$] and P$_{CO_2}$. Hence

$$[CO_2] = 0.03 \ P_{CO_2} \qquad (4)$$

The hydration of CO$_2$ is slow; at 38° C about 5 sec are required for half equilibration.[20, 29] The reverse reaction, dehydration of H$_2$CO$_3$, is about 500 times faster since the equilibrium constant is the ratio of the backward and forward rate constants. The uptake of CO$_2$ in the tissues and its loss in the lungs are dependent on the rapidity of the hydration and dehydration reactions of CO$_2$ since blood spends only a few seconds in the tissue and lung capillaries. These kinetic considerations led to the discovery of an enzyme, carbonic anhydrase (CA), by Roughton which catalyzes the reaction

$$CO_2 + H_2O \overset{CA}{\rightleftharpoons} H_2CO_3$$

Carbonic anhydrase is found in high concentrations in red blood cells, ion secreting tissues and neuroglia. In blood, CA speeds up the reactions sufficiently to make diffusion the factor limiting CO$_2$ exchange in tissue and lungs.

IONIZATION OF CARBONIC ACID. Carbonic acid dissociates rapidly to form hydrogen ions and bicarbonate ions: H$_2$CO$_3$ = H$^+$ + HCO$_3^-$. The equilibrium condition is

$$K_{H_2CO_3} = \frac{[H^+] \, [HCO_3^-]}{[H_2CO_3]} \qquad (5)$$

The value of K$_{H_2CO_3}$ in dilute solutions is 1.6×10^{-4} mols per liter at 38° C.[20] Hence, carbonic acid is almost completely dissociated in the body where [H$^+$] = 4×10^{-8} mols per liter. It is more convenient to write the equilibrium equation for the overall reaction with [H$_2$CO$_3$] replaced by [CO$_2$] in equation (5). The reaction and equilibrium condition are

$$CO_2 + \dot{H}_2O = H_2CO_3 = H^+ + HCO_3^-$$

and

$$K_a' = \frac{[H^+] \, [HCO_3^-]}{[CO_2]} \qquad (6)$$

In plasma at 38° C, K$_a'$ = 8×10^{-7} mols per liter.[31, 60, 61] For example, arterial [CO$_2$] is calculated above as $0.03 \times 40 = 1.2$ mmols per liter = 1.2×10^{-3} mols per liter. The concentration of hydrogen ions in plasma is 40 nmols per liter = 4×10^{-8} mols per liter. Thus [HCO$_3^-$] = K$_a'$[CO$_2$]/[H$^+$] = $8 \times 10^{-7} \times$

$(1.2 \times 10^{-3}/4 \times 10^{-8}) = 2.4 \times 10^{-2}$ mols per liter = 24 mmols per liter, the normal value of bicarbonate concentration in plasma.

The equilibrium condition can be put into an even more usable form by substituting [CO$_2$] = 0.03 P$_{CO_2}$ into equation (6):

$$K_a' = \frac{[H^+] \, [HCO_3^-]}{0.03 \ P_{CO_2}} \qquad (7)$$

With this equation any one of the three important physiologic variables, [H$^+$], [HCO$_3^-$] and P$_{CO_2}$ can be calculated if the other two are known. As shown below, the acid-base status of the body can be estimated with reasonable accuracy from measurement of these quantities in the blood and from other clinical information.

MEASUREMENT OF pH, Pa$_{CO_2}$ and [HCO$_3^-$]. The hydrogen ion concentration, [H$^+$], (more correctly H$^+$ ion activity, a$_H$), of blood can be calculated from equation (7) if P$_{CO_2}$ and [HCO$_3^-$] are known. Also [H$^+$] can be calculated from the pH as measured with a glass membrane electrode which is porous only to H$^+$ and thus a potential proportional to the logarithm of the concentration (actually activity) ratio of H$^+$ develops across the glass membrane. The partial pressure of CO$_2$ in the blood can be estimated from alveolar P$_{CO_2}$, measured directly with a P$_{CO_2}$ electrode or calculated from equation (7) if pH and [HCO$_3^-$] are known. Bicarbonate concentration can be estimated from the total CO$_2$ content of a sample of blood plasma and either the pH or P$_{CO_2}$ by using equation (7).

The total CO$_2$ concentration of a solution, [Total CO$_2$], is defined as the sum of [CO$_2$] and [HCO$_3^-$] where [CO$_2$] includes [H$_2$CO$_3$]. [Total CO$_2$] can be measured by adding an excess of strong acid to a sample of known volume, shaking vigorously and collecting and measuring the volume of CO$_2$ liberated. The strong acid drives the reaction, H$^+$ + HCO$_3^-$ $\rightleftharpoons$ H$_2$CO$_3$ $\rightleftharpoons$ H$_2$O + CO$_2$ far to the right because of the high [H$^+$]. For example if [H$^+$] is increased to 0.1 mol per liter and P$_{CO_2}$ is reduced to that of air (0.3 mm Hg), then, from equation (7), [HCO$_3^-$] = 7.2×10^{-5} mmols per liter, negligibly smaller compared with the normal plasma value of 24 mmols per liter.

Estimation of [Total CO$_2$] is simple and is done routinely in the clinic. Since [Total CO$_2$] = [CO$_2$] + [HCO$_3^-$], then [HCO$_3^-$] =

$[\text{Total CO}_2] - [\text{CO}_2] = [\text{Total CO}_2] - 0.03$ Pco_2. Substituting in equation (7) gives

$$K_a' = \frac{[\text{H}^+]([\text{Total CO}_2] - 0.03\ \text{Pco}_2)}{0.03\ \text{Pco}_2} \quad (8)$$

Usually, $[\text{Total CO}_2]$ and $p\text{H}$ are measured and Pco_2 is calculated. For example, suppose that measurements on a sample of plasma give $p\text{H} = 7.3$ and $[\text{Total CO}_2] = 25.5$ mmols per liter. Since $p\text{H} = -\log[\text{H}^+]$, $[\text{H}^+] = 10^{-p\text{H}} = 10^{-7.3} = 10^{0.7} \times 10^{-8} = 0.5 \times 10^{-8}$ mols per liter $= 50$ nmols per liter. Solving equation (8) for $0.03\ \text{Pco}_2$ gives

$$0.03\ \text{Pco}_2 = \frac{[\text{Total CO}_2]}{K_a'/[\text{H}^+] + 1} = 25.5/(800/50 + 1)$$
$$= 25.5/11 = 1.5 \text{ mmols per liter.}$$

Hence, $\text{Pco}_2 = 1.5/0.03 = 50$ mm Hg and $[\text{HCO}_3^-] = [\text{Total CO}_2] - 0.03\ \text{Pco}_2 = 25.5 - 1.5 = 24$ mmols per liter.

THE HENDERSON-HASSELBALCH EQUATION. The equilibrium condition for the CO_2 system is usually written in logarithmic form with $[\text{H}^+]$ replaced by $p\text{H}$ because $p\text{H}$ is the quantity measured and for other reasons given above. The equilibrium condition, $K_a' = [\text{H}^+][\text{HCO}_3^-]/0.03\ \text{Pco}_2$, is transformed into the Henderson-Hasselbalch equation for the CO_2 system by taking the logarithm of both sides,

$$\log K_a' = \log [\text{H}^+] + \log \frac{[\text{HCO}_3^-]}{0.03\ \text{Pco}_2}$$

Substituting $p\text{H} = -\log [\text{H}^+]$, defining $p\text{K}_a' = -\log K_a'$ and rearranging terms gives

$$p\text{H} = p\text{K}_a' + \log \frac{[\text{HCO}_3^-]}{0.03\ \text{Pco}_2} \quad (9)$$

$K_a' = 8 \times 10^{-7}$, so $p\text{K}_a' = -\log 8 \times 10^{-7} = -\log 8 - \log 10^{-7} = -0.9 + 7 = 6.1$. Thus the Henderson-Hasselbalch equation for plasma at $38°$ C is

$$p\text{H} = 6.1 + \log \frac{[\text{HCO}_3^-]}{0.03\ \text{Pco}_2} \quad (10)$$

For example, in normal plasma, $[\text{HCO}_3^-] = 24$ mmols per liter and $[\text{CO}_2] = 0.03\ \text{Pco}_2 = 0.03 \times 40 = 1.2$ mmols per liter. Hence $[\text{HCO}_3^-]/0.03\ \text{Pco}_2 = 24/1.2 = 20$. The logarithm of 20 is the sum of the log of 2 and the log of 10; $\log 20 = \log (2 \times 10) = \log 2 + \log 10 = 0.3 + 1 = 1.3$; $p\text{H} = 6.1 + 1.3 = 7.4$, the normal $p\text{H}$ of plasma.

The Henderson-Hasselbalch equation shows clearly the effects of Pco_2 on the acidity of a solution. An increase in Pco_2 adds carbonic acid to the solution and titrates it in the acid direction (lower $p\text{H}$). Since $[\text{H}_2\text{CO}_3]$ is proportional to Pco_2 the actual amount of acid added and buffered cannot be calculated solely from the change in Pco_2; if the solution is well buffered, considerable acid is added by any increase in Pco_2; i.e., CO_2 continues to dissolve and hydrate to H_2CO_3 as the H^+ are taken up by the buffer. On the other hand, adding noncarbonic acid to a solution with constant Pco_2 causes little $p\text{H}$ change; most of the excess H^+ ions combine with HCO_3^- and leave the solution as CO_2.

The Buffering Action of a Bicarbonate–Carbonic Acid Solution at Fixed Pco_2. CONCENTRATION OF ADDED ACID. The titration curves of a bicarbonate solution at various fixed Pco_2 values are shown in Figure 27–4. These curves are determined in much the same ways as those of a phosphate buffer: A strong, noncarbonic acid such as HCl is added to the bicarbonate solution which is equilibrated with a gas of fixed Pco_2 by vigorous bubbling. After each addition of nonvolatile acid, a sample of the solution is taken and its $p\text{H}$ is measured. (Since the Pco_2 is known, $[\text{HCO}_3^-]$ can be calculated.) The titration curve is obtained by plotting the amount of acid added against the resulting $p\text{H}$ as was done in Figure 27–2. However, the amount of noncarbonic acid added is equal to the decrease in $[\text{HCO}_3^-]$ (plus the negligible increase in $[\text{H}^+]$); the vast majority of added H^+ ions combine with bicarbonate to form carbonic acid and disappear into the gas phase as carbon dioxide. Therefore, the decrease in $[\text{HCO}_3^-]$ closely approximates the amount of acid added; a plot of $[\text{HCO}_3^-]$ against $p\text{H}$ gives an upside down titration curve of the CO_2 system, as shown in Figure 27–4. It must be emphasized that this is true only if the solution has no other buffer substances, i.e., if other nonbicarbonate buffers were present, some of the added H^+ would combine with them and reduce the $p\text{H}$ change and, more importantly, not all of the H^+ would combine with HCO_3^-; the amount of acid added would be greater than the decrease in bicarbonate. The estimation of the amount of acid added to solutions containing protein buffers and bicarbonate can be estimated from the $p\text{H}$, $[\text{HCO}_3^-]$ and buffer value of the solution by a back-titration method described below.

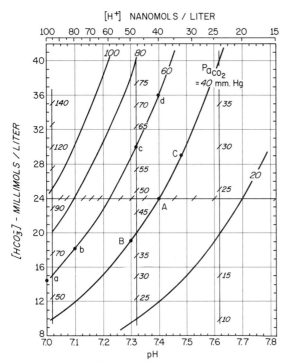

[H⁺]   NANOMOLS / LITER

Figure 27–4 The [HCO_3^-]-pH diagram for blood plasma. [H^+] is shown across the top. Pa_{CO_2} isobars are calculated from the Henderson-Hasselbalch equation [equation (10)] with Pa_{CO_2} held constant at the values attached to each curve. Several features of this graph aid understanding of its characteristics: (i) Any isobar can be superimposed on any other isobar by sliding the whole curve horizontally. (ii) At fixed pH, Pa_{CO_2} is proportional to [HCO_3^-]. The vertical lines at pH's of 7.02, 7.32 and 7.62 have slanting scribe lines indicating the Pa_{CO_2} at those points; the proportionality factors are 4, 2 and 1 respectively. (iii) Along any horizontal line, Pa_{CO_2} is proportional to [H^+]; at [HCO_3^-] = 24 mmols per liter, Pa_{CO_2} = [H^+] if Pa_{CO_2} is given in mm Hg and [H^+] in nmols per liter. The slanting scribe marks along the [HCO_3^-] = 24 mmols per liter line indicate the Pa_{CO_2} at the point as read off the [H^+] scale at the top. (iv) A Pa_{CO_2} isobar is the titration curve of a HCO_3^-–H_2CO_3 solution with Pa_{CO_2} held fixed. The slope of the curve represents the buffer value at that point and is 2.3 [HCO_3^-]. Point A is the normal value for human plasma, points B and C the changes caused by adding 5 mmols of strong acid and strong base respectively. See text for significance of small letters.

TITRATION CURVES AT FIXED P_{CO_2}.

At P_{CO_2} = 40 mm Hg and [HCO_3^-] = 24 mmols per liter, pH = 7.40 as calculated above. This is plotted as point A in Figure 27–4. Adding 5 mmols of strong acid per liter of solution reduces [HCO_3^-] by nearly 5 to 19 mmols per liter. Hence pH = 6.1 + log 19/1.2 = 6.1 + log 15.8 = 6.1 + 1.2 = 7.30 (point B, Fig. 27–4). A negligible fraction of the H^+ ions added remains in the ionic

form. After the addition, [H^+] = 10^{-pH} = $10^{-7.3}$ = $10^{-(8-0.7)}$ = $10^{-8} \times 10^{0.7}$ = 5×10^{-8} mols per liter = 50 nmols per liter, whereas 5,000,000 nmols per liter of H^+ ions were added. This value of [H^+] can also be obtained by reading from the scale at the top of Figure 27–4.

Similarly, adding 5 mmols of strong base per liter raises [HCO_3^-] to 29 mmols per liter and pH = 6.1 + log (29/1.2) = 7.48 (point C, Fig. 27–4). Other points for P_{CO_2} = 40 mm Hg are calculated and plotted in the same way and a smooth curve is drawn through them to make the P_{CO_2} 40 mm Hg *isobar*. The same procedure was used to construct the titration curves (isobars) for P_{CO_2} = 20, 60, 80 and 100 mm Hg as shown in Figure 27–4.

ESTIMATION OF P_{CO_2} ON [HCO_3^-]-pH GRAPHS. As shown in Figure 27–4, P_{CO_2} isobars are concave upward. Solving the Henderson-Hasselbalch equation (10) for [HCO_3^-] gives

$$[HCO_3^-] = 0.03 \ P_{CO_2} \ 10^{(pH - 6.1)} \qquad (11)$$

showing that these curves are exponential for fixed P_{CO_2}. Study of this and equation (10) reveals three properties of P_{CO_2} isobars which make it relatively easy to estimate the P_{CO_2} at points falling between the isobars shown on the graph:

(i) At constant pH, *i.e.*, along any vertical line in Figure 27–4, P_{CO_2} is proportional to [HCO_3^-]. In particular if pH = 7.32, P_{CO_2} (in mm Hg) = 2 × [HCO_3^-] (in mmols per liter) as shown by the vertical line at pH = 7.32 in Figure 27–4. The slanting marks show 5 mm Hg intervals. Comparison with the left ordinate scale shows that P_{CO_2} is twice [HCO_3^-], *e.g.*, at pH = 7.32 and [HCO_3^-] = 20 mmols per liter, P_{CO_2} = 40 mm Hg. Similarly, at pH = 7.02, P_{CO_2} = 4[HCO_3^-]; and at pH = 7.62, P_{CO_2} = [HCO_3^-]. (Note that these proportionality factors depend on the units used.)

(ii) At a fixed [HCO_3^-], P_{CO_2} is proportional to [H^+]. This is most easily seen from the equilibrium condition, $K_a' \times 0.03 \ P_{CO_2}$ = [H^+] × [HCO_3^-]. In particular, along the horizontal line at [HCO_3^-] = 24 mmols per liter, P_{CO_2} (in mm Hg) = [H^+] (in nmols per liter) as shown by the slanting marks along the [HCO_3^-] = 24 mmols per liter line.

(iii) Any P_{CO_2} isobar can be superimposed on any other isobar by sliding the whole curve horizontally, *e.g.*, the 80 mm Hg isobar is everywhere 0.3 pH to the left of the 40 mm Hg isobar in Figure 27–4.

LOGARITHMS AND THE CALCULATION OF pH. The logarithm to the base ten of a number is the power to which 10 must be raised to give the number. More operationally, for every positive number there is another number that is its logarithm that can be calculated by a fixed rule. This rule is the means for translating between pH and a_H. A table of number pairs for translating back and forth can easily be made up by remembering that $\log 2 = 0.3$. Since $\log AB = \log A + \log B$ and $\log A/B = \log A - \log B$ and $\log 2 = 0.3$, then $\log 4 = \log 2 \times 2 = \log 2 + \log 2 = 0.3 + 0.3 = 0.6$, $\log 8 = 0.9$, $\log 16 = 1.2$, $\log 1.6 = \log 16/10 = \log 16 - \log 10 = 1.2 - 1.0 = 0.2$, $\log 5 = \log 10/2 = \log 10 - \log 2 = 1.0 - 0.3 = 0.7$, $\log 2.5 = \log 10/4 = 0.4$ and $\log 1.25 = \log 2.5/2 = 0.4 - 0.3 = 0.1$. With these values, a P_{CO_2} isobar can be calculated with sufficient accuracy for most purposes. For example, let $P_{CO_2} = 60$ mm Hg; $[CO_2] = 0.03 \times 60 = 1.8$ mmols per liter. Take 8, 10, 16 and 20 times 1.8, *i.e.*, 14.4, 18, 28.8 and 36 mmols per liter. The logarithms of the ratio $[HCO_3^-]/0.03\ P_{CO_2}$ for each value are 0.9, 1.0, 1.2 and 1.3 and the corresponding pH's are simply 6.1 plus the log ratios, giving 7.0, 7.1, 7.3 and 7.4 respectively. These values are plotted as points a,b,c,d in Figure 27–4, sufficient to draw the $P_{CO_2} = 60$ mm Hg isobar with considerable accuracy.

BUFFER VALUE OF BICARBONATE SOLUTIONS AT CONSTANT P_{CO_2}. Figure 27–4 shows that a solution containing HCO_3^- is an excellent buffer if P_{CO_2} is held constant. The buffer value at any point is, by definition, the slope of a P_{CO_2} isobar at that point. Since the curves are exponential the buffer value increases in direct proportion to $[HCO_3^-]$ and is independent of P_{CO_2} or pH. Calculation shows that buffer value = $2.3\ [HCO_3^-]$, *e.g.*, buffer value = 55 sl for $[HCO_3^-] = 24$ mmols per liter. In other words, the molar buffer value is 2.3 per pH unit, a value 4 times that of an ordinary buffer (0.575 per pH unit). The fourfold improvement in buffer value is due to the fact that P_{CO_2} and thus $[H_2CO_3]$ are held constant. In an ordinary buffer, adding acid increases the concentration of HA and decreases that of A^- simultaneously; at constant P_{CO_2} only the $[HCO_3^-]$ changes; $[H_2CO_3]$ stays constant.

EFFECT OF DILUTION. As mentioned earlier, diluting a solution containing a buffer has practically no effect on the pH. However, this is not true of the bicarbonate-carbonic acid solution at fixed P_{CO_2}; $[H_2CO_3]$ is constant but $[HCO_3^-]$ is decreased and hence pH falls; *e.g.*, if the volume of solution were doubled by adding water, $[HCO_3^-]$ would be halved and pH would thus fall by 0.3 (Fig. 27–4). Adding water is equivalent to adding acid and re-

moving water is equivalent to adding base, the amounts per liter being equal to the change in $[HCO_3^-]$. Changes in the bodily water balance thus have acid-base effects.

TITRATION OF BLOOD AND EXTRACELLULAR FLUID *IN VITRO*

The foregoing sections, "Buffers and Buffering" and "The Carbon Dioxide System," describe the physical-chemical process of buffering of H^+ ions in laboratory conditions (*in vitro*). This section shows how the principles of buffering are used to estimate the acid-base status of the extracellular fluid (*cf.* Chap. 26, Vol. II) from measurements of the pH and P_{CO_2} (or [Total CO_2]) of blood.

In the situations described above where a solution is titrated with acid or alkali, the system consisting of the solution and the acid/alkali to be added is called a closed system because there is no interchange of matter between the system and the rest of the universe, *i.e.*, the experimenter completely controls the amount of acid/alkali added to the solution. This is certainly not the case for the extracellular fluid (ECF); it is an open system because acid/alkali can be added to the ECF by the lungs, kidneys, gut, cells and bone as well as by the clinician attempting therapy (intravenous infusion, control of inspired P_{CO_2}). Nevertheless, it is useful to discuss the titration of extracellular fluid as a closed system, because knowledge of the buffer properties of the ECF permits an estimate, from the pH and P_{CO_2} of blood, of how much acid/alkali has been added to the ECF by pathological conditions in, for example, the lungs, kidneys or gut. This estimate of added acid/alkali is necessary for rational therapy.

Extracellular fluid consists of the blood and the interstitial fluid, the fluid outside the blood vessels which is in diffusion equilibrium with the blood and in direct contact with cells. It is meaningful to regard the extracellular fluid as a buffer system because it consists of a fairly well-stirred water solution containing among other substances H_2CO_3-HCO_3^- and the acid and conjugate-base forms of hemoglobin and plasma proteins. It is not possible to take a sample of interstitial fluid from a patient because of the tight packing of cells; hence the buffer value of extracellular fluid is calculated from direct measurements on blood, indirect measurements of the volume of the interstitial fluid and the assumption that this fluid contains a negligible concentration of protein and other nonbicarbonate buffers.

If an unknown amount of noncarbonic acid is added to the ECF by a pathological process (e.g., diarrhea, kidney disease), then it is possible to calculate the amount of acid added from the buffer value of the ECF and from the deviations of pH and bicarbonate concentration from normal. Thus knowledge of the buffering value of the ECF of a patient is the key to diagnosis of acid-base disturbances.

Blood and extracellular fluid can be titrated either with carbonic acid (by changing the P_{CO_2}) or with a nonvolatile (noncarbonic) acid. Any respiratory malfunction which alters arterial P_{CO_2} titrates the ECF with carbonic acid. Many normal and pathological conditions cause addition of noncarbonic acid to the ECF. For example, heavy exercise causes lactic acid production in cells, much of which appears in the ECF; acute vomiting, in effect, adds alkali to the ECF because of the loss of HCl from the stomach and prolonged vomiting and diarrhea add acid because of loss of alkaline secretions from the small intestine; impaired renal excretion of acids adds acid to the ECF. When the ECF is titrated with carbonic acid, the added acid is buffered only by the protein buffers; the carbonic acid–bicarbonate system does not act as a buffer in this circumstance. On the other hand, noncarbonic acid is buffered both by the protein and H_2CO_3-HCO_3^- buffer systems. It is easier to understand and explain the titrations of ECF with carbonic acid than with noncarbonic acid and so CO_2 titration is considered first. It is also simpler to consider first the titration of blood rather than of extracellular fluid.

Titration of Blood with CO_2. An increase in Pa_{CO_2}, the partial pressure of CO_2 in arterial blood, titrates the blood (and ECF) in the acid direction. The concentration of added acid (amount of acid added to a liter of blood) depends not only on the change in P_{CO_2} but on the buffer value of the blood. For every carbonic acid molecule that has its H^+ ion taken up by a buffer, one HCO_3^-

ion appears in the blood. The reaction with hemoglobin, Hb^- is

$$H_2CO_3 + Hb^- \rightleftharpoons HHb + HCO_3^-$$

It follows, then, that the concentration of acid added to a buffer solution by a change in P_{CO_2} is equal to the change in $[HCO_3^-]$ so that the appropriate graphical coordinates for CO_2 titration curves are $[HCO_3^-]$ vs. pH such as in Figure 27–4. The P_{CO_2} titration of a sample of oxygenated blood and its various buffer components are shown in Figure 27–5. The curves are approximately linear over the range of pH's shown (7.0 to 7.7). The procedure for obtaining a curve is to equilibrate a sample of blood (in vitro) with a particular P_{CO_2}, measure its pH and repeat for other P_{CO_2} values. Since P_{CO_2}

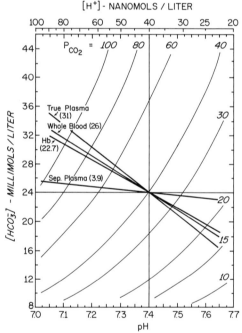

Figure 27–5 CO_2 titration curves of a sample of blood in vitro and its components. Acid is added or removed by increasing or decreasing Pa_{CO_2}. The amount of acid added is equal to the increase in $[HCO_3^-]$ because for every H^+ formed by the dissociation of H_2CO_3 a HCO_3^- also appears and nearly all the H^+ ions appearing combine with buffers. The negative of the slope of the titration curve, the buffer value (in sl) is shown in parentheses. True plasma buffer value (31 sl) is higher than that of whole blood because of the Donnan equilibrium which forces much of the HCO_3^- formed in the cell into the plasma. True plasma is plasma equilibrated with red cells and then separated for analysis.

and pH are known, $[HCO_3^-]$ can be calculated or the values can be plotted directly on Figure 27–5 by finding the point where the isobar for the given P_{CO_2} crosses the vertical line through the measured pH.

BUFFER VALUE OF PLASMA PROTEINS. The buffer value of a solution containing more than one kind of buffer is simply the sum of the individual buffer values. The buffer value of blood is almost entirely due to hemoglobin and plasma proteins. The expected buffer value of plasma can be calculated from the amount of protein. Since plasma proteins are heterogeneous, buffer value is measured in millimols per gram times pH (gram-pH) rather than millimols per millimol times pH. According to Van Slyke and co-workers,[59] the buffer value of plasma proteins is 0.1 mmols of acid per gram-pH; that of hemoglobin is about 0.18 mmol per gram-pH. These values have nearly the same ratio as the histidine contents of plasma proteins and hemoglobin.[20] A liter of plasma contains about 70 g of protein, so buffer value = 0.1×70 = 7.0 sl. Since a liter of blood contains 0.55 liter of plasma, the contribution of plasma to the buffer value of blood is 0.55×7 = 3.9 sl. The line labeled Sep. Plasma in Figure 27–5 shows the CO_2 titration curve of the plasma in 1 liter of blood. It can be seen that plasma proteins contribute only about one-sixth of the total buffer value of blood.

TITRATION OF HEMOGLOBIN WITH CO_2.[61] A liter of red cells contains 334 g or $334/16.7 = 20$ mmols of hemoglobin; a liter of blood contains 0.45 liter of red cells and $0.45 \times 20 = 9$ mmols of hemoglobin. The buffer value of hemoglobin in a liter of blood water is thus $3 \times 9 = 27$ sl, i.e., the measured molar buffer value of hemoglobin is for a liter of water rather than a liter of solution. Since a liter of blood contains only 0.84 liter of water, the actual buffer value of hemoglobin in a liter of blood is $0.84 \times 27 = 22.7$ sl. This buffer value is shown by the titration curve labeled Hb in Figure 27–5. The CO_2 titration curve of whole blood has a slope approximately equal to the sum of the slopes of the titration curves for plasma proteins and hemoglobin, $3.9 + 22.7 = 26.6$ sl (line labeled whole blood in Fig. 27–5).

TITRATION OF TRUE PLASMA WITH CO_2.

The titration of blood with CO_2 is complicated by the presence of red blood cells. Red cell membranes are highly permeable to anions and so HCO_3^- and Cl^- are equilibrated with the transmembrane potential of about -10 mV. The negative potential means that HCO_3^- ion concentration is lower inside cells than outside (ratio of about 0.7). When the P_{CO_2} of blood is increased, most of the excess H^+ ions added are buffered by hemoglobin in red cells and a majority of the HCO_3^- ions formed when H^+ ions combine with Hb, exchange for Cl^- ions in the plasma (the Cl^--shift). The result is that the increase in HCO_3^- concentration in plasma is greater than expected from the concentration of added acid and the buffer value of blood; $[HCO_3^-]$ is correspondingly lower inside the red cells. Thus, if the bicarbonate concentration of plasma, $[HCO_3^-]_p$, is made the abscissa, the titration curve is steeper than that of whole blood, having a slope of 31 sl (true plasma curve, Fig. 27–5). To repeat, this is because the negative transmembrane potential of red blood cells makes plasma bicarbonate concentration higher than that in the red cells. Plasma equilibrated at some P_{CO_2} with red cells present and then separated for analysis of $[HCO_3^-]$ is called *true plasma* as compared with *separated plasma*, which is separated first and then titrated with CO_2 and analyzed.

The buffer value of true plasma is used in estimating the change in plasma pH caused by a change in Pa_{CO_2}; the buffer value of whole blood is used in estimating the amount of noncarbonic acid added to the blood. Of course, buffer value depends on the concentrations of hemoglobin and plasma protein in blood.

CONCENTRATION OF ACID ADDED BY A CHANGE IN P_{CO_2}. Study of Figure 27–5 shows that the concentration of carbonic acid added to a solution by a particular change in P_{CO_2} depends on the buffer value of the solution being titrated. For example, a change of P_{CO_2} from 40 to 80 mm Hg increases bicarbonate concentration of true plasma from the normal value of 24 to 31 mM, an increase of 7 mM, whereas the same change in P_{CO_2} increases $[HCO_3^-]$ of separated plasma from 24 to 25.2 mM, an increase of only 1.2 mM. The greater the buffer value of the solution, the greater is the concentration of carbonic acid added by the change in P_{CO_2}. If the solution contains no buffer, a change in P_{CO_2} produces a negligible change in $[HCO_3^-]$, e.g., increasing P_{CO_2} from 40 to 80 mm Hg reduces pH from 7.4 to 7.1; $[H^+]$ increases from 400 to 800 nM and $[HCO_3^-]$ must increase by an equal amount. The increase, 400 nM or 0.0004 mM, raises $[HCO_3^-]$ from 24 to 24.0004 mM. Of course, the greater the buffer concentration is, the smaller the change in pH produced by a change in P_{CO_2}.

The reason for this behavior of the CO_2 system at equilibrium is that an increase in P_{CO_2} correspondingly increases H_2CO_3 concentration and hence the product $[H^+] \times [HCO_3^-]$. If buffer is present, some of the newly formed H^+ ions combine with buffer, leaving HCO_3^- in place of buffer anion, more H_2CO_3 dissociates to keep the $[H^+] \times [HCO_3^-]$ product constant and more CO_2 hydrates to maintain $[H_2CO_3]$. The result is that $[H^+]$ is lower and $[HCO_3^-]$ higher with buffer present, i.e., H_2CO_3 is a strong acid at body pH's. The constant P_{CO_2} implies that there is a very large store of CO_2 to maintain $[H_2CO_3]$ constant no matter how much H^+ combines with buffer.

CARBAMINOHEMOGLOBIN FORMATION.[20, 53] Titrating protein solutions with carbon dioxide is complicated by the direct combination of CO_2 with free end amino groups of hemoglobin and plasma proteins to form a carbamino compound. The reaction is of the form $R - NH_2 + CO_2 \rightleftharpoons R - NHCOO^- + H^+$. Carbamic acids ($R - NHCOOH$) are fairly strong and completely dissociated at normal body pH. Carbaminohemoglobin forms only if the free amino group is in the uncharged form ($R - NH_2$ rather than $R - NH_3^+$) and the reaction is between the end amino group and dissolved CO_2, not bicarbonate or carbonic acid. An appreciable fraction of the [Total CO_2] of the blood is carbamino compounds of hemoglobin (about 1 mmol per liter of blood). The concentration of these compounds depends on the concentrations of dissolved CO_2 (hence on Pa_{CO_2}) and of uncharged free amino groups in the solution (hence on the pH). An increase in Pa_{CO_2} tends, on the one hand, to increase the concentration of carbamino compounds by increasing $[CO_2]$ and, on the other hand, to decrease their concentration by decreasing pH, which

promotes formation of charged free amino groups according to the reaction, $R - NH_2 + H^+ \rightleftharpoons R - NH_3^+$, which is forced to the right. In living animals, the concentration of carbamino compounds is practically independent of Pa_{CO_2}.

Oxygenated hemoglobin contains about 0.1 mmol and reduced hemoglobin about 0.3 mmol of carbamino hemoglobin per mmol.[23] The change in amount of carbamino hemoglobin with the oxygenation of the solution means that some CO_2 (up to 30 per cent) is transported from the tissues to the lungs in this form. However, the role of carbamino compounds in the buffer value of the blood is practically negligible; the combination of CO_2 with hemoglobin releases H^+ which must be buffered just as those produced by dissociation of carbonic acid must be. The only difference with respect to acid-base behavior between the two is that bicarbonate ions penetrate the erythrocyte membrane easily whereas carbaminohemoglobin is confined to the cell because of the large size of the hemoglobin molecule.

Titration of Blood with Noncarbonic Acid. Adding noncarbonic acid to blood in vitro with P_{CO_2} held constant titrates both the protein and carbonic acid–bicarbonate buffer systems in the acid direction. Since part of the added H^+ combines with hemoglobin and plasma proteins and part with HCO_3^- to form CO_2, the amount of added acid is greater than the decrease in bicarbonate concentration by the concentration of added acid which is buffered by proteins. This is different from a pure bicarbonate solution as described in connection with Figure 27–4. Since the state of dissociation of the various ionizable groups on protein molecules depends on pH but not on Pa_{CO_2}, the amount of acid added can be determined by back-titrating to the original pH by reducing the P_{CO_2}. As P_{CO_2} is reduced, carbonic acid is removed from the solution, H^+ ions dissociate from proteins, combine with HCO_3^- and disappear as CO_2. When the pH has returned to its original value, the proteins have returned to their original ionization state. At the original pH, the total change in $[HCO_3^-]$ of blood must equal the amount of acid added. In other words, since none of the added H^+ ions are now combined with the protein buffers, all of it must have combined with HCO_3^-

and disappeared from the system in the form of CO_2. For this to happen, however, the P_{CO_2} must be reduced, i.e., back-titrated in the alkaline direction.

TITRATION OF PLASMA WITH HCL. The back-titration method of estimating the excess noncarbonic acid concentration is illustrated in Figure 27–6. Consider first the simpler case in which 1 liter of separated plasma is titrated in vitro. Assume that 8

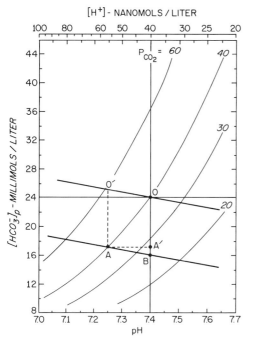

[H⁺] - NANOMOLS / LITER

Figure 27–6 Titration of plasma with noncarbonic acid. Method of estimating amount of noncarbonic acid added to a liter of plasma. Adding 8 mmols of strong noncarbonic acid such as HCl titrates the plasma from O to A (Pa_{CO_2} held fixed at 40 mm Hg). Position of point A is determined by back-titrating along titration curve of plasma proteins to the original pH by lowering Pa_{CO_2} and thus removing carbonic acid. Slope of the titration curve is independent of Pa_{CO_2} and of pH. Thus a titration curve, AB, is constructed through A with a slope of −7 sl (buffer value of plasma). The intersection of this line with original pH (point B) determines concentration of noncarbonic acid added. Since the charge on the plasma protein is the same at B as at O (pH's are the same), all of the added acid must have combined with HCO_3^- and disappeared from the solution as CO_2. The distance OB is thus 8 mmols per liter, the concentration of noncarbonic acid added. The distance OA' is the concentration of the original noncarbonic acid buffered by HCO_3^- at constant Pa_{CO_2} and A'B the amount buffered by titration of the plasma proteins in the acid direction; i.e., A'B = −7 (pH_A − pH_B) = 7 × 0.15 = 1.05 mmols per liter. Note that O'A = OB.

mmols of noncarbonic acid (HCl) are added to a liter of plasma whose initial pH = 7.4 and $[HCO_3^-] = 24$ mM and P_{CO_2} is held constant at 40 mm Hg by vigorous bubbling of a gas having a P_{CO_2} of 40 mm Hg through the solution. The HCl titrates the plasma along the $P_{CO_2} = 40$ mm Hg isobar toward lower pH and lower $[HCO_3^-]$ (from 0 to A in Fig. 27–6). Part of the added, completely dissociated acid reacts with bicarbonate to form carbonic acid, which is converted to CO_2 and enters the gas phase. The rest of the acid reacts with the plasma proteins. The number of millimoles per liter of added H^+ buffered by plasma protein is determined by back-titrating to the original pH by lowering the P_{CO_2}. The resulting reduction in $[H_2CO_3]$ also reduces the $[H^+]$ and $[HCO_3^-]$. The buffer value of plasma is not affected by the addition of noncarbonic acid so the titration curve is simply shifted to the left and down (AB, Fig. 27–6). At normal pH (point B), the concentration of HCO_3^- removed must equal the concentration of noncarbonic acid added since the charge on the protein buffers depends only on the pH and it is normal. Thus point B must be 8 mmols per liter directly below 0. The vertical distance OA' represents the concentration of acid originally buffered by HCO_3^-, and A'B is that buffered by plasma protein. Also the greater the buffer value of proteins, the less the change in the $[HCO_3^-]$ of plasma produced by a given concentration of added acid. Note that when noncarbonic acid is added, the $[HCO_3^-]$ *decreases*, whereas when carbonic acid is added, $[HCO_3^-]$ *increases*.

TITRATION OF BLOOD WITH HCL. The procedure for determining the concentration of noncarbonic acid added to whole blood is identical with that used for separated plasma but is complicated by the unequal distribution of HCO_3^- between red blood cells and plasma described above. Since measurements of pH, P_{CO_2} and [Total CO_2] are made on plasma, a correction must be made for the excess concentration of HCO_3^- in plasma over that in red cells. In theory, the best procedure for determining concentration of added acid is to plot $[HCO_3^-]$ of whole blood on the ordinate and to use the buffer value of whole blood for the CO_2 back titration to the original pH. In practice it is easier to keep the measured value, $[HCO_3^-]_p$, on the ordinate, and use the buffer value of true plasma to calculate the concentration of acid added to plasma and then to correct this value for the unequal HCO_3^- distribution. This method is described after some preliminary considerations.

The buffer values of both whole blood and true plasma are nearly directly proportional to the concentration of hemoglobin in blood. As the hemoglobin concentration increases, so does the fraction of HCO_3^- outside cells and the difference between the buffer values of whole blood and separated plasma. Since hemoglobin concentration varies considerably between patients, direct measurements of hemoglobin concentration of blood or buffer value of blood or true plasma are advisable in acid-base disorders. Siggaard-Andersen[56] has described a simple and reliable method of estimating buffer value graphically from measurements of pH at two known P_{CO_2} values. He also gives a graphic means of estimating the amount of extra noncarbonic base or acid (negative of his Base Excess) in the blood by using the buffer value and original pH. The concentration of added noncarbonic acid in blood can be estimated as follows: [Excess noncarbonic acid] (mM) $= (1 - 0.00125 [HbO_2]) \times$ change in $[HCO_3^-]$ of plasma at pH = 7.4 (distance OB, Fig. 27–7), where hemoglobin concentration, $[HbO_2]$, is in grams per liter. For $[HbO_2] =$ 150 g per liter (normal), the correction factor is 0.81, i.e., the buffer value of normal whole blood is 0.81 of that for true plasma, $0.81 \times 31 = 25$ sl, as described above.

The use of this correction factor is illustrated by an example: Suppose an unknown amount of noncarbonic acid is added to a sample of blood maintained at $P_{CO_2} =$ 40 mm Hg. After the addition, the plasma is found to have a pH = 7.27 and $[HCO_3^-]_p =$ 18 mM (point A, Fig. 27–7). Hemoglobin concentration is normal. What is the concentration of added noncarbonic acid in millimoles per liter?

Since hemoglobin is normal, the buffer value of true plasma is 31 sl. A titration curve having a slope of −31 sl is drawn through point A (line AB, Fig. 27–7). Point B is determined as follows. The pH change is $7.4 - 7.27 = 0.13$. The decrease in $[HCO_3^-]$ is $0.13 \times 31 = 4.0$ mM and since at point A $[HCO_3^-] = 18$ mM, at point B $[HCO_3^-] = 18 - 4 = 14$ mM and pH = 7.40.

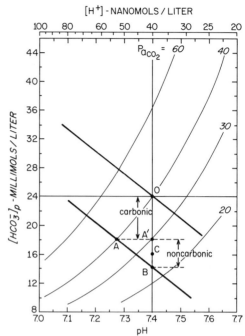

Figure 27–7 Estimation of noncarbonic acid excess in fully oxygenated whole blood. See legend to Figure 27–6. Most correct estimate is OC = 0.81 OB = 8.1 mmols per liter. Note that a much larger fraction of the added noncarbonic acid was buffered by protein buffers of blood (A′B) than was the case for separated plasma. Hence, the fall in pH was only 0.13 instead of 0.15 as in Figure 27–6. The portion of the added acid labeled carbonic (OA′) is that buffered by HCO_3^- and that labeled noncarbonic (A′B) is the portion taken up by the protein buffers. Both are overestimated by 1/0.81 (see text).

Since the pH is normal at point B, all the added acid has disappeared as CO_2, and the apparent concentration of added acid is thus $24 - 14 = 10$ mM. However, this concentration must be multiplied by the factor 0.81 to correct for the lower concentration of HCO_3^- in red cells than in plasma: 0.81×10 mM = 8.1 mM is the estimated concentration of added noncarbonic acid and is represented by point C, Figure 27–7. If the hemoglobin concentration is lower than normal, the correction factor is smaller than 0.81. The method illustrated in Figure 27–7 is for use with fully oxygenated blood only. Other initial values must be used otherwise.

The estimation of the concentration of noncarbonic acid added to blood is of little fundamental significance since noncarbonic acid from any source is added to the extracellular fluid rather than just to the blood (see below). However, knowledge of the procedure is frequently necessary for interpreting measurements made on blood samples.

Titration of Extracellular Fluid *in Vivo*. CO$_2$ TITRATION. It is not possible to obtain a sample of extracellular fluid from a patient, so titrating ECF *in vitro* with CO$_2$ is a thought experiment. Since interstitial fluid contains little or no protein, its buffer value is negligibly small and most of the buffer substances in ECF are contained in the blood. The CO$_2$ titration curve of the whole extracellular fluid can be calculated from the buffer value of 1 liter of blood and from the total volumes of blood and ECF. The buffer value of 1 liter of well-mixed ECF is less than that of blood by the factor by which interstitial fluid dilutes protein buffers of blood. A typical value for the blood volume of a 70 kg man is 5.3 liters, 2.4 liters of red cells and 2.9 liters of plasma. Inulin space, the readily diffusible extracellular volume, is about 12 liters, and thus total extracellular volume, including red cells, is $12 + 2.4 = 14.4$ liters (Chap. 26, Vol. II). The 5.3 liters of blood contain $5.3 \times 9 = 47.7$ mmols of hemoglobin, and 2.9 liters of plasma contain $70 \times 2.9 = 203$ g of plasma protein. These amounts are contained in 14.4 liters of extracellular fluid so that a liter contains $47.7/14.4 = 3.3$ mmols of hemoglobin and $203/14.4 = 14$ g of plasma protein. The molar buffer value of hemoglobin is 3.0 per pH, and so the buffer value due to hemoglobin in a liter of extracellular fluid is $3.0 \times 3.3 = 10$ sl and that due to plasma protein is $0.1 \times 14 = 1.4$ sl. The predicted buffer value of extracellular fluid is approximately $10 + 1.4 = 11.4$ sl. The correction for the unequal transmembrane distribution of HCO$_3^-$ ions is small because of the small fraction of the extracellular fluid volume occupied by red cell water. Experimental evidence showing that the buffer value of ECF is about 11 sl is given in the next section on CO$_2$ titration of the whole body.

NONCARBONIC ACID TITRATION OF ECF. The concentration of noncarbonic acid added to the extracellular fluid is estimated in exactly the same way as described for plasma in connection with Figure 27–6, using a buffer value of 11 sl for the CO$_2$ back-titration. No correction like that described for whole blood (Fig. 27–7) is necessary.

TITRATION OF THE WHOLE BODY WITH CARBON DIOXIDE

The whole body can be titrated with CO$_2$ by controlling the ventilation rate or the fraction of CO$_2$ in the inspired air; the buffer value of the body can be estimated from the resulting changes in [HCO$_3^-$]$_p$ and pH. Such data are commonly called a CO$_2$ uptake curve when [HCO$_3^-$] or [CO$_2$] is plotted against P$_{CO_2}$. The resulting graph is curved and difficult to interpret; the corresponding pH-[HCO$_3^-$] plot is linear and simple to interpret. CO$_2$ titration data are obtained by having a volunteer hyperventilate for a few minutes, thus reducing the P$_{A_{CO_2}}$ and Pa$_{CO_2}$ and titrating the body in the alkaline direction, or, more frequently, by having the subject breathe 5 to 10 per cent CO$_2$ for a time, thereby titrating his blood in the acid direction.[19] Arterial blood samples are taken at appropriate times. More recently, volunteers[6] have entered chambers where P$_{CO_2}$ is held fixed and CO$_2$ titration curves are obtained for times up to an hour. Dogs have been subjected to the same procedure for days or weeks.[18, 50]

Unlike well-stirred solutions *in vitro*, no single unique CO$_2$ titration curve is obtained when the whole body is titrated with CO$_2$. The values of plasma pH and [HCO$_3^-$] found following a change to a new P$_{I_{CO_2}}$ (pressure of inspired CO$_2$) depend markedly on the elapsed time between the change in P$_{I_{CO_2}}$ and the taking of the blood sample.[10, 69] Several processes with progressively lengthening time courses are involved, and the slope of the titration curve changes in time until all the processes involved have reached new steady-state (unchanging) values in weeks or even months. The body is said to be in a *transient* acid-base state if pH, [HCO$_3^-$]$_p$, and/or Pa$_{CO_2}$ are changing in time. The known processes giving rise to transients in pH and [HCO$_3^-$] of plasma following an increase in P$_{I_{CO_2}}$ are: (i) CO$_2$ transport to tissues. Time is required for the circulation to deliver enough CO$_2$ to the extracellular fluid and cells to raise their P$_{CO_2}$ values to their steady-state values somewhat above Pa$_{CO_2}$. (ii) Transmembrane H$^+$ transfer. Active transport processes transfer H$^+$ from the intracellular to the extracellular fluid or vice versa. (iii) Bone carbonate release. (iv) Increased rate of acid excretion by the kidneys.

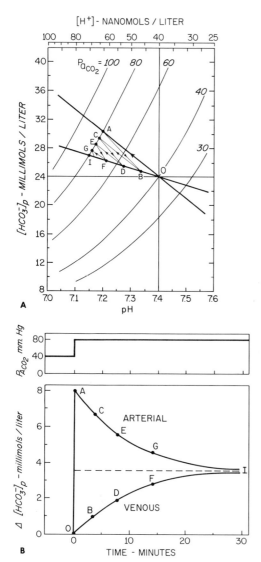

Figure 27–8 The extracellular $[HCO_3^-]$ circulatory mixing transient following a change in PA_{CO_2} from 40 to 80 mm Hg. *A*, Changes in plasma $[HCO_3^-]$ and pH of a segment or "slug" of blood in successive passes through the lungs and tissues. At the time PA_{CO_2} was increased, the segment was in the lungs and was titrated from O to A along the true plasma titration curve. When the segment reached the tissues it gave up most of its excess CO_2, $[H^+]$ and plasma $[HCO_3^-]$ to the extracellular space which is still at point O. This transfer moves the segment of blood to point B and titrates the extracellular fluid from O to B. On returning to the lungs, the blood segment is titrated from B to C, parallel to OA. This process continues, C to D, D to E, E to F, and so forth until the blood, extracellular fluid and intracellular fluid reach a new steady state at point I. Note that point A is on the titration curve of true plasma; point I is on a titration curve having a slope of −11.4 sl, the calculated value for extracellular fluid. (See Fig. 27–9.) *B, upper:* Time course of change in PA_{CO_2}; *lower:* Time course of the changes, $\Delta[HCO_3^-]_p$, in the bicarbonate concentration of arterial and venous blood. The letters on the

Transients in Body CO_2 Stores.[21,40,41,57] RATE OF CO_2 TRANSPORT TO TISSUES. The time course of CO_2 uptake by the lungs, blood, interstitial fluid and tissues, following a sudden increase in PI_{CO_2} is a complicated process, having several components. The approximate time courses of these changes in arterial and venous blood and in the whole body are shown in Figures 27–8 and 27–10. Figure 27–8A illustrates the process of CO_2 transport from the blood to the tissues following a stepwise increase in PA_{CO_2}. Following the sudden increase in PA_{CO_2}, the CO_2 content of blood in the pulmonary vein likewise increases suddenly. After a few tens of seconds, the time required for the high CO_2 blood to reach the tissue capillaries, the blood gives up most of its extra CO_2 (mainly HCO_3^-) to the interstitial fluid, or does not pick up its normal load of CO_2 in the tissues. The venous blood returning from the tissues has a P_{CO_2} about equal to that of the tissues and possibly less than that of the arterial blood. The low CO_2 venous blood picks up another load of CO_2 in the lungs and carries it to the tissues, gradually filling up the tissues with CO_2. This process is illustrated in Figure 27–8A as movements on the $[HCO_3^-]_p$-pH diagram and the time course is shown in Figure 27–8B.

The time necessary to reach a new steady state in a tissue thus depends on the ratio between tissue volume, V, and blood flow, $\dot{V}_b$, and on the buffer value of the tissue. If the same amount of CO_2 were delivered to the tissues every minute, it would take about $V/\dot{V}_b$ minutes to fill up the tissue; *e.g.*, for the whole body, V = 43.4 liters and $\dot{V}_b$, cardiac output, is about 5 liters per min, so 43.5/5 = 8.7 min. However, the amount delivered to the tissues decreases in proportion as tissue $[HCO_3^-]$ and P_{CO_2} build up. Analysis shows that 63 per cent of the final amount is delivered in 8.7 min, 63 per cent of the remainder in the next 8.7 min and so on; *i.e.*, this is an exponential process with a time constant of 8.7 min. Unfortunately, this calculation is much too simple; the

curves refer to corresponding points in part A. The arterial blood is presumed equilibrated with PA_{CO_2} and the venous blood with the extracellular fluid. All measurements must be made in fully oxygenated blood to be comparable. Curves are exponentials with time constant = 10 min; actual curves are more complicated.

blood flow to various tissues and organs varies considerably, depending on bodily activity. The major determinant of equilibration time is blood flow in muscle, which constitutes nearly half the body mass. In addition, the uptake of CO_2 by muscle cells may be delayed by the slow hydration of CO_2 to H_2CO_3 because muscle does not contain carbonic anhydrase. Blood flow to muscle is much lower than to most other tissues; at rest the time constant for filling muscle with CO_2 is about 30 min. The time constant for most other tissues is about 2 min.[21,40,41,57] Blood flow to muscle may be halved during anesthesia and may increase severalfold in exercise. Hence the duration of the whole body transient state depends primarily on muscle blood flow. Normal body CO_2 production, about 0.23 liter per min, does not affect these conclusions; however, increased CO_2 production due to exercise does produce a transient.

TITRATION CURVE OF EXTRACELLULAR FLUID.[6,9] The length of time necessary for interstitial fluid to reach steady state values of pH, $[HCO_3^-]$ and P_{CO_2}, 10 to 60 min, is a measure of how long it takes for blood and interstitial fluid to mix. This mixing time, though long compared with *in vitro* experiments, is short compared with the time courses of the other processes listed above. Thus, it is expected that measurements made on blood samples taken 10 to 60 min after a change in PA_{CO_2} will give values that fall close to the calculated titration curve for extracellular fluid; at this time ECF is fairly well mixed and transmembrane H^+ transfer, carbonate release from bone and renal H^+ excretion processes are so slow that they have probably had a negligible effect on ECF bicarbonate concentration. Figure 27–9 shows that this expectation is correct. Measurements of pH, $[HCO_3^-]_p$, and PA_{CO_2} were made on samples of arterial blood drawn from volunteers who breathed room air, 5 per cent or 10 per cent CO_2.[6] It is seen that the average values (larger filled circles) fall along a titration curve having a slope of −11.8 sl, surprisingly close to the value of −11.4 sl calculated above from blood buffer value and volume and volume of extracellular fluid.

TRANSIENTS IN TOTAL BODY CO_2. Figure 27–10 shows the approximate time course of changes in the total CO_2 content

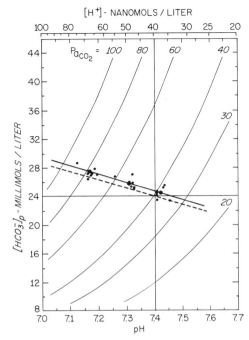

Figure 27–9 CO_2 titration curve of extracellular fluid. Small points, individual measurements made on seven volunteer subjects; large points, average values for controls (breathing room air, right point) and for inspired CO_2 concentrations of 5 per cent (center) and 10 per cent (left). Arterial blood samples were drawn before and 10 to 60 minutes after an increase in PI_{CO_2}. PA_{CO_2} depends on PI_{CO_2} and on alveolar ventilation. Solid line drawn through large points has a slope of −11.8 sl; dashed line is the calculated titration curve of extracellular fluid and has a slope of −11.4 sl. (Data of Brackett, Cohen and Schwartz, *New Engl. J. Med.*, 1965, 272, 6–12.)

of the body in a person at rest following a step change in PI_{CO_2}. Note that the ordinate is total CO_2 content of the body (except for bone) rather than bicarbonate concentration of plasma used in several previous figures. The initial rapid change is due to CO_2 uptake in blood; the successively slowing time course thereafter is due to extracellular and cellular uptake, the limiting factor being the blood flow to the various tissues and the rate of CO_2 hydration in muscle.[41] It must be emphasized that changes in blood flow to muscle drastically alter this curve and the probable time course must be considered when estimating the acid-base status of a patient. Note that there are CO_2 transients when noncarbonic acid is added to the ECF (*e.g.*, by cellular metabolism, renal malfunction, intravenous infusion by a physician); some time is re-

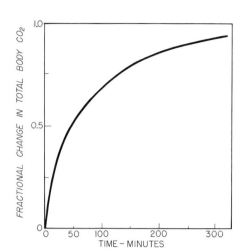

Figure 27–10 Approximate time course of changes in total body CO_2 stores following a step increase in PI_{CO_2}. *Ordinate:* Fraction of final change in total CO_2 stores of body. Over the time course shown, the change in CO_2 consists of dissolved CO_2, HCO_3^- and H_2CO_3. (After Fahri and Rahn, *Anesthesiology*, 1960, *21*, 604–614; and Sullivan *et al.*, *Amer. J. Physiol.*, 1964, *206*, 887–890.)

quired for the CO_2 formed by combination of added H^+ with HCO_3^- to be carried from the tissues to the lungs.

CO_2 Titration Curves of Plasma as a Function of Time. Clearly, from the foregoing, nothing such as a unique whole body titration curve exists; the apparent buffer value of plasma varies with time after a change in PA_{CO_2} and the time of taking the sample must be stated. As pointed out by Fahri and Rahn,[21] taking account of the CO_2 transient explains the wide variations in measured buffer values obtained from whole body CO_2 titrations.[6, 9, 18, 19]

Considerations based on Figures 27–8 and 27–10 lead to the predictions that CO_2 titration curves based on arterial blood sampling, as is commonly done, will have buffer values that change with time in the manner illustrated in Figure 27–11: Immediately after a step change in PA_{CO_2} the slope of the titration curve will be near that of true plasma. The buffer value will then decrease steadily until it reaches the value for extracellular fluid in 10 to 60 min. These predictions are supported by experimental data. Davenport[19] took arterialized blood samples from volunteers 2 to 4 min after the beginning of hyperventilation or CO_2 inhalation. The experimental points fall on a titration curve having a buffer value of

about 27 sl (curve labeled 2 to 4 min, Fig. 27–11). Brackett and co-workers[6] took serial arterial blood samples from volunteers for periods up to an hour after they had entered a chamber containing air with an elevated PCO_2. The major part of the CO_2 transient is over in about 10 min but there are slow changes throughout the hour. Their data are plotted in Figure 27–9 for comparison with the calculated buffer value of extracellular fluid and also in Figure 27–11 (curve labeled 10 to 60 min, ECF). The buffer value is 11.8 sl, almost exactly that

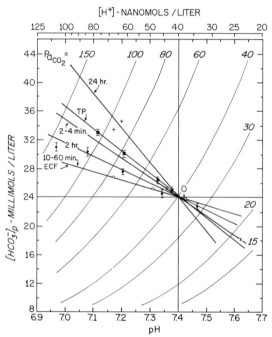

Figure 27–11 CO_2 titration curves of the whole body, plotted on the $[HCO_3^-]$-pH diagram, at various times after a sudden increase in PA_{CO_2} (initial $[HCO_3^-]$ and pH adjusted so lines go through point 0). True plasma (TP, ■) and extracellular fluid (ECF) titration curves are shown for reference. Measurements were made on true arterial plasma. At 2 to 4 min after PA_{CO_2} change, buffer value had fallen from 31 to 27 sl (X); at 10 to 60 min to 11.6 sl (0), the buffer value of extracellular fluid. Thereafter buffer value increases again due to transmembrane H^+ transfer, carbonate release from bone and renal compensation. In dogs, buffer value is 18.6 sl after 0.5 to 2.0 hours (●), and approximately 48 after 24 hours (+). Reasons for early changes are illustrated in Figure 27–8. (2 to 4 min curve after Davenport, *The ABC of acid-base chemistry*, Chicago, University of Chicago Press, 1958; true plasma and 10 to 60 min curves after Brackett *et al.*, *New Engl. J. Med.*, 1965, *272*, 6–12; 0.5 to 2 hour curve after Cohen *et al.*, *J. clin. Invest.*, 1964, *43*, 777–786; 24 hour points after Polak *et al.*, *J. clin. Invest.*, 1961, *40*, 1223–1237.)

calculated above for extracellular fluid. Since these were normal, conscious volunteers, muscle blood flow was reasonably high and the CO_2 transient was probably over in less than 45 min.

Titration curves obtained at times greater than 1 hour should have the same slope as that of extracellular fluid unless other processes are operating which remove H^+ or add HCO_3^- ions to the ECF. As mentioned above, three such processes are known (there may be others): (i) transmembrane transfer of H^+, (ii) release of carbonate from bone and (iii) renal H^+ ion excretion. Since renal compensation is slow—requiring many hours to many days to reach completion—any changes in buffer value in a matter of hours are due to transmembrane transfer or carbonate release. Cohen and co-workers[18] measured the CO_2 uptake of dogs after 0.5 to 2.0 hours in an atmosphere containing a fixed P_{CO_2}. The slope of the CO_2 titration curve is nearly -19 sl (Fig. 27-11, 2 hour curve), considerably greater than that of the extracellular fluid of a dog (about 10 sl). Hence, some of the H^+'s appearing in the extracellular fluid because of the increased P_{CO_2} have penetrated the cell membrane or have combined with carbonate from bone ($2H^+ + CO_3^= \rightleftharpoons H_2CO_3 \rightleftharpoons H_2O + CO_2$). After 1 day, $[HCO_3^-]$ has increased so that the buffer value is about 50 sl (24 hour curve, Fig. 27-11), but how much of this is due to increased renal acid excretion is not certain.[50] The available evidence indicates that approximately half of the acid added to extracellular fluid leaves it over a period of several hours.

The three rather slow processes described above which remove excess H^+ ions from the ECF are similar to chemical buffering in their effects on the CO_2 titration curve but actually are active, physiological responses of the body to combat the acid-base unbalance. These *compensatory* responses introduce the concept of the *primary* cause (pathology) leading to an acid-base disorder and the various bodily regulatory mechanisms which act to restore the acid-base state toward normal. The next section describes the terminology used to describe acid-base disorders and compensatory responses. The following sections describe present knowledge of the known compensatory mechanism: transmembrane H^+ transfer and carbonate release from

bone. Renal H^+ secretion mechanisms are described in Chapter 25, Volume II.

COMPENSATORY MECHANISMS

Primary Cause and Compensatory Response. Three quantities are needed to specify an acid-base disturbance: pH, P_{CO_2} and the extra noncarbonic acid (proportional to the change in plasma $[HCO_3^-]$ at normal pH). In specifying clinical acid-base disturbances, it is useful to have terms describing the direction of changes in these quantities from normal. Hence the range of pH, $[HCO_3^-]_p$, and P_{CO_2} values found in normal, healthy people must be known. In addition, the *primary* disturbance or pathologic condition that *caused* the acid-base change must be distinguished from the further changes which may occur in the body to *compensate* for the original change.

TERMINOLOGY AND NORMAL VALUES. *Acidosis* is defined as any condition in which the plasma pH is less than normal for that individual. The "normal" range of arterial blood pH is 7.35 to 7.45 (Fig. 27-12).[3, 19] Thus, any pH below 7.35 is clearly an acidosis. *Alkalosis* is defined as an arterial blood pH greater than 7.45. A condition in which arterial CO_2 pressure is greater than normal is usually accompanied by a fall in pH and for this reason is often called "respiratory acidosis." However, the noncommittal term *hypercapnia*, meaning raised Pa_{CO_2}, is more accurate. Since hypercapnia can occur simultaneously with alkalosis, "respiratory acidosis" is misleading. Similarly, *hypocapnia* refers to a condition of lower than normal Pa_{CO_2}. The range of Pa_{CO_2} values found in humans at sea level is 35 to 48 mm Hg (Fig. 27-12).

No satisfactory term for an excess (or deficit) of noncarbonic acids in the body has been generally accepted. The most common term is "metabolic acidosis" as distinguished from respiratory acidosis. However, "metabolic acidosis" is sometimes found when the pH is above normal, so the term acidosis is misleading. Noncommittal terms such as "nonvolatile" or "noncarbonic" seem to offer advantages. However, neither of these words lends itself to simple characterization of the condition resulting from an excess or deficit of non-

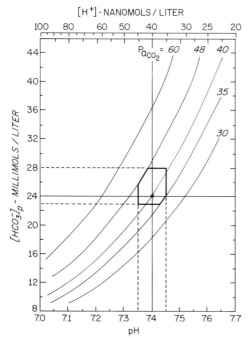

[H⁺] - NANOMOLS / LITER

Figure 27–12 Normal acid-base range in humans living at sea level. Any point within the hexagon is within the range of variation of values measured in healthy, normal people. People living at higher altitudes have significantly different values. (After Davenport, *The ABC of acid-base chemistry*. Chicago, University of Chicago Press, 1958.)

volatile or noncarbonic acids; "hypernoncarbonosis" and "hypernoncarbonicity" are clumsy. The terms "metabolic acidosis" and "alkalosis" are well established and will be used occasionally, but usually such changes will be termed excess (or deficit) of noncarbonic acids or noncarbonic acid excess or deficit. Siggaard-Andersen's term[56] "base-excess" suffers from the same difficulty (although "excess basicity" has been used). In addition, the most common acid-base imbalances are acidoses in which there may be a negative base excess.

COMPENSATION. Most data indicate that pH or $[H^+]$ is the most important factor in determining the bodily response to acid-base unbalances. In addition to the buffers of the extracellular space, four other mechanisms may act either to limit further the changes in pH, or, in the absence of permanent disability, to restore pH to normal. (i) H^+ are transferred across tissue cell membranes in a direction which changes blood pH toward normal. (ii) A deviation in plasma pH acts on the respiratory center

to change alveolar ventilation and PA_{CO_2} so as to reduce the deviation. This is called *respiratory compensation* (of a primary noncarbonic acid excess or deficit) and is usually not complete, *i.e.*, does not restore pH to normal. (iii) Carbonate is released from bone. (iv) Renal excretion of noncarbonic acid or base increases to eliminate an excess of noncarbonic acid or base or to compensate for the pH changes due to an abnormal Pa_{CO_2}. The latter is called renal compensation (of a primary hyper- or hypocapnia). Again it must be emphasized that the crux of clinical diagnosis is to distinguish between the cause of the disturbance, *e.g.*, diabetes or emphysema, and the compensatory response, *e.g.*, hyperventilation or renal excretion of noncarbonic acid (retention of alkali).

Interactions Between Cells and Extracellular Fluid. An acid load added to the extracellular fluid is handled by the body in at least three ways: (i) Buffering by the extracellular fluid, (ii) transfer of acid from extracellular to intracellular fluid, and (iii) neutralization by bone carbonate. The relative contributions of these three are only approximately known and probably vary with the nature of the acid-base disturbance. The acid-base state of the extracellular fluid is susceptible to fairly direct measurement but may not be an accurate reflection of the overall H^+ ion balance of the cells and ECF. For example, it is probable that muscle cells[2, 34] respond actively, in some circumstances, to ameliorate extracellular acid-base disturbances by taking up some of the excess H^+ in ECF, while other types of cells such as brain[44] maintain their internal pH's relatively constant by expelling excess H^+ into the ECF. Knowledge of the processes which determine cellular acid-base state and the exchanges of H^+/HCO_3^- between cells and ECF is essential for understanding the bodily responses to acid-base imbalances and forms part of the basis for rational therapy.

The acid-base state of a cell (or any fluid) is specified if any two of the three quantities pH, $[HCO_3^-]$ and P_{CO_2} are known for the cell fluid. Cellular acid-base state is the resultant of three separate processes:

(i) Cellular metabolic activity. A change in the rate of production of the acids of intermediary metabolism changes the acid-base state of a cell. For example, an increase in the lactic acid production rate (such as occurs in heavy exercise[46]) increases steady-state lactic acid concentration, which titrates cell buffers in the acid direction, reduces bicarbonate concentration and decreases pH. The role of these processes in determining the cell's acid-base state has been investigated extensively in the past few years.[1, 44, 55] A facultative increase in metabolic acid production in cells seems to be the main mechanism whereby the body compensates (restores plasma pH to normal) for alkalosis, either metabolic or respiratory (i.e., an alkalosis caused by either a deficit of noncarbonic acid or hypocapnia). Acidosis of either type (hypercapnic or noncarbonic acid excess) causes a decrease in the cellular concentrations of metabolic acids but this is insufficient to restore pH to normal. During hypercapnia, transmembrane transport of HCO_3^- plays an important role in reducing the change in pH_i of brain cells,[44] but bone carbonate release and renal excretion of H^+ appear to be the major processes acting to move ECF pH toward normal.

(ii) Partial pressure of carbon dioxide. Since CO_2 in the hydrated form, H_2CO_3, is an acid, an increase in P_{CO_2} titrates cell buffers in the acid direction.

(iii) Transmembrane transfer of HCO_3^- and/or H^+ ions. Acid-base interactions between cells and the extracellular fluid are mediated largely by fluxes, both active and passive, of appropriate substances through the membrane. In a system such as the body where CO_2 equilibrates rapidly, the entry of a H^+ ion into a cell is exactly equivalent to the entry of a H_2CO_3 molecule plus the exit of a HCO_3^- ion.

H^+/HCO_3^- fluxes are probably both passive and active. There is little information on the sizes of these fluxes, but there is little doubt that H^+ and HCO_3^- are not distributed at equilibrium across cell membranes, can penetrate them, and thus that part of a cell's metabolic energy production is used to maintain an active extrusion of H^+ (or uptake of HCO_3^-). The factors which regulate H^+ pumping are poorly understood, but even this rudimentary understanding is helpful in comprehending acid-base regulatory mechanisms.

INTRACELLULAR pH.[15, 65] H^+ are not equilibrated across muscle cell membranes; indirect[2, 64] and direct measurements[16]

agree that intracellular pH of muscle is about 7.0 (6.8 to 7.2) whereas the pH would be 5.9 if H^+ were equilibrated with a transmembrane potential of -90 mV. The inside negative potential difference across the membrane acts to drive the positively charged H^+ from outside to inside and HCO_3^- from inside to outside. If both were equilibrated, the external-to-internal concentration ratios would be like that of Cl: $[HCO_3^-]_o/[HCO_3^-]_i = [H^+]_i/[H^+]_o = [Cl^-]_o/[Cl^-]_i = 116$ mM$/3.9$ mM $= 30$. The measured ratio is $[H^+]_i/[H^+]_o = 10^{pH_o - pH_i} = 10^{7.4-7.0} = 2.5$ (range 1.6 to 4); i.e., internal $[H^+]$ is about 2.5 times external $[H^+]$ and external $[HCO_3^-]$ is 2.5 times internal $[HCO_3^-]$. *Despite the fact that $[H^+]$ in a muscle cell is greater than that outside, the tendency for them to diffuse out because of this concentration difference is far outweighed by the tendency for them to diffuse in because of the transmembrane voltage. Similarly, there is a net tendency for outward diffusion of HCO_3^-.*

The disequilibrium of H^+ and HCO_3^- across the cell membrane indicates that either the membrane is impermeable to both these ions or that H^+ are actively transported out of the cell, or HCO_3^- into the cell or both. The available evidence indicates that the membrane is permeable to both H^+ and HCO_3^- (see below). CO_2 penetrates all membranes easily because of its high lipid solubility. Since H^+ and HCO_3^- are not equilibrated and are permeable the disequilibrium must be maintained by an active transport system which utilizes metabolic energy.

ESTIMATION OF INTRACELLULAR pH.[15, 64, 65] There are three ways of estimating intracellular pH, (pH_i): (i) Injecting pH-sensitive dyes into the cytoplasm, (ii) measuring the distribution of a weak acid or base between the tissue and interstitial fluid on the assumption that the membrane is relatively impermeable to the ionic form, (iii) direct measurements with an intracellular glass membrane H^+-sensitive electrode. Direct measurements indicate that intracellular pH is about 7.0.

The distribution pattern of a weak acid between plasma and tissues provides an estimate of the over-all pH of a tissue or of the whole body. This technique has recently received great emphasis because of the discovery by Waddell and Butler[64] that 5,5-dimethyl-2,4-oxazolidinedione (DMO) is almost ideal for this type of measurement. The principle is that the dissociation of a weak acid depends on the [H$^+$] of the medium it is dissolved in. DMO has pK' of 6.13, so that the ratio of ionized to nonionized DMO at plasma pH of 7.43 is 20 to 1, whereas this ratio in a cell at pH of 7.03 would be only 8 to 1. If the cell membrane is permeable to the acid (nonionized) form and relatively impermeable to the anionic form, intracellular pH can be calculated from measurements of tissue and plasma DMO concentrations, interstitial volume and plasma pH.

Carbonic acid can also be used to estimate pH_i. Measurements of tissue and plasma [Total CO_2], Pco_2 and the plasma pH permit the calculation of pH_i from the Henderson-Hasselbalch equation. This method has been used extensively but is uncertain because CO_2 may combine directly with intracellular proteins to form carbamino groups and because cellular Pco_2 is an unknown amount higher than arterial Pco_2. Generally speaking, pH_i values calculated in this manner are somewhat higher than those obtained with DMO and intracellular pH electrodes.

The exact range of intracellular pH values is not known accurately, but the general agreement of many different methods in the hands of many different investigators strongly indicates that H$^+$ are not equilibrated in cells;[15, 22, 64, 65] the observed concentration is about 10 times lower than expected from the membrane potential and external concentration.

TRANSMEMBRANE FLUXES OF H$^+$ AND HCO$_3^-$. The net passive fluxes of H$^+$ and HCO$_3^-$ through the membrane determine the rate of the active transport process necessary to maintain internal pH constant. The desired quantity is the sum of the two fluxes; an influx of one H$^+$ is equivalent to an efflux of one HCO$_3^-$ and an influx of one H$_2$CO$_3$. The net flux of an ion depends directly on the membrane's permeability to it and the sum of the electrical and concentration gradients acting on the ion. Since muscle transmembrane potential and concentrations of H$^+$, HCO$_3^-$ and H$_2$CO$_3$ on both sides of the membrane are known, fluxes can be calculated if the permeability is known.

The passive permeability of muscle cell membranes to H$^+$ ions (P_H) is high, about 10^3 times K$^+$ permeability (P_K),[71] and the permeability to HCO$_3^-$ is about 0.1 of P_K.[70] Despite the large value of P_H, the concentrations of H$^+$ ions in cells and bathing fluids are so much lower than those of HCO$_3^-$ that the calculated H$^+$ flux is con-

siderably smaller than the calculated HCO$_3^-$ flux. The size of the passive bicarbonate flux is such that several hours are required to reach a new steady state following a change in active flux induced, for example, by a change in external pH.[2, 34]

POSSIBLE NATURE OF THE H$^+$ ION PUMP. In order to maintain intracellular pH at a nonequilibrium steady state, there must be an active efflux of H$^+$ (or influx of HCO$_3^-$) equal to the passive efflux of HCO$_3^-$. The metabolic energy requirement for the pump is small compared to that of the Na$^+$-K$^+$ pump which maintains cellular Na$^+$ and K$^+$ ion concentrations far from equilibrium. Little is known of the mechanism of the pump or of the factors that regulate its rate. Although the mechanism may well be the same in all cells, it is probable that the absolute pumping rate and the factors regulating it depend, in part, on the type of cell (e.g., muscle, nerve, liver).

There are at least three general possible mechanisms of H$^+$ pumping (Fig. 27–13): (i) A coupled pump where H$^+$ ions "hitch" a ride on the Na$^+$-K$^+$ pump; (ii) a completely separate H$^+$/HCO$_3^-$ pump; (iii) a sodium gradient pump where a Na$^+$ for H$^+$ exchange carrier uses the Na$^+$ gradient to power H$^+$ extrusion.

Coupled H$^+$ Pump. It is possible that H$^+$ ions compete with Na$^+$ and/or K$^+$ ions for the binding sites on the coupled Na$^+$-K$^+$ pump which maintains the steady-state ion concentrations in cells. If this were so, H$^+$ ions would be actively transported out in place of Na$^+$ and/or actively transported in, taking the place of K$^+$. A Na$^+$-K$^+$ exchange pump with the Na$^+$ and K$^+$ legs having high but different affinities for H$^+$ ions would pump H$^+$ both into and out of the cell at rates directly dependent on the external and internal H$^+$ ion concentrations, respectively. Such a pump can account for the observed steady state of Na$^+$, K$^+$ and H$^+$ (Fig. 27–13C). The primary reasons for considering this peculiar model of a pump (i.e., it pumps H$^+$ ions both ways) are (i) the transmembrane shifts in Na$^+$ and K$^+$ which occur in acid-base disturbances are simply explained by this model (see below), and (ii) chemically speaking, it is entirely possible that H$^+$ ions *do* compete with Na$^+$ and K$^+$ for carrier sites.

Separate H$^+$ Pump. Since the maintenance of intracellular pH near its normal

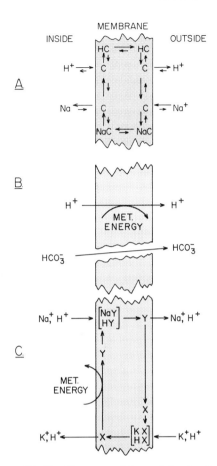

Figure 27–13 Three possible models of active H^+ ion pumping. *A,* Sodium gradient pump. A carrier, M, can bind both Na^+ and H^+ and can cross the membrane only in combined form MH^+ or MNa^+. Low internal Na^+ concentration, maintained by a Na^+–K^+ pump, means that M will carry Na^+ into the cell and H^+ out of the cell. *B,* Simple H^+ pump with an accompanying leak of HCO_3^-. Leakage of H^+ is assumed to be negligible. *C,* Two-way H^+ transport via the coupled Na^+-for-K^+ pump. Low internal $[H^+]$ is maintained because affinity of H^+ for the Na^+ carrier, Y, is about 2.5 times greater than the affinity of H^+ for the K^+ carrier, X.

value of about 7 is essential to cell function, it may well be that there is a mechanism for pumping H^+ out of cells separately from the Na^+-K^+ pump. This is illustrated in Figure 27–13*B* where an active efflux of H^+, driven by metabolic energy, balances the net outward leak of HCO_3^-. There is little evidence for or against the existence of such a pump.

Since the Na^+-K^+ pump is detected biochemically as a Na^+-K^+ activated, ouabain inhibited ATPase, the H^+ pump might be detected biochemically as an ATPase whose activity is affected by pH, $[HCO_3^-]$ and/or PCO_2. A bicarbonate activated ATPase is found in the microsomal fraction (membrane fragments) of tissue homogenates of the acid secreting cells of the stomach[36, 66] and from salivary glands which secrete alkali.[35] It is not certain that this ATPase activity occurs in the cell surface membranes, but if it does, then it is possibly a reflection of H^+/HCO_3^- pumping machinery. In order to test this possibility, it is necessary to correlate the bicarbonate activated ATPase activity with H^+ pumping rate in intact cells, under a variety of conditions. Unfortunately, there are no well-developed methods of estimating H^+ pumping rates in cells.

Another approach is to ascertain whether or not the apparent variations of H^+ pumping rate with external acid-base state (pH, $[HCO_3^-]$, PCO_2) are consistent with a H^+ pump that is activated by $[HCO_3^-]$. For example, analysis of the data[2, 34] on the pH of excised skeletal muscle under various conditions indicates that the rate of active H^+ influx depends directly on the partial pressure of CO_2 and inversely on the H^+ ion concentration of the external medium or directly with $[HCO_3^-]$, a finding consistent with a bicarbonate activated ATPase. The functional effect is that muscles take up H^+ ions from extracellular fluid when its pH falls and PCO_2 stays constant, *i.e.,* an excess of noncarbonic acid (metabolic acidosis). This behavior does not apply to brain cells which closely regulate their pH.[55]

Sodium Gradient H^+ Pump. There is good evidence that the Ca^{++} concentration of cell fluid is maintained at less than 10^{-6} M by a pump mechanism that derives its energy from the Na^+ concentration gradient which, in turn, is maintained by the Na^+-K^+ pump. The probable mechanism is a "carrier" molecule, M, confined to the membrane, which has affinities for both Ca^{++} and Na^+ and which can move through the membrane only if combined with a Ca^{++} or 2 Na^+'s. The reaction is $CaM + 2Na^+ \rightleftharpoons Na_2M + Ca^{++}$. When the carrier is at the inside surface of the membrane, the fraction of carrier molecules that combine with Ca^{++} depends on the ratio $[Ca^{++}]_i/[Na^+]_i^2$ and the fraction of M's combined with Ca^{++} at the outside surface is proportional to $[Ca^{++}]_o/[Na^+]_o^2$. If intracellular calcium concentration should rise (as a result of contractile activity, for example), then the fraction of M's at the inside surface in the form CaM will increase and there will be an increased efflux of Ca^{++} and influx of Na^{++}. The transport of some amino acids and sugars into cells is also powered by the Na^+ gradient. Thus, a H^+-ion pump powered by the Na^+-ion concentration gradient is a reasonable possibility. There is little or no evidence either for or against this possibility, but the apparent inhibition of H^+ extrusion by an increase in external $[H^+]$ does follow naturally.

BUFFER VALUE OF MUSCLE. Active transport of H^+ obscures the role of intracellular buffers in maintaining intracellular pH. If the cell membrane were impermeable to H^+ and HCO_3^-, then an increase in P_{CO_2} would titrate cell buffers in the acid direction and the slope of the intracellular titration curve would be a measure of the intracellular buffer value. The addition of noncarbonic acid to the extracellular fluid would have no effect on cell pH and vice versa, and cellular metabolic acid production would be buffered only by cell buffers.

Since there are active and passive fluxes of H^+ and HCO_3^-, the steady-state cell pH depends more on the characteristics of the pump than on the buffer value of the intracellular fluid; the duration of the transient state depends both on intracellular buffer value and the H^+ ion pumping rate. For example, a sudden increase in P_{CO_2} titrates cell (and extracellular) buffers in the acid direction and the immediate changes in cell $[HCO_3^-]$ and pH depend only on the buffer value of intracellular fluid. The fall in cell pH probably stimulates H^+ pumping rate in muscle cells but not enough to balance the increased leakage of bicarbonate resulting from the increase in internal $[HCO_3^-]$. Thus cell bicarbonate concentration slowly falls and consequently so does the slope of the "titration" curve. The buffer value of the cell fluid thus determines how much HCO_3^- is made available for delivery to the extracellular space by an increase in P_{CO_2} and thus determines the extent that muscle cells can help ameliorate an extracellular acidosis.

The buffer value of intracellular fluid is not accurately known and certainly varies from one tissue to another but can be estimated in two ways: (i) from the amount of buffer material found in cells, e.g., the sum of buffer values of phosphate, protein and other buffer materials determined by direct chemical analysis; (ii) titration of the tissue with CO_2.

Calculating tissue buffer value directly is hampered by lack of knowledge of the buffer values of tissue proteins. The buffer value is unlikely to be as high as that of hemoglobin in terms of unit weight. Since histidine is the main buffer in protein at body pH, buffer value is roughly proportional to histidine content. From this it is estimated that the minimum buffer value of muscle protein is about 15 sl. Phosphate, inorganic and organic, is high in muscle, 40 to 70 mmols per liter. All of these phosphate compounds have pK's of about 6.5 except creatine phosphate with a pK of 4.5. Creatine phosphate thus has nearly zero buffer value at cell pH values whereas the others have nearly maximal molar buffer values, 0.45 per pH. Noncreatine phosphate concentration is about 40 mEq per liter, so buffer value is about $40 \times 0.45 = 18$ sl. Anserine and carnosine contribute about $15 \times 0.575 = 9$ sl. The total calculated buffer value is $15 + 18 + 9 = 42$ sl. This value is somewhat smaller than the measured buffer value of muscle homogenates, about 50 sl.[32]

There are considerable data on the CO_2 titration curves of muscle in situ, but the results of different investigators differ.[22,32,64] The data follow a general trend which can be reasonably well approximated by a straight line with a buffer value of 15 sl.

This apparent buffer value is so low that it cannot be accounted for even if it is assumed that muscle is the soure of all the HCO_3^- appearing in the extracellular fluid. For example, the increase of extracellular buffer value from 11 to 19 sl over 2 hours as shown in Figure 27–12 would decrease the buffer value of muscle from the calculated value of 42 down to 30 sl. (The reduction in buffer value of muscle is greater than the increase in the whole body buffer value because of the smaller slope of the P_{CO_2} isobars at the lower $[HCO_3^-]$ of muscle.) The most likely explanation of this discrepancy is that other tissues, particularly brain, extrude H^+ into the extracellular fluid to maintain intracellular pH nearly constant despite the increased P_{CO_2}. Muscle would then take up not only H^+ added to the extracellular space by the increase in P_{CO_2} but also that added to the extracellular space by tissues which extrude H^+.

Bone Carbonate Release. In 1917, Van Slyke obtained the first evidence that sources outside the extracellular fluid add HCO_3^- to it in a noncarbonic acidosis. He infused HCl in dogs and found that the change in plasma pH was considerably less than expected from the quantity of HCO_3^- and other buffers in the extracellular fluid. It was generally assumed for many years thereafter that the extra HCO_3^- found in the ECF after a few hours came from cells, primarily muscle, and that the contribution from bone, if any, is on a much longer time scale (days). However, recent findings indicate that release of carbonate from bone is the predominant source of alkali for neutralizing excess noncarbonic acid added to the extracellular fluid even at times as short as 1 or 2 hours. The early carbonate release is in the form of Na_2CO_3 while the later, maintained release is as $CaCO_3$. It has long been recognized that chronic noncarbonic acidosis causes bone demineralization (loss of $CaCO_3$).

Burnell and Teubner[11, 12] infused about 10 mmols of HCl per kg of body weight into dogs over a 7 hour period and measured the regular acid-base parameters and the changes in bone carbonate, sodium and calcium. In these circumstances, muscle cell pH does not change appreciably; hence, most of the added acid probably remained in the extracellular fluid. Since extracellular fluid volume is about one-fifth of body weight, 10 mmols of HCl per kg body weight is 50 mmols per liter of ECF. This is more than twice the initial HCO_3^- concentration (about 20 mM). Nevertheless, at the infusion rate used (about 5 mmols HCl/hour per liter of ECF) HCO_3^- concentration and pH remained steady for 6 hours at about 5 mM and 7.15, respectively. The concentration of HCl neutralized by the buffer in ECF is equal to the fall in $[HCO_3^-]$ at constant pH (*cf.* Figs. 27–6 and 27–7) and is $(23-5) + 11$ sl $\times (7.4 - 7.15) = 18 + 2.7 \simeq 21$ mM. Although 50 mM HCl was neutralized only 21 mM was neutralized by ECF. The remaining 30 mmols per liter of HCl was neutralized by alkali not initially present in the extracellular fluid. Burnell showed that the excess HCl was neutralized by Na_2CO_3 released from bone: The measured decrease in bone Na_2CO_3 was found

to be approximately equal to 30 mmols per liter of ECF. Further, since the rate of acid infusion was 5 mmols/hour per liter of ECF, this represents the maximal rate of release of carbonate from bone, at least in the conditions of this experiment.

Release of carbonate from bone thus appears to be the predominant source of alkali for neutralizing excess noncarbonic acid in the ECF at times as short as a few hours.

The role of bone in alleviating hypercapnia (respiratory) acidosis is more obscure. In chronic hypercapnia where the P_{CO_2} is less than about 65 mm Hg, pH is restored to near normal values by renal compensatory mechanisms which increase plasma $[HCO_3^-]$ and there is no apparent need to mobilize carbonate from bone. It is not known whether or not carbonate is transiently released from bone during the early stages of hypercapnia (a few hours to a few days) to help relieve the acidosis until renal compensation is completed.

The total body store of bone carbonate is large, about 50 times the amount of HCO_3^- in cells and extracellular fluid. However, in an open system such as the body, if a substance is continually lost, the store, no matter how large, will be eventually used up. For example, neutralization by bone carbonate of noncarbonic acid added to the ECF by an acute disturbance such as temporarily uncontrolled diabetes or diarrhea makes "sense" in that the use of a trivial amount of bone carbonate restores ECF pH and $[HCO_3^-]$ to normal. On the other hand, use of bone carbonate to neutralize a chronic excess of noncarbonic acid, such as occurs in renal disease, causes gradual demineralization and weakening of the bone as the carbonate store is used up. The immediate threat to life posed by the chronic acidosis is avoided at the expense of bone carbonate loss which is a much more long term threat to life.

Wills[67] proposes that the primary function of the parathyroid gland is in regulating acid-base balance rather than in regulating calcium.

REGULATION OF BODY ACID-BASE STATE

The preceding section described the mechanisms whereby the body can act to ameliorate acid-base unbalances. This concluding section "puts it all together" by giving descriptions of typical examples of acid-base disturbances, the body's compensatory responses to them and the factors controlling the degree of response of the compensatory mechanisms. The mechanisms called into action depend on the severity of the unbalance, on the elapsed time since its onset and on the type of unbalance, carbonic or noncarbonic. It must be emphasized, once again, that interpretation of the acid-base state from measurements of plasma pH, HCO_3^- and P_{CO_2} requires a good understanding of the time courses of the various compensatory responses; transient conditions must be distinguished from steady-state conditions.

Compensatory Responses to Hypercapnia and Hypocapnia.[4, 7, 42] Consider the sequence of events occurring following a sudden but maintained decrease in alveolar ventilation so that $P_{A_{CO_2}}$ is increased to a higher steady value. The first, most rapid transient is in the alveolar $P_{A_{CO_2}}$, followed by the CO_2 mixing transient described above, the length of time (1/2 to 2 hours) necessary for the circulation to "fill" the tissues with more CO_2. The CO_2 titration curve has a slope of about 11 sl at this time, the value expected for extracellular fluid. Over the next several days, active responses (transmembrane H^+ transfer, renal H^+ secretion and possibly bone carbonate release) add further HCO_3^- to the ECF, thus increasing the plasma pH toward normal along the new P_{CO_2} isobar. The stimulus for initiating these compensatory responses is probably the fall in ECF pH. A new steady state is reached when plasma $[HCO_3^-]$ has increased to the value dictated by the maximal rate at which the kidneys can reabsorb bicarbonate.

TRANSMEMBRANE ION EXCHANGE. The sequence of changes in ECF produced by hyper- and hypocapnia have been studied by Pitts and co-workers.[27, 48, 49] They estimated extracellular and plasma volumes and measured plasma concentrations of the principal cations and anions in nephrectomized dogs before and during radical alterations of $P_{a_{CO_2}}$ induced by CO_2 breathing or hyperventilation. These data permit the calculation of total extracellular ion content. The nephrectomy was done to avoid complications due to renal compensation and, therefore, the results reflect only

the contributions of transmembrane H^+ transfer and carbonate release from bone. The results of one of their experiments are plotted in Figure 27–14.

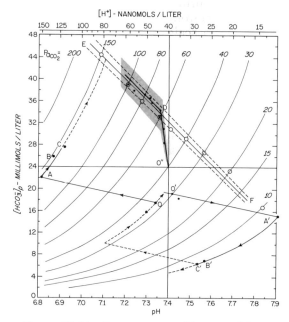

Figure 27–14 Compensation of hypo- and hypercapnia by transmembrane H^+ transfer, bone carbonate release and renal H^+ secretion. Curves OABC and O'A'B'C' show extent of transmembrane H^+ transfer plus bone carbonate release in anesthetized, nephrectomized dogs. OA and O'A' are expected titration curves of extracellular fluid. B and C are actual values 1 and 2 hours respectively after increasing $P_{A_{CO_2}}$ to 145 mm Hg by increasing $P_{I_{CO_2}}$. H^+ which left extracellular fluid were replaced by Na^+ and K^+. Primed points have same time significance with respect to start of hyperventilation which reduced $P_{A_{CO_2}}$ to 6 mm Hg. Most of the rapid compensatory fall in $[HCO_3^-]_p$ (A' to C' in 2 hours) is due to appearance of lactic acid in extracellular fluid; most of the remainder is due to uptake of Na^+ and loss of H^+ from cells. Actual sequence of experiment was O'B'C' OABC; O differs from O' because of insufficient recovery time from hyperventilation (dashed lines). Line FDE is limit of HCO_3^- reabsorption by kidneys, i.e., maximum attainable $[HCO_3^-]$ in plasma, and thus represents maximum possible renal compensation of hypercapnia. Curve O''DE and surrounding band shows extent of renal compensation of hypercapnia in patients with chronic lung disease. This shows that, chronically, there is complete compensation, i.e., restoration of normal pH if $P_{a_{CO_2}}$ is 65 mm Hg or less. For higher $P_{a_{CO_2}}$ values, compensation is limited by maximum reabsorptive ability of the kidney as shown by the coincidence of the curves, O''DE, and line FDE for pH less than about 7.35. (After Giebisch et al., J. clin. Invest., 1955, 34, 231–245; Pitts, Harvey Lect., 1952–53, 48, 172–209; Pitts, Physiology of the kidney and body fluids. Chicago, Year Book Medical Publishers, 1963; Rector et al., J. clin. Invest., 1960, 39, 1706–1721; and Refsum, Clin. Sci., 1964, 27, 407–415.)

The acid-base consequences of changing the composition of the inspired gas from room air to 20 per cent CO_2 and 80 per cent O_2 are shown by the curve OABC in Figure 27-14. Point O is the control value for this particular dog and point A is the hypothetical result of titrating extracellular fluid along a slope of approximately −10 sl to the extremely high Pa_{CO_2} value of 145 mm Hg. Point A would have been reached about one half-hour after the start of CO_2 breathing if no transmembrane or other transfer of H^+ had occurred. The blood $[HCO_3^-]$ had increased more than expected from the ECF titration curve in 1 hour and had reached point B. There was a continuing rise in $[HCO_3^-]_p$ during the next hour and the plasma had moved approximately along the $Pa_{CO_2} = 145$ mm Hg isobar to point C after 2 hours. Note that this rise in $[HCO_3^-]$ moves plasma pH toward normal and intracellular pH away from normal. Actually, part of the added HCO_3^- may have come from bone, but the relative contributions of bone and muscle cells to the increased $[HCO_3^-]$ of plasma are not known for hypercapnia.

The expected increase in plasma $[HCO_3^-]$ is the product of the buffer value of the extracellular fluid and the pH change induced by the increase in Pa_{CO_2}. This can be seen to be about 5 mmols per liter due to a pH decrease of about 0.5. The volume of the extracellular fluid was estimated at 6.55 liters so that $6.55 \times 4 = 26$ mmols of HCO_3^- appeared in the extracellular fluid as a result of the buffering of added carbonic acid by plasma proteins and hemoglobin. Since most carbonic acid was buffered in the red cells and the HCO_3^- formed diffused mainly into the plasma and interstitial fluid, an equal amount of Cl^- must have entered the red cells (chloride shift). Although the results varied considerably in these experiments, approximately this amount of Cl^- disappeared from the extracellular fluid. In addition to the 26 mmols of HCO_3^- resulting from buffering, another 46 mmols appeared over 2 hours (C, Fig. 27-14). The Na^+ content of extracellular fluid increased by 40 mmols and K^+ content by 9 mmols, a total of 49 mmols, close to the change of 46 mmols in HCO_3^-. There was, however, a consistent surplus of $Na^+ + K^+$ over HCO_3^-. This was attributed to an increase in extracellular phosphate concentration. These results can be explained qualita-

tively by the two-way transport of H^+ via the Na^+–K^+ pump postulated above (Fig. 27-13C). The net outflux of HCO_3^- from cells is attributed to the net influx of H^+ because the $[H^+]$ in the extracellular fluid increases more than $[H^+]$ in the intracellular fluid along with a consequent greater increase in the amount of H^+ carried inward on the K^+ leg of the pump. The increased activity of the inward leg would also increase the activity of the outer leg and increase efflux of Na^+. The result is a net exit of Na^+ due to the increased pump activity and a net retention of K^+ in the extracellular fluid due to the increased use of the inward leg by H^+ and a net movement of H^+ into cells. It is more difficult to explain this mixed outflux of Na^+ and K^+ in terms of a H^+ pump–HCO_3^- leak model (Fig. 27-13B) or the Na^+ gradient pump (Fig. 27-13A). Findings of this type are an experimental basis for postulating the H^+, Na^+–K^+, H^+ pump but are no more than suggestive.

Pitts and co-workers[27, 48, 49] also studied the redistribution of ions in the extracellular fluid during extreme hyperventilation. The results of one such experiment are illustrated in Figure 27-14 by the curve O'A'B'C'. The letters have the same time significance as the curve OABC in CO_2 breathing. The most noteworthy feature of the experiment is the high rate at which the plasma $[HCO_3^-]$ fell following the onset of hyperventilation. Balance studies showed that most of the fall in HCO_3^- was due to a large efflux of lactic acid, presumably in the un-ionized, acid form, from the cells and a decrease in phosphate. In addition, there were modest losses of Na^+ and much smaller losses of K^+ from the extracellular fluid. The Cl^- content increased about as expected from the titration of blood buffers in the alkaline direction. The exchange of cell H^+ for plasma Na^+ can be explained along the lines described above for hypercapnia: The fall in Pa_{CO_2} titrates the extracellular fluid further in the alkaline direction than it does the cell fluid. Hence, the number of H^+ entering on the inward leg of the pump decreases more than the number leaving on the outward leg. The decrease in $[H^+]_o$ slows the pump so that there is a net inward leakage of Na^+ and little change in net K^+ flux because the decrease in the rate of the inward leg is due entirely to the fall in $[H^+]_o$.

CONTROL OF RENAL H^+ SECRETION. As

mentioned previously (Chap. 25, Vol. II), the H$^+$ and HCO$_3^-$ concentrations in urine are determined by active cation exchange processes centered around a Na$^+$-for-K$^+$ exchange pump which presumably also pumps H$^+$. The concern here is not the mechanics of the process but the factors controlling the rates of H$^+$ secretion and HCO$_3^-$ reabsorption.

Secretion of H$^+$ can reduce the pH of the luminal fluid to about 4.4; the amount of H$^+$ appearing in the urine depends on the amount of buffer in urine and on the rate of ammonia secretion. Phosphate is the only buffer of any consequence in urine, but the rate of phosphate filtration is relatively constant. The increased renal H$^+$ secretion of hypercapnia and acidosis is due to increased ammonia production. Ammonia is formed in the kidney by metabolism and the neutral form distributes at equal concentrations in the cells and lumen. When the luminal fluid's pH is reduced by the secretion of H$^+$, the ammonia takes up some of the H$^+$ to become NH$_4^+$. The more acid the solution, the greater the concentration of NH$_4^+$ and the greater the rate of H$^+$ excretion.

In either noncarbonic or chronic hypercapnic (respiratory) acidosis, several days to more than a week are needed to reach a new steady state despite a three- to tenfold increase in acid excretion after a few days. Thus a noncarbonic acid excess of 10 mmols per liter of extracellular fluid means that there is a total of 10 mEq per liter × 14.3 liters = 143 mmols of excess acid in the ECF plus an unknown amount of excess acid in the body cells. A typical rate of acid secretion is about 1.5 mEq per hour in normal man and 15 mEq per hour in chronic noncarbonic acid excess. In a normal man at least 143/1.5 = 96 hours = 4 days would be required to excrete the excess acid in the ECF. Additional time would be required to excrete the excess acid in cells and thus restore normal H$^+$ balance. However, acid production would increase during this period and a somewhat shorter time would be adequate. This is in addition to the normal metabolic production of noncarbonic acid by body processes which must also be secreted. (See section below on fecal excretion of alkali.)

HCO$_3^-$ REABSORPTION IN CHRONIC HYPERCAPNIA. The kidneys respond differently to chronic hypercapnia and to noncarbonic acidosis. In the latter, renal acid secretion, if normal or near normal, restores [HCO$_3^-$] and pH to normal, whereas in hypercapnic acidosis, renal compensation increases [HCO$_3^-$] above normal but, depending on the Pa$_{CO_2}$, may or may not restore pH to normal, i.e., may not completely compensate the acidosis (cf. Fig. 27–14, line O″DE). As plasma [HCO$_3^-$] increases, so does the filtered load of HCO$_3^-$ and H$^+$ secretion in the kidney is devoted increasingly to reabsorption of HCO$_3^-$ rather than to acidification of the urine. The degree of renal compensation in chronic hypercapnia is, in some cases, limited by the maximum HCO$_3^-$ reabsorptive capacity of the kidneys, which in turn is determined by the maximal rate of H$^+$ secretion in both the proximal and distal tubules. This limiting bicarbonate concentration is reached when the filtered load of HCO$_3^-$ equals the amount reabsorbed and the plasma level remains constant at its maximum possible value.

The kinetics of renal reabsorption of HCO$_3^-$ and secretion of H$^+$ are best presented in terms of the amount of HCO$_3^-$ reabsorbed per liter of glomerular filtrate—i.e., concentration of HCO$_3^-$ in the reabsorbate—rather than in terms of the amount reabsorbed per minute. When expressed in this way—[HCO$_3^-$] of reabsorbate plotted against [HCO$_3^-$] of plasma—HCO$_3^-$ reabsorption shows a rather sharp threshold; the [HCO$_3^-$] of reabsorbate is equal to the [HCO$_3^-$] of filtrate; i.e., all is reabsorbed, until plasma [HCO$_3^-$] (increased by HCO$_3^-$ infusion keeping Pa$_{CO_2}$ normal) reaches about 26 mmols per liter.[49] At higher plasma [HCO$_3^-$], reabsorbate [HCO$_3^-$] is constant; i.e., under normal circumstances, the maximum HCO$_3^-$ reabsorption is 26 mmols per liter of filtrate. However, this maximum value is directly dependent on Pa$_{CO_2}$: the higher the Pa$_{CO_2}$, the greater the reabsorbate [HCO$_3^-$] and thus the higher the steady-state value of plasma bicarbonate concentration. The relationship between steady-state [HCO$_3^-$] and Pa$_{CO_2}$ is a straight line when plotted on the [HCO$_3^-$]-pH diagram. Line EDF in Figure 27–14 shows that the maximum plasma [HCO$_3^-$] varies directly with the pH, the apparent buffer value being 38 sl. The kind of mechanism giving rise to this type of reabsorptive behavior is not known.

Regardless of the mechanism, the dependence of maximum plasma $[HCO_3^-]$ on P_{CO_2} limits renal compensation for the acidosis of primary hypercapnia. If the plasma $[HCO_3^-]$ necessary to restore plasma pH to normal is less than the maximum plasma $[HCO_3^-]$, then compensation is complete; i.e., pH returns to normal. This is illustrated by the band, O″DE in Figure 27–14, which represents the range of steady-state changes in $[HCO_3^-]_p$ found in a large number of patients with chronic hypercapnia due to a pathologic lung condition such as emphysema.[51] Despite considerable variation among patients, the average renal compensation falls almost exactly on the line of maximum plasma $[HCO_3^-]$ for Pa_{CO_2} values greater than 65 mm Hg (DE). Below 65 mm Hg, pH is restored practically to normal (O″D). The correlation between these two sets of data (FDE and O″DE) is clearly the result of the same function, maximum renal HCO_3^- reabsorptive ability.

The extracellular fluid titration curve drawn through O″ parallel to OA in Figure 27–14, and the maximum renal compensation curve, O″DE, delimit the area in which the plasma $[HCO_3^-]$ and pH of patients with primary hypercapnia will be found. Points not reasonably close to O″DE probably represent transient states if kidney function is normal. Points below the extracellular fluid titration curve and to the left of $Pa_{CO_2} = 40$ mm Hg represent a combination of primary hypercapnia and primary noncarbonic acid excess.

COMPENSATION OF CHRONIC HYPOCAP-NIA. Chronic hypocapnia (hyperventilation) is infrequent but occurs in some forms of hysteria, drug medication (heavy salicylate therapy) and heart disease in which arterial oxygen concentration is sufficiently low to provide a significant hypoxic drive to ventilation. Clinically the extent of renal compensation in chronic hypocapnia is usually complete.[56]

Compensatory Responses to Noncarbonic Acid Imbalance. The CO_2 titration curve of a solution gives a measure of the protein buffers present. The pH changes due to the addition of noncarbonic acid to a solution containing CO_2–HCO_3^- are much smaller than those resulting from adding an equal amount of carbonic acid because noncarbonic acids are buffered by both HCO_3^- and proteins, whereas carbonic acid is buffered only by proteins. The buffer value of the extracellular fluid at fixed Pa_{CO_2} is the sum of the buffer values of HCO_3^- and proteins that due to HCO_3^- is 2.3 $[HCO_3^-]_p = 2.3 \times 24 = 55$ sl, that to proteins is 11 and total buffer value is $55 + 11 = 66$ sl. In cells $[HCO_3^-]$ is about 10 mmols per liter and the contribution of cell protein and phosphate buffers — about twice that of HCO_3^- — is dominant.

The mechanisms whereby the body deals with an excess of noncarbonic acid in the blood were described above. This section first describes the possible sources of an excess or deficit of noncarbonic acid in the ECF and then explains the factors determining the extent of resulting respiratory compensation.

SOURCES OF NONCARBONIC ACID IN THE ECF. Three categories of noncarbonic acid production can be distinguished: (i) Sulfur- and phosphorus-containing substances in the diet which are metabolized to sulfuric and phosphoric acids. These are eventually excreted by the kidneys but must be buffered during their stay in the body. (ii) In some circumstances, organic acid products of intermediary metabolism accumulate in substantial quantities and must be buffered until excreted or until metabolized to CO_2 and water. For example, lactic acid is produced in great quantities in heavy muscular exercise when energy output temporarily exceeds oxygen supply. The lactic acid diffuses out of the cells in the un-ionized, acid form and produces a temporary excess of metabolizable noncarbonic acid in the extracellular fluid. Clinically more important is the acidosis of uncontrolled diabetes; metabolic intermediates accumulating in the ECF may produce coma and death. In both cases, restoring normal conditions removes the excess acid quickly by converting it to CO_2, which is excreted rapidly. The body's response to mineral acids is different from that to lactic acid, because the latter distributes through body water and is metabolized while mineral acids are confined to the ECF and are not metabolized.[45] (iii) Hydration. Excess fluid intake dilutes all the solutes of the body water. The consequent reduction in $[HCO_3^-]$ causes a corresponding increase in $[H^+]$ if Pa_{CO_2} is constant. However, there will be a compensatory hypocapnia. This type of acidosis is more rapidly corrected by excretion of the excess water than by renal secretion of the apparent excess H^+. The acidosis of hydration is a chemical equilibrium effect since the total body H^+

stores have not changed. Dehydration has the opposite effects. However, compensatory responses complicate the picture.[38]

Fecal Excretion of Alkali.[14] Some evidence indicates that there is a steady loss of alkali in the feces, presumably arising from loss of some of the alkaline secretions of the small intestine. Loss of alkali in feces is, of course, equivalent to adding acid to the body fluids. The rate of loss of alkali in the feces is not accurately known but appears to be about the same as the normal rate of secretion of acid in the urine, indicating that the diet is approximately neutral. The loss of alkali in feces serves to emphasize the importance of detailed balance studies where *all* sources and sinks of H^+ are accounted for, *e.g.*, feces, urine, bone. Camien and colleagues[14] argue strongly that the term "potential acidity" should be used instead of H^+ when speaking of acid-base balance in the body. The factors affecting and/or controlling the amount of alkali lost in the feces have not been adequately delineated.

RESPIRATORY COMPENSATION OF NON-CARBONIC ACID UNBALANCES. In acid-base disturbances, respiratory minute volume and renal excretion of acid are regulated, by appropriate mechanisms, in such a way as to change pH toward normal, *i.e.*, the maintenance of pH near normal is more important for adequate body function than maintenance of a Pa_{CO_2} or $[HCO_3^-]$. As described above, the renal response to chronic hypercapnia is to increase plasma $[HCO_3^-]$ far above its normal value, thereby restoring pH to near normal values. On the basis that pH is the quantity primarily regulated, it would be expected that, in conditions of noncarbonic unbalance, respiratory minute volume is regulated in such a way as to change pH toward normal. This is the case but compensation is not complete, *i.e.*, pH is not restored to normal.

As pointed out in Chapter 23, Volume II, alveolar ventilation is determined by the $[H^+]$ of blood, Pa_{CO_2} and Pa_{O_2}. Respiratory minute volume is increased by increases in $[H^+]$ and Pa_{CO_2} and by a decrease in Pa_{O_2}. Excess noncarbonic acid increases plasma $[H^+]$, which stimulates respiration, reduces Pa_{CO_2} and changes $[H^+]$ toward normal. This compensatory response will not restore $[H^+]$ to normal because the reduction in Pa_{CO_2} decreases the drive to the respiratory center; *e.g.*, if compensation were complete, plasma $[H^+]$ would be normal and total drive to the respiratory center would be subnormal, due to the lowered Pa_{CO_2}. Gray[30] developed an equation describing the dependence of the alveolar ventilation ratio on $[H^+]$ and Pa_{CO_2} and this equation can be used to predict the degree of respiratory compensation for any degree of noncarbonic acid unbalance. Since alveolar P_{CO_2} is inversely proportional to alveolar ventilation, Gray's equation can be written

$$\frac{Pa_{CO_2} \text{ (Normal)}}{Pa_{CO_2} \text{ (Exper.)}} = \text{ventilation ratio}$$
$$= 0.22\,[H^+] + 0.262 Pa_{CO_2} - 18 \quad (12)$$

where $[H^+]$ is in nmols per liter and Pa_{CO_2} in mm Hg. This equation can be used to calculate Pa_{CO_2} as a function of plasma $[H^+]$ if alveolar and plasma P_{CO_2}'s are assumed to be equal. This relationship is plotted on the $[HCO_3^-]$-pH diagram in Figure 27–15, together with representative data from the clinical literature[8, 54] showing that the "Gray line" predicts the extent of compensation of chronic noncarbonic acid excess with reasonable accuracy. As a rule of thumb, compensation is about halfway between the $Pa_{CO_2} = 40$ mm Hg isobar and the $pH = 7.4$ vertical line.

The significance of the respiratory compensation curve ("Gray line") can be grasped from study of the lines OA and AB in Figure 27–15. Suppose that 10 mmols of noncarbonic acid per liter of ECF are infused rapidly via a vein (*i.e.*, a noncarbonic acid excess of 10 mM). Assume for the sake of simplicity that the acid is added so fast that arterial P_{CO_2} remains constant. In this case the ECF is titrated along the $Pa_{CO_2} = 40$ isobar to point A. The blood reaching the respiratory center has a subnormal pH and normal P_{CO_2}; hence the respiratory center is stimulated, minute volume is increased, and the ECF is titrated by the fall in Pa_{CO_2} along the 11 sl line from A toward B. As pH increases and Pa_{CO_2} falls, the drive to the respiratory center decreases (see Gray's equation above). Point B, on the Gray line, is the point of balance between the stimulatory effect of the subnormal pH and the inhibitory effect of the subnormal Pa_{CO_2} such that alveolar P_{CO_2} remains constant at a subnormal value. The concentration of ex-

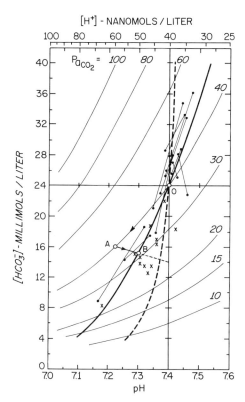

Figure 27–15 Extent of respiratory compensation of noncarbonic acidosis. Solid line, the "Gray" line, extent of compensation predicted from Gray's equation relating alveolar ventilation to plasma [H+] and Pa_{CO_2}. Dashed line, predicted compensation on assumption that Pa_{CO_2} term drops out after 24 hours because of compensatory changes in cerebrospinal fluid [HCO₃⁻]. Curve OAB illustrates how these points may be reached; excess noncarbonic acid titrates from O to A; hyperventilation due to stimulation of respiration by decreased pH titrates extracellular fluid along its buffer curve to B, a point on the Gray line. Solid dots connected by lines show the acid-base status of patients with normal renal function (upper group) and chronic renal insufficiency (lower group) before (lower) and after (upper) taking large doses of NaHCO₃ for 3 days. In general the changes tend to parallel the "Gray" line. Crosses show the acid-base status of another group of patients with chronic renal insufficiency. (After Bradley and Semple, *J. Physiol. (Lond.)*, 1962, *160*, 381–391 (•). Schwab, *Klin. Wschr.*, 1962, *40*, 765–772 (X).)

cess noncarbonic acid is estimated by extending the CO_2 titration line AB until it reaches normal pH (dashed line, Fig. 27–15). The [HCO₃⁻] here is 14 mM, 10 mM below the normal value of 24 mM.

Respiratory compensation for a deficit of noncarbonic acid is usually rather less than predicted in the chronic state[52] and depends on the method of induction of the alkalosis

in volunteers.[28] Compensation is clearly present in shorter-term disturbances produced by infusion or ingestion of NaHCO₃ (Fig. 27–15).[8] This short-term respiratory compensation (hypoventilation) is usually somewhat less than predicted from Gray's equation in part because appreciable hypoventilation reduces Pa_{O_2}, which may stimulate respiration. Hypoxic drive to respiration is probably the main reason for relatively small hypoventilatory compensation for chronic noncarbonic alkalosis.

It takes time to alter alveolar and plasma P_{CO_2} values following a change in alveolar ventilation so that there is a lag in respiratory compensation of a noncarbonic acid unbalance. The time course of respiratory compensation is described above; the main component is the time required for circulation to alter the body's stores of carbon dioxide—30 min to a few hours. Thus, compared to the week-long time scale of renal compensation of a carbonic acid unbalance, respiratory compensation of a noncarbonic acid unbalance is relatively fast.

Adaptation to Chronically Altered Pa_{CO_2}. It was pointed out in Chapter 23, Volume II, that the effect of Pa_{CO_2} on respiration is mediated through a [H+]-sensitive area in the fourth ventricle which is bathed by cerebrospinal fluid. However, the change in [H+] due to altered Pa_{CO_2} is compensated for in about 24 hours by an appropriate change in cerebrospinal [HCO₃⁻]. On this basis, the degree of respiratory compensation for noncarbonic acid excess should increase over 24 hours, at which time [H+] of plasma should be the main factor determining the respiratory minute volume (dashed line in Fig. 27–15) and there may be a 24-hour component of the respiratory compensation transient. However, there is little clinical evidence for a greater degree of respiratory compensation in chronic—as compared with acute—noncarbonic acid excess (*cf.* Fig. 27–15).

USING THE [HCO₃⁻]ₚ-pH DIAGRAM

The principles developed in this chapter can best be summarized by reviewing the deductions that can be made about acid-base status of a patient from the position of a point on the [HCO₃⁻]ₚ-pH diagram. In addition to defining the direction of the acid-base unbalance—*e.g.*, acidosis or alkalosis, hyper- or hypocapnia (excess or deficit of carbonic acid), excess or deficit of noncarbonic acid—the location of the point helps to distinguish between the primary pathologic condition and any compensatory changes and to determine whether or not a steady state has been reached. More specifically, the answers to several questions are wanted: (i) Is the primary cause of the acid-base disturbance carbonic or noncarbonic? (ii) Have there been compensatory changes? If so, to what extent? (iii) Is the patient in a steady state of unbalance? If not, what is the phase of the transient state, *e.g.*, circulatory mixing transient, bone carbonate release or renal? (iv) What is the total unbalance in the extracellular fluid? (v) Are corrective measures indicated, *e.g.*, intravenous infusion of alkali or acid?

Examples. The manner and extent to which these answers are obtainable from the $[HCO_3^-]_p$-pH diagram can best be illustrated by specific examples, but all available clinical information should be used in reaching a decision. The following examples illustrate how the maximum information can be obtained from the position of a point on the $[HCO_3^-]_p$-pH diagram (Fig. 27–16) and the limitations and uncertainties which hamper judgment.

Point A in Figure 27–16 ($pH = 7.34$, $[HCO_3^-] = 31.5$ mmols per liter) represents acidosis (slight), significant excess of carbonic acid (hypercapnia, $Pa_{CO_2} = 60$ mm Hg) and a significant deficit of noncarbonic acid. (i) The high Pa_{CO_2} and a $[HCO_3^-]$ above the ECF buffer line means that the primary pathologic condition is carbonic acid excess. The cause of the hypercapnia must be determined by other means, such as lung function tests. (ii) Considering the variability of renal compensatory responses and the point's nearness to the steady-state, maximal renal compensation line, this is likely a complete renal compensation of the primary hypercapnia. (iii) The patient is likely in a steady state, having a chronic excess of carbonic acid and deficit of noncarbonic acid. (iv) There is a compensatory noncarbonic acid deficit of 7 mmols per liter of extracellular fluid (vertical distance between A and ECF, the extracellular fluid titration curve). (v) No acid-base corrective measures are indicated, since the patient is in a compensated steady state. Therapy would be directed toward improving lung function.

Point B ($pH = 7.23$, $[HCO_3^-] = 32.5$ mmols per liter, $Pa_{CO_2} = 80$ mm Hg) also represents a case of hypercapnia and acidosis. (i) The point is above the extracellular buffer line, so the primary disturbance is an excess of carbonic acid. (ii) This excess is partially compensated by bone carbonate release, transmembrane H^+ transfer and renal acid excretion. (iii) The point is well away from the region of maximal renal compensation, so the patient is in a transient state. There are at least two ways that the blood could have reached point B: It may have been reached a few days after the onset of respiratory insufficiency, renal compensation not yet having had time to reach completion. On the other hand, point B may have been reached from point A about 30 min after a sudden worsening of lung function, increase in Pa_{CO_2} and titration of the extracellular fluid in the acid direction. Since point B is on the extracellular titration curve through A, the deficit of noncarbonic acid at points A and B is 7 mmols per liter. It is not possible to determine which of these two possibilities is the more correct one from the data given, but might be from the patient's history. In either case, the patient is in a transient state with respect to renal compensation and the blood will move up the $Pco_2 = 80$ mm Hg isobar to the maximum renal compensation curve over a period of days, provided that Pco_2 remains constant. (iv) The noncarbonic acid deficit of 7 mmols per liter is a partial renal compensation of the carbonic acid excess. (v) Therapy would center around improving respiratory function.

Point B' ($pH = 7.43$, $[HCO_3^-]_p = 30.7$ mM, $Pa_{CO_2} = 48$ mm Hg) shows the approximate position of the plasma about 1 hour after restoration of lung function to normal. The compensatory deficit of noncarbonic acid now becomes a primary noncarbonic acid deficit (metabolic acidosis), and so the blood approaches the respiratory compensation steady-state curve (RC, Figure 27–16) along an 11 sl titration curve (BB', Fig. 27–16) parallel to the ECF curve. If respiratory function remained normal, the excess bicarbonate would be rapidly (many hours) lost in the urine and the point would approach normal along the Gray line (B'O, Fig. 27–16).

Point C ($pH = 7.21$, $[HCO_3^-]_p = 26.6$ mmols per liter) falls near the extracellular fluid titration curve, so the disturbance is

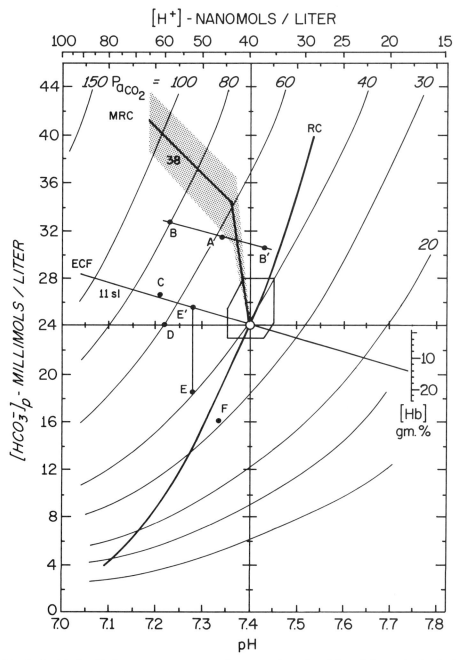

Figure 27–16 $[HCO_3^-]_p$-pH diagram summarizing the principles of acid-base balance in the body. A point on this diagram permits a number of conclusions to be drawn concerning a patient's acid-base status and permits an estimate of his total extracellular fluid H^+ unbalance. Steady-state or chronic compensation lines for hypercapnia (maximum renal compensation line, MRC) and excess noncarbonic acid (respiratory compensation, Gray line, RC) are reproduced from previous figures. Points near these lines indicate that the patient is in a steady state of acid-base unbalance. Points not close to these lines indicate a transient state. The hexagon in the middle indicates the normal range of variations; this range always must be taken into account in diagnoses. Amount of excess noncarbonic acid in the extracellular fluid can be estimated by multiplying the excess per liter (vertical distance between point and the ECF line) by the volume of extracellular fluid as estimated from body weight. Buffer value of ECF depends on blood hemoglobin concentration. Scale labeled [Hb] allows approximate correction of the slope of the ECF titration curve for hemoglobin concentration of blood. Corrected titration curve is obtained by marking hemoglobin concentration (in grams per cent) on [Hb] scale and drawing a straight line through this point and normal (pH = 7.4, $[HCO_3^-]_p$ = 24 mM). Significance of points A-F is described in text. A similar diagram is given by Refsum, *Scand. J. clin. lab. Invest.*, 1971, *27* (Suppl. 118), 8–11.

uncompensated primary hypercapnia. Therapy would be directed at improving respiratory function. This is a fairly improbable point since it would be found only about an hour after onset of hypercapnia (movement along extracellular buffer line) or in acute increase of noncarbonic acid on top of a previously compensated hypercapnia (movement down $P_{CO_2} = 70$ mm Hg isobar).

Point D ($pH = 7.22$, $[HCO_3^-]_p = 24$ mmols per liter) is a combined carbonic and noncarbonic acidosis: (i) The primary disturbances are hypercapnia (hypoventilation) and noncarbonic acid excess. (ii) No compensatory changes have occurred. (iii) This is almost certainly an acute, transient state. (iv) Noncarbonic acid excess amounts to 2 mmols per liter. (v) This state could be brought about, for example, by gastrointestinal upset in a patient with emphysema where considerable amounts of alkali are lost from the intestinal tract.

Point E ($pH = 7.28$, $[HCO_3^-]_p = 18.3$ mmols per liter) is acidosis with excess noncarbonic acid and normal P_{CO_2}. No compensatory changes have occurred and the patient is in a transient state or is a hyporeactor with respect to respiratory compensation. An alternate possibility is that the patient's blood normally has a rather high P_{CO_2}, say 48 mm Hg, and that a Pa_{CO_2} of 40 mm Hg represents a substantial respiratory compensation for the primary noncarbonic acidosis. If kidney function is normal, this point represents a transient state of excess noncarbonic acid, which is excreted by the kidneys during the next several days provided there is no continuous extra source of the excess acid, such as in uncontrolled diabetes. In some instances, giving an intravenous infusion of $NaHCO_3$ or sodium lactate might be desirable to neutralize some of the excess noncarbonic acid, the amount to be given depending on the total noncarbonic acid excess in the body. However, on the basis of present knowledge, it is not possible to estimate accurately how much of the excess acid has been neutralized by bone carbonate or gone into cells. A safe and simple procedure is to neutralize only the excess noncarbonic acid in the ECF. The amount of $NaHCO_3^-$ to be administered is equal to the noncarbonic acid excess per liter of extracellular fluid times the number of liters of extracellular fluid. At point E, this is 7 mmols per liter (distance EE', Fig.

27–16) $\times$ 14 liters extracellular fluid = 100 mmols of $NaHCO_3$ for a 70 kg man. Estimates such as this are subject to large errors due to uncertainty about ECF volume. Since about half of the excess noncarbonic acid appearing in ECF is neutralized by bone carbonate, administration of 100 mmol of bicarbonate will not overcorrect the disturbance. (See also reference 43.)

Point F ($pH = 7.33$, $[HCO_3^-]_p = 16$ mmols per liter) represents about the same condition as E except that respiratory compensation is clearly substantial, about as expected from the Gray line. This is a steady state of respiratory compensation and, except in kidney disease, a transient state of renal compensation. Renal H^+ excretion will restore the blood to normal over several days. As excess noncarbonic acid is eliminated from the body, the blood will "move" along the Gray line toward normal, since the respiratory steady state is achieved in about an hour and hence will always be near the steady state value during the renal transient.

The same principles and considerations are used in interpreting points on the alkaline side of normal.

The large variations in the apparent buffer value of blood following a sudden change in arterial P_{CO_2} (Fig. 27–8) indicate that caution is required in obtaining arterial blood samples from conscious patients; hyperventilation induced in a patient by the sight of a hypodermic needle and syringe approaching his arm could cause marked changes in the pH and $[HCO_3^-]$ of such samples. The obtaining of arterialized capillary blood samples is less traumatic and may avoid this difficulty. Venous blood samples are much more representative of the actual acid-base status of the tissue from which the blood comes, but the complication of correcting for the low oxygen tension and the difficulty in obtaining mixed venous blood samples (right atrium) make this procedure unattractive in the clinic.

Six Areas of the $[HCO_3^-]_p$-pH Diagram. The $[HCO_3^-]_p$-pH diagram can be divided into several regions. A point to the left of $pH = 7.4$ is acidosis; to the right is alkalosis. Any P_{CO_2} greater than 40 is hypercapnia and less than 40 is hypocapnia. The body is in a steady state of acid-base balance at any point on the maximum renal compensa-

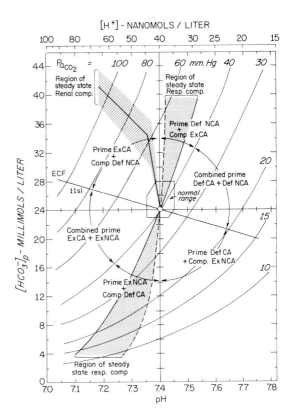

Figure 27–17 Six diagnostic regions of the $[HCO_3^-]_p$-pH diagram. The $pH = 7.40$ vertical line, $Pa_{CO_2} = 40$ mm Hg isobar and the ECF titration curve divides the $[HCO_3^-]_p$-pH, plane into six regions, denoted by curved arrows. The primary cause and compensatory response, if any, are shown for each region. Abbreviations: ExNCA = excess of noncarbonic acid (metabolic acidosis). ExCA = excess of carbonic acid (hypercapnia, respiratory acidosis). DefNCA = deficit of noncarbonic acid (metabolic alkalosis). DefCA = deficit of carbonic acid (hypocapnia, respiratory alkalosis). Comp = compensatory or compensation. Prime = primary cause or pathology.

tion line and in a steady state of respiratory function anywhere along the Gray line but is in a transient state of renal acid secretion unless there is a pathologic renal condition. Points not near these lines represent different stages in the respiratory and renal system transients.

The $[HCO_3^-]_p$-pH diagram is divided into six regions by three lines (Fig. 27–17): (i) The vertical line through $pH = 7.40$ separates acidosis from alkalosis. More accurately, the vertical strip between $pH = 7.35$ and 7.45 separates acidosis from alkalosis. (ii) The strip between $Pa_{CO_2} = 35$ and 48 mm Hg separates an excess of carbonic acid (hypercapnia, hypoventilation). (iii) The ECF titration curve (11 sl line through normal) separates an excess of noncarbonic acid (metabolic acidosis, negative base excess) from a deficit of noncarbonic acid (metabolic alkalosis, positive base excess). The six regions thus formed are named in Figure 27–17 according to the primary functional disorder and compensatory response, if any. The normal range and maximum renal and respiratory compensation curves are

also shown. The plotting of acid-base data from blood on the $[HCO_3^-]_p$-pH graph permits (i) diagnosis of the primary cause; (ii) determination of the degree of compensatory response, transient or steady-state; and (iii) estimation of the concentration of noncarbonic and carbonic acids in the extracellular fluid; and (iv) provides a rational basis for therapy. Thus, a reasonably accurate picture of the acid-base status of a patient can be obtained by considering the factors which cause the blood to reach the state represented by a point on the $[HCO_3^-]_p$-pH diagram.

PROBLEMS

1. A 1 liter solution contains 10 mmols per liter of a weak acid whose dissociation constant is 40 nmols per liter. Initially, 1 mmole of the weak acid is in the form HA (acid) and 9 mmols in the form A^- (conjugate base). Concentrated HCl of negligible volume is added to the solution in 1 mmol increments until a total of 8 mmols have

been added. Calculate the $[H^+]$ initially and after each addition.

$$K_a' = \frac{[H^+]\,[A^-]}{[HA]}$$

Assume change in $[A^-]$ = amount added acid. Tabulate the results and make a plot of the titration curve of the solution in the following three ways: (i) Amount of added acid (mmols) on the abscissa and corresponding $[H^+]$ on the ordinate. (ii) pH on the abscissa and amount of added acid on the ordinate. (iii) Measure the maximum slope of the titration curve and compare with the expected value of 5.75 slykes.

2. Derive the Henderson-Hasselbalch equation, $pH = pK + \log [A^-]/[HA]$, where $pK = -\log K$, from the weak acid equilibrium condition $[H^+] = K[HA]/[A^-]$.

3. For carbonic acid, the Henderson-Hasselbalch equation is

$$pH = pK_a' + \log [HCO_3^-]/aP_{CO_2}$$

For human plasma, $pK_a' = 6.1$ and $a = 0.03$ mmols/liter per mm Hg. Calculate and plot the titration curve for a solution containing 24 mmols per liter of $NaHCO_3$ and 125 mmols per liter of $NaCl$ (equivalent to plasma) with P_{CO_2} held constant at 40 mm Hg. Note that the amount of added strong acid (HCl) is equal to the negative of the change in $[HCO_3^-]$ because every H^+ added combines with a HCO_3^- to form H_2CO_3 and thence CO_2 except for the negligible few that remain in the ionized form.

4. On a $[HCO_3^-]_p$-pH diagram (e.g., Fig. 27–16) draw CO_2 titration curves through $pH = 7.4$ and $[HCO_3^-] = 24$ for (i) true plasma (31 slykes), (ii) whole blood (26 slykes), and (iii) extracellular space (11 slykes) for a normal person.

5. Complete the accompanying table. Sufficient accuracy for $[HCO_3^-]_p$, P_{CO_2}, total $[CO_2]$ and excess or deficit on noncarbonic acid values can be obtained by plotting the points on the $[HCO_3^-]_p$-pH diagram (e.g., on Fig. 27–16) and reading off the desired values by interpolation. Assume $[Hb] = 15.0$ g per cent. (Buffer slope = 11 slykes.) Under "diagnosis" give the cause of the acid-base abnormality in each case, the compensatory response, if any, and the extent of completion of the compensatory response, i.e., transient or steady-state. (See table below.)

6. A 70 kg man ingests 12.5 g of $NaHCO_3$. Estimate and plot plasma pH and $[HCO_3^-]_p$ (assume that the $NaHCO_3$ is entirely absorbed and remains entirely in the extracellular fluid):

   (i) After absorption but before any respiratory compensation.

   (ii) After any respiratory compensation is completed.

   (iii) Draw the probable path followed by the blood as the $NaHCO_3$ is excreted by the kidney.

7. The following measurements were obtained from true plasma of a normal person:

	pH	Pa_{CO_2}	$[HCO_3^-]_p$
Hyperventilation	7.62	20	
Normal	7.42	40	

Calculate $[HCO_3^-]_p$ and the buffer value of the true plasma. Estimate about how long after the change in respiration the blood samples were taken and give reasons for your answer.

8. A 60 kg patient comes into the hospital with acute asthma. Hemoglobin concentration is normal. Pa_{CO_2} is found to be 65 mm Hg. What arterial plasma bicarbonate concentration would be expected (i) 2 min, (ii) 30 min, (iii) 2 hours, and (iv) 5 days

pH	PA_{CO_2} (MM HG)	$[HCO_3^-]$ (MMOL/L)	TOTAL $[CO_2]$ (MMOL/L)	EXCESS OR DEFICIT OF NONCARBONIC ACID	"DIAGNOSIS"
A. 7.45	20				
B. 7.34	60				
C. 7.21			28		
D. 7.28	40				
E.	50		33		
F. 7.33	32				
G. 7.60	25				

pH	Pa_{CO_2} (MM HG)	$[HCO_3^-]p$ (MMOL/L)	TOTAL CO_2 (MMOL/L)	EXCESS (+) OR DEFICIT (−) OF NONCARBONIC ACID (MMOL/L)	"DIAGNOSIS"
A. 7.45	20	13.5	14.1	10	Partially comp. hypocap., renal transient
B. 7.34	60	32	33.8	−7	Partially comp. hypercap., nearly steady-state
C. 7.21	67	26	28	0	Hypercap., no comp., transient, (see point E', Fig. 27–16)
D. 7.28	40	18.5	19.7	7	Excess NCA, no comp., transient (see point E, Fig. 27–16)
E. 7.43	50	31.5	33	−6.5	Deficit of NCA; resp. comp., steady-state
F. 7.33	32	16	17	9	Excess NCA with partial resp. comp., steady-state
G. 7.60	25	24	25	−2.5	Combined hypocap. and deficit of NCA, transient

Note: Italicized numbers were given in the original problem.

after the onset of the attack? Assume Pa_{CO_2} remains steady at about 65 mm Hg. Explain the reasons for the changes in $[HCO_3^-]_p$.

ANSWERS

1. (i) The graph should be identical in shape to that in Figure 27–1A. The value of $[H^+]$ is 40 nmols per liter when 4 mmols of acid have been added so that $[A^-]/[HA] = 1$ and $[H^+] = K$.

(ii) The graph should be identical in shape to that in Figure 27–2. Since $pK = -\log K = 7.4$, $pH = 7.4$ when added acid = 4 mmols per liter.

(iii) The expected slope of the line tangent to the titration curve is the maximum molar buffer value times total buffer concentration = 0.575×10 mmols per liter = 5.75 slykes.

2. The Henderson-Hasselbalch equation is derived for the CO_2-$[HCO_3^-]$ system in equations (8) to (10).

3. The plot is the $Pa_{CO_2} = 40$ mm Hg isobar.

4. (i) At $pH = 7.0$, $\triangle pH = 7.0 - 7.4 = -0.4$. For buffer value = −31 slykes, $\triangle[HCO_3^-]_p = (-31)(-0.4) = 12.1$ mM. $[HCO_3^-] = 12.1 + 24 = 36.1$ mM. Plot point 7.0, 36.1 mM and draw straight line through this point and normal (7.4, 24 mM).

(ii) For whole blood $\triangle[HCO_3^-] = (-26)$

$(-0.4) = 10.4$ mM and $[HCO_3^-] = 10.4 + 24 = 34.4$ mM at $pH = 7.0$.

(iii) For ECF $\triangle[HCO_3^-] = (-11)(-0.4) = 4.4$ mM, $[HCO_3^-] = 28.8$ at pH 7.0.

5. Answer to question 5 is table shown above.

6. Assume volume of ECF = 14 L

Molecular weight of $NaHCO_3 = 84$; 12.5 g

$NaHCO_3 \times \dfrac{1 \text{ mol}}{84 \text{ g } NaHCO_3} = 0.14$ mols;

$\dfrac{0.14 \text{ M}}{14 \text{ L}} = 10$ mM/L of HCO_3^- added to ECF. This is deficit of noncarbonic acid, so $[HCO_3^-]$ at $pH = 7.4 = 24 + 10 = 34$ mmol/L.

(i) To find $[HCO_3^-]$ at $Pa_{CO_2} = 40$ mm Hg, draw 11 slyke line through 7.4, 34 mmol/L and go along 11 sl line to $Pa_{CO_2} = 40$ mm Hg isobar:

$$[HCO_3^-] = 33, pH = 7.53.$$

(ii) After respiratory compensation

$P_{CO_2} \simeq 50$ mm Hg, $pH \simeq 7.47$, $[HCO_3^-] = 33.5$.

(iii) The point follows the Gray line toward normal values as excess bicarbonate is excreted.

7.

	pH	Pa_{CO_2}	$[HCO_3^-]_p$
Hyperventilation	7.62	20	20 mM
Normal	7.42	40	25 mM

Note: Italicized numbers were given in the original problem. Buffer value = $\dfrac{25 - 20}{7.62 - 7.42}$

= 25 slykes.

Samples were taken a few minutes after change in respiration: At zero time $\beta = 31$ and is approaching 11 slykes in 30 to 60 min because of time required for blood to equilibrate with ECF.

8. (i) 28 mM per liter; (ii) 26 mM per liter; (iii) 26 − 30 mM per liter; (iv) 37 mM per liter. The early fall (i) to (iii) results from dilution of plasma by ECF. The late rise (iv) is the result of renal H^+ secretion and HCO_3^- reabsorption.

REFERENCES

1. Adler, S., Anderson, B., and Zemotel, L. Metabolic acid-base effects on tissue citrate content and metabolism in the rat. *Amer. J. Physiol.*, 1971, *220*, 986–992.

2. Adler, S., Roy, A., and Relman, A. S. Intracellular acid-base regulation. I. The response of muscle cells to changes in CO_2 tension or extracellular bicarbonate concentration. *J. clin. Invest.*, 1965, *44*, 8–20.

3. Altman, P. L., and Dittmer, D. S., eds. *Blood and other body fluids*. Washington, D.C., Federation of American Societies for Experimental Biology, 1961.

4. Arbus, G. S., Hebert, L. A., Levesque, P. R., Etsten, B. E., and Schwartz, W. B. Characterization and clinical application of the "significance band" for acute respiratory alkalosis. *New Engl. J. Med.*, 1969, *280*, 117–123.

5. Blumentals, A. S., ed. Symposium on acid-base balance. *Arch. intern. Med.*, 1965, *116*, 647–742.

6. Brackett, N. C., Jr., Cohen, J. J., and Schwartz, W. B. Carbon dioxide titration curve of normal man. Effect of increasing degrees of acute hypercapnia on acid-base equilibrium. *New Engl. J. Med.*, 1965, *272*, 6–12.

7. Brackett, N. C., Jr., Wingo, C. F., Orhan, M., and Solano, J. T. Acid-base response to chronic hypercapnia in man. *New Engl. J. Med.*, 1969, *280*, 124–130.

8. Bradley, R. D., and Semple, S. J. G. A comparison of certain acid-base characteristics of arterial blood, jugular venous blood and cerebrospinal fluid in man, and the effect on them of some acute and chronic acid-base disturbances. *J. Physiol. (Lond.)*, 1962, *160*, 381–391.

9. Brown, E. B., Jr. *In vivo* and *in vitro* carbon dioxide dissociation: Vertebrates, Part 1. CO_2 dissociation curves. In: *Biological handbooks, respiration and circulation*, P. L. Altman and D. S. Ditmars, eds. Bethesda, Maryland, Federation of American Societies for Experimental Biology, 1971.

10. Brown, E. B., Jr. Whole body buffer capacity. In: *Ion homeostasis of the brain*, B. K. Siesjo and S. C. Sørensen, eds. Copenhagen, Munksgaard, 1971.

11. Burnell, J. M., and Teubner, E. J. Changes in bone sodium and carbonate in metabolic acidosis and alkalosis in the dog. *J. clin. Invest.*, 1971, *50*, 327–331.

12. Burnell, J. M., and Teubner, E. J. Personal communication.

13. Butler, J. N. *Ionic equilibrium, a mathematical approach.* Reading, Mass., Addison-Wesley, 1964.

14. Camien, M. N., Simmons, D. H., and Gonick, H. C. A critical reappraisal of "acid-base" balance. *Amer. J. clin. Nutr.*, 1969, *22*, 786–793.

15. Caldwell, P. C. Intracellular pH. *Int. Rev. Cytol.*, 1956, *5*, 229–277.

16. Caldwell, P. C. Studies on the internal pH of large muscle and nerve fibres. *J. Physiol. (Lond.)*, 1958, *142*, 22–62.

17. Christensen, H. N. *Neutrality control in the living organism.* Philadelphia, W. B. Saunders Co., 1971.

18. Cohen, J. J., Brackett, N. C., Jr., and Schwartz, W. B. The nature of the carbon dioxide titration curve in the normal dog. *J. clin. Invest.*, 1964, *43*, 777–786.

19. Davenport, H. W. *The ABC of acid-base chemistry*, 4th ed. Chicago, University of Chicago Press, 1969.

20. Edsall, J. T., and Wyman, J. *Biophysical chemistry*, Vol. 1. New York, Academic Press, 1958.

21. Farhi, L. E., and Rahn, H. Dynamics of changes in carbon dioxide stores. *Anesthesiology*, 1960, *21*, 604–614.

22. Fenn, W. O. Carbon dioxide and intracellular homeostasis. *Ann. N.Y. Acad. Sci.*, 1961, *92*, 547–558.

23. Ferguson, J. K. W. Carbamino compounds of CO_2 with human haemoglobin and their role in the transport of CO_2. *J. Physiol. (Lond.)*, 1936, *88*, 40–55.

24. Forster, R. E., Edsall, J. T., Otis, A. B., and Roughton, F. J. W., eds. *CO_2: chemical, biochemical, and physiological aspects.* Washington, D.C., Scientific and Technical Information Division, Office of Technology Utilization, National Aeronautics and Space Administration, 1969.

25. Frisell, W. R. *Acid-base chemistry in medicine.* New York, The Macmillan Company, 1968.

26. German, B., and Wyman, J., Jr. The titration curves of oxygenated and reduced hemoglobin. *J. biol. Chem.*, 1937, *117*, 533–550.

27. Giebisch, G., Berger, L., and Pitts, R. F. The extrarenal responses to acute acid-base disturbances of respiratory origin. *J. clin. Invest.*, 1955, *34*, 231–245.

28. Goldring, R. M., Cannon, P. J., Heinemann, H. O., and Fishman, A. P. Respiratory adjustment to chronic metabolic alkalosis in man. *J. clin. Invest.*, 1968, *47*, 188–202.

29. Gray, B. A. The rate of approach to equilibrium in uncatalyzed CO_2 hydration reactions: The theoretical effect of buffering capacity. *Resp. Physiol.*, 1971, *11*, 223–234.

30. Gray, J. S. *Pulmonary ventilation and its physiological regulation.* Springfield, Ill., Charles C Thomas, 1950.

31. Hastings, A. B., Sendroy, J., Jr., and Van Slyke, D. D. Studies of gas and electrolyte equilibria in blood. XII. The value of pK' in the Hen-

derson-Hasselbalch equation for blood serum. *J. biol. Chem.*, 1928, 79, 183–192.

32. Heisler, N., and Piiper, J.   The buffer value of rat diaphragm muscle tissue determined by PCO_2 equilibration of homogenates. *Resp. Physiol.*, 1971, 12, 169–178.

33. Hills, A. G.   *Acid-base balance chemistry, physiology, pathophysiology.* Baltimore, Williams and Wilkins Co., 1973.

34. Izutsu, K. T.   Intracellular pH, H ion flux and H ion permeability coefficient in bullfrog toe muscle. *J. Physiol. (Lond.)*, 1972, 221, 15–27.

35. Izutsu, K. T., and Siegel, I. A.   A microsomal HCO_3^--stimulated ATPase from the dog submandibular gland. *Biochim. biophys. Acta (Amst.)*, 1972, 284, 478–484.

36. Kasbekar, D. K., and Durbin, R. P.   An adenosine triphosphatase from frog gastric mucosa. *Biochim. biophys. Acta (Amst.)*, 1965, 105, 472–482.

37. Kildeberg, P.   *Clinical acid-base physiology.* Baltimore, Williams and Wilkins Co., 1968.

38. Libermann, I. M., and García-Pierce, H.   Quantitative displacement of blood acid-base equilibrium in acutely water depleted dogs. *Pflügers Arch. ges. Physiol.*, 1971, 330, 51–60.

39. Masoro, E. J., and Siegel, P. D.   *Acid-base regulation: Its physiology and pathophysiology.* Philadelphia, W. B. Saunders Co., 1971.

40. Matthews, C. M. E., Laszlo, G., Campbell, E. J. M., Kibby, P. M., and Freedman, S.   Exchange of $^{11}CO_2$ in arterial blood with body CO_2 pools. *Resp. Physiol.*, 1968, 6, 29–44.

41. Matthews, C. M. É., Laszlo, G., Campbell, E. J. M., and Read, D. J. C.   A model for the distribution and transport of CO_2 in the body and the ventilatory response to CO_2. *Resp. Physiol.*, 1968, 6, 45–87.

42. McNicol, M. W., and Campbell, E. J. M.   Severity of respiratory failure. Arterial blood gases in untreated patients. *Lancet*, 1965, 336–338.

43. Mellemgaard, K., and Astrup, P.   The quantitative determination of surplus amounts of acid or base in the human body. *Scand. J. clin. lab. Invest.*, 1960, 12, 187–199.

44. Messeter, K., and Siesjö, B. K.   The intracellular pH' in the brain in acute and sustained hypercapnia. *Acta physiol. scand.*, 1971, 83, 210–219.

45. Mithoefer, J. C., and Karetzky, M. S.   Comparative effects of organic and mineral acid on acid-base balance and gas exchange. *J. lab. clin. Med.*, 1968, 72, 924–932.

46. Osnes, J.-B., and Hermansen, L.   Acid-base balance after maximal exercise of short duration. *J. appl. Physiol.*, 1972, 32, 59–63.

47. Perutz, M. F.   The hemoglobin molecule. *Proc. Roy. Soc.*, 1969, B 173, 113–149.

48. Pitts, R. F.   Mechanisms for stabilizing the alkaline reserves of the body. *Harvey Lect.*, 1954, 48, 172–209.

49. Pitts, R. F.   *Physiology of the kidney and body fluids.* Chicago, Year Book Medical Publishers, 1963.

50. Polak, A., Haynie, G. D., Hays, R. M., and Schwartz, W. B.   Effects of chronic hypercapnia on electrolyte and acid-base equilibrium. I. Adaptation. *J. clin. Invest.*, 1961, 40, 1223–1237.

51. Refsum, H. E.   Acid-base status in patients with chronic hypercapnia and hypoxaemia. *Clin. Sci.*, 1964, 27, 407–415.

52. Roberts, K. E., Poppell, J. W., Vanamee, P., Beals, R., and Randall, H. T.   Evaluation of respiratory compensation in metabolic alkalosis. *J. clin. Invest.*, 1956, 35, 261–266.

53. Roughton, F. J. W.   Transport of oxygen and carbon dioxide. *Handb. Physiol.*, 1964, sec. 3, vol. I, 767–825.

54. Schwab, M.   Das Säure-Basen-Gleichgewicht im arteriellen Blut und Liquor cerebrospinalis bei chronischer Niereninsuffizienz. *Klin. Wschr.*, 1962, 40, 765–772.

55. Siesjo, B. K., Folbergrova, J., and MacMillan, V.   The effect of hypercapnia upon intracellular pH in the brain, evaluated by the bicarbonate–carbonic acid method and from the creatine phosphokinase equilibrium. *J. Neurochem.*, 1972, 19, 2483–2495.

56. Siggaard-Andersen, O.   *The acid-base status of the blood.* Baltimore, Williams and Wilkins Co., 1964.

57. Stoddart, J. C.   The effects of voluntarily controlled alveolar hyperventilation on carbon dioxide excretion. *Quart. J. exp. Physiol.*, 1967, 52, 369–381.

58. Van Slyke, D. D.   On the measurement of buffer values and on the relationship of buffer value to the dissociation constant of the buffer and the concentration and reaction of the buffer solution. *J. biol. Chem.*, 1922, 52, 525–570.

59. Van Slyke, D. D., Hastings, A. B., Hiller, A., and Sendroy, J., Jr.   Studies of gas and electrolyte equilibria in blood. XIV. The amounts of alkali bound by serum albumin and globulin. *J. biol. Chem.*, 1928, 79, 769–780.

60. Van Slyke, D. D., and Sendroy, J., Jr.   Studies of gas and electrolyte equilibria in blood. XV. Line charts for graphic calculations by the Henderson-Hasselbalch equation, and for calculating plasma carbon dioxide content from whole blood content. *J. biol. Chem.*, 1928, 79, 781–798.

61. Van Slyke, D. D., and Sendroy, J., Jr.   Studies of gas and electrolyte equilibria in blood. XVII. The effect of oxygenation and reduction on the carbon dioxide absorption curve and the pK' of whole blood. *J. biol. Chem.*, 1933, 102, 505–519.

62. Van Slyke, D. D., Sendroy, J., Jr., Hastings, A. B., and Neill, J. M.   Studies of gas and electrolyte equilibria in blood. X. The solubility of carbon dioxide at 38° in water, salt solution, serum, and blood cells. *J. biol. Chem.*, 1928, 78, 765–799.

63. Van Slyke, D. D., Wu, H., and McLean, F. C.   Studies of gas and electrolyte equilibria in the blood. V. Factors controlling the electrolyte and water distribution in the blood. *J. biol. Chem.*, 1923, 56, 765–849.

64. Waddell, W. J., and Butler, T. C.   Calculation of intracellular pH from the distribution of 5,5-dimethyl-2,4-oxazolidinedione (DMO). Application to skeletal muscle of the dog. *J. clin. Invest.*, 1959, 38, 720–729.

65. Waddell, W. J., and Bates, R. G.   Intracellular pH. *Physiol. Rev.*, 1969, 49, 285–329.

66. Wiebelhaus, V. O., Sung, C. P., Helander, H. F., Shah, G., Blum, A. L., and Sachs, G. Solubilization of anion ATPase from necturus oxyntic cells. *Biochim. biophys. Acta (Amst.)*, 1971, *241*, 49–56.

67. Wills, M. R. Fundamental physiological role of parathyroid hormone in acid-base homeostasis. *Lancet*, 1970, 802–804.

68. Winters, R. W., Engel, K., and Dell, R. B. *Acid-base physiology in medicine. A self-instruction program.* Westlake, Ohio, The London Co., 1967.

69. Woodbury, J. W. Regulation of pH. In: *Physiology and Biophysics*, 19th ed. T. C. Ruch and H. D. Patton, eds. Philadelphia, W. B. Saunders Co., 1965.

70. Woodbury, J. W., and Miles, P. R. Anion conductance of frog muscle membranes: One channel, two kinds of pH-dependence. *J. gen. Physiol.*, 1973, *62*, 324–353.

71. Woodbury, J. W., White, S. H., Mackey, M. C., and Weakly, J. N. High membrane H^+ permeability of frog skeletal muscle. *Proc. int. Union Physiol. Sci.*, 1968, 7, 472.

CHAPTER 28 THE URINARY BLADDER*

by THEODORE C. RUCH

The function of the bladder is of great concern to urologists, physiatrists and neurologists and of some concern to neurophysiologists. Conflicting concepts, unphysiologic terminology, ignorance of fundamental anatomy and physiology of such structures as the "internal sphincter," and the intermixture of mechanical, pathologic and neural factors—all contribute to making bladder function a complex subject.

Innervation of Bladder. The motor nerve supply to the bladder and its sphincters is derived from both divisions of the autonomic nervous system and from the somatic nervous system (Fig. 28–1). Afferent and efferent fibers are conducted in the same nerves. Recent research indicates that the innervation is considerably more complicated than parasympathetic vs. sympathetic innervations (see intramural plexus, below).

PARASYMPATHETIC. The preganglionic fibers supplying the bladder issue from the spinal cord mainly at S_3, although there is a small contribution from S_2, S_4, or both, established in man by root blocks.[31] This outflow accounts for bladder disturbance in lesions of the conus medullaris or the cauda equina. The preganglionic fibers traverse the *pelvic* nerves (nervi erigentes) and

the inferior hypogastric plexus, intermingling with sympathetic fibers, to synapse with postganglionic neurons in the ganglion clumps lodged in the bladder wall. These fibers constitute the efferents for the detrusor muscle and the internal sphincter.

SYMPATHETIC. The preganglionic outflow is from the upper lumbar and lower thoracic segments of the spinal cord. Fibers from the lumbar prevertebral ganglia and the preaortic nerve plexus descend along the abdominal aorta to form, at its bifurcation, the "presacral nerve"— more properly the superior hypogastric plexus— which, in turn, divides into two strands, the hypogastric nerves. The hypogastric nerves, coursing along the anterior surface of the sacrum, join the inferior hypogastric plexus and are distributed to the bladder. Some have attributed a detrusor relaxing or filling function to the sympathetic fibers as opposed to the parasympathetically mediated emptying function. Urologically and physiologically, the bladder fulfills two

---

*Bors, E., and Comarr, E., *Neurological urology.*[3a] Boyarsky, S., ed., *Neurogenic bladder.*[4] Kuru, M., *Physiol. Rev.*, 1965, *45*, 426–494.[36] Langworthy, O. R., et al., *Physiology of micturition.*[38] McLellan, F. C., *The neurogenic bladder.*[43] Pedersen, E., ed., *Acta neurol. scand.*, Symposium, 1966, *42*, Suppl. 20, 1–186.[51] Ruch, T. C., *Handb. Physiol.*, 1960, sec. 1, vol. II, 1207–1223.[56]

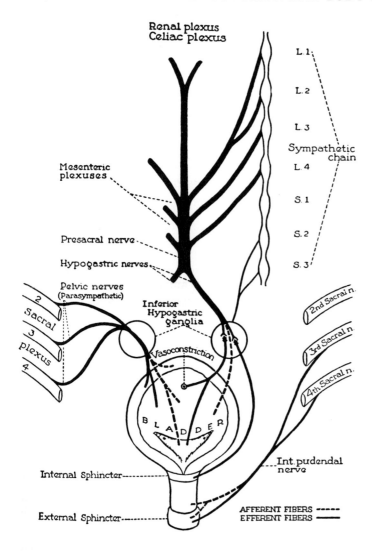

Renal plexus
Celiac plexus

L. 1
L. 2
L. 3
Sympathetic chain
L. 4
S. 1
S. 2
S. 3

Mesenteric plexuses
Presacral nerve
Hypogastric nerves
Pelvic nerves (Parasympathetic)
Inferior Hypogastric ganglia
Sacral 3 plexus 4
Vasoconstriction
2nd Sacral n.
3rd Sacral n.
4th Sacral n.
BLADDER
Internal Sphincter
Int. pudendal nerve
External Sphincter
AFFERENT FIBERS -----
EFFERENT FIBERS ——

Figure 28–1 Efferent and afferent pathways to bladder and sphincters. Postganglionic parasympathetic neurons are omitted. (From McLellan, *The neurogenic bladder.* Springfield, Ill., Charles C Thomas, 1939.)

opposite functions with the same musculature. The sympathetic fibers to the internal sphincter (motor) probably prevent reflux into the bladder during ejaculation.

SOMATIC. Somatic efferent fibers are confined to the external sphincter and prostatic urethra, which they reach by way of the third and fourth anterior roots and the pudendal nerve.

AFFERENT. The afferent fibers for the micturition reflex (and for the sense of bladder fullness) traverse the pelvic nerve. Painful impulses from the bladder dome are conducted by the hypogastric nerves; those from the trigone by the pelvic nerve. Sensory impulses from the urethra traverse the pudendal nerve. The afferent terminals are unencapsulated and mainly found in the tissue around muscle fascicles.[21a]

INTRAMURAL PLEXUSES. Many of the perplexing results of nerve and pharmacological stimulation of the bladder and the role of the autonomic divisions become understandable in light of histochemical analysis of the intramural plexuses of the bladder. Following upon the development of the fluorescence method of displaying adrenergic synapses, Hamberger and Norberg[30] demonstrated adrenergic and nonadrenergic (presumably cholinergic) cell bodies in the intramural bladder ganglia. Adrenergic terminals surround not only small adrenergic cell bodies but also the large nonadrenergic cell bodies. These basket-like terminations (not actually in contact with the cell body) suggest that the two autonomic divisions interact at a neural level rather than or as well as at the muscle level.

This type of analysis was extended further by Elbadawi and Schenk[20, 21] who found ganglion cells with short axons widely distributed within the walls of the bladder; these were adrenergic and cholinergic in about equal numbers, the latter with axons ending on the muscle cells in an almost one-to-one relationship. Postganglionic neurons synapse with these short neurons or terminate directly on the muscle cells. In addition, cell bodies are found in an extraganglionic site in the muscular coat. Thus there are three plexuses: extrinsic ganglia along the pelvic and hypogastric nerves as well as the pelvic or extramural plexus (sympathetic and parasympathetic), an intramural plexus and a third or extraganglionic site in the muscular coat.

It is not certain that this arrangement means a three- or a two-neuron chain; Elbadawi and Schenk favor the former. The more significant point is that, using histological procedures for both adrenergic and cholinergic neurons, they found a "cross-over" relationship so that a postganglionic neuron (adrenergic or cholinergic in respect to its terminals on muscle cells) may be influenced by both adrenergic (presumably sympathetic) and cholinergic (presumably parasympathetic) preganglionic neurons. Because noradrenalin inhibits ganglionic transmission, the preganglionic sympathetics could block discharge of parasympathetic postganglionic neurons, and adrenergic sympathetic fibers could inhibit sympathetic synapses which are cholinergic. These recent histochemical findings may explain several puzzling aspects of bladder physiology such as the biphasic response to acetylcholine. They mean that sections or stimulations of the two autonomic divisions are unphysiological evidence. But, perhaps most important, they mean that the sympathetic division, through its inhibitory relation to cholinergic phenomena, may have a greater role in bladder neurophysiology than has been suspected. Whether the postganglionic innervation of muscle cells of the bladder is cholinergic in all species has been questioned.[1]

Clinical and Laboratory Study of Bladder. The methods commonly used for clinical and laboratory study of the bladder are (i) retrograde or indirect cystometry (catheter passed through the urethra and connected with a manometer); (ii) the new method of direct cystometry[47] (needle penetrating the bladder wall), which yields information about urethral resistance as well as detrusor power (Fig. 28–2); (iii) flow rates and cast distance of urine; (iv) cystography with x-ray opaque substances; (v) electromyography of bladder or sphincters; (vi) radio capsule telemetry;[26] (vii) cystoscopy, a viewing of the bladder wall which yields a surprising amount of functional information; (viii) the recording of times and volumes of micturition; and (ix) the reaction to Urecholine. A relatively recent development is the combination of several of the aforementioned techniques simultaneously.

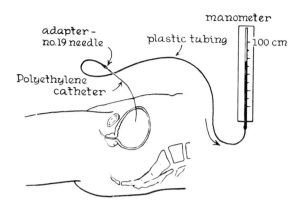

Figure 28–2   *Upper,* The method of direct cystometry. *Lower,* The range of voiding pressures occurring in patients having obstruction of the urinary passage. (From Murphy and Schoenberg, *J. Urol. (Baltimore),* 1960, *84,* 106–110.)

Recently, urodynamic or bioengineering analysis of several simultaneous parameters, particularly pressures and flow rates, has led to estimates of outflow resistance and more sophisticated concepts which are in the process of refinement conceptually and methodologically.[31a]

CYSTOMETRY. The cystometer is used to determine a pressure–volume curve of the bladder by means of a fluid (or air) source under pressure, a manometer, either isotonic or isovolumetric-isometric (a strain gauge), and a catheter connected by a three-way stopcock. The intravesical pressure should be measured while the bladder is connected with the manometer and disconnected from the filling reservoir, and

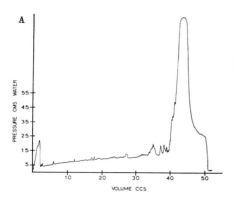 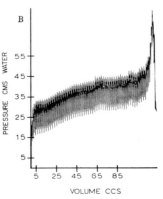

Figure 28-3 *A*, Air cysto-metrogram, and *B*, water cysto-metrogram, both at flow rate of 37 ml per minute and French 22 Foley catheter. (From Bradley *et al.*, *J. Urol. (Baltimore)*, 1968, *100*, 451–455.)

after the fluid has ceased to flow. Fluid is introduced into the bladder either steadily (at a slow rate) or in volume increments spaced a few minutes apart. This may be done through the recording catheter, a suprapubic or a double catheter. Pressure is plotted against volume to yield a cysto-metrogram. Rhythmic tonus waves, the threshold, vigor and duration of the micturition reflex, and subjective sensations are all important and are noted on the record.

Much of the detail of the conventional cystometric records is known to be due to artifacts, especially when applied to small laboratory animals such as the cat. The resistance to flow at the orifice or along the lumen of the catheter (size and length of tubing) and the viscosity of the bladder wall itself are the principal causes of arti-facts, and both increase with the incre-mental method, *i.e.*, with intermittent high inflow rates. This initial pressure rise has erroneously been considered to be a reflex phenomenon. Substitution of air for water greatly reduces viscous drag so that a rapid inflow rate (37 ml per min, 22 French catheter) could be used and a cystometro-gram could be obtained in one or two minutes from the canine bladder.[7] As shown in Figure 28–3, the same procedure with water yields an abnormal cystometrogram. However, such rapid filling roles are not ordinarily used, and recording pressure just before the next increment avoids the instrumental errors.

As shown in Figure 28–4, the cystometro-gram or pressure–volume curve typically consists of two parts, with the first segment

(I) divided again into two parts:* (i) an initial rise when the first increment is added (segment Ia), which is probably due to pressure of the fluid above the bladder neck and to intra-abdominal pressure;[60] (ii) segment Ib, a prolonged, nearly flat

---

*When it is necessary to distinguish between the two components of segment I, segments Ia and Ib can be used. This terminology replaces that introduced by the author.[56]

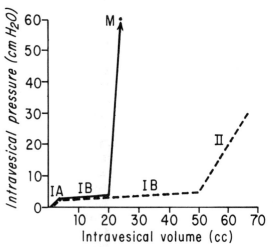

Figure 28-4 Schematic cystometrograms. *M* indi-cates peak pressure during micturition contraction. *Segment IA*, or the initial rise, is segment from zero to first point of inflection. *Segment IB*, or the tonus limb, begins at the first inflection point and either ends at micturition contraction or, in the absence of micturi-tion reflex, continues into *segment II*, the ascending limb. This transition is made angular for emphasis. (From Tang and Ruch, *Amer. J. Physiol.*, 1955, *181*, 249–257.)

segment (0 to 10 or 15 cm H_2O), the steepness of which is held to reflect detrusor "tonus;"* (iii) in a normally innervated bladder a sharp rise of pressure ensues, indicating a micturition reflex contraction of some strength (150 cm). The threshold for micturition is, then, the volume of filling (stretch) just previous to the reflex. When reflex micturition is wanting, the micturition pressure rise is replaced by a rapid rise. Segment II probably represents the passive stretch of elastic connective tissue of the vesical wall. The cystometrogram therefore reflects two bladder phenomena: (i) the amount of bladder tonus and (ii) the threshold and strength of the micturition contraction of the detrusor. The third problem of micturition is, of course, the response and control of the sphincters.

Bladder Hypotonus. A normal bladder holds its contents under very low pressure, usually less than 10 cm H_2O. Moreover, it holds increasing volumes of fluid with little increase in intravesical pressure. (Ureteral discharge proceeds without interference, and hydroureter is thus prevented.) In clinical parlance, a steeper than normal curve is termed "hypertonic," and a flatter than normal curve, "hypotonic." The clinical significance of the slope of the curve, *i.e.*, of bladder tonus, has long been confusing because it is so nonspecific and ambiguous diagnostically. The role of infection and inflammation and clinical spasticity in the genesis of hypertonicity and the cumulative effects of distension in producing hypertonicity need to be kept in mind in conclusions drawn from patients and chronic animal preparations alike. Many writers,[17, 24, 43, 45, 59] directly or by implication, consider bladder tonus a reflex phenomenon. They liken it to the stretch reflex ("tonus") of skeletal muscle or ascribe it to a "peripheral reflex" served by an intramural plexus of the bladder wall. However, a myotatic reflex yields a steadily increasing reflex contraction to increasing

stretch (see Chap. 8, Vol. I), whereas the bladder accommodates a new volume *with little persisting increase in pressure.* The somatic reflex analogy would be the lengthening reaction (see Chap. 8, Vol. I), but no such inhibitory reflex has been demonstrated. The flatness of the initial limb of the normal cystometrogram is sometimes accounted for by assuming that the stretch reflex arcs are activated at low volumes but are "inhibited" from the brain. These are all "neurogenic" theories of bladder tonus. The author's view, the myogenic theory, is that the reaction to stretch and supposed accommodation to increasing volumes are largely physical properties of the smooth muscle and connective tissue of the bladder wall, and are entirely nonreflex and nonneural in origin. These two hypotheses lead to quite different interpretations of the neurogenic disturbances of bladder function. The first emphasizes the overactivity or release of tonic bladder reflexes, spinal or peripheral, to explain "hypertonic" bladder behavior. The second emphasizes the physical state of the bladder wall to increased nociceptive reflexes as a result of infection and of changes secondary to interferences with the micturition reflex, *i.e.*, stretching or shrinking. More refined methods may prove both mechanisms are involved.

Nesbit *et al.*,[49] Tang and Ruch[62] and Carpenter and Root[13] studied the effects of anesthesia, spinal cord or root section and autonomic blocking agents on segments Ib of the cystometrogram in human patients and experimental animals. Whereas all abolish micturition, none reduces bladder tonus, which indicates that it is not reflex in nature (Fig. 28–5, center and right). In the intervals between spontaneous rhythmic contractions, no efferent or afferent impulses are detected in sacral parasympathetic neurons or nerves.[29] Sabetian[58] instilled a powerful mucosal irritant (tincture of capsici) into the bladder and found no change in bladder tone. Furthermore, as shown in Figure 28–5, brain stem transections which greatly augment (release) limb myotatic reflex tonus and the excitability of the micturition reflex fail to influence bladder tonus.[63] This also is true of patients with greatly overactive micturition reflexes caused by cortical lesions.[1]

---

*The term "tonus," recognized by physiologists as a cloak for ignorance, is used here operationally to mean "that which is reflected in the pressure–volume curve of the bladder." It is a generic term and must be qualified when the mechanisms of specific examples are known.

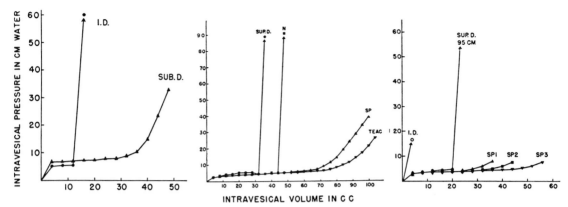

Figure 28–5 Cystometrograms obtained in cats after various neural lesions of brain stem or spinal cord and after autonomic blocking agents *(TEAC)*. Long arrows represent micturition and point to dot or square giving pressure generated. Threshold is next volume to right of point where arrows originate. Note the three phases of cystometrograms seen when micturition is absent, *e.g., SUB.D.* and *SP.* The abbreviations *I.D.*, *SUB.D.* and *SUP.D.* mean inter-, sub- and supracollicular decerebration; *SP1, SP2* and *SP3*, successive determinations after spinal transection. (After Tang and Ruch, *Amer. J. Physiol.*, 1955, *181*, 249–257.)

The absence of supraspinal control of bladder tonus has been confirmed by Edvardsen[18] by direct cystometry. However, he describes an increase in bladder tonus from interrupting a hypothetical continuously active inhibitory pathway with afferents in the sacral posterior roots and an efferent limb in the hypogastric nerves. This was operative only at higher bladder volumes, making the segment Ib rise less sharply before than after sectioning posterior roots. Since the cystometrogram in most experiments rose more sharply only at higher bladder volumes, the hypothesized reflex in these experiments may be a nociceptive reflex rather than a reflex adaptation of the bladder wall to larger volumes. Nociceptive fibers from the bladder reach the spinal canal through the lumbar as well as sacral posterior roots.

Even death or deep anesthesia is ineffective. In changing segment Ib the one factor which will alter the initial limb of the cystometrogram is the stretch produced by the accumulation of urine or that involved in a cystometric determination when the bladder is unprotected by the micturition reflex, as after spinal cord or pelvic nerve section. Successive decreases in bladder tonus occur after each of a series of cystometric determinations (Fig. 28–5, right and 28–16) is carried into segment II. Changes which might be ascribed to a drug or a sacral nerve section may actually be due simply to the stretch involved in the control cystometric determination. These hysteresis-like effects occur in the isolated bladder (Fig. 28–6) and

are accompanied by a reduction in the force of contraction at equal volumes caused by the viscous slack from previous stretching.[12]

Gjone[24] has attacked the concept that vesical tonus is non-neurogenic by reporting changes in the Ib segment of the cystometrogram consequent to denervation, especially sympathectomy. Division of the parasympathetic supply depressed the Ib segment somewhat and sympathectomy elevated it. In fact, after a parasympathectomy-induced depression, sympathectomy elevated segment Ib to nearly normal levels,

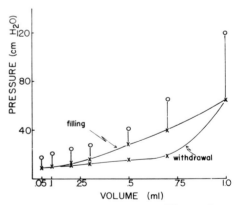

Figure 28–6 Cystometrogram on filling and passive emptying the isolated rat bladder. The open circles indicate the pressure immediately after the volume increment; the X's indicate the pressure before the next increment. (From Carpenter, *Invest. Urol.*, 1968, 6, 273–283.)

i.e., the end state was total deprivation of extrinsic nerve supply. Several unphysiological elements were involved in Gjone's experiments: anesthesia, injection into the bladder of large volumes, more often and within a space of 2 sec, which could damage muscle fibers and stimulate nociceptors, and finally, the surgical interventions. In early experiments, exposure of the bladder led to erroneous results and stimulation of pain receptors might well activate a maintained sympathetic discharge in Gjone's experiments. In any case, the effect of the neurectomies was to shift the cystometrogram up or down, in no way changing its shape, *i.e.*, the sympathetic nervous system seems to exert some kind of continuous relaxing effect on the bladder, the origin of the discharge being unknown. That a sympathetic discharge inhibitory to the bladder is not dependent upon nociceptive stimulation of surgical trauma is indicated by a slight depression of segment Ib by injection of isoproterenol (a beta-adrenergic stimulator)[5] in relatively intact dogs. These pharmacological effects may or may not be important clinically,[6, 27] but there remains the important physiological point of whether, as many investigators still believe, the sympathetic nervous system plays any part in the so-called accommodation of the bladder to increasing volumes or whether this is a matter of physics.

PHYSICAL ANALYSIS OF BLADDER TONE. A physical analysis of bladder tone confirms the aforementioned physiologic analysis and reveals a common interpretative error in cystometry.[60, 67] If the response of the bladder to increasing volume is thought of in terms of mural tension rather than intravesical pressure, any need to postulate a reflex or other "vital" explanation for the flatness of segment Ib of the cystometrogram disappears. Analysis shows that the tension in the bladder walls is steadily mounting as volume increases. The same rise has been made for the stomach, especially the muscular pyloric antrum (Chap. 1, Vol. III). This is easily visualized if you consider two points on the bladder as connected by a string. Even when the pull (tension) on a long string is very high, little pressure is exerted upon your finger when you attempt to bend the string. The difference between pressure and mural tension derives from the law of Laplace.

Consider the bladder as a perfect sphere composed of two hemispheres as in Figure 28–7

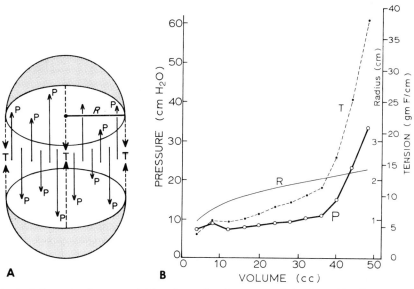

A **B**

Figure 28–7 *A*, Schematic diagram which aids in the derivation of the law of Laplace. *B*, Plot of radius (R) and intravesical pressure (P) showing that mural tension (T) rises even when the pressure curve is relatively flat. (Data from Tang, *J. Neurophysiol.*, 1955, *18*, 583–595. *A* after Winton and Bayliss, *Human physiology.* Boston, Little, Brown and Co., 1955.)

derived from Winton and Bayliss.[67] The pressure in the bladder as recorded in the cystometrogram is acting to force the two hemispheres apart. This total force must be P (force per unit area) times the area of the imaginary circular plane separating the two hemispheres, or $P \cdot \pi R^2$. This total force must be just balanced by the forces in the vesical wall at the junction of the two hemispheres. If these forces are of magnitude T per unit length of wall, then T times the circumferences ($T \cdot 2\pi R$) is the total force in the vesical wall holding the hemispheres together. This permits writing the equation of Laplace for a sphere:

$$T \cdot 2\pi R = P\pi R^2 \text{ or } P = 2(T/R).$$

The variables in this equation are plotted along with the cystometrogram in Figure 28–7B. The radius, R, is derived from the volume, V, assuming a spherically shaped bladder, from the geometric relation

$$V = (4/3) \pi R^3 \text{ or } R = \sqrt[3]{(3/4\pi)\ V}.$$

The Pressure, P, is that which is directly measured when obtaining the cystometrogram. The tension, T, is calculated from the values of P and R using the law of Laplace as derived above. Tension can be expressed also as a function of pressure and volume by combining the two preceding relations:

$$T = (P/2) \cdot \sqrt[3]{(3/4\pi)\ V} = 0.31 \cdot P \cdot \sqrt[3]{V}.$$

In the initial segment of these curves (below about 35 cc volume), equivalent to segment Ib of Figure 28–4 the tension and the radius both increase but at an approximately proportional rate, so that the pressure remains relatively constant. This increase in tension with stretch would be expected in any elastic material, whether living or inanimate, which obeys Hooke's law; thus, it is unnecessary to postulate an active "receptive relaxation" to explain a nearly flat cystometrogram. At higher volumes (above 35 cc volume, equivalent to segment II), the tension increases much more rapidly than the radius, leading to a steep rise in pressure. This increase in tension may simply be a consequence of exceeding the elastic range, since apparently the mural tissue passively becomes "stiffer" at large extensions. This interpretation is consistent with the hysteresis effect described above in neurally inactive bladders.

From this, it is clear that more information could be obtained about the mechanical state of the bladder tissue by plotting tension as a function of extension (or radius inferred from volume) rather than using the conventional cystometrogram plotting pressure vs. volume. The graphs in Figure 28–7B are derived from measurements of cat bladder; however, results from the human bladder are essentially similar, as confirmed by Hinman and Miller.[32] In application of the law of Laplace to bladder emptying, it is more complicated, since radius is continually changing and the law describes passive rather than active tension of musculature contraction.

It is apparent that the flatness of segment Ib of the cystometrogram requires no explanation in terms of reflex inhibitory relaxation. At the higher volumes of segment II, the sharply rising tension and stress mean that elastic elements are no longer approximating Hooke's law or, more probably, that some inelastic connective tissue element such as collagen has come into play.

Bladder Hypertonus. In extreme form hypertonus is manifested by an infected and fibrotic decentralized bladder (Fig. 28–8) or in the so-called uninhibited bladder, which differs by exhibiting strong micturition contractions; both show a so-called "climbing type" of cystometrogram. Convincing evidence that so-called "hypertonus" is also a physical rather than a reflex or neural phenomenon was given by Veenema et al.[65] and was analyzed further by Dahl.[16] If the bladder is maintained in an empty state by leading the ureters elsewhere than the bladder, a "climbing" type

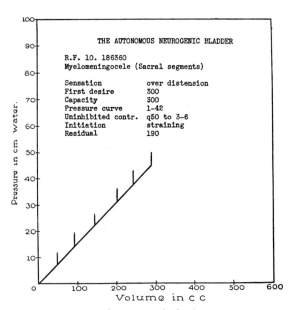

Figure 28–8 The type of climbing tonus curve caused by decentralization and resultant infection of the bladder. The vertical lines represent small micturitions. (From McLellan, *The neurogenic bladder.* Springfield, Ill., Charles C Thomas, 1939.)

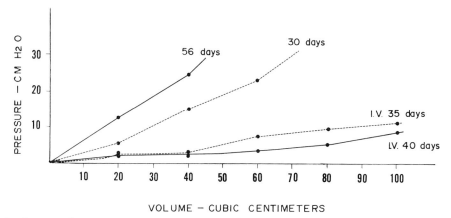

Figure 28–9 Sequential cystometrograms of empty bladder after diversion of the ureters to the ileum. (From Dahl, *Invest. Urol.*, 1969, 7, 160–167.)

(hypertonicity) cystometrogram results (Fig. 28–9). Thus the "climbing" type of curve seen in patients may be a physical change caused by shrinkage of the bladder rather than any type of neurogenic hypertonus. In the experiments on dogs, the "climbing" type of curve remained after general or spinal anesthesia[65] and was unaltered by spinal cord section or section of the cauda equina dividing sensory and motor nerves to the bladder. When the bladder again received urinary input by a ureteroileovesical union, periodic emptying was restored and the cystometrogram became progressively flatter.[16] Schmaelzle et al.[58a] found bladder capacity to be reduced to an average of 12.5 per cent of initial value by urinary diversion and recovered to about 80 per cent six weeks after restoration. While shrunken, a second determination showed some hysteresis effect but nothing suggesting the breaking of a fibrotic contracture; nor was the microscopical appearance significantly altered. The degree of infection is undoubtedly a variable. The abnormality of the vesical wall is demonstrable biochemically.[8] The bladder wall of dogs was infiltrated uniformly with a sclerosing agent which also induced bladder sepsis. The contractile force to electrical stimulation was reduced, segment Ib of the cystometrogram was converted to the climbing type (slope increased 2 to 10 times over the normal) and the collagen of the muscle was increased. The latter would explain the "hypertonic" cystometrogram. In neither of these studies were cystometrograms rapidly repeated, which would afford some clue to the role of fibrosis, although the reversibility bears on this point.

The results of these experiments suggest that the "climbing" type of curve, like that seen in patients, can be a rapidly developing physical change caused by shrinkage of the bladder rather than any type of neurogenic hypertonus. These experiments imitate the shrinkage of the bladder caused by small, frequent micturitions. Thus, a series of studies suggests that abnormalities in segment Ib of the cystometrogram are the result of physical changes in the bladder wall, the "flat" type of curve being caused by stretch and the "climbing" type by shrinkage. Neither is due to an underactive or overactive hypothetical accommodation reflex. However, as seen above, extrinsic neural influences on the Ib limb now seem to have been proved. The effect of the structural changes in the bladder wall on micturition and possible nociceptive reflexes will be discussed later.

Rhythmicity.[52, 53] Observed with an isometric cystometer, the cat's bladder, containing 12 to 20 ml of fluid, commences to contract rhythmically. The rate is slow, one contraction every 2 or 3 min; the duration is 5 to 15 sec and the amplitude 1 to 2 cm H_2O. Like bladder tonus, decerebration, spinal transection, lumbosacral rhizotomy or mural ganglionectomy did not abolish this rhythmicity, nor did ganglionic blocking agents (TEAC or atropine). Plum and Colfelt[53] concluded, therefore, that the rhythmic contractions were myogenic. The phenomenon is also termed autonomous

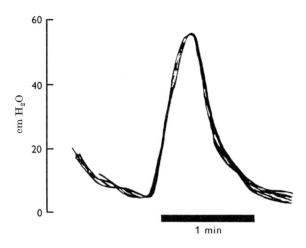

Figure 28–10 Six superimposed records of "spontaneous" rhythmical bladder contractions developing the same peak contractions. (From McPherson, *J. Physiol. [Lond.]*, 1966, *185*, 185–196.)

rhythm. Gjone[24] disputes the myogenic origin of the vesical rhythm on the grounds that a sympathectomy enhances it whereas pelvic nerve or sacral root section largely abolishes it. However, it is clear that Gjone and Plum and Colfelt were observing entirely different vesical rhythms. It seems that one is, in fact, myogenic or at least peripherally generated and one, the second, has all the properties of a spinal reflex. The first type was confirmed by La Grange[37] by applying strain gauges over the bladder. He recorded low force, high frequency irregular contractions out of phase as recorded in different gauges. These were local (nonpropagated and independent of extrinsic innervation and either myogenic or originating in the neural plexus). McPherson[44] recorded a second type of rhythmic contraction from the cat's bladder, occurring at remarkably regular intervals varying in different experiments from 45 to 120 sec. They are much more powerful than the first type, having amplitudes of 20 to 60 cm of water. The shapes of successive pressure waves are virtually identical, as shown in Figure 28–10. In a chloralose-anesthetized cat, the second type of wave was altered. After section of the spinal cord, only small (> 10 cm H_2O) and irregular waves were recorded, and they were not altered by nerve stimulation (Fig. 28–11). de Groat and Ryall[29] observed both types of rhythmic contractions in the same experiment, large waves (60 cm H_2O) repeated as low as three per min, with smaller waves occurring in the intervening periods at four to six per min. The large wave increased in rate (to six per min) when the baseline pressure was increased. The large contractions were preceded by action potentials in parasympathetic neurons and in

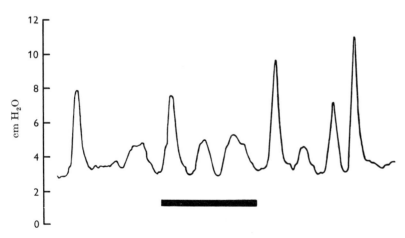

Figure 28–11 Contractions of the cat's bladder two hours after spinal transection of the spinal cord at T3. Black bar equals one minute. (After McPherson, *J. Physiol. [Lond.]*, 1966, *185*, 185–196.)

bladder afferents, regarded as secondary to the contraction. Transection of the spinal cord abolished the large contractions, leaving low amplitude contractions, which were not accompanied by action potentials in the pelvic nerve neurons and were unaffected by complete denervation of the bladder. However, 7 to 38 days after spinal transection, large contractions returned and were accompanied by action potentials in the parasympathetic neurons. La Grange's[37] observation of rhythmic bladder contractions by strain gauges on the serosal surface suggest that Gjone and McPherson were recording two different modes of rhythm, depending on extrinsic innervation. Under any second category he describes (i) low amplitude, short duration, high frequency contractions, and (ii) high amplitude, long duration and low frequency contractions. Both involve the bladder wall nearly simultaneously and coordinate reflexly from the spinal cord. As stated above, there are two types of bladder rhythms, one myogenic and one (perhaps with two subvarieties) dependent upon and influenced by spinal afferents and descending pathways. The latter are undoubtedly aborted micturition reflexes. The small

contractions may set up afferent discharge which "triggers" the larger ones and when the volume of fluid in the bladder is great enough, they fuse into a full-fledged micturition.

In man, rhythmicity is influenced by the higher levels of the nervous system and exhibits the same behavior in response to drugs and functional denervation and the same accelerated tempo and increased amplitude prior to micturition as in the cat.

Micturition. Micturition involves the contraction of the detrusor muscle of the bladder, the shaping or funneling of the bladder outlet and the opening and closing of the sphincters. The detrusor contraction is a spinal stretch reflex subject to inhibition and facilitation from higher centers. Micturition is the longest lasting episodic but nonrhythmical reflex such as breathing, or tonic, such as postural reflexes. Its initiation is voluntary, but its maintenance is not; is this maintenance spinal or peripheral? The requisite receptors for the initiation and sustaining of micturition have been demonstrated electrophysiologically in nerves of the bladder. Figure 28–12 shows single, slowly adaptive, distension-sensitive sen-

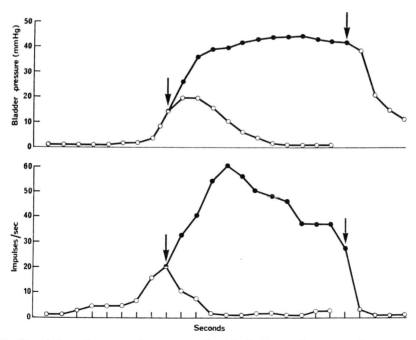

Figure 28–12 Correlation of pressure (upper record) with discharge of receptors (lower record) induced by occluding urethral cannula. Open circles indicate urethral cannula open; filled circles indicate cannula closed between the first and second arrows. (From Iggo, *Acta neuroveg. (Wien)*, 1966, 28, 121–134.)

sory end organs, some coming into play early and some later in the stretch which was very rapid. They adapt rapidly and could not sustain contraction. Stretch- (tension-) related receptors "in series" with the muscle fibers and hence violently stimulated by contraction have been demonstrated electrophysiologically.[33] They are found in the perifascicular connective tissue which is tied closely to and hence in series with the muscle bundles.[21a] These receptors reflexly increase and sustain bladder contraction (positive feedback). If tension is increased by obstructing outflow, these receptors fire even more actively and their reflex effect would tend to overcome resistance. Another factor in sustaining contractions may be Barrington's second reflex, initiated by passage of fluid through the external urethra, a somatovisceral reflex via the pudendal nerve. Why does the contraction stop when the bladder is empty? Three possible factors contribute: (i) adaptation of receptors, (ii) a recurrent or Renshaw inhibition[28] and (iii) afferents in the pelvic nerve that are inhibitory to parasympathetic neuron excitation. Starting and maintaining micturition is such a problem that stopping it has received little attention.

Rhythmic contractions of myogenic origin (see above) can build up into weak and brief emptying contractions. Conway and Bradley[15] plotted the spread of activity in the dog's bladder reflexively excited. The contraction spreads at the rate of about 1 cm per sec, begins in the midline and concomitantly stretches the urethrovesical junction and the intervening bladder wall which later contracts. The midline dome contraction spreads laterally. Although voluntary micturition begins with a relaxation of perineal musculature, the smooth muscle of the bladder is under direct voluntary control and micturition can occur without contraction of striated muscle.[17, 40]

The external sphincter (m. compressor urethrae) can, especially in the male, be powerfully contracted voluntarily and stop the flow of urine during detrusor contraction. However, the external sphincter cannot be opened voluntarily; it opens by a reflex inhibition of tonic pudendal nerve discharge from afferent fibers originating in the bladder and urethra.

The external urethral sphincter, examined electromyographically,[23] behaves much like the external anal sphincter which is recorded from in the urologic clinic because of its accessibility.* It exhibits a tonic activity which increases in frequency as the bladder is filled and ceases several seconds before micturition starts. This inhibition of tone is a reflex originating in the bladder wall[9a] and carried out by the sacral segments facilitated from the brain stem; it is very sensitive to spinal shock. Attempting to force fluid through the distal urethra by increasing pressure causes a progressively stronger contraction of the sphincter muscle. The stimulus for this activity is mainly the stretching of the distal urethral wall. The sphincter can withstand a perfusion pressure of 50 cm H_2O, much higher than the pressure at which flow starts in micturition; the stretch reflex of the sphincter, like that of the detrusor, may be hyperactive. Thus, spasticity of the external sphincter and the pelvic floor and depression of the reflex arc between bladder and sphincter are factors in neurogenic bladder dysfunction in man; the pudendal nerve is sometimes sectioned to decrease the resistance of the external sphincter.

That the internal sphincter is a gate opened by the pelvic nerve or the sympathetic nerve fibers or by impulses traversing a neural plexus has been directly disproved. Lapides,[39] from experiments based on hydrodynamic principles, supports the view that the internal sphincter opens purely mechanically as a result of contractions of the detrusor muscle and its extensions. The internal sphincter is not truly a sphincter in the literal sense, since the fibers which enclose its opening have no annular organization, arising instead from the inner longitudinally running detrusor muscle fascicles, especially the anterior ones which continue longitudinally into the urethra, forming an internal layer.

Woodburne[68] notes that "to stress the internal longitudinal layer of the urethra is not to deny that other components of the bladder wall continue in spiraling and circular directions external to the longitudinal coat as an external layer of the urethra." For example, the external

---

*Such records have disclosed a new clinical syndrome, a sphincter contraction rather than relaxation during micturition (dysergy), occurring in some patients with spinal injury.[55]

longitudinal fibers of the posterior wall spiral to end in the anterior parts of the inferior urethra.

No true annular sphincter at the bladder neck is seen. The circular and oblique bundles of fibers might serve as a weak sphincter. However, such bundles arch toward and then *away* from the neck and could open rather than close the proximal urethra. The bladder neck does, however, contain "an exceedingly rich collection of elastic fibers taking a circular direction even in the longitudinal bundles." These elastic fibers could serve to close the urethral passage.

The wall tension opposing distension is exerted by the mechanical properties (tonicity) of the smooth muscle and elastic fibers. That the sphincter resists bladder pressure successfully follows from the law of Laplace, $P = T/R$ (see above), since the radius of the urethra is much smaller than that of the bladder and hence the urethral wall exerts a much greater pressure. Stretching (dilation), which reduces wall tension of the sphincter, causes incontinence. Shortening the urethra diminishes and stretching it longitudinally increases resistance to flow through it. Lapides[39] has shown in dogs that a tubular "neourethra" fashioned from the fundus wall if 4 cm long restrains urine when the bladder is at rest and opens during micturition. There can be no question of inhibitory relaxation of such artificial sphincters. This is strong evidence that the urethra opens and closes by an active contraction of the longitudinal muscle. As shown in Figure 28–13, Lapides visualizes the bladder, which ordinarily resembles a chemical flask (Florence), as becoming pear-shaped during micturition because contraction of the longitudinal muscles makes the sphincter wider and shorter and allows intravesicular pressure to overcome the pressure exerted by the sphincter walls.

From the passive nature of the internal sphincter action, Bradley *et al.*[10] developed an apparatus for doing this by burying a radio frequency receiver beneath the skin. The receiver is activated with an external radio frequency source of power. However, many problems remain to be solved.[9b]

CLINICAL CORRELATIONS.[3, 3a, 17, 34, 38, 43, 45] *Tabetic Bladder Dysfunction.* The most severe neurogenic disturbance of bladder function is that occurring clinically with tabes dorsalis or experimentally after section of the sacral posterior roots. Progressive enlargement and overflow incontinence commence immediately and persist; the micturition reflex is in complete and

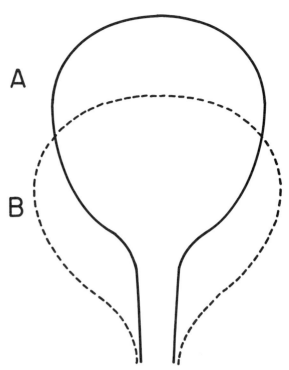

Figure 28–13 Outline of the bladder wall before (A) and just after (B) a micturitional contraction of the detrusor, showing that the critical proximal centimeters of the urethra have increased in diameter (because longitudinal muscle has contracted, making the distance between two points less and obliterating the angle between body and urethra). (After Lapides, *J. Urol. (Baltimore)*, 1958, *80*, 341–353.)

enduring abeyance. Voluntary micturition is possible only with great effort. The micturition reflex suffers the same fate as the stretch reflex in deafferented skeletal muscle. The segment Ib of the pressure—volume curve is lower and flatter than normal, and the segment II shifts to the higher volumes. Without good reason, this is commonly ascribed to loss of a tonic stretch reflex. Actually, the hypotonus simply represents changes in the property of the smooth muscle secondary to prolonged dilatation. The primary difficulty, therefore, is the absence of the micturition reflex which normally protects the bladder muscle from stretching. Exactly the same situation exists during the initial stages of spinal shock following transection (Fig. 28–5, right).

Decentralized or Autonomous Bladder. Lesions of the conus medullaris, cauda equina or pelvic nerve destroy entirely the parasympathetic efferent as well as the pelvic afferent connections of the bladder. The bladder is thus decentralized except for the sympathetic efferent and afferent connections which may depress bladder activity (see above). The behavior of the bladder is conditioned by the smooth muscle and by whatever neural control is effected through the mural plexus.[17, 38, 43, 45] Initially no active micturition contractions are elicited by effort or in cystometric examination. There is overflow incontinence. Later, a remarkable change occurs; small, brief waves of contraction occur spontaneously or in reaction to stretch or increased intra-abdominal pressures. The larger waves are accompanied by opening of the sphincters, but are inadequate in tension and duration for effective micturition. This results in escape of only small amounts of fluid, leaving residual urine. Such contractions, and particularly the partially coordinated action of the sphincter and detrusor, are often held to mean that the mural plexus serves a "peripheral" reflex. Several facts are difficult to square with the assumption of an independently active peripheral plexus; e.g., no such activities follow interruption of the spinal reflex arcs for micturition by posterior root section. The bladder wall undergoes great hypertrophy after decentralization, which may alter the initial limb of the cystometrogram. Bors[3a] uses the term lower motor neuron type for this condition and upper motor neuron type for the type described in the next section. The latter is indefensible both in neurology and in urology because there are many types of descending neurons affecting the bladder reflexes as well as skeletal reflexes.

Automatic Bladder of Spinal Transection. As noted, the initial stage of acute spinal shock following transection of the spinal cord above the sacral region is like that following decentralization. No micturition reflex is elicitable; after a period of retention, overflow incontinence supervenes. The retention apparently represents failure of the sequential detrusor-sphincter action. In favorable cases an automatic or "reflex" bladder is established, sometimes with a large residual urine resulting from the weakness and brevity of the micturition reflex; at other times the micturition reflex seems hyperirritable. The lack of voluntary control and knowledge of the time of micturition can be partly circumvented by using sensory stimulations or bladder pressure to precipitate micturition. Some victims of spinal injury or of descending pathways in the brain develop what is termed *hypertonic cord bladder*[45] or a *reflex neurogenic bladder*[43] or commonly the "uninhibited cord bladder." The Ib segment of the cystometrogram rises sharply, quite strong micturition contractions occur with small degrees of filling (100 ml), and residual urine is small. Smaller, spontaneous, regularly occurring detrusor contractions are also present. The possibility of an irritative basis as in cystitis must be taken into account. Clinically the patient complains of increased frequency of micturition and an increased urgency to the point that voluntary control cannot be maintained for

more than a very few minutes, leading to embarrassing incontinence. This is commonly given an oversimplified interpretation in terms of a powerful spinal micturition reflex seen in the human infant which with further development of the brain comes under the control of an "upper motor neuron," an explanation analogous to the explanation of spasticity of skeletal muscles, one which is now considered to be an oversimplification in view of the multiplicity of descending influence, facilitatory as well as inhibitory, affecting the motor neuron. There are many rather than *one* upper motor neurons (see below) affecting the excitability of sacral parasympathetic neurons.[57] The period of absent or weak micturition contractions after spinal transection negates the idea of a powerful spinal reflex which needs to be curbed by the brain. More probably micturition is a long-circuited reflex executed through and controlled through various brain structures. However, another question is just what type of bladder reflex is overactive and what is the role of cystitis.

Physiology of the Overactive Bladder. In discussing the overactive neurogenic bladder dysfunction in man and animals, it was suggested that "this may simply be caused by overactivity of the micturition reflex induced by stretch receptors. Another possibility, which should be explored, is that the bladder possesses a second mode of contraction, allied to nociceptive somatic reflexes which are favored in the spinal state and which are phasic rather than sustained. Such nociceptive reflexes could be operative both in cystitis and after spinal cord lesions."[56]

Highly suggestive evidence that there are two modes of micturition contraction was brought out[29] in a previous discussion of rhythmicity. Two modes of reflex response in parasympathetic neurons, early and late, followed stimulation of the central end of a branch of the pelvic nerve. The early response was produced by high threshold afferent fibers. In chronic spinal cats (28 to 38 days) brief contractions, but stronger than the spontaneous ones, could be induced not only by introduction of small volumes (0.4 to 2 ml) into the bladder, but also by tactile stimulation of the perineal region, strong pressure on a foot pad and electrical stimulation of the central end of a pelvic nerve filament. The latter caused early responses in the pelvic nerve and EPSP's in parasympathetic neurons; the late response was never observed in chronic

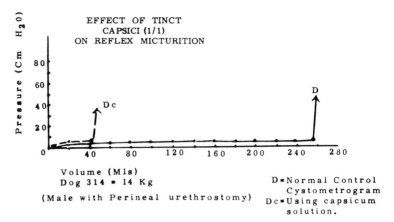

EFFECT OF TINCT
CAPSICI (1/1)
ON REFLEX MICTURITION

Figure 28–14 Effect of mucosal irritant on reflex micturition. D, normal control. Dc, after infusion of bladder with tincture of capsicum. (From Sabetian, *Brit. J. Urol.*, 1965, 37, 417–423.)

D = Normal Control
Cystometrogram
Dc = Using capsicum
solution.

spinal animals. The early response in the chronic spinal animal lacks the sustained character of the second response when spinal cord is intact. In a general way these experiments point to a dual mode of response from bladder afferents but do not identify the receptor source.

Normal micturition contraction is unquestionably induced by receptors in the wall of the bladder, and the mucosa plays little or no role. Sabetian[58] has shown that local anesthesia of the mucosa leaves a perfectly normal cystometrogram and micturition in experimental animals. However, as shown earlier by Plum and Colfelt,[53] if the mucosa is irritated chemically (tincture of capsicum), the volume threshold is reduced dramatically (in Fig. 28–14, from 260 to 40 ml). The micturition and return to baseline in these acute experiments could be called normal. However, adding to these the structural changes from irritation and from a continued shrinkage of the bladder from frequent micturition and the rising tonus limb, the frequent micturition of an "uninhibited" or spastic bladder could be envisaged. The hypothesis of a "mucosal micturition" reflex is supported by several fragments of evidence: (i) the stimulus is mucosal and nociceptive; (ii) severe vasomotor and sudomotor effects occur from distension of the bladder in spinal men; (iii) the bladder has an "in series" arrangement of receptors for true micturition to guarantee the continuance of micturition (nociceptive reflexes tend to be phasic); and (iv) in its higher control, micturition has many resemblances to postural (stretch reflexes), whereas (see below) nociceptive reflexes are hyperactive

in spinal animals. The hypothetical second or nociceptive mode of micturition must play a part in the paraplegic bladder. This does not mean that a neurogenic bladder dysfunction based on hyperactive mural engendered micturition does not occur. However, before this conclusion is reached, the external sphincter must be considered. If in the paraplegic the limbs are spastic, the external sphincter and pelvic floor would also be spastic, leading to short, aborted micturitions, and clinically this seems to be the obstacle to bladder training but obviously not to the overactive bladder exhibited by retrograde cystometry. With the sphincter mechanism ruled out, it will be shown in the following sections that overactivity (and hyperactivity) of the detrusor reflex micturition could be expected from brain lesions. Such hyperactivity cannot be ruled out as a result of spinal cord lesions if extensor spasticity of the limbs occurs. The hypothesis would then become that there are two modes of bladder contraction, one originating from mucosal receptors and one from muscle receptors, and these will imitate the changes in limb flexor and extensor reflexes in the way they are affected by transection of the spinal cord. Failure of the extensor stretch reflex compares with failure of micturition; release of flexor reflexes compares with overactive nociceptive micturition contractions, with consequent physical changes in the bladder wall. In the decerebrate preparation the extensor reflex overactivity (decerebrate rigidity) parallels the overactivity of the stretch receptor induced micturition. Lapides et al.[39a] used Banthine (methantheline bromide), a ganglionic blocking agent, to distinguish an uninhibited bladder and small capacity inflamed bladder. In the former, contractions disappear and bladder capacity increases; in the latter they persist. Lapides *et al.* concluded that they are local, noncholinergic, myogenic contractions. This does not disprove the existence of a second type of reflex, only that it is not cholinergic and that sympathetic preganglionic fibers synapse with cholinergic postganglionic fibers.[20, 21, 29] The parallel between limb reflexes and bladder reflexes will be discussed further in the following section.

Encephalic Control of Micturition. Micturition, a stretch reflex of the bladder, like the stretch reflex of the skeletal muscle,[59, 61, 63] is subject to control, facilitatory and inhibitory, from several levels of the nervous system. These are summarized in Figure 28–15, right. For both skeletal and vesical reflexes the basic arc is spinal and, since they both fail for a period after spinal transection, they both must depend on the brain for facilitation. One origin of the facilitatory impulses for the bladder, as stated by Barrington[2] and others, lies in the anterior pontine region in the location indicated in Figure 28–15. Section above this level, as in the classic intercollicular decerebration, results in extremely hyperactive micturition reflexes[61] (Fig. 28–5, left). Cooling or focal lesions at this level or section just below it result in a "spinal" cystometrogram, *i.e.*, complete failure of micturition because a

facilitatory area has been destroyed (Fig. 28–5, left). The micturition reflex is thus comparable to decerebrate rigidity, although one is phasic and the other postural. This pontine area is a "center" rather than a way station from higher centers, since it facilitates after levels above it are removed. It is presumed to receive a long circuited input from the stretch afferents of the bladder.

The extremely low threshold of the micturition reflex in the intercollicular decerebrate preparation (I.D. in Fig. 28–5) indicates that areas above this level have a net inhibitory effect, and this is borne out experimentally. A transhypothalamic section preventing the cerebral cortex, basal ganglia and anterior hypothalamus from acting on the sacral micturition reflex arc lowers the threshold (T.D. in Fig. 28–16), proving that the structures above have a strong net tonic inhibitory effect. A section through the anterior midbrain (SUP. D.) causes a rise in threshold, proving the existence of a posterior hypothalamic facilitatory center. If the experiment begins with the same supracollicular decerebration (SUP. D. in left of Fig. 28–16), the same high threshold is obtained, and after an intercollicular decerebration a few millimeters caudally a low threshold is obtained (see Fig. 28–5), proving the existence of a midbrain inhibitory center (M in Fig. 28–15).

There exists therefore a tonic influence from at least four levels of the nervous system, and successive levels are alternately

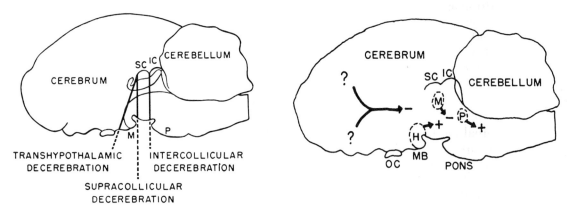

Figure 28–15 Schematic sagittal section of cat's brain, showing, on left, transections employed in studying brain stem control of micturition. Diagram at right shows locus of areas (*H,M,P*), as determined by transections and Horsley-Clarke lesions, which facilitate and inhibit micturition: + means facilitation and − means inhibition of micturition reflex; *SC* and *IC*, superior and inferior colliculi; *M* and *MB*, midbrain and mammillary bodies; *OC*, optic chiasm. (From Tang, *J. Neurophysiol.*, 1955, *18*, 583–595, and Tang and Ruch, *J. comp. Neurol.*, 1956, *106*, 213–245.)

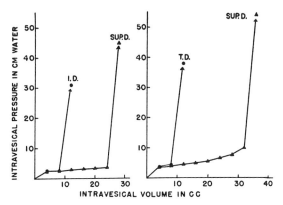

Figure 28-16 Cystometrograms after transections of the brain stem establishing a posterior hypothalamic (right) and a midbrain (left) brain stem micturition center. For planes of transection and abbreviations see Figure 28-15, and for explanation see text. (From Tang and Ruch, *Amer. J. Physiol.*, 1955, *181*, 249–257.)

inhibitory and facilitatory (see Fig. 28-17).

Edvardsen[18] has confirmed the aforementioned findings and adds the information that the supraspinal control is not exercised through the sympathetic nervous system. He does, however, argue that sympathetic reflexes interact to determine micturition threshold, the sympathectomized cats giving the lower thresholds with a given level of transection. He assumes that the threshold occurs at a constant pressure and by sympathectomy this level is obtained at smaller volumes because the pressure volume curve is said to rise more rapidly (see above).

Because the stretch reflex of the skeletal muscles and the vasoconstrictor neurons

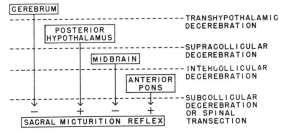

Figure 28-17 A summarizing diagram indicating the net facilitatory or inhibitory action of various levels of the nervous system deduced from surgical procedures shown at the right. For simplicity, the diagram does not take into account the possibility that the descending pathways from the higher structures terminate on lower ones, including the bulbar reticular inhibitory and facilitatory areas of Kuru[36] and coworkers. (From Tang, *J. Neurophysiol.*, 1955, *18*, 583–595.)

of the sympathetic nervous system are tonically facilitated from medullary levels, a similar tonic facilitation of vesical reflexes would be predicted, but is not revealed by transection experiments. Stimulation recording and degeneration experiments by Kuru[36] and co-workers, sympathetically reviewed by Boyarsky and Ruskin,[6] indicate that a lateral reticular spinal tract carries facilitatory impulses from the medulla, pons and mesencephalon. Only the second of these were found by Tang and Ruch,[63] the mesencephalic level being inhibitory. A medial (and posterior) bulbar area was said to be a "vesico-inhibitory" center acting by way of the ventral reticular tract and was described by Kuru. These postulated bulbar areas, if tonic, are exactly balanced or have no ascending afferent input because a section below the anterior pontine micturition facilitatory area is equivalent to a spinal transection. A possibility is that pontine and higher centers exert their influence in part through bulbar facilitatory and inhibitory areas of Kuru which would then be "way stations" rather than "centers."

To summarize, the micturition reflex is tonically influenced from at least three levels of the brain stem, and successive levels are alternately inhibitory and facilitatory. The reflex is also influenced by two suprasegmental structures, the cerebellum and the cerebral hemispheres.

Stimulation experiments as early as 1941[14] indicated that the anterior lobe of the cerebellum exerts both excitatory and inhibitory effects on the bladder through one of the brain stem regions described above, since the effects occur after removal of the cerebral cortex.[11] In the most recent study,[9] stimulation of the fastigial nucleus inhibited the micturition reflex in a supracollicular decerebrate cat. This is another parallel with the stretch reflex of limb muscles which become greatly hyperactive when the anterior lobe of the cerebellum is included in a decerebration. A supracollicular decerebration does not decide whether fastigial nucleus inhibits the pontine facilitatory area or inhibits the mesencephalic inhibitory area.

Considerable clinical information[1a] has placed a sensory "center" for the control of micturition (and defecation) in the superior frontal gyrus at the midline and on the adjoining medial surface, as shown in Figure

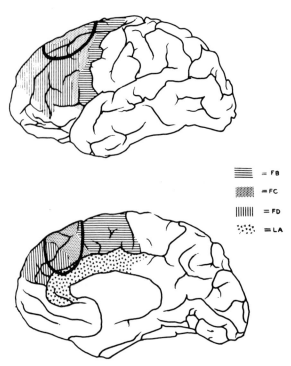

=== = FB

▓▓▓ = FC

|||||| = FD

⋰⋰⋰ = LA

Figure 28–18 The area of the human cerebral cortex concerned with initiation and inhibition of micturition (superimposed on cytoarchitectural areas of Von Economo). (From Andrew and Nathan, *Brain*, 1964, 87, 233–262.)

28–18. Pathologic and surgical damage to this area may end the desire to micturate and reduce the awareness that micturition is imminent. The ability to restrain micturition or to stop it, once begun, is also im-

paired. Cystometric curves resemble very closely those seen after decerebration in cats and suggest that the bladder reflex is freed from inhibition. The micturition reflex threshold is as low as 100 ml and incontinence occurs not as an overflow phenomenon but through overactivity of the unopposed subcortical facilitating centers. As pointed out above, after a transection leaving the posterior hypothalamus intact, the micturition reflex is overactive, which is consistent with these clinical observations.

That many areas of the cerebral cortex are probably involved in micturition is shown by stimulation experiments (Fig. 28–19). These areas are sensorimotor areas I and II, anterior cingulate, orbital and piriform regions.[25]

In view of the multiple cortical areas and brain stem levels involved in bladder function, it is not surprising that the spinal pathways are in doubt. One is posterolateral and superficial.[35] A recent description[22] places a pathway lateral and deep (Fig. 28–20).

The higher control of the micturition reflex is obviously far more complex than the disinhibition of a spinal reflex from damage to the "upper motor neuron" described in the clinical literature[43, 45] which ignores some studies on man.[66] The representations of the bladder above the brain stem have proliferated, *i.e.*, the amygdala,[19] basal ganglia[41, 42] and thalamus.[41, 46] Recorded isometrically in unanesthetized cats it was first observed[42] that stimulation of the globus pallidus inhibited the bladder

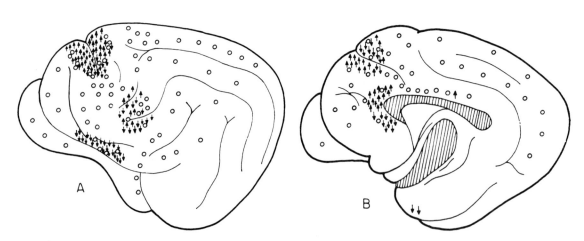

Figure 28–19 The cat's brain in lateral *(A)* and medial *(B)* views, showing four cortical loci where electrical stimulation causes bladder contraction,↑, or relaxation,↓, or no response, ○. Note that some of the areas are divided into an inhibitory and an excitatory zone; the area on the orbital surface is purely inhibitory. (From Gjone and Setekleiv, *Acta physiol. scand.*, 1963, 59, 337–348.)

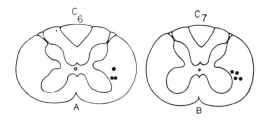

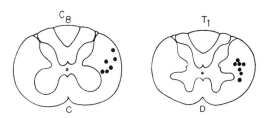

Figure 28-20 Location of the vesicopressor pathway (dots) at four spinal cord levels. (From Foley, *Proc. Soc. exp. Biol. Med. (N.Y.)*, 1970, *133*, 25–29.)

contractions of the second type described above (there was no change in tone). Coincident with a spontaneous reflex contraction, increased firing of single pallidal units preceded contraction and occurred again during relaxation with decreased firing during the contraction.[54] In later experiments[41] other subcortical ganglia proved even more powerful inhibitors, with the red nucleus and substantia nigra stimulation being most effective and the subthalamic nucleus and N. ventralis lateralis of the thalamus being less so. These observations are consistent with the occurrence of bladder disorders, especially heightened bladder activity, in association with Parkinsonism. Thalamotomy, according to Murnaghan,[46] tends to worsen the condition. Whether, as suggested by clinical studies, these areas exert a tonic effect on micturition needs clarification.

As early as 1935, Watts and Uhle[66] described the cystometrograms in a series of patients with brain tumors. Some exhibited a sharply rising cystometrogram accompanied by urgency to micturate at very small volumes (50 ml) and involuntary micturition, the classic uninhibited neurogenic bladder dysfunction. The cystometrograms of another group were flat, with the desire to void not appearing in some patients until the bladder filling reached 600 to 700 ml; some had difficulty in initiating micturition, but some had no history of urinary disturbances. In a general way, these observations are consistent with the findings of Gjone and Setekleiv[25] of facilitatory as well as cortical inhibitory areas. A more definite correlation is difficult.

The multiplicity of cortical and subcortical influences on the bladder makes micturition more understandable in familar neurophysiologic terms. In fact, no better illustration of the similarity between the higher control of somatic and autonomic reflexes can be found than that obtained by comparing the micturition reflex with the myotatic reflex of the limb extensor muscles after lesions at different levels of the nervous system.* Although we have come to expect a representation of autonomic functions at various levels of the nervous system, it is surprising that such a simple act as micturition, which can only start, continue and stop, should have and presumably require such manifold representations.

The neural mechanism for efficient storage and expulsion of urine consists of the sacral spinal reflex arcs, the anterior pons, the midbrain, the posterior hypothalamus and the cerebral cortex. The neural mechanism which determines when and where this mechanism will operate, the decision or the trigger, a matter of considerable importance to the individual, is the contribution of the cerebral cortex.

Cardiovascular Effects of Bladder Activity.[64] Bladder distension and emptying in spinal men mysteriously have a significant effect on the heart, the vasculature and the sweat glands well above the level of transection. The effects on blood pressure occur when the nervous system is intact though largely antagonized by the baroreceptor reflexes. The neural mechanism of these effects in man is not worked out. They are not exactly correlated with pain, because cauda equina lesions will eliminate the "viscero-vascular" reflex from the bladder but not all pain which from the bladder dome is served by the lumbar sympathetic nerves. This, the logical candidate for a diffuse pathway, seems to be ruled out.

---

*However, certain differences must be kept in mind; *e.g.*, neither the spinal cord nor the brain controls any reflex bladder tone because none exists, whereas the tone of somatic muscles originates entirely in spinal reflexes and suprasegmental mechanisms.

The receptors for this reflex are located throughout the bladder wall, but the response to tension applied to the trigone is larger. In the cat, the afferent pathway is mainly the pelvic nerve but also the hypogastric nerves. With increasing distension the order of vascular changes is vasoconstriction (veins as well as arterials), rise in blood pressure, tachycardia and cardiac arrhythmias. Arguments that the vascular effects are not allied to pain receptors do not take into account the dual pathway of pain impulses from the bladder.

The effects upon blood pressure and pulse rate have been so marked that they are incorporated in an instrumental test for the diagnosis of neurogenic bladder dysfunction.[55]

Psychosomatic Aspects. Because the "resting" experimental animal shows no change in bladder tonus when its neural axis is sectioned at various levels does not mean that the bladder is not subject to phasic influences from higher centers. Students commonly observe that impending examinations may cause urinary frequency. Experimentally, psychological stimuli such as those involved in a psychiatric interview have been shown to cause the bladder to contract more strongly than do some physiologic stimuli. Although cortical and subcortical centers do not tonically influence bladder tone, electrical stimulation of these centers causes the bladder to contract. Psychological stimuli may be likened therefore to direct brain stimulation. Shame, guilt and embarrassment inhibit the act of micturition by a conditioned or learned response.

REFERENCES

1. Ambache, N., and Alboozar, M. Non-cholinergic transmission by postganglionic motor neurones in the mammalian bladder. *J. Physiol. (Lond.),* 1970, *210*, 761–783.

1a. Andrew, J., and Nathan, P. W. Lesions of the anterior frontal lobes and disturbances of micturition and defecation. *Brain,* 1964, *87*, 233–262.

2. Barrington, F. J. F. The relation of the hind-brain to micturition. *Brain,* 1921, *44*, 23–53.

3. Bors, E. Neurogenic bladder. *Urol. Surv.,* 1957, 7, 177–250.

3a. Bors, E., and Comarr, E. *Neurological urology; physiology of micturition, its neurological disorders and sequelae.* Baltimore, University Park Press, 1971.

4. Boyarksy, S., ed. *Neurogenic bladder.* Baltimore, Williams and Wilkins Co., 1967.

5. Boyarksy, S., Labay, P., Gregg, R., and Levie, B. Pharmacologic studies of the nature of the sympathetic nerves of the urinary bladder. *Paraplegia,* 1968, *6*, 137–150.

6. Boyarsky, S., and Ruskin, H. Physiology of the bladder. In: *Urology,* Vol. I. M. F. Campbell and J. H. Harrison, eds. Philadelphia, W. B. Saunders Co., 1970.

7. Bradley, W., Carren, S., Shapero, R., and Wolfson, J. Air cystometry. *J. Urol. (Baltimore),* 1968, *100*, 451–455.

8. Bradley, W., Chou, S., Markland, C., and Swaiman, K. Biochemical assay technique for estimation of bladder fibrosis. *Invest. Urol.,* 1965, 3, 59–64.

9. Bradley, W. E., and Teague, C. T. Cerebellar influence on the micturition reflex. *Exp. Neurol.,* 1969, *23*, 399–411.

9a. Bradley, W. E., and Teague, C. T. Electrophysiology of pelvic and pudendal nerves in the cat. *Exp. Neurol.,* 1972, *35*, 378–393.

9b. Bradley, W. E., Timm, G. W., and Chou, S. N. A decade of experience with electronic simulation of the micturition reflex. *Urol. int. (Basel),* 1971, *26*, 283–303.

10. Bradley, W. E., Wittmers, L. E., and Chou, S. N. An experimental study of the treatment of the neurogenic bladder. *J. Urol. (Baltimore),* 1963, *90*, 575–582.

11. Bruhn, J. M., Foley, J. O., Emerson, G. M., and Emerson, J. D. Autonomic response to cerebellar stimulation in the decorticate cat. *Amer. J. Physiol.,* 1961, *201*, 700–702.

12. Carpenter, F. G. Motor responses of bladder smooth muscle in relation to elasticity and fiber length. *Invest. Urol.,* 1968, *6*, 273–283.

13. Carpenter, F. G., and Root, W. S. Effect of parasympathetic denervation on feline bladder function. *Amer. J. Physiol.,* 1951, *166*, 686–691.

14. Connor, G. J., and German, W. J. Functional localization within the anterior cerebellar lobe. *Trans. Amer. neurol. Ass.,* 1941, *67*, 181–186.

15. Conway, C. J., and Bradley, W. E. Measurement of spread of excitation in the urinary detrusor muscle during reflex induction. *J. Urol. (Baltimore),* 1969, *101*, 533–539.

16. Dahl, D. S. Reversible vesical hypertonicity, a consequence of the chronic empty state. *Invest. Urol.,* 1969, 7, 160–167.

17. Denny-Brown, D., and Robertson, E. G. On the physiology of micturition. *Brain,* 1933, *56*, 149–190. *Idem:* The state of the bladder and its sphincters in complete transverse lesions of the spinal cord and cauda equina. *Ibid.* 397–463.

18. Edvardsen, P. Nervous control of urinary bladder in cats. I. The collecting phase. *Acta physiol. scand.,* 1968, *72*, 157–171. *Idem:* II. The expulsion phase. *Ibid.* 172–182. *Idem:* III Effects of autonomic blocking agents in the intact animal. *Ibid.* 183–193. *Idem:* IV. Effects of autonomic blocking agents on responses to peripheral nerve stimulation. *Ibid.* 234–247.

19. Edvardsen, P., and Ursin, T. Micturition thresholds in cat with amygdala lesions. *Exp. Neurol.,* 1968, *21*, 495–501.

20. Elbadawi, A., and Schenk, E. A. Dual innerva-

tion of the mammalian urinary bladder: A histochemical study of the distribution of cholinergic and adrenergic nerves. *Amer. J. Anat.*, 1966, *119*, 405–427.

21. Elbadawi, A., and Schenk, E. A. A new theory of the innervation of the bladder musculature. Part 2, The innervation apparatus of the ureterovesical junction. *J. Urol. (Baltimore)*, 1971, *105*, 368–371. *Idem*: Part 3, Postganglionic synapses in uretero-vesico-urethral autonomic pathways. *Ibid.* 372–374.

21a. Fletcher, T. B., and Bradley, W. E. Afferent nerve endings in the urinary bladder of the cat. *Amer. J. Anat.*, 1970, *128*, 147–158. See also *J. comp. Neurol.*, 1969, *136*, 1–20.

22. Foley, A. L. A descending vesicopressor pathway in the monkey. *Proc. Soc. exp. Biol. (N.Y.)*, 1970, *133*, 25–29.

23. Garry, R. C., Roberts, T. D. M., and Todd, J. K. Reflexes involving the external urethral sphincter in the cat. *J. Physiol. (Lond.)*, 1959, *149*, 653–665.

24. Gjone, R. Peripheral autonomic influence on the motility of the urinary bladder in the cat. I. Rhythmic contractions. *Acta physiol. scand.*, 1965, *65*, 370–377. *Idem*: II. "Tone." *Ibid.* 1966, *66*, 72–80. *Idem*: III. Micturition. *Ibid.* 1966, *66*, 81–90.

25. Gjone, R., and Setekleiv, J. Excitatory and inhibitory bladder responses to stimulation of the cerebral cortex in the cat. *Acta physiol. scand.*, 1963, *59*, 337–348.

26. Gleason, D. M., and Lattimer, J. K. A miniature radio transmitter which is inserted into the bladder and which records voiding pressures. *J. Urol. (Baltimore)*, 1962, 87, 507–509.

27. Gregg, R. A., Boyarsky, S., and Labay, P. Blocking of beta adrenergic receptors in the urinary bladder using sotalol. *Sth. med. J. (Bgham., Ala.)*, 1969, *62*, 1366–1373.

28. de Groat, W. C., and Ryall, R. W. Recurrent inhibition in sacral parasympathetic pathways to the bladder. *J. Physiol. (Lond.)*, 1968, *196*, 579–591.

29. de Groat, W. C., and Ryall, R. W. Reflexes to sacral parasympathetic neurons concerned with micturition. *J. Physiol. (Lond.)*, 1969, *200*, 87–108.

30. Hamberger, B., and Norberg, K.-A. Adrenergic synaptic terminals and nerve cells in bladder ganglia of the cat. *Int. J. Neuropharmacol.*, 1965, *4*, 41–45.

31. Heimburger, R. F., Freeman, L. W., and Wilde, N. J. Sacral nerve innervation of the human bladder. *J. Neurosurg.*, 1948, 5, 154–164.

31a. Hinman, F., Jr., Boyarsky, S., Pierce, J. M., Jr., and Zinner, N. R., eds. *Hydrodynamics of micturition.* Springfield, Ill., Charles C Thomas, 1971.

32. Hinman, F., Jr., and Miller, E. R. Mural tension in vesical disorders and ureteral reflux. *Trans. Amer. Ass. Genitourin. Surg.*, 1963, 55, 13–20.

33. Iggo, A. Physiology of visceral afferent systems. *Acta neuroveg (Wien)*, 1966, *28*, 121–134.

34. Juul-Jensen, P. Neurological bladder dysfunction—cystometry in diagnosis and treatment. *Acta neurol. scand.*, 1962, *38*(Suppl. 3), 113–130.

35. Kerr, F. W. L., and Alexander, S. Descending autonomic pathways in the spinal cord. *Arch. Neurol. (Chic.)*, 1964, *10*, 249–261.

36. Kuru, M. Nervous control of micturition. *Physiol. Rev.*, 1965, *45*, 426–494.

37. La Grange, R. G. Peripheral autonomic regulation of the canine urinary bladder: Model proposal. *Invest. Urol.*, 1971, *9*, 64–81.

38. Langworthy, O. R., Kolb, L. C., and Lewis, L. G. *Physiology of micturition.* Baltimore, Williams & Wilkins Co., 1940.

39. Lapides, J. Structure and function of the internal vesical sphincter. *J. Urol. (Baltimore)*, 1958, *80*, 341–353.

39a. Lapides, J., and Dodson, A. I., Jr. Observations on effect of methantheline (Banthine) bromide in urological disturbances. *Arch. Surg.*, 1953, *66*, 1–9.

40. Lapides, J., Sweet, R. B., and Lewis, L. W. Role of striated muscle in urination. *J. Urol. (Baltimore)*, 1957, 77, 247–250.

41. Lewin, R. J., Dillard, G. V., and Porter, R. W. Extrapyramidal inhibition of the urinary bladder. *Brain Res.*, 1967, *4*, 301–307.

42. Lewin, R. J., and Porter, R. W. Inhibition of spontaneous bladder activity by stimulation of the globus pallidus. *Neurology (Minneap.)*, 1965, *15*, 1049–1052.

43. McLellan, F. C. *The neurogenic bladder.* Springfield, Ill., Charles C Thomas, 1939.

44. McPherson, A. The effects of somatic stimuli on the bladder in the cat. *J. Physiol. (Lond.)*, 1966, *185*, 185–196.

45. Munro, D. The cord bladder—its definition, treatment and prognosis when associated with spinal cord injuries. *New Engl. J. Med.*, 1936, *215*, 766–777.

46. Murnaghan, G. F. Neurogenic disorders of the bladder in Parkinsonism. *Brit. J. Urol.*, 1961, *33*, 403–409.

47. Murphy, J. J., and Schoenberg, H. W. Observations on intravesical pressure changes during micturition. *J. Urol. (Baltimore)*, 1960, *84*, 106–110.

48. Nesbit, R. M., and Baum, W. C. Cystometry: Its neurologic diagnostic implication. *Neurology*, 1954, *4*, 190–199.

49. Nesbit, R. M., and Lapides, J. Tonus of the bladder during spinal "shock." *Arch. Surg.*, 1948, *56*, 138–144.

50. Nesbit, R. M., Lapides, J., Valk, W. W., Sutler, M., Berry, R. L., Lyons, R. H., Campbell, K. N., and Moe, G. K. The effects of blockade of the autonomic ganglia on the urinary bladder in man. *J. Urol. (Baltimore)*, 1947, 57, 242–250.

51. Pedersen, E., ed. The neurogenic bladder. *Acta neurol. scand.*, 1966, *42*(Suppl. 20), 1–186.

52. Plum, F. Autonomous urinary bladder activity in normal man. *Arch. Neurol. (Chic.)*, 1960. *2*, 497–503.

53. Plum, F., and Colfelt, R. H. The genesis of vesical rhythmicity. *Arch. Neurol. (Chic.)*, 1960. *2*, 487–496.

54. Porter, R. W. A pallidal response to detrusor contraction. *Brain Res.*, 1967, *4*, 381–383.

55. Roussan, M., Abramson, A. S., D'Oranzio, G., and Boyarsky, S. Changes mediated by the somatic and autonomic nervous systems in response

to bladder filling: A method of study. In: *The neurogenic bladder.* S. Boyarsky, ed. Baltimore, Williams and Wilkins Co., 1967.

56. Ruch, T. C. Central control of the bladder. *Handb. Physiol.,* 1960, sec. 1, vol. II, 1207–1223.

57. Ruch, T. C., and Tang, P. C. The higher control of the bladder. In: *The neurogenic bladder.* S. Boyarsky, ed. Baltimore, Williams and Wilkins Co., 1967.

58. Sabetian, M. The role of the vesical mucosa in reflex micturition. *Brit. J. Urol.,* 1965, *37,* 417–423. *Idem*: The genesis of bladder tone. *Ibid.* 424–432.

58a. Schmaelzle, J. F., Cass, A. S., and Hinman, F., Jr. The effect of disuse and restoration of function on vesical capacity. *J. Urol. (Baltimore),* 1969, *101,* 700–705.

59. Sherrington, C. S. Postural activity of muscle and nerve. *Brain,* 1915, *38,* 191–234.

60. Smith, R. F., Watson, M. R., and Ruch, T. C. Unpublished observations.

61. Tang, P. C. Levels of brain stem and diencephalon controlling micturition reflex. *J. Neurophysiology,* 1955, *18,* 583–595.

62. Tang, P. C., and Ruch, T. C. Non-neurogenic basis of bladder tonus. *Amer. J. Physiol.,* 1955, *181,* 249–257.

63. Tang, P. C., and Ruch, T. C. Localization of brain stem and diencephalic areas controlling the micturition reflex. *J. comp. Neurol.,* 1956, *106,* 213–245.

64. Taylor, D. E. M. Afferent pathways and efferent mechanisms in the bladder viscero-vascular reflex. *Quart. J. exp. Physiol.,* 1968, *53,* 262–272.

65. Veenema, R. J., Carpenter, F. G., and Root, W. S. Residual urine, an important factor in interpretation of cystometrograms. An experimental study. *J. Urol. (Baltimore),* 1952, *68,* 237–241.

66. Watts, J. W., and Uhle, C. A. W. Bladder dysfunction in cases of brain tumor: A cystometric study. *J. Urol. (Baltimore),* 1935, *34,* 10–30.

67. Winton, F. R., and Bayliss, L. E. *Human physiolgy.* Boston, Little, Brown & Co., 1955.

68. Woodbourne, R. T. Anatomy of the bladder. In: *The neurogenic bladder.* S. Boyarsky, ed. Baltimore, Williams and Wilkins Co., 1967.

INDEX

Page numbers in *italic* type refer to illustrations.